Lecture Notes in Computer Science

Vol. 270: E. Börger (Ed.), Computation Theory and Logic. IX, 442 pages. 1987.

Vol. 271: D. Snyers, A. Thayse, From Logic Design to Logic Programming. IV, 125 pages. 1987.

Vol. 272: P. Treleaven, M. Vanneschi (Eds.), Future Parallel Computers. Proceedings, 1986. V, 492 pages. 1987.

Vol. 273: J.S. Royer, A Connotational Theory of Program Structure. V, 186 pages. 1987.

Vol. 274: G. Kahn (Ed.), Functional Programming Languages and Computer Architecture. Proceedings. VI, 470 pages. 1987.

Vol. 275: A.N. Habermann, U. Montanari (Eds.), System Development and Ada. Proceedings, 1986. V, 305 pages. 1987.

Vol. 276: J. Bézivin, J.-M. Hullot, P. Cointe, H. Lieberman (Eds.), ECOOP '87. European Conference on Object-Oriented Programming. Proceedings. VI, 273 pages. 1987.

Vol. 277: B. Benninghofen, S. Kemmerich, M.M. Richter, Systems of Reductions. X, 265 pages. 1987.

Vol. 278: L. Budach, R.G. Bukharajev, O.B. Lupanov (Eds.), Fundamentals of Computation Theory. Proceedings, 1987. XIV, 505 pages. 1987.

Vol. 279: J.H. Fasel, R.M. Keller (Eds.), Graph Reduction. Proceedings, 1986. XVI, 450 pages. 1987.

Vol. 280: M. Venturini Zilli (Ed.), Mathematical Models for the Semantics of Parallelism. Proceedings, 1986. V, 231 pages. 1987.

Vol. 281: A. Kelemenová, J. Kelemen (Eds.), Trends, Techniques, and Problems in Theoretical Computer Science. Proceedings, 1986. VI, 213 pages. 1987.

Vol. 282: P. Gorny, M.J. Tauber (Eds.), Visualization in Programming. Proceedings, 1986. VII, 210 pages. 1987.

Vol. 283: D.H. Pitt, A. Poigné, D.E. Rydeheard (Eds.), Category Theory and Computer Science. Proceedings, 1987. V, 300 pages. 1987.

Vol. 284: A. Kündig, R.E. Bührer, J. Dähler (Eds.), Embedded Systems. Proceedings, 1986. V, 207 pages. 1987.

Vol. 285: C. Delgado Kloos, Semantics of Digital Circuits. IX, 124 pages. 1987.

Vol. 286: B. Bouchon, R.R. Yager (Eds.), Uncertainty in Knowledge-Based Systems. Proceedings, 1986. VII, 405 pages. 1987.

Vol. 287: K.V. Nori (Ed.), Foundations of Software Technology and Theoretical Computer Science. Proceedings, 1987. IX, 540 pages. 1987.

Vol. 288: A. Blikle, MetaSoft Primer. XIII, 140 pages. 1987.

Vol. 289: H.K. Nichols, D. Simpson (Eds.), ESEC '87. 1st European Software Engineering Conference. Proceedings, 1987. XII, 404 pages. 1987.

Vol. 290: T.X. Bui, Co-oP A Group Decision Support System for Cooperative Multiple Criteria Group Decision Making. XIII, 250 pages. 1987.

Vol. 291: H. Ehrig, M. Nagl, G. Rozenberg, A. Rosenfeld (Eds.), Graph-Grammars and Their Application to Computer Science. VIII, 609 pages. 1987.

Vol. 292: The Munich Project CIP. Volume II: The Program Transformation System CIP-S. By the CIP System Group. VIII, 522 pages. 1987.

Vol. 293: C. Pomerance (Ed.), Advances in Cryptology – CRYPTO '87. Proceedings. X, 463 pages. 1988.

Vol. 294: R. Cori, M. Wirsing (Eds.), STACS 88. Proceedings, 1988. IX, 404 pages. 1988.

Vol. 295: R. Dierstein, D. Müller-Wichards, H.-M. Wacker (Eds.), Parallel Computing in Science and Engineering. Proceedings, 1987. V, 185 pages. 1988.

Vol. 296: R. Janßen (Ed.), Trends in Computer Algebra. Proceedings, 1987. V, 197 pages. 1988.

Vol. 297: E.N. Houstis, T.S. Papatheodorou, C.D. Polychronopoulos (Eds.), Supercomputing. Proceedings, 1987. X, 1093 pages. 1988.

Vol. 298: M. Main, A. Melton, M. Mislove, D. Schmidt (Eds.), Mathematical Foundations of Programming Language Semantics. Proceedings, 1987. VIII, 637 pages. 1988.

Vol. 299: M. Dauchet, M. Nivat (Eds.), CAAP '88. Proceedings, 1988. VI, 304 pages. 1988.

Vol. 300: H. Ganzinger (Ed.), ESOP '88. Proceedings, 1988. VI, 381 pages. 1988.

Vol. 301: J. Kittler (Ed.), Pattern Recognition. Proceedings, 1988. VII, 668 pages. 1988.

Vol. 302: D.M. Yellin, Attribute Grammar Inversion and Source-to-source Translation. VIII, 176 pages. 1988.

Vol. 303: J.W. Schmidt, S. Ceri, M. Missikoff (Eds.), Advances in Database Technology – EDBT '88. X, 620 pages. 1988.

Vol. 304: W.L. Price, D. Chaum (Eds.), Advances in Cryptology – EUROCRYPT '87. Proceedings, 1987. VII, 314 pages. 1988.

Vol. 305: J. Biskup, J. Demetrovics, J. Paredaens, B. Thalheim (Eds.), MFDBS 87. Proceedings, 1987. V, 247 pages. 1988.

Vol. 306: M. Boscarol, L. Carlucci Aiello, G. Levi (Eds.), Foundations of Logic and Functional Programming. Proceedings, 1986. V, 218 pages. 1988.

Vol. 307: Th. Beth, M. Clausen (Eds.), Applicable Algebra, Error-Correcting Codes, Combinatorics and Computer Algebra. Proceedings, 1986. VI, 215 pages. 1988.

Vol. 308: S. Kaplan, J.-P. Jouannaud (Eds.), Conditional Term Rewriting Systems. Proceedings, 1987. VI, 278 pages. 1988.

Vol. 309: J. Nehmer (Ed.), Experiences with Distributed Systems. Proceedings, 1987. VI, 292 pages. 1988.

Vol. 310: E. Lusk, R. Overbeek (Eds.), 9th International Conference on Automated Deduction. Proceedings, 1988. X, 775 pages. 1988.

Vol. 311: G. Cohen, P. Godlewski (Eds.), Coding Theory and Applications 1986. Proceedings, 1986. XIV, 196 pages. 1988.

Vol. 312: J. van Leeuwen (Ed.), Distributed Algorithms 1987. Proceedings, 1987. VII, 430 pages. 1988.

Vol. 313: B. Bouchon, L. Saitta, R.R. Yager (Eds.), Uncertainty and Intelligent Systems. IPMU '88. Proceedings, 1988. VIII, 408 pages. 1988.

Vol. 314: H. Göttler, H.J. Schneider (Eds.), Graph-Theoretic Concepts in Computer Science. Proceedings, 1987. VI, 254 pages. 1988.

Vol. 315: K. Furukawa, H. Tanaka, T. Fujisaki (Eds.), Logic Programming '87. Proceedings, 1987. VI, 327 pages. 1988.

Vol. 316: C. Choffrut (Ed.), Automata Networks. Proceedings, 1986. VII, 125 pages. 1988.

Vol. 317: T. Lepistö, A. Salomaa (Eds.), Automata, Languages and Programming. Proceedings, 1988. XI, 741 pages. 1988.

Vol. 318: R. Karlsson, A. Lingas (Eds.), SWAT 88. Proceedings, 1988. VI, 262 pages. 1988.

Vol. 319: J.H. Reif (Ed.), VLSI Algorithms and Architectures – AWOC 88. Proceedings, 1988. X, 476 pages. 1988.

Vol. 320: A. Blaser (Ed.), Natural Language at the Computer. Proceedings, 1988. III, 176 pages. 1988.

Vol. 321: J. Zwiers, Compositionality, Concurrency and Partial Correctness. VI, 272 pages. 1989.

Vol. 322: S. Gjessing, K. Nygaard (Eds.), ECOOP '88. European Conference on Object-Oriented Programming. Proceedings, 1988. VI, 410 pages. 1988.

Vol. 323: P. Deransart, M. Jourdan, B. Lorho, Attribute Grammars. IX, 232 pages. 1988.

Lecture Notes in Computer Science

Edited by G. Goos and J. Hartmanis

380

J. Csirik J. Demetrovics
F. Gécseg (Eds.)

Fundamentals of Computation Theory

International Conference FCT '89
Szeged, Hungary, August 21–25, 1989
Proceedings

Springer-Verlag
New York Berlin Heidelberg London Paris Tokyo Hong Kong

Editors

János Csirik
Férenc Gécseg
József Attila Tudományegyetem Bolyai Intézete
Aradi vértanúk tere 1, H-6720 Szeged, Hungary

János Demetrovics
Computer and Automation Institute
Hungarian Academy of Sciences
P.O. Box 63, H-1502 Budapest, Hungary

CR Subject Classification (1987): D.2.4, E.5, F, I.1.2, I.2–3

ISBN 3-540-51498-8 Springer-Verlag Berlin Heidelberg New York
ISBN 0-387-51498-8 Springer-Verlag New York Berlin Heidelberg

Printed in Germany

Printing and binding: Druckhaus Beltz, Hemsbach/Bergstr.
2145/3140-543210 – Printed on acid-free paper

PREFACE

This volume constitutes the proceedings of the conference on Fundamentals of Computation Theory held in Szeged, Hungary, August 21-25, 1989. The conference is the seventh in the series of the FCT conferences initiated in 1977 in Poznan-Kornik, Poland. The conference was organized by the Attila József University (Szeged) with cooperation of the Computer and Automation Institute of the Hungarian Academy of Sciences.

The papers in this volume are the texts of invited addresses and shorter communications falling in one of the following sections:

Efficient Computation by Abstract Devices: Automata, Computability, Probabilistic Computations, Parallel and Distributed Computing

Logics and Meanings of Programs: Algebraic and Categorical Approaches to Semantics, Computational Logic, Logic Programming, Verification, Program Transformations, Functional Programming

Formal Languages: Rewriting Systems, Algebraic Language Theory

Computational Complexity: Analysis and Complexity of Algorithms, Design of Efficient Algorithms, Algorithms and Data Structures, Computational Geometry, Complexity Classes and Hierarchies, Lower Bounds

The shorter communications were selected on March 21 and 22, 1989 at the Program (Selection) Committee Meeting in Szeged from the large number of papers submitted to FCT '89.

Thanks are due to the members of the Program Committee for their work in evaluating the submitted papers, to the members of the Organizing Committee for their hard job in all organizational matters as well as to all referees of FCT '89.

Szeged, August 1989

János Csirik   János Demetrovics   Ferenc Gécseg

CONFERENCE COMMITTEES

PROGRAM COMMITTEE

G. Ausiello, J. Berstel, L. Budach, R.G. Bukharajev, Phan Dinh Dieu, P. van Emde Boas, the late A.P. Ershov, F. Gécseg, J. Gruska, J. Hartmanis, J. Heintz, G. Hotz, K. Indermark, H. Jürgensen, M. Karpinski, L. Lovász, O.B. Lupanov, G. Mirkowska, A. Mostowski, A. Pultr, J.H. Reif, G. Rozenberg, J. Sakarovitch, A. Salomaa, E. Szemerédi, H. Thiele, I. Wegener, Wu Wen-Tsün

ORGANIZING COMMITTEE

J. Demetrovics (Chairman), J. Csirik (Secretary), Z. Ésik (Secretary), M. Bartha, Gy. Horváth, J. Virágh

CONFERENCE CHAIRMAN

Ferenc Gécseg

REFEREES OF FCT '89

Ablayev, F.M.
Adámek, J.
Albrecht, A.
America, P.H.M.
Arnolds, A.
Asveld, P.R.J.
Badouel, E.
Banachowski, L.
Barendregt, H.P.
Becker, B.
Bilardi, G.
Bloom, S.L.
Boanon, L.
Brandenburg, F.J.
Briot, J.-P.
Chlebus, B.S.
Chrétienne, P.
Chytil, M.
Courcelle, B.
Culik, K. II
Dahlhaus, E.
Darondeau, P.
Dassow, J.
Dauchet, M.
Delporte-Gallet, C.
Diekert, V.
Diks, K.
Ďuriš, P.
Ehrig, P.
Engelfriet, J.
Enikeev, A.I.
Grigorieff, S.
Habel, A.
Haddad, S.
Hansel, G.
Harju, T.
Hoffmann, F.
Hromkovič, J.
Huttenlocher, D.
Jędrzejowicz, J.
Jung, H.
Karhumäki, J.
Kleijn, H.C.M.
Klop, J.W.
Kluźniak, F.
Kolla, R.
Korec, I.
Kreowski, H.-J.
Krob, D.
Kuchen, H.
Kudlek, M.
Kusnetsov, S.E.
Lehmann, T.
Lescanne, P.
Linna, M.
Loeckx, J.
Lukaszewicz, W.
Machi, A.
Marek, I.
Mazoyer, J.
Mehlhorn, K.
Meinel, C.
Ch. Meyer, J.-J.
Molitor, P.
Müller, F.
Nait Abdallah, M.A.
Nasirov, I.R.
Niwiński, D.
Nurmeev, N.N.
Oberschelp, W.
Ollagnier, J.M.
Orlowska, E.
Pelletier, M.
Perrot, J.-F.
Pin, J.E.
Privara, I.
Ranjan, D.
Reutenauer, C.
Rovan, B.
Rudak, L.
Ruohonen, K.
Ružička, P.
Salimov, F.I.
Salwicki, A.
Samitov, R.K.
Schinzel, B.
Schupp, P.E.
Seidl, H.
Sieber, K.
Skowron, A.
Solovjev, V.D.
Spaniol, O.
Stabler, E.P. Jr.
Štěpánek, P.
Štura, J.
Szepietowski, A.
Teitelbaum, T.
Thomas, W.
Treinen, R.
Validov, F.I.
Vitányi, P.M.B.
Vogler, H.
Waack, S.
van Westrhenen, S.C.
Wiedermann, J.
Wolf, G.

Contents

ON WORD EQUATIONS AND MAKANIN'S ALGORITHM

Habib Abdulrab, Jean-Pierre Pécuchet
Laboratoire d'Informatique de Rouen et LITP.
Faculté des Sciences, B.P. 118, 76134 Mont-Saint-Aignan Cedex †
E.m.: mcvax!inria!geocub!abdulrab mcvax!inria!geocub!pecuchet

ABSTRACT. — We give a short survey of major results and algorithms in the field of solving word equations, and describe the central algorithm of Makanin.

Introduction

An algebra equipped with a single associative law is a **semigroup**. It is a **monoid** when it has a unit. The free monoid generated by the set A (also called **alphabet**) is denoted by A^*. Its elements are the **words** written on the alphabet A, the neutral element being the empty word denoted by 1. The operation is the concatenation denoted by juxtaposition of words. The **length** of a word w (the number of letters composing it) is denoted by $|w|$. For a word $w = w_1 \ldots w_n$, with $|w| = n$, we denote by $w[i] = w_i$ the letter at the ith position. The number of occurrences of a given letter $a \in A$ in a word w, will be denoted by $|w|_a$.

In this terminology, the term algebra (in the sense of [Fag Hue], [Kir]) built on a set of variables V, a set C of constants, and a set of operators constituted of an associative law, is nothing else than the free monoid $T = (V \bigcup C)^*$ over the alphabet of **letters** $L = V \bigcup C$.

A unifier of two terms $e_1, e_2 \in T$ is a monoid morphism $\alpha : T \longrightarrow T$ (i.e. a mapping satisfying $\alpha(mm') = \alpha(m)\alpha(m')$ and $\alpha(1) = 1$), leaving the constants invariant (i.e. satisfying $\alpha(c) = c$ for every $c \in C$) and satisfying the equality $\alpha(e_1) = \alpha(e_2)$.

The pair of words $e = (e_1, e_2)$ is called an **equation** and the unifier α is a **solution** of this equation.

A solution $\alpha : T \longrightarrow T'$ **divides** a solution $\beta : T \longrightarrow T''$ if there exists a **continuous** morphism $\theta : T' \longrightarrow T''$ (i.e. satisfying $\theta(x) \neq 1$ for every x) such as $\beta = \alpha\theta$. We also say that α is **more general** than β. A solution α is said to be **principal** (or **minimal**) when it is divided by no other but itself (or by an **equivalent** solution, i.e. of the form $\alpha' = \alpha\theta$ with θ, an isomorphism).

The two main problems concerning systems of equations are the existence of a solution, and the computation of the set of minimal solutions (denoted by μCSU_A in [Fag Hue]). All these problems reduce to the case of a single equation, as by [Alb Law] every infinite system of equations is equivalent to one of its finite subsystems, and a finite system can be easily encoded in a single equation [Hme].

---

† This work was also supported by the Greco de Programmation du CNRS and the PRC Programmation Avancée et Outils pour l'Intelligence Artificielle.

The study of properties and structure of the set of solutions of a word equation was initiated by Lentin and Schützenberger ([Len Sch], [Len]) in the case of constant-free equations ($C = \emptyset$).

In particular, Lentin shows that every solution is divided by a unique minimal one and gives a procedure (known as the **pig-pug**) allowing to enumerate the set of minimal solutions. This procedure extends without difficulty to the general case of an equation with constants (cf. [Plo], [Pecl]). The minimal solutions are obtained as labels of some paths of a graph. When this graph is finite, as in the case when no variable appears more than twice, we obtain a complete description of all solutions.

The problem of the existence of a solution was first tackled by Hmelevskii who solved it in the case of three variables [Hme], then by Makanin who solved the general case [Makl]. He gave an algorithm to decide whether a word equation with constants has a solution or not.

This paper is divided into two parts. The first one will be devoted to a brief presentation of the pig-pug method which gives, for simple cases, the most efficient unification algorithm. The rest of this paper will be devoted to Makanin's Algorithm [Makl] as it is implemented by Abdulrab [Abdl]. In order to keep a reasonable size to this paper, most of the proofs will be omitted.

1. The pig-pug

In the remaining part of this paper, we assume without loss of generality, that the alphabets of variables $V = \{v_1 \dots v_n\}$ and of constants $C = \{c_1 \dots c_m\}$ are finite and disjoint. We make the convention to represent the variables by lower-case letters, as $x, y, z \dots$, and the constants by upper-case letters as $A, B, C \dots$. We call **length** of an equation $e = (e_1, e_2)$ the integer $d = |e_1 e_2|$.

The **projection** of an equation e over a subset Q of V is the equation obtained by "erasing" all the occurrences of $V \setminus Q$. Consequently, an equation has 2^n projections $(\Pi_Q e_1, \Pi_Q e_2)$ where $\Pi_Q : (V \bigcup C)^* \longrightarrow (Q \bigcup C)^*$ is the projection morphism.

One easily proves the following proposition which reduces the research of a solution to that of a continuous one.

Proposition 1.1 *An equation e has a solution iff one of its projections has a continuous solution.* ∎

The pig-pug method consists in searching for a continuous solution α in the following manner: it visits the lists $e_1[1], \dots, e_1[|e_1|]$ and $e_2[1], \dots, e_2[|e_2|]$ of symbols of e from left to right and at the same time, one tries to guess how their images can overlap. At each step, one makes a non deterministic choice for the relative lengths of the images of the first two symbols $e_1[1]$ et $e_2[1]$. According to the choice made :

$$|\alpha(e_1[1])| < |\alpha(e_2[1])|, \quad |\alpha(e_1[1])| = |\alpha(e_2[1])|, \quad |\alpha(e_1[1])| > |\alpha(e_2[1])|$$

one applies to the equation one of the three substitutions to variables :

$$e_2[1] \leftarrow e_1[1]e_2[1], \quad e_2[1] \leftarrow e_1[1], \quad e_1[1] \leftarrow e_2[1]e_1[1].$$

The process is repeated until the **trivial** equation (1,1) is obtained.

The application of the pig-pug method to all the projections of an equation gives a graph labeled by the three previous transformations. The nodes of the graph are the transformed equations.

The following result, a proof of which can be found in [Pec1], shows that this graph enumerates all the minimal solutions.

Theorem 1.2 [Len] *The set of minimal solutions of a word equation is given by the labels of the paths linking the root to the trivial equation in the pig-pug graph.* ∎

One can consider the graph associated with an equation e, as a deterministic automaton M, in which :

1) the alphabet is given by the projections and the variable substitution.
2) the states of the automaton are the vertices of the graph.
3) the final state is $(1, 1)$.

The language accepted by this automaton, designated $L(M)$, is precisely the set of all the minimal solutions of e, that is, a complete set of minimal unifiers.

When the graph is finite, $L(M)$ is a regular language. But in the general case the automaton is infinite. And $L(M)$ in not necessarily regular, and not even a context-free language, (for example, this is the case of the automaton M associated with the equation $e = (xAAyx, AxxBz)$).

Note that $L(M)$ is not necessarily minimal, in the sense that, equivalent minimal unifiers can be generated within $L(M)$.

In the general case, the pig-pug's graph will be infinite. However one can always decide the existence of a solution by:

Theorem 1.3 [Mak2] *One can construct a recursive function F such that, if an equation of length d has a solution, then there exists one in which the length of the components of the solutions are bounded by $F(d)$.* ∎

The only known function F is that derived from Makanin's algorithm that we will see now. Another reason for the study of this algorithm is that it leads to a better pruning of the graph, and is more efficient than the pig-pug method in some cases.

2. Length equations

Before describing Makanin's algorithm, we introduce the notion of length equations which is related to integer programming.

First note that if $Card(C) = 0$, the equation e has necessarily the trivial solution $\alpha(v) = 1$, for every $v \in V$. Consequently, we can assume that $Card(C) > 0$. An equation is called **simple** if $card(C) = 1$. Such equations are related to integer equations in the following way:

Let e be a simple equation, consider the commutative image e' of e:

$$(v_1^{p_1} \ldots v_n^{p_n} c_1^{p_{n+1}}, v_1^{q_1} \ldots v_n^{q_n} c_1^{q_{n+1}})$$

The linear diophantine equation: $(p_1-q_1)v_1'+\ldots+(p_n-q_n)v_n' = (q_{n+1}-p_{n+1})$. is called the **length equation** associated with e.

The isomorphism between $c_1{}^*$ and $(N,+)$ gives the following correspondence:

Proposition 2.1 *There is a bijective correspondence between the solutions of a simple equation and the non-negative integer solutions of its length equation.* ▮

With every equation e we associate the simple equation e' obtained by the substitution of all the constants of e by c_1. The length equation associated with e' is by definition that associated with e.

The following necessary condition is easily shown:

Proposition 2.2 *If an equation e admits a solution, then its length equation admits a non-negative integer solution.* ▮

Thus solving simple equations reduces to integer programming. Next we shall see how Makanin's algorithm can solve non-simple equations.

3. Equation with scheme and position equation

In this section, we describe two basic notions appearing in Makanin's algorithm: the notion of an equation with scheme, and that of a position equation. We show how to compute a position equation from an equation with scheme.

3.1. Equation with scheme

Obviously, there are many possible ways of choosing the positions of the symbols of e_1 according to those of the symbols of e_2. For example, the following diagrams illustrate such possibilities for the equation $e = (AyB, xx)$.

```
*__A__*________y__*__B__*      *__A__*__y__*__B__*      *__A__*__y__*__B__*
*__________x__*________x__*    *__x__*________x__*      *________x__*__x__*
```

Informally, a scheme applicable to an equation $e = (e_1, e_2)$ indicates how to locate the positions of the symbols of e_1 according to those of e_2 in a possible solution of e.

Formally, a **scheme** is a word $s \in\ = \{=,<,>\}^*\ =$, that is a word over the alphabet $\{=,<,>\}$ beginning and ending with the letter $=$.

A scheme s is called **applicable** to an equation $e = (e_1, e_2)$ if the following conditions are satisfied:

1) $|s|_< + |s|_= = |e_1| + 1$.
2) $|s|_> + |s|_= = |e_2| + 1$.

where $|s|_\phi$, is the number of occurrences of ϕ in s.

The left and right boundaries of a symbol t (denoted by $lb(t)$ and $rb(t)$) in a scheme s applicable to e are the integers of the interval $[1\ |s|]$, defined in the following way:

If $t = e_1[n]$ then $lb(t)$ is the length of the prefix of s whose length is equal to n over the alphabet $\{=, <\}$, and $rb(t)$ is the length of the prefix of s whose length is equal to $n+1$ over $\{=, >\}$. The definition in the case $t = e_2[n]$ is obtained from the previous one by exchanging $<$ and $>$.

An **equation with scheme** is a 6-tuple (V, C, e, s, lb, rb), where e is an equation over the alphabet of variables V and the alphabet of constants C, s is a scheme applicable to e, lb and rb are left and right boundary maps.

3.2. Informal presentation of an example

An equation with scheme will now be transformed into a new object over which further processes will be applied. Before we give the formal definitions in the next sections, we introduce here the different notions and notations via an example.

Consider the equation with scheme

$$(\{A, B\}, \{x, y, z\}, AxyBz = zzx, =<><=<=, lb, rb)$$

corresponding to the following diagram:

```
*__A__*_________x__*__y__*__B__*__z__*
*__________z__*________z__*__________x__*
=     <     >     <     =     <     =
```

This equation with scheme will be transformed into a so called position equation. This object inherits the seven boundaries of the equation with scheme and of all occurrences of constants, but variables will be treated in a special manner.

Variables with single occurrence, like y, will disappear.

Other occurrences of variables will be renamed in order to avoid the growth of the equations appearing in the pig-pug method. The renamed occurrences of a similar variable will be associated via a symmetrical binary relation on variables (called duality relation) or a positional equivalence, depending on the number of these occurrences, as shown below:

For a two occurrence variable, like x, the two occurrences will be renamed x_1 and x_2 and associated in the duality relation. That is, $x_1 = dual(x_2)$ and $x_2 = dual(x_1)$.

For a m occurrence variable u, with m greater than two, like z, $2m-2$ renamed variables will be generated as follows :

1) $m-1$ renamed variables (here z_1 and z_2) will receive the place of the first occurrence of u, (here the first occurrence of $z = e_2[1]$).

2) $m-1$ other renamed variables (here z_3 and z_4) will receive the place of the $m-1$ other occurrences of u, (here, $e_2[2]$ and $e_1[5]$).

3) each variable of the first set is in duality with one of the second set, (here, we have $dual(z_1) = z_3, dual(z_3) = z_1, dual(z_2) = z_4, dual(z_4) = z_2$).

So, the duality relation gives a sort of "link" between the first $m-1$ occurrences associated with u, and the last $m-1$ occurrences. Since $z_1 = z_2$ because they have the same position, the equality of all the renamed variables $z_1 = z_2 = z_3 = z_4$ is obtained.

We then obtain an object which can be illustrated by the following diagram:

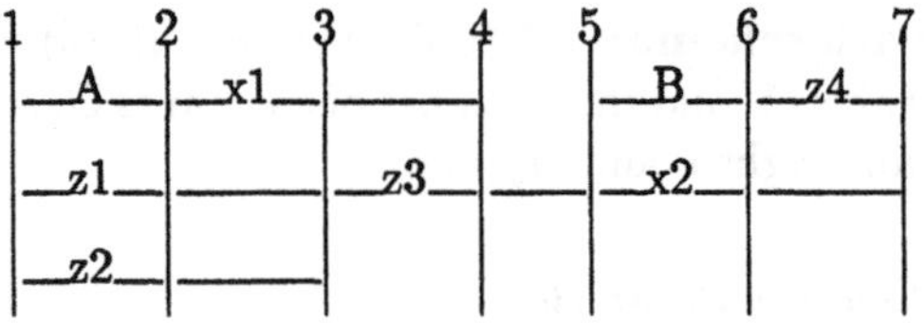

As in the pig-pug method, this position equation E will be transformed by pointing out the leftmost element (here the occurrence of A) for substituting it into the largest leftmost element (here the first largest leftmost variable z_1). The difference is that we put A at the beginning of the dual of z_1 (i.e. z_3), rather than substitute A in every variable associated with z_1. Note that, there are two ways to put A as a prefix of z_3. Either A takes all the segment between 3 and 4, or a part of this segment.

In order to avoid any loss of information during this move, a link is created between old and new positions of A in the form of a list called connection. This transformation gives rise to the two new position equations E' and E'' represented below and corresponding to the two possible positions of A relatively to the boundary 4. The link or connection (2 z_1 4) means that the prefix of z_1 ending at boundary 2 is equal to the prefix of its dual (i.e. z_3) ending at boundary 4.

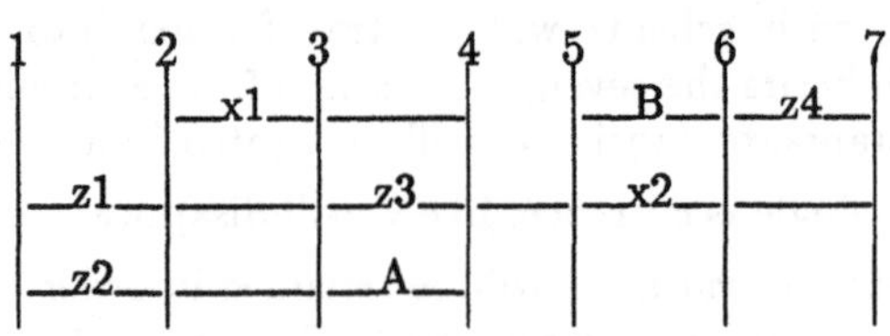

(2 z1 4)

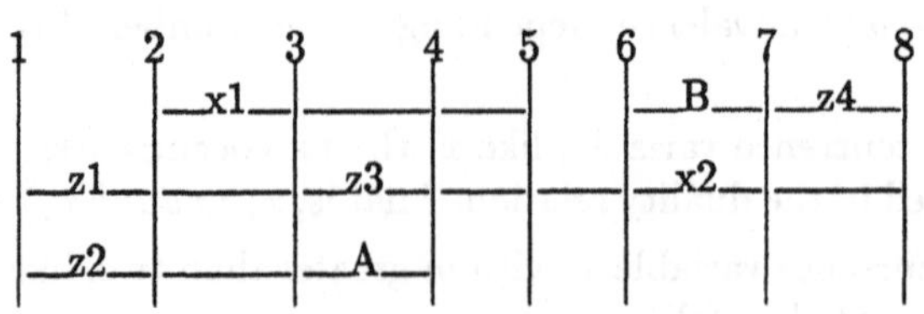

(2 z1 4)

Intuitively, if a position equation has a **solution**, then one can replace its variables by some words such that the positional constraints and the duality relation are satisfied. There is an intimate relationship between the existence of a solution of a position equation and the existence of solutions of the position equations resulting from its transformation. In the previous equation E, A is a prefix of z_1 and $z_1 = dual(z_1) = z_3$. So, A is also a prefix of z_3, and one of the two position equations E' and E'' has a solution. Conversely, if E' or E'' has a solution, so has E.

The moves used in the transformations of position equations will ensure that the number of variable occurrences remains bounded. But the growth of equation length appearing in the pig-pug method will now reappear in the growth of connection length. So, what did we win? The fact that one can prove that position equations with "long" connections can be eliminated, even while in the pig-pug method we have no direct proof that "long" equations can be deleted.

3.3. Position equation

Let us now give the formal definitions.

The notion of position equation given here is a constraint version of the notion of generalized equation introduced by Makanin [Mak1].

Formally, a position equation E is given by the following data:

Definition 3.3.1 :

a- An alphabet of **constants** $C = \{c(1)\dots c(r)\}, (r \geq 0)$.

b- A list of **variables** $X = x(1)\dots x(2n), (n \geq 0)$.

c- An involution **dual** : $X \longrightarrow X$ exchanging each variable base with its dual; i.e. $x = dual(dual(x))$, for all $x \in X$.

d- A list called **Bases** consisting of $2n+m$ elements, $(m \geq r)$. The $2n$ first elements are the variables of X. The other bases are the occurrences of the constants of C (each constant having at least one occurrence in E).

e- A nonempty strictly ascending integer list $\{1, \dots, \$\}$ called **Boundaries**.

f- Two mappings
left_boundary, right_boundary : $Bases \longrightarrow Boundaries$.
satisfying the following two conditions:

$left\_boundary(x) < right\_boundary(x)$, for all $x \in X$.

$right\_boundary(w) = successor(left\_boundary(w))$ for each constant base w.

The boundaries of E are divided into two parts: the **essential** and the **inessential** boundaries. The set of essential boundaries contains all the left and right boundaries of the bases of E.

g- A finite (possibly empty) set of **connections**. Each connection is a list of the form $(p, y(1), .., y(k), q)$ where p and q are two boundaries and $y(1)\dots y(k) (k \geq 1)$ are variable bases.

Each connection satisfies some conditions described in [Pecl].

The unique difference between this notion and that of generalized equation of Makanin is that the boundaries are *totally* ordered and that the right boundary of a constant is the successor of its left boundary.

The bases with left boundary equal to 1 are called the **leading bases**. The first variable of X with left boundary equal to 1 and with greatest right boundary is called the **carrier**.

4. Transformation of position equations

As seen in the example of 3.2, every position equation will be submitted to a transformation, which in the general case, consists in selecting the longest variable x with left boundary equal to 1 (i.e. the carrier), and in transferring all the other bases with left boundary equal to 1 under $dual(x)$.

More precisely, five different cases will occur corresponding to the five types of position equations listed below: (we assume that the boundary list $\{1,\ldots,\$\}$ is the interval [1 \$])

Type 1 : The position equation E has no carrier.

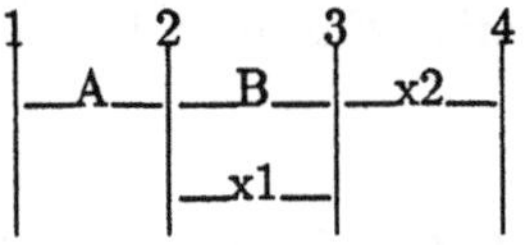

Type 2 : E has a carrier v, there is no other leading base, and the right boundary of v is equal to 2.

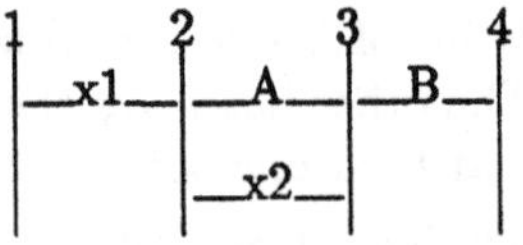

Type 3 : E has a carrier v, there are no other leading base, the right boundary of v is greater than 2 and equal to the second minimal essential boundary.

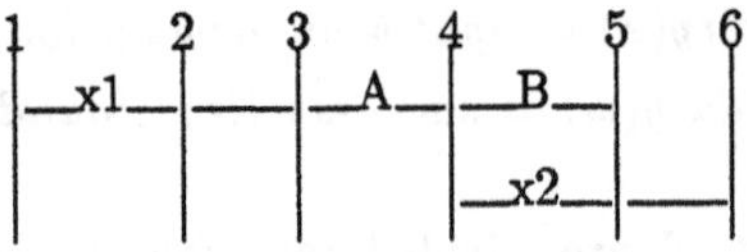

(2 x1 5)

Type 4 : E has a carrier v, there is no other leading base, the right boundary of v is greater than 2 and greater than the second minimal essential boundary.

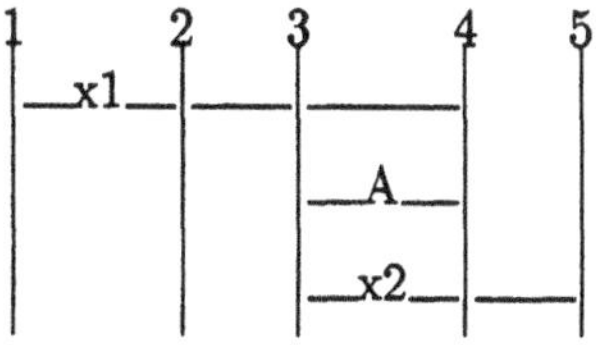

(2 x1 4)

<u>*Type* 5</u> : E has a carrier and other leading bases.

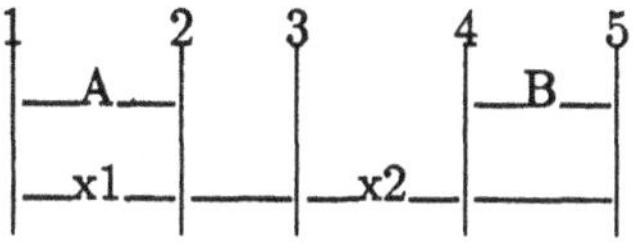

The transformation of a position equation E leads to a set of transformed position equations denoted by $T(E)$. Let us now describe more precisely what is this transformation in one of these cases.

The transformation of a position equation E of Type 5 consists in transferring every leading base w under $dual(v)$, where v is the carrier. Each possibility gives a new position equation. If w is a variable base the list of the connections of a transformed position equation $E' \in T(E)$ is obtained in the following way:

a- Each occurrence of w in each connection of E is replaced by v, w and each occurrence of $dual(w)$ is replaced by $dual(w), dual(v)$:

$(\ldots, w, \ldots, dual(w), \ldots) \longrightarrow (\ldots, v, w, \ldots, dual(w), dual(v), \ldots)$

b- Let p be the right boundary of w in E, and l' its right boundary in E'. The new connection (p, w, l') is added to E'. In addition, each connection with concluding boundary equal to p is transformed into:

$(....p) \longrightarrow (...., v, l')$.

Example :

The set $T(E)$ computed from the position equation of Type 5 given in this section consists in the following two position equations:

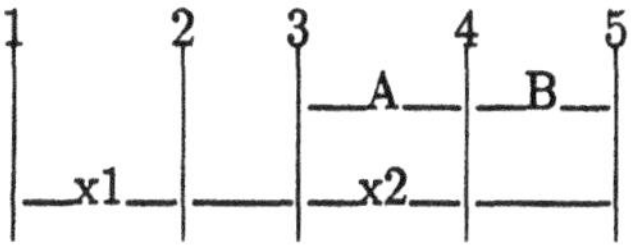

(2 x1 4)

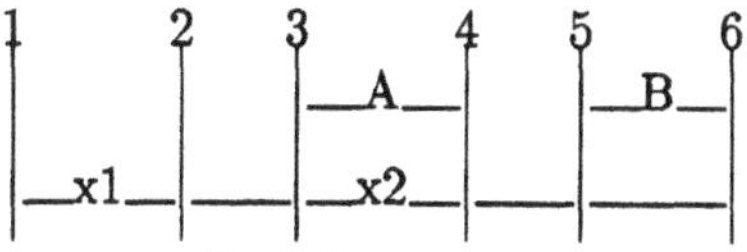

(2 x1 4)

5. Makanin's algorithm

Let e be an equation in L^*. We start by a general description of the algorithm:

If the length equation associated with e has no non-negative integer solution then, by proposition 2.1, it is clear that e has no solution. Otherwise if the equation e is simple then, by proposition 2.2, e has a solution. Otherwise the algorithm develops a tree level by level. The tree is denoted by $\mathcal{A}$ and its levels by $L_i (i \geq 0)$.

The first level L_0 contains the position equations computed from the schemes applicable to the projections of e. It is important to observe that it is the only step where new variable bases will be generated. Their number will then remain bounded.

The step from level L_i to level L_{i+1} is based on the two fundamental operations of normalization and transformation.

Each position equation E of L_i is transformed as seen in 4 into a set denoted by $T(E)$ of position equations from which the non admissible (cf. [Abd1]) ones are deleted.

If any position equation E' of this set is not normalized (cf. [Pec1]) it is replaced by a set $N(E')$ of normalized position equations computed from E'. The result of these transformation and normalization process to every position equation of L_i leads to a set $N(T(L_i))$ of admissible and normalized position equations.

The next level L_{i+1} will be deduced from $N(T(L_i))$ by deleting a certain number of position equations.

We first identify position equations which differ only by a renaming of bases or of boundaries, and we delete every position equation already occurring at a previous level.

We then also delete all position equations whose maximum length of connections is greater than a number $K(d)$ (cf. [Mak1]), depending only on the length $d = |e|$ of the original equation e. This is the central point which ensures the finiteness of the developed tree $\mathcal{A}$ and makes the algorithm to halt.

The development of levels is repeated until we obtain an empty level, in which case the equation e has no solution, or a level containing a simple position equation, in which case e has a solution.

This leads to the following algorithm.

Makanin's algorithm:

Input : an equation e of L^*

Output : YES if e admits a solution, NO otherwise.

1- If the length equation associated with e has no non-negative integer solution then
END: NO.

2- If the equation is simple then
END: YES.

3- $i \leftarrow 0$.

4-

$$L_i \leftarrow \bigcup_{j=1}^{2^{Card(V)}} e(Pj)$$

where P_j is the j-th projection of e, and $e(Pj)$ is the set of all the admissible and normalized position equations, computed from the schemes applicable to Pj.

5- Loop:

5-a If $L_i = ()$ then
END: NO.

5-b If L_i contains a simple equation then
END: YES.

5-c $i \leftarrow i + 1$.

5-d $L_i \leftarrow$ the set of all admissible and normalized position equations resulting from the transformation and the normalization of the elements of L_{i-1}, the elimination of all the already developed position equations and of the position equations containing any connection whose length is greater than $K(|e|)$.

Remark :

This version of the algorithm differs from the original one of [Mak1] by a strengthening of normalization conditions introduced in [Pec1] and the elimination of the already developed position equations introduced in [Abd1].

Conclusion

In this paper, we gave a general description of Makanin's algorithm which allows to decide whether a word equation has a solution or not.

When a word equation has a solution, one can use the pig-pug method to compute the set of all the minimal solutions of e. This set is presented as a language accepted by a deterministic automaton (not necessarily finite).

New concepts, such as equations with schemes and position equations are introduced in our description of Makanin's algorithm. A new presentation permitting to all its steps to be illustrated graphically, is described in this paper. The fundamental point is that our version has been effectively implemented.

Some improvements concerning the reduction of the number of types of position equations, and the the elimination of the equivalence test between the levels of $\mathcal{A}$, (this enables us to reduse the complexity of the algorithm by one exponential), are given here. Other technical results concerning our work in this area can be found in [Abd1], [Abd2], [Pec1].

References:

[Abd1] H. Abdulrab : Résolution d'équations sur les mots: étude et implémentation LISP de l'algorithme de Makanin. (Thèse), University of Rouen (1987). And Rapport LITP 87-25, University of Paris-7 (1987).

[Abd2] H. Abdulrab : Solving word equations. to appear in RAIRO d'Informatique théorique.

[Alb Law] M.H. Albert, J. Lawrence : A proof of Ehrenfeucht's conjecture. Theoretical Computer Science 41 p. 121-123, (1985).

[Fag Hue] F. Fages, G. Huet : Complete sets of unifiers and matchers in equational theories. Theoretical Computer Science, 43, p. 189-200, (1986).

[Hme] Yu. I. Hmelevskii : Equations in free Semigroups. Trudy Mat. In st. Steklov, 107, (1971).

[Kir] C. Kirchner : From unification in combination of equational theories to a new AC-Unification algorithm. Proc. of the CREAS, Austin, Texas. (1987).

[Len] A. Lentin : Équations dans le monoïde libre, Gauthier-Villars, Paris, (1972).

[Len Sch] A. Lentin, M.P. Schutzenberger : A combinatorial problem in the theory of free monoids. Proc. of the University of north-Carolina, p. 128-144, (1967).

[Mak1] G.S. Makanin : The problem of solvability of equations in a free semigroup. Mat. Sb. 103(145) (1977) p. 147-236 English transl. in Math. USSR Sb. 32, (1977).

[Mak2] G.S. Makanin : Equations in a free group. Math. USSR Izvestiya Vol. 21, No 3, (1983).

[Pec1] J.P. Pécuchet : Équations avec constantes et algorithme de Makanin. (Thèse), University of Rouen, (1981).

[Plo] G. Plotkin : Building-in equational theories. Machine Intelligence 7, p. 73-90, (1972) .

Complexity classes with complete problems between P and NP-C

Carme Àlvarez, Josep Díaz* and Jacobo Torán
Departament de Llenguatges i Sistemes Informàtics
Universitat Politècnica de Catalunya
Pau Gargallo 5, 08028 Barcelona

Abstract

We study certain language classes located between P and NP that are defined by polynomial time machines with bounded amount of nondeterminism. We observe that these classes have complete problems, and find characterizations of the classes using robust machines with bounded access to the oracle, and in terms of nondeterministic complexity classes with polylog running time. We also study the relationship of these classes to P and NP.

1. Introduction and basic definitions.

If $P \neq NP$, are there classes with "natural" complete problems between P and the class of NP-complete problems (NP-C)? It is well known [La 75], that if $P \neq NP$ there are sets of intermediate complexity, lying between P and NP-complete. (See also chapter 7 of [Ba,Di,Ga 88]). As a matter of fact, several classes of sets have been proposed, which include the class P and do not seem to be complete for NP; among these classes we could mention the class UP of NP languages that can be accepted by nondeterministic polynomial-time machines with unique accepting paths, [Va 76], the class FewP of languages recognized by nondeterministic Turing machines with a bounded number of accepting paths, [Al 85],[Ca,He 89], [Kö,Sc,To,To 89], the class R of languages accepted by polynomially clocked probabilistic machines which have zero error probability for inputs not in the language, and error probability bounded by $\varepsilon < \frac{1}{2}$ for strings in the language [Gi 77], and the class ZPP of languages which are recognized in polynomial time by Las Vegas type algorithms [Gi 77], (see also chapter 6 of [Ba,Di,Ga 88]). A common characteristic of these classes is the fact that the existence of complete problems for any of them is not known [Ha,He 86].

In this work, we study a hierarchy of classes that have "natural" complete problems, and appear to be strictly between P and NP-complete. These classes arise considering

---

* The research of this author was supported by CIRIT grant EE87/2

problems that can be solved in polynomial time using a small amount of nondeterminism:

Definition. For any $i \geq 1$, let

$$\beta_i = \{L \subseteq \Sigma^* | \exists A \in P, \exists c \in \mathbb{N}, \forall x \in \Sigma^* x \in L \Leftrightarrow \exists y, |y| \leq c\lceil \log^i(|x|)\rceil \text{ and } \langle x, y\rangle \in A\}$$

Recall, that from the classical characterization of NP in terms of quantifiers we have:

$$NP = \{L \subseteq \Sigma^* | \exists A \in P, \exists k, \ \forall x \in \Sigma^* x \in L \Leftrightarrow \exists y, |y| \leq |x|^k \text{ and } \langle x, y\rangle \in A\}$$

The class β_i is therefore the class of problems which can be recognized in polynomial time by a nondeterministic machine which for any input w, uses at most an "$O(\log^i(|w|))$ amount of nondeterminism". Notice that from the definition it follows:

$$P = \beta_1 \subseteq \beta_2 \subseteq \ldots \subseteq \beta_i \subseteq \ldots NP$$

We define the Bounded-Nondeterminism hierarchy as $BND = \bigcup_{i \geq 1} \beta_i$.

This hierarchy is not new, it appeared in [Ki,Fi 84] illustrating the issue of relativizations separating P from NP. In the article by Kintala and Fisher, some oracles separating every class in the BNL hierarchy are found. We shall discuss this issue in the next sections. Also the hierarchy is briefly mentioned in [Ba 84], [Sc,Bo 84], and [Xu,Do,Bo 83]. We should remark that none of the mentioned papers goes into a deep study of the hierarchy itself or the relations between this class and some other structural classes placed in the "no men's land" between P and NP-complete.

The BND hierarchy seems to be strict although the result is hard to prove since it implies P≠NP.

In this article we show that all the classes in BND have complete problems, which are obtained as a simple modification of certain complete problems for P. We also obtain (section 3) a characterization of the classes in BND in terms of one way robust Turing machines, showing that the small amount of nondeterminism defining the classes can be simulated by bounding the number of questions to the oracle of a deterministic robust machine. We observe an unusual phenomenon related with the β-classes and the robust machines, presenting a relativized world in which β_k helps (in the sense of [Ko 87]) a class strictly bigger than β_k.

Although the complete problems for β_k are simple variations of complete problems for P, the class seems to be much closer to NP than to P and many of the results obtained for the class NP can be translated to the classes β_k.

2. Complete problems in β_k.

Let us start by showing the existence of complete problems for β_k. In order to enumerate the machines recognizing languages in β_k, we define a machine model for this class.

Definition: For every k, a β_k-machine is a nondeterministic polynomial time machine equiped with a clock, such that for a certain constant $c > 0$ the machine works as follows:

> With input x, the machine first writes nondeterministically $c\lceil \log^k(|x|)\rceil$ symbols on one of the working tapes. After this, the machine works deterministically, and if it has to take a nondeterministic decision, the machine follows the path given by the written nondeterministic string.

It should be clear that these machines recognize the languages in β_k, and that for every k, there exists a recursive enumeration of the β_k machines. We can suppose that machine M_i with an input of length n can make $i \cdot \lceil \log^k(n)\rceil$ nondeterministic choices. From this enumeration, we define in a canonical way the following β_k-complete language

$$L_k = \{\langle i, x, 0^n\rangle | M_i \text{ accepts } x \text{ in } n \text{ or less steps }\},$$

where, as we have said, M_i is a nondeterministic machine using $i \cdot \lceil \log^k(|x|)\rceil$ nondeterministic choices.

This complete language gives us a base from which to define more "natural" complete problems for these classes. The complete problems are found by introducing a certain amount of "controlled nondeterminism" into P-complete problems. For example, consider the following variations of the circuit value problem, [La 75], and of the generator problem, [Jo,La 76], which are β_k-complete:

β_k-CVP: **Input:** A pair $\langle x, y\rangle$ such that $x \in \{0,1\}^*$ and y encodes a boolean circuit with $|x| + \lceil \log^k(|x|)\rceil$ input gates.

Question: Is there a string $z \in \{0,1\}^*$ of length $\lceil \log^k(|x|)\rceil$ such that the circuit encoded by y with input xz outputs a 1?

β_k-GEN: **Input:** A set X of elements; a binary operation $*$ defined on X and explicitly given by a table and defined in such a way that in $\log^k(|X|)$ cases the operation $*$ can have two possible values; a subset $S \subseteq X$, and a distinguished element x of X.

Question: Is there a way to define $*$ unambiguously so that x is contained in the smallest subset of X which contains S and is closed under the given operation $*$?

Theorem 2: For any $k \geq 1$, β_k-CVP and β_k-GEN are β_k-complete.

It is clear that the problems are in β_k. The rest of the proof of the theorem is basically the same as the hardness proof (for P) for the problems CVP and GEN presented in [La 75] and [Jo,La 76]. We refer to the mentioned articles for the details.

Observe that the above two problems are self-reducible. In following sections, we will make use of this fact. Adding controlled amounts of nondeterminism to other P-complete problems, it is not hard to obtain other problems complete for the β-classes; on the other hand, it is interesting to observe that by restricting nondeterminism in NP-complete problems (for example SAT with a polylogarithmic number of variables) we obtain problems that do not seem complete for the β-classes.

3. Robust machines and the β-classes.

In this section we give a characterization of the β-classes using robust machines, i.e. oracle machines that recognize the same language independent of the oracle they use. This kind of machines were first introduced in [Sc 85], and also studied in [Ko 87], and [Ha,He 87]. For a survey on the area we refer to [Sc 88].

Definition: For any language L, $L \in P_{1-\text{help}}$ if and only if there is a deterministic polynomial time bounded machine M and an oracle A such that:
i/ For every $x \in L$, $x \in L(M, A)$ and
ii/ For every oracle B and every $x \notin L$, $x \notin L(M, B)$.

Definition: Let K be a language class. For any language L, $L \in P^{K}_{1-\text{help}}$ if and only if there is a deterministic polynomial time bounded machine M and an oracle $A \in K$ such that:
i/ For every $x \in L$, $x \in L(M, A)$ and
ii/ For every oracle B and every $x \notin L$, $x \notin L(M, B)$.

By bounding the access to an NP oracle, we can characterize the β-classes in terms of robust machines. Between square brackets, we indicate the amount of oracle queries allowed.

Theorem 3: For every $k \geq 2$,
i/ $P_{1-\text{help}}^{\text{NP}\ [\log^k]} = P_{1-\text{help}}^{\beta_k\ [\log^k]} = \beta_k$
ii/ $\beta_k \subseteq P^{\beta_k}_{1-\text{help}}$

Proof: (Sketch) i/ The inclusions from left to right follow from the fact that an accepting computation path from a deterministic machine that makes $\log^k$ questions to the oracle can be found guessing $\log^k$ bits; since the machine is robust, if a certain input is not in the language recognized by the machine, none of the possible paths that can be guessed for this input will accept.

The first inclussion from right to left is straightforward. For the other inclusion, given a language $L \in \beta_k$ which for a certain polynomial time predicate P satisfies:

$$L = \{x \mid \exists y, |y| \leq \lceil \log^k(|x|) \rceil\ \langle x, y \rangle \in P\} \quad \text{let}$$

$$Pref(L) = \{\langle x, y \rangle \ \mid \ \exists y'\ |y| + |y'| = \lceil \log^k(|x|) \rceil \ \text{ and } \ \langle x, yy' \rangle \in P\}.$$

Clearly Pref(L) is also in β_k. Any set L in β_k can be semi-decided following the usual methods by a deterministic robust machine doing binary search in oracle Pref(L). This proves also result ii/. □

Result ii/ only seems to hold in one direction; consider a language L in β_k and the language $L' = \{\langle a_1, a_2, \ldots, a_n \rangle \ \mid \ a_i \in L \text{ for } i = 1 \ldots n\}$. Clearly L' is in $P^{\beta_k}_{1-\text{help}}$ since in order to check that a given string w is in L', the robust machine only needs to ask the oracle for the witnesses of all the substrings in w, and then check that they are correct.

This can be done in polynomial time with a "good" oracle in β_k. On the other hand, L' does not seem to be in β_k since in order to check the validity of an input, $n \log^k$ nondeterministic bits are needed.

This is a strange phenomenon since all the classes that have appeared in the literature helping some other class in the sense of one-way robust machines, can only help themselves or some of their subclasses, ($P^{NP}_{1-\text{help}} = NP$, $P^{P}_{1-\text{help}} = P$, $P^{BPP}_{1-\text{help}} \subseteq R$, [Sc 85], [Ko 87]), and β_k seems to help a class strictly greater than itself. It would be very hard to prove that β_k is strictly contained in $P^{\beta_k}_{1-\text{help}}$ since it would inmediately imply $P \neq NP$, but we still can prove the result in a relativized world. In the next theorem we find a relativization where β_k is strictly contained in $P^{\beta_k}_{1-\text{help}}$. First we define relativized robust machines.

Definition: For any language L and any oracle B, $L \in (P_{1-\text{help}})^B$ if and only if there is a deterministic polynomial time bounded machine M and an oracle A such that:
i/ For every $x \in L$, $x \in L(M, A \oplus B)$ and
ii/ For every oracle D and every $x \notin L$, $x \notin L(M, D \oplus B)$.

Intuitively a relativized robust machine is just a deterministic machine with an oracle which is the join of two sets. Questions to one of the sets in the join are always answered correctly, but the questions to the other set have to be verified by the machine. Following the techniques in [Sc 85], [Ko 87] it is not hard to see that for any oracle B, $(P_{1-\text{help}})^B = NP^B$.

Definition: Let K be a language class that can be relativized. For any language L and any oracle B, $L \in (P^{K}_{1-\text{help}})^B$ if and only if there is a deterministic polynomial time bounded machine M and an oracle $A \in K^B$ such that:
i/ For every $x \in L$, $x \in L(M, A \oplus B)$ and
ii/ For every oracle D and every $x \notin L$, $x \notin L(M, D \oplus B)$.

Observe that in the above definition we also relativize the helping class. The motivation for this is that if we give more power to the robust machine by letting it access "two" oracles, we also have to give more power to the helping class or it will not be able to help the machine. It is also easy to prove that for any oracle B, $(P^{NP}_{1-\text{help}})^B = NP^B$.

Theorem 4: For any $k \geq 1$, there is an oracle B such that $\beta_k^B \neq (P^{\beta_k}_{1-\text{help}})^B$.

Proof: For any set B, consider the test language

$$L_B = \{0^n \mid \text{there are at least } n \text{ elements of length } \lceil \log^k(n) \rceil \text{ in } B\}.$$

i/ For every set B, $L_B \in (P^{\beta_k}_{1-\text{help}})^B$. Consider the set

$$\begin{aligned} B' = &\{0^n\#u \mid \exists v\ |u| + |v| = \lceil \log^k(n) \rceil \text{ and } uv \in B\} \cup \\ &\{0^n\#x\#u \mid |x| = \lceil \log^k(n) \rceil \text{ and } \exists v\ |u| + |v| = \lceil \log^k(n) \rceil,\ x \in B,\ uv \in B,\ x < uv\} \end{aligned}$$

Clearly $B' \in \beta_k^B$ since in order to check that a given input is in the set, we only need to guess $\log^k$ bits, at the most, and then make at the most two questions to B.

Informally we describe now a deterministic one-way robust machine that recognizes L_B using oracle B': With input 0^n, by binary search in the oracle find the smallest string x_1 of length $\lceil \log^k(n) \rceil$ in B (smallest x_1 of length $\lceil \log^k(n) \rceil$ such that $0^n\#x_1 \in B'$). Then use x_1 to get the second smallest string of size $\lceil \log^k(n) \rceil$ in B' (smallest x_2 such that $0^n\#x_1\#x_2 \in B'$), next, use x_2 as the key to get the following one in lexicographical order, x_3, (smallest x_3 such that $0^n\#x_2\#x_3 \in B'$) and so on, until we have n strings. Finally, check that all of them are in B by directly asking oracle B.

ii/ There is a set B such that $L_B \notin \beta_k^B$.

Let $M_1, M_2 \ldots$ be an enumeration of β_k-nondeterministic oracle machines, with the computation time of M_s bounded by polynomial p_s and its nondeterminism bounded by $s\lceil \log^k \rceil$. We will construct set B in stages:

stage 0: $B_0 := \emptyset$; $n_0 := 0$

stage s: Let n_s be the smallest m such that m is a power of 2, $s \log^k(m) > p_i(n_i)$ for $i < s$, and

$$\frac{\binom{2^{\log^k(m)}}{m}}{\binom{p_s(m)}{m}} > 2^{s\log^k(m)}.$$

If $0^{n_s} \in L(M_s, B_{s-1})$ then $B_s := B_{s-1}$
else

1. If there are n_s strings $w_1, \ldots w_{n_s}$ of length $\log^k(n_s)$ such that $0^{n_s} \notin L(M_s, B_{s-1} \cup \{w_1, \ldots w_{n_s}\})$ then let $B_s := B_{s-1} \cup \{w_1, \ldots w_{n_s}\}$.
2. If the condition in 1 is not true then there is an integer r, $(r < n_s)$, and r strings of length $\log^k(n_s)$, $w_1 \ldots w_r$ such that $0^{n_s} \in L(M_s, B_{s-1} \cup \{w_1, \ldots w_r\})$. Let $B_s = B_{s-1} \cup \{w_1, \ldots, w_r\}$.

end (of stage s).

Let $B = \bigcup_s B_s$. Following similar arguments as in [Ba,Gi,So 75] it is not hard to check that $L_B \notin \beta_k^B$. It is only left to show that the assertion in 2 is true.

Claim: For every n_s, if for every sequence of n_s strings $\{w_1, \ldots w_{n_s}\}$ of length $\log^k(n_s)$, $0^{n_s} \in L(M_s, B_{s-1} \cup \{w_1, \ldots w_{n_s}\})$ then there is an integer r, $r < n_s$, and r strings $\{w_1, \ldots w_r\}$ of length $\log^k(n_s)$, such that $0^{n_s} \in L(M_s, B_{s-1} \cup \{w_1, \ldots w_{n_r}\})$.

Proof of the claim: Given n_s, we will call a string of $s\log^k(n_s)$ bits, forcing (determinizing) the computation of M_s, a computation path of M_s. Observe that there are at the most $2^{s\log^k(n_s)}$ computation paths, and that this is independent of the oracle we use.

Suppose that the claim is not true, then for every tuple of n_s strings $\{w_1, \ldots w_{n_s}\}$ of length $\log^k(n_s)$, in every accepting path of machine M_s with input 0^{n_s} and oracle

$B_{s-1} \cup \{w_1 \dots w_{n_s}\}$ exactly n_s of the words of length $\log^k(n_s)$ queried to the oracle are answered "yes" (there are only n_s words of this length in the oracle). But then for every tuple $\{w_1, \dots w_{n_s}\}$ there is an accepting path that contains exactly the strings $\{w_1 \dots w_{n_s}\}$ answered "yes".

We will say that a tuple of n_s strings $\{w_1 \dots w_{n_s}\}$ "is included" in a computation path if every element from the tuple $\{w_1, \dots w_{n_s}\}$ is queried by M_s with input 0^{n_s} and oracle $B_{s-1} \cup \{w_1, \dots w_{n_s}\}$. Observe that although in a computation path the range of oracle queries that can be made can be exponential (the machines are non-adaptive), the number of tuples of n_s strings "included" in a computation path is at the most $\binom{p_s(n_s)}{n_s}$. The explanation for this is that a tuple included in the computation path is forced by choosing which n_s queries, from the $p_s(n_s)$ oracle queries that can be made by M_s, are answered "yes".

But then for every tuple $\{w_1, \dots w_{n_s}\}$ there is an accepting path that contains exactly the mentioned strings answered "yes".

From the above two facts follows that the number of computation paths of M_s that accept 0^n for some oracle $B_{s-1} \cup \{w_1, \dots w_{n_s}\}$ is greater or equal than the number of different tuples $\{w_1, \dots w_{n_s}\}$, divided by the number of tuples that can be "included" in a computation path. But this number is

$$\frac{\binom{2^{\log^k(n_s)}}{n_s}}{\binom{p_s(n_s)}{n_s}},$$

which is strictly greater than $2^{s \log^k(n_s)}$, the number of possible computation paths of M_s with input 0^{n_s}. This is a contradiction, and therefore assertion 2 is true. □

4. Relationship between the nondeterministic polylog classes and the BND hierarchy.

Let us introduce the class NPL of languages which can be recognized in polylog time by nondeterministic machines. This class has been recently considered in [Im,Na 88]. We shall use NPL to characterize the β-classes. To define non-trivially the NPL class, the machine model used must have indirect access to the input bits.

Definition: An NPL-machine is a canonical nondeterministic Turing machine with polylog running time (i.e., for a certain constant $k > 0$, the machine makes at the most $\lceil \log^k(n) \rceil$ steps, being n the input size), which has a special query tape where positions of the input can be written, and a special query state. When the machine enters the query state, and a position i is written in the query tape, the machine writes down in the tape the i-th symbol of the input string.

The class NPL is the class of languages recognized by NPL-machines.

It follows from the definition that a NPL-machine can only access a polylog number of input bits.

We can consider the canonical extension of NPL-machines to oracle NPL-machines. The problem that arises when one considers the normal definition of an oracle machine is that in the case of NPL machines, there is too little time to write things on the oracle tape. This fact creates strange restrictions, for example, it is not true that an oracle NPL-machine with oracle A can recognize the set A. In order to avoid these problems, we will assume that initially the oracle tape contains input x. The oracle queries have then the pattern $\langle x, y\rangle$, being x the input and y a string of polylog length with respect to x.

From the way the classes NPL and BND have been defined, it seems natural to think of an relation between them. It is known that NPL is strictly included in BND since using the restriction of the NPL-machines to read the input, it is not hard to define a language in P which is not in NPL.

The following result gives us a characterization of BND in terms of NPL-machines, showing that both classes are the same if we let them query an oracle in P (observe that BND does not get any extra help from a P oracle).

Theorem 5: $BND = NPL^P$.

Proof: From right to left, let $L \in NPL^P$, then there exists a constant $k > 0$ and a set $B \in DPL^P$ such that

$$L = \{x \mid \exists y, |y| \leq \log^k(|x|), \text{ and } \langle x, y\rangle \in B\}$$

where DPL^P is the class of languages recognized by deterministic polylog machines with oracle in P.
Since $B \in DPL^P$ and $DPL^P \subseteq P^P$ we get $B \in P$, and by the way β_k has been defined we get that $L \in \beta_k$.

To prove the converse, let us consider the set $L \in BND$. There exists a constant k and a set A in P such that

$$L = \{x \mid \exists y, |y| \leq \lceil \log^k(|x|)\rceil, \text{ and } \langle x, y\rangle \in A\}$$

Let us consider the oracle NPL-machine M which with oracle A recognizes L. M on input x guesses a string y of length less than or equal to $\log^k(|x|)$. Then M accepts x iff $\langle x, y\rangle \in A$. But as $A \in P$ and M is a NPL-machine then $L \in NPL^P$. □

5. Similarities between the β-classes and NP.

Although we have seen that β_k-complete problems can be obtained with small modifications from P-complete problems, we will see in this section several results which indicate that the classes β_k share many of the properties of the class NP. We will begin by re-stating in terms of β_k an old result of Adleman [Ad 79] which says that the difference between P and NP is the ability of the NP-machines to manufacture randomness. In the next theorem, we prove that this difference also exists between P and β_k, which in a certain sense means that the ability to manufacture randomness is independent of the amount of nondeterminism used by the machine. Our proof follows directly a similar one in [He,We 89]. We introduce first two definitions:

Definition: Given a β_k machine M and an input x, a *certificate of M on x* is an accepting path of $M(x)$.

Definition: Let M_U be a universal Turing machine, $z \in \Sigma^*$, and $f, g : \mathbb{N} \longrightarrow \mathbb{N}$. Define the *Kolmogorov complexity relative to string z* as

$$K[f(n), g(n)|z] = \{x \mid \exists y,\ |y| \leq f(|x|) \wedge M_u(y\#z) \text{ prints } x \text{ within } g(x) \text{ steps}\}.$$

We say that a string x is Kolgomorov simple relative to z if there is a constant c such that $x \in K[c \log n, n^c|z]$.

Theorem 6: $P = \beta_k$ iff β_k has Kolmogorov simple certificates relative to the input.

Proof. If $P = \beta_k$, by using the self-reducibility of β_k-GEN we can find certificates for any instance in the language. In the other direction, there are only polynomially many simple strings relative to the input and we can produce all of them in polynomial time, and check if at least one of them is a true certificate. □

Besides this similarity between the sets accepted by machines using a polynomial amount of nondeterminism and a polylog amount of nondeterminism, there are other common characteristics; one of them is that β_k cannot have complete tally sets unless $P = \beta_k$. This result is analogous to the well known result of Berman about the non existence of NP-complete tally sets if $P \neq NP$. The result is given in the following theorem, whose proof is almost the same as to the one given by Berman [Be 78], but using the set β_k-GEN, shown previously to be complete in β_k. The details of the proof are left to the reader.

Theorem 7: If there exists a co-sparse and β_k-complete set, then $P = \beta_k$.

As a corollary to this last theorem we can state,

Corollary: If there exists a tally β_k-complete set, then $P = \beta_k$.

It is an open question whether the result holds also for sparse sets. Mahaney's proof for sets in NP [Ma 80], cannot be carried over to the β-classes since it needs a massive use of nondeterminism.

It is also interesting to compare the ability of β_k-machines to compress queries to an oracle in NP, with the ability of NP-machines to do the same thing. It is well known that an NP-machine with access to an orace in NP can be simulated by another machine that queries the oracle just once ($\mathrm{NP}^{\mathrm{NP}} = \mathrm{NP}^{\mathrm{NP}[1]}$). We present results in this line, for the class β_k. Recall that between the square brackets we indicate the number of oracle queries allowed.

Theorem 8: For every k, and $j \geq k$

i/ $\beta_k^{\mathrm{NP}[\log^k(n)]} = \beta_k^{\mathrm{NP}[2]}$

ii/ $\beta_k^{\mathrm{NP}[\log^{k+j}(n)]} = \beta_{k+j}^{\mathrm{NP}[2]}$

Proof: We sketch i/. ii/ is analogous. The direction from right to left is obvious; for the other one, a β_k-machine M making $\log^k$ queries to an oracle in NP can be simulated by another machine M' making just 2 queries to another NP oracle in the following way: M' guesses a computation path of M, and a list of oracle answers from the oracle ($O(\log^k(n))$ bits). With this information M' computes the list of oracle queries corresponding to the guessed answers and the guessed computation path, and needs just two queries to the new oracle to check that all the queries corresponding to a (guessed) positive answer are in the old oracle, and all the queries corresponding to a negative answer are not in the old oracle. Since NP is closed under disjunctive and conjunctive truth table reducibilities, the new oracle is also in NP. □

The β-classes share other properties with NP, for instance, it is easy to show that they are closed under polynomial time many-one reducibility. Moreover, they do not seem to be closed under polynomial time Turing reducibility (it is not hard to find a relativization in which these classes are not closed under this reducibility).

However, the β-classes are closed under polynomial time disjunctive reducibilities and do not seem to be closed under conjunctive reducibility (again this conjecture can be proved in a relativized world with the technique used in the proof of theorem 4).

Final remarks.

We have presented a hierarchy of β-classes, which are located between P and NP-complete, and which have complete problems. We have characterized these classes in terms of robust machines, introducing the concept of relativized robust machine.

We also characterized the β-classes using nondeterministic polylog time complexity classes, which share some of the properties of the β-classes. The relationship between the β-classes and other complexity classes that lie between P and NP, like R or ZPP, is an interesting open question since any result in this direction will clarify the relationship between nondeterminisim and probabilism. We do not know either whether the β-classes lie within Schöning's Low or High hierarchies (see [Sc 86]), but we conjecture that the classes are not included in either of them.

References

[Ad 79] L. Adleman: Time, Space and Randomness. Tech Report MIT/LCS/TM-131, MIT 1979.

[Al 85] E. Allender: *Invertible functions.* Ph.D. dissertation, Georgia Institute of Technology, 1985.

[Ba,Gi,So 75] I. Baker, J. Gill, R. Solvay: Relativization of the P=?NP question. *SIAM J. Comput.*, 4, 1975, 431–442.

[Ba 84] J.L. Balcázar: *En torno a oraculos que ciertas maquinas consultan, y las espantables consecuencias a que ello da lugar* Ph.D. dissertation, FIB. 1984.

[Ba,Di,Ga 88] J.L. Balcázar, J.Diaz, J.Gabarró: *Structural Complexity*, (Vol.1) 1988, Springer-Verlag.

[Be 78] P. Berman: Relationships between density and deterministic complexity of NP-complete languages. *Proceedings 5th ICALP*, 1978, 63–72.

[Bo 74] R. Book: Tally languages and complexity classes. *Information and Control* , 1974, 186–193.

[Ca,He 89] J. Cai, L. Hemachandra: On the power of parity. *Symp. Theor. Aspects of Comput. Sci.* 1989, 229–240.

[Gi 77] J. Gill: Computational complexity of probabilistic Turing machines. *SIAM J. Comput.*, **6**, 1977, 675–695.

[Ha,He 86] J. Hartmanis, L. Hemachandra: Complexity classes without machines: on complete languages for UP. *13th. ICALP*, 1986, 123–135.

[Ha,He 87] J. Hartmanis, L. Hemachandra: One-way functions, robustness, and the non-isomorphism of NP-complete sets. *2nd Structure in Complexity Theory Conference*, 1987, 160–175.

[He,We 89] L. Hemachandra, G. Wechsung: Using randomness to characterize the complexity of computation. To appear in *IFIP* 89.

[Im,Na 88] R. Impagliazzo, M. Naor: Decision Trees and Downward Closures. *3rd. Structure in Complexity Theory Conference*, 1988, 29–38.

[Jo,La 76] N. Jones, W. Laaser: Complete problems for deterministic polynomial time. *Theoretical Computer Science*, **3**, 1976, 105–118.

[Ki,Fi 84] C. Kintala, P .Fisher: Refining nondeterminism in relativized complexity classes. *SIAM J. Comput.*, **13**, 1984, 329–337.

[Ko 87] K. Ko: On helping by robust oracle machines. *Theoretical Computer Science*, **52**, 1987, 15–36.

[Kö,Sc,To,To 89] J. Köbler, U. Schöning, S. Toda, J. Torán: Turing machines with few accepting computations and low sets for PP. To appear in *4th Structure in Complexity Theory Conference*, 1989.

[La 75] R. Ladner: On the structure of polynomial time reducibility. *J.ACM*, **22** 1975, 155–171.

[Ma 80] S. Mahaney: Sparse complete sets for NP: solution of a conjecture of Berman and Hartmanis. In *Proceedings IEEE Symposium on Foundations of Computer Science*, 1980, 54–60.

[Sc,Bo 84] U. Schöning, R. Book: Immunity, relativizations, and nondeterminism. *SIAM J. Comput.*, **9**, 1984, 46–53.

[Sc 85] U. Schöning: Robust algorithms: a different approach to oracles. *Theoretical Computer Science*, 40, 1985, 57–66.

[Sc 86] U. Schöning: *Complexity and Structure.* Lecture Notes in Computer Science, 1986, Springer-Verlag.

[Sc 88] U. Schöning: Robust oracle machines. *Proceedings 13th Math. Fundations of Computer Science*, 1988, 93–107.

[Va 76] L. Valiant: Relative complexity of checking and evaluating. *Information Processing Letters*, **5**, 1976, 20–23.

[Xu,Do,Bo 83] M. Xu, J. Doner, R. Book: Refining nondeterminism in relativizations of complexity classes. *JACM* **30**, 1983, 677-685.

Interpretations of synchronous flowchart schemes

Miklós Bartha
Bolyai Institute, A. József University, Szeged, Hungary 6720

1 Introduction

Synchronous flowchart schemes are algebraic models of synchronous systems, described earlier by Leiserson and Saxe [11] by means of so called finite edge-weighted directed multigraphs. In our presentation a synchronous flowchart scheme is like an ordinary one (see e.g. [7] or [2]), but its edges have a nonnegative integer, the delay number (or weight, see [11]) associated with them. The reader is referred to [3] for the exact definition of synchronous schemes, which were called (ambiguously) systolic in that paper. Since all the functional elements (cf. [11]) of a synchronous system can be constructed from simple operations, we can assume that the "boxes" of the corresponding synchronous scheme are labelled by a (singly) ranked alphabet Σ, and an interpretation of the scheme is fixed by a pointed Σ-algebra (i.e. a $\Sigma_\perp$-algebra, where $\perp$ is a distinguished constant symbol). In the computation process modelled by such a synchronous Σ-scheme the boxes represent the operations prescribed by their labels, and the delay number n associated with an edge e is implemented by n consecutive registers placed along e.

We can associate a finite state structural Mealy automaton (cf. [10]) A with each synchronous Σ-scheme F in the following way. A state of A is some assignment of values to all the registers of F. With each clock tick, A maps the current state - current input pair into a new state - current output pair. The transition function of A is specified by the "global" behavior of the interconnected boxes in F.

If the delay number of an edge in F happens to be zero, then no register stops the signals along the corresponding interconnection. To avoid feedback of rippling, we assume that in every cycle of F there exists at least one edge which has strictly positive delay number.

The reader is assumed to be familiar with [3], where an equational axiomatization was given for synchronous flowchart schemes. The semantics of synchronous schemes can be defined in feedback theories (introduced also in [3]). The variety of feedback theories is in close connection with that of iteration theories (cf. [7]), which has many relevant subclasses well-known from the literature. To enlight feedback theories from the point of view of semantics, in this paper we define a large subclass of this variety, the class of feedback algebras induced by ordinary pointed Σ-algebras. It will be shown that feedback algebras provide an adequate semantics to the step-by-step behavior of synchronous flowchart schemes.

Although this paper is logically a sequel to [5], the reader is not assumed to be familiar with that work, because its main construction (the F^{∞}-construction) is summarized in Section 3.

2 Preliminaries

Our algebras defining the syntax and semantics of synchronous schemes are sorted by $N \times N$ (N denotes the set of all nonnegative integers), and they are equipped with the operations *composition* ($\cdot$), *sum* ($+$), *feedback* ($\uparrow$ or $\Uparrow$) and constants 1, 0, x, ϵ, 0_1. This type of algebras is denoted by S, and the subtype of S not containing feedback is denoted by D. The S-operations are the simplest ones by which we can construct schemes from boxes and mappings (i.e. interconnections between the boxes). The most elementary mappings are those represented by our constants, namely $1 : 1 \to 1$, $0 : 0 \to 0$, $\epsilon : 2 \to 1$, $0_1 : 0 \to 1$, which are unique of their sorts, and the transposition $x : 2 \to 2$. The reader is referred to [2,3] for a detailed explanation of the types S and D. However, since we shall be dealing with algebraic theories (theories, for short), we shall use also *tupling* ($\langle \ldots \rangle$) as a derived operation. For the representation of theories by D-algebras, iteration theories by S-algebras and vice versa, the reader is referred again to [2,3]. The notation and terminology is adopted without any change from these works, as it was done in the other two papers [4,5] written on the subject, too. Thus, for example, by writing that f is a morphism $p \to q$ we mean that f is an element of sort (p, q) in some D-(S-) algebra. When dealing with tuples of the form $\langle f_1, \ldots, f_n \rangle$ we usually omit the restriction of each f_i being a morphism $p_i \to q$ for a fixed $q \in N$. Rather, we assume that q is known from the context, and require only that $f_i : p_i \to q_i$ with $q_i \leq q$ for each $i \in [n] (= \{1, \ldots, n\})$. In such cases one should substitute $f_i + 0_{q-q_i}$ for f_i in mind. For a morphism $f : p \to q$ and $n \in N$, $n \otimes f$ stands for $\sum_{i=1}^{n} f : n \cdot p \to n \cdot q$.

We shall often be working in such S-algebras simultaneously which have their underlying sets in common. In doubtful expressions the appropriate algebra (in which the operations are to be performed) will be indicated as a subscript of the uncertain operations. In addition, the feedback operation in (ordinary) flowchart scheme algebras and iteration theories (but only in these algebras) will always be denoted by $\Uparrow$, to underline the different behavior of feedback in them.

Given a ranked alphabet Σ, T_Σ and $T(\Sigma)$ denote the free Σ-algebra and the free algebraic theory generated by Σ, respectively. If $Y = \{y_1, \ldots, y_n, \ldots\}$ is a set of variable symbols, then $T_\Sigma^{(\infty)}(Y)$ denotes the set of (infinite) Σ-trees over Y, as usual. It is known that $T(\Sigma)^{(\infty)}(p, q)$ consists of p-tuples of trees belonging to $T_\Sigma^{(\infty)}(Y_q)$ ($Y_q = \{y_1, \ldots, y_q\}$). If A is a Σ-algebra, then Pol A denotes the clone algebra of A, i.e. the algebraic theory of all polynomials over A.

Base morphisms (algebraic constants) of sort $p \to q$ will be treated as mappings of $[p]$ into $[q]$ in our algebras. The following mappings will be distinguished throughout the paper:

n is a shorthand for $n \otimes 1$.

$\kappa(n,p) : p \cdot n \to n \cdot p$. This permutation rearranges p blocks of length n to n blocks of length p, i.e. $\kappa(n,p)$ takes $(j-1)\cdot n + i$ $(j \in [p], i \in [n])$ to $(i-1)\cdot p + j$.

$\alpha\#s$. If $\alpha : r \to r$ is a permutation and s is a sequence $(n_1, \ldots, n_r)$ of nonnegative integers with $n = \sum_{i=1}^{r} n_i$, then $\alpha\#s : n \to n$ is the "blocked performance" of α on s, i.e. $\alpha\#s$ sends $j + \sum_{i=1}^{k} n_i$, where $j \in [n_{k+1}]$ to the number $y + j$, where y is the sum of numbers n_i such that $\alpha(i) < \alpha(k+1)$ (see also [7]).

3 The F^∞-construction

The F^∞-construction [5] is a standard inverse limit construction by which every pointed algebraic theory T can be extended to a feedback theory $F^\infty T$ in a natural way. (Recall that a theory is pointed if it is equipped with a further constant $\perp : 1 \to 0$.) The construction proceeds as follows.

First we define a pointed theory $n\text{-res}^* T$ for every $n \in N$ such that $n\text{-res}^* T(p,q) = T(n \cdot p, n \cdot q)$. A morphism $F : p \to q$ in $n\text{-res}^* T$ will be represented in the form $F = \langle f_1, \ldots, f_n \rangle$, where $f_i : p \to n \cdot q$ in T for each $i \in [n]$. Composition is performed in $n\text{-res}^* T$ exactly like in T, the sum of $F = \langle f_1, \ldots, f_n \rangle : p_1 \to q_1$ and $G = \langle g_1, \ldots, g_n \rangle : p_2 \to q_2$ is

$$F + G = \langle f_1 + g_1, \ldots, f_n + g_n \rangle : p_1 + p_2 \to q_1 + q_2,$$

and for any constant symbol c, $c_{n\text{-res}^* T} = n \otimes c_T$. Lemma 2 in Section 4 shows that $n\text{-res}^* T$ is a theory, and T can be embedded into $n\text{-res}^* T$ by the injection $n\otimes : f \mapsto n \otimes f, f \in T$. In the sequel we shall be interested in the subtheory $n\text{-res}\, T \subseteq n\text{-res}^* T$ for which $\langle f_1, \ldots, f_n \rangle \in n\text{-res}\, T$ iff $f_i \in T(p, i \cdot q)$ for each $i \in [n]$.

The tuple $\langle f_1, \ldots, f_n \rangle \in n\text{-res}\, T$ expresses the step-by-step behavior of a synchronous system (with q input and p output ports) in which the functional elements are represented by morphisms in T. Indeed, $f_i \in T(p, i \cdot q)$ can be considered as the function which specifies the output appearing on the output ports in the ith step (clock cycle), depending on the inputs that have arrived to the input ports in the ith and all the previous steps.

The second step of the construction is to make $n\text{-res}\, T$ a feedback theory. Let $F = \langle f_1, \ldots, f_n \rangle : 1 + p \to 1 + q$ be a morphism in $n\text{-res}\, T$, and let Γ be the doubly ranked alphabet consisting of the symbols $\gamma_i : 1 + p \to i \cdot (1+q)$, $i \in [n]$ and $\perp : 1 \to 0$. To define $\uparrow F$, consider the Γ-flowchart scheme $\mathcal{F}$:

$$\Uparrow^{n-1} ((n-1+0_1+n\cdot p)\cdot \kappa(n,2)\#(1^n, p^n)\cdot \langle \gamma_1, \ldots, \gamma_n \rangle \cdot \kappa(2,n)\#(1,q)^n \cdot (\perp + n - 1 + n \cdot q))$$

(see also Fig. 1 in the case $n = 3$). Observe that $\mathcal{F}$ is loop-free (i.e. it is a DΓ-scheme, see [3]) so it is possible to evaluate it in T by the valuation $\gamma_i = f_i$ and $\perp = \perp_T$. Moreover, it is easy to see that the resulting morphism $\langle f'_1, \ldots f'_n \rangle : p \to q$ is in $n\text{-res}\, T$. We define $\uparrow F$ to be this morphism $\langle f'_1, \ldots, f'_n \rangle$. The meaning of the expression defining $\uparrow F$ is obvious: the first input-output port pair of the "system" F becomes a register. The morphism f'_i specifies the sequence of outputs appearing on the remaining p output ports in the ith step, assuming that in the jth step ($j \leq i$) the input arrives to the first input port of F from the first output port of F, providing its output computed in the $(j-1)$th step. For $j - 1 = 0$ this output is taken to be $\perp$.

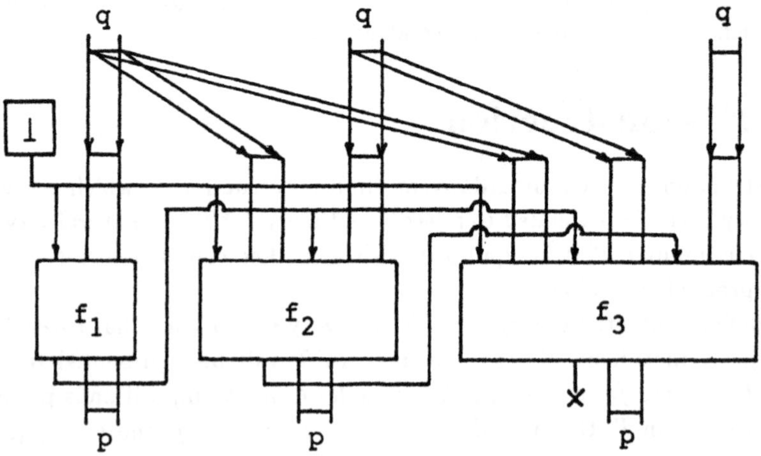

Figure 1

To characterize the algebraic properties of feedback in n-res T, construct a sequence of expressions $\mathbf{f}^0, \mathbf{f}^1, \ldots, \mathbf{f}^m, \ldots : 1 + p \to q$ over the variable $\mathbf{f} : 1 + p \to 1 + q$ in the following way:

— $\mathbf{f}^0 = (1 + p) \otimes \perp + 0_q$;

— $\mathbf{f}^{i+1} = \mathbf{f} \cdot \langle (\nabla + 0_p) \cdot \mathbf{f}^i, q \rangle$ for $i \geq 0$,

where $\perp$ and ∇ denote the terms $\uparrow \epsilon$ and $\uparrow x$, respectively.

Remark 1 It is important to note that each $\mathbf{f}^i$ becomes a theory expression if we treat $\perp$ and ∇ as constants in it. We shall need this observation in the proof of Theorem 3 below.

The proofs of the following two statements need a considerable amount of computation, therefore we omit them here.

Lemma 1 *The identity*

$$\uparrow \mathbf{f} = (0_1 + p) \cdot \mathbf{f}^n \text{ for } \mathbf{f} : 1 + p \to 1 + q$$

is valid in n-res T.

Theorem 1 n-res T *is a feedback theory for every* $n \in N$.

Finally, we define a homomorphism $\phi_n : (n+1)$-res $T \to n$-res T for every $n \in N$. ϕ_n is the trivial map which "forgets the result of the computation in the last clock cycle", i.e. for $F = \langle f_1, \ldots, f_{n+1} \rangle$, $\phi_n(F) = \langle f_1, \ldots, f_n \rangle$. Let $F^\infty T$ denote the limiting S-algebra of the inverse chain $(n\text{-res}\, T, \phi_n), n \in N$. By Theorem 1, $F^\infty T$ is a feedback theory. Moreover, it is an extension of T (due to the embeddings $n\otimes : T \to n\text{-res}\, T$), and the equation $\uparrow \epsilon = \bot_T$ holds in $F^\infty T$ (regarding T as a subtheory of $F^\infty T$). $F_1^\infty T$ will denote the smallest sub-S-algebra of $F^\infty T$ containing T. If $\psi : T \to T'$ is a homomorphism between pointed theories, then using Lemma 1, ψ can be extended to a homomorphism $\psi_n (= n\text{-res}\, \psi) : n\text{-res}\, T \to n\text{-res}\, T'$ between feedback theories. Thus, we can define $\psi_\infty (= F^\infty \psi) : F^\infty T \to F^\infty T'$ to be the unique homomorphism for which the diagram below commutes. In this way F^∞ becomes a functor from the category of pointed theories to the category of feedback theories.

$$\begin{array}{ccccccccc}
T & \xleftarrow{\phi_1} \cdots \xleftarrow{\phi_{n-1}} & n\text{-res}\, T & \xleftarrow{\phi_n} & (n+1)\text{-res}\, T & \xleftarrow{\phi_{n+1}} & \cdots & F^\infty T \\
\downarrow \psi_1 = \psi & & \downarrow \psi_n & & \downarrow \psi_{n+1} & & & \downarrow \psi_\infty = F^\infty \psi \\
T' & \xleftarrow{\phi'_1} \cdots \xleftarrow{\phi'_{n-1}} & n\text{-res}\, T' & \xleftarrow{\phi'_n} & (n+1)\text{-res}\, T' & \xleftarrow{\phi'_{n+1}} & \cdots & F^\infty T'
\end{array}$$

Remark 2 Note that the above diagram does not assure that, in the category of feedback theories, $F^\infty \psi$ is a unique morphism $F^\infty T \to F^\infty T'$ extending ψ in general. Consequently, $F_1^\infty T$ need not be the free feedback theory generated by T (as a partial feedback theory).

4 Interpretations of synchronous schemes

Let Σ be a *nonempty* ranked alphabet, fixed for the rest of the paper. $\mathsf{Sf}(\Sigma)$ will denote the S-algebra of synchronous Σ-schemes (SΣ-schemes, see [3]). The following two definitions are adopted from [5].

Definition 1 Let $A = (A, \Sigma)$ be a pointed Σ-algebra. $F_1^\infty(\text{Pol}\, A)$ is called the feedback algebra induced by A.

Definition 2 An interpretation of SΣ-schemes is a pointed Σ-algebra A. The semantics of an SΣ-scheme under interpretation A is the image of the scheme by the unique homomorphism

$$\|\ \|_A : \mathsf{Sf}(\Sigma) \to F_1^\infty(\text{Pol}\, A)$$

determined by the mapping $\sigma \mapsto \sigma_A, \sigma \in \Sigma$.

Theorem 2 *$T_{\Sigma_\bot}$, the free pointed Σ-algebra is a Herbrand interpretation of* SΣ*-schemes, i.e. for any two* SΣ*-schemes F, G:*

$$\|F\|_{T_{\Sigma_\bot}} = \|G\|_{T_{\Sigma_\bot}} \text{ iff } \|F\|_A = \|G\|_A \text{ under all interpretations } A.$$

Proof. By assumtion, $\Sigma_\perp$ contains at least two symbols, hence $\mathrm{Pol}\,T_{\Sigma_\perp}$ is not trivial. It is well-known that in this case $\mathrm{Pol}\,T_{\Sigma_\perp} \cong T(\Sigma_\perp)$. Consequently, there exists a unique homomorphism $|\ |_A : \mathrm{Pol}\,T_{\Sigma_\perp} \to \mathrm{Pol}\,A$ (of pointed theories) extending the mapping $\sigma \mapsto \sigma_A, \sigma \in \Sigma$. By the F^∞-construction, $|\ |_A$ induces a homomorphism $F^\infty|\ |_A$ between the feedback algebras induced by $\mathrm{Pol}\,T_{\Sigma_\perp}$ and $\mathrm{Pol}\,A$, respectively, showing that $\|\ \|_A$ always factors through $\|\ \|_{T_{\Sigma_\perp}}$.

Note, however, that if $\Sigma = \emptyset$, then $\mathrm{Pol}\,T_{\Sigma_\perp}$ becomes the trivial pointed theory. Since in this case the homomorphism $|\ |_A$ does not exist for every A, Theorem 2 does not hold either.

$F_1^\infty(\mathrm{Pol}\,T_{\Sigma_\perp})$ might be called the feedback algebra *freely induced* by Σ. In the sequel we are going to prove that this algebra is isomorphic to the free feedback theory $\mathrm{Ft}(\Sigma)$ *generated* by Σ. Equivalently, we show that $F_1^\infty(T(\Sigma_\perp)) \cong \mathrm{Ft}(\Sigma)$, which is true even if $\Sigma = \emptyset$.

Let Δ also be a ranked alphabet. Recall from [9] that a deterministic (and totally defined) finite state top down tree transducer (*dfst*) $\mathcal{T}$ from Σ to Δ is specified by:

(i) A finite nonempty set of states, which we choose $[n]$ for some $n \in N$ for convenience;

(ii) A set of rewriting rules, which contains for each $i \in [n]$ and $\sigma \in \Sigma$ a rule of the form

$$i\sigma(y_1, \ldots, y_k) \to t_i(i_1 y_{j_1}, \ldots, i_l y_{j_l}),$$

where $\sigma \in \Sigma_k$, and the right-hand side of the above rule represents a tree in $T_\Delta([n] \times Y_k)$ as it is described in [9].

$\mathcal{T}$ operates in the well-known way, inducing a function (tree transformation) from T_Σ into T_Δ (or, equivalently, a function from $T_\Sigma(Y)$ into $T_\Delta([n] \times Y)$). Note that we omitted the initial state from the definition, because we do not need it for our present purposes.

To characterize the transformation induced by $\mathcal{T}$ as a homomorphism between algebraic theories, we use the trick of Arnold and Dauchet [1,8]. The trick is in fact the encoding of a state-variable pair iy_j to a single variable $y_{(j-1)\cdot n+i}$, so that the collection of the rules concerning $\sigma \in \Sigma_k$ can be condensed into a tuple $\langle t_1, \ldots, t_n \rangle \in T(\Delta)(n, n \cdot k)$. Thus, $\mathcal{T}$ defines a (rank preserving) mapping from Σ into $n\text{-dil}\,T(\Delta)$ (see [8] for the details), which can be extended uniquely to a homomorphism $\tau : T(\Sigma) \to n\text{-dil}\,T(\Delta)$. τ can be viewed as the algebraic representation of the transformation induced by $\mathcal{T}$. Recall from [1] that for any magmoid M, $n\text{-dil}\,M$ is the magmoid in which $n\text{-dil}\,M(p,q) = M(n \cdot p, n \cdot q)$, and the operations (composition and sum) are performed exactly like in M, furthermore, $1_{n\text{-dil}\,M} = n_M$ and $0_{n\text{-dil}\,M} = 0_M$. It is easy to extend the construction (functor) n-dil to algebraic theories (D-algebras in our sense) by defining

$$x_{n\text{-dil}\,T} = (x\#(n,n))_T,\ \epsilon_{n\text{-dil}\,T} = \langle n,n \rangle_T \text{ and } (0_1)_{n\text{-dil}\,T} = (0_n)_T$$

in a theory T. It was (implicitly) proved in [1] that in this way $n\text{-dil}\,T$ becomes a theory.

Lemma 2 *The D-algebras $n\text{-res}^*\,T$ and $n\text{-dil}\,T$ are isomorphic.*

Proof. Let $\xi_n : n\text{-dil}\,T \to n\text{-res}^*\,T$ be the mapping which sends $f \in n\text{-dil}\,T(p,q)(= T(n \cdot p, n \cdot q))$ to $\kappa(n,p)^{-1} \cdot f \cdot \kappa(n,q)$. An easy computation (left to the reader) shows that ξ_n is a homomorphism. The inverse of ξ_n is ς_n, which sends $f \in n\text{-res}^*\,T(p,q)$ to $\kappa(p,n)^{-1} \cdot f \cdot \kappa(q,n)$. (In fact the pair (ξ_n, ς_n) determines a natural isomorphism between the functors n-dil and n-res*.)

Let $\mathcal{T}$ be a *dfst* from Σ to Δ over the states $[n]$. Adding the rewriting rules $i\bot \to \bot, i \in [n]$, $\mathcal{T}$ is extended to a *dfst* from $\Sigma_\bot$ to $\Delta_\bot$. Moreover, making use of the well-known tree ordering (cf. [6,12]), $\mathcal{T}$ can be extended to a *dfst* on infinite trees "by continuity" (see e.g. [6]). In this case $\mathcal{T}$ determines a (continuous) homomorphism of $T^\infty(\Sigma)$ into $n\text{-dil}\,T^\infty(\Delta)$.

We define a *dfst* $\mathcal{T}_n$ from Σ_∇ to Σ for each positive integer n in the following way ($\Sigma_\nabla = \Sigma \cup \{\nabla\}$ with ∇ having rank 1). The set of states of $\mathcal{T}_n$ is $[n]$, and its rewriting rules are:

$i\sigma(y_1, \ldots, y_k) \to \sigma(iy_1, \ldots, iy_k)$ for every $i \in [n], \sigma \in \Sigma_k$;

$1\nabla(y_1) \to \bot$ and $(i+1)\nabla(y_1) \to iy_1$ for $i \in [n-1]$.

Let $\tau_n : T^\infty(\Sigma_\nabla) \to n\text{-dil}\,T^\infty(\Sigma)$ denote the homomorphism determined by $\mathcal{T}_n$ in the above sense, furthermore, let τ_0 be the unique homomorphism of $T^\infty(\Sigma_\nabla)$ into the trivial algebraic theory.

The following lemma will take a crutial part in the proof of Theorem 4. It expresses that for each $n \in N$, every infinite tree in $T^\infty(\Sigma_\nabla)$ can be converted in a homomorphic way to a series of n trees in $T^\infty(\Sigma)$, which belongs to $n\text{-res}\,T^\infty(\Sigma)$.

Lemma 3 *For every $n \in N$ and $t : p \to q$ in $T^\infty(\Sigma_\nabla)$,*

(i) $\xi_n(\tau_n(t)) \in n\text{-res}\,T^\infty(\Sigma)$;

(ii) $\phi_n(\xi_{n+1}(\tau_{n+1}(t))) = \xi_n(\tau_n(t))$.

Proof. The first statement is obvious by the definition of $\mathcal{T}_n$. According to the characterization of $\mathcal{T}_n$ by τ_n, a state-variable pair iy_j is encoded as $y_{(j-1)\cdot n+i}$, which causes a difference between the interpretation of iy_j by τ_{n+1} and τ_n. Using the n-res "philosophy", however, iy_j is encoded to $y_{(i-1)\cdot q+j}$, *independently* of n. It remains to observe that the isomorphism ξ_n does exactly the desired conversion from the n-dil type representation of the transform of t by $\mathcal{T}_n$ to the n-res type one.

Further on we denote the homomorphism $\xi_n \circ \tau_n$ by ω_n (note that $(\xi_n \circ \tau_n)(t) = \xi_n(\tau_n(t))$). Since ξ_n and τ_n are continuous, ω_n is continuous as well.

Lemma 4 *For every $n \in N$ and $t : 1 \to q$ in $T^\infty(\Sigma_\nabla)$,*

$$\omega_{n+1}(\nabla(t)) = \langle \bot, \omega_n(t) \rangle_{T^\infty(\Sigma)}.$$

Proof. By definition, $\omega_{n+1}(\nabla) = \langle \bot, y_1, \ldots, y_n \rangle_{T^\infty(\Sigma)}$. Thus,

$$\begin{aligned} \omega_{n+1}(\nabla(t)) &= \omega_{n+1}(\nabla) \cdot \omega_{n+1}(t) = \langle \bot, y_1, \ldots, y_n \rangle \cdot \omega_{n+1}(t) \\ &= \langle \bot, \phi_n(\omega_{n+1}(t)) \rangle = \langle \bot, \omega_n(t) \rangle \quad \text{(Lemma 3)}. \end{aligned}$$

Let $t : 1 + p \to 1 + q$ be a morphism in $T^\infty(\Sigma_\nabla)$, and define the trees $t^0, t^1, \ldots, t^n, \ldots$ by applying the polynomials $\mathbf{f}^0, \mathbf{f}^1, \ldots, \mathbf{f}^n, \ldots$ constructed in Section 3 on t.

Lemma 5 *If $n \leq m$, then $\omega_n(t^m) = \omega_n(t^n)$.*

Proof. Induction on m. The base step $m = 0$ is trivial. For $m \geq 1$ the case $n = 0$ is again trivial. If $n \geq 1$ as well, then

$$\begin{aligned}
\omega_n(t^m) &= \omega_n(t \cdot \langle(\nabla + 0_p) \cdot t^{m-1}, q\rangle) && \text{(def.)}\\
&= \omega_n(t) \cdot \omega_n(\langle\nabla((1 + 0_p) \cdot t^{m-1}), q\rangle)\\
&= \omega_n(t) \cdot \langle\langle\bot, \omega_{n-1}(1 + 0_p) \cdot \omega_{n-1}(t^{m-1})\rangle_{T_1}, \omega_n(q)\rangle_{T_2} && \text{(Lemma 4)}\\
&\quad (T_1 = T^\infty(\Sigma), T_2 = n\text{-res}\, T^\infty(\Sigma))\\
&= \omega_n(t) \cdot \langle\langle\bot, \omega_{n-1}(1 + 0_p) \cdot \omega_{n-1}(t^{n-1})\rangle_{T_1}, \omega_n(q)\rangle_{T_2} && \text{(ind.)}
\end{aligned}$$

The same result is obtained if we perform the above computation starting from $\omega_n(t^n)$, showing that $\omega_n(t^m) = \omega_n(t^n)$.

$T^\infty(\Sigma_\nabla)$ can be given the structure of an iteration (even rational) theory in the well-known way (see e.g. [12]). Considering $T^\infty(\Sigma_\nabla)$ as the equivalent S-algebra (cf. [2]), we introduce the operation $\uparrow$ on $T^\infty(\Sigma_\nabla)$ by

$$\uparrow f = \Uparrow (f \cdot (\nabla + q)) \text{ for } f : 1 + p \to 1 + q.$$

Let θ be the congruence relation of $T^\infty(\Sigma_\nabla)$ induced by the equation $\Uparrow (\epsilon \cdot \nabla) = \bot$. It is easy to see that the S-algebra $T^\infty_\nabla(\Sigma)$, which is the same as $T^\infty(\Sigma_\nabla)/\theta$ with the only exception that $\Uparrow$ is replaced in it by $\uparrow$, is a feedback theory. Moreover, it was proved in [4] that $\mathrm{Ft}(\Sigma)$ (the free feedback theory generated by Σ) is isomorphic to the sub-S-algebra of $T^\infty_\nabla(\Sigma)$ generated by Σ.

Theorem 3 *ω_n defines a homomorphism between the feedback theories (i.e. S-algebras) $T^\infty_\nabla(\Sigma)$ and n-res $T^\infty(\Sigma)$.*

Proof. By the definition of $\mathcal{T}_n$, if $t \equiv t'$ (θ), then $\tau_n(t) = \tau_n(t')$. Hence it is enough to prove that if $t : 1 + p \to 1 + q$ is a morphism in $T^\infty(\Sigma_\nabla)$, then

$$\omega_n(\uparrow t) = \uparrow (\omega_n(t)).$$

According to the definition of iteration in $T^\infty(\Sigma_\nabla)$ (cf. [12]), $\uparrow t$ is the supremum of the directed chain $(0_1 + p) \cdot t^m, m \geq 0$.

$$\begin{aligned}
\omega_n(\uparrow t) &= \omega_n(\sqcup_m((0_1 + p) \cdot t^m))\\
&= \sqcup_m(\omega_n(0_1 + p) \cdot \omega_n(t^m)) && \text{(continuity)}\\
&= \omega_n(0_1 + p) \cdot \omega_n(t^n). && \text{(Lemma 4)}
\end{aligned}$$

Observe that

$$\omega_n(\nabla) = \langle\bot, y_1, \ldots, y_{n-1}\rangle = (\nabla)_{n\text{-res}\, T^\infty(\Sigma)},$$

$$\omega_n(\bot) = \langle\bot, \ldots, \bot\rangle = (\bot)_{n\text{-res}\, T^\infty(\Sigma)}.$$

Since ω_n is a homomorphism between algebraic theories, by Remark 1 we have:

$$\begin{aligned}
\omega_n(0_1 + p) \cdot \omega_n(t^n) &= ((0_1 + p) \cdot (\omega_n(t))^n)_{n\text{-res}\, T^\infty(\Sigma)}\\
&= \uparrow (\omega_n(t)) && \text{(Lemma1)}
\end{aligned}$$

Theorem 4 $\mathrm{Ft}(\Sigma) \cong F_1^\infty T(\Sigma_\perp)$.

Proof. By Theorem 3, ω_n is a homomorphism of $T_\nabla^\infty(\Sigma)$ into n-res $T^\infty(\Sigma)$, and by Lemma 3, $\omega_n = \phi_n \circ \omega_{n+1}$. It follows that there exists a unique homomorphism $\omega_\infty : T_\nabla^\infty(\Sigma) \to F^\infty T^\infty(\Sigma)$ such that $\omega_n = \phi_n^\infty \circ \omega_\infty$, where ϕ_n^∞ is the standard homomorphism of $F^\infty T^\infty(\Sigma)$ into n-res $T^\infty(\Sigma)$. To prove that ω_∞ is injective, let t and t' be two different morphisms (tuples of infinite trees) of any sort $p \to q$ in $T^\infty(\Sigma_\nabla)$. Then there must be a node u in t and a node u' in t' with the following property. u and u' are contained in the same (say ith) component of t and t', and the labelled paths leading to these nodes from the root of the ith component of t and t' differ only in the last label. Let n be the number of ∇-nodes along the longest common prefix of these two paths. It is evident that $\omega_{n+1}(t)$ and $\omega_{n+1}(t')$ should already be different. Observe that in the trees of $T^\infty(\Sigma_\nabla)$ belonging to $\mathrm{Ft}(\Sigma)$ every infinite branch is scattered by an infinite number of ∇-nodes. Consequently, if t is a morphism in $\mathrm{Ft}(\Sigma)$, then $\omega_n(t)$ is finite for every $n \in N$, i.e. it belongs to n-res $T(\Sigma_\perp)$ already. Thus, the restriction of ω_∞ to $\mathrm{Ft}(\Sigma)$ provides the required isomorphism.

References

[1] A. Arnold and M. Dauchet, Théorie des magmoïdes, *RAIRO Inform. Théor.* **12** (1978), 235–257 and **13** (1979), 135–154.

[2] M. Bartha, A finite axiomatization of flowchart schemes, *Acta Cybernet.* **2** (1987), 203–217.

[3] M. Bartha, An equational axiomatization of systolic systems, *Theoret. Comput. Sci.* **55** (1987), 265–289.

[4] M. Bartha, Connections between feedback and iteration theories, submitted for publication.

[5] M. Bartha, Towards a mathematical theory of synchronous logical circuits and systolic networks, submitted for publication.

[6] J. Bilstein and W. Damm, Top down tree transducers for infinite trees, Proceedings, 6th Colloquium on Trees in Algebra and Programming (CAAP' 81), *Lecture Notes in Computer Science* **112** (1981), 117–134.

[7] S. L. Bloom and Z. Ésik, Axiomatizing schemes and their behaviors, *J. Comput. System Sci.* **31** (1985), 375–393.

[8] M. Dauchet, Transductions de forêts — Bimorphismes de magmoïdes, Thèse d'Etat, Université de Lille, 1977.

[9] J. Engelfriet, Bottom up and top down tree transformations, a comparison, *Math. Systems Theory* **9** (1975), 198–231.

[10] F. Gécseg and I. Peák, Algebraic Theory of Automata, Akadémiai Kiadó, Budapest, 1972.

[11] C. E. Leiserson and J. B. Saxe, Optimizing synchronous systems, *J. VLSI Comput. Systems* **1** (1983), 41–67.

[12] J. B. Wright, J. W. Thatcher, E. G. Wagner, and J. A. Goguen, Rational algebraic theories and fixed-point solutions, *in* "Proceedings, 17th IEEE Symposium on Foundations of Computer Science, Houston, Texas, 1976," pp. 147–158.

GENERALIZED BOOLEAN HIERARCHIES AND BOOLEAN HIERARCHIES OVER RP (Conference Abstract[1])

ALBERTO BERTONI – Università degli Studi di Milano

DANILO BRUSCHI – Università degli Studi di Milano and University of Wisconsin

DEBORAH JOSEPH – University of Wisconsin

MEERA SITHARAM – University of Wisconsin

PAUL YOUNG – University of Washington and University of Wisconsin

1. INTRODUCTION.

In this paper we study the complexity hierarchy formed by taking the Boolean closure of sets in RP, the class of sets decidable in random polynomial time. This hierarchy, which we call the *Boolean hierarchy over RP* and denote by RBH, is analogous to the difference hierarchy for *r.e.* sets studied by Ershöv ([Er 68a 68b 69]) and to the Boolean hierarchy over NP studied by ([BuHa 88], [CaHe 86], [Ka 88], [KöSchWa 87], [Wa 86], [Wa 88], [WeWa 85]). RBH lies above RP and below BPP. It is of particular interest because so little is known about sets that might be in $BPP - RP$. In fact, since Adleman and Huang ([AdHu 87]) showed that primality testing lies in RP, (and hence in ZPP), there have been no natural candidates for the class $BPP - RP$. Thus, the examples we give for the Boolean hierarchy over RP should help renew the belief that the classes BPP and RP are truly different.

Our study begins with a uniform proof of the following metatheorem:

- Boolean hierarchies over arbitrary complexity classes satisfy the same definitional equivalences as the Boolean hierarchy over NP. Moreover, many of the properties of polynomial time machines that make a constant number of queries to an oracle in NP (adaptively and non-adaptively) generalize to machines that query an oracle from an arbitrary complexity class, C. For instance, C could be Random Polynomial time, (RP), Unique Polynomial time, *FewP,* the nondeterministic exponential time classes $NEXP^{linear}$ and $NEXP^{poly}$, or the recursively enumerable sets, (RE).

---

[1] This abstract presents in brief form the authors' work on generalized Boolean hierarchies and Boolean hierarchies over RP. [BBJSY 89] contains a complete presentation of this work, and copies will be available at the conference and at the University of Wisconsin address below. This work was supported in part by the Ministero della Pubblica Istruzione, through "Progetto 40%: Algoritmi e Strutture di Calcolo," the National Science Foundation under grant DCR-8402375, and by the Wisconsin Alumni Research Foundation under a Brittingham Visiting Professorship. The last four authors' 1988-89 address is: Computer Sciences Department, University of Wisconsin, 1210 West Dayton St., Madison, WI 53706, U.S.A.

We establish this very general theorem in Section 2 by generalizing many of the theorems for the Boolean hierarchy over *NP* to Boolean hierarchies over fairly arbitrary complexity classes C. In addition, we give a careful treatment of *extended* Boolean hierarchies over fairly arbitrary classes. Some of the proofs for the hierarchy over *NP* generalize trivially. However, some of the proofs over *NP* use the existence of complete sets or other special properties of *NP*, and these proofs require modification for classes such as *RP*.

Having proved that Boolean hierarchies over most complexity classes have similar definitional invariants and relationships to bounded query classes, in Section 3 we focus our attention on the Boolean hierarchy over *RP*. We study its structural properties and give several natural examples of sets in the hierarchy:

- Fermat showed that every integer can be represented as the sum of at most four integer squares. From recent work of Bach, Miller, Rabin, and Shallit ([BaMiSh 86], [RaSh 88]), it follows that the set $\{n : n$ is perfect and is the sum of two integer squares$\}$ is in *RP*, and that the set $\{n : n$ is perfect and is the sum of three integer squares$\}$ is in *RP*. Combining these results we can see that the set $\{n : n$ is perfect and is the sum of three integer squares but not a sum of two integer squares$\}$ is in the second level of the Boolean hierarchy over *RP*. Similarly, $\{n : n$ is perfect and is the sum of four integer squares but not a sum of three integer squares, or n is perfect and is the sum of two integer squares$\}$ is in the third level of the Boolean hierarchy over *RP*.
- Schwartz proved that checking whether a polynomial p represented as a straight line program is identically zero is in *coRP*, ([Sch 80]). We show that this problem is equivalent to checking whether two straight line programs compute the same natural number and thus that the latter problem is also in *coRP*. However, the problem of deciding whether the polynomial represented by a straight line program is a monomial seems not to be in either *RP* or co*RP*, although we can show that it is in the second level of the Boolean hierarchy over *RP*. In fact, we show that if this latter problem is in *RP* or *coRP*, then the problem of checking whether straight line programs are identically zero and the problem of checking whether two straight line programs compute the same natural number are both in *ZPP*.

In Section 3.2 we briefly explore how truth-table reducibility from self-reducible sets in *NP* together with membership in *BPP* can force sets from the Boolean hierarchy over *NP* into the Boolean hierarchy over *RP*.

One would like to prove that the Boolean hierarchy over *RP* is a proper hierarchy and satisfies the set containments "which it should" with respect to the polynomial time hierarchy, other probabilistic classes, and the Boolean hierarchy over *NP*. This goal is obviously too ambitious, so in Section 4 we give various oracle constructions including:

- There is an oracle, X, relative to which the obvious proper containments which one *expects* to hold for Boolean hierarchies over *RP* and for *RP vs NP* do hold with respect to X.

This at least shows that no simple proof should show that the Boolean hierarchy over *RP* does not behave as we would "expect" it to behave.

1.1 . BASIC DEFINITIONS AND NOTATION.

Definition 1.1. A set S is in RP *if there is a nondeterministic polynomial time Turing machine, M, and a positive fraction $1/c$ such that*

$$x \in S \Rightarrow M(x) \text{ accepts on at least } 1/c \text{ of its paths}$$
$$x \notin S \Rightarrow M(x) \text{ rejects on all of its paths.}$$

The machine M is called an RP-***acceptor*** for S, and its acceptance criterion is said to have ***one-sided*** error probability. The complement class of RP is denoted $coRP$, and the intersection of RP and $coRP$ is the class ZPP. Obviously from the above definition, RP is contained in NP.

The class BPP, or Bounded error Probabilistic Polynomial time, is given by

Definition 1.2. $S \in BPP$ *if there exists a nondeterministic polynomial time Turing machine, M, and positive fractions $1/c, 1/d$, with $(1-(1/d)) < 1/c$, such that*

$$x \in S \Rightarrow M(x) \text{ accepts on at least } 1/c \text{ of its paths}$$
$$x \notin S \Rightarrow M(x) \text{ rejects on at least } 1/d \text{ of its paths.}$$

The machine M is called a BPP-***acceptor*** for S, and its acceptance criterion is said to have a ***two-sided*** error probability. Notice from this definition that BPP is closed under complements, and that RP and $coRP$ are subsets of BPP. Following work of Sipser, Gacs and Lautemann have independently shown that BPP is contained in the $\Sigma_2^P \cap \Pi_2^P$ level of the Meyer-Stockmeyer hierarchy, ([La 83], [Si 83]).

Ko ([Ko 82]) and Zachos ([Za 86]) introduced a polynomial time hierarchy over RP (denoted RH) which is analogous to the Meyer-Stockmeyer hierarchy over NP. Zachos showed that the entire random polynomial time hierarchy, RH, is contained in BPP. Recalling our motivation for studying the structure of $BPP - RP$, it should be noted that there are no known natural candidates for sets in $\Sigma_2^{RP} - RP$. This is partly because Σ_2^{RP}-acceptors have awkward acceptance criteria which natural problems do not seem to satisfy. In fact, Zachos and Heller have attempted to prove that RH collapses to $\Sigma_2^{RP} \cap \Pi_2^{RP}$ $(= \Delta_2^{RP})$.

In Section 2 we construct regular and extended Boolean hierarchies over RP, all of which lie between RP and Δ_2^{RP}.

2. PROPERTIES OF GENERALIZED BOOLEAN HIERARCHIES.

In this section we lay the foundations for discussing both regular and extended Boolean hierarchies over quite general complexity classes, and over RP in particular.

Definition 2.1.

i) For each integer k suppose that h_k is a k-ary Boolean function. For any set Y, let χ_Y denote the characteristic function of Y. Then for any collection of sets C

$$h_k[C] =_{def} \{ S : \exists C_1, \ldots, C_k \in C \ [\chi_S(x) = h_k(\chi_{C_1}(x), \ldots, \chi_{C_k}(x))] \}.$$

ii) The **regular Boolean hierarchy** *as defined over NP by Cai and Hemachandra ([CaHe 86]) is denoted by* $NP[k] =_{def} h_k[NP]$, *where*

$$h_k = (...((x_1 \wedge \neg x_2) \vee x_3)\ldots) \vee x_{k-1}) \wedge \neg x_k \quad \text{for } k \text{ even, and}$$
$$h_k = (...((x_1 \wedge \neg x_2) \vee x_3)\ldots) \wedge \neg x_{k-1}) \vee x_k \quad \text{for } k \text{ odd.}$$

iii) The **regular Boolean hierarchy** *over an arbitrary collection of sets C is denoted by* $C[k] =_{def} h_k[C]$, *where* h_k *is as in (i) and (ii).*

Note that other base sequence of functions $\lambda k h_k$ for building Boolean hierarchies have been proposed by Hausdorff, by Köbler and Schöning, and by Wechsung and Wagner. Several of these were shown by Hausdorff to give equivalent hierarchies for any base collection of sets C, and *all of these proposed definitions have been shown to be equivalent for regular Boolean hierarchies over the base class NP*, ([KöSchWa 87]).

Boolean hierarchies are known to have an intimate connection with "bounded query" classes. For bounded query classes, $P_{tt}^{C}[k]$ denotes the class of sets that are reducible to some set in C via some bounded-truth-table which has at most k variables, and $P_{T}^{C}[k]$ denotes the class of languages recognizable by a polynomial time Turing machine which, on input x, makes at most k queries to an oracle in C. For example, Beigel has briefly considered Boolean hierarchies and bounded query classes over arbitrary collections of sets, C, and with *no* assumptions on C has proved the following results, ([Be 87], Section 4.1, Theorems 17 and 18), for all integers k:

1. $P_{tt}^{C}[k] \subseteq P_{T}^{C}[k] \subseteq P_{tt}^{C}[2^k - 1]$.
2. $C[k] \cup coC[k] \subseteq P_{tt}^{C}[k] \subseteq C[2^k(2k+3)] \cap coC[2^k(2k+3)]$.[2]

On the other hand, by restricting the base class C to be NP, one gets much sharper results:

3. $P_{T}^{NP}[k] = P_{tt}^{NP}[2^k - 1]$, ([Be 87]).
4. $NP[k] \cup coNP[k] \subseteq P_{tt}^{NP}[k] \subseteq NP[k+1] \cap coNP[k+1]$, ([Wa 88]).

These examples show that by placing *some* restrictions on the base classes C which one uses in building Boolean hierarchies over C, one should get stronger results and more control over the Boolean hierarchy. We will return to this point after we discuss extended Boolean hierarchies.

The class, Θ_2^P, of sets polynomial time truth-table reducible to NP, (which by 3 above is $P_T^{NP}[O(log)]$) has received increasing attention as an important subclass of Δ_2^P. For example, in [BHW 89] it is shown that $\Theta_2^P \subseteq PP$. Obviously, for the regular Boolean hierarchy over NP, $\cup_k NP[k] \subseteq \Theta_2^P$, but since only *bounded* polynomial time truth-tables are involved in building $\cup_k NP[k]$, it is very unlikely that $\cup_k NP[k] = \Theta_2^P$. Hence, since they should run all the way through Θ_2^P, it is reasonable to try to build *extended* Boolean hierarchies over NP (and over other complexity classes such as RP) by using fairly *arbitrary* polynomial time truth-tables.

---

[2] Beigel's proof is actually given not for the Boolean hierarchy $C[k]$ over C based on the Cai-Hemachandra definition in 2.1, but rather on the Köbler-Schöning definition, which is based on the parity function.

In [Wa 88], Wagner gives one definition for building extended Boolean hierarchies through Θ_2^P. His idea, which is mathematically very elegant but to our taste not philosophically appealing, is to base the definition directly on sets which encode the "mind-changes" of Boolean formulas over NP. The underlying theory which justifies his definition is worked out very carefully for *regular* Boolean hierarchies over NP and is quite elegant, ([KöSchWa 87], [Wa 88]). Cai and Hemachandra, on the other hand, have suggested building extended Boolean hierarchies over NP based on their particular sequence h_k of Definition 2.1. Roughly, their idea for defining a set S in an extended Boolean hierarchy over NP is as follows: to define the characteristic function, χ_S, of the set S, given x one computes one of their formulas $h_{t(x)}$ of Definition 2.1 (ii) via a polynomially computable function t, and then one gets $\chi_S(x)$ by substituting into $h_{t(x)}$ the characteristic functions of sets which are uniformly reducible to SAT in polynomial time. By the results of [Wa 88] these approaches are equivalent for the *regular* Boolean hierarchy over NP. Although the machinery for proving the equivalence of these definitions for regular Boolean hierarchies over NP should work for extended hierarchies over NP, to the best of our knowledge proofs have never been published. Furthermore, since RP probably does not have complete sets, once one considers Boolean hierarchies over RP rather than over NP, it is not clear what sets should be substituted for the complete set SAT if one uses the Cai and Hemachandra approach in building hierarchies over RP.

Most generally, for arbitrary complexity classes C, one should build sets S in the *extended Boolean-hierarchy over* C as follows: given x compute an *arbitrary* truth-table T_x *using whatever computational power is appropriate for the class* C, and define $\chi_S(x)$ by substituting into T_x the characteristic functions of any *infinite* sequence of sets from C which can be *uniformly* obtained, again using whatever computational power is appropriate for the class C.

Using such a very general definition for building extended Boolean hierarchies and building on the work described in Wagner's very useful survey, [Wa 88], we prove very general theorems which apply to Boolean hierarchies over fairly arbitrary complexity classes. In adapting the work in [Wa 88], we must overcome the fact that we work with base classes, C, in which there may be no nice enumerations of the members of the class, there may be no complete sets, and "existential guessing" may not be freely available. Basically, we prove the following

Metatheorem 2.2. *All regular Boolean hierarchies over fairly arbitrary complexity classes, C, satisfy the same definitional equivalences as the regular Boolean hierarchy over NP. Furthermore, at all finite levels and at least at lower infinite levels, all extended Boolean hierarchies over these same base classes satisfy the same definitional equivalences. Moreover, the proofs of these equivalences show that many of the properties that relate the number of parallel and sequential calls to an oracle in NP generalize to machines that query an oracle from these more general complexity classes. Classes, C, for which these results apply not only include classes like NP and RP, but also such classes as Uniform RNC, $FewP$, the nondeterministic exponential time classes $NEXP^{linear}$ and*

$NEXP^{poly}$, RE, classes in the polynomial hierarchy, and classes in the arithmetic hierarchy. For NP, $NEXP^{linear}$, $NEXP^{poly}$, and RE the basic equivalences hold at all levels of the extended Boolean hierarchy.

Full statements and proofs may be found in [BBJSY 89]. We emphasize here that *all* of the specific results cited in 2.2 are corollaries of quite general, *uniform,* theorems.

3. THE BOOLEAN HIERARCHY OVER RP.

In Section 3 we focus our attention on the Boolean hierarchy over *RP*. In Section 3.1 we give examples of sets in the Boolean hierarchy over *RP*, and in Section 3.2 we investigate when self-reducibility and membership in *BPP* can force sets in *BH* to lie in *RH*.

The regular Boolean hierarchy over *RP* was defined by Definition 2.1, (iii). In light of Theorem 2.2, the definition of the regular Boolean hierarchy over *RP* remains invariant if one uses a variety of other standard Boolean functions in Definition 2.1. The regular Boolean hierarchy over *RP* can also be characterized using either machine-acceptors or the probabilistic quantifiers introduced in [Za 86]; these characterizations are given in [BBJSY 89].

3.1 . EXAMPLES OF LANGUAGES IN RBH.

Here we give natural examples of languages which lie in the Boolean hierarchy over *RP*. These languages are of particular interest because they represent natural problems which are in *BPP* but do not seem to lie in *RP*.

In [SoSt 77] Solovay and Strassen showed that primality testing is in *coRP*. Recently Adleman and Huang, ([AdHu 87]), have shown that primality testing is also in *RP*, and hence primality testing is now known to be in *ZPP*. As a consequence problems such as determining whether a number is perfect and determining whether a number is a Carmichael number that were previously conjectured to be in $BPP - RP$ are now known to be in *RP*. However, we can use the fact that the set of perfect numbers is in *RP* to give examples of problems that are in the Boolean hierarchy over *RP* but seem not to be in *RP*.

Our first examples use Fermat's well-known result that every integer can be represented as the sum of at most four integer (possibly zero) squares. Combining Fermat's result with recent work of Bach, Miller, Rabin, and Shallit ([BaMiSh 86], [RaSh 88]) shows that the set $\{n : n$ is perfect and is the sum of two integer squares$\}$ is in *RP* and that the set $\{n : n$ is perfect and is the sum of three integer squares$\}$ is in *RP*. These results together give that

1. $\{n : n$ is perfect and is the sum of three integer squares but not a sum of two integer squares $\}$ is in $RP[2]$.
2. $\{n : n$ is perfect and is the sum of four integer squares but not a sum of three integer squares $\} \cup \{n : n$ is perfect and is the sum of two integer squares $\}$ is in $RP[3]$.

Currently there are no known techniques that place either of these two sets in *RP*.[3]

[3] Of course, we should also point out that it is not known that there are infinitely many perfect numbers (although this is widely believed to be the case), and even if there are infinitely many perfect numbers either of the sets described above could be empty.

A second, and rather different, collection of problems involving polynomials that are represented by straight line programs can also be shown to be in RBH. Schwartz has shown that checking whether a polynomial p, represented as a straight line program, is identically zero is a $coRP$ problem ([Sch 80]). More formally,

Definition 3.1.1. *A straight line program Φ is an ordered sequence of instructions $I_1, \cdots, I_n$ such that an individual instruction I_k has one of the following forms:*

$x_k \leftarrow x_j + x_s$; $x_k \leftarrow x_j - x_s$; $x_k \leftarrow x_j x_s$; $x_k \leftarrow 0$; $x_k \leftarrow 1$;

where x_k, x_j, x_s are indexed variables and $j, s < k$.

For a straight line program Φ, we will let x_n denote the variable with greatest index contained in Φ, let N_Φ denote the number contained in the variable x_n at the end of the computation of Φ, and let $l(\Phi)$ denote the number of instructions of Φ. If we allow instructions of the form: $x \leftarrow z$; where z is a formal variable, then Φ describes a polynomial $P_\Phi(z)$ of degree at most $2^{l(\Phi)}$.

With this notation we can now formally describe our next set of problems.

Problem 3.1.1 (Schwartz).
Instance : A straight line program, Φ_1, with a formal variable.
Question: Is P_{Φ_1} identically null?

Problem 3.1.2.
Instance : Two straight line programs, Φ_1 *and* Φ_2.
Question: Is N_{Φ_1} equal to N_{Φ_2}?

As mentioned above, Schwartz has shown that Problem 3.1.1 is in $coRP$. Problem 3.1.2 can obviously be reduced to Problem 3.1.1, so it too is in $coRP$. A somewhat less obvious proof, ([BBJSY 89]), shows that Problem 3.1.1 can be reduced to Problem 3.1.2.

Next we consider a problem involving straight line programs that is in RBH, but seems not to be in RP or $coRP$. To state the problem, for a polynomial $p(z) = \sum_{k=0}^{n} a_k p^k$, we will denote the coefficient of z^k in $p(z)$ by $[z^k]p(z)$.

Problem 3.1.3.
Instance : A straight line program, Φ, with a formal variable.
Question: Does Φ represent a monomial? I.e., does there exist exactly one k such that $[z^k]P_\Phi$ is not equal to 0?

Using the following lemma we will show that Problem 3.1.3 is in $RP[2]$.

Lemma 3.1.1. *For each continuous function p, the following two sentences are equivalent:*

1. $\exists\alpha\exists\beta[p(x) = \alpha x^\beta$ *and* $\alpha \neq 0]$, and
2. $\forall x[p(1)p(x^2) = p^2(x)$ *and* $p(1) \neq 0]$.

Now given a straight line program Φ which represents a polynomial $P_\Phi(z)$, to verify whether $P_\Phi(z)$ is a monomial we do the following:

1. Compute a straight line program Φ_1 such that $N_{\Phi_1} = P_\Phi(1)$;
2. Compute a straight line program Φ_2 such that $P_{\Phi_2}(x) = N_{\Phi_1} P_\Phi(z^2) - P_\Phi^2(z)$;
3. If $N_{\Phi_1} \neq 0$ and $P_{\Phi_2} = 0$, then accept; else reject.

Given Φ, Steps 1 and 2 can be computed in polynomial time. Step 3 can be computed with one query to an RP oracle and with one independent query to a coRP oracle. Thus Problem 3.1.3 is in $RP[2]$.

Next we give some evidence that it is unlikely that Problem 3.1.3 is either in RP or in $coRP$.

Theorem 3.1.3. *If Problem 3.1.3 is in $RP \cup coRP$, then Problems 3.1.1 and 3.1.2 are each in ZPP.*

To prove this it is adequate to show that if Problem 3.1.3 is in RP (or coRP), then Problem 3.1.2 and its complement are in RP (and coRP). Notice that verifying that the number N_Φ, represented by the straight line program Φ, is not equal to zero is equivalent to verifying that N_Φ is a monomial. In fact: $N_\Phi \neq 0$ *if and only if* N_Φ is a monomial, since every integer greater than zero is a monomial.

On the other hand verifying that N_Φ is equal to zero, is equivalent to verifying that the program $\Phi_1 = 1 + \Phi * z$ is a monomial (since Φ_1 is a monomial only if $\Phi * z$ is null). In fact: P_{Φ_1} is a monomial *if and only if* $N_\Phi = 0$.

This shows Problem 3.1.2 and its complement are in RP, and thus Problem 3.1.2 is in ZPP. The same reasoning is applied for the case in which Problem 3.1.3 can be solved with a coRP algorithm.

We should also comment that by generalizing Problem 3.1.1. to multivariate polynomials we have obtained problems that lie in higher levels of the Boolean hierarchy over RP. (See [BBJSY 89] for details.)

3.2 . SELF-REDUCIBLE SETS, BPP AND RBH.

In this section, we study the relationship between BPP and the Boolean hierarchy over RP by extending a result of Ko: "If a disjunctive self-reducible set in NP is also in BPP, then it must be in RP." ([Ko 82]).

We too start with the class of disjunctive self-reducible sets in NP. For each set, S, in this class and for each level, $NP[k]$, of the Boolean hierarchy over NP, we construct a canonical set, $S_k \in NP[k]$, such that membership of S_k in BPP forces S_k into $RP[k]$. Furthermore, S_k will be complete for a reasonable subset, $C_S[k] \subseteq NP[k]$, and the membership of S_k in BPP will force all of $C_S[k]$ into $RP[k]$.

Definition 3.2.1. *Let h_k be as in Definition 2.1. For any set S in NP, let*

$$C_S =_{def} \{Y : Y \leq_m^P S\}, \text{ and}$$

$$S_k =_{def} \{\langle x_1, \ldots, x_k\rangle : h_k(\chi_S(x_1), \ldots, \chi_S(x_k))\}.$$

Theorem 3.2.2.

1.) $C_S \subseteq NP$, (and so $C_S[k] \subseteq NP[k]$).

2.) $S_k \in NP[k]$.

3.) S_k is $\leq_m^P$-complete for $C_S[k]$.

4.) S disjunctive self-reducible and $S_k \in BPP$ implies that $C_S[k] \subseteq RP[k]$.

Theorem 3.2.3.

1.) *Let $<^P_{ptt}$ denote positive polynomial time truth-table reducibility. Then Theorem 3.2.2 holds with $<^P_{ptt}$ replacing $\leq^P_m$ in Definition 3.2.1 and Theorem 3.2.2.*

2.) *Theorem 3.2.2 holds with $\leq^P_T$ replacing $\leq^P_m$ in Definition 3.2.1 and Theorem 3.2.2, provided we further restrict C_S to disjunctive self-reducible members of NP.*

The proofs of Theorems 3.2.2 and 3.2.3 directly use Ko's result together with Zachos's result, ([Za 86]), that BPP is closed under $\leq^P_T$ reductions.

4. ORACLE SEPARATION RESULTS.

Ideally we would like to prove that the Boolean hierarchy over RP is a proper hierarchy and behaves "as it should" with respect to the polynomial time hierarchy, other probabilistic classes, and the Boolean hierarchy over NP. This goal is obviously too ambitious, so we settle for an oracle construction. Moreover, we are almost always working within BPP, and it is known, ([BeGi 81]), that with respect to a random oracle $BPP^X = P^X$ with probability one – this in spite of strong intuitions that in the "real world" $BPP \neq P$. Thus, even proving that with respect to a random oracle X the Boolean hierarchy over RP behaves "as it should" with probability one is too ambitious. Our more modest goal here is to investigate what may happen in relativized worlds. More specifically we construct a single oracle, X, relative to which the obvious proper containments which one *expects* to hold in the "real world" at least hold with respect to the oracle. This demonstrates that no simple proof should show that the real RBH world does *not* behave as we expect. (Of course we also prove that it is possible to get other oracles which witness the existence of what we presume to be "pathological" structures.) We state the most interesting of our results in the remainder of this section.

Our most important result, Theorem 4.1, summarizes the way we believe most proper containments should appear and gives a single oracle relative to which these containments are indeed proper.

Theorem 4.1. *There exists an oracle X such that:*

1.) *The regular Boolean hierarchy over RP and the regular Boolean hierarchy over NP are proper infinite hierarchies, and the regular Boolean hierarchy over RP is properly included in PP.*

2.) *For extended Boolean hierarchies over NP and over RP, the preceding results all hold at least through levels up to $O(\log|x|)$.*

3.) *For extended Boolean hierarchies over NP and over RP the preceding results all hold above $O(\log|x|)$ except possibly the proper containment in PP.*

Our remaining theorems give oracles which witness the existence of possibly pathological Boolean hierarchies over NP and over RP. Theorem 4.3 and Theorem 4.4 are obtained using results of [Ra-82] where oracles D and E are constructed such that $P^D \neq RP^D = NP^D$.

Theorem 4.2. *There exists an oracle A such that the Boolean hierarchy over RP is interwined with the Boolean hierarchy over NP, and the hierarchies extend to the $k-th$ level where they collapse together with it $PSPACE^A$.*

Theorem 4.3. *There exists an oracle B such that the Boolean hierarchy over RP and the Boolean hierarchy over NP coincide and are proper infinite hierarchies.*

Theorem 4.4. *There exists an oracle C such that the Boolean hierarchy over RP collapses while the Boolean hierarchy over NP has infinite levels.*

Finally we note that since Rackoff also proved his results for *UP*, we can use similar techniques to obtain the above theorems for Boolean hierarchies over *UP* instead of over *RP*.

Again, in [GeGr 86] it is proved that there exists an oracle F such that $P^F \neq RP^F \neq UP^F \neq NP^F$. Modifying this construction, one can prove:

Theorem 4.5. *There exists a single oracle G such that the regular Boolean hierarchies over RP, UP, and NP all extend through infinite levels.*

The techniques for these constructions are similar to those used by Baker, Gill and Solovay, [BaGiSo 75], as extended by Cai and Hemachandra to work on machines which characterize the Boolean hierarchy over *NP*. However, the fact that we deal with probabilistic machines requires some changes in these techniques. Specifically, we have to overcome the following difficulties: (1) the set of oracle probabilistic machines is not enumerable and may change as the oracle changes, (2) the subtree of accepting (or rejecting) computations of an *RP* machine may contain exponentially many queries to a given oracle.

To realize our constructions we generalize Rackoff's notion of a critical string, ([Ra 82]), for diagonalizing against single *RP* machines to the notion of a critical set of strings which works on k-tuples of *RP* machines (as the machines of the Boolean hierarchy over *RP* are).

Complete statements and proofs of these results may be found in [BBJSY 89]. For the regular Boolean hierarchy over *RP* many of these separations can also be witnessed by *strong* separations, that is by sets which are *immune* for *RP*, and also immune for other low classes of *RBH*, ([BJY 89]).

5. BIBLIOGRAPHY.

[Ad 65] J. Addison, "The method of alternating chains," *Symp Theor Models,* North-Holland, (1965), 1-16.

[AdHu 87] L. Adleman and M. A. Huang, "Recognizing primes in random polynomial time," *ACM Symp Theory Computing,* (1987), 462-469.

[BaMiSh 86] E. Bach, G. Miller and J. Shallit, "Sums of divisors, perfect numbers and factoring," *SIAM J Computing,* 15, (1986), 1143-1154.

[BaGiSo 75] T. Baker, J. Gill and R. Solovay, "Relativizations of the $P = NP$ question," *SIAM J Computing,* 4 (1975), 431-442.

[Be 87] R. Beigel, "Bounded queries to *SAT* and the Boolean hierarchy," *preprint,* (1987).

[BHW 89] R. Beigel, L. Hemachandra and G. Wechsung, "On the power of probabilistic polynomial time: $P^{NP[log]} \subseteq PP$," *Structure in Complexity Conference,* (1989), to appear.

[BeGi 81] C. Bennet and J. Gill, "Relative to a random oracle A, $P^A \neq NP^A \neq coNP^A$ with probability one," *SIAM J Computing,* 10 (1981), 96-113.

[BBJSY 89] A. Bertoni, D. Bruschi, D. Joseph, M. Sitharam and P. Young, "Generalized Boolean hierarchies and Boolean hierarchies over *RP*," *Univ Wisconsin, CS Dept Tech Report,* 809 (1989), 1-50.

[BJY 89] D. Bruschi, D. Joseph and P. Young, "Strong separations for the Boolean hierarchy over *RP*," *Univ Wisconsin, CS Dept Tech Report,* 847 (1989), 1-12.

[BuHa 88] S.Buss and L. Hay, "On truth-table reducibility to *SAT* and the difference hierarchy," *Structure in Complexity Conference,* (1988), 224-233.

[CGHHSWW 88] J. Cai, T. Gundermann, J. Hartmanis, L. Hemachandra, V. Sewelson, K. Wagner, and G. Wechsung, "The Boolean hiearchy I: structural properties," *SIAM J Comput,* 6 (1988), 1232-1252.

[CGHHSWW 89] J. Cai, T. Gundermann, J. Hartmanis, L. Hemachandra, V. Sewelson, K. Wagner, and G. Wechsung, "The Boolean hiearchy II: applications," *SIAM J Comput,* 7 (1989), 95-111.

[CaHe 86] J. Cai and L. Hemachandra, "The Boolean hierarchy: hardware over *NP*," *Structure in Complexity Conference,* (1986), 105-124.

[Er 68a] Y. Ershŏv, "A hierarchy of sets, I," *Algebra and Logic,* 7 (1968), 25-43.

[Er 68b] Y. Ershŏv, "A hierarchy of sets, II," *Algebra and Logic,* 7 (1968), 15-47.

[Er 69] Y. Ershŏv, "A hierarchy of sets, III," *Algebra and Logic,* 9 (1969), 20-31.

[GeGr 86] J. Geske and J. Grollmann, "Relativizations of unambiguous and random polynomial time classes," *SIAM J Computing,* 15 (1986), 511-519.

[GuWe 86] T. Gundermann and G. Wechsung, "Nondeterministic Turing machines with modified acceptance," *Proc MFCS,* L N CS, 233 (1986), 396-404.

[Ha 78] F. Hausdorff, *Set Theory,* Chelsea, 3rd ed., 1978.

[HiZa 84] P. Hinman and S. Zachos, "Probabilistic machines, oracles and quantifiers," *Proc Recursion Theory Week,* L. N. Math 1141 (1984), 159-192.

[Ka 88] J. Kadin, "The polynomial time hierarchy collapses if the Boolean hierarchy collapses," *Structure in Complexity Conference,* (1988), 278-292.

[KöSchWa 87] J. Köbler, U. Schöning and K. Wagner, "The difference and truth-table hierarchies for *NP*," *RAIRO,* 21 (1987), 419-435.

[Ko 82] K. Ko, "Some observations on probabilistic algorithms and *NP*-hard problems," *Inform Proc Letters,* 14 (1982), 39-43.

[La 83] C. Lautemann, "*BPP* and the polynomial hierarchy," *Inform Proc Letters,* 17 (1983), 215-217.

[Pu 65] H. Putnam, "Trial and error predicates and a solution to a problem of Mostowski," *J Sym Logic,* 30 (1965), 49-57.

[Ra 80] M. Rabin, "Probabilistic tests for primality," *J Number Theory,* 12 (1980), 128-138.

[RaSh 88] M. Rabin and J. Shallit, "Randomized algorithms in number theory," To appear *Comm Pure Appl Math.*

[Ra 82] C. Rackoff, "Relativized questions involving probabilistic algorithms," *JACM,* 29 (1982), 261-268.

[Sch 80] J. Schwartz, "Fast probabilistic algorithms for the verification of polynomial identities," *JACM,* 27 (1980), 701-717.

[Si 83] M. Sipser, "A complexity theoretic approach to randomness," *Proc Symp Theory Comput,* (1983), 330-335.

[SoSt 77] R. Solovay and V. Strassen, "A fast Monte-Carlo test for primality," *SIAM J Comput* 6 (1977), 84-85; [Erratum: 7 (1978), 118].

[Wa 86] K. Wagner, "More complicated questions about maxima and minima, and some closure properties of *NP*," *ICALP,* L N Comp Sc, 226 (1986), 434-443.

[Wa 88] K. Wagner, "Bounded query computations," *Structure in Complexity Conference,* (1988), 260-277.

[WeWa 85] G. Wechsung and K. Wagner, "On the Boolean closure of *NP*," *Proc Conf Fundament Comput Theory,* L N Comp Sc, 199 (1985), 485-493.

[ZaHe 84] S. Zachos and H. Heller, "A decisive characterization of *BPP*," *Information and Control,* 69 (1986), 125-135.

[Za 86] S. Zachos, "Probabilistic quantifiers, adversaries and complexity classes: an overview," *Structure in Complexity Conference,* (1986), 383-400.

The Equational Logic of Iterative Processes

Stephen L. Bloom
Department of Computer Science
Stevens Institute of Technology
Hoboken, NJ 07030

Abstract

We describe a formalization of the equational logic of flowchart algorithms which has been shown to apply to a large class of other iterative processes connected with the theory of computation.

The last time I addressed this audience was at a meeting dedicated to the memory of Calvin Elgot. At that time, I reviewed Elgot's contributions to the algebraic analysis of flowchart algorithms.[1] In a sense, the current talk is a continuation of the previous one.

At that 1981 FCT Conference I met a native of Szeged, Zoltán Ésik. He had been interested in Elgot's papers and he told me about some of his own results. Since that time, he and I have been working together. Perhaps at the present occasion we will find other collaborators. At any rate, in this talk I hope to explain what we have been working on, and why we think it is a fundamental topic.

To my never ending surprise, many people find the formalism of algebraic theories a formidable barrier. So, for the moment, I'll follow the advice of Ernie Manes and try a version not found in our earlier papers.[2]

Suppose that C is a category with finite coproducts. You will recall that if X and Y are objects in C , then the coproduct $X+Y$ of X and Y is a C-object equipped with two coproduct injections

$$in_X : X \to X+Y;\ in_Y : Y \to X+Y$$

with the following universal property. If $f : X \to Z$ and $g : Y \to Z$ are any two C morphisms, there is a unique morphism $\langle f,\ g\rangle : X+Y \to Z$ such that

$$in_X \cdot \langle f,\ g\rangle = f \text{ and } in_Y \cdot \langle f,\ g\rangle = g.$$

---

[1] Mainly, the papers [9] and [10].

[2] Thanks also to my colleagues Klaus Sutner and Ralph Tindell for helpful comments.

The morphism $\langle f, g\rangle$ is called the source pairing of f and g. C will also have an initial object, say 0, and any object X is isomorphic to $X+0$. We write $0_Z : 0 \to Z$ for the unique morphism from the initial object[3] to the object Z. If $f : X \to Y$ and $g : Z \to W$ we write $f \oplus g : X + Z \to Y + W$ for $\langle f \cdot in_Y,\ g \cdot in_W\rangle$. For example, in the category **Pfn** of sets and partial functions, the coproduct object of two sets is their disjoint union; the empty set is the initial object.

Elgot focused attention on the *iteration equation for* f, when $f : X \to X + Y$. The iteration equation for f is the fixed point equation in the variable $\xi : X \to Y$:

$$\xi \;=\; f \cdot \langle \xi,\ 1_Y\rangle.$$

1_Y denotes the identity morphism on Y. A special case is when Y is the initial object so that, up to isomorphism, $f : X \to X$. In this case, the iteration equation for f is

$$\xi \;=\; f \cdot \xi,$$

where $\xi : X \to 0$.

Elgot formalized those categories in which "most" morphisms have a unique solution to their iteration equation. These categories were called *iterative theories.* When the iteration equation for f has a unique solution, it is denoted $f^\dagger$ and called the *iterate of f.* (Further technical requirements are needed to get iterative theories, but we need not be concerned with them here.) Such categories contain a complete set of operations for the control structure of flowchart algorithms – composition, case statements (via coproducts) and a vector iteration operation, and thus can serve as target categories for a denotational semantics of flowchart algorithms.

One important thing to notice is that the operation $f \mapsto f^\dagger$ is a partial operation, only defined when the iteration equation for f has a unique solution.
I'll describe two examples. One familiar situation in which the iteration equation arises is in the following matrix theory Mat_S. In this category, the objects are the natural numbers. A morphism $n \to p$ in Mat_S is an n by p matrix with entries in the semiring S of all subsets of words on an alphabet X. Matrix multiplication is the category composition. It is a pleasant fact that the coproduct of, say, 1 and 2 is 3, and the coproduct injections are the following two matrices:

$$in_X \;=\; \begin{bmatrix} 1 & 0 & 0 \end{bmatrix} : 1 \to 3$$

$$in_Y \;=\; \begin{bmatrix} 0 & 1 & 0 \\ 0 & 0 & 1 \end{bmatrix} : 2 \to 3.$$

In a similar way, one can show that the coproduct of n and m is $n + m$.

If we write a morphism $n \to n + p$ as

$$f \;=\; [A \;\; B]$$

[3] We will usually neglect to distinguish among isomorphic objects, or naturally isomorphic functors.

where A is n by n and B is n by p, then the iteration equation for f becomes

$$\xi = A\xi + B,$$

the equation associated with the description of the behavior of finite automata, as well as with the definition of rational power series. Here, if the entries of the matrix A are sets of words which do not contain the empty word, then there is a unique solution of the iteration equation for f. (This fact is sometimes called *Arden's Lemma.*)

One more preliminary example. Let Σ be a ranked set. If Y is a finite set, a Σ-Y *tree* is a finite or infinite rooted tree t whose internal vertices (of outdegree $k > 0$) are labeled with an element in Σ of rank k; the leaves of t are labeled by elements in Σ of rank 0 or by elements in the set Y. The category $\Sigma\,\mathrm{TR}$ has finite sets as objects. A morphism $f : X \to Y$ is a function from X to Σ-Y trees. The composite of $f : X \to Y$ with $g : Y \to Z$ is the function whose value at $x \in X$ is the Σ-Z tree obtained by attaching a copy of the tree yg to each leaf labeled y of the Σ-Y tree xf. Here (up to an isomorphism) $\{1\} + \{1\} = \{1,2\}$. The interesting trees in this category are those trees $[n] \to [p]$ obtained by unfolding flowchart schemes (with vertices labeled by letters in Σ) with n begin vertices and p exit vertices, for some $n, p \geq 0$. We use the notation $[n]$ for the set consisting of the first n positive integers. If $f : [n] \to [n+p]$ is the unfolding of a scheme, then the unique solution to the iteration equation for f is the unfolding of the scheme $[n] \to [p]$ obtained from f by identifying the first n exits with the corresponding begin vertices. The subtheories of trees which come from unfolding schemes have computational importance, since two flowchart schemes will have the same stepwise behaviors for all interpretations of their atomic letters iff they unfold to the same tree.

The meticulous reader will have noticed that in our two examples the iteration equation for certain morphisms will not have a unique solution. In the matrix theory case, if the empty word does occur in some of the entries in the matrix A, there will be many solutions. And if $\mathbf{1}_X : X \to X$ takes $x \in X$ to the Σ-X tree which consists only of a leaf labeled x, then any function which takes $x \in X$ to a tree with no leaves will be a solution to the iteration equation for $\mathbf{1}_X$, namely

$$\xi = \mathbf{1}_X \cdot \xi.$$

Iteration theories were introduced to avoid worrying about special cases. An **iteration theory** is a category C with finite coproducts, equipped with an operation of *iteration* which maps each morphism $f : X \to X + Y$ to a morphism $f^\dagger : X \to Y$. Most importantly, this operation must satisfy all of the equations satisfied by † in each tree theory $\Sigma\mathrm{TR}$, for all Σ. (Of course Y can be recovered from $X + Y$ so that the operation $f \mapsto f^\dagger$ is meaningful.)

The original motivation for this definition was to focus attention on the equational properties of the control structures of flowchart algorithms: composition, case statements and iteration. Further, by insisting that the iteration operation is total, we have the opportunity to use standard equational logic, rather than worry about the logic of partial operations.

In hindsight, two additional reasons for the usefulness of the concept *iteration theory* can be given. I tend to use one, and Ésik prefers the other. The first reason is this. There are many categories, not closely related to flowchart algorithms, with a fixed point operation which has all of the equational properties of flowchart algorithms. Knowledge of the iteration theory identities helps understand these structures. I will describe some of these structures immediately below. The second reason is that the formalism of iteration theories is precisely the one needed to solve a system of fixed point equations of the form:

$$\begin{aligned} x_1 &= f_1(x_1, \ldots, x_n, \ldots, x_{n+p}) \\ &\vdots \\ x_n &= f_n(x_1, \ldots, x_n, \ldots, x_{n+p}) \end{aligned}$$

Some examples of iteration theories.

- Partial functions. Let Ω be a collection of sets closed under disjoint union. Let C be the category whose objects are the sets in Ω and whose morphisms $X \to Y$ are all partial functions from X to Y. This is an example of an ordered theory, where $f \leq g : X \to Y$ if the graph of f is contained in the graph of g. We define the iteration operation on the partial function

  $$f : X \to X + Y$$

  as the least solution of the iteration equation for f. It follows that $xf^\dagger = y$ iff there is a finite sequence

  $$x = x_0, x_1, \ldots, x_k$$

  with $x_i f = x_{i+1} \in X$ for $i < k$, and with $x_k f = y$. Then C is an iteration theory, and this is the only way to define the operation $f \mapsto f^\dagger$ so that C is an iteration theory. (See [4].)

- Theories of *sequacious functions* are iteration theories. These theories were introduced by Elgot in [9] to mirror the stepwise behavior of algorithms with an "external state" X, where X is a fixed set. The objects in the theory Seq_X are (finite) sets. A morphism $A \to B$ in Seq_X is a sequacious function

  $$f : X^+ \times A \cup X^\infty \to X^+ \times B \cup X^\infty.$$

  The set of finite and infinite sequences of elements of X is denoted X^∞. A sequacious function f is one which satisfies the following conditions.

  1. if $f(x, a) = (u, b)$ then $u = xu'$, for some $u' \in X^*$;
  2. if $f(x, a) = w \in X^\infty$ then $w = xw'$, for some $w' \in X^\infty$;
  3. if $f(x, a) = (u, b)$, then $f(vx, a) = (vu, b)$, all $v \in X^*$;
  4. if $f(x, a) = w \in X^\infty$ then $f(vx, a) = vw$, all $v \in X^*$;
  5. $f(w) = w$, $w \in X^\infty$.

Note that a sequacious function $A \to B$ is determined by its values on pairs of the form $(x,a) \in X \times A$. If $f : A \to A+B$ is *positive* in the sense that the length of $f(x,a)$ is at least 2, then there is a unique solution to the iteration equation for f; the iteration operation can be extended to all sequacious functions so that Seq_X is an iteration theory. (See [4].)

- The rational theories and ω-continuous ordered theories studied by the ADJ group (see e.g. [18] and [16]), are iteration theories. These theories have the property that their hom-sets $T(X,Y)$ are equipped with a partial order for which certain sequences have a least upper bound. Enough least upper bounds exist to define iteration as a least upper bound of approximations, imitating Scott's well known construction.

- ω-functors on a category of ω-categories.[4] An ω-category is one with an initial object $\perp$ in which all diagrams of the shape

$$a_0 \xrightarrow{f_0} a_1 \xrightarrow{f_1} \ldots a_n \xrightarrow{f_n} a_{n+1} \to \ldots$$

have a colimit. An ω-functor between ω-categories preserves these colimits. Suppose that Ω is a collection of ω-categories closed under finite products. The iteration theory $Th_\omega(\Omega)$ has the categories in Ω as objects; a morphism $C \to D$ is an ω-functor

$$F : D \to C.$$

Note the reversal of direction. The theory composition of $F : X \to Y$ and $G : Y \to Z$ is functor composition in the opposite order:

$$F \cdot G \;=\; Z \xrightarrow{G} Y \xrightarrow{F} X.$$

Thus, coproducts in the theory turn into products in Ω. For a functor

$$F : D \times C \to C$$

and an object x in D, let $F_x : C \to C$ be the endofunctor which takes

$$\alpha : u \to v$$

in C to

$$F(x,u) \xrightarrow{F(1_x,\alpha)} F(x,v)$$

in C. For each object x in D, consider the diagram

$$\perp \to \perp F_x \xrightarrow{g_1} \perp F_x^2 \xrightarrow{g_2} \ldots$$

[4]The previous footnote applies most strongly here.

in which the first arrow $\perp \to \perp F_x$ is uniquely determined by the initiality of $\perp$, and g_{n+1} is obtained by applying F_x to g_n, for $n \geq 0$. For each such diagram, we choose a particular colimit

$$\perp F_x^n \longrightarrow xF^\dagger,\ n \geq 0,$$

whose vertex is $xF^\dagger$, by definition. The object $xF^\dagger$ is determined only up to isomorphism. But once a choice of a colimit diagram has been made for each object in D, we can extend the definition of $F^\dagger$ to morphisms in D in a unique way so that it becomes an ω-functor. Ésik and I have shown in [5] that each of the theories $Th_\omega(\Omega)$ is an iteration theory. Further, in the case that each category in Ω is an "effective cpo" [17], the subcategory of $Th_\omega(\Omega)$ whose morphisms are the computable maps is an iteration theory.

What is the point? As shown nicely in the work of Lehmann, Plotkin and Smyth, [13], [15], the fixed point equations used to specify circular data types can be solved using ω colimits in appropriate categories. One example is a standard equation describing "stacks of elements in A":

$$S \ = \ 1 + A \times S.$$

Let C be the ω-category of sets, and let

$$F : C \times C \to C$$

be the ω-functor taking the pair of sets A, X to

$$(A, X)F \ = \ 1 + A \times X,$$

where 1 is a one point set. Since $F^\dagger$ is a solution of the iteration equation,

$$AF^\dagger \ = \ 1 + A \times (AF^\dagger).$$

Thus, if $AF^\dagger$ is not too large, perhaps $AF^\dagger$ is the set of stacks we intend when writing the above fixed point equation. Indeed, since $F^\dagger$ is defined as the above colimit, $AF^\dagger$ is an "initial F_A algebra" in the terminology of Adamek [1] and can be shown to be isomorphic to the collection of all finite sequences of elements of A.

What is gained? Since the ω-functors form an iteration theory, we know that they satisfy many identities. Thus it is possible to simplify or prove equivalent different descriptions of circular data types. We give one short example of this possibility. Let $\omega \bullet x$ denote a copower of the object x with itself ω times, and suppose that we are trying to solve the fixed point equation

$$v \ = \ G(v) \tag{1}$$

in an ω-category C, where on objects v,

$$G(v) \ = \ 1 + \omega \bullet (a + v) \times v$$

for a fixed object a in $\mathcal{C}$. Consider the functor $\mathcal{C}^2 \to \mathcal{C}$ defined on objects by

$$F_1(u,v) = a + u + v$$

It is not hard to see that for any object v,

$$F_1^\dagger(v) = \omega \bullet (a + v),$$

so that

$$G(v) = F_2(F_1^\dagger(v), \mathbf{1}(v)),$$

where

$$F_2(v) = 1 + u \times v.$$

We use a special case of what we call the *pairing identity* for iteration theories:

$$\langle F_1,\ F_2\rangle^\dagger = \langle F_1^\dagger \cdot H^\dagger,\ H^\dagger\rangle$$

where

$$H(v) = 1 + (\omega \bullet (a + v)) \times v.$$

It follows that to solve (1) we may solve simultaneously the system of two fixed point equations

$$\begin{aligned} u &= a + u + v \\ v &= 1 + u \times v \end{aligned}$$

and then take the second component of the solution.

As one further example, we mention that, as will be shown in a forthcoming paper, modulo bisimilarity Milner's synchronization trees also form an iteration theory.

I think now we may justly say that iteration theories occur quite frequently in computer science. It seems that if one looks hard enough at an iterative process, one finds an iteration theory. Thus it is useful to know what information can be derived from the fact that a particular category is an iteration theory. By definition, an iteration theory satisfies at least the identities valid for the tree theories. But what are these identities?

At least three sets of axioms for iteration theories have been found. I will mention here only two. The first group was obtained by Ésik in [11] and consists of the following four axioms:

- *the left zero identity:*

$$(0_X \oplus f)^\dagger = f,$$

for $f : X \to Y$;

- *the right zero identity:*

$$(f \oplus 0_Z)^\dagger = f^\dagger \oplus 0_Z;$$

for $f : X \to X + Y$;

- *the pairing identity:*

$$\langle f, g\rangle^\dagger = \langle f^\dagger \cdot \langle h^\dagger, \mathbf{1}_Z\rangle, h^\dagger\rangle,$$

where $f : X \to X + Y + Z$, $g : Y \to X + Y + Z$ and h is defined by

$$h := g \cdot \langle f^\dagger, \mathbf{1}_{Y+Z}\rangle : Y \to Y + Z$$

- *the commutative identity:* This identity is the most difficult to state, let alone understand! We call any morphism formed from the coproduct injections by means of source pairing a *base morphism*. Now suppose that $Y = Y_1 + \ldots + Y_m$, $m \geq 1$, is a coproduct of the m objects Y_i. Suppose that $f : X \to Y + Z$ and that $\rho : Y \to X$ is a base epimorphism. Lastly suppose that for each $i = 1, \ldots, m$, $\rho_i : Y \to Y$ is a base morphism such that $\rho_i \cdot \rho = \rho$. Then, writing in_i for in_{Y_i}, the commutative identity is

$$\langle in_1 \cdot \rho \cdot f \cdot (\rho_1 \oplus \mathbf{1}_Z), \ldots, in_m \cdot \rho \cdot f \cdot (\rho_m \oplus \mathbf{1}_Z)\rangle^\dagger = \rho \cdot (f \cdot (\rho \oplus \mathbf{1}_Z))^\dagger.$$

The second group of axioms was found by Ştefănescu in [12]. In addition to the commutative identity just mentioned, there are four others:

- *the fixed point identity:*

$$f^\dagger = f \cdot \langle f^\dagger, \mathbf{1}_Y\rangle,$$

for $f : X \to X + Y$.

- *the parameter identity:*

$$(f \cdot (\mathbf{1}_X \oplus g))^\dagger = f^\dagger \cdot g,$$

for $f : X \to X + Y$, $g : Y \to Z$.

- *the composition identity:*

$$(f \cdot g)^\dagger = f \cdot (g \cdot (f \oplus \mathbf{1}_Z))^\dagger,$$

for $f : X \to Y$, $g : Y \to X + Z$.

- *the double dagger identity:*

$$f^{\dagger\dagger} = (f \cdot (\langle \mathbf{1}_X, \mathbf{1}_X\rangle \oplus \mathbf{1}_Y)^\dagger,$$

for $f : X \to X + X + Y$.

Each of these identities can be thought of as a method to solve a system of fixed point equations. For example, the pairing identity shows that the simultaneous solution of a system of n equations can be reduced to the solution of n single fixed point equations. A special case of the commutative identity shows that the system of equations

$$\begin{aligned} x_1 &= f(z_1, z_2, y) \\ x_2 &= f(z_3, z_4, y) \end{aligned}$$

where $z_j \in \{x_1, x_2\}$, $j = 1, \ldots, 4$, has the solution $x_1 = x_2 = x$, where x is the solution to

$$x = f(x, x, y).$$

I should mention that most of the iteration theories which seem to occur naturally have what we call a *functorial dagger*. This property, discovered by Arbib and Manes in [2], is the following implication:

$$\begin{array}{ccc} X & \xrightarrow{f} & X+Y \\ \rho\Big\downarrow & & \Big\downarrow \rho\oplus\delta \\ Z & \xrightarrow[g]{} & Z+W \end{array} \quad \Rightarrow \quad \begin{array}{ccc} X & \xrightarrow{f^\dagger} & Y \\ \rho\Big\downarrow & & \Big\downarrow \delta \\ Z & \xrightarrow[g^\dagger]{} & W \end{array}$$

i.e. the right hand side commutes if the left hand side does. (We can restrict the implication to the case that ρ is a base epimorphism.) It is easy to show that this implication implies the commutative identity, and can thus be used to help show that a particular theory is an iteration theory.

We can use the dagger identities in the context of matrix theories to get an answer to the question that was asked explicitly by Conway in [8] and implicitly in [14]: What are appropriate axioms to impose on the Kleene *-operation? We answer the question in the following framework. Suppose that we have a theory of all matrices over a semiring S which is an iteration theory, as in the first example above. Thus, to any n by $n+p$ matrix $f = [A\ B]$ the dagger operation assigns an n by p matrix $f^\dagger$ in such a way that all the axioms for iteration theories are valid. Consider the n by $2n$ matrix of the form

$$f = [A\ \mathbf{1}_n]$$

where A is n by n and $\mathbf{1}_n$ is the identity matrix. Define the operation

$$A \mapsto A^*$$

on the square matrices over S by

$$A^* := [A\ \mathbf{1}_n]^\dagger.$$

We can prove that the star operation on square matrices will satisfy the following identities:

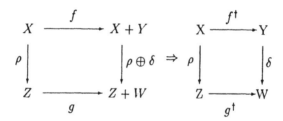

$$\begin{aligned} a^* &= 1_n + aa^*; \\ (ab)^* &= 1_n + a(ba)^*b; \\ (a+b)^* &= (a^*b)^*a^* \end{aligned}$$

as well as the following inductive equation:

$$\begin{bmatrix} a & b \\ c & d \end{bmatrix} = \begin{bmatrix} (a+bd^*c)^* & a^*b(d+ca^*b)^* \\ d^*c(a+bd^*c)^* & (d+ca^*b)^* \end{bmatrix}$$

where a is n by n, b is n by 1, c is 1 by n and d is 1 by 1. (This is an incomplete list of equational properties of *. Others are implied by the commutative identity.) The operation * is not necessarily idempotent. Indeed, the initial matrix iteration theory consists of all matrices over the following linearly ordered semiring S_0, whose elements in order are:

$$0, 1, 2, ..., 1^*, (1^*)^2, (1^*)^3, ..., 1^{**}.$$

Addition and multiplication on the integers are the standard operations; otherwise, the operations are given by:

$$\begin{aligned} x + y &= \max\{x, y\}, \text{ if } x \text{ or } y \geq 1^*; \\ n(1^*)^k &= (1^*)^k, \text{ if } n, k \geq 1; \\ (1^*)^n(1^*)^p &= (1^*)^{n+p}; \\ x1^{**} = 1^{**}x &= 1^{**}, \text{ if } x \neq 0. \end{aligned}$$

The star operation on S_0 is defined by:

$$x^* = \begin{cases} 1 & \text{if } x = 0; \\ 1^* & \text{if } x = 1; \\ 1^{**} & \text{otherwise.} \end{cases}$$

If one insists that addition is idempotent and that $1^* = 1$, then the corresponding initial matrix theory consists of all matrices over the two element semiring with $1 = 1 + 1 = 1^*$. See [6] for further information.

Finally, let me only mention that one can get models of propositional dynamic logic in iteration theories, and do partial correctness logic as well (see [3] and [7]). In fact, consideration of just how to justify standard Hoare logic in iteration theories has led to a system of partial correctness logic which applies to all flowchart programs, not just the while-programs. But this is a long story, and will require another talk.

References

[1] J. Adamek. Free algebras and automata realizations in the language of category theory. *Comm. Math. Univer. Carolinae*, pages 589–602, 1974.

[2] M.A. Arbib and E. Manes. Partially additive categories and flow-diagram semantics. *J. Algebra*, (1):203–227, 1980.

[3] S.L. Bloom and Z. Ésik. Floyd–Hoare logic in iteration theories. Submitted for publication.

[4] S.L. Bloom and Z. Ésik. Varieties of iteration theories. *SIAM Journal of Computing*, 17(5):939–966, 1988.

[5] S.L. Bloom and Z. Ésik. Equational logic of circular data type specification. *Theoretical Computer Science*, 63(3):303–331, 1989.

[6] S.L. Bloom and Zoltán Ésik. Matrix and matricial iteration theories. Submitted for publication.

[7] Stephen L. Bloom. A note on guarded theories. Lecture Notes in Computer Science. Springer-Verlag. To appear.

[8] John Conway. *Regular Algebras and Finite Machines*. Chapman and Hall, London, 1971.

[9] Calvin C. Elgot. Monadic computation and iterative algebraic theories. In J. C. Shepherson, editor, *Logic Colloquium 1973, Studies in Logic*, Volume 80. North Holland, Amsterdam, 1975.

[10] Calvin C. Elgot. Structured programming with and without goto statements. *IEEE Transactions on Software Engineering SE-2*, (1):41–54, 1976.

[11] Z. Ésik. Identitites in iterative and rational theories. *Computational Linguistics and Computer Languages*, 14:183–207, 1980.

[12] Gh. Ştefănescu. On flowchart theories: Part I. The deterministic case. *Journal of Computers and System Science*, (35):163–191, 1987.

[13] D. Lehmann and M.B. Smyth. Algebraic specification of data types: a synthetic approach. *Mathematical Systems Theory*, 14:97–139, 1981.

[14] E. G. Manes. Assertional categories. Volume 298 of *Lecture Notes in Computer Science*, pages 85–120. Springer-Verlag, 1987.

[15] M.B. Smyth and G.D. Plotkin. The category theoretic solution of recursive domain equations. *SIAM Journal of Computing*, 11(4):761–783, 1982.

[16] E.G. Wagner, J.W. Thatcher, and J.B. Wright. Programming languages as mathematical objects. Volume 64 of *Lecture Notes in Computer Science*, pages 84–101. Springer–Verlag, 1978.

[17] Klaus Weihrauch. *Computability*. EATCS Monographs on Theoretical Computer Science. Springer–Verlag, Berlin, Heidelberg, New York, Tokyo, 1987.

[18] J.B. Wright, J.W. Thatcher, J. Goguen, and E.G. Wagner. Rational algebraic theories and fixed–point solutions. Proceedings 17th IEEE Symposium on Foundations of Computing, pages 147–158, Houston, Texas, 1976.

The Distributed Bit Complexity of the Ring: From the Anonymous to the Non-anonymous Case

Hans L. Bodlaender *

Shlomo Moran **

Manfred K. Warmuth ***

ABSTRACT:

The *distributed bit complexity* of an asynchronous network of processors is a lower bound on the worst case bit complexity of computing any non-constant function of the inputs to the processors [MW]. This concept attempts to capture the amount of communication required for any useful computation on the network.

The aim of this kind of research is to characterize networks by their bit complexity. In [MW] Moran and Warmuth studied the bit complexity of a ring of n processors under the assumption that all the processors in the ring are identical (anonymous [ASW]), i.e. all processors run the same program and the only parameter to the program is the input of the processor. It was shown that for anonymous rings it takes $\Omega(n \log n)$ bits to compute any non-constant function. We would like to release the assumption that the processors are anonymous by allowing each of the processors to have a distinct identity (which becomes a second parameter to the program). If the set of possible identities grows double exponentially in n, then by a simple reduction to the anonymous case one can show that the lower bound holds as well [MW].

In this paper we show that the $\Omega(n \log n)$ bit lower bound for computing any non-constant function holds even if the set of possible identities is very small, that is, $n^{1+\varepsilon}$, for any positive ε.

1. INTRODUCTION.

Consider a distributed network of processors that communicate asynchronously along the links of the network. Each processor receives an external input. The processors compute a function of the total input configuration of the network and then terminate. The aim of this line of research is to establish lower bounds on the amount of resources (complexity) required to compute any non-constant function on a given network. The lower bound must hold for any algorithm that computes the non-constant function.

As a complexity measure of an algorithm we use the worst case message and bit complexity over all possible input configurations, all possible choices (if any) for the identities of the processors and all possible delay times of the communication links of the network. The minimum worst case message and bit complexity for computing any non-constant function on the network is called the (asynchronous) *distributed message* and *bit complexity,* respectively, of the network [MW]. In this paper we give results about the distributed bit complexity of a ring of n processors.

---

* Department of Computer Science, University of Utrecht, 3508 TA Utrecht, the Netherlands. Part of this work was done while this author was at the Laboratory for Computer Science at MIT being supported by a grant from the Netherlands Organization for the Advancement of Pure Research (Z.W.O.).

** Department of Computer Science, the Technion, Haifa 32000, Israel. This research was supported in part by Technion V.P.R. Funds - Wellner Research Fund, and by the Foundation for Research in Electronics, Computers and Communications, administrated by the Israel Academy of Sciences and Humanities. Part of this work was done while this author was visiting the University of California at Santa Cruz being supported by ONR grant N00014-86-K-0454.

*** Department of Computer and Information Sciences, University of California, Santa Cruz, CA 95064. This author gratefully acknowledges the support of ONR grant N00014-86-K-0454. Part of this work was done while this author was visiting the Technion, Israel Institute of Technology, supported by the Wolberg fund.

In general we seek to investigate how the distributed bit complexity depends on the network. Intuitively, this complexity increases with the amount of "symmetry" given in the network which is determined by the two following factors.

a) The topology of the network: To compute the function value, each processor needs to get some global information about the input configuration. Topological symmetry in the network makes it hard to gather global information without causing a large number of messages/bits to be sent.

b) The degree of "anonymity" between the processors. Three cases are considered in the literature: In the first the processors are anonymous (i.e. they have no id's). In the second each processor has a distinct id. The id's can be used to break symmetry in the network and cut down message traffic. In a third case the network has one distinguished processor (the *leader*) who can coordinate the computation.

In the case where the network has a leader then this leader can coordinate the computation. It can first instigate messages that collect the global information that is necessary to compute the non-constant function. Secondly, it will broadcast the function value to the network. The bit complexity of such an algorithm can be linear in the size (number of edges + number of vertices) of the network. Note that in some networks the topology of the network distinguishes a small number of processors. For example in the star network the central processor is naturally the leader and in a chain the two end processors can coordinate the computation. The definition of distributed complexity is mainly useful for highly symmetric networks, such as the ring, the hypercube or the torus. Each of these networks is used in practice.

The distributed complexity of a network gives a theoretical limit of the capabilities of the network. It can be used to guide the choice of networks used in practice. Previous research and this paper concentrate on the distributed complexity of the ring of n processors. It leads to an interesting case study as to how the processors may avoid symmetry.

The distributed bit complexity of a ring of n anonymous processors is $\Theta(n \log n)$ [MW]. This holds for arbitrary ring size and input size. However the lower bound proof of $\Omega(n \log n)$ in [MW] assumes that the processors are anonymous or that the processors have distinct identities but the set of possible identities is very large ($\Omega(n\, 2^{n^n})$, which is unrealistically large).

If the set of distinct identities is very small, i.e. in $\{1,2, \cdots ,n+c\}$, for some constant c, then it is easy to compute non-constant functions in $O(n)$ bits (use processor with identities in $\{1,2, \cdots ,c+1\}$ as leaders). We show in this paper that the approach of giving processors distinct identities from a reasonably large set does not help to break the $\Omega(n \log n)$ lower bound.

Theorem 1. Let f be any non-constant function on Σ^n for some arbitrary alphabet Σ, and let AL be any asynchronous algorithm that computes f on ring of n processors, labeled with n distinct identities chosen from a set X of size at least $n^{1+\varepsilon}$, for some $\varepsilon > 0$. Then the worst case bit complexity of AL is $\Omega(n \log n)$.

Thus in highly symmetric networks such as the ring the only way to compute non-constant functions in $O(n)$ bits is essentially to pre-elect a leader. However, electing a leader costs $\Omega(n \log n)$ messages even if the set of possible identities is only of size cn, for any constant $c>1$ [PKR, B1].

The proof of Theorem 1 is constructive: Given an algorithm AL that computes a non-constant function on a ring of n processors, we use cut-and-paste techniques to construct a computation of AL in which $\Omega(n \log n)$ bits are sent. cut-and-paste method was first used in [MW] for the anonymous case, where the processors have no id's. When the processors have id's then cutting and pasting is more involved, since in the final ring constructed all processors must have distinct id's from a small set of possible id's ($O(n^{1+\varepsilon})$). The key idea is to iteratively apply cutting and pasting to many lines of processors in parallel. A case analysis shows that in the final set of lines there must be a line that can be embedded in a ring of n processors with distinct id's, and a computation of AL on this line requires $\Omega(n \log n)$ bits.

Note that the lower bound of Theorem 1 does not depend on the size of the input alphabet and the size of the ring. In contrast, it has been shown that that the distributed message complexity of the anonymous ring *does* depend on both these sizes. Specifically, large input alphabets, as well as small non-divisors of the ring size, can reduce the distributed message complexity of the anonymous ring [ASW, MW, DG, B2].

The main future challenge will be to determine the distributed bit complexity of other networks, such as the hypercube. Recently, the distributed bit complexity of the torus was shown to be $\Theta(n)$ [BB]. The proofs for the case when the processors have identities are expected to be more involved (see this paper) than in the anonymous case. Some techniques developed in this paper for rings are likely to be applied to other networks. Also, it is an interesting open problem to determine whether the lower bound of Theorem 1 can be extended to the case where the set of distinct identities is of size cn, for some constant c.

Probabilistic ways to break symmetry are studied in [AAHK] and the "probabilistic" distributed bit complexity of anonymous rings is shown to be $\Theta(n\sqrt{\log n})$. It would be interesting to know whether this bound remains if the processors have distinct identities in a reasonably large range.

2. DEFINITIONS AND BASIC RESULTS.

A *processor* consists of an input letter and an id. Let Σ be the input alphabet, which is allowed to be arbitrary large. Let X be a set of id's, and suppose that $|X| \geq n^{1+\varepsilon}$, for some constant $\varepsilon > 0$. Let $\sigma = \sigma_1 \cdots \sigma_n \in \Sigma^n$ be an input word and $x = (x_1, \cdots, x_n)$ be a sequence of n distinct identities taken from X. A *ring configuration* $R(\sigma, x)$ consists of processors $p_1, \cdots, p_n$, where p_i is connected by a link to $p_{i (mod\ n)+1}$, (for $i = 1, \cdots, n$), and processor p_i has input σ_i and id x_i. A *line configuration* $L(\sigma, x)$ is defined similarly, except that there is no link between p_1 and p_n; informally, a line configuration can be viewed as a ring configuration in which there is an infinite delay on the link connecting p_1 and p_n. The *size* of a ring configuration (or a line configuration) is the number of processors in the configuration. We use $\cdot$ to denote the concatenation of two sequences. Thus, for line configurations $L_1 = L(\tau, x)$ and $L_2 = L(\omega, y)$, $L_1 \cdot L_2$ denotes the line configuration $L(\tau\omega, xy)$.

Consider an execution of some distributed algorithm on a ring or a line configuration. The *history* of a link e in this execution, denoted by $h(e)$, is the sequence of messages delivered on e with their directions. Formally, $h(e) = (d_1 m_1 d_2 m_2 \cdots d_s m_s)$, where d_i is either R (for right) or L (for left), and $m_i \in \{0,1\}^+$ is the ith message delivered on e, in direction d_i. A message is considered delivered when it is accepted and in case of a tie, messages from the left are delivered before message from the right. The *length* of a history is the number of characters in it. Note that the length of a history of a link is at most twice the number of bits delivered on the link. In [MW] a similar notion of history was defined for processors instead of links.

A *history sequence* of a line configuration $L(\sigma, x)$ is a sequence $H(L) = (p_1, h_1, p_2, h_2, \cdots, h_{n-1}, p_n)$, where the p_i's are the processors in L and h_i is the history of the link connecting p_i and p_{i+1}. *A segment of size m* of the history sequence $H(L)$ defined above is a sequence $(p_i, h_i, \cdots, h_{i+m-2}, p_{i+m-1})$ $(i+m-1 \leq n)$. The *length* of a segment of a history sequence is the sum of the lengths of the histories of its links.

Let $H_1 = (p_1,h_1, \cdots ,h_{n-1},p_n)$ and $H_2 = (q_1,h'_1, \cdots ,h'_{m-1},q_m)$ be two segments of history sequences, and let h be a history. Then $H_1 \cdot h \cdot H_2$ denotes the segment $(p_1,h_1, \cdots ,h_{n-1},p_n,h,q_1,h'_1, \cdots ,h'_{m-1},q_m)$. Finally, we say that a history sequence H of a line configuration L is *produced* by the algorithm AL if there is an execution of AL on L with history H.

A basic tool in our proof is converting executions of AL on ring configurations to executions on line configurations and vice-versa. Since the output value of any execution of AL on a ring configuration $R(\sigma,x)$ must be $f(\sigma)$ for all possible delay times of the asynchronous links, we may choose particular delay times for the proofs. The basic delay strategy [MW], here called *semi-synchronized execution*, is an execution in which internal computation at a processor takes no time and links are either *blocked* (very large delay) or are *synchronized* (it takes exactly one time unit to traverse the link); each unblocked link may become blocked at any time, and once it becomes blocked it remains so indefinitely.

Consider an execution of AL on $R(\sigma,x)$ which is (fully) synchronized (no link is blocked), and assume that this execution terminates in less than t time units. Without loss of generality, let $t = nk$ for some integer k. As in [MW], we associate the *canonical line configuration* $D(\sigma,x) = L(\sigma^{2k},x^{2k})$ with the ring configuration $R(\sigma,x)$. The $2t = 2nk$ processors in $D(\sigma,x)$ are denoted by $p_{1,1},p_{2,1}, \cdots ,p_{n,1},p_{1,2}, \cdots ,p_{n,k},p'_{1,1}, \cdots ,p'_{n,k}$. Note that, by definition, processors $p_{i,j}$ and $p'_{i,j}$ have identity x_i and input σ_i. Informally, $D(\sigma,x)$ consists of $2k$ copies of the $R(\sigma,x)$ that were cut at the link $p_n - p_1$ and then concatenated to one line of $2kn$ processors. Thus, processors $p_{i,j}$ and $p'_{i,j}$ in $D(\sigma,x)$ correspond to the processor p_i in the jth and $(k+j)^{th}$ copies of $R(\sigma,x)$.

Let $D = D(\sigma,x)$ be a canonical line configuration. A *canonical execution* of AL on D is a semi-synchronized execution in which for $i = 1, \cdots ,t$, the ith leftmost link and the ith rightmost links of D are blocked at time i (by the definition of line configuration, the link connecting $p_{1,1}$ and $p'_{n,k}$ is blocked at time 0). Finally, the *canonical history sequence* of D, to be denoted by $H(D)$, is the history sequence produced by to the canonical execution of AL on D. The relation between a synchronized execution on a ring configuration $R(\sigma,x)$ and a canonical execution on the corresponding line configuration $D(\sigma,x)$ is given by the following:

Lemma 2.1: [MW]. In a canonical execution of $D(\sigma,x)$, both $p_{n,k}$ and $p'_{1,1}$ output $f(x)$ and terminate. □

The processors $p_{n,k}$ and $p'_{1,1}$ will be called the *center processors* of $D(\sigma,x)$.

Lemma 2.2 [MW]: Let $D = D(\sigma,x)$ be the canonical line configuration of $R = R(\sigma,x)$, and let $H = H(D)$. The bit complexity of the synchronized execution of AL on R is bounded from below by half of the history length of any segment of size n of H.

Proof: Any n consecutive histories in H are prefixes of the histories of the corresponding edges in a synchronized execution on the ring configuration $R(\sigma,x)$. The result now follows from the observation that the length of the history of a link is at most twice the number of bits delivered on it. □

Lemma 2.3 [MW]: Let $W_1,...,W_k$ be k distinct words over an alphabet of size $r > 1$. Then there is a word W_i s.t. $|W_i| \geq \log_r(k/2)$ (for $1 \leq i \leq k$) and $|W_1|+|W_2|+ \cdots +|W_k|>(k/2)\log_r(k/2)$.

Proof: Represent the W_i with an r-ary tree, s.t. each W_i corresponds to a path from the root to an internal node or a leaf of the tree. In the tree each leaf is responsible for some W_i. Assume the overall length of the words W_i is minimized. Then in the corresponding tree all internal nodes except possible one internal node of maximum level have degree r. Hence at least half of the nodes are leaves. The lemma is implied by the fact that the average height of the leaves in an r-ary tree with v leaves is at least $\log_r v$. □

The above two lemmas imply the following.

Lemma 2.4: Let $R = R(\sigma,x)$ be a ring configuration, and let $D = D(\sigma,x)$. If there are n consecutive histories in a $H(D)$ that contain at least $\frac{1}{12}n$ distinct histories, then the algorithm *AL* for computing f requires $\Omega(n \log n)$ bits in the worst case. □

In view of the above lemma, we assume the following for the rest of the paper.

Assumption Q: There is no canonical history sequence of *AL* in which n consecutive histories contain more than $\frac{1}{12}n$ distinct histories.

In the below proofs we manipulate existing history sequences to create new ones. Two basic rules are used in our manipulations, which are presented below.

Rule 1: Let $H = (q_1,h_1, \cdots ,q_i,h_i,q_{i+1}, \cdots ,h_{m-1},q_m)$ be a history sequence that is produced by an execution E on a line $L = L(\sigma,x)$, and let h'_i be any prefix of h_i. Then there are histories $h'_1, \cdots ,h'_{i-1}\ h'_{i+1}, \cdots ,h'_{m-1}$, where h'_j is a prefix of h_j, such that the history sequence $H' = (q_1,h'_1, \cdots q_i,h'_i,q_{i+1}, \cdots ,h'_1,q_m)$ is produced by some execution E' of *AL* on L.

Proof: Consider the execution E of *AL* on L. On a certain moment during this execution, the history of the i'th link will be h'_i. Then stop this execution by blocking all links. Clearly, for each j, $1 \le j \le m$, the history of the j'th link will be a prefix of h_j. □

Rule 2: Let $H = H_1 \cdot h \cdot H_2$ and $H' = H'_1 \cdot h \cdot H'_2$ be two history sequences, corresponding to executions E and E', respectively, of *AL* on line configurations $L = L_1L_2$ and $L' = L'_1L'_2$, where h is the history of the links connecting L_1 with L_2 and L'_1 with L'_2. Then the history sequence $\hat{H} = H_1 \cdot h \cdot H'_2$ is produced by an execution $\hat{E}$ of *AL* on the line configuration $\hat{L} = L_1L'_2$.

Proof: $\hat{E}$ is obtained by alternating the execution E, restricted to L_1, and E' restricted to L'_2. Suppose $h=(d_1m_1, \cdots ,d_sm_s)$. Let e be the link between L_1 and L'_2. Then, for i from 1 to s: if $d_i=R$, then do a part of execution E on L_1, until message m_i is sent on link e, else do a part of execution E' on L'_2, until message m_i is sent on link e. The resulting execution $\hat{E}$ of *AL* produces the history sequence $H_1 \cdot h \cdot H'_2$ on $\hat{L} = L_1L'_2$. □

One specific way in which Rule 2 above will be used is the *maximal shrinking*: Consider a history sequence $H = H_1 \cdot h \cdot H_2 \cdot h \cdot H_3$ that corresponds to an execution of *AL*; then the history sequence $H' = H_1 \cdot h \cdot H_3$ also corresponds to some execution of *AL*. By repeated applications of this operation, any segment of a history sequence can be *shrunk* to a segment in which all the histories are distinct (A similar technique was also used in [MW]). If a history segment H is produced by some execution of *AL* on the corresponding line of processors, then $E(H)$ denotes one such execution.

3. CONSTRUCTING FAMILIES OF CONTRADICTING COMPUTATIONS.

Let τ and ω be two input configurations s.t. $f(\tau) \neq f(\omega)$. We will construct two families of history sequences $F(\tau)$ and $F(\omega)$ that correspond to executions of AL on τ and ω, resp. The identities of the processors in the history sequences in $F(\tau)$ and $F(\omega)$ will belong to sets X_τ and X_ω, resp., where $X_\tau \cap X_\omega = \Phi$, $X_\tau \cup X_\omega = X$ and $0 \leq |X_\tau| - |X_\omega| \leq 1$. $F(\tau)$ and $F(\omega)$ will have similar properties. The description and construction of $F(\tau)$ is given below:

Each history sequence H in $F(\tau)$ will satisfy the following properties:

Property (a): There is an actual execution of AL, $E(H)$, that produced H.

Property (b): H can be written as $H = LE \cdot h_L \cdot LI \cdot h_C \cdot RI \cdot h_R \cdot RE$, where LE, LI, RI and RE are segments of history sequences connected by links having histories h_L, h_C and h_R, satisfying the following:

Property (b1): There is a canonical line configuration $D = D(\tau, x)$, where $x \in X_\tau{}^n$, such that $LI \cdot h_C \cdot RI$ is a segment of the canonical history sequence $H(D)$. Moreover, the rightmost processor of LI and the leftmost processor of RI (which are connected by a link with history h_C) are the center processors of D, and hence they output $f(\tau)$.

Property (b2): All the identities of the processors in $LE \cdot RE$ are distinct from each other and no id of a processor in $LE \cdot RE$ is also an id of a processor in $LI \cdot h_C \cdot RI$.

The segments LE and RE of H, as well as the histories h_L and h_R, may be empty. The segment $LI \cdot h_C \cdot RI$ will be called the *inner part* of H.

To construct the set $F(\tau)$, we construct a list $F = (F_0, \cdots, F_i, \cdots)$, where each F_i is a set of history sequences. F_0 is the empty set, and F_{i+1} is obtained by applying one of the operations (i) - (iii) below to F_i. Eventually we get a set F_N to which none of these operations is applicable; this F_N is $F(\tau)$.

(i) *ADD*: Assume that there are n identities in X_τ that do not appear in any history sequence in F_i, and let $x = (x_1, \cdots, x_n)$ be a sequence of such identities. Let $D = D(\tau, x)$ be the canonical line configuration of $R(\tau, x)$. F_{i+1} is obtained by adding $H(D)$, the canonical history sequence of D, to F_i. (Note that $H(D)$ can be written as $LI \cdot h_C \cdot RI$, where h_C is the history of the link connecting the center processors of $D(\tau, x)$).

(ii) *LEFT-JOIN* : This operation replaces two history sequen H and H' in F_i by their *LEFT-JOIN*, which is the history sequence $\hat{H}$ defined below; the resulting set is F_{i+1}. The property of $\hat{H}$ which we need for our proof is that its inner part is shorter than the inner parts of both of H and H'. The definition of this operation follows:

Let $H = LE \cdot h_L \cdot LI \cdot h_C \cdot RI \cdot h_R \cdot RE$ and $H' = LE' \cdot h_L' \cdot LI' \cdot h_C' \cdot RI' \cdot h_R' \cdot RE'$ be in F_i. The operation *LEFT−JOIN* can be applied to the histories H and H' iff they satisfy the following conditions:

(a) The identities appearing in H are distinct from those appearing in H'.

(b) The size of the inner part of H is at least as large as the size of the inner part of H'.

(c) There is a history h such that $LI = LI_1 \cdot h \cdot LI_2$ and $LI' = LI_1' \cdot h \cdot LI_2'$, where the size of LI_1 is at most n.

(d) LE is of size at most $\frac{1}{12}n$.

The *LEFT-JOIN* of H and H' is the history sequence $\hat{H} = \hat{LE} \cdot h \cdot LI_2' \cdot h_C' \cdot RI' \cdot h_R' \cdot RE'$, where $\hat{LE}$ is obtained by performing a maximal shrinking on the segment $LE \cdot h_L \cdot LI_1$ (Note that all the identities in $LE \cdot h_L \cdot LI_1$ are distinct and different from those of $LI_2' h_C' RI' h_R' RE'$.)

(iii) *RIGHT-JOIN*: This operation is defined similarly to *LEFT-JOIN*.

For the family $F(\tau)$ to exist, we need the following lemma.

Lemma 3.1: Let $\mathbf{F} = (F_0, \cdots, F_i, \cdots)$ be a sequence of sets of history sequences such that $F_0 = \Phi$ and F_{i+1} is obtained by applying one of the operations (i) - (iii) above to F_i. Then $\mathbf{F}$ is finite.

Proof: Recall the size of the canonical histories is $2t$. Define the *cost* $C(H)$ of a history sequence $H = LE \cdot h_L \cdot LI \cdot h_C \cdot RI \cdot h_R \cdot RE$ to be 3^{2t-s}, where s is the size of $LI \cdot h_C \cdot RI$; the cost $C(F)$ of F is defined as the sum of the costs of the history sequences in it. Since each id can occur in at most one history sequence in F_i, it follows that F_i contains at most $n^{1+\varepsilon}$ history sequences. Hence $C(F_i)$ is bounded from above by $n^{1+\varepsilon} \cdot 3^{2t}$. We now show that for all i, $C(F_{i+1}) \geq C(F_i) + 1$, thus proving the lemma.

If F_{i+1} is obtained from F_i by the operation *ADD*, then $C(F_{i+1}) = C(F_i) + C(H(D)) = C(F_i) + 1$. Suppose F_{i+1} is obtained from F_i by a *LEFT-JOIN* operation (The case of a *RIGHT-JOIN* operation is similar). Now note that the size of the inner part of $\hat{H}$, the segment $LI_2' \cdot h_C' \cdot RI'$, is smaller than the size of the inner part of H' and thus also smaller than the size of the inner part of H. This implies that $C(\hat{H}) \geq 3C(H')$ and $C(\hat{H}) \geq 3C(H)$, so $C(F_{i+1}) = C(F_i) + C(\hat{H}) - C(H) - C(H') \geq C(F_i) + 1$. □

Let $F = F(\tau)$ be a set F_N satisfying the above lemma. Some basic properties of history sequences in F for are given in the next lemma.

Lemma 3.2: For $H = LE \cdot h_L \cdot LI \cdot h_C \cdot RI \cdot h_R \cdot RE$ be in $F(\tau)$ the following properties hold:

(a) Both LE and RE are of size at most $\frac{1}{6}n$.

(b) Any segment of size $\leq n$ in $LI \cdot h_C \cdot RI$ contains at most $\frac{1}{12}n$ distinct histories.

(c) Any identity occurring in H does not occur in any other history sequence in F.

(d) There is an execution $E(H)$ of AL.

Proof:

(a) When a new history is constructed by a $LEFT-JOIN$ operation, then the LE-segment of the new history is produced by maximally shrinking an LE-segment of size at most $\frac{1}{12}$ and part of an $LI-$segment of size at most n. The latter part is contained in a canonical history sequence. Therefore by Assumption Q it contains at most $\frac{1}{12}$ distinct histories. We conclude that after maximal shrinking the new LE-segment has size at most $\frac{1}{6}n$. A similar argument shows that the RE-segments of histories produced by a $RIGHT-JOIN$ are at most of size $\frac{1}{6}n$.

(b) This follows from Assumption Q and the fact that there exists a canonical line configuration that contains $LI \cdot h_c \cdot RI$ as a segment.

(c) This follows directly from the construction of $F(\tau)$.

(d) This follows from the construction of $F(\tau)$ and the correctness of Rule 2. □

4. PROOF OF THEOREM 1.

Let F be $F(\sigma)$ for $\sigma \in \{\tau,\omega\}$. We call a history sequence of F *finished* if a certain condition (defined below) holds. We then show in the Main Lemma that the existence of such *finished* history sequences in both $F(\tau)$ and $F(\omega)$ implies the $\Omega(n \log n)$ lower bound. Finally we prove by a counting argument that if either $F(\tau)$ or $F(\omega)$ does not contain a finished history sequence and the id set is large enough ($n^{1+\varepsilon}$), then the same lower bound must hold.

A history sequence $H = LE \cdot h_L \cdot LI \cdot h_C \cdot RI \cdot h_R \cdot RE$ in F is called *left unfinished (right unfinished)* if LI (RI) is of size at least $\frac{1}{12}n$ and LE (RE) is of size at most $\frac{1}{12}n$. A history sequence is called *finished* iff it is neither left unfinished nor right unfinished.

Main Lemma: If both $F(\tau)$ and $F(\omega)$ contain a finished history sequence then the message complexity of AL is $\Omega(n \log n)$.

Proof: If $H = LE \cdot h_L \cdot LI \cdot h_C \cdot RI \cdot h_R \cdot RE$ is finished then at least one of the following must hold:

(ir1) Both LI and RI have size smaller than $\frac{1}{12}n$.

(ir2) Both LE and RE have size larger than $\frac{1}{12}n$.

(ir3) The size of RE is larger than $\frac{1}{12}n$ and the size of LI is smaller than $\frac{1}{12}n$.

(ir4) The size of RI is smaller than $\frac{1}{12}n$ and The size of LE is larger than $\frac{1}{12}n$.

To prove the lemma it is suffices to show that if both $F(\tau)$ and $F(\omega)$ contain a history sequence satisfying one of the properties (ir1)-(ir4) above, then the bit complexity of AL is $\Omega(n \log n)$. This is done in the next 3 lemmas.

Lemma 4.1: If a history sequence H in $F(\tau)$ satisfies (ir1), then no history sequence in $F(\omega)$ satisfies (ir1), and vice versa.

Proof: Assume that the lemma is false. Then there is a history sequence $H_\tau = LE \cdot h_L \cdot LI \cdot h_C \cdot RI \cdot h_R \cdot LE$ in $F(\tau)$ that satisfies (ir1). So the size of the segment $LI \cdot h_C \cdot RI$ is at most $\frac{1}{6}n$ and all identities in this segment (and thus in all of H_τ) are distinct). By Lemma 3.2(a) the size of LE and of RE is at most $\frac{1}{6}n$. Clearly the total size of H_τ is at most $\frac{1}{2}n$. By Property (b1), in the execution $E(H_\tau)$ the center processors output $f(\tau)$. Similarly, there is a history sequence H_ω in $F(\omega)$ that has size at most $\frac{1}{2}n$ and in the execution $E(H_\omega)$ the center processors output $f(\omega)$.

Observe that all id's occurring in H_τ and H_ω are distinct. Thus it is possible to embed the line configurations of these history sequences into a ring configuration $R = R(\sigma,x)$ of size n: Concatenate the processors of H_τ and H_ω and close the line to a ring of size n by adding the needed number of processors. Now repeat both executions $E(H_\tau)$ and $E(H_\omega)$ on the corresponding segments of R (block all links not contained in the two segments). We get an execution of AL on a ring of size n with distinct identities taken from X in which some processors output $f(\tau)$ and another output $f(\omega)$. This is a contradiction. □

In view of the above lemma, we may assume without loss of generality, that $F(\tau)$ does not contain a history sequence satisfying (ir1). In the sequel we denote $F(\tau)$ by F and show that no history in F satisfies any of the conditions (ir2) to (ir4), unless the lower bound holds. Clearly, this completes the proof of the Main Lemma.

Lemma 4.2: If a history sequence H in F satisfies (ir2), then the bit complexity of AL is $\Omega(n \log n)$.

Proof: Let $H = LE \cdot h_L \cdot LI \cdot h_C \cdot RI \cdot h_R \cdot RE$ be a history sequence satisfying (ir2). Assume first that $LI \cdot h_C \cdot RI$ is of size at most n. Then all the identities in H are distinct; moreover, by (ir2) the number of distinct histories in both LE and RE is at least $\frac{1}{12}n$, and by Lemma 3.2 (a) this number is at most $\frac{1}{6}n$. By Assumption Q, the number of distinct histories in $LI \cdot h_C \cdot RI$ is at most $\frac{1}{12}n$. Thus, by applying maximal shrinking to $LI \cdot h_C \cdot RI$ we get a history sequence $\hat{H}$ whose size is at most $\frac{5}{12}n$, and which contain at least $\frac{1}{12}n$ distinct histories. By Lemma 2.3, the length of $\hat{H}$ is $\Omega(n \log n)$. Since $\hat{H}$ is of size smaller than n, it can be embedded in a ring configuration R of size n. Thus, we have an execution of AL on R whose bit complexity is $\Omega(n \log n)$.

We are left with the case where the size of the inner part of H is larger than n.

Let $p = p_{i,j}$ be the leftmost processor in LI, and let q be the rightmost processor in H which has the same identity as p (note that q is either $p_{i,j}$ or $p'_{i,j}$ for some j). Let h be the history of the link to the left of q. Then H can be written as $LE \cdot h_L \cdot H1 \cdot h \cdot H2 \cdot h_R \cdot RE$. Observe that all id's in LE and $H2 \cdot h_R \cdot RE$ are distinct. Let $\overline{H2}$ be the segment produced by maximal shrinking from $H2$. Replace $H2$ by $\overline{H2}$ in H to get the history sequence $\overline{H} = LE \cdot h_L \cdot H1 \cdot h \cdot \overline{H2} \cdot h_R \cdot RE$. By the definition of q, the segment $H2$ is of size at most n. Thus by Assumption Q it contains at most $\frac{1}{12}n$ distinct histories and hence $\overline{H2}$ is of size at most $\frac{1}{12}n$.

By Property (b1) of the history sequences in $F(\tau)$ there is a canonical line configuration D, that contains $h_L \cdot LI \cdot h_C \cdot RI$ as a consecutive subsequence, and in which p and q have the same identity. This implies that h_L is a prefix of h or vice versa, so assume that h_L is a prefix of h. We use this to produce a history sequence of size less than n that contains LE. Then we use the fact that LE has at least $\frac{1}{12}$ distinct histories to derive the lower bound (In the symmetric case h is a prefix of h_L and one can construct a history sequence of size less than n that contains RE). By using Rule 1 on h and h_L we get a history sequence $H' = LE' \cdot h_L' \cdot H1' \cdot h_L \cdot H2' \cdot h_R' \cdot RE'$ that is produced by some execution of AL. By using Rule 2 on H and H' (with $h = h_L$) we get a history sequence $\hat{H} = LE \cdot h_L \cdot H2' \cdot h_R' \cdot RE'$ that is produced by another execution of AL. In $\hat{H}$ all the identities are distinct. The sizes of LE and RE' are at most $\frac{1}{6}n$, and the size of $H2'$ is at most $\frac{1}{12}n$ and thus the total size of $\hat{H}$ sums to at most $\frac{5}{12}n$. As above we embed $\hat{H}$ in a ring of size n and run the execution $E(\hat{H})$ on the segment $\hat{H}$ of the ring. $\hat{H}$ contains the segment LE which has at least $\frac{1}{12}$ distinct histories. Thus by Lemma 2.3 the length of LE is $\Omega(n \log n)$ and this completes the proof of the lemma. □

Lemma 4.3: If a history sequence H in F satisfies (ir3) or (ir4), then the bit complexity of AL is $\Omega(n \log n)$.

Proof: This proof follows the same outline of the proof of Lemma 4.2, and is omitted from this version. □

This completes the proof of the Main Lemma. □

The Main Lemma aboves implies that if both $F(\tau)$ and $F(\omega)$ contain a finished history sequence, then Theorem 1 holds. Thus, in order to complete the proof of Theorem 1, it is suffices to prove the following:

Lemma 4.4: If $F(\tau)$ or $F(\omega)$ contains only unfinished history sequences, then the bit complexity of AL is $\Omega(n \log n)$.

Proof: We prove the lemma for $F = F(\tau)$. First observe that there are at most $n-1$ identities in X_τ that do not occur in any history sequence H in F (otherwise Operation (i) is applicable to F, in contrast with the definition F). Since each history sequence in F contains less than $2n$ distinct identities, there are at least $M = \frac{|X_\tau| - n + 1}{2n}$ distinct history sequences in F. At least half of the history sequences of F are either right or left unfinished. Without loss of generality, assume that there are at least $K = \frac{1}{2}M$ distinct history sequences in F that are left unfinished. Note that $K = n^{\varepsilon'}$ for some $\varepsilon' > 0$.

Let $H_1, \cdots, H_K$ be the history sequences in F which are left unfinished. Then the left inner part of each H_i is of size at least $\frac{1}{12}n$. Let Q_i be the set of the histories of the first $\frac{1}{12}n$ links of the inner part of H_i. By the definition of F, the operation *LEFT-JOIN* can be applied to no pair of these H_i's, and hence the sets Q_i are disjoint. Let l_i be the length of the minimal length history in Q_i, and let j be such that $l_j = \max\{l_i : 1 \le i \le K\}$. By Lemma 2.3, $l_j = \Omega(\log K) = \Omega(\log n)$, which means that the inner part of Q_j contains a history segment of size $\frac{1}{12}n$ and of length $\Omega(n \log n)$. By Lemma 2.2, the length of this segment is a lower bound on the bit complexity of the synchronized computation of the ring R that corresponds to the inner part of H_j. This complete the proof of the lemma, and hence the proof of Theorem 1. □□□

REFERENCES

[AAHK] K. Abrahamson, A. Adler, L. Higham and D. Kirkpatrick, "Randomized Function Evaluation on a Ring," Tech. Rep. 87-20, Dept. of Comp. Sc., Univ. of British Columbia, Vancouver, Canada, 1987.

[ASW] C. Attiya, M. Snir and M. K. Warmuth, "Computing on an anonymous ring," J. ACM, October 1988.

[B1] H. L. Bodlaender, "New lower bound techniques for distributed leader finding and other problems on rings of processors," Technical Report RUU-CS-88-18, University of Utrecht, 1988, to appear in Theor. Comp. Sc.

[B2] H. L. Bodlaender, unpublished note.

[BB] P. W. Beame and H.L. Bodlaender, "Distributed Computing on Transitive Networks: The Torus", proceedings STACS 89, p. 294-303, 1989.

[DG] P. Duris and Z. Galil, "Two lower bounds in asynchronous distributed computation," proceedings FOCS 87, p. 326-330, 1987.

[MZ] Y. Mansour and S. Zaks, "On the bit complexity of distributed computations in a ring with a leader," Information and Computation, Vol. 75, No. 2, 1987, pp. 162-177.

[MW] S. Moran and M. Warmuth, "Gap theorems for distributed computation," proceedings PODC, p. 131-140, 1986.

[PKR] J. Pachl, E. Korach and D. Rotem, "Lower bounds for distributed maximum-finding algorithms," JACM 31, pp. 905-918, 1984.

THE JUMP NUMBER PROBLEM FOR BICONVEX GRAPHS AND RECTANGLE COVERS OF RECTANGULAR REGIONS

Andreas Brandstädt
Sektion Mathematik, Friedrich-Schiller-Universität Jena
Universitätshochhaus, DDR-6900 Jena

Abstract.
Let $P=(V, \leq_P)$ be a finite partially ordered set (poset) with $|V|=n$ and let $L=(l_1,\ldots,l_n)$ be a linear extension of P. The pair (l_i,l_{i+1}), $1\leq i\leq n-1$, is a jump of P in L if $l_i \not\leq_P l_{i+1}$. The jump number problem is the problem of finding the minimum number of jumps in any linear extension of a given poset P. It is known that for posets P_1,P_2 with the same comparability graph also the jump numbers of P_1 and P_2 coincide and that for chordal bipartite graphs the jump number decision problem is NP-complete.
We show in this paper that the jump number of biconvex graphs (a subclass of chordal bipartite graphs) can be determined in polynomial time using several reformulations of the problem and a duality relation between rectangle independent point sets and rectangle covers of rectangular regions known from a result of Chaiken/Kleitman/Saks/Shearer and Franzblau/Kleitman. This solves the jump number problem for biconvex graphs by means of computational geometry. Furthermore for bipartite permutation graphs (a subclass of biconvex graphs) the rectangle cover approach yields a greedy solution which is faster than the dynamic programming solution given by Steiner/Stewart. An optimal rectangle cover can be determined in linear time using the geometric description of the region or the defining permutation.

1. Definitions and preliminary results

$P=(V, \leq_P)$ is a partially ordered set (poset) if $\leq_P$ is reflexive, transitive and antisymmetric.
$L=(V, \leq_L)$ is a linear order if L is a poset in which every pair of elements is related w.r.t. $\leq_L$.

A linear order L is a linear extension of a poset P if for all $u,v \in V$ $u \leq_P v$ implies $u \leq_L v$. In a linear extension $L=(l_1 \dots l_n)$ of P the pair (l_i, l_{i+1}), $1 \leq i \leq n-1$, is called a step if $l_i \leq_P l_{i+1}$ and otherwise is called a jump.

Let s(P,L) denote the number of steps of P in L and let j(P,L) denote the number of jumps in L. Obviously $j(P,L)+s(P,L)=n-1$. The jump number j(P) of the finite poset P is

$j(P)=\min\{j(P,L):$ **L is a linear extension of** $P\}$ and

the step number s(P) of P is

$s(P)=\max\{s(P,L):$ L is a linear extension of $P\}$.

The jump number problem is the following

<u>Problem 1.</u>

Given a finite poset P, find j(P)./In the following we recall some reformulations of Problem 1 given in /15/ (Problem 2 to Problem 5) and continue this sequence of reformulations by Problem 6. By a theorem of /9/,/11/ the jump number of a poset depends only on the corresponding comparability graph $G_P=(V,E_P)$ with $\{x,y\} \in E_P$ iff (x,y) or (y,x) is contained in the transitive closure of $\leq_P$. Therefore in the following we study the jump number problem for comparability graphs instead of posets. Let j(G) be the jump number of any poset having G as its comparability graph.

Throughout this paper it will be assumed that G is connected (otherwise one can determine j(G) by investigating the components of G).

The first reformulation of Problem 1 is

<u>Problem 2.</u>

Given a comparability graph G, find j(G)./For bipartite graphs it is known that the jump number decision problem

JUMP NUMBER = $\{(G,k):$ G is a bipartite graph and $j(G) \leq k\}$

is NP-complete (/13/).

We recall now the definition of the following hierarchy of subclasses of bipartite graphs:

1. G=(V,E) is <u>chordal bipartite</u> if each cycle of length at least 6 has a chord (i.e. an edge between two nonconsecutive vertices of the cycle).
2. For a graph G=(X,Y,E) an ordering of X has the adjacency property if for each $y \in Y$ $N(y)=\{x: \{x,y\} \in E\}$ consists of vertices which are consecutive in the ordering of X (N(y) forms an interval in the ordering of X).

   A bipartite graph is <u>convex</u> if there is an ordering of X or Y which fulfills the adjacency property.
3. A bipartite graph is <u>biconvex</u> if there are orderings of X and of Y which both fulfill the adjacency property.

4. Define permutation graphs as in /6/:
   Let $\pi=(i_1 \dots i_n)$, $\{i_1,\dots,i_n\}=\{1,\dots,n\}$, be a permutation and $G_\pi=(\{1,\dots,n\},E_\pi)$ with $\{i,j\}\in E_\pi$ iff $(i-j)(\pi^{-1}(i)-\pi^{-1}(j))<0$ (/6/). A graph $G=(V,E)$ is a permutation graph iff there is a permutation π of $\{1,\dots,|V|\}$ such that G is isomorphic to G_π. A graph G is a <u>bipartite permutation</u> graph if G is bipartite and G is a permutation graph.

In /2/ the following containments of these four classes are mentioned: bipartite permutation $\subset$ biconvex $\subset$ convex $\subset$ chordal bipartite.
As usually a subset M of edges, $M\subseteq E_P$, is a matching if for any two edges $\{u,v\}$, $\{x,y\}\in M$ $\{u,v\}\cap\{x,y\}=\emptyset$.
An M-alternating cycle for a matching M is an even-length cycle in which every second edge is from M.
A matching M is an alternating-cycle-free matching if it contains no M-alternating cycle.
Chaty and Chein (/4/) proved that for bipartite posets P Problem 1 is equivalent to the following

<u>Problem 3.</u>

Find the maximum size of an alternating-cycle-free matching in G_P.
(in fact the maximum size of an alternating-cycle-free matching in G_P is equal to the maximum step number of P).
In /12/ it is observed that for chordal bipartite graphs Problem 2 is equivalent to

<u>Problem 4.</u>

Find the maximum size of an alternating-C_4-free matching in G_P. The corresponding decision problem is shown to be NP-complete in /12/ for chordal bipartite graphs.
In /15/ Problem 4 is solved in linear time using a dynamic programming approach for bipartite permutation graphs.
The dimension of a poset P, denoted by dim P, is defined as the minimum number of linear extensions whose intersection is P. Thus a poset is 2-dimensional iff it is the intersection of 2 linear extensions.
It is well-known that 2-dimensional posets are exactly the posets whose comparability graphs are the permutation graphs. Bipartite permutation graphs are characterized in /1/,/14/:

<u>Theorem 1</u> (/1/,/14/):

Let $P=(X,Y,\leq_P)$ be a bipartite poset. Then dim $P\leq 2$ iff there is an ordering $x_1,x_2,\dots,x_m$ of X and $y_1,y_2,\dots,y_n$ of Y such that if $x_i\leq_P y_j$ and $x_k\leq_P y_l$ with $i<k$ and $l<j$ then $x_i\leq_P y_l$ and $x_k\leq_P y_j$ also hold. Such an ordering will be called a strong ordering.

Corollary 1:

Let $P=(X,Y,\leq_P)$ be a 2-dimensional bipartite poset with a strong ordering for X and Y. For every $y_i \in Y$ let L_i (R_i) denote the leftmost (rightmost) predecessor of y_i with respect to the strong ordering.

Then a) For every $y_i \in Y$ the predecessor set $P(y_i)=[L_i,R_i]$ is an interval on X

b) The left and right end points of these intervals form a nondecreasing sequence in the strong ordering i.e. $L_1 \leq L_2 \leq \ldots \leq L_n$ and $R_1 \leq R_2 \leq \ldots \leq R_n$.

This neighbourhood structure is essentially used in /15/.

For a bipartite poset $P=(X,Y,\leq_P)$ define the adjacency matrix $A(P)=(a_{ij})$ by $a_{ij}=1$ if $x_j \leq_P y_i$ and 0 otherwise. This matrix can be interpreted also as a set of grid points with a point (i,j) in the set iff $a_{ij}=1$.

By corollary 1 for 2-dimensional bipartite posets the corresponding set of grid points has the shape shown in Figure 1:

Figure 1: Typical shape for regions A(P) of 2-dimensional bipartite posets (bipartite permutation graphs):

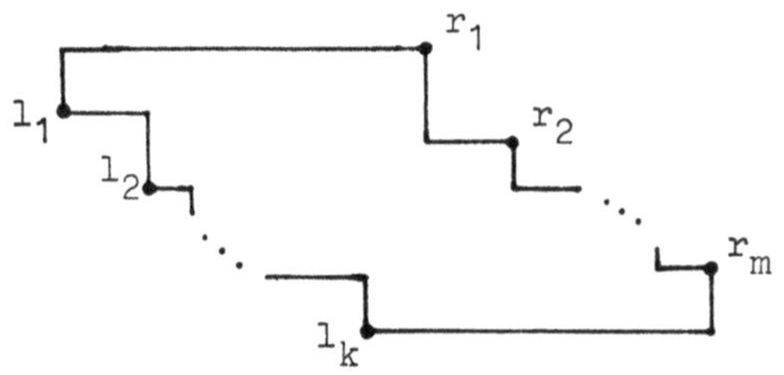

(a 2-staircase normal form (2-s.n.f.))

Note that the region A(P) is completely described by the sequences $l_1,\ldots,l_k$ and $r_1,\ldots,r_m$. We denote this by $A(P)=A(l_1,\ldots,l_k;r_1,\ldots,r_m)$. For an edge set $E' \subseteq E_P$, $G_P=(X,Y,E_P)$, let $A(E')$ be the corresponding set of grid points i.e. $(i,j) \in A(E')$ iff $\{x_i,y_j\} \in E'$ and let $A(G_P)=A(E_P)$. Obviously M is a matching in G_P iff M contains at most one point of each row or column of $A(G_P)$. M is an alternating-C_4-free matching in G_P iff M is a matching in G_P with the property: if $(i,j),(k,l) \in A(M)$ with $(i,j) \neq (k,l)$ then (i,l) or (k,j) are not in A(M).

Obviously the maximum size of an alternating-C_4-free matching in G_P does not depend on permutations of rows and columns in A(P) and thus one can consider a normal form of A(P).

For grid points (i,j),(k,l) let

$\hat{R}((i,j),(k,l)) = \{(i,j),(k,l),(i,l),(k,j)\}$ if $i \neq k$ and $j \neq l$ and
$\hat{R}((i,j),(k,l)) = \{(i,j),(k,l)\}$ otherwise.
For $a,b \in A$ a and b are A-independent if $\hat{R}(a,b) \nsubseteq A$. For $S \subseteq A$ S is A-independent if for all $a,b \in S$, $a \neq b$, a and b are A-independent.
Note that a maximum alternating-C_4-free matching in G_P corresponds to a maximum A-independent set of grid points in A(P) for bipartite posets P. Let $i(A) = \max \{ |S| : S \subseteq A$ and S is A-independent$\}$.
Thus the next reformulation of Problem 1 is

Problem 5.

Given a finite poset P, find i(A(P)).

2. Jump number for biconvex graphs

A set of grid points A is convex in rows (convex in rows and columns) if for all $x,y \in A$ on one row (on one row or column also the interval between x and y is contained in A. A is biconvex if it is convex in rows and columns. Let $R=(R_i)_{i=1,\dots,k}$ be a set of rectangles $R_i \subseteq A$.

R covers A if $\bigcup R = \bigcup_{i=1}^{k} R_i \supseteq A$.

Let $r(A) = \min\{|R| : R$ is a set of rectangles in A which covers $A\}$.
For $a,b \in A$ let a,b be rectangle-independent in A if a and b are not in a common rectangle in A. $S \subseteq A$ is rectangle-independent in A if for all $a,b \in S$ a,b are rectangle-independent in A.
Let $i_R(A) = \max\{ |S| : S \subseteq A$ and S is rectangle-independent in $A\}$.
It is clear that $i(A) = i_R(A)$ for biconvex regions A and there are convex but not biconvex regions for which $i(A) \neq i_R(A)$. In the following we restrict ourselves to biconvex regions A. In /8/ it is shown that for regions A which are convex in rows $i_R(A) = r(A)$ and can be determined in $O(n^2)$ steps where n is the length of the description of A. For biconvex regions the equality was shown already in /3/. This yields the reformulation of Problem 1 for biconvex graphs:

Problem 6.

For biconvex region A, find r(A).
According to /8/ we have

Theorem 2.

For biconvex graphs the jump number problem is solvable in polynomial time.

Proof: Evidently for a biconvex graph G=(X,Y,E) A(G) is convex in rows and columns for orderings of X and Y which fulfill the adjacency property. In order to solve Problem 2 for G we can solve Problem 6 using the algorithm of /8/. If for $G=(X,Y,E)$ $n=|X|+|Y|$ then the length of the description of $A(G)$ is linear in n and therefore the time bound for Problem 6 is $O(n^2)$ as in the algorithm of /8/. □

3. Jump number for bipartite permutation graphs

Assume that $G_\pi=(X,Y,E_\pi)$ is a bipartite permutation graph and $A=A(G_\pi)$ is in 2-staircase normal form (2-s.n.f.)
We describe a simple greedy algorithm for this case which was discovered independently also by H. Jung and H. Fauck (Berlin) (/7/,/10/) (which were motivated by computational geometry). /7/ contains an algorithm which simultaneously constructs a minimum rectangle cover of A and a maximum A-independent set of points in A.
For a 2-s.n.f. $A=A(l_1,\ldots,l_k:r_1,\ldots,r_m)$ let $R(l_i,r_j)$ denote the rectangle given by the lowest leftmost point l_i and the highest rightmost point r_j if this rectangle is nonempty.

Algorithm 1:

Input : $A=A(l_1,\ldots,l_k;r_1,\ldots,r_m)$ - a 2-s.n.f.
Output: a) $R_1,\ldots,R_l$ - a minimum set of rectangles in A which covers A.
b) $r(A)=l$.

1. Let $R_1=R(l_1,r_1)$ and $j:=1$.

Repeat

2. If $R_j=R(l_{i_j},r_{k_j})$ and $i_j<k$, $k_j<m$ then

2.1 $R_{j+1}=R(l_{i_j+1},r_{k_j+1})$ if r_{k_j+1} is at most one row lower than l_{i_j} and l_{i_j+1} is at most one column to the right of r_{k_j}

2.2 $R_{j+1}=R(l_{i_j+1},r_{k_j})$ if r_{k_j+1} is at least two rows lower than l_{i_j}

2.3 $R_{j+1}=R(l_{i_j},r_{k_j+1})$ if l_{i_j+1} is at least two columns to the right of r_{k_j}

3. If $R_j=R(l_{i_j},r_{k_j})$ and ($i_j=k$ or $k_j=m$) then

3.1 If $i_j=k$ then $R_{j+1}=R(l_k,r_{k_j+1})$

3.2 If $k_j=m$ then $R_{j+1}=R(l_{i_j+1},r_m)$

4. $j:=j+1$ and $R_j:=R_{j+1}$;

until $R_j=R(l_k,r_m)$.

Figure 2: The cases 2.1, 2.2 and 2.3 of Algorithm 1:

2.1

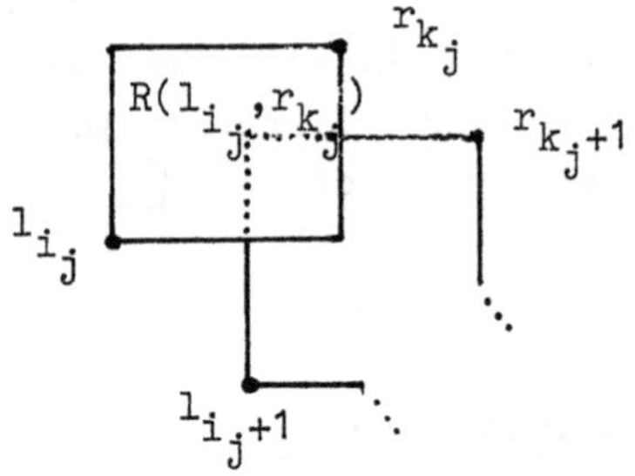

2.2:

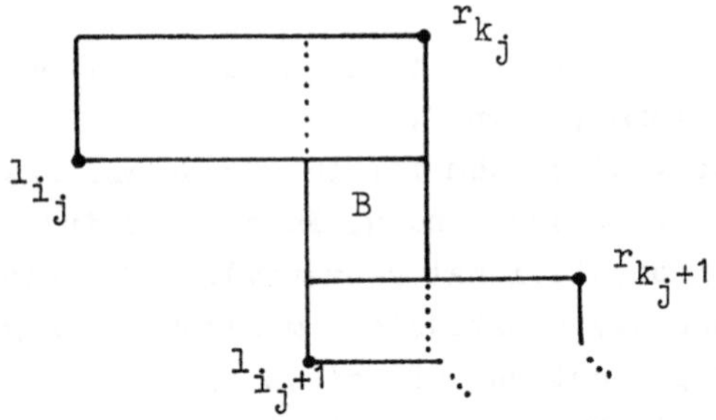

2.3:

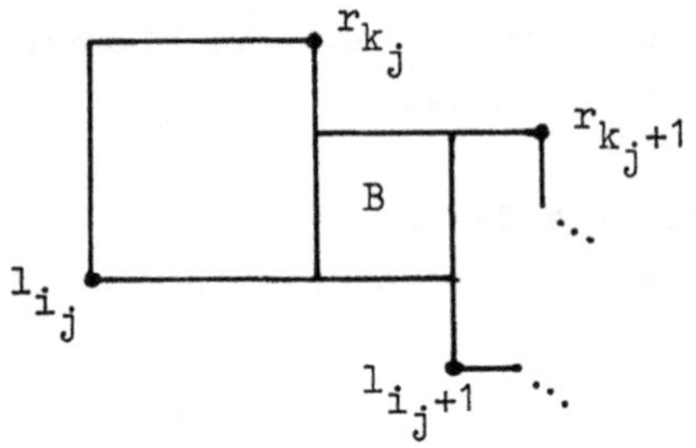

For the case of a region A in 2-s.n.f. which is a subset of the plane R^2 instead of a set of grid points the algorithm can be easily modified in order to get a minimum rectangle cover of A.

Theorem 3:

Algorithm 1 determines a minimum rectangle cover of $A=A(l_1,\dots,l_k; r_1,\dots,r_m)$ within $O(k+m)$ steps.

Theorem 4:

If for a bipartite permutation graph G_π the graph is described as a 2-s.n.f. then the jump number problem for G_π can be solved by Algorithm 1 counting the minimum number of rectangles within $O(k+m)$ steps.

Example 1:

Let $\pi = (3\ 4\ 1\ 7\ 2\ 8\ 10\ 5\ 6\ 9)$ and $X=\{3,4,7,8,10\}$, $Y=\{1,2,5,6,9\}$.

$A(G_\pi)$:

	1	2	5	6	9
3	x	x			
4	x	x			
7		x	x	x	
8			x	x	
10			x	x	x

$l_1=(4,1)$, $l_2=(7,2)$, $l_3=(10,5)$; $r_1=(3,2)$, $r_2=(7,6)$, $r_3=(10,9)$.
The result of Algorithm 1 is
$R_1=R((4,1),(3,2))$, $R_2=R((7,2),(7,6))$, (case 2.1),
$R_3=R((10,5),(7,6))$ (case 2.2), $R_4=R((10,5),(10,9))$ (case 3.1)
Algorithm 1 yields time linear in the description of the area whereas the dynamic programming of /15/ is linear only in the number of vertices and edges of G.
If G_π is described by its permutation π then also Algorithm 1 can be performed for π since $l_1,\dots,l_k$ and $r_1,\dots,r_m$ can be simply recognized in π , and it can also be recognized whether case 2.1, 2.2, 2.3, 3.1, 3.2 is fulfilled.
Theorem 5:
If a bipartite permutation graph is given by its permutation $\pi=(i_1\dots i_n)$ then in O(n) steps the jump number of G_π can be determined.

4. Rectangle graphs and independent grid points

For a region A of grid points define $G_A=(A,E_A)$ with $\{x,y\}\in E_A$ iff $\hat{R}(x,y)\subseteq A$ for $x,y\in A$.
Clearly the A-independent point sets in A correspond to independent sets of vertices in G_A.
Now assume that A is in 2-s.n.f. For grid points x,y define the partial order $\leq_1$ by $x \leq_1 y$ if x is not lower than **y** and not to the right of y. Obviously $\leq_1$ is a partial order.
Proposition 1:
If A is in 2-s.n.f. then each independent set of points in A is a chain w.r.t. $\leq_1$.
Proposition 2:
C is a maximal clique in G_A iff C is a maximal rectangle in A.
For 2-s.n.f. A the graph G_A is perfect as we will show. For biconvex regions the same assertion does not hold:

Example 2:

A:

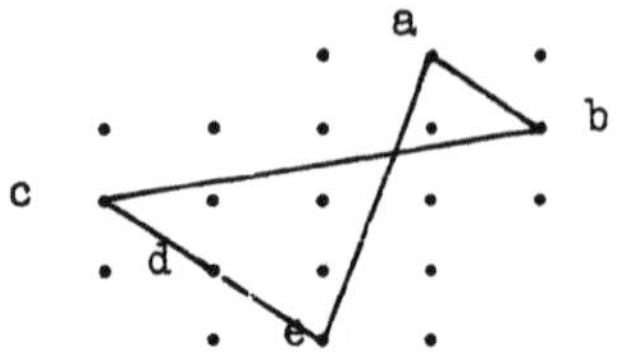

(a,b,c,d,e) is an induced C_5 in G_A and thus G_A is not perfect.
Let $\overline{G}_A = (A, \overline{E}_A)$ i.e. $\{x,y\} \in \overline{E}_A$ if $\hat{R}(x,y) \nsubseteq A$.

Proposition 3:

If $\{x,y\} \in \overline{E}_A$ then $x \leq_1 y$ or $y \leq_1 x$.
Let us orient all edges of E_A corresponding to $\leq_1$ from the upper left to the lower right point of the edge.

Proposition 4:

For 2-s.n.f. A the graph $\overline{G}_A$ is transitively orientable.
Thus G_A is a co-comparability graph for 2-s.n.f. A and these graphs are known to be perfect.
The perfectness of G_A together with Proposition 2 gives another proof that for 2-s.n.f. A $i(A) = r(A)$ holds but for biconvex A this approach does not work with G_A as the corresponding graph as Example 2 shows. It would be interesting whether one can define suitable other graphs corresponding to A such that the equality $i(A)=r(A)$ follows from the perfectness of certain graphs.

Open Problem: Determine the complexity of the jump number problem for convex graphs.

Acknowledgement:

This work was stimulated by discussions with M.M. Sysło (Wrocław) and D. Kratsch (Jena) concerning the jump number problem for convex graphs. Thanks also to H. Jung (Berlin) and H. Fauck (Berlin) for the references on computational geometry.

References:

/1/ A. Brandstädt, D. Kratsch, On the restriction of some NP-complete graph problems to permutation graphs, Report N/84/80, Friedrich-Schiller-Universität Jena, appeared in FCT'85, LNCS 199, 53-62

/2/ A. Brandstädt, J. Spinrad, L. Stewart, Bipartite permutation graphs are bipartite tolerance graphs, Congressus Numerantium Vol. 58, 1987, 165-174

/3/ S. Chaiken, D.J. Kleitman, M. Saks, J. Shearer, Covering Regions by Rectangles, SIAM J. Alg. Discr. Math. 1981, 394-410

/4/ G. Chaty, M. Chein, Ordered matchings and matchings without alternating cycles in bipartite graphs, Utilitas Mathematica 16, 1979, 183-187

/5/ M. Chein, M. Habib, The jump number of dags and posets: an introduction, Ann. Discr. Math. 9, 1980, 189-194

/6/ S. Even, A. Pnueli, A. Lempel, Permutation graphs and transitive graphs, J. ACM 19, 1972, 400-410

/7/ H. Fauck, Ein optimaler sequentieller und ein paralleler Algorithmus zur Konstruktion minimaler Überdeckungen einfacher, rechtwinkliger, monotoner Polygone in der Ebene durch Rechtecke, Diploma thesis 1988, Humboldt-Universität Berlin

/8/ D. Franzblau, D.J. Kleitman, An Algorithm for Constructing Regions with Rectangles: Independence and Minimum Generating Sets for Collections of Intervals, 16th STOC 1984, 167-174

/9/ M. Habib, Comparability invariants, in Ordres: Description et Roles, ed. M. Pouzet and D. Richard, 371-386, North-Holland 1984

/10/ H. Jung, H. Fauck, personal communication

/11/ R.H. Möhring, Algorithmic aspects of comparability graphs and interval graphs, in Graphs and Orders, ed. I. Rival, 41-101, D. Reidel Publ. Co. 1985

/12/ H. Müller, Alternating-cycle-free matchings in chordal bipartite graphs, 1988, submitted to Order

/13/ W.R. Pulleyblank, On minimizing setups in precedence constrained scheduling, to appear in Discr. Appl. Math.

/14/ J. Spinrad, A. Brandstädt, L. Stewart, Bipartite permutation graphs, Discr. Appl. Math. 18 (1987), 279-292

/15/ G. Steiner, L. Stewart, A linear time algorithm to find the jump number of 2-dimensional bipartite partial orders, Order 3 (1987), 359-367

/16/ L. Stewart, Permutation Graph Structure and Algorithms, Ph. D. thesis, University of Toronto, 1985

/17/ M. M. Sysło, Minimizing the jump number for partially ordered sets: A graph-theoretic approach, Order 1 (1984) 7-19

/18/ M. M. Sysło, A graph-theoretic approach to the jump number problem, in Graphs and orders, ed. I. Rival, 185-215 D. Reidel Publ. co. 1985

Recent Developments in the Design of Asynchronous Circuits

J.A. Brzozowski and J.C. Ebergen

Computer Science Department
University of Waterloo
Waterloo, Ont., Canada N2L 3G1

Abstract

Some recent developments in the design of asynchronous circuits are surveyed. The design process is considered in two parts. First, the communication behaviour of the component to be designed is formally specified and this specification is decomposed into a network of basic components. Second, the basic components are realized using gate circuits.

In the first part of the design process we use trace theory to reason about all possible sequences of events. Components are specified by regular-expression-like programs, called commands, whose semantics is based on directed trace structures. We formalize the concepts of speed-independent and delay-insensitive circuits in the context of a network of basic components.

In the second part we use switching theory for the analysis of gate circuits. Three different delay models are discussed: the feedback-delay, the gate-delay, and the gate-and-wire-delay model. The last two models correspond to speed-independent and delay-insensitive circuits, respectively. We point out that networks of components are commonly operated in the 'input-output mode' (where inputs may change as soon as outputs have responded to a previous input change), whereas gate circuits are usually operated in the 'fundamental mode' (where the entire gate circuit must stabilize before another input change is permitted).

We note that delay-insensitive gate circuits are unlikely to exist for most basic components. For this reason, it is important that analysis and design methods are developed using bounded-delay models.

1 Introduction

In recent years a number of important results have been obtained in the area of asynchronous circuits. The purpose of this paper is to describe some of these key developments and to refer to others which, for lack of space, cannot be discussed properly here.

Before we present the new results, we briefly emphasize the increasing importance of asynchronous circuits in the rapidly changing world of computer technology. As computer systems become more and more distributed, it is more difficult to achieve proper communication and synchronization among all the parts in such systems. Each of these parts is usually an independently clocked (i.e. synchronous) system. The synchronization of such systems involves many timing problems, some of which —like the metastability problem[6] — are of a fundamental nature. It is believed that proper design techniques for asynchronous circuits can alleviate these problems considerably.

For many years asynchronous circuits have been studied using Boolean algebra as the main formalism. These studies started with Huffman[13] and Muller and Bartky[19]; in the latter work the name *speed-independent circuit* was coined. Several different models[2, 3, 11, 23] were applied in order to describe and verify circuit behaviour as accurately as possible. It was only recently that a unifying theory was found in which the differences and similarities among these models could be explained[5].

A somewhat different approach to the design of asynchronous circuits was advocated by Molnar et al. in the Macro Modules project[7] from which the term *delay-insensitive circuit* evolved and, more recently, by Seitz[25] who coined the term *self-timed system.* Ideas expressed by Molnar and Seitz have influenced researchers at Eindhoven University of Technology[10, 20, 21, 27, 29] where a formalism, called *trace theory*, was developed for the design of such circuits. A similar formalism was used recently by Dill[8] for automatic verification of speed-independent circuits.

Martin[14, 16] uses the language of *Communicating Sequential Processes* (CSP)[12] to specify the behaviours of components to be designed. Such a CSP specification can then be compiled into a self-timed circuit. Many interesting circuits have been designed in this way, culminating with a fast asynchronous microprocessor[15]. Techniques similar to Martin's were also applied at Philips Research[1] where unexpectedly good results have been obtained.

Another demonstration of the usefulness of asynchronous circuits was given by Ivan Sutherland in his 1988 Turing Award lecture[28], where he shows how special types of asynchronous circuits, called *micropipelines*, can be used conveniently in the design of many fast processing components.

The design of asynchronous circuits is an attractive area of research, in particular because it lies on the boundary of theory and practice. On the one hand it contains simple and elegant mathematics, from formal language theory to semantics. On the other hand it is practical: many circuits have been used in actual designs and exhibit an unexpectedly good performance and robustness. Furthermore, such circuits are particularly well-suited for implementing parallel computations.

Although asynchronous circuits have been studied for many years now, the new approaches and major breakthroughs make us believe that this field is still very young and that more results are to be expected in the near future. Some difficult problems, however, still remain. We discuss some of these problems in the next sections.

2 The Producer and Consumer Paradigm

To illustrate the design of asynchronous circuits, we discuss a simple example starting with a behavioural specification and ending with a gate-level implementation. The necessary terminology and notation will be developed along the way. The producer-consumer setting of the example is due to Dijkstra[9]; the final circuit is a special case of a micropipeline[28].

Consider a 'producer' that outputs data items to be stored in a buffer and a 'consumer' that removes such items from the same buffer. Let a and b denote the production and consumption of data items, respectively. The producer and consumer act independently of each other; consequently, together they might generate any sequence of a's and b's. This may lead to unacceptable situations, however: the producer may cause an overflow of the buffer —which is assumed to be finite— and the consumer may cause an underflow. Consequently we need to design a controller which ensures that no items are produced if the buffer is full and none are consumed if the buffer is empty.

To keep the example simple, assume that the buffer has 2 places. Then the set of allowed sequences of productions and consumptions is the language accepted by the (incomplete) finite automaton defined by the state graph of Figure 1, where the initial state (corresponding to the empty buffer) is designated by an incoming arrow, and all the states are accepting states. Examples of allowed sequences are: ϵ (the empty trace), a, aa, ab, aab, $aabb$, etc.

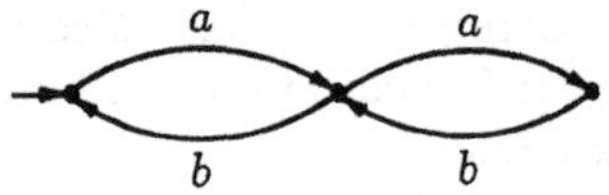

Figure 1: State graph for buffer controller.

In order to ensure that only allowed sequences occur, the controller of Figure 2 will send an 'acknowledge' signal to the producer and one to the consumer. Consider first the producer side and assume that the buffer is not full. The producer supplies input a notifying the controller that an item is being stored. After some delay, the controller will respond with an output p, informing the producer that another item can be stored. For convenience, signals that are inputs (respectively outputs) to the controller will be identified by ? (respectively !).

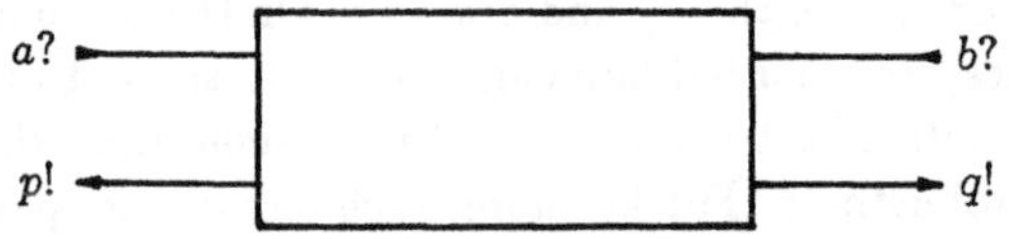

Figure 2: Buffer controller.

When the consumer and the controller are considered in isolation, the communication protocol between the two consists of an alternation of input a and output p, starting with a. Thus, all allowed behaviours are defined by the state graph of Figure 3(a). Similarly, the protocol between the controller and consumer is as shown in Figure 3(b). Initially the buffer is empty. When it becomes non-empty, the controller informs the consumer of this fact by sending output q. The consumer then removes an item while sending input b to the controller. This then repeats.

In the following we give a formal specification of the buffer controller using trace theory; the material from here to Section 7 is based on [10, 20, 27, 29]. Finite sequences of symbols are called *traces* and sets of such sequences, together with the indication which symbols are inputs and which are outputs, are described by *directed trace structures*. Formally,

(a) (b)

Figure 3: (a) Producer interface (b) Consumer interface.

a directed trace structure is a triple $< A, B, X >$, where A is the *input alphabet*, B is the *output alphabet*, and X is a set of traces constructed from symbols in $A \cup B$. A trace structure is called *regular* when its trace set is a regular set. Regular directed trace structures can be represented by expressions called *commands*, which are similar in many ways to regular expressions. Commands are defined inductively. The atomic commands ϵ, $a?, a!$, and $!a?$ represent the atomic trace structures $< \emptyset, \emptyset, \{\epsilon\} >$, $< \{a\}, \emptyset, \{a\} >$, $< \emptyset, \{a\}, \{a\} >$, and $< \{a\}, \{a\}, \{a\} >$, respectively. For commands E, $E0$, and $E1$, the expressions $E0; E1$ (concatenation), $E0|E1$ (union), $[E]$ (repetition), and $\mathbf{pref}E$ (prefix-closure) are also commands. Let $\mathbf{i}E, \mathbf{o}E$, and $\mathbf{t}E$ denote the input alphabet, output alphabet, and trace set of the directed trace structure represented by command E. The directed trace structures represented by $E0;E1$, $E0 \,|\, E1$, $[E]$, and $\mathbf{pref}E$ are defined by

$$\begin{aligned}
E0; E1 &= < \mathbf{i}E0 \cup \mathbf{i}E1, \mathbf{o}E0 \cup \mathbf{o}E1, (\mathbf{t}E0)(\mathbf{t}E1) >, \\
E0 \mid E1 &= < \mathbf{i}E0 \cup \mathbf{i}E1, \mathbf{o}E0 \cup \mathbf{o}E1, \mathbf{t}E0 \cup \mathbf{t}E1 >, \\
[E] &= < \mathbf{i}E, \mathbf{o}E, (\mathbf{t}E)^* >, \text{ and} \\
\mathbf{pref}E &= < \mathbf{i}E, \mathbf{o}E, \{t_0 \mid (\exists t_1 :: t_0 t_1 \in \mathbf{t}E)\} >,
\end{aligned}$$

where concatenation of sets is denoted by juxtaposition and * denotes Kleene's closure. (Here, we use the same notation for the command and the language defined by the command.) With the above definitions, the communication behaviours between producer and controller and between consumer and controller can be described by $\mathbf{pref}[a?; p!]$ and $\mathbf{pref}[q!; b?]$ respectively.

3 Parallel Composition and Synchronization

The complete communication behaviour of the controller of Figure 2 is specified by a proper synchronization of the two communication protocols; the overall protocol must ensure that the number of items contained in the buffer is always at most two and at least zero. In order to describe this proper co-operation between the two sides of the controller, we introduce a new operation on directed trace structures called *weaving*.

Formally, the weave $E0 \| E1$ of two directed trace structures represented by the commands $E0$ and $E1$ is defined by

$$\begin{aligned}
E0 \| E1 \;=\; & < \mathbf{i}E0 \cup \mathbf{i}E1, \mathbf{o}E0 \cup \mathbf{o}E1, \\
& \{t \in (\mathbf{a}E0 \cup \mathbf{a}E1)^* \mid t\!\downarrow\!\mathbf{a}E0 \in \mathbf{t}E0 \wedge t\!\downarrow\!\mathbf{a}E1 \in \mathbf{t}E1\} > .
\end{aligned}$$

Here, $\mathbf{a}E = \mathbf{i}E \cup \mathbf{o}E$ and $t\!\downarrow\!B$ denotes the projection of trace t on alphabet B, i.e. the trace from which all symbols not in B have been deleted. Informally, a weave of two specifications represents all behaviours that are in accordance with each of the two specifications.

As an example, consider the two commands $E0 = \mathbf{pref}[a?; c!]$ and $E1 = \mathbf{pref}[b?; c!]$. According to the above definitions of weaving we have

$$\mathbf{i}(E0\|E1) = \{a, b\},\ \mathbf{o}(E0\|E1) = \{c\},\ \text{and } \mathbf{t}(E0\|E1) = \{\epsilon, a, b, ab, ba, abc, bac, ..\}.$$

Alternatively, this directed trace structure can be represented by $\mathbf{pref}[a?\|b?; c!]$.

Notice that, in a weave, common symbols must match. One could also say that weaving expresses 'parallel co-operation with synchronization on common symbols.' There are two special cases of weaving $E0$ and $E1$: if $\mathbf{a}E0 \cap \mathbf{a}E1 = \emptyset$, weaving amounts to interleaving or shuffle; if $\mathbf{a}E0 = \mathbf{a}E1$, weaving amounts to intersection.

Returning to the buffer, we give a specification for the communication behaviour of the controller using weaving and projection. For this purpose we introduce a so-called *internal symbol* $!x?$ in the two communication protocols to achieve proper synchronization. This internal symbol is 'projected away' after weaving. The complete communication behaviour of the controller is given by

$$E = (\mathbf{pref}[a?; !x?; p!] \parallel \mathbf{pref}[!x?; q!; b?]) \downarrow \{a, p, b, q\}.$$

Because of the synchronization on the common (internal) symbol $!x?$, there are always at most two and at least zero items in the buffer. To see this, notice that, because of the first command in the weave, $0 \le \#_a t - \#_x t \le 1$ for each trace t in $\mathbf{t}E$, where $\#_a t$ denotes the number of a's in t. Similarly, because of the second command in the weave, $0 \le \#_x t - \#_b t \le 1$. Consequently, we have $0 \le \#_a t - \#_b t \le 2$. Moreover, each trace t with this property is also contained in $\mathbf{t}E$.

A command equivalent to E is the following:

$$E1 \;=\; \mathbf{pref}(a?; [(p!; a?)\|(q!; b?)]).$$

Both commands define the same trace structure. From this last command it is readily verified that the communication behaviour specified by E can be represented also by the state graph of Figure 4.

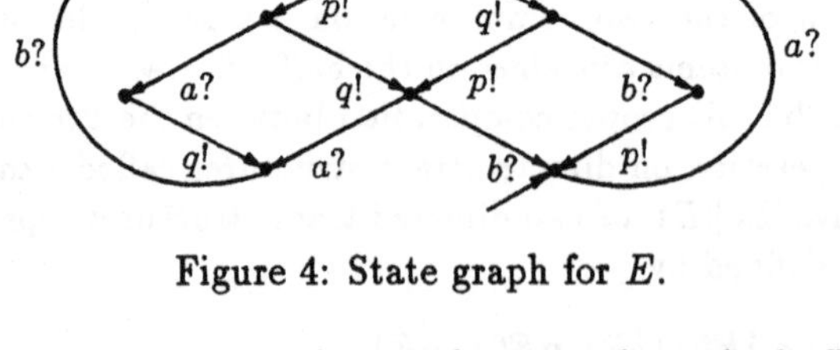

Figure 4: State graph for E.

The command $E1$ does not generalize to the case where the buffer has $n > 2$ places. The state graph of Figure 4 does generalize, but its size is exponential in n. The command E generalizes easily. For example, for $n = 4$ the command becomes

$$(\mathbf{pref}[a?; !x1?; p!] \parallel \mathbf{pref}[!x1?; !x2?] \parallel \mathbf{pref}[!x2?; !x3?] \parallel \mathbf{pref}[!x3?; q!; b?]) \downarrow \{a, p, b, q\}.$$

One can show that, for each trace t in the above command, we have $0 \leq \#_a t - \#_b t \leq 4$[27].

The length of the generalized version of the command E is linear in n. This illustrates that commands may be preferred to state graphs in case parallel operation and synchronization are involved.

4 Specification and Implementation

Thus far, we have specified the communication behaviour of the buffer controller by means of the rather abstract notion of a directed trace structure, which can be represented by a command. Such a specification can be interpreted not only in a formal mechanistic way, but also in physical terms like voltage transitions on wires. We explain these interpretations by means of the specification for the so-called *(Muller) C-element* (named after Muller[19]).

Formally, a C-element is specified by the command **pref**$[a?\|b?; c!]$. It represents a basic component with three terminals: inputs a and b, and output c. Its schematic is shown in Figure 5. Initially, the environment for this component produces communication actions

C-element	**pref**$[a?\|b?; c!]$
WIRE	**pref** $[a?; b!]$
IWIRE	**pref** $[b!; a?]$

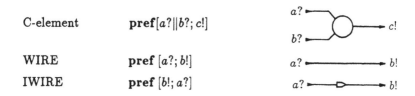

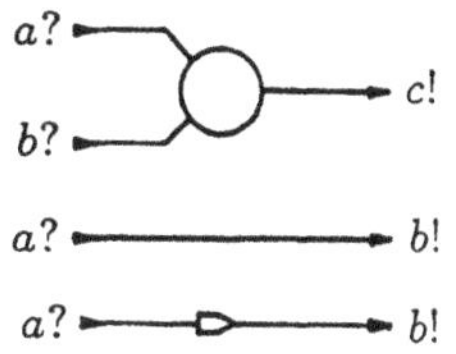

Figure 5: C-element, WIRE, and IWIRE.

(at terminals) a and b; then the component will respond with a communication action (at terminal) c. Only after c has been received may the environment produce the next communication actions a and b, after which the component will respond with c again, etc. We call this mode of operation —where outputs may be generated only after certain inputs have occurred and where next inputs may be generated only after certain outputs have occurred— the *input-output mode* of operation.

In a physical interpretation of the C-element, the symbols a, b, and c stand for voltage transitions at the corresponding terminals. A voltage transition can be high-going or low-going; both transitions are denoted by the same symbol. Input transitions are caused by the environment and output transitions are caused by the circuit.

With this physical interpretation in mind we can construct the state graph of Figure 6 for the behaviour of the C-element; the states in this graph are represented by the voltage levels at terminals a, b, and c respectively. Initially all voltages levels are 0. For convenience, n-tuples like a, b, c and $0, 0, 0$ are written as abc and 000, etc. The unstable states are represented by dashed circles.

In Figure 5 specifications of two other basic components, viz., the WIRE and the IWIRE, are given as well. The WIRE component has two terminals. It first receives input a and then responds with output b. According to the input-output mode of operation, the environment may produce the next input a only after it has received the output b, after

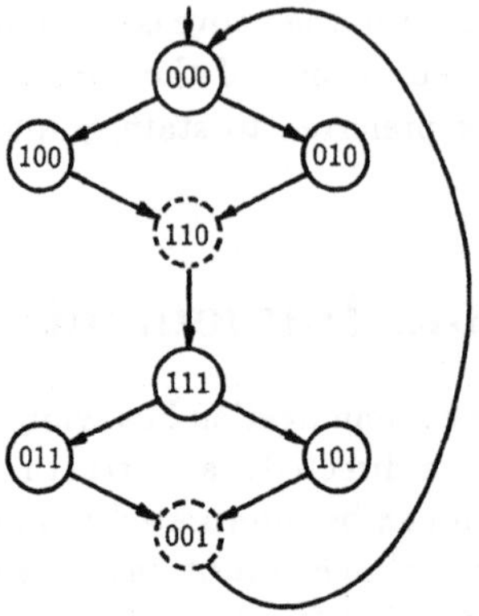

Figure 6: State graph for C-element.

which the WIRE will respond with output b again, etc. The IWIRE can be seen as an initialized WIRE: it starts by producing an output; after this, its behaviour is the same as that of the WIRE. The IWIRE is denoted by an open arrowhead in the schematic.

5 Decomposition

Given a specification of the communication behaviour of a component, like command E for the controller, we would like to 'decompose' this component into a network of some 'basic' components. In other words, we would like to find a network of basic components that produces the outputs as specified in E, if the environment produces the inputs as specified in E.

We first illustrate the concept of decomposition by means of the network of Figure 7. This network consists of a connection of two WIRE components, one IWIRE, and

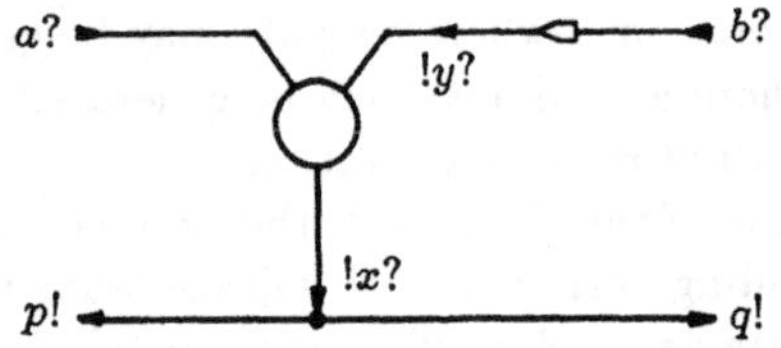

Figure 7: A network of basic components.

one C-element. Their respective specifications are $E_1 = \mathbf{pref}[x?;p!]$, $E_2 = \mathbf{pref}[x?;q!]$, $E_3 = \mathbf{pref}[y!;b?]$, and $E_4 = \mathbf{pref}[a?\|y?;x!]$. We show that the controller specified by

$$E = (\mathbf{pref}[a?;!x?;p!] \parallel \mathbf{pref}[!x?;q!;b?]) \downarrow \{a,p,q,b\}$$

can be decomposed into E_1, E_2, E_3, and E_4. This is expressed by $E \rightarrow (E_1, E_2, E_3, E_4)$, where the network of the components E_1, E_2, E_3 and E_4 is denoted by (E_1, E_2, E_3, E_4).

In order to take the environment of the network into account as well, we take the *reflection* of E in which we interchange the role of component and environment. More formally, the reflection of a directed trace structure represented by command E, is denoted by $\overline{E}$ and defines the directed trace structure $\overline{E} = <\mathbf{o}E, \mathbf{i}E, \mathbf{t}E>$, where the inputs and outputs are interchanged. In our example, $\overline{E}$ specifies when its outputs a and b are produced; these form the inputs of the network (E_1, E_2, E_3, E_4). Instead of considering the network (E_1, E_2, E_3, E_4) and its environment as specified in E, we consider the network $(E_0, E_1, E_2, E_3, E_4)$ from now on, where $E_0 = \overline{E}$.

Formally, in order to prove that the network of Figure 7 behaves as specified in E, we need to demonstrate that four conditions hold for the network $(E_0, E_1, E_2, E_3, E_4)$. The first condition requires that there be no dangling inputs or outputs, i.e. that every input be connected to an output and vice versa. In formula:

$$(\cup i : 0 \le i < 5 : \mathbf{o}E_i) = (\cup i : 0 \le i < 5 : \mathbf{i}E_i). \tag{1}$$

If (1) holds, we say that the network $(E_0, E_1, E_2, E_3, E_4)$ is *closed.*

The second condition is that no outputs of distinct components are connected to each other. In formula

$$\mathbf{o}E_i \cap \mathbf{o}E_j = \emptyset \quad \text{for } 0 \le i, j < 5 \wedge i \ne j. \tag{2}$$

When (2) holds, we say that the network is *free of output interference.* If (1) and (2) hold, then every symbol is an output of only one component in the network.

Conditions (1) and (2) are conditions on the structure of the network. They are formulated in terms of the alphabets of the directed trace structures. The next two conditions are behavioural conditions; they are phrased in terms of the trace sets and the alphabets.

The third condition prescribes that the input-output mode of operation may not be violated for any component in the network. We can simulate the network by generating traces of symbols, representing joint behaviours of the components in the network. Formally, we construct the trace set X of all joint behaviours as follows. Initially, $X = \{\epsilon\}$. Choose a trace t, symbol z, and index $i, 0 \le i < 5$, such that $t \in X \wedge z \in \mathbf{o}E_i \wedge tz\downarrow \mathbf{a}E_i \in \mathbf{t}E_i$ holds (i.e. after joint behaviour t, component E_i can produce output z). If for all $j, 0 \le j < 5$ we have $tz \downarrow \mathbf{a}E_j \in \mathbf{t}E_j$, (component j can accept z, i.e. its input-output mode of operation is not violated), then we add tz to X. Otherwise, we stop the simulation and say that the network has *computation interference.* Our third condition is

$$\text{The network is free of computation interference.} \tag{3}$$

Testing for computation interference can be done by an algorithm involving a finite state graph[8, 10]. One can verify that (3) is satisfied for the network $(E_0, E_1, E_2, E_3, E_4)$. The trace set X that can be generated for this network can be represented by

$$X = \mathbf{t}(\mathbf{pref}[!a?;!x?;!p?] \parallel \mathbf{pref}[!y?;!x?;!q?;!b?]).$$

The only difference between X and $\mathbf{t}E$ is the symbol y which we introduced in the decomposition as the output of the IWIRE and input of the C-element.

The fourth condition is that every trace of the component specified (here E) may also occur in the simulation we described above (excluding symbols not in $\mathbf{a}E$) and that only such traces may occur. In formula:

$$X \downarrow \mathbf{a}E = \mathbf{t}E. \tag{4}$$

If (4) is satisfied we say that the network behaves as specified. Condition (4) is satisfied by the network of Figure 7 as well.

The decomposition of Figure 7 generalizes to a decomposition for the controller of the n-place buffer. The network for this decomposition is given in Figure 8 for $n = 4$. Notice that the synchronization on common symbols is realized by the C-elements.

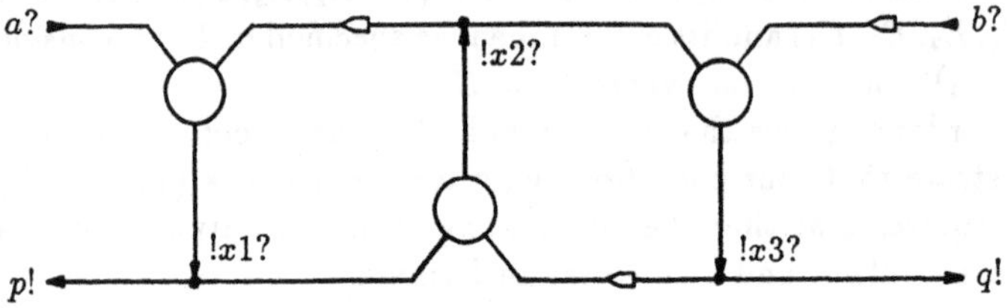

Figure 8: Decomposition for 4-place buffer controller.

The formalization of decomposition as given above is taken from [10]. The verification of the proof obligations mentioned above can be automated. Dill has designed a verifier that checks whether conditions (1) through (3) hold [8]. Such a verification method, however, is proportional to the number of states in the global state graph, which can grow as the product of the numbers of states of the components. For this reason, it is essential that theorems be developed that allow for a more efficient design or verification of a decomposition.

6 DI Decomposition

In the previous section we gave a formal definition of decomposition in terms of trace structures. The physical interpretation of decomposition is intended to correspond to the realization of a circuit by a network of sub-circuits. These sub-circuits may have arbitrary, non-negative response times. The communications between the sub-circuits, however, are assumed to be instantaneous. Thus, a circuit obtained by means of decomposition can be called a *speed-independent circuit*, i.e. its correctness is independent of any delay in the response times of the components.

In practice, the sub-circuits are connected to each other by means of wires, which may have unspecified delays. Such delays may affect the correctness of the circuit. If the correctness of the circuit is independent of any delays in the response times of components *and* connection wires, then we call such a circuit a *delay-insensitive circuit*.

While a speed-independent circuit is formally described by means of a decomposition, a delay-insensitive circuit is formally described by means of a *DI decomposition*. A DI decomposition is a decomposition in which all connection wires between the components are taken into account. Formally, these connection wires are represented by WIRE components and connect components with each other through an intermediate boundary as exemplified in Figure 9.

We give a brief description of a delay-insensitive circuit. For more details the reader is referred to [10]. First, we define the *enclosure* $enc(E_1)$, i.e. the component enclosed by the intermediate boundary, by renaming the symbols in the command E_1 to their 'localized' versions. The collection of WIRE components connecting the enclosure $enc(E_1)$ with its

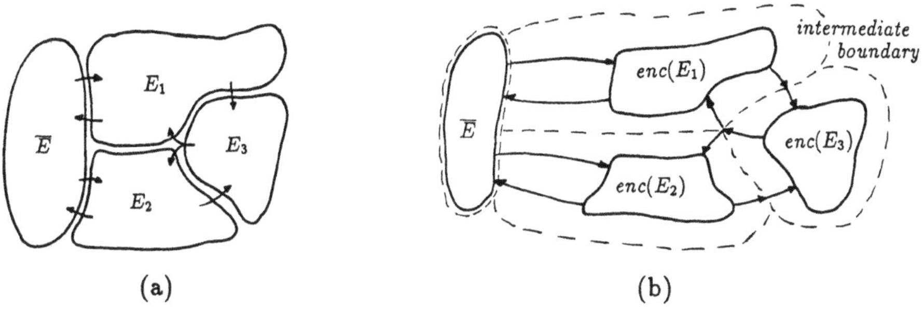

Figure 9: (a) Decomposition. (b) DI Decomposition.

intermediate boundary is denoted by $Wires(E_1)$. E_2, E_3, and E_4 are treated similarly. We say that the components E_1, E_2, E_3, and E_4 form a DI decomposition of component E, denoted by $E \stackrel{DI}{\rightarrow} (E_1, E_2, E_3, E_4)$ if and only if

$$E \rightarrow (i : 0 \leq i < 5 : enc(E_i),\ Wires(E_i)\ ). \tag{5}$$

In general, DI decompositions are more difficult to derive and verify than decompositions because of all the (connection) WIRE components. The two decompositions are equivalent, however, if all the constituent components are so-called DI components. A component E is called a DI component, if

$$E \rightarrow (enc(E),\ Wires(E)).$$

This property formalizes that the communication behaviour between component and environment is insensitive to wire delays. Verification of the DI property reduces to verifying that the network $(\overline{E},\ Wires(E), enc(E))$ is free of computation interference. The basic components C-element, WIRE, and IWIRE, for example, are DI components.

Since all basic components of the decomposition of Figure 7 are DI components, this decomposition is a DI decomposition, i.e. (5) holds. Accordingly, the circuit of Figure 7 represents not only a speed-independent circuit but also a delay-insensitive circuit. The same reasoning holds for Figure 8.

The idea of formalizing delay-insensitivity using a characterization of a DI component originates from Molnar[18]. Udding was the first to give a rigorous formulation of the DI property in terms of directed trace structures[29].

Shannon showed that any switching function can be realized by a gate circuit with only a finite number of gate types[26]. Similarly, we can ask ourselves 'Can any DI component be decomposed into a network of components chosen from a finite basis of DI components?' In [10] it is shown that, indeed, any regular DI component can be so decomposed. Consequently, such a decomposition is also a DI decomposition. The C-element, WIRE, and IWIRE component are members of such a basis. Other components are, for example, the XOR (or MERGE) component specified by $\mathbf{pref}[(a?|b?); c!]$, the TOGGLE component specified by $\mathbf{pref}[a?; b!; a?; c!]$, and an arbiter-like component.

If a component is specified by a command satisfying a certain syntax, then its decomposition can be described as a syntax-directed translation into a network of basic (DI) components[10]. Another attractive property of this translation is that the number of basic components in the final network is proportional to the length of the command. We also mention, however, that the decompositions obtained thus may not be optimal.

7 Realizations of Components by Gate Circuits

Having decomposed a component to be designed into a network of basic components such as C-elements and WIREs, one is faced with the problem of realizing the basic components. Here, we assume that this is to be done using logic gates; space limitations prevent us from discussing other types of realizations, such as those based on the commonly used MOS technology[30].

We introduce a number of ideas related to the design of asynchronous gate circuits by using the example of the C-element. The input-output behaviour of the C-element of Figure 5 has been described by the state graph of Figure 6. We assume that the inputs a and b can only change one at a time. The following illustrates a frequently used approach to gate circuit design. Construct a combinational gate circuit with inputs a, b, and c, and output C. The output C gives the 'excitation' of c, i.e. the next value that the (sequential) circuit output c should assume, if the present values of the inputs and the output are given by a, b, and c. From Figure 6 we observe that the output c should become 1 if $a = 1$ and $b = 1$. Once the output c becomes 1, it should remain 1 as long as $a = 1$ or $b = 1$. Thus, we have $C = ab + (a+b)c = ab + ac + bc$, where ab denotes the AND function and $a + b$ denotes the OR function of a and b. A gate circuit corresponding to this expression is shown in Figure 10, where the rectangle between C and c represents a delay. The presence of such a delay is implicitly assumed when we talk about the present

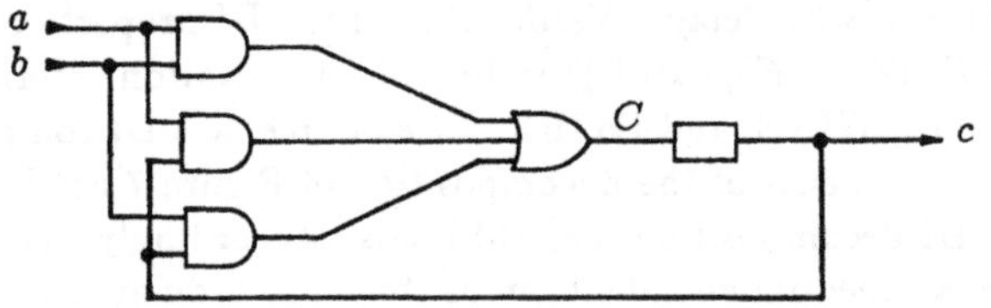

Figure 10: Gate circuit for C-element.

value c of the output and the excitation C to which the output is tending to change. Since C is assumed to be a Boolean function of a, b, and c, the value of C is computed from those of a, b, and c without delay.

Our design of the buffer controller began with a high-level specification of the controller component that led to the decomposition of the component in terms of some basic components. Furthermore, this decomposition has the important property that the network behaviour is independent of the delays in basic components and wires. Thus, the decomposition is delay-insensitive under the assumption that the environment co-operates, i.e. that the input-output mode of operation is used. A natural question now arises: Can each basic component be realized by a delay-insensitive *gate* circuit? In particular, is the circuit in Figure 10 for the C-element delay-insensitive? We consider such questions in the next two sections.

8 Fundamental Mode versus Input-Output Mode

Classical switching theory[17] assumes that gate circuits operate in *fundamental mode.* This means that the environment of the gate circuit co-operates in such a way that it produces a next input only after the entire circuit has stabilized. This is a much more restricted environment than the one in the input-output mode of operation, where an input is allowed to change as soon as the output changes, but possibly before all the gates in the circuit have had the chance to stabilize.

For simplicity, we first consider the question whether the circuit of Figure 10 is speed-independent. In other words, we assume that each gate has a delay and that wires have no delays. This model is also called the *gate-delay model.* The gate-delay model for the circuit of Figure 10 is shown in Figure 11. This model is more realistic than the one of Figure 10, where there is only one delay before the output c. Suppose that the circuit of

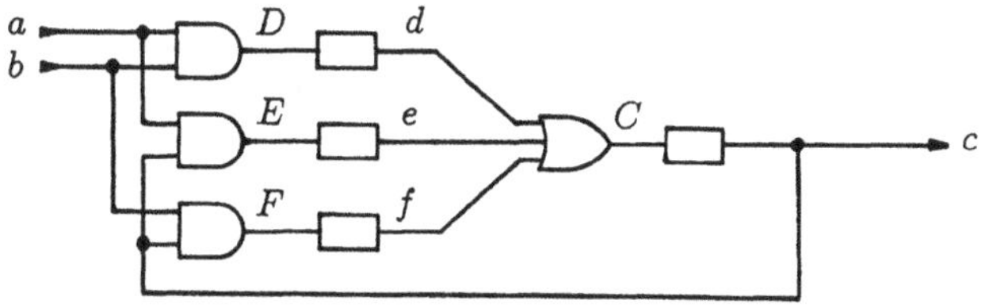

Figure 11: Gate-delay model.

Figure 11 is started in state $ab = 00, cdef = 0000$. When a changes to 1, the excitation remains $CDEF = 0000$, i.e. the circuit is stable and no further changes take place. Next, suppose b changes to 1, i.e. we reach $ab = 11, cdef = 0000$. Now $D = 1$, and the other excitations remain 0. After some time, d becomes 1 and changes C to 1. Eventually, c, e, and f also become 1 and the circuit stabilizes in $ab = 11, cdef = 1111$. Altogether, we have verified that, starting with $ab = 10$ and $cdef = 0000$, the change to $ab = 11$ results reliably in the state $cdef = 1111$, when fundamental mode is assumed.

Now consider the same transitions from the stable state $ab = 00, cdef = 0000$ when ab becomes 10 and then 11 in the input-output mode. The following sequence of events is possible, where underlined entries represent unstable gates:

ab	$cdef$	
00	0000	stable initial state
10	0000	input a changes, state is still stable
11	$0\underline{0}00$	input b changes
11	$\underline{0}100$	output d changes
11	$11\underline{0}\underline{0}$	output c changes

In the input-output mode, the environment may change input b again now. Thus, the following is possible:

10	$1\underline{1}\,\underline{0}0$	input b changes
10	$\underline{1}0\,\underline{0}0$	output d changes (before output e)
10	$00\,00$	output c changes (before output e)

The final state reached in this transition is the stable state $ab = 10, cdef = 0000$. Consider the signals a, b, and c only; we have just shown above that the following trace is possible: $t = abcbc$. This is not in accordance with the C-element specification, which requires that all the traces must be in the trace set of $\mathbf{pref}[a?||b?;c!]$.

Altogether, we have shown that a circuit behaving properly in fundamental mode may not behave properly in input-output mode. On the other hand, one can view fundamental mode operation as an input-output mode operation with 'slowly' changing inputs. If all gate outputs have become stable, then the circuit outputs, being outputs of some gates, have also reached their new values. Hence, operating a circuit in fundamental mode does not violate any input-output mode principles.

9 Fundamental Mode Analysis

We have seen in the last section that fundamental mode operation may differ from input-output mode operation. However, we do not reject the fundamental mode approach for two reasons. First, if it is impossible to realize a circuit specification to operate correctly in fundamental mode, then it is certainly impossible to do this in input-output mode. Therefore, fundamental mode realizability is a necessary condition for input-output mode realizability. Second, very little work has been done on input-output mode analysis, whereas much is known about the fundamental mode approach. In this section we briefly summarize the known results concerning fundamental mode analysis, including some very recent findings.

The first question that arises when one is choosing a model for a gate circuit is what assumptions are to be made about the presence of delays in the circuit. Three different models have been used in the past. The first one is the *feedback-delay model*, where one chooses a set of wires in the circuit with the property that cutting all these wires removes all the loops in the circuit —thus making the resulting circuit a combinational one. One then associates a delay with each wire in this set. An example of such a model is the circuit of Figure 10, where only one delay is assumed. This model was introduced by Huffman[13].

The second model is the *gate-delay model* in which a delay is associated with each gate. This corresponds to the concept of speed-independence described in Section 5 for a decomposition of a component into a network of basic components. This model was used by Muller and Bartky[19]. An example of this approach is the circuit of Figure 11.

The third model, the *gate-and-wire-delay model*, corresponds to the concept of delay-insensitivity described in Section 6 for DI decomposition. Such a model was used implicitly or explicitly by many authors; see, for example, [3]. To illustrate this model, consider Figure 11; here one would have to add a delay in the wire from input a to the top AND gate and also one from input a to the middle AND gate, etc. Altogether, ten additional wire delays have to be introduced.

Having selected a delay model for a given circuit, we can associate a state variable with each delay and find the excitation function for that variable, i.e. the Boolean function that specifies the value to which that variable is tending to change. For example, for the single state variable c in Figure 10, the excitation is $C = ab + ac + bc$. For Figure 11, we have four excitation functions: $D = ab$, $E = ac$, $F = bc$, and $C = d + e + f$.

In order to cover all three delay models, we will refer to a set of state variables and their excitation functions as a *network*. An *internal state* of a network is a tuple of binary values assigned to the state variable tuple, say y, and an *input state* is a tuple of binary values assigned to the input variable tuple, say x. For example, for the network of Figure 11 we have $x = (a, b)$ and $y = (c, d, e, f)$. A *total state* is a pair (input state, internal state). A state variable is *stable* in a given total state if its value in that state is equal to its excitation in that state. A total state is stable if all of its state variables are stable. In fundamental mode analysis we start with a stable total state $(x0, y0)$ of a network and then change the input to $x1$ and keep it at that value until the circuit 'has had a chance to stabilize.' Only then is the input allowed to change again. Of course, not all such transitions result in a single stable state. If more than one stable state can be reached or if an oscillation occurs, the behaviour is considered improper.

In general, the new state $(x1, y0)$ is unstable. If there is only one variable unstable, then eventually that variable must change, and a unique next total state is reached. If two variables are unstable, either can change first, or both can change at the same time. Thus, there are three possible next states. The situation where two or more variables are unstable is called a *race*. A commonly used *race model* —dating back to Huffman[13], but formalized by Brzozowski and Yoeli[3]— assumes that, in any unstable state, any subset of the set of unstable variables may 'win the race,' i.e. change to its corresponding excitation state. This model has been called the GMW (general multiple winner) race model.

It turns out that, if one uses the GMW race model, then each of the three delay models (feedback, gate, gate-and-wire) may yield different results[5]. The most accurate (and realistic) model of the three is the gate-and-wire-delay model. Unfortunately, the computation time for this model is exponential in the number of state variables. Recently, however, it has been shown [4] that the results for the gate-and-wire-delay model can be obtained also by an efficient method called *ternary simulation* introduced by Eichelberger[11]. Moreover, it was shown in [5] that, when a different race model —the so-called XMW model— is used, all three delay models yield the same results as ternary simulation. Thus, in the XMW race model one is permitted to use the feedback-delay model without losing any information. The XMW model uses ternary algebra based on the three values 0, 1, and $\times$, where the last value corresponds to an 'uncertain' signal.

While the XMW analysis (in any delay model) or, equivalently, the GMW analysis in the gate-and-wire-delay model give useful results, these results are frequently pessimistic in the sense that they include timing problems which are very unlikely to occur in practice. These analysis models are indeed so pessimistic that only very few sequential circuit behaviours can be realized in delay-insensitive fashion[23]. For example, Seger has shown that there does not exist a delay-insensitive gate circuit realizing a modulo-2 counter.

The situation is even worse for the input-output mode operation. We conjecture, that such commonly used circuits as the set-reset latch and the C-element do not have delay-insensitive gate circuit implementations. In fact, it appears that no non-trivial sequential behaviours have such realizations.

10 Bounded-Delay Models

The results mentioned in the previous section are rather discouraging because the basic components needed for delay-insensitive decomposition cannot be designed in delay-insensitive fashion from gates. What then is the solution to this dilemma? The practical answer is that we have to make some assumptions about the sizes of delays in circuit elements and wires, i.e. we are led to some type of *bounded-delay model.* Such a model has been used informally for many years. See, for example, [22]. A simple example of such an approach is the following. First, design a gate circuit using Huffman's feedback variable approach (as illustrated by the example in Figure 10). Then introduce a sufficiently large delay in the output to make sure that all the gates and wires in the circuit stabilize before the new output value reaches the output terminal. Such an approach will work if each delay has an upper bound.

We illustrate the bounded-delay approach with the circuit of Figure 12 for the C-element. The unlabeled rectangles and the thin ovals represent the gate and wire delays,

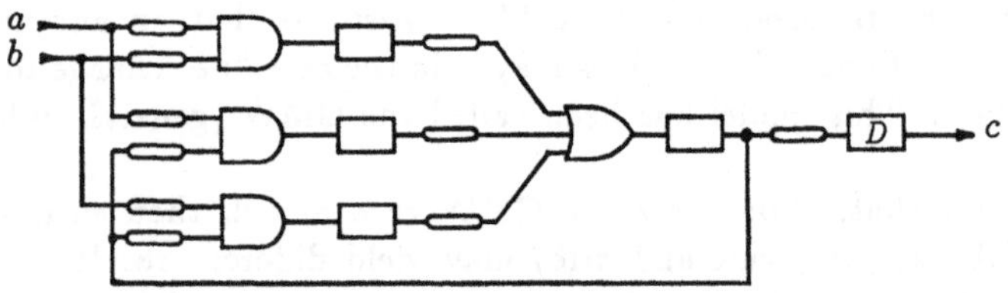

Figure 12: Bounded-delay model

respectively. The delay element labeled D is added by the designer to 'slow down' the output. Suppose we know that all wire delays are at most one time unit, and that all gate delays are at most two time units. One can verify that the circuit will behave properly in the input-output mode, if the output delay D is at least four time units. Notice that the presence of the output delay forces the input-output mode operation to become identical to the fundamental mode operation of the circuit.

It is an open problem whether there exist general systematic design techniques for circuits operating in some appropriate bounded-delay model. In fact, the analysis of circuits under the bounded-delay assumption is far from trivial. Some new results have been obtained by Seger who has shown that bounded-delay analysis can be done efficiently[24].

11 Concluding Remarks

By means of a simple example we have illustrated some of the recent developments in the design of asynchronous circuits. Because of space limitations, we have not been able to discuss important results obtained by others. In particular, we would like to mention the recent developments made by A.J. Martin. The interested reader will find an extensive overview in[16].

References

[1] C. van Berkel, C. Niessen, M. Rem, R. Saeijs, VLSI Programming and Silicon Compilation: a Novel Approach from Philips Research, *Proceedings of IEEE International Conference on Computer Design 1988, (ICCD '88)*, 1988.

[2] J.A. Brzozowski and M. Yoeli, *Digital Networks*, Prentice-Hall, Englewood Cliffs, New Jersey, 1976.

[3] J.A. Brzozowski and M. Yoeli, On a Ternary Model of Gate Networks, *IEEE Transactions on Computers*, Vol. C-28, pp. 178-183, 1979.

[4] J.A. Brzozowski and C-J. Seger, A Characterization of Ternary Simulation of Gate Networks, *IEEE Transactions on Computers*, Vol. C-36, pp. 1318-1327, 1987.

[5] J.A. Brzozowski and C-J. Seger, A Unified Framework for Race Analysis of Asynchronous Networks, *Journal of the ACM*, Vol. 36, pp. 20-45, 1989.

[6] T.J. Chaney and C.E. Molnar, Anomalous Behavior of Synchronizer and Arbiter Circuits, *IEEE Transactions on Computers*, Vol. C-22, pp. 421-422, 1973.

[7] W.A. Clark and C.E. Molnar, Macromodular Computer Systems, in *Computers in Biomedical Research*, Vol. IV, (R. Stacy and B. Waxman, eds.), Academic Press, New York, 1974.

[8] D.L. Dill, Trace Theory for Automatic Hierarchical Verification of Speed-Independent Circuits, in *Advanced Research in VLSI, Proceedings of the Fifth MIT Conference*, (J. Allen and F. Leighton, eds.), MIT Press, pp. 51-68, 1988.

[9] E. W. Dijkstra, Hierarchical Ordering of Sequential Processes, *Acta Informatica*, Vol. 1, pp. 115-138, 1971.

[10] J. C. Ebergen, *Translating Programs into Delay-Insensitive Circuits*, CWI Tract 56, Centre for Mathematics and Computing Science, Amsterdam, 1989.

[11] E.B. Eichelberger, Hazard Detection in Combinational and Sequential Switching Circuits, *IBM Journal of Research and Development*, Vol. 9, pp. 90-99, 1965.

[12] C.A.R. Hoare, Communicating Sequential Processes, *Communications of the ACM*, Vol. 21, pp. 666-677, 1978.

[13] D.A. Huffman, The Synthesis of Sequential Switching Circuits, in *Sequential Machines: Selected Papers*, (E.F. Moore ed.), Addison-Wesley, Reading Massachusetts, pp. 3-62, 1964, First appeared in the *J. Franklin Inst.*, Vol. 257, pp. 161-190, 1954.

[14] A. J. Martin, Compiling Communicating Processes into Delay-Insensitive VLSI Circuits, *Distributed Computing*, Vol. 1, pp. 226-234, 1986.

[15] A. J. Martin et al., The Design of an Asynchronous Microprocessor, in *Advanced Research in VLSI, Proceedings of the Decennial Caltech Conference on VLSI*, (C.L. Seitz ed.), 1989.

[16] A. J. Martin, Programming in VLSI: From Communicating Processes to Delay-Insensitive Circuits, in *UT Year of Programming Institute on Concurrent Programming*, (C.A.R. Hoare ed.), Addison-Wesley, 1989.

[17] E.J. McCluskey, *Introduction to the Theory of Switching Circuits*, McGraw-Hill Book Company, New York, 1965.

[18] C.E. Molnar, T.P. Fang and F.U. Rosenberger, Synthesis of Delay-Insensitive Modules, in *Proceedings 1985, Chapel Hill Conference on VLSI*, (H. Fuchs ed.), Computer Science Press, pp.67-86, 1985.

[19] D. E. Muller and W.S. Bartky, A Theory of Asynchronous Circuits, *Proceedings of an International Symposium on the Theory of Switching*, Vol. 29 of the *Annals of the Computation Laboratory of Harvard University*, Harvard University Press, Cambridge, Mass., pp. 204-243, 1959.

[20] M. Rem, Concurrent Computations and VLSI Circuits, in *Control Flow and Data Flow: Concepts of Distributed Computing*, (M. Broy ed.), Springer-Verlag, pp. 399-437, 1985.

[21] M. Rem, Trace Theory and Systolic Computations, in *Proceedings PARLE, Parallel Architectures and Languages Europe*, Vol. 1, (J.W. de Bakker, A.J. Nijman and P.C. Treleaven eds.), Springer-Verlag, pp. 14-34, 1987.

[22] F. Rosenberger, C. Molnar, T. Chaney, and T-P. Fang, Q-modules: Internally Clocked Delay-Insensitive Modules, *IEEE Transactions on Computers*, Vol. 37, pp.1005-1018, 1988.

[23] C-J. Seger, *Models and Algorithms for Race Analysis in Asynchronous Circuits*, Ph. D. Thesis, Department of Computer Science, University of Waterloo, Research Report CS-88-22, 1988.

[24] C-J. Seger, The Complexity of Race Detection in VLSI Circuits, in *Advanced Research in VLSI, Proceedings of the Decennial Caltech Conference on VLSI*, (C.L. Seitz ed.), pp. 335-350, 1989.

[25] C.L. Seitz, System Timing, in *Introduction to VLSI Systems*, C. Mead and L. Conway, Addison-Wesley, pp. 218-262, 1980.

[26] C. E. Shannon, A Symbolic Analysis of Relay and Switching Circuits, *Trans. AIEE*, pp. 731-723, 1938.

[27] J. L.A. van de Snepscheut, *Trace Theory and VLSI Design*, Lecture Notes in Computer Science 200, Springer-Verlag, 1985.

[28] I. E. Sutherland, *Micropipelines, The 1988 Turing Award Lecture*, to appear in CACM.

[29] J. T. Udding, A Formal Model for Defining and Classifying Delay-Insensitive Circuits and Systems, *Distributed Computing*, Vol. 1, pp. 197-204, 1986.

[30] N. Weste and K. Eshragian, *Principles of CMOS VLSI Design A Systems Perspective*, Addison Wesley, 1985.

New Simulations between CRCW PRAMs

Bogdan S. Chlebus[*], *Krzysztof Diks*[*], *Torben Hagerup*[†] and *Tomasz Radzik*[‡]

[*] Instytut Informatyki, Uniwersytet Warszawski, PKiN, p. 850, 00–901 Warszawa, Poland.

[†] Fachbereich Informatik, Universität des Saarlandes, D–6600 Saarbrücken, West Germany.

[‡] Computer Science Department, Stanford University, Stanford, California 94305.

Abstract: This paper is part of a continued investigation of the relative power of different variants of the CRCW PRAM with infinite global memory. The models that we consider are the standard PRIORITY and COMMON PRAMs, together with the less well-known COLLISION$^+$ and TOLERANT PRAMs. We describe several new results for the simulation of an n-processor PRIORITY PRAM on weaker machines:

(1) on an n-processor TOLERANT PRAM: Slowdown $O(\sqrt{\log n})$;

(2) on an n-processor COLLISION$^+$ PRAM: Slowdown $O(\log\log n \log^{(3)} n)$;

(3) on a COMMON PRAM with kn processors ($k \leq \log n/2$): Slowdown $O(\log n/(k \log(\log n/k)))$;

(4) on a TOLERANT PRAM with kn processors ($2 \leq k \leq \log n$): Slowdown $O(\log n/\log k)$;

(5) on a randomized n-processor COLLISION$^+$ PRAM: Expected slowdown $O(\log\log n)$.

1. Introduction

The PRAM is one of the most popular models of parallel computers. A PRAM consists of a collection of sequential processors numbered $1, \ldots, p$ and operating synchronously on a global memory, which in this paper will be assumed infinite. Various PRAMs have been introduced, differing in the conventions regarding concurrent reading and writing, i.e., attempts by several processors to access the same memory cell in the same step. CRCW (concurrent-read concurrent-write) PRAMs allow simultaneous reading from as well as simultaneous writing to each cell. Simultaneous writing is not immediately logically meaningful, and various different rules for the

---

[†] Supported by the Deutsche Forschungsgemeinschaft, SFB 124, TP B2, VLSI Entwurfsmethoden und Parallelität.

[‡] Part of the research was carried out while the author was at Instytut Informatyki, Uniwersytet Warszawski.

resolution of write conflicts have been introduced and used in concrete algorithms. The conflict resolution rules of interest to us are:

COMMON [K]: All processors writing to a given cell in a given step must be writing the same value, which then gets stored in the cell;

TOLERANT [GR]: If more than one processor attempts to write to a given cell in a given step, then the contents of that cell do not change;

COLLISION$^+$ [CDHR]: If the processors attempting to write to a given cell in a given step all attempt to write the same value, then that value gets stored in the cell; if at least two values differ, a special collision symbol is stored in the cell;

ARBITRARY [SV]: If several processors simultaneously attempt to write to a given cell, then one of them succeeds and writes its value, but there is no rule assumed to govern the selection of the successful processor;

PRIORITY [G]: If several processors simultaneously attempt to write to a given cell, then the lowest-numbered processor among them succeeds.

This paper continues the investigation, begun in [CDHR], of the relative power of different CRCW PRAMs.

For brevity, a CRCW PRAM working according to the COMMON (TOLERANT, etc.) rule will be called a COMMON (TOLERANT, etc.). The number of processors of a particular machine is indicated by a postfixed integer in parentheses (e.g., PRIORITY(n) denotes a PRIORITY PRAM with n processors). A simulation by one PRAM M_2 of a single step of another PRAM M_1 is a computation by M_2 that changes the state of M_2's global memory exactly as the single step under consideration changes the state of M_1's global memory (i.e., it implements the same state transition). As a technical requirement, we assume the simulating machine M_2 to have an additional infinite global memory that can be used to hold the variables of the simulation (if an upper bound on the space used by M_1 is easily computable, an upper segment of a single memory can be used for the same purpose). We say that M_2 simulates M_1 with *slowdown* T if at most T steps of M_2 are needed to simulate a single step of M_1.

As observed in [CDHR], some relations between the CRCW PRAM models defined above are obvious. For instance, ARBITRARY is stronger than COLLISION$^+$ in the sense that COLLISION$^+$(n) can be simulated on ARBITRARY(n) with slowdown $O(1)$. If we express this fact as "COLLISION$^+$ $\leq$ ARBITRARY", then the following partial ordering is easy to establish:

$$\begin{matrix} \text{COMMON} & & \\ & \leq & \\ & & \text{COLLISION}^+ \leq \text{ARBITRARY} \leq \text{PRIORITY}. \\ & \leq & \\ \text{TOLERANT} & & \end{matrix}$$

Grolmusz and Ragde [GR] proved that COMMON is not comparable with TOLERANT in the sense of this relation "$\leq$".

In this paper we contribute several new efficient algorithms simulating PRIORITY (the strongest commonly used CRCW PRAM) on the weaker models COMMON, TOLERANT and COLLISION$^+$. New and previously known simulation results are summarized in Table 1.

Simulated machine	Simulating machine	Slowdown	Reference
	Any(n)	$O(\log n)$	(sorting)
	COMMON(n)	$O\left(\frac{\log n}{\log\log n}\right)$	[FRW2]
	COMMON($n \log n$)	$O(1)$	[CDHR]
	COMMON(kn) ($1 \le k \le \log n/2$)	$O\left(\frac{\log n}{k\log(\log n/k)}\right)$	new
PRIORITY(n)	TOLERANT(n)	$O(\sqrt{\log n})$	new
	TOLERANT(kn) ($2 \le k \le n$)	$O\left(\frac{\log n}{\log k}\right)$	new
	COLLISION$^+$(n)	$O((\log\log n)^2)$	[CDHR]
	COLLISION$^+$(n)	$O(\log\log n \log^{(3)} n)$	new
	COLLISION$^+$(n) (randomized)	$O(\log\log n)$ (expected)	new
	ARBITRARY(n)	$O(\log\log n)$	[CDHR]

Table 1: Upper bounds for inter-CRCW PRAM simulations.

There is only one significant lower bound corresponding to the entries in the table: Ragde, Szemerédi, Steiger and Wigderson [RSSW] proved a lower time bound of $\Omega(\sqrt{\log n})$ on COMMON(n) for a problem that can be solved in constant time on TOLERANT(n). Assuming certain additional restrictions on the size of the global memory, the communication between processors or the way in which the input is provided, further separation results have been obtained (cf. [FRW1, FRW2, LY]).

Consider a simulated machine M_1 with n processors and a simulating machine M_2 with at least n processors. We shall always identify the processors of M_1 with n processors of M_2, called the *main processors*. Every write step of M_1 conceptually partitions its processors and, by extension, the main processors of M_2 into groups: Two processors are in the same group exactly if they attempt to write to the same memory cell (create an extra group for processors that do not attempt to write to any cell). As in [FRW1], we call these groups *colour classes* and say that two processors are of the same colour iff they are in the same colour class. In this paper, M_1 will always be a PRIORITY PRAM, so that the write attempt of a processor of M_1 is successful exactly if the processor is the lowest-numbered processor, called the *winner*, within its colour class. Hence the task of the simulation algorithm is to determine, using the available write conflict resolution rule, the winner of each colour.

In the interest of readability, we will pretend as a descriptional device that each processor P of the simulating machine M_2 at each step is located in a particular global memory cell, which we denote by *Loc*(P). In addition to its usual integer contents, each memory cell can contain arbitrarily many processors; a cell that contains at least one processor is said to be *non-empty*. Once a processor has computed the address of some global memory cell in one of its local registers, it can move to that cell. We call a machine with these properties an *allocated PRAM*. Its realization in terms of standard PRAMs is obvious: Each processor has a dedicated local register whose contents is interpreted as the current location of the processor, and every move by the processor corresponds to a change of the value of this register.

Definition: The *Find-First problem* of size n is as follows: Given in the memory of an allocated PRAM an array $A[1..n]$, each cell of which is empty or contains exactly one processor, find the lowest-numbered non-empty cell of A (i.e., after the computation, each processor initially in A must know whether initially it had any processors in A strictly to its left).

The Find-First problem is closely related to that of determining the winner of each colour in the simulation of PRIORITY: For each possible colour γ, take an array $A_\gamma[1..n]$ and move each main processor P of the simulating machine to the cell $A_\gamma[i]$, where γ is the colour of P and i is its number. For each colour, this constructs an instance of the Find-First problem whose solution determines the winner of that colour. We hence have

Proposition 1: Suppose that an allocated PRAM of some type τ can solve Find-First problems of size n in time T. Then PRIORITY(n) can be simulated with slowdown $T+O(1)$ on a (standard) PRAM of type τ.

In the following section we describe solutions to the Find-First problem on COMMON, TOLERANT and COLLISION$^+$ and state the corresponding simulation results. Section 3 is devoted to a randomized simulation of PRIORITY(n) on COLLISION$^+$(n).

In most of the paper we ignore questions of rounding and assume expressions to be integer-valued whenever this is convenient. The changes needed to take rounding into account are easy, but would obscure the exposition. We also omit details of the computation of addresses, mainly of cells representing nodes in trees, and of the allocation of space to subproblems solved recursively.

The variable n always denotes the number of processors of a simulated machine and the size of Find-First problems. By a b-ary tree, where $b \geq 2$, we will mean a complete, ordered tree with n leaves and degree b of each internal node. We assume the leaves of the tree to be numbered $1,\ldots,n$ from left to right. A b-ary tree *over* an array $A[1..n]$ is a b-ary tree whose ith leaf, for $i=1,\ldots,n$, coincides with $A[i]$. The height of a b-ary tree is $\log n/\log b$.

2. Deterministic simulations

Fich, Ragde and Wigderson [FRW2] proved that Find-First problems of size n (there called "leftmost prisoner problems") can be solved on COMMON(n) in time $O(\log n/\log\log n)$. We show how to speed up the solution if each of the processors defining the problem, in the sequel called the *main processors* (note how this agrees with the use of the term in Section 1), has at its disposal k additional auxiliary processors, where k is some integer with $1 \le k \le \log n/2$.

Let a Find-First problem be defined by some array $A[1..n]$, and let T be a b-ary tree over A, where $b = 2^{\log n/k}$. We say that a leaf of T is *occupied* exactly if it contains at least one processor. An internal node of T is occupied iff some leaf in its subtree is occupied. An occupied node is called a *local winner* if it is either the root of T or the leftmost occupied son of its father. The *overall winner* is the leftmost occupied leaf in T. The following proposition is obvious (see Fig. 1):

Proposition 2: An occupied leaf is the overall winner if and only if all its ancestors in T are local winners.

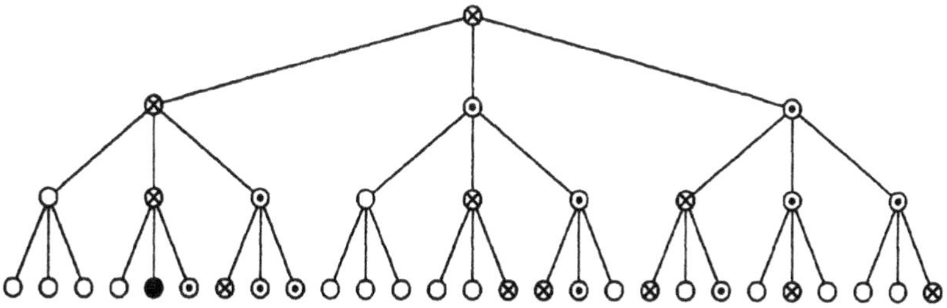

Fig. 1. An illustration of Proposition 2. ⊙ denotes an occupied node, ⊗ a local winner, and ● the global winner.

Observe that the overall winner yields the solution to the Find-First problem defined by A. Using the above proposition, we can determine the overall winner on COMMON as follows:

(1) Recall that each main processor P is located in some leaf $Loc(P)$ of T. Distribute its associated auxiliary processors among the ancestors of $Loc(P)$, one processor to each ancestor.

(2) Each auxiliary processor P marks $Loc(P)$ as occupied.

(3) Compute the local winner among the sons of each occupied internal node. To do this, execute the Find-First algorithm of [FRW2].

(4) Compute the overall winner as the unique leaf in T whose proper ancestors are all local winners. This requires the computation of a k-way OR by each set of k auxiliary processors associated with some main processor.

All steps except (3) can be executed in constant time on COMMON, and because step (3) can be executed in time $O(\log b/\log\log b)$ [FRW2], where $b = 2^{\log n/k}$, we have the following

Lemma 1: If each main processor has k additional auxiliary processors, where $1 \le k \le \log n/2$, then Find-First problems of size n can be solved on COMMON in time $O(\log n/(k \log(\log n/k)))$.

Remark: Lemma 1 was discovered independently by Ragde [R], who also has a matching lower bound.

Combining Lemma 1 with Proposition 1, we obtain:

Theorem 1: PRIORITY(n) can be simulated on COMMON(kn), where $1 \le k \le \log n/2$, with slowdown $O(\log n/(k \log(\log n/k)))$.

The Find-First algorithm of [FRW2] does not work on TOLERANT. For this reason, a simulation on TOLERANT(kn), where $2 \leq k \leq n$, should employ the auxiliary processors in a different way that we now describe.

Take a k-ary tree T over the input array. For each main processor P, assign its k auxiliary processors to the k leaves in T with the same father as $Loc(P)$. Then use the algorithm of [GR] for the computation of OR on TOLERANT to determine the local winner within each group of leaves with the same father.

In the next step, only main processors located in leaves that are local winners (together with their auxiliary processors) move to the fathers of their current locations. Continuing in this fashion, the overall winner is determined in $O(\log n/\log k)$ steps.

Lemma 2: If each main processor has k additional auxiliary processors, where $2 \leq k \leq n$, then Find-First problems of size n can be solved on TOLERANT in time $O(\log n/\log k)$.

Theorem 2: PRIORITY(n) can be simulated on TOLERANT(kn), where $2 \leq k \leq n$, with slowdown $O(\log n/\log k)$.

We now return to the Find-First problem without auxiliary processors, but would like to apply the algorithms developed above. The idea is to eliminate from the original problem a large number of processors that are known to not be the winner, and then to use these as auxiliary processors of the remaining processors.

Definition: The *Partitioning problem* of size n is as follows: Given in the global memory of an allocated PRAM an array $A[1..n]$ containing at least two processors, but with no cell containing more than one processor, colour each processor in A either black or white in such a way that the number of white processors is at least one, but does not exceed the number of black processors.

The Partitioning problem was introduced in [CDHR], where it was shown that Partitioning problems of size n can be solved in time $O(\log\log n)$ on COLLISION$^+$. The same algorithm works on TOLERANT, but we here show a more general result. Assume that each main processor has k additional processors, where $2 \leq k \leq \log n$. The following algorithm then solves the Partitioning problem on TOLERANT in time $O(\log\log n/\log k)$ (cf. Fig. 2):

(1) If $n < 4$, then colour the leftmost processor white and colour the remaining processors black. Otherwise execute steps (2)–(6).

(2) Let T be a b-ary tree over the input array, where $b = 2^{\lceil \log n/k \rceil}$, and distribute the auxiliary processors of each main processor P among the at most k ancestors of $Loc(P)$.

Each node of T now is empty or contains one or more processors. A non-empty node is called *single* if it contains only one processor; otherwise it is called *full*. Observe that each non-empty leaf of T is a single node.

(3) Each auxiliary processor P determines the status (single or full) of $Loc(P)$.

(4) Each main processor P moves to the single ancestor of $Loc(P)$ of maximum height.

Step (4) is easy since on the path from a non-empty leaf to the root of T, there is exactly one change from single to full nodes.

(5) Each main processor P checks whether some processor is located at a brother of $Loc(P)$. If this is the case, then the father of $Loc(P)$ is called an *active* node. Otherwise P colours itself black.

(6) For each active node v, solve recursively the Partitioning problem defined by the main processors located in the sons of v.

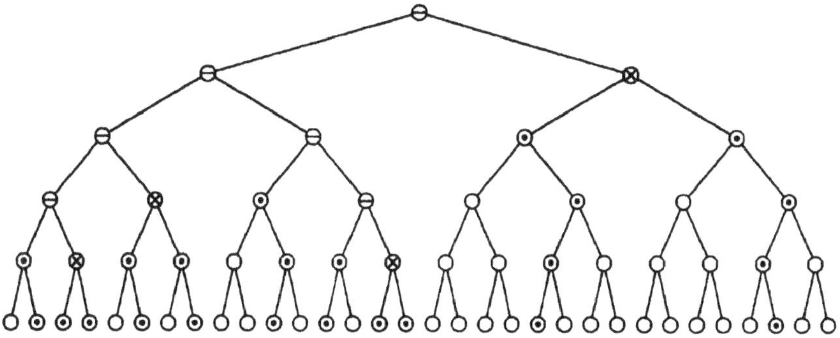

Fig. 2. A Partitioning problem and the corresponding classification of tree nodes. ⊙ denotes a single node, ⊗ an active node, and ⊖ a non-active full node.

To prove the correctness of the algorithm, observe that whenever a processor is coloured white, it has a mate that is coloured black (see step (1)). Hence the number of white processors does not exceed the number of black processors. On the other hand, at least one processor is coloured white. This is clear for $n < 4$, and for $n \geq 4$ it suffices by induction to show the existence of at least one active node. But each full node of maximum depth in T clearly is an active node.

Steps (1)–(5) of the algorithm can be executed in constant time on TOLERANT. Since each recursive call reduces the logarithm of the problem size by a factor of k, the total execution time is $O(\log\log n/\log k)$.

Lemma 3: If each main processor has k additional auxiliary processors, where $2 \leq k \leq \log n$, then Partitioning problems of size n can be solved on TOLERANT in time $O(\log\log n/\log k)$.

The algorithm given has two interesting properties:

(a) Suppose that we want to solve the Partitioning problem again, but now only for the white processors, which we call *survivors*. Then we can associate with each survivor a total of $2k$ auxiliary processors; it suffices that each white processor takes over the auxiliary processors of its black mate. If we repeatedly solve the Partitioning problem, the number of auxiliary processors per survivor hence doubles in each Partitioning step.

(b) If we want to apply the algorithm for the Partitioning problem to the solution of Find-First problems, it seems that we must guarantee that the winner of the Find-First problem always survives. Unfortunately, the winner may be coloured black in step (5). We remedy this situation by maintaining, during the whole computation, a function F from the set of main processors to the set of cells of the input array, such that the following invariant holds after each step:

Let P be the leftmost survivor that is either uncoloured or white, and let L be the overall winner of the original Find-First problem. Then $F[P] = L$. (*)

Initially, $F[P] = Loc(P)$ for all main processors P, and the invariant is obviously satisfied. In order to preserve its validity, we modify step (1) to colour white the processor with the leftmost F value, and we add at the end of step (5) the following instructions designed to "save" L, whenever it is about to "disappear", by copying it from a processor that has just coloured itself black in the present step (5) (let us call such a processor *recent black*) to at least one uncoloured processor:

(a) Each recent black processor P moves to the father of $Loc(P)$ and marks it by $F[P]$.

(b) Each recent black processor P together with its auxiliary processors checks whether $F[P]$ is strictly to the left of all F values written at proper ancestors of $Loc(P)$. If this is the case, then P notifies all processors at proper ancestors of $Loc(P)$ that their F values are not leftmost. Such processors are said to die. On the other hand, if P finds at some proper ancestor of $Loc(P)$ an F value that is equal to or to the left of $F[P]$, then P itself dies.

(c) Each uncoloured main processor Q checks with its auxiliary processors whether there is a live processor P at some proper ancestor of $Loc(Q)$. There is at most one such processor P. If P exists and $F[P]$ is to the left of $F[Q]$, then Q sets $F[Q] := F[P]$.

Now we can show how to apply the above algorithm to speed up the solution of Find-First problems on TOLERANT and COLLISION$^+$. Without loss of generality we assume that initially each main processor has two auxiliary processors.

The solution of the Find-First problem on TOLERANT is as follows: First carry out $\log\log n$ Partitioning steps in order to get $\log n$ auxiliary processors for each survivor. By Lemma 3, this can be done in time $O((\log\log n)^2)$. Then execute $\sqrt{\log n}$ further Partitioning steps. Again by Lemma 3, this takes time $O(\sqrt{\log n})$. Now we can assume that each survivor has $2^{\sqrt{\log n}}$ auxiliary processors, and Lemma 2 allows us to find the winner P of the Find-First problem defined by the survivors in constant time. Finally, by the invariant, $F[P]$ gives the solution to the original Find-First problem, and we have:

Theorem 3: PRIORITY(n) can be simulated on TOLERANT(n) with slowdown $O(\sqrt{\log n})$.

Remark: Theorem 3 answers a question posed in [GR], namely whether there is a simulation with slowdown $o(\log n)$ of PRIORITY(n) on TOLERANT(n).

The solution of Find-First problems on COLLISION$^+$ is very similar: Carry out $\log\log n$ Partitioning steps in order to get $\log n$ auxiliary processors for each survivor, and then finish in constant time by Theorem 1. In each Partitioning step we can use auxiliary processors gathered in previous steps. Hence in the ith Partitioning step, each survivor has 2^i auxiliary processors, and by Lemma 1, the ith Partitioning step can be executed in time $O(\log\log n/i)$, giving a total execution time of

$$O\Big(\sum_{i=1}^{\log\log n} \frac{\log\log n}{i}\Big) = O(\log\log n \log^{(3)} n).$$

Theorem 4: PRIORITY(n) can be simulated on COLLISION$^+$(n) with slowdown $O(\log^{(2)} n \log^{(3)} n)$.

3. A randomized simulation.

This section outlines a randomized simulation of PRIORITY(n) on COLLISION$^+$(n). As usual, the task of the simulating COLLISION$^+$ is to solve one Find-First problem for each colour class defined by a write step of the simulated PRIORITY. The simulation is always correct; only its execution time is a random variable. To our knowledge, it is the first randomized simulation of one PRAM model on another (except for simulations based on randomized sorting). An important ingredient is the following

Lemma 4 [CDHR] : If each main processor has $\log n$ additional auxiliary processors, then Find-First problems of size n can be solved on COMMON (and therefore on COLLISION$^+$) in constant time.

Before we state the simulation algorithm itself, we introduce a useful subroutine. Let us say that m processors $P_1, \ldots, P_m$ *scatter* over a cells if they carry out the following computation: For $i = 1, \ldots, m$, P_i chooses a random number z_i in the range $1 \,..\, a$, sets $C[z_i] = 0$, where $C[1 \,..\, a]$ is some array, then (simultaneously with the remaining processors) writes its processor number to $C[z_i]$, and finally reads $C[z_i]$. As a result of the scattering, each processor P_i knows if it *collides*, i.e., if $z_j = z_i$ for some $j \neq i$. If P_i does not collide, it moves to $C[z_i]$.

An overview of the simulation algorithm looks as follows: We first draw a random sample B of expected size $\Theta(n/\log n)$ from the set of all n processors. The size of B is chosen so that using Lemma 4 and all available processors, we can with high probability find the winner of each colour class within B in constant time. Next each processor compares itself with the winner in B of the same colour (if any). If it loses to the winner in B, i.e., its number is larger, then it clearly cannot be the overall winner, and we may forget about its write request. In this way the number of processors that must still be taken into account is significantly reduced, and large colour classes tend to lose more of their elements. In fact, we can prove that with high probability, all colour classes are of size $(\log n)^{O(1)}$ after this step.

We next scatter the elements of each colour class over an array that is also of size $(\log n)^{O(1)}$, but much bigger that the expected size of colour classes. Processors that do not collide can be dealt with in $O(\log\log n)$ deterministic time in a straightforward manner. On the other hand, the number of colliding processors with high probability is $O(n/\log n)$, so that we again can use Lemma 4.

Each application of Lemma 4 to a set U of processors of size $O(n/\log n)$ poses a problem that was ignored above: It is not sufficient to have enough auxiliary processors; these must also be associated with the elements of U. This forces us to compute B in a rather roundabout fashion: We first obtain a random sample A of expected size $O(n/\log n)$. Then the elements of A scatter over an array of size larger than $|A|$, but still $O(n/\log n)$, and B is taken as the set of processors in A that do not collide. Since the elements of B can be given distinct integers of size $O(n/\log n)$ (their position in the array), the allocation of auxiliary processors to the elements of B is trivial.

In the second application of Lemma 4, the colliding processors are numbered consecutively, but only locally within ranges of processor indices of size $(\log n)^{O(1)}$ (a global numbering would be too slow). One can show that the (random) distribution of the colliding processors is sufficiently even that with high probability, each range contains $\Theta(\log n)$ auxiliary processors for each colliding processor in the range. Hence the allocation of auxiliary processors can be done locally within ranges in $O(\log\log n)$ deterministic time.

Theorem 5: For any constant α, one step of a PRIORITY(n) can be simulated by $O(\log\log n)$ steps of a randomized COLLISION$^{+}(n)$ with probability at least $1 - n^{-(\log n)^{\alpha}}$.

Proof: Omitted for lack of space.

References

[CDHR]: B. S. Chlebus, K. Diks, T. Hagerup and T. Radzik: "Efficient Simulations between Concurrent-Read Concurrent-Write PRAM Models". 13th Symposium on Mathematical Foundations of Computer Science (1988), Springer Lecture Notes in Computer Science 324, 231–239.

[FRW1]: F. E. Fich, P. Ragde and A. Wigderson: "Relations Between Concurrent-Write Models of Parallel Computation". *SIAM Journal on Computing* **17** (1988), 606–627.

[FRW2]: F. E. Fich, P. Ragde and A. Wigderson: "Simulations Among Concurrent-Write PRAMs". *Algorithmica* **3** (1988), 43–51.

[G]: L. M. Goldschlager: "A Universal Interconnection Pattern for Parallel Computers". *Journal of the ACM* **29** (1982), 1073–1086.

[GR]: V. Grolmusz and P. Ragde: "Incomparability In Parallel Computation". Proceedings, 28th Annual Symposium on Foundations of Computer Science (1987), 89–98.

[K]: L. Kučera: "Parallel Computation and Conflicts in Memory Access". *Information Processing Letters* **14** (1982), 93–96.

[LY]: M. Li and Y. Yesha: "Separation and Lower Bounds for ROM and Nondeterministic Models of Parallel Computation". *Information and Computing* **73** (1987), 102–128.

[R]: P. Ragde, personal communication, 1988.

[RSSW]: P. Ragde, W. Steiger, E. Szemerédi and A. Wigderson: "The Parallel Complexity of Element Distinctness is $\Omega(\sqrt{\log n})$". *SIAM Journal on Discrete Mathematics* **1** (1988), 399–410.

[SV]: Y. Shiloach and U. Vishkin: "An $O(\log n)$ Parallel Connectivity Algorithm". *Journal of Algorithms* **3** (1982), 57–67.

ABOUT CONNECTIONS BETWEEN SYNTACTICAL AND COMPUTATIONAL COMPLEXITY. (1)

Jean-Luc Coquidé, Max Dauchet, Sophie Tison
LIFL (URA 369-CNRS)
Université de Lille-Flandres-Artois.
UFR IEEA, 59655 VILLENEUVE D 'ASCQ Cedex FRANCE.
e mail: dauchet@frcitl71.bitnet

ABSTRACT: We prove in this paper three kinds of results:
1/ It is undecidable whether a term rewriting system preserves recognizability.
2/ ground term rewriting systems with recognizable control are as powerfull as general rewriting systems.
3/ Influence on decidability and complexity of termination of syntactical restrictions on term rewriting systems reduced to one or two rules.

INTRODUCTION.

The three results investigate connection between syntactical restriction and computational complexity or decision problems.

It is well known that there is unfortunatly no general "continuous" or "practical" connection between both point of view. For example, any Prolog program can be reduced to one rule and three facts (and so any problem about this class is undecidable) but any termination problem reduced to a rule and one fact is decidable (Dauchet & all (1988), Devienne & Lebègue (1986,1988). But by an other way, there are some "positive" results. For example, tree automata and recognizable (or rational) tree languages are the most famous classes (Gecseg & Steinby , 1984) in our area.

The problem is important because every time we get a "good" syntactical class, we get investigations tools and concepts from a theoretical point of view, and we get syntactical tools for software engineering from a practical point of view.

Our first and second results are somewhat "negative" one, because they prove that restrictions which seem very very strong are not sufficient to get decidable classes (part 1/) or are not computational restrictions at all (part 2/).

1/In the first part we solve a problem stated by Courcelle (1988).

We prove that it is undecidable whether a rewriting system "preserves" recognizability or not;-we mean if the corresponding equivalence relation satures any recognizable forest in a recognizable too forest-; In particular, B.Courcelle proved that in a theory (F,E) the equational sets are recognizable iff E satisfies this property ; We know already that ground

(1) The preparation of this paper was supported in part by the "GRECO de Programmation" and the PRC "Mathématiques et Informatique". Part of the ESPRIT Basic Research Action " Algebraic and Syntactic Methods in Computer Sciences".

rewriting systems (as any system whose equations are bilinear and separated) preserve recognizability ; However, there are many other cases where systems satisfie this property; Unfortunately -but as expected- this property is undecidable.

2/ In the second part we present a method to simulate a term rewriting system by a ground term rewriting system with rational control that proves this notion of control completely suppresses the hierarchy for the term rewriting systems.

The notion of control we study here is quite different from the usual conditional rewriting (Kaplan & Jouannaud(1987)). Conditional rewriting formalizes usual concepts of programmation. Our notion of control is closely related to concurrence cooperation and communication notions(see example 4). The conclusion of this part is that " rational control is much more powerfull than it seems".

3/ We recently proved that termination is undecidable in the one rewriting rule case (Dauchet, 1988), even if this rule is left linear, complete and non-ambiguous case (Dauchet, 1989). That means that we cannot decide if computations always terminate, even with a term rewriting system reduced to one rule, and even if no variable occurs twice in the left-hand side, no variable is erased and even if there is no overlapping possible between two occurrences of the left-hand side. Here, we draw a survey of decidability results about termination for a lot of "small classes" of term rewriting systems).

We suppose the reader is familiar with tree automata (Gecseg & Steinby , 1984) and, for the last part, with term rewriting systems- also called term rewriting- (see for example Dershowitz (1987).)

Part 1. **It is undecidable whether a rewriting system "preserves" recognizability**

NOTATIONS : Given an alphabet V, V* denotes the monoïd generated by V, i.e. the set of words on V; for a word u,$\tilde{u}$ denotes the mirror of u. Given a set X of variables and a gradued set F, $T_F(X)$ denotes the term algebra over X; Elements of $T_F(X)$ are called terms or trees;T_F denotes the set of ground terms - i.e. terms without variables-; $|t|$ denotes the depth of t; If W is a subset of F where any letter is of arity one, we note sometimes T(W,X) as W*; Similary, uv stands for u(v) when all the symbols occurring are of arity 0 or 1. A rewriting rule is a pair of terms denoted $l \rightarrow r$; A rewriting system is a set of rewriting rules; Given a set of rules R,the relations $\rightarrow, \leftrightarrow, \xrightarrow{*}, \xleftrightarrow{*}$ are defined as usual, i.e. for a set of terms F, $R_{\leftrightarrow}(F) = \{t / \exists u \in F, t \xleftrightarrow{*} u\}$; We suppose the reader is familiar with the notions of recognizability.

PROOF:

In order to prove our result, we associate with any Post Problem PCP a rewriting system R and we prove that there is a solution to PCP iff R "preserves" recognizability:

Let PCP = (V,n,α,β) be a Post correspondence problem, where V is an alphabet,$n \geq 1$, and $\alpha = (\alpha_1,...,\alpha_n)$, $\beta = (\beta_1,...,\beta_n)$; For any word $u = u_1...u_n$ in V*, α_u (resp. β_u) denotes the word $\alpha_{u_1} ...\alpha_{u_n}$ (resp.$\beta_{u_1} ...\beta_{u_n}$).

Let Δ be an alphabet defined by $\Delta = W \cup I \cup \{k, r, eng, a\}$, where W is obtained by associating with each letter of V a letter of arity 1, I is obtained by associating with each integer between 1 and n a letter of arity 1, k is a new symbol of arity 2, r and eng new symbols of arity 3, a a new symbol of arity 0;

Let us now define a finite rewriting system R on by the following rules divided in eight metarules:

1: $k(x,y) \rightarrow k(i(x),\alpha_i(y)) \quad \forall i \in I$;

2: $k(i(x),y) \rightarrow (a,i(x),y) \quad \forall i \in I$;

3: $r(x,i(y),\beta_i(z)) \rightarrow r(i(x),y,z) \quad \forall i \in I$;

4: $r(x,a,a) \rightarrow eng(a,a,a)$;

5: $eng(x,y,z) \rightarrow eng(l(x),y,z)$ / $eng(x,l(y),z)$ / $eng(x,y,l(z)) \quad \forall l \in W \cup I$;

6: $eng(a,y,z) \rightarrow eng(eng(a,a,a),y,z)$ /$eng(y,eng(a,a,a),z)$ / $eng(y,z,eng(a,a,a))$;

7: $eng(u(x),v(y),a) \rightarrow k(u(x),v(y))$ with either:

. $u \in \Delta$, $v \in \Delta$, ($u \notin I \cup \{a\}$ or $v \notin W \cup \{a\}$);

. $u = i$, $|v| \leq |\alpha_i|$, $\alpha_i \notin vW^*$;

. $u = a$, $v \in \Delta$, $v \neq a$;

8: $eng(u(x),v(y),z) \rightarrow r(u(x),v(y),z)$ with either:

. $u \in \Delta$, $u \notin I \cup \{a\}$, $v = \varepsilon$; or:

. $u = v = a$; or:

. $v \in \Delta$, $v \notin I \cup \{a\}$, $u = \varepsilon$;

Nota: In the rule 4, the variable x occurs only at the left side; It may be surprizing when we reverse the rule; But actually, this rule may be replaced by the rules:

. $r(i(x),a,a) \rightarrow r(x,a,a)$;

. $r(a,a,a) \rightarrow eng(a,a,a)$;

Now, we can prove the following theorem:

Theorem: there is a solution to PCP iff the closure of any recognizable set of terms by $\overset{*}{\leftrightarrow}_R$ is recognizable.

Proof: Let F_0 be the recognizable set of trees whose root is of arity 2 or 3;

Lemma 1: If there exists a solution to PCP, $R_{\leftrightarrow}(k(a,a)) = F_0$.

Proof of the lemma 1:

.First, as the root of members of rules of R are always of arity 2 or 3, $R_{\leftrightarrow}(k(a,a))$ is obviously included in F_0; Conversely, let us prove we can obtain any tree t of F_0 from $k(a,a)$, by induction on nocc[t] the number of occurrences of r,k,eng in t:

<u>-nocc[t] = 1:</u>

First case: root(t) = eng;

. $eng(a,a,a)$ may be obtained from $k(a,a)$; actually by hypothesis, there is some u in I^+ such that $\alpha_u = \beta_u$; Now:

-by the rule 1, we have $k(a,a) \rightarrow k(ua,\alpha_u a)$;

-by the rule 2, $k(ua,\alpha_u a) \rightarrow r(a,ua,\alpha_u a) = r(a,ua,\beta_u a)$;

-by the rule 3, $r(a,ua,\beta_u a) \rightarrow r(\tilde{u}a,a,a)$;

-by the rule 4, we obtain *eng*(a,a,a);

. Now, any tree of root *eng* and where other letters are of arity 1 or 0 may be obtained from *eng*(a,a,a) by the rules 5;

Second case: root(t) is *r* or *k*; we prove now that if we may obtain any tree t of root *eng* with nocc[t] inferior or equal to n, we may obtain any tree t of root *k* or *r* with nocc[t] = n:

. root(t) is *k*; Let *k*(u,v) be such a tree;

-either u = ma, v = na, with $n = \alpha_m$; then *k*(u,v) is obtained directly from *k*(a,a) by the rules 1;

-either not; then $u = u_1(u_2)$, $v = v_1(v_2)$, with $v_1 = \alpha_{u_1}$ and $(\text{root}(u_2) \notin I \cup \{a\}$ or $(u_2 = i, v_2 \notin \alpha_i.T(\Delta))$ or $(u_2 = a, v_2 \neq \varepsilon))$. Then, nocc[$eng(u_2,v_2,a)$] is equal to n, and so $eng(u_2,v_2,a)$ may be obtained from k(a,a) by hypothesis; As $eng(u_2,v_2,a) \rightarrow k(u_2,v_2)$ by the rule 6, and $k(u_2,v_2) \rightarrow k(u,v)$ by the rule 1, *k*(u,v) may be obtained too!;

.root(t) is *r*; Let *r*(u,v,w) be such a tree;

-either u belongs to I^+a; So $u = u_1(a)$; as shown above, we may obtain $k(\tilde{u}_1(v),\beta_{\alpha_1}(w))$; Then $k(\tilde{u}_1(v),\beta_{\alpha_1}(w)) \rightarrow r(a,\tilde{u}_1(v),\beta_{\alpha_1}(w))$ by the rule 2 and $r(a,\tilde{u}_1(v),\beta_{\alpha_1}(w)) \xrightarrow{*} r(u,v,w)$ by the rule 3 (applied several times);

-either u = *a*; then, either v doesn't belong to $I\Delta^*$ and $eng(a,v,w) \rightarrow r(a,v,w)$ by the rule 8, either v belongs to $I\Delta^*$ and *r*(a,v,w) is obtained from *k*(v,w) by the rule 2;

-either u doesn't belong to I^*a; so, $u = u_1(l(u_2))$, with u_1 in I^* and l in $\Delta \backslash (I \cup \{a\})$; Then nocc[$eng(l(u_2),\tilde{u}_1(v),\beta_{\alpha_1}(w))$] = n; By the rule 8 $eng(l(u_2),\tilde{u}_1(v),\beta_{\alpha_1}(w)) \rightarrow r(l(u_2),\tilde{u}_1(v),\beta_{\alpha_1}(w)))$; by several applications of the rule 3, $r(l(u_2),\tilde{u}_1(v),\beta_{\alpha_1}(w)) \xrightarrow{*} r(u,v,w)$;

-<u>Induction step:</u> Let us now suppose we may obtain any tree of F_0 with nocc[t] inferior to n(n>1); We prove now we may obtain any tree t of F_0 with nocc[t] equal to n; by what preceedes, we have just to prove it for trees whose root is eng: Let $t = eng(t_1,t_2,t_3)$ such a tree; We may suppose without lost of generality that t contains an occurrence of *r*,*eng* or *k*; So, $t_1 = u_1(v_1)$ with u_1 in $(I \cup W)^*$ and v_1 in F_0; As nocc[$eng(a,t_2,t_3)$] is inferior to n, $eng(a,t_2,t_3)$ is in the class of *k*(a,a) by induction hypothesis; now $eng(a,t_2,t_3) \rightarrow eng(eng(a,a,a),t_2,t_3)$ by the rule 6; By induction hypothesis, *eng*(a,a,a) is equivalent to v_1; So, $eng(v_1,t_2,t_3)$ is equivalent to *k*(a,a); As $eng(v_1,t_2,t_3) \rightarrow$ t by the rules 5, the proof is complete;

q.e.d

Lemma 2: If there is a solution to PCP,$R_{\leftrightarrow}(F)$ is recognizable for any recognizable set of terms F.

Proof of the lemma 2: Let F be a recognizable set of terms; Then F can be seen as $F_1 \cup F_2$, with F_1 the part of F where no letter of arity 2 or 3 occurs, F_2 the part of F where any tree contains an occurrence of *r*,*eng* or *k*; F_1 and F_2 are obviously recognizable, and $R_{\leftrightarrow}(F) = R_{\leftrightarrow}(F_1) \cup R_{\leftrightarrow}(F_2)$; As any member of rule contains an occurrence of *r*,*eng* or *k*, $R_{\leftrightarrow}(F_1) = F_1$; Now, let $F_3 = h(F_2)$ where h applies any letter of arity 2 or 3 on b, and F_4 the set of terms obtained from F_3

by substituing any b by F_0; Obviously, $R_{\leftrightarrow}(F_2) = F_4$ and F_4 is recognizable. So, $R_{\leftrightarrow}(F) = F_1 \cup F_4$ and is recognizable; q.e.d.

Lemma 3: If there is no solution to PCP,$R_{\leftrightarrow}(k(a,a))$ is not recognizable .

Proof of the lemma 3: Let us define two set of terms G_1, G_2 by:

$G_1 = \{ k(ua,va) \,/\, v = \alpha_u,\ u \in I^* \}$

$G_2 = \{ r(ua,va,wa) \,/\, \beta_{\tilde{u}} w = \alpha_{\tilde{u}v}, uv \neq \varepsilon,\ uv \in I^* \}$

Let G be $G_1 \cup G_2$; we are going to prove G is $R_{\leftrightarrow}(k(a,a))$; Any tree of G_1 may be obtained obviously from $k(a,a)$ by the rules 1; Now let us prove any tree of G_2 may be obtained from a tree of G_1:

-$r(ua,va,wa)$ may be obtained from $r(a,\tilde{u}va,\beta_{\tilde{u}}wa)$ by the rules 3 if u belongs to I^*;

-as uv is not empty, $r(a,\tilde{u}va,\beta_{\tilde{u}}wa)$ may be obtained from $k(\tilde{u}va,\beta_{\tilde{u}}wa)$ by the rules 2; Furthermore as $\beta_{\tilde{u}}w = \alpha_{\tilde{u}v}$, $k(\tilde{u}va,\beta_{\tilde{u}}wa)$ belongs to G_1, q.e.d.;

So, any tree of G belongs to $R_{\leftrightarrow}(k(a,a))$; Let us now prove the reverse inclusion; as $k(a,a)$ belongs to G, it is sufficient to prove $\{t / \exists s \in G,\ t \rightarrow s \text{ or } t \leftarrow s\}$ is included in G; Let us study what happens for each metarule applied:

-1: $t \leftarrow s$ by the rule 1, with s in G: so $t = k(iua,\alpha_i\alpha_u a)$ for some i in I and t belongs to G;

-1 bis: Similary, if $t \rightarrow s$ by the rule 1, with s in G, t belongs to G;

-2: $t \leftarrow s$ by the rule 2 with s in G: so $s = r(a,v,w)$ with $w = \alpha_v$ and $t = k(v,w)$ belongs to G;

-2 bis: proof as for 2;

-3: $s = r(ua,iva,\beta_i wa) \rightarrow t = r(iua,va,wa)$; As s belongs to G, $\beta_{\tilde{u}}\beta_i w = \alpha_{\tilde{u}iv}$; so $\beta_{i\tilde{u}}w = \alpha_{i\tilde{u}v}$ and t belongs to G;

-3 bis: similar;

-4: $t \leftarrow s$ by the rule 4 with s in G is impossible: actually, s should be of the form $r(ua,va,wa)$ with $v = w = \varepsilon$; So, by the condition implied by the definition of G, u is not empty and $\alpha_u = \beta_u$, i.e. there is some solution to PCP!;

-4 bis: $t \rightarrow s$ by the rule 4 with s in G is impossible, as the right member's root of a rule 4 is *eng*;

-5,5 bis,6,6 bis,7: impossible!;

-7 bis: $t \rightarrow s$ by the rule 7 with s in G is impossible: actually let $s = k(u(s_1),v(s_2))$; as s belongs to G, if u belongs to Δ either u belongs to I, either u = a; if u = a, v = a too and the rule 7 may not be applied; if u = i, $v(s_2) = \alpha_i(s_3)$ and the rule 7 may not be applied too;

-8: let $s = r(u(s_1),v(s_2),w(s_3))$ in G; by the definition of G, u or v is different from a, and if u(resp.v) is a letter, u(resp. v) belongs to I; So, the rule 8 may not be applied;

-8 bis: impossible! q.e.d.

Now, the proof of the theorem s immediate: actually, G_1 is obviously not recognizable; as G_1 is the intersection of G with the recognizable set of trees whose root is *r*, G is not recognizable too; So, if there is no solution to PCP, $R_{\leftrightarrow}(k(a,a))$ is not recognizable; So, by lemma 2, $R_{\leftrightarrow}(F)$ is recognizable for any recognizable set of terms F, iff there is a solution to PCP; as the exixtence of solution for a PCP is undecidable, **it is undecidable whether R "preserves" recognizability.**

Part 2: **Ground term rewriting systems with recognizable control are as powerful as general rewriting systems.**

Definition:
Rules with rational control are similar to contextual rewriting rules; A rule is of the form: $F :: l \rightarrow r$ where F is a recognizable set of terms and l and r are ground terms. The meaning of a rule is as usual, but applications of the rule are restricted to terms in F only.

example:

$not(a_1 \text{ or } \alpha_1) :: a_0 \rightarrow a_1$
$a(a_2) :: a(a_1) \rightarrow \alpha_1$
$a(\alpha_2) :: a(\alpha_1) \rightarrow a_1$
$b(a_1,a_2) \rightarrow b'$
$b(a_2,a_1) \rightarrow b'$

$not(a_2 \text{ or } \alpha_2) :: a_0 \rightarrow a_2$
$\alpha_1 :: a(a_2) \rightarrow \alpha_2$
$a_1 :: a(\alpha_2) \rightarrow a_2$
$b(\alpha_1,\alpha_2) \rightarrow b'$
$b(\alpha_2,\alpha_1) \rightarrow b'$

then if $n \geq p$: $b(a^n(a_0),a^p(a_0)) \xrightarrow{*} b(a^{n-p}(\alpha),\alpha')$ with $(\alpha,\alpha') = (a_1,a_2)$ or (a_2,a_1) or (α_1,α_2) or (α_2,α_1).

Though the restriction of ground rules was very strong, we prove that the adjunction of rational control allows to simulate every trs with the particular systems we have got.

Simulation:

We need the following special symbols:
ζ is a binary function symbol, *copy* a constant symbol

We pose : $\hat{t} = \zeta(copy\ , t)$

$\hat{T}_\Sigma = \{\hat{t} \,/\, t \in T_\Sigma\}$

Δ_r is the set of special symbols used during the simulation of the rule r

$\hat{R}_r$ is the ground trs controlled rationnally which simulates the rule r.

Lemma: $\forall (t,u) \in (T_\Sigma)^2$

$[\, t \underset{r}{\rightarrow} u \,] \Leftrightarrow [\, \hat{t} \underset{\hat{R}}{\rightarrow} t_1 \overset{*}{\underset{\hat{R}}{\rightarrow}} t_2 \underset{\hat{R}}{\rightarrow} \hat{u}, \quad (t_1,t_2) \in (T_{\Sigma \cup \Delta_r})^2 , \ (\hat{t},\hat{u}) \in (\hat{T}_\Sigma)^2$

and *property (Q)*: all the rules other than the first are controlled by a recognizable set of terms every tree of which contains an element of Δ_r]

Sketch of proof:

1) We copy the tree at the place *copy*; that copy is realized from left to right and ascending back to the root of the tree which corresponds to the postfixed notation in the form of a list structure.

example: b_1 will be copied in the form:

```
        b1
       /  \
     b2    b3
     /\    /\
   a1 a2  b4 a5
          / \
        a3   a4
```

```
 ------------------------------------
 |   |   |   |   |   |   |   |   |
 a1  a2  b2  a3  a4  b4  a5  b3  b1
```

2) We simulate a Turing machine by ground system with rational control. Let a Turing machine T of the form :

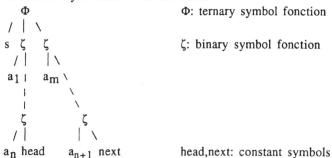

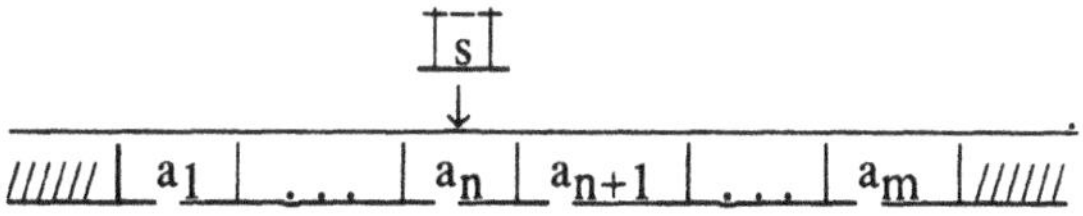

s denotes the state, $a_1,\ldots,a_m$ denote the contents of the tape.
T is simulated by a term T' of the form:

```
      Φ                          Φ: ternary symbol fonction
    / | \
   s  ζ   ζ                      ζ: binary symbol fonction
     /|   |\
   a1 |   am \
      |       \
      |        \
      ζ         ζ
     /|         | \
   an head   an+1 next           head,next: constant symbols
```

We need the following special symbols:

1) s_a denotes the fact that T is about to perform the state transition labelled by s,a;

2) the symbols head', next' and $next_a$ are needed in order to mark intermediate states and to perform moves of the read/write head;

The simulation works as follows:

1)enter a new transition: s <u>and</u> ζ(a,head) :: s $\rightarrow$ s_a

2)perform the instruction:

α) move left:

s_a <u>and</u> head :: next $\rightarrow$ ζ(a,next')
s_a <u>and</u> next' :: ζ(a,head) $\rightarrow$ head'
s_a <u>and</u> head' :: next' $\rightarrow$ next

β) move right:

s_a <u>and</u> head :: ζ(b,next) $\rightarrow$ $next_b$
s_a <u>and</u> $next_b$ :: head $\rightarrow$ ζ(b,head')
s_a <u>and</u> head' :: $next_b$ $\rightarrow$ next

γ) write "b" on field: $s_a :: \zeta(a,head) \rightarrow \zeta(b,head')$

3) take successor state(t): head' <u>and</u> next :: $s_a \rightarrow t$

$t :: head' \rightarrow head$

3) We recopy the tree; the recopy is realized from right to left and descending in agreement with 1).

Simulation theorem: $\forall (t,u) \in (T_\Sigma)^2 \; [\, t \xrightarrow[R]{*} u \,] \Leftrightarrow [\, \hat{t} \xrightarrow[\hat{R}]{*} \hat{u} \,]$ with $\hat{R} = \cup \hat{R}_r$

sketch of proof:

It is sufficient to observe that if we associate to every rule r the trs $\hat{R}_r$, the property (Q) allows not to mix the derivations.

Conclusion: This result means that every usual derivation can be simulated by a ground trs with rational control. But the number of steps is really increasing. It would be interesting to study simulations which respect classes of complexity (up to a constant, up to a polynomial relation).

Part 3 **Termination and halting of special term rewriting systems.**

In this last part, we study the decidability of the termination or of the halting for some simple classes of term rewriting systems (trs).

We suppose the reader is familiar with trs (see Dershowitz, 1987).

A trs *terminates* iff there is no infinite computation.

A trs *halts for a term t* iff there is no infinite computation starting from t. (So, a trs terminates iff it halts for every term).

Figure 1 draws the frontier between decidability and undecidability of these problems across a syntactical classification of trs reduced to one or two rules.

Let us recall that termination (and halting) are undecidable for a left linear rule (Dauchet 1989). But we get decidability in special cases.

Let us introduce the definition of deterministic.

Definition: A rule is *non-overlapping* (Dershowitz, 1987, def. 23), i.e. "non-ambiguous", i.e. "without critical pairs", i.e. the left-hand side does not overlap a non-variable proper subterm of itself.

Example: If the left hand side is A(A(x)), the rule is overlapping.

Definition: Let R: l→ r be a rewriting rule. We will say that R is *deterministic* iff it satisfies the two following properties:

(R'1) R is *non-overlapping*

(R'2) There is at the most a non-variable subterm r' of r unifiable with l (it is said that l *overlaps* r) or, excusively, at most a non-variable subterm l' of l unifiable with r (it is said that r overlaps l) (as usually, variables of different occurrences of rules are considered disjoint).

The following rewriting rules are deterministic.

Example 1: some like-program schemes...

F(s(x)) → if zero(x) then 1 else F(x)

Example 2: some Prolog-like programs...
P(+(x,y),z)) → P(x,+(y,z))

Counter-example: A(x,A(by,z)) → A(ax, A(x,y))
is not deterministic there are two different overlaps between the left-hand side and the right-hand side.

Proposition: Termination is decidable for a non overlapping and deterministic rewriting rule.

Conjectures:

(1) Termination and halting problems are decidable in low degree polynomial time for a linear right-constant-free term rewriting rule of rank ≥ 2. (A term is said *constant-free* if no constant occurs in it; a rewriting rule l→r is *right-constant-free* if r is is constant free.)

(2) Termination and halting problems are decidable in low degree polynomial time for a linear term rewriting rule of rank ≥ 2.

(3) The termination and halting problems for a linear term rewriting rule can be reduced to the same problem for a word rewriting rule (i.e. a semi-Thue rule).

(4) The termination and halting problems for a word rewriting rule are decidable in low degree polynomial time.

Obviously, (1) is a particular case of (2); (3)+(4) imply (2). But we think that a proof of (2) will be much more tedious than a proof of (1); roughly, (2) and (3) may be of same complexity but (4) seems much more difficult. It is obvious that (3)+(4) imply decidability in the case of a linear term rewriting rule.

Such results will be interesting because it is always interesting to get efficient algorithms and because our mind is that solving (4) requires a new kind of pumping lemma, and pumping lemma are a crucial tool for a lot of decidability problems.

The following result, which is a particular case of (1), illustrates the conjecture (1) and points the claim that (1) is easier than (2)!

Claim: *Termination problem is decidable in linear time for a linear right-constant-free term rewriting rule of rank 2.*

Sketch of proof: b(x,y) denotes a letter of rank 2; x and y denote the two variables; u,u',v,v',w,w',m,m' are concatenations of symboles of rank one (identified to words).

Let l → r be the rule. It is easy to check that the rule does not terminate iff

i/ l = l(x,y) = u(b(v(x),w(y)) and r = u'(l(v'(x),w'(y)))

or

ii/ l = l(m(x),y) and r = u'(l(y,m(m'(x))))

figure 1: classification of"small" trs

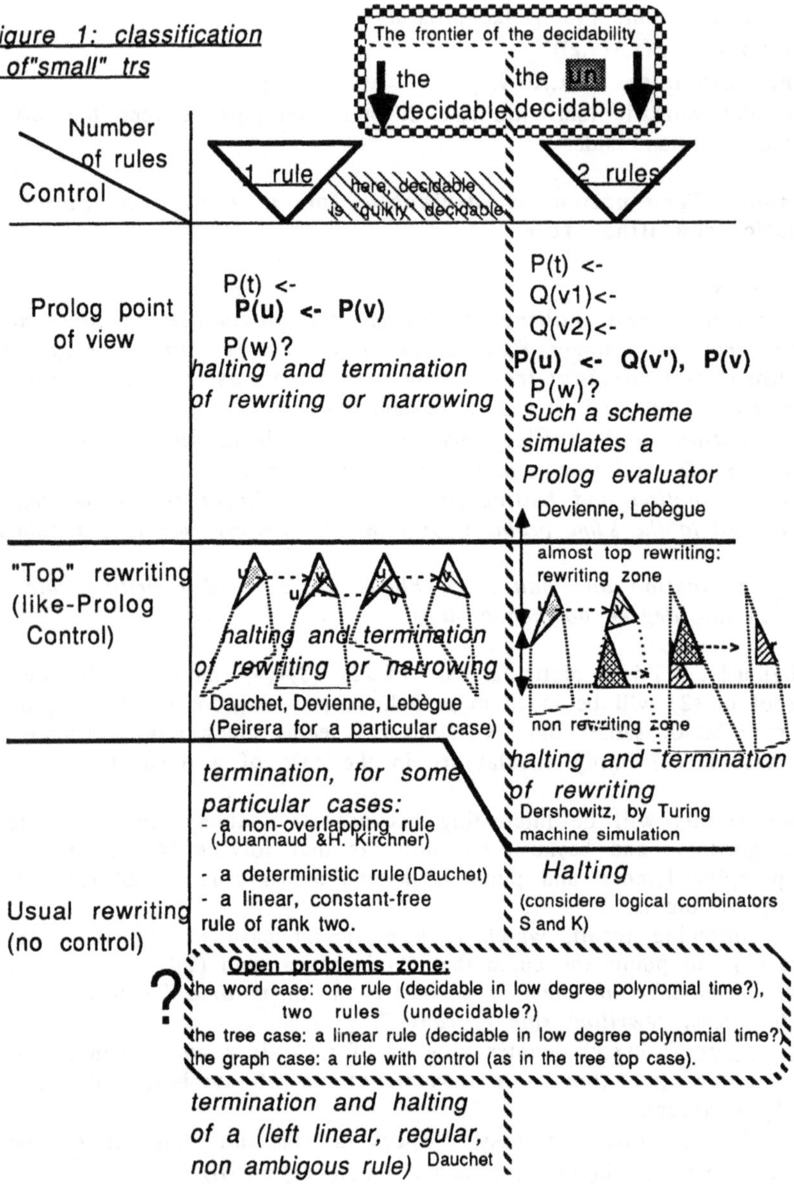

REFERENCES

Arnold,A & Dauchet,M. Theorie des magmoïdes. *RAIRO Informatique théorique* 12 pp. 235-257 and 13 pp. 135-154 (1979)

Book, R.V. (1987). Thue Systems as Rewriting Systems. *J. Symbolic computation,* 3 p. 39-68.

Courcelle,B. On recognizable sets and automata (*to appear*)

Dauchet,M. (1989). Simulation of Turing Machines by an only left-linear rewriting rule. *R.T.A. 89, North Carolina, USA.*

Dauchet,M & De Comité,F .A gap between linear and non linear term rewriting systems. *Lectures Notes in Computer Science 256 (1987) Springer Verlag* pp. 95-104

Dauchet,M., Devienne Ph.&Lebègue P. (1988). Décidabilité de la terminaison d'une règle de réécriture en tête, *Journées AFCET-GROPLAN, Bigre+Globule* 59, p. 231-237.

Dauchet,M.,Heuillard,Lescanne & Tison,S. Decidability of the confluence of ground term rewriting systems.*Rapport INRIA 675 and LICS'87.*

Dauchet,M.& Tison,S .Tree automata and decdability in ground term rewriting systems.*FCT'85,Lectures Notes in Computer Science* vol. 199,pp80-84 (1985)

Dauchet, M. (1988). Termination of rewriting is undecidable in the one-rule case. *MFCS 1988, Carlsbad. Lec. notes Comp. Sci.* 324, p. 262-268.

Dershowitz, N. (1987). Termination. *J. Symbolic computation,* 3 p.69-116.

Devienne Ph. & Lebègue P. (1986). Weighted graphs, a tool for logic programming, *CAAP 86, Nice, Springer Lec. notes Comp. Sci.* 214, p.100-111.

Devienne Ph. (1988). Weighted Graphs, a tool for expressing the Behaviour of Recursive Rules in Logic Programming. *To appear in proceeding of FGCS' 88, Tokyo.*

Gecseg,F & Steinby,M. Tree automata. *Akadémiaï Kiado,Budapest* 1984

Ginsburg,S & Spanier,E.H. Control sets on grammar.*MST 2* (1968) pp. 157-178

Greibach,S. Control sets on contex-free grammar forms. *JCSS 15* (1977) pp. 35-98

Huet, G. & Lankfork D.S. (1978). On the uniform halting problem for term rewriting systems. *Rapport Laboria* 283, INRIA.

Huet, G. & Oppen D. C. (1980). Equations and rewriting rules: A survey, *in R. V. Book, ed., New York: Academic Press.* Formal Language Theory: Perspectives and Open Problems, pp. 349-405.

Jouannaud J. P. (1987). Editorial of *J. Symbolic computation,* 3, p.2-3.

Jouannaud J.P. & H. Kirchner (1984). Construction d'un plus petit ordre de simplification. *R.A.I.R.O. Informatique Théorique/ Theoretical Informatics,* 18-3, pp. 191-207.

Kaplan,S & Jouannaud,J.P. Conditional term rewriting system. *Procedings 1987. Springer -Verlag LNCS*

Zhang,H & Remy,J.L. Contextual Rewriting. *Lectures notes in Computer Science* Vol 202 pp. 46-62 (1985).

Completeness in Approximation Classes

Pierluigi Crescenzi
Dipartimento di Informatica e Sistemistica, Via Buonarroti 12, 00185 Roma Italy
Alessandro Panconesi
Department of Computer Science, Cornell University - Ithaca NY 14850, USA *

ABSTRACT

We introduce a formal framework for studying approximation properties of NP optimization (NPO) problems. The classes we consider are those appearing in the literature, namely the class of approximable problems within a constant ϵ (APX), the class of problems having a Polynomial-time Approximation Scheme (PAS) and the class of problems having a Fully Polynomial-time Approximation Scheme (FPAS). We define natural approximation preserving reductions and obtain completeness results for these classes. A complete problem in a class can not have stronger approximation properties unless P=NP. We also show that the degree structure of NPO allows intermediate degrees, that is, if P≠NP, there are problems which are neither complete nor belong to a lower class.

1. Introduction

The widespread belief that NP-complete problems cannot be solved by polynomial time algorithms made researchers look for different strategies than exact resolution in order to deal with these problems. Since many of the most important NP-complete problems are *the recognition version* of an optimization problem, it is natural to ask the following question: "Can we devise polynomial time algorithms always finding solutions *close* to the optimum? "

Several results are known on the approximability or non-approximability of the so called NP Optimization problems (optimization problems whose recognition version is in NP) when the quality of the approximation is measured by the relative error. In particular four classes have been identified:

i) Problems which are not approximable in polynomial time unless P=NP;
ii) APX: problems which are approximable within some fixed relative error $\epsilon > 0$;
iii) PAS: problems which can be approximated within *any* ϵ by algorithms having as input an instance x and ϵ. Such algorithms are called Polynomial-time Approximation Schemes, their complexity must be polynomial in $|x|$.
iv) FPAS: problems which can be approximated by Polynomial-time Approximation Schemes whose running time is polynomial in both the size of the input and $1/\epsilon$. These algorithms are called Fully Polynomial-time Approximation Schemes.

In spite of some remarkable attempts (see [PM], [AMP], [Kr], [KS]) the reasons why a problem is approximable or not are still not clear. In this paper we are interested in the problem of determining *lower-bounds* concerning the approximability of NPO problems, that is we would like to develop techniques that would allow to prove statements like "If P≠NP, then problem F is not in PAS" or "If P≠NP, then problem F is not in APX" and so on. Generally this kind of results has been obtained via polynomial reductions mapping an NP-complete problem into the given optimization problem and showing that the approximability of the latter would imply the former to be in P. For the class FPAS Garey and Johnson [GJ] have developed an alternative approach based on the notion of *Strong NP-completeness*; an optimization problem whose recognition version is Strong NP-complete is not in FPAS. Roughly speaking, a problem is Strong NP-complete if its NP-completeness does not depend on the presence of large weights in the input instance. Here, "large" means non-polynomial in the length of the instance. In order to derive similar criteria for

* This work was done when the author was at Dipartimento di Informatica e Sistemistica, Via Buonarroti 12, 00185 Roma Italy

the other classes, it seems natural to try to define useful concepts of completeness. This approach has been taken by Orponen and Mannila who show completeness results in NPO [OM].

In the present paper we continue along the same line of research. We introduce natural approximation preserving reductions and show completeness the existence of complete problems both in the class APX and in the class PAS.

Perhaps surprisingly, in spite of the fact that APX $\supseteq$ PAS and the reductions we use are very natural, reducibility in PAS is not a refinement of that in APX.

Besides, one of the most relevant results in the paper shows that in all classes APX, PAS, and NPOthere exist intermediate problems which are neither complete nor in the lower class. This answers a question posed in [OM] on the existence of incomplete problems in the approximation classes. The significance of this result can be explained in the following way; usually a problem F in a class, say NPO, is proved not to be in a lower class, say APX, by proving a statement like "If F $\in$APX, then P=NP". The existence of incomplete problems shows that the proof of non approximability does not imply completeness in NPOand similarly for the other classes.

Thus our notion of completeness captures a deeper level of structure than the notion of NP-completeness. In fact, an NP-complete problem, when considered in its optimization version, can be approximable or not, complete or incomplete (in our sense).

The paper is organized as follows. In section 2 we formally define the classes of approximable and non-approximable problems and we introduce special kinds of reductions and hence of completeness. In section 3 we prove completeness results; as a by-product we are able to show that the reducibility in PAS is not a specialization of that in APX. In section 4 we show the existence of incomplete problems with the well known delayed diagonalization technique.

2. A Formal Framework For Optimization Problems

Every known NPO problem F can be characterized by the following objects:

- $I_F \subseteq \Sigma^*$ is the space of *input instances*, it is recognizable in polynomial time;
- a polynomial predicate $\pi_F(x, y)$ and a polynomial $q_F(n)$ define the set of *feasible solutions*

  $$D_F = \{(x, y) \mid x \in I_F \wedge \pi_F(x, y) = \text{TRUE} \wedge \mid y \mid \leq q_F(\mid x \mid)\};$$

  where the set

  $$D_F(x) = \{y \mid (x, y) \in D_F\}$$

  is the set of feasible solutions on input x;
- $f_F : D_F \longrightarrow Z^+$ is the *objective function*, it is computable in polynomial time.

The *optimum on input* x (max or min) of the optimization problem F is

$$f_F^*(x) = opt\{f_F(x, y) \mid y \in D_F(x)\}.$$

Without any loss in generality we can assume that the predicate π_F incorporates the test "$x \in I_F$?", so we can define

Definition 1. *An NPO-problem F is a triple $F = (q_F, \pi_F, f_F)$ where: i) $q_F(n)$ is a polynomial; ii) $\pi_F(x, y)$ is a polynomial-time decidable predicate; iii) $f_F : D_F \longrightarrow Z^+$ is a polynomial-time computable function.*

Definition 1 allows to associate to any NPO problem F a non-deterministic Turing machine (NTM) N_F defined as follows:

NTM FOR F:
guess $y \in \{0, 1\}^{q_F(|x|)}$
if $\pi_F(x, y)$ = FALSE then ABORT
output $f_F(x, y)$

The above machine has the property that, for every x, $y \in D_F(x)$ iff there is a computation path with output $f_F(x, y)$. This characterization will turn out to be useful for proving completeness.

Let us make an example in order to explain the previous definitions.

Max-CLIQUE: the set of input instances I_{CLQ} is the set of all (strings encoding) undirected graphs $G = (V, E)$. The polynomial predicate $\pi_{CLQ}(G, G')$ is "Is G a graph and G' a clique of G?". Thus $D_{CLQ}(G)$ is the set of all cliques contained in G. Finally, the objective function is $f_{CLQ}(G, G') = \| V' \|$. The NTM for Max-CLIQUE is the following

NTM for Max-CLIQUE:
guess a subgraph $G' = (V', E')$ of G
if G is not a graph or G' is not a clique then ABORT
output $\| V' \|$

In the sequel we will consider only maximization problems. Nevertheless all our results can be shown to hold with respect to minimization problmes also.

To define classes of approximable NPO problems we need the notion of relative error; the following is a widely used definition (see [PS])

Definition 2. *Let F be an* NPO *problem. Given $x \in I_F$, for any $y \in D_F(x)$ the quality of y with respect to F is*

$$Q(F, y) = \begin{cases} 0, & \text{if } f_F^*(x) = 0; \\ \frac{f_F^*(x) - f_F(x,y)}{f_F^*(x)}, & \text{otherwise.} \end{cases}$$

We can now define classes of approximable problems. With $(0,1)_Q$ we indicate the set $(0,1) \cap Q^+$.

Definition 3. *An* NPO *problem F is in* APX *if there exists an $\epsilon \in (0,1)_Q$ and a polynomial DTM T such that, for any $x \in I_F$: $T(x) \in D_F(x)$ and $Q(F, T(x)) \leq \epsilon$.*

Definition 4. *T is a polynomial-time approximation scheme (pas) for F if, for every input (x, ϵ) such that $x \in I_F$: i) $(x, T(x, \epsilon)) \in D_F$; ii) T's complexity is $h_T(x, \epsilon)$ where h_T is polynomial in $| x |$ and arbitrary in $1/\epsilon$.*

The definition of h_T allows to have complexities like $| x |^{1/\epsilon}$ or $k^{1/\epsilon} | x |$, which arise in practice (see for example [HS], [PS]).

Definition 5. *An* NPO *problem F is in* PAS *if there exists a pas T_F for it.*

Definition 6. *An* NPO *problem F is in* FPAS *if there exists a pas T_F for it whose complexity is $h_T(x, \epsilon) = p(x, 1/\epsilon)$ where p is a polynomial in both $| x |$ and $1/\epsilon$.*

Clearly, the following inclusions hold NPO$\supseteq$ APX $\supseteq$ PAS $\supseteq$ FPAS; it is also well known that these inclusions are strict in the hypothesis P$\neq$NP.

In order to introduce the notion of completeness for our classes we now define three kinds of reductions between problems. All of them are refinements of the following

Definition 7. *Given $F, G \in$ NPO a NPO-reduction from F to G is a triple t_1, t_2, c where:*
- *$t_1 : I_F \longrightarrow I_G$ is polynomially computable;*
- *$t_2(x, y)$ is a polynomially computable function such that $y \in D_G(t_1(x)) \Rightarrow t_2(x, y) \in D_F(x)$;*
- *$c : (0,1)_Q \longrightarrow (0,1)_Q$.*

Roughly speaking, if we want to map F into G we use t_1 to go from F to G and t_2 to go back. In order to map approximated solutions into approximated solutions we have to specify the role of c.

The first reduction, called A-reduction, is needed to introduce completeness in NPO.

Definition 8. *Let F,G be two* NPO *problems; F is said to be A-reducible to G, in symbols $F \leq_A G$, if there exists a NPO-reduction from F to G such that, for any $y \in D_G(t_1(x))$:*

$$Q(G,y) \leq \epsilon \Longrightarrow Q(F,t_2(x,y)) \leq c(\epsilon).$$

It is easy to show that the previous definition satisfies the following fact.

FACT. *If F is A-reducible to G and* $G \in$ APX, *then* $F \in$ APX.

Definition 9. *A problem* $G \in$ NPO *is* NPO-*complete if, for any* $F \in$ NPO, $F \leq_A G$.

Analogously, we define reductions for the classes APX and PAS. The only difference between the reduction in NPO and that in APX concerns the role of the mapping c:

Definition 10. *Let F,G be two* NPO *problems; F is said to be P-reducible to G, in symbols* $F \leq_P G$, *if there exists a NPO-reduction from F to G such that, for any* $y \in D_G(t_1(x))$:

$$Q(G,y) \leq c(\epsilon) \Longrightarrow Q(F,t_2(x,y)) \leq \epsilon.$$

Definition 11. *A problem* $G \in$ APX *is* APX-*complete if, for any* $F \in$ APX, $F \leq_P G$.

In [OM] reductions equivalent to $\leq_A$ and $\leq_P$ are defined but completeness with respect to the latter is not proved.

To define completeness in PAS we need to modify the function c substantially

Definition 12. *Let F,G be two* NPO *problems; F is said to be F-reducible to G, in symbols* $F \leq_F G$, *if there exist 3 functions* t_1, t_2, c *such that*

i) t_1, t_2 *are as in definition 7,*

ii) $c: (0,1)_Q \times I_F \longrightarrow (0,1)_Q$

iii) for any $y \in D_G(t_1(x))$: $Q(G,y) \leq c(\epsilon, x) \Longrightarrow Q(F,t_2(x,y)) \leq \epsilon$,

iv) c's time complexity is $p(1/\epsilon, |x|)$, *p a polynomial, and*

v) c's value is $1/q(1/\epsilon, |x|)$, *q a polynomial.*

Definition 13. *A problem* $G \in$ PAS *is* PAS-*complete if, for any* $F \in$ PAS, $F \leq_F G$.

The definitions above satisfy the following facts.

- $F \leq_P G$ and $G \in$ PAS $\Rightarrow F \in$ PAS, and
- $F \leq_F G$ and $G \in$ FPAS $\Rightarrow F \in$ FPAS.

The following fact can also be easily proved.

FACT. *The defined reductions are reflexive and transitive.*

The reductions we defined are quite natural, nevertheless they are related in a strange way. As APX $\supseteq$ PAS it would seem reasonable to assert "$F \leq_F G$ implies $F \leq_P G$"; in other words, the F-reduction should be definable as a P-reduction with some additional constraint. But this is not the case; in the next section we will show that, surprisingly, *any* APX problem is F-reducible to a PAS-complete problem.

3. APX-completeness

The aim of this section is to show the APX-completeness of the following problem, BOUNDED SAT (BSAT):

INSTANCE: a boolean formula φ with variables $x_1, \ldots, x_n$ of weights $w_1, \ldots, w_n$ and a separate weight W. The weights must satisfy

$$\sum_{i=1}^{n} w_i \leq 2W$$

PROBLEM: maximize the following function, defined on the assignments of φ

$$f(y) = \begin{cases} W, & \text{if } \varphi(y) = \text{FALSE}; \\ \sum_{i=1}^{n} w_i x_i, & \text{otherwise.} \end{cases}$$

To understand the result we first consider the case of NPO-completeness.

For a class that has a machine representation a common way of showing completeness is to define a universal machine for the class. This method also works for NPO. If we consider tuples like $X = (x, N_F, 0^k)$, where $N_F = (q_F, \pi_F, f_F)$, we can define a "universal" problem U_N where $\pi_U(X, y)$ is "simulate $\pi_F(x, y)$ for k steps; if k is too little reject", and $f_U(X, y)$ is "simulate $f_F(x, y)$ for k steps; if k is too little output 0". The universal problem is then $U_N = (| X |, \pi_U, f_U)$. Clearly, if k is big enough to simulate non-deterministically every branch of N_F we have that U_N is *exactly the same* as F. As a consequence, if we define $t_1(x) = (x, N_F, 0^k)$, $k = p_F(| x |)$ where p_F is N_F's polynomial time bound, and t_2 as the identity function we can easily show that $Q(U_N, y) = Q(F, t_2(y))$; NPO-completeness follows.

However, much more can be proved; by modifying Cook's proof of the NP-completeness of SAT slightly it is possible to prove NPO-completeness for a weighted version of SAT. By modifying other NP-completeness proofs, natural problems can be proved NPO-complete. For example, the following theorem is from [OM].

Theorem 1. *TSP and 01-INTEGER PROGRAMMING are* NPO*-complete.*

In APX the situation is considerably different because we do not have a machine model to simulate. $F \in$ APX if there is a NTM for it *and* a polynomial time T approximating F. The right idea seems to consider tuples like $(x, \epsilon, T, N_F, 0^k)$ where T is an ϵ-approximating algorithm for N_F. In doing so we have to face two kinds of problems: i) we cannot know in advance whether T approximates N_F; ii) since a problem in APX can be approximated by any $\epsilon \in (0, 1)$ we have to "map" all the ϵ's into one fixed ϵ_0.

We start by defining our "universal" problem U_A by means of a non-deterministic algorithm; from its definition it will be clear that U_A is an NPO problem.

The inputs for the algorithm are of the form $X = (x, \epsilon, T, N_F, 0^k)$ where

- T is a polynomial-time TM,
- N_F is an NPO problem, that is a triple $N_F = (q_F, \pi_F, f_F)$,
- 0^k is a padding of k 0's, ϵ is a rational in $(0, 1)$, and x is an input for T and N_F.

On input $X = (x, \epsilon, T, N_F, 0^k)$ the machine for U_A performs the following non-deterministic algorithm. The algorithm is divided into two parts: the first, the TRUNK, is deterministic while the second, the BRANCHES, is non deterministic.

DEFINITION OF U_A

TRUNK. For k steps do the following: if $\pi_F(x, T(x))$ = TRUE then set $t = f_F(x, T(x))$. If k is too small abort computation (X is not a valid input instance).

BRANCHES. Simulate *non-deterministically* k steps of N_F. If k is too small output the value $A(X) + t$, $A(X) + \min\{t/(1-\epsilon), f_F(x, y)\}$ otherwise ($y \in D_F(x)$). The value $A(X)$ is a non-negative, polynomially computable function whose value will be specified later.

Notice that if we perform the Trunk, U_A outputs at least the value $A(X) + t$; moreover $u_A^*(X)$, U_A's optimum value, is bounded in terms of t. In the next lemma we show how $A(X)$ and t are used to show APX membership of U_A.

Lemma 1. *U_A is in* APX.

PROOF. In this lemma we specify the value of $A(X)$, which appears in the above definition of U_A.

We want show that the following is an 1/2-approximated algorithm for U_A: if k is big enough to perform the Trunk, output $T(x)$.

What we have to show is that $Q(U_A, T(x)) \leq 1/2$.

Let $a = A(X) + t$ and $b = A(X) + t/(1-\epsilon)$. By definition of U_A we have $a \leq u_a^*(X) \leq b$ and hence

$$Q(U_A, T(x)) \leq \frac{b-a}{b} = t\,\frac{\epsilon}{A(X)(1-\epsilon)+t} \leq \epsilon$$

In order to define $A(X)$ we distinguish two cases

i) $[\epsilon \leq 1/2]$. We set $A(X) = 0$, we have $Q(U_A, T(x)) \leq \epsilon \leq 1/2$.

ii) $[\epsilon > 1/2]$. In this case $A(X)$ is "responsible" for the ratio $(b-a)/b$ to be less than or equal to $1/2$.

We define $A(X)$ in such a way that

$$\frac{b-a}{b} = t\frac{\epsilon}{A(X)(1-\epsilon)+t} = \frac{1}{2}$$

that is

$$A(X) = t\frac{2\epsilon-1}{1-\epsilon}$$

Thus $A(X)$ is polynomially computable in $|X|$ and we know it is always the case that $Q(U_A, T(x)) \leq 1/2$. QED

Lemma 2. *U_A is* APX-*complete.*

PROOF. Let F be any problem in APX. Then, there must be a pair T, δ such that T δ-approximates F. Let p_T, p_F be the polynomial bounds of T and N_F. We have to exhibit a P-reduction $F \leq_P U_A$. Given x define $k = \max\{p_F(|x|), p_T(|x|)\}$, and

$$t_1(x) = (x, \delta, T, N_F, 0^k) = X$$

The size of k is big enough to simulate N_F, hence $D_{U_A}(x) = D_F(x)$; as a consequence we can define t_2 as $t_2(x, y) = y$.

By our choice of k and *since T δ-approximates F* we have $u_A^*(X) \leq A(X) + f_F^*(x)$ and hence

$$Q(U_A, y) = \frac{f^*(x) - f(x,y)}{A(X) + f^*(x)}$$

From the definition of P-reduction we have to find a function $c(\epsilon)$ that for all ϵ's satisfies

$$Q(U_A, y) \leq c(\epsilon) \Longrightarrow Q(F, y) \leq \epsilon$$

Suppose then

$$Q(U_A, y) = \frac{f_F^*(x) - f_F(x,y)}{A(X) + f_F^*(x)} \leq c(\epsilon).$$

We want to define $c(\epsilon)$ in order to satisfy the above implication.

From the definition of $A(X)$ we consider two cases. If $A(X) = 0$ it is enough to set $c(\epsilon) = \epsilon$, otherwise (we simplify the last inequality)

$$\begin{aligned} Q(F, y) &\leq c(\epsilon)\left(\frac{A(X)}{f_F^*(x)} + 1\right) \\ &= c(\epsilon)\left(\frac{t}{f_F^*(x)}\frac{2\delta-1}{1-\delta} + 1\right) \\ &\leq c(\epsilon)\frac{\delta}{1-\delta} \end{aligned}$$

Then, in order to have $Q(F, y) \leq \epsilon$ we set

$$c(\epsilon) = \frac{1-\delta}{\delta}\epsilon$$

Since $(1-\delta)/\delta$ is a constant of the reduction process the above is a valid P-reduction. QED

We can now state

Theorem 2. BSAT *is* APX-*complete.*

PROOF. BSAT is clearly in APX. To prove completeness we show $U_A \leq_P$ BSAT. Let T_A be the $1/2$-approximated algorithm for U_A and N_A be the NTM for U_A. We introduce some conventions. Without any loss in generality we can assume that N_A has a write-only output tape whose generic cell will be denoted as v_i; with y we will indicate a computation path of N_A, and with $N_A(X, y) = v$ we will mean that $N_A(X)$ has value v along computation path y. Since T_A $1/2$-approximates U_A we know that only a bounded amount of output cells will be used, more specifically at most $v_0, \ldots, v_m$ can possibly be used where

$$v = \sum_{k=0}^{m} 2^{m-k} v_k \leq 2T_A(X)$$

We are now ready to use (a modification of) Cook's proof of the NP-completeness of SAT to show $U_A \leq_P$ BSAT. To construct the instance of BSAT we first set the value $W = T_A(X)$. This can be done in polynomial time. By Cook's proof we know that there exists a boolean formula φ, constructable in polynomial-time, such that

$$N_A(X,y) = v \iff \varphi(X',y',v') = \text{TRUE}$$

where X', y', v' are variables obtained from X, y, v according to Cook's construction, in particular there is exactly one v_i' for each tape cell v_i. We put the weights only on the v_i''s:

$$w(v_i') = w_i = 2^{m-i},\ 0 \le i \le m$$

Since

$$\sum_{i=0}^{m} w_i \le 2W$$

what we constructed is a valid BSAT instance. If we define

$$t_2(X,(X',y',v')) = \begin{cases} T_A(X), & \text{if } \varphi(X',y',v') = \text{FALSE} \\ y, & \text{otherwise} \end{cases}$$

we have $Q(U_A, t_2(X,(X',y',v'))) = Q(\text{BSAT},(X',y',v'))$. QED

4. PAS-Completeness

According to the same line of the preceeding section we will show that the following problem, LINEAR BSAT (LBSAT), is PAS-complete

INSTANCE: a boolean formula φ with variables $x_1, \ldots, x_n$ of weights $w_1, \ldots, w_n$ and a separate weight W. The weights must satisfy

$$\sum_{i=1}^{n} w_i \le \left(1 + \tfrac{1}{n-1}\right) W.$$

PROBLEM: maximize the following function defined on the assignment of φ

$$f(y) = \begin{cases} W, & \text{if } \varphi(y) = \text{FALSE} \\ \sum_{i=1}^{n} w_i x_i, & \text{otherwise} \end{cases}$$

We can modify the definition of U_A to get a complete problem for PAS, let us call it U_P. U_P is defined in the same way as U_A except that in the Trunk, on input $X = (x, \delta, T, N_F, 0^k)$, $T(x,\delta)$ is simulated instead of $T(x)$. In order to show membership in PAS for U_P we have to modify the function $A(X)$, this is done in the next

Lemma 3. *U_P is in* PAS.

PROOF. We have to show that U_P has a polynomial-time approximation scheme, that is an algorithm T that given any ϵ is able to ϵ-approximate U_P in time $h_T(X,\epsilon)$ where h_T is polynomial in $|X|$ and arbitrary in $1/\epsilon$. The way of obtaining this is to define $A(X)$ in such a way that if we want to approximate U_P with an error smaller than $1/|X|$ then $1/\epsilon$ is so big to allow a deterministic simulation of U_P in polynomial-time.

As in Lemma 1 let $b = A(X) + t/(1-\delta)$ and $a = A(X) + t$.

PAS-ALGORITHM FOR U_P

Input: $X = (x, \delta, T, N_F, 0^k)$ and ϵ' the *wanted approximation.*

If k is big enough to perform the Trunk compute $Q = (b-a)/b$, abort otherwise (X is not a valid input).

If $Q \le \epsilon'$ then output $T(x,\delta)$ otherwise simulate U_P on input X *deterministically* and print an optimal solution y^*, i.e. $u_P(X,y^*) = u_P^*(X)$.

We now show that for any input (X,ϵ') this algorithm approximates U_P within ϵ' in time polynomial in $|X|$ and exponential in ϵ'. Notice that, similarly to the U_A case, $Q(U_P, T(x,\delta)) \le Q$. There are two possibilities; the first is when $Q \le \epsilon'$. In this case, since $Q(U_P, T(x,\delta)) \le Q$ and the complexity of the Trunk is linear, the algorithm... works.

The second case is when $Q > \epsilon'$. We shall define $A(X)$ to manage this case properly. A few calculations show that $Q = \delta t/(A(X)(1-\delta) + t)$; if we define $A(X)$ in such a way that $Q = 1/|X|$, that is

$$A(X) = \frac{t(\delta|X|-1)}{1-\delta},$$

we have that $Q > \epsilon' \Rightarrow 1/\epsilon' > |X|$. $A(X)$ is polynomially computable.

Notice now that to simulate U_P deterministically we need $k2^k \leq | X | 2^{|X|} < | X | 2^{1/\epsilon'}$ steps. In other words, if $Q > \epsilon'$ we can simulate U_P deterministically in time polynomial in $| X |$ and exponential in $1/\epsilon'$. Hence our algorithm is a polynomial-time approximation scheme. QED

Theorem 3. *U_P is* PAS*-complete.*

PROOF. Given any $F \in$ PAS we have to exhibit an F-reduction witnessing $F \leq_F U_P$. Let T_F be the PAS algorithm for F and p_F, h_T the complexities of N_F and T_F. Define $t_1(x) = (x, 1/2, T_F, N_F, 0^k)$ where $k = \max\{p_F(| x |), h_T(1/2, | x |)\}$. Notice the complexity given by h_T; by our choice of $\delta = 1/2$, k is a polynomial value in $| x |$ no matter what h_T is (any other fixed rational in $(0,1)$ would be as good as $1/2$). The same choice implies that $A(X) = t(| X | -2)$. The other two parts of the F-reduction are defined as

- $t_2(x, y) = y$; recall that by our choice of k we have enough time to simulate both T_F and N_F and hence $D_{U_P}(X) = D_F(x)$;
- $c(x, \epsilon) = \epsilon/(| X | -1)$. This is a valid definition for c since both its complexity and its value are polynomial in $1/\epsilon$ and $| X |$.

We have now to show that, for every ϵ,

$$Q(U_P, y) \leq c(x, \epsilon) \Longrightarrow Q(F, t_2(x, y)) \leq \epsilon.$$

Simple calculations yield

$$Q(U_P, y) = \frac{f_F^*(x) - f_F(x,y)}{A(X) + f_F^*(x)};$$

Then saying that

$$Q(U_P, y) \leq c(x, \epsilon)$$

is equivalent to say that

$$\begin{aligned} Q(F, t_2(x, y)) &= \frac{f_F^*(x) - f_F(x, y)}{f_F^*(x)} \\ &\leq c(x, \epsilon)\left(\frac{A(X)}{f_F^*(x)} + 1\right) \\ &\leq c(x, \epsilon)\left(\frac{A(X)}{t} + 1\right) \end{aligned}$$

Substituting in the last expression the values of $A(X)$ and $c(x, \epsilon)$ yields our conclusion, namely $Q(F, t_2(x, y)) \leq \epsilon$. QED

Corollary 1. LBSAT*is* PAS*-complete.*

PROOF. We only show LBSAT $\in$ PAS, $U_P \leq_F$ LBSAT can be shown using exactly the same argument of Theorem 2.

To show LBSAT $\in$ PAS is very easy. The definition of LBSAT is such that the value W always insures an error less than or equal to $1/n$. If an approximation $\epsilon < 1/n$ is wanted then $2^{1/\epsilon} > 2^n$ and we can find the optimal assignment deterministically in polynomial-time. QED

Theorem 3 has another interesting corollary.

Definition 14. *The closure of a class $C \subseteq NPO$ with respect to a reduction $\leq_R$ is the set*

$$\mathcal{C}(C, \leq_R) = \{F \mid \exists G \in C \text{such that} F \leq_R G\}.$$

This is a restatement in our framework of the notion of closure of a class of languages with respect to a reduction.

Corollary 2. $\mathcal{C}$(PAS, $\leq_F$) *includes* APX.

PROOF. Observe the role of $\delta = 1/2$ in the proof of theorem 3; as already pointed out any other rational in $(0,1)$ could be used. In fact, if $F \in$ APX can be approximated within some ϵ_F the proof of Theorem 3 with $\delta = \epsilon_F$ shows that F is F-reducible to U_P. QED

The intuitive reason for this to happen is that the function c in the definition of P-reduction must be independent of $|x|$ in order to respect time bounds. On the other hand, in the F-reduction c can be polynomially dependent on $|x|$. For this very same reason we have that the inclusion of Corollary 2 is proper. To see this, pick $F \in NPO - APX$ with objective function whose value is either 0 or 1 (for example, the characteristic function of SAT) and F-reduce it to any PAS problem with objective function bounded in terms of $p(|x|)$, p a polynomial (for example a suitable version of U_P); the reduction is possible since the definition of F-reduction allows $c(x, \epsilon) = 1/p(|x|)$.

5. The Existence Of Intermediate Degrees

A natural theoretical question to ask in our framework is whether, assuming P≠NP, incomplete problems can exist; by using delayed diagonalization (see [La,Ho]) we are able to show two different versions of this fact, namely

Theorem 4. *If P≠NP, there is a* NPO *problem* INT *such that:*

- INT $\notin$ APX,
- INT *is not* NPO-*complete, and*
- INT *is* APX-*hard via P-reductions.*

Theorem 5. *If P≠NP, there is a* NPO *problem* INT' *such that:*

- INT' $\notin$ APX,
- INT' *is not* NPO-*complete, and*
- INT' *is not* APX-*hard via P-reductions.*

These and the following theorems are interesting because they explain the difference there is between showing that a given problem is not approximable (or, more in general, that it does not have some approximation property) and showing that it is complete. In particular they show that NPO-completeness is not a consequence of the proof of statements like "if $F \in$ APX then P=NP" usually used to show non-approximability of problems. Moreover, Theorem 5 shows that there are problems with a quite counterintuitive characteristic; even if they are strong enough to be non-approximable they are not able to represent "weaker" problems, namely those in APX.

PROOF OF THEOREM 4. Let T_i, ϵ_j, A_k be enumerations of polynomial DTM's, rationals in $(0,1)$ and A-reductions respectively. Then let U_N be an NPO-complete problem informally introduced at the beginning of section 2; since it is possible to reduce every $F \in$ NPO to U_N in such a way that $Q(U_N, y) = Q(F, t_2(x, y))$ we have that U_N is also APX-hard with respect to P-reductions. Finally, let U_A and T_A the APX-complete problem of section 2 with its $1/2$-approximated algorithm.

The theorem relies on the possibility of definining the new problem INT partly as U_N and partly as U_A. INT is defined as U_N in order to be non approximable and as U_A in order to be incomplete. More precisely, consider the following "even" predicates C_{2k}'s:

$C_{2k}(x)$ = TRUE iff $k = (i, j)$ and $T_i(x)$ does not approximates $U_N(x)$ within ϵ_j.

Because of the non-approximability of U_N (we suppose P≠NP) we know that $C_{2k}(x)$ = TRUE infinitely often.

Similarly, consider the "odd" predicates C_{2k+1}'s

$C_{2k+1}(x)$ = TRUE iff $A_k = (t_{1k}, t_{2k}, c_k)$ *fails* at x, that is
$$Q(U_N, t_{2k}(x, T_A(t_{1k}(x)))) > c_k(1/2).$$

If $C_{2k+1}(x)$ were false almost everywhere U_N would be approximable within $c_k(1/2)$ by the the polynomial algorithm $t_2(x, T_A(t_1(x)))$ plus a finite table. Notice that if $C_{2k+1}(x)$ = TRUE x is a witness that U_N cannot be A-reduced to U_A via A_k.

In order to define INT we need a polynomial TM $D(x)$ such that

range$(D) = 1^*$,
$\forall k\, \exists x : C_{2k}(x) = \text{TRUE} \wedge D(x) = 1^{2k}$,
$\forall k\, \exists x : C_{2k+1}(x) = \text{TRUE} \wedge D(x) = 1^{2k+1}$.

The proof that such a machine exists is in a forthcoming lemma. We now show that D is what we need in order to conclude the theorem.

We define

$$\text{INT}(x) = \begin{cases} U_N(x), & \text{if } |D(x)| \text{ is even;} \\ U_A(x), & \text{otherwise.} \end{cases}$$

We first show that no T_i can claim to approximate INT within ϵ_j that is, INT $\notin$ APX. Since *rangeD* $= 1^*$ we have that there is an x such that: *i)* $C_{2k}(x) = \text{TRUE}$, where $k = (i, j)$; *ii)* on x INT is defined the same as U_N. In other words, x is a witness that T_i does not ϵ_j-approximates U_N; on that x INT and U_N are defined the same.

Similarly, INT cannot be NPO-complete. If it were it would be $U_N \leq_A$ INT via some A_k, but this is impossible since for each k there is x such that: *i)* A_k fails at x; *ii)* on x INT is defined to be the same as U_A.

Finally, INT is APX-hard via P-reductions because both U_N and U_A are and we can tell which of the two INT(x) is in polynomial-time by running $D(x)$. QED

Lemma4. *There is a polynomial-time computable function $D(x)$ such that*

range$(D) = 1^*$,
$\forall n\, \exists x : C_n(x) = \text{TRUE} \wedge D(x) = 1^n$.

PROOF. We inductively define points x_k and z_k :

i) $x_0 = \min\{x \mid C_0(x) = \text{TRUE}\}$; z_0 is the time needed to find x_0 by following this procedure: test $C_0(x)$ for increasingly larger x's.

ii) $x_k = \min\{x \mid C_k(x) \wedge x > z_{k-1}\}$; z_k is the time needed to find x_k by following this procedure: test $C_k(x)$ for increasingly larger x's provided that $x > z_{k-1}$.

Our machine D acts as follows: on input x for $|x|$ steps look for $z_0, z_1, \ldots$; let z_k be the largest, output 1^{k+1}.

The claim follows from the fact that there are infinitely many z_k's and that for each k we have $z_k < x_{k+1} < z_{k+1}$ and by observing that $D(x) = 1^{k+1}$ whenever $z_k \leq x < z_{k+1}$. QED

The proof of theorem 5 is analogous to that of theorem 4, and hence it is omitted.

By similar arguments we can show this very general theorem

Theorem 6. *Let $\mathcal{X}, \mathcal{Y}$ be two classes among* NPO, APX, PAS, FPAS*with $\mathcal{Y} \subseteq \mathcal{X}$. Given $G \in \mathcal{Y}$ and H $\mathcal{X}$-complete it is possible to build F which is not $\mathcal{X}$-complete but belongs to $\mathcal{X} - \mathcal{Y}$. Moreover it is possible to make it either $\mathcal{Y}$-hard or not.*

6. Concluding remarks and open questions

We defined natural approximation preserving reductions and proved the completeness in the classes APXand PASfor weighted versions of SAT. We also illustrated a certain "pathology" of the F-reduction with respect to the A-reduction: the closure of PASwith respect to the F-reduction strictly includes APX.

Then we proved the existence of incomplete problems. These results illustrate the rich structure of NPO: problems which are NP-complete in their recognition version behave differently in their original optimization form with respect to both approximation properties and completeness in the approximation classes.

Some interesting questions can now be asked. Are there natural complete problems in APXand PAS? Similarly, are there natural incomplete problems? In this latter case, "natural" has a somewhat wider interpretation than in the former question: we ask if it is possible to show, under some reasonable assumptions, the existence of incomplete problems without making use of diagonalization, in particular whether some weighted version of SAT can be proved incomplete.

Acknowledgements

We wish to thank Giorgio Ausiello and Pekka Orponen for many helpful discussions and suggestions.

References

[AMP] G.Ausiello - A. Marchetti Spaccamela - M. Protasi *"Toward a unified approach for the classification of NP-complete problems"* TCS 12 1980

[Co] S.A. Cook *"The complexity of theorem-proving procedures"* Proc. 3rd ACM STOC 1971

[DL] W.Fernandez De La Vega - G.S. Lueker *"Bin Packing can be solved within* $1 + \epsilon$ *in linear time"* Combinatorica 1 1981

[GJ] M.R.Garey - D.Johnson *"Computers and Intractability: a guide to the theory of NP-completeness"* Freeman, San Francisco 1979

[Ho] S.Homer *"On Simple And Creative Sets in NP"* TCS 47 1986

[HS] D.Hochbaum - D.Shmoys *"Using Dual Approximation Algorithms for Scheduling problems: theoretical and practical results"* JACM 34 1987

[IK] O.H. Ibarra - C.E. Kim *"Fast approximation for the Knapsack and Sum of Subset problems"* JACM 22 1975

[Jo] D.Johnson *"Approximation algorithms for combinatorial problems"* Proc. 5th ACM STOC 1973

[Kr] M.W. Krentel *"The Complexity of Optimization Problems"* PhD Thesis, Cornell University May 87

[KS] B.Korte - R. Schrader *"On the existence of fast approximation schemes"* NON LINEAR PROGRAMMING 4 1981

[La] R. Ladner *"On The Structure Of Polynomial Time Reducibility"* JACM 22 1975

[OM] P.Orponen - H.Mannila *"On approximation preserving reductions: complete problems and robust measures"* Technical Report, University of Helsinki 1987

[PM] A.Paz and S.Moran *"NP-optimization problems and their approximation"* Proc. 4th ICALP 1977

[PS] C.Papadimitriou - K.Steiglitz *"Combinatorial Optimization: Algorithms and Complexity"* Prentice-Hall, Englewood Cliffs, New Jersey 1982

[Sh] S.K. Sahni *"Algorithms for scheduling indepent tasks"* JACM 23 1976

SEPARATING COMPLETELY COMPLEXITY CLASSES RELATED TO POLYNOMIAL SIZE Ω-DECISION TREES

Carsten Damm, Christoph Meinel

Sektion Mathematik
Humboldt-Universität zu Berlin
DDR-1086 Berlin, PF 1297

ABSTRACT.

In proving exponential lower and polynomial upper bounds for parity decision trees and collecting similar bounds for nondeterministic and co-nondeterministic decision trees we completely separate the complexity classes related to polynomial size deterministic, nondeterministic, co-nondeterministic, parity, and alternating decision trees.

INTRODUCTION.

One of the hardest problems in complexity theory is to separate complexity classes, especially those of feasible problems. Although complexity classes like L , NL , or P are identified and investigated for a long time, no considerable success was achieved up to now in separating these classes. However, a most promising approach for doing this is to investigate the corresponding nonuniform complexity classes. These nonuniform classes can be described by means of circuitry-based computation devices such as Boolean circuits, formulas, decision trees, or branching programs to which combinatorial techniques and counting arguments can be applied more directly than to the very complex Turing machines. Indeed, first exponential lower bounds could be proved for some restricted types of Boolean circuits and branching programs. Besides exponential lower bounds for monotone circuits [Ra85,An85], and for bounded depth circuits [Ya85,Ha86], exponential lower bounds were obtained for decision trees and for read-once-only branching programs [Bu85,We84,A&86]. Moreover, by means of certain exponential lower and polynomial upper bounds it could be proved in [KMW88] that nondeterministic (or, equivalently, disjunctive), co-nondeterministic (or, equivalently, conjunctive), as well as alternating branching programs are strictly more powerful than deterministic ones under the read-once-only restriction. However, almost nothing is known for parity branching programs which represent the remaining fifth computing concept due to a classification given in [Me88].

In order to settle this question we start in the following with the investigation of tree-structured Ω-branching programs, $\Omega \subseteq \mathbb{B}_2$, which are known in the case $\Omega = \emptyset$ as decision trees . After classifying Ω-decision trees, $\Omega \subseteq \mathbb{B}_2$, into the five types of ordinary, disjunctive, conjunctive, parity, and alternating decision trees we establish strong differences in the computational power of all these five types. Of course, the lower bounds for read-once-only ordinary, disjunctive, and conjunctive branching programs and the upper bounds for read-once-only alternating branching programs apply also to the corresponding ordinary, disjunctive, conjunctive, and alternating decision trees, which can be thought of as special read-once-only branching programs. But proving certain exponential lower and polynomial upper bounds for parity decision trees we confirm for the first time the assumption that the parity computation concept differs from the well-known concepts of deterministic, nondeterministic, co-nondeterministic, and alternating computations.

The paper is organized as follows. In Section 1 , after introducing and classifying Ω-decision trees (Theorem 1) we characterize their computational power in terms of certain bounded depth circuit models (Proposition 1). Then, in Section 2 , considering various functions we derive a variety of lower and upper bounds for the different types of Ω-decision trees (Lemmas 1 through 3). In the final third section we completely separate the complexity classes defined by polynomial size ordinary, nondeterministic (i.e. disjunctive), co-nondeterministic (i.e. conjunctive), parity, and alternating decision trees (Theorem 2). Considering the nondeterministic, co-nondeterministic and parity computation concepts we prove in terms of Ω-decision trees that none of them can be dominated by one or two of the remaining concepts. Finally, it is shown that alternating decision trees are stronger than nondeterministic, co-nondeterministic and parity decision trees together.

1. Polynomial Size Ω-Decision Trees.

A *decision tree* (*DT*) is a directed binary tree, whose leaves are labelled by Boolean constants and whose inner nodes are labelled by Boolean variables taken from a set $X = \{x_1, \ldots, x_n\}$. An Ω-*decision tree* T is a decision tree some of whose inner nodes are labelled by Boolean functions $\omega \in \Omega$ from a set $\Omega \subseteq \mathbb{B}_2$ of 2-argument Boolean functions instead of Boolean variables. The Boolean values assigned to the leaves of T extend recursively to Boolean values associated with all nodes of T in the following way: if the successor nodes v_0 , v_1 of a node v of T carry the Boolean values δ_0 , δ_1 and if v is labelled by a Boolean variable x_i we associate with v the value δ_0 (δ_1) iff $x_i = 0$ $(x_i = 1)$. If v is labelled by a Boolean function ω then we associate with v the value $\omega(\delta_0, \delta_1)$. T is said to *accept* (*reject*) an input $w \in \{0,1\}^n$ if the root of T associates with 1 (0) under w . The function computed by an Ω-decision tree T is the Boolean function that takes on the value 1 on those inputs accepted by T.

Besides the *depth* (i.e., the length of a longest path in an Ω-decision tree T) the *size* ,denoted by $Size(T)$, of an Ω-decision tree T is the most important complexity measure. It is defined to be the number of inner nodes of T . By $\Omega\text{-}DT(f)$ we denote the minimal size of a decision tree that computes the function f :

$$\Omega\text{-}DT(f) = \min \{Size(T) \mid T \text{ is an } \Omega\text{-decision tree for } f\}.$$

Ω-decision trees with $\Omega = \{\vee\}, \{\wedge\}, \{\oplus\}$ and $\{\vee, \wedge\}$ are called *disjunctive*, *conjunctive*, *parity* and *alternating* decision trees, respectively.

Via the usual correspondence between binary languages $A \subseteq \{0,1\}^*$ and sequences $\{f_n\}$ of Boolean functions $f_n \in \mathbb{B}_n$, namely

$$w \in A \quad \text{iff} \quad f_{|w|}(w) = 1 ,$$

a sequence $\{T_n\}$ of Ω-decision trees is said to *accept a language* $A \subseteq \{0,1\}^*$ if, for all $n \in \mathbb{N}$, T_n computes the characteristic function of the n-th *restriction* A^n of A, $A^n = A \cap \{0,1\}^n$.

By $\mathcal{P}_{\Omega\text{-}DT}$, $\Omega \subseteq \mathbb{B}_2$, we denote the class of languages acceptable by (sequences of) *polynomial size* Ω-decision trees.

In order to compare the computational power of different Ω-decision trees we introduce the computational equivalence of decision trees. An Ω-decision tree T and an Ω'-decision tree T' ,

Ω, $\Omega' \subseteq \mathbb{B}_2$, are said to be *computationally equivalent* if they accept the same set and if their sizes coincide, to within a constant factor.

In full analogy to the classification of Ω-branching programs given in [Me88] we can prove the following theorem which completely classifies Ω-decision trees.

THEOREM 1.

For each Ω-decision tree, $\Omega \subseteq \mathbb{B}_2$, there is a computationally equivalent Ω'-decision tree with

$$\Omega' = \emptyset \ , \ \Omega' = \{\vee\} \ , \ \Omega' = \{\wedge\} \ , \ \Omega' = \{\oplus\} \ , \ or \ \ \Omega' = \{\vee,\wedge\} \ . \ \blacksquare$$

Since the sizes of computationally equivalent Ω-decision trees coincide, to within constant factors, due to the classification result of Theorem 1 each polynomial size Ω-decision tree is computationally equivalent to an ordinary, a disjunctive, a conjunctive, a parity, or an alternating decision tree of polynomial size.

COROLLARY 1.

There are at most five complexity classes related to polynomial size Ω-decision trees, $\Omega \subseteq \mathbb{B}_2$:

$$\mathcal{P}_{DT} \ , \ \mathcal{P}_{\{\vee\}-DT} \ , \ \mathcal{P}_{\{\wedge\}-DT} \ , \ \mathcal{P}_{\{\oplus\}-DT} \ \ \text{and} \ \ \mathcal{P}_{\{\vee,\wedge\}-DT} \ .$$

These five classes are related in the following manner:

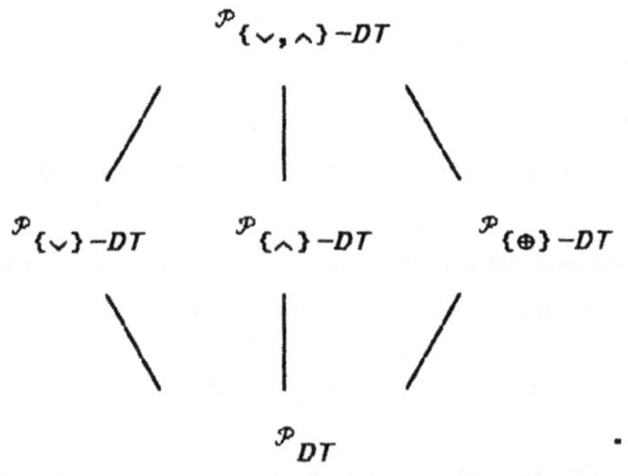

. ■

The various decision tree complexities of a Boolean function are related to certain depth 2 formula sizes of this function. Recall, a *disjunctive normal form* over the set of variables $\{x_1, x_2, \ldots, x_n\}$ is the disjunction of conjunctions over the corresponding set of literals (i.e., the set of variables and their negations). Similarily a *conjunctive normal form* is a conjunction of disjunctions over the literals. In analogy we define a *parity normal form* to be the parity over conjunctions of literals. Note the difference between a parity normal form of a Boolean function f and its description by means of a polynomial over $\mathbb{F}_2$: The polynomial representation is unique (since it disallows negations of variables) whereas the parity normal form is not.

The appropriate complexity measure of such normal form representations of a Boolean function is their *length*, which is defined to be the number of literals occurring in it. Let us denote by $DNF(f)$, $CNF(f)$, and $PNF(f)$ the minimal length of a disjunctive, of a conjunctive, and of a parity normal form of f , respectively. As usual, we denote by $L(f)$ the fan-in 2 formula size of f .

We can add the following inequalities relating *DT*-complexities and certain formula sizes to those given by Wegener in [We84].

PROPOSITION 1.

Let $f \in \mathbb{B}_n$ *be a Boolean function. Then it holds*

(i) $DNF(f) \leq n \cdot \{\vee\}\text{-}DT(f)$ *and* $\{\vee\}\text{-}DT(f) \leq 2 \cdot DNF(f)$,

(ii) $CNF(f) \leq n \cdot \{\wedge\}\text{-}DT(f)$ *and* $\{\wedge\}\text{-}DT(f) \leq 2 \cdot CNF(f)$,

(iii) $PNF(f) \leq n \cdot \{\oplus\}\text{-}DT(f)$ *and* $\{\oplus\}\text{-}DT(f) \leq 2 \cdot PNF(f)$,

(iv) $L(f) \leq 3 \cdot \{\wedge,\vee\}\text{-}DT(f)$ *and* $\{\wedge,\vee\}\text{-}DT(f) \leq 2 \cdot L(f)$.

PROOF :

(i) The first inequality relies on the fact, that an disjunctive decision tree T computes 1 when given the input vector x iff there is an accepting path in T consistent with x . On the other hand accepting paths directly correspond to conjunctions. Observing that in directed trees there is a bijection between maximal paths and sinks the inequality follows. The second inequality is simply proved by simulating the conjunctions in a disjunctive normal form by deterministic decision trees and combining the results by an binary tree all of whose inner nodes are labelled with $\vee$.

(ii) and (iii) are proved similar to (i). For (ii) one uses duality.

(iv) The first inequality results from a standard technique introduced by Wegener [We84], which uses the selection function *sel* in order to connect decision tree complexity and formula size.

The second inequality is obvious. Just replace the input gates by single node decision trees. ∎

If we denote by $\mathcal{P}_{DNF}$, $\mathcal{P}_{CNF}$, $\mathcal{P}_{PNF}$ the complexity classes of all languages with polynomial size disjunctive, conjunctive, and parity normal forms we can restate Proposition 1 as follows.

COROLLARY 2.

(i) $\mathcal{P}_{\{\vee\}-DT} = \mathcal{P}_{DNF}$,

(ii) $\mathcal{P}_{\{\wedge\}-DT} = \mathcal{P}_{CNF}$,

(iii) $\mathcal{P}_{\{\oplus\}-DT} = \mathcal{P}_{PNF}$,

(iv) $\mathcal{P}_{\{\wedge,\vee\}-DT} = \mathcal{NC}^1$. ■

($\mathcal{NC}^1$ denotes the class of languages accepted by polynomial size (fan-in 2) formulas.)

Let us only remark that Wegener [We84] proved

$$DT(f) \geq DNF(f) + CNF(f)$$

which now can be interpreted as a relation between the (ordinary) decision tree complexity of a Boolean function f , on the one hand, and its disjunctive and conjunctive decision tree complexity on the other hand.

PROPOSITION 2.

Let f be a Boolean function. Then it holds

$$DT(f) \geq \frac{1}{2} \cdot (\{\vee\}\text{-}DT(f) + \{\wedge\}\text{-}DT(f)) . \blacksquare$$

2. LOWER AND UPPER BOUNDS

Let maj_n , $par_n \in \mathbb{B}_n$ and $t_{n,n-1} \in \mathbb{B}_{n^2}$ be defined as follows. The *majority function* maj_n takes on value 1 on those input vectors $x \in \{0,1\}^n$ that contain at least $\lceil n/2 \rceil$ ones. The *parity function* par_n takes on value 1 on those input vectors x that contain an odd number of ones. Finally, $t_{n,n-1}$ is the function defined on n^2 variables arranged in an $n \times n$ matrix that takes on value 1 on those inputs x that contain at least n-1 ones in each row. It is well-known that all the above-mentioned functions can be computed by polynomial size Boolean formulas and consequently by polynomial size alternating decision trees.

LEMMA 1.

(i) *There is a conjunctive decision tree of size* $O(n^3)$ *which computes* $t_{n,n-1}$.

(ii) *Each disjunctive decision tree which computes the function* $t_{n,n-1}$ *is of size* $2^{\Omega(n)}$.

(iii) *Each parity decision tree which computes the function* $t_{n,n-1}$ *is of size* $2^{\Omega(n\log n)}$.

PROOF.

Claim (i) follows from Proposition 1 and the observation that $t_{n,n-1}(x) = 1$ if and only if in each row of x any pair of entries contains at least one value 1.

Claim (ii) : Since $t_{n,n-1}$ is monotone $DNF(t_{n,n-1})$ can be estimated by the number of prime implicants of $t_{n,n-1}$ which is n^n .

Claim (iii) can be obtained by means of Razborov's approximation technique [Ra86] for circuits of bounded depth. Due to Proposition 1 it suffices to prove the lower bound $PNF(t_{n,n-1}) = 2^{\Omega(n)}$. The method makes extensive use of the isomorphism

$$\mathbb{B}_n \cong \mathbb{F}_2[x_1, x_2, \ldots, x_n]/(x_i = x_i^2) .$$

In the sequel we do not distinguish at all between the function and the polynomial. Let $\mathbb{P}(d)$, $d \in \mathbb{N}$, be the subspace of all polynomials of degree not larger than d . For $f, g \in \mathbb{B}_n$, $G \subseteq \mathbb{B}_n$ let

$$\rho(f,g) = \#\{a \in \{0,1\}^n \mid f(a) \neq g(a)\}$$

be the usual Hamming distance for Boolean functions, and let

$$\rho(f,G) = \min_{g \in G} \rho(f,g) .$$

By $L_k^*(f)$, $k \geq 1$, we denote the minimum number of gates in a circuit computing f over the unbounded fan-in basis $\{\wedge,\oplus\}$ which has at most k levels of $\wedge$-gates. Since, obviously, $PNF(f) \geq L_1^*(f)$ we restrict our considerations to the case $k = 1$.

For $k = 1$ Lemma 1 of [Ra86] states:

(1) *If* $f \in \mathbb{B}_n$ *and* $r \geq 1$ *then* $L_1^*(f) \geq \rho(f,\mathbb{P}(r)) \cdot 2^{r-n}$.

In [Ra86] the main idea behind estimating $\rho(f,\mathbb{P}(r))$ is to associate a so-called *intersection matrix* to f the *rank* of which (over $\mathbb{F}_2$) yields a lower bound for $\rho(f,\mathbb{P}(r))$ under certain conditions. More precisely, let $A \subseteq \{0,1\}^n$ be a set of input vectors. For $x, y \in A$ let $x \wedge y = (x_1 \wedge y_1, x_2 \wedge y_2, \ldots, x_n \wedge y_n)$. $M^A(f)$ is the $\#A \times \#A$ matrix with entry $f(x \wedge y)$ in row x and column y .

In [Ra86,Pa86] the following property of the linear map M^A is proved

(2) *If for all* $x, y \in A$ *the number of ones in* $x \wedge y$ *exceeds* r *then, for any* $f \in \mathbb{B}_n$ *, it holds*

$$\rho(f, P(r)) \geq \operatorname{rank} M^A(L(f)) .$$

Here L is the linear isomorphism

$$L : \mathbb{B}_n \longrightarrow \mathbb{B}_n \quad , \; L(f)(x) = \bigoplus_{y \leq x} f(y).$$

For $t_{n,n-1}$ we choose

$A_n = \{x \in \{0,1\}^n |\ x$ contains exactly $n - 1$ ones in each row$\}$.

It suffices to compute $L(t_{n,n-1})$ merely on those inputs that appear as intersection $x \wedge y$ of two inputs in A_n. Since $t_{n,n-1}$ is monotone $L(t_{n,n-1})$ vanishes on inputs outside of $t^{-1}_{n,n-1}(1)$. If $t_{n,n-1}(z) = 1$ for $z \in A_n$ then $t_{n,n-1}(y) = 1$ for some $y \leq z$ if and only if $y = z$. Hence for $x, y \in A_n$ $L(f)(x \wedge y) = 1$ iff $f(x \wedge y) = 1$ iff $x = y$. That means $\operatorname{rank} M^{A_n}(t_{n,n-1}) = \# A_n = n^n$. On the other hand the number of ones in $x \wedge y$ is at least $n(n - 1)$ for $x, y \in A_n$. Combining (1) and (2) one gets

$$PNF(t_{n,n-1}) \geq n^n / 2^{2n-1} = 2^{\Omega(n \log n)}$$

the desired lower bound.

LEMMA 2.

(i) *Each disjunctive decision tree which computes the parity function* par_n *is of size* $2^{\Omega(n)}$.

(ii) *Every conjunctive decision tree which computes the parity function* par_n *is of size* $2^{\Omega(n)}$.

(iii) *There is a linear size parity decision tree which computes the parity function* par_n .

PROOF.

Claims (i) and (ii) are consequences of Proposition 1 and the well-known lower bounds $DNF(par_n) = CNF(par_n) = n \cdot 2^{n-1}$ for the length of disjunctive and conjunctive normal forms for the parity function.

Claim (iii) is obvious. ■

LEMMA 3.

(i) *Each disjunctive decision tree which computes the majority function* maj_n *is of size* $2^{\Omega(n)}$.

(ii) *Each conjunctive decision tree which computes the majority function* maj_n *is of size* $2^{\Omega(n)}$.

(iii) *Each parity decision tree which computes the majority function* maj_n *is of size* $2^{\Omega(n)}$.

PROOF.

Claims (i) to (iii) follow from Proposition 1 and the fact that majority is not reducible to parity by means of polynomial size, constant depth circuits [Ra86]. ∎

3. Separation Results

Theorem 1 and the lower and upper bounds derived in the last section allow to determine exactly the relations between the Ω-decision tree complexity classes, a fact which seldomly happens in complexity theory.

THEOREM 2.

(i) *All the complexity classes* $\mathcal{P}_{DT}$, $\mathcal{P}_{\{\vee\}-DT}$, $\mathcal{P}_{\{\wedge\}-DT}$, $\mathcal{P}_{\{\oplus\}-DT}$ *and* $\mathcal{P}_{\{\wedge,\vee\}-DT}$ *defined by polynomial size* Ω*-decision trees are different from each other.*

(ii) *The classes* $\mathcal{P}_{\{\vee\}-DT}$, $\mathcal{P}_{\{\wedge\}-DT}$, *and* $\mathcal{P}_{\{\oplus\}-DT}$ *are incomparable. Moreover neither one nor the union of two of them contains one of the others.*

(iii) *The union of the classes* $\mathcal{P}_{DT}$, $\mathcal{P}_{\{\vee\}-DT}$, $\mathcal{P}_{\{\wedge\}-DT}$, *and* $\mathcal{P}_{\{\oplus\}-DT}$ *does not exhaust all of* $\mathcal{P}_{\{\wedge,\vee\}-DT}$.

PROOF.

First, one observes that $\mathcal{P}_{DT}$, $\mathcal{P}_{\{\oplus\}-DT}$ and $\mathcal{P}_{\{\wedge,\vee\}-DT} = \mathcal{NC}^1$ are closed under complement, and that $\mathcal{P}_{\{\vee\}-DT}$ and $\mathcal{P}_{\{\wedge\}-DT} =$ co-$\mathcal{P}_{\{\vee\}-DT}$ are not (Lemma 1).

Claim (i) and (ii) can be obtained from the various lower and upper bounds given in Lemmas 1 and 2.

Finally, claim (iii) is implied by Lemma 3 and by the fact that $maj_n \in \mathcal{P}_{\{\vee,\wedge\}-DT}$. ∎

We conclude by drawing a picture which shows the relations of complexity classes related to polynomial size ordinary, disjunctive, conjunctive, parity, and alternating decision trees.

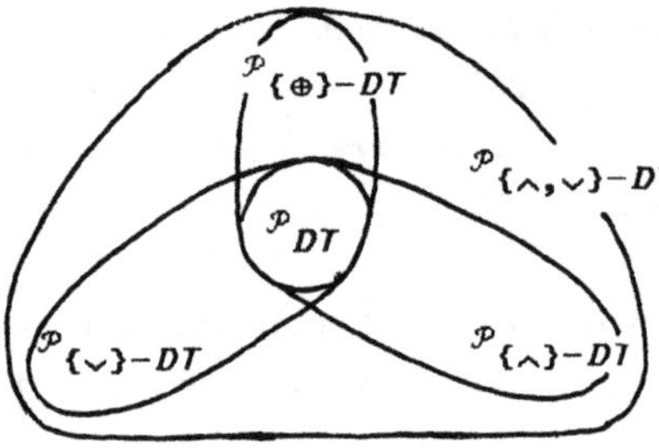

Acknowledgements

We would like to thank Matthias Krause for some very insightful discussions. Thanks are also due to Ingo Wegener for providing us with various material concerning the problems under consideration.

References

[A&86] M.Ajtai, L.Babai, P.Hajnal, J.Komlos, P.Pudlak, V.Rdl, E.Szemeredi, G.Turán: Two lower bounds for branching programs, Proc. 18th ACM STOC, 30-38.

[An85] A.E.Andreev: On a method of obtaining lower bounds for the complexity of individual monotone functions, Dokl. Akad. Nauk SSSR 282/5, 1033-1037.

[Bu85] L.Budach: A lower bound for the number of nodes in a decision tree. EIK 21 (1985), 221-228.

[FSS81] Furst, Saxe, M. Sipser: Parity, circuits and the polynomial time hierarchy, Proc. 22th IEEE FOCS, 1981, 260-270.

[Ha86] J.Hastad: Improved lower bounds for small depth circuits, Proc. 18th ACM STOC (1986), 6-20.

[KMW88] M.Krause, Ch.Meinel, S.Waack: Separating the eraser Turing machine classes L, NL, co-NL and P, Proc. MFCS'88, LNCS 324, 405-413.

[Me88] Ch.Meinel: The power of polynomial size Ω-branching programs, Proc. STACS'88 Bordeaux, LNCS 294, 81-90.

[Pa86] M.S.Paterson: On Razborov's result for bounded depth circuits over {⊕,∧}, manuscript.

[Ra85] A.A.Razborov: A lower bound for the monotone network complexity of the logical permanent, Matem. Zametki 37/6 (1985).

[Ra86] A.A.Razborov: Lower bounds on the size of circuits of bounded depth over the basis {∧,⊕}, preprint, Steklov Inst. for Math., Moscow 1986, (see also Mat. Zam. 41, no.4 (1987), 598-607) (in Russian).

[We84] I.Wegener: Optimal decision trees and one-time-only branching programs for symmetric Boolean functions, Inf. and Control, Vol. 62, Nos. 2/3 (1984), 129-143.

[We88] I.Wegener: On the Complexity of Branching Programs and Decision Trees for Clique Functions, JACM Vol. 35, No. 2 (1988), 461-471.

[Ya85] A.C.Yao Separating the polynomial-time hierarchy by oracles, Proc.26th IEEE FOCS (1985), 1-10.

ON PRODUCT HIERARCHIES OF AUTOMATA

P. Dömösi[1], Z. Ésik[2,3,4], B. Imreh[2]

1) Dept. of Mathematics, L. Kossuth University,
Debrecen, Egyetem tér 1., 4010, Hungary
2) Bolyai Institute, A. József University,
Szeged, Aradi v. tere 1., 6720, Hungary
3) Institute for Informatics, Technical University,
München, Arcisstr. 21, D – 8000 München 2, FRG

1. Introduction

The problem of finding economic realizations of a given automaton by a product of "simple" automata is of interest both from theoretical and practical point of view. In its general form, all the components of a Gluškov-type product of automata are fed back to one another. There have been two major approaches in decreasing the direct interconnection complexity of the Gluškov-type product. The hierarchy of α_i-products is due to Gécseg [11] and contains one particular product notion for each nonnegative integer i. The main characteristic of the α_i-product is that the length of the feedback cycles is bounded by i. Thus the α_0-product, the bottom of the hierarchy, coincides with the familiar cascade composition or loop-free product, cf. [19, 26]. Another hierarchy, the ν_i-hierarchy appears in [5], where i is now a positive integer. In a ν_i-product, each component is fed back to at most i of the component automata.

The aim of the present paper is to provide a systematic comparison of the inherent realization capacities of the above product notions. We treat both homomorphic and isomorphic realization. The emphasis is on the ν_i-hierarchy compared with the α_i-hierarchy. The section on α_i-products is included for the reader's convenience.

2. Basic notions and notation

An *automaton* is a triple $\mathcal{A} = (A, X, \delta)$ with finite nonempty sets A (state set), X (input set), and transition $\delta : A \times X \to A$ that extends to words in X^* as usual. We write $u^{\mathcal{A}}$ for the state transformation induced by a word $u \in X^*$. The *characteristic semigroup* $S(\mathcal{A}) = \{u^{\mathcal{A}} | u \in X^+\}$ consists of those transformations induced by nonempty words.

A class $\mathcal{K}$ of automata is a nonempty class. We say that $\mathcal{K}$ satisfies the *Letičevskiĭ criterion* if there exist $\mathcal{A} = (A, X, \delta) \in \mathcal{K}$, a state $a \in A$, input letters $x_1, x_2 \in X$ and words $u_1, u_2 \in X^*$ with $\delta(a, x_1) \neq \delta(a, x_2)$ and $\delta(a, x_1u_1) = \delta(a, x_2u_2) = a$. If $\mathcal{K}$ does not satisfy the Letičevskiĭ criterion but we have $\delta(a, x_1) \neq \delta(a, x_2)$ and $\delta(a, x_1u) = a$ for some automaton $\mathcal{A} = (A, X, \delta) \in \mathcal{K}$, where $a \in A$, $x_1, x_2 \in X$ and $u \in X^*$, then

[4] Research supported by Alexander von Humboldt Foundation

$\mathcal{K}$ satisfies the *semi-Letičevskiĭ criterion.* Otherwise we say that $\mathcal{K}$ does not satisfy the Letičevskiĭ criteria.

Let $\mathcal{A}_t = (A_t, X_t, \delta_t)$, $t = 1, \ldots, n$, $n \geq 1$, be automata and take a finite nonempty set X. Given a system of feedback functions

$$\varphi_t : A_1 \times \ldots \times A_n \times X \to X_t, \quad t = 1, \ldots, n,$$

the *general product* (or Gluškov-type product, cf. [18]) of $\mathcal{A}_1, \ldots, \mathcal{A}_n$ with respect to X and $\varphi = (\varphi_1, \ldots, \varphi_n)$ is the automaton $\mathcal{A} = \mathcal{A}_1 \times \ldots \times \mathcal{A}_n(X, \varphi) = (A, X, \delta)$ with $A = A_1 \times \ldots \times A_n$ and

$$\delta((a_1, \ldots, a_n), x) = (\delta_1(a_1, x_1), \ldots, \delta_n(a_n, x_n)),$$

$$x_t = \varphi_t(a_1, \ldots, a_n, x), \quad t = 1, \ldots, n,$$

for all $(a_1, \ldots, a_n) \in A$ and $x \in X$.

Let $\mathcal{A}$ be the above general product and set $N = \{1, \ldots, n\}$. A function $\gamma : N \to P(N)$ into the power set of N is a *neighbourhood function* of $\mathcal{A}$ if each feedback function $\varphi_t(a_1, \ldots, a_n, x)$ is independent of any state variable a_s with $s \notin \gamma(t)$. The concept of a general product with a neighbourhood function is essentially the same as the notion of a network of automata as defined e.g. in [25]. If $\mathcal{A}$ has a neighbourhood function γ with $\gamma(t) \subseteq \{1, \ldots, t + i - 1\} \cap N$ $(1 \leq t \leq n)$, where $i \geq 0$ is a fixed integer, then $\mathcal{A}$ is called an α_i*−product* of the $\mathcal{A}_t$. If for a positive integer i there exists a neighbourhood function γ of $\mathcal{A}$ with $|\gamma(t)| \leq i$ $(1 \leq t \leq n)$, then $\mathcal{A}$ is a ν_i*−product*. Taking $i = 0$ we obtain the notion of the *quasi-direct product*.

Let $\mathcal{K}$ be a class of automata. We shall use the following notations:

$$\begin{aligned}
\mathcal{P}_g(\mathcal{K}) &:= \text{all general products of automata from } \mathcal{K};\\
\mathcal{P}_{\alpha_i}(\mathcal{K}) &:= \text{all } \alpha_i - \text{products of automata from } \mathcal{K};\\
\mathcal{P}_{\nu_i}(\mathcal{K}) &:= \text{all } \nu_i - \text{products of automata from } \mathcal{K};\\
\mathcal{S}(\mathcal{K}) &:= \text{all subautomata of automata from } \mathcal{K};\\
\mathcal{H}(\mathcal{K}) &:= \text{all homomorphic images of automata from } \mathcal{K};\\
\mathcal{I}(\mathcal{K}) &:= \text{all isomorphic images of automata from } \mathcal{K}.
\end{aligned}$$

We write g−product for the general product and q−product for the quasi-direct product. Let β refer to any type of the product defined above. A class $\mathcal{K}$ is *homomorphically (isomorphically)* β*−complete* if $\mathcal{H}\mathcal{S}\mathcal{P}_\beta(\mathcal{K})$ $(\mathcal{I}\mathcal{S}\mathcal{P}_\beta(\mathcal{K}))$ is the class of all automata. The operator $\mathcal{P}_{1\beta}$ is defined by $\mathcal{P}_{1\beta}(\mathcal{K}) = \{\mathcal{A}(X,\varphi) | \mathcal{A} \in \mathcal{K},\ \mathcal{A}(X,\varphi) \text{ is a } \beta-\text{product}\}$. Thus $\mathcal{P}_{1\beta}(\mathcal{K})$ consists of all first β−powers of automata from $\mathcal{K}$. Notice that $\mathcal{P}_{1g}(\mathcal{K}) = \mathcal{P}_{1\alpha_1}(\mathcal{K})$ and $\mathcal{P}_{1q}(\mathcal{K}) = \mathcal{P}_{1\alpha_0}(\mathcal{K})$.

3. α_i−products

The first result shows that the α_i−products form a proper hierarchy with respect to isomorphic realization.

3.1. Theorem [20] There is a class $\mathcal{K}$ such that $\mathcal{I}\mathcal{S}\mathcal{P}_{\alpha_i}(\mathcal{K}) \subset \mathcal{I}\mathcal{S}\mathcal{P}_{\alpha_{i+1}}(\mathcal{K})$ for all $i \geq 0$.

For homomorphic realization the situation is quite different. To see this, we recall a classical result of Letičevskiĭ.

3.2. Theorem [23] A class $\mathcal{K}$ of automata is homomorphically $g-$complete if and only if it satisfies the Letičevskiĭ criterion.

It is proved in [7] that the very same holds for the α_2-product. Another proof is given in [9].

3.3. Theorem [7] A class $\mathcal{K}$ is homomorphically α_2-complete if and only if it satisfies the Letičevskiĭ criterion.

In view of Theorem 3.3, we might think that $\mathcal{H}\mathcal{S}\mathcal{P}_g(\mathcal{K}) = \mathcal{H}\mathcal{S}\mathcal{P}_{\alpha_2}(\mathcal{K})$ holds for every class $\mathcal{K}$. This fact has been proved in [10], providing a full description of $g-$varieties, i.e., classes of the form $\mathcal{H}\mathcal{S}\mathcal{P}_g(\mathcal{K})$.

3.4. Theorem [10] Suppose that $\mathcal{K}$ satisfies the semi-Letičevskiĭ criterion. Then $\mathcal{H}\mathcal{S}\mathcal{P}_g(\mathcal{K}) = \mathcal{H}\mathcal{S}\mathcal{P}_{\alpha_1}(\mathcal{K})$. If $\mathcal{K}$ does not satisfy the Letičevskiĭ criteria then already $\mathcal{H}\mathcal{S}\mathcal{P}_g(\mathcal{K}) = \mathcal{H}\mathcal{S}\mathcal{P}_{\alpha_0}(\mathcal{K})$.

3.5. Corollary [10] $\mathcal{H}\mathcal{S}\mathcal{P}_g(\mathcal{K}) = \mathcal{H}\mathcal{S}\mathcal{P}_{\alpha_2}(\mathcal{K})$ holds for every class $\mathcal{K}$.

Let $\mathcal{Z}_2$ be the automaton obtained from the cyclic group Z_2 by letting it act on itself. The Krohn- Rhodes decomposition, see [1, 6], implies that $\mathcal{H}\mathcal{S}\mathcal{P}_{\alpha_0}(\{\mathcal{Z}_2\})$ is the class of all permutation automata whose characteristic semigroup is a group of order 2^n, for some $n \geq 0$. Since $\mathcal{P}_{1\alpha_1}(\{\mathcal{Z}_2\}) = \mathcal{P}_{1g}(\{\mathcal{Z}_2\})$ contains an isomorphic copy of each two-state automaton, we see that $\mathcal{H}\mathcal{S}\mathcal{P}_{\alpha_0}(\{\mathcal{Z}_2\}) \subset \mathcal{H}\mathcal{S}\mathcal{P}_{\alpha_0}\mathcal{P}_{1\alpha_1}(\{\mathcal{Z}_2\}) = \mathcal{H}\mathcal{S}\mathcal{P}_{\alpha_1}(\{\mathcal{Z}_2\})$. In fact an automata $\mathcal{A}$ belongs to $\mathcal{H}\mathcal{S}\mathcal{P}_{\alpha_1}(\{\mathcal{Z}_2\})$ if and only if every subgroup of $S(\mathcal{A})$ is of order 2^n, for some $n \geq 0$, again by the Krohn-Rhodes decomposition. On the other hand, by Theorem 3.3, $\mathcal{H}\mathcal{S}\mathcal{P}_{\alpha_2}(\{\mathcal{Z}_2\})$ is the class of all automata. We have thus proved the following result.

3.6. Proposition [11] There is a class $\mathcal{K}$ with $\mathcal{H}\mathcal{S}\mathcal{P}_{\alpha_0}(\mathcal{K}) \subset \mathcal{H}\mathcal{S}\mathcal{P}_{\alpha_1}(\mathcal{K}) \subset \mathcal{H}\mathcal{S}\mathcal{P}_{\alpha_2}(\mathcal{K})$.

4. ν_i-products

For homomorphic realization, the α_i-hierarchy collapses at $i = 2$. If a class $\mathcal{K}$ satisfies the semi-Letičevskiĭ criterion then we already have $\mathcal{H}\mathcal{S}\mathcal{P}_g(\mathcal{K}) = \mathcal{H}\mathcal{S}\mathcal{P}_{\alpha_1}(\mathcal{K})$, and even $\mathcal{H}\mathcal{S}\mathcal{P}_g(\mathcal{K}) = \mathcal{H}\mathcal{S}\mathcal{P}_{\alpha_0}(\mathcal{K})$ holds if $\mathcal{K}$ does not satisfy the Letičevskiĭ criteria. On classes with the Letičevskiĭ criterion and on those not satisfying the Letičevskiĭ criteria, the ν_i- products behave in the same way, while on classes satisfying the semi-Letičevskiĭ criterion they behave quite differently. In fact they form a proper hierarchy. The exact results are formulated by the next three theorems.

4.1. Theorem [4] A class $\mathcal{K}$ of automata is homomorphically ν_3-complete if and only if it satisfies the Letičevskiĭ criterion.

4.2. Theorem [2] There exists a class $\mathcal{K}$ satisfying the semi-Letičevskiĭ criterion and such that $\mathcal{HSP}_{\nu_i}(\mathcal{K}) \subset \mathcal{HSP}_{\nu_{i+1}}(\mathcal{K}) \subset \mathcal{HSP}_{\alpha_0}(\mathcal{K})$ hold for all $i \geq 1$. Consequently also $\mathcal{ISP}_{\nu_i}(\mathcal{K}) \subset \mathcal{ISP}_{\nu_{i+1}}(\mathcal{K})$, for all $i \geq 1$.

4.3. Theorem [15] If $\mathcal{K}$ does not satisfy the Letičevskiĭ criteria then $\mathcal{HSP}_g(\mathcal{K}) = \mathcal{HSP}_{\nu_1}(\mathcal{K})$.

It should be noted that the existence of a class $\mathcal{K}$ with $\mathcal{ISP}_{\nu_i}(\mathcal{K}) \subset \mathcal{ISP}_{\nu_{i+1}}(\mathcal{K})$, all $i \geq 1$, has already been shown in [16], where also isomorphically ν_i-complete classes have been characterized.

The proof of Theorem 4.1 is a technically very much involved direct construction that we plan to publish elsewhere. A result that improves on both Theorem 4.3 and the second part of Theorem 3.4 will be given below.

The results given thus far completely describe the relations between the α_i-products and the relations between the ν_i-products, both for homomorphic and isomorphic realization. Theorem 4.2 also shows that none of the ν_i-products is "more general" than any of the α_i-products.

4.4. Corollary Given integers $i \geq 0$ and $j \geq 1$, there is a class $\mathcal{K}$ with $\mathcal{HSP}_{\alpha_i}(\mathcal{K}) \not\subseteq \mathcal{HSP}_{\nu_j}(\mathcal{K})$. Consequently, also $\mathcal{ISP}_{\alpha_i}(\mathcal{K}) \not\subseteq \mathcal{ISP}_{\nu_j}(\mathcal{K})$.

The rest of this section is devoted to showing that neither the α_0-product nor the α_1-product is more general than any of the ν_j-products. (Here the restriction is due to the fact that, by Corollary 3.5, we of course have $\mathcal{HSP}_{\nu_j}(\mathcal{K}) \subseteq \mathcal{HSP}_{\alpha_2}(\mathcal{K})$ for all $j \geq 1$ and for any class $\mathcal{K}$.) But first we need some new concepts.

Given an automaton $\mathcal{A} = (A, X, \delta)$, define $\mathcal{A}^+ = (A, S(\mathcal{A}), \delta')$ by $\delta'(a, u^{\mathcal{A}}) = \delta(a, u)$, for all $a \in A$ and $u \in X^+$. Thus $\mathcal{A}^+$ is just the transformation semigroup corresponding to the automaton $\mathcal{A}$. We set $\mathcal{K}^+ = \{\mathcal{A}^+ | \mathcal{A} \in \mathcal{K}\}$ for a class $\mathcal{K}$ of automata.

Take any product notion defined in Section 2 and call it the β-product. Let $\mathcal{P}_\beta$ be the corresponding operator. If $\mathcal{K}$ is a class of automata we denote $\mathcal{P}^+_\beta(\mathcal{K}) = \mathcal{P}_\beta(\mathcal{K}^+) = \mathcal{P}_\beta\mathcal{P}_{1q}(\mathcal{K}^+)$. Any automaton in $\mathcal{P}^+_\beta(\mathcal{K})$ is termed a β^+-product of automata in $\mathcal{K}$. Thus the α_0^+-product is a counterpart of the wreath product of transformation semigroups, see [1, 6]. In a similar way we define $\mathcal{S}^+(\mathcal{K}) = \mathcal{SP}_{1q}(\mathcal{K}^+)$. We have $\mathcal{A} \in \mathcal{HS}^+(\{\mathcal{B}\})$ for two automata $\mathcal{A}$ and $\mathcal{B}$ if and only if the transformation semigroup corresponding to $\mathcal{A}$ divides the transformation semigroup corresponding to $\mathcal{B}$, see [6]. A class $\mathcal{K}$ is homomorphically (isomorphically) $s-\beta^+$-complete if $\mathcal{HS}^+\mathcal{P}^+_\beta(\mathcal{K})$ $(\mathcal{IS}^+\mathcal{P}^+_\beta(\mathcal{K}))$ is the class of all automata.

Homomorphically $s-\alpha_0^+$-complete classes are fully characterized by the Krohn–Rhodes decomposition, see [1, 6]. For isomorphically $s-\alpha_0^+$-complete classes as well as isomorphically and homomorphically $s-\alpha_i^+$- complete classes with $i \geq 1$ we refer to [12, 14]. It turns out that, for $i = 0, 1$, there exists no finite isomorphically $s-\alpha_i^+$-complete class of automata. The same holds for homomorphically $s-\alpha_i^+$-complete classes with $i = 0, 1$. The references [5, 3] provide full descriptions of isomorphically and homomorphically $s-\nu_i^+$-complete classes. For related results see also [8, 13, 21, 22, 24]. Here we only recall a consequence of the characterization of isomorphically $s-\nu_i^+$-complete classes.

4.5. Theorem [5] The automaton Z_2 is isomorphically $s - \nu_1^+$–complete.

Since every transformation in the semigroup of Z_2 is already induced by an input letter, we even have that $IS^+P_{\nu_1}^+(\{Z_2\}) = IS^+P_{\nu_1}(\{Z_2\})$ is the class of all automata. Theorem 4.5 reveals the fact that, as opposed to Theorem 4.2, already the ν_1–product is very powerful in some cases: every transformation semigroup divides the transformation semigroup corresponding to a ν_1–power of Z_2.

We now turn back to comparing the α_i–products and the ν_i–products.

4.6. Proposition There is a class K such that $ISP_{\nu_j}(K) \not\subseteq ISP_{\alpha_i}(K)$ for all $i \geq 0$ and $j \geq 1$.

Indeed, let $K = \{Z_2\}$. Supposing $ISP_{\nu_j}(K) \subseteq ISP_{\alpha_i}(K)$ we also have $P_{\nu_j}(K) \subseteq ISP_{\alpha_i}(K)$, so that $P_{\nu_j}^+(K) \subseteq ISP_{\alpha_i}(K)$ because of the special structure of Z_2. But then

$$IS^+P_{\nu_1}^+(K) \subseteq IS^+P_{\nu_j}^+(K) \subseteq IS^+ISP_{\alpha_i}(K) \subseteq IS^+IS^+P_{\alpha_i}^+(K) = \\ = IS^+P_{\alpha_i}^+(K).$$

By Theorem 4.5 this implies that $K = \{Z_2\}$ is a finite isomorphically $s - \alpha_i^+$–complete class, contradiction.

The last statement of this section is proved in a similar way.

4.7. Proposition There is a class K such that we have $HSP_{\nu_j}(K) \not\subseteq HSP_{\alpha_i}(K)$ for $i = 0, 1$ and all $j \geq 1$.

5. $\alpha_0 - \nu_i$–products

An $\alpha_0 - \nu_i$*–product* is an α_0–product that is also a ν_i–product. Thus e.g. an $\alpha_0 - \nu_1$–product is a product $A = A_1 \times \ldots \times A_n(X, \varphi)$ with the additional property that for each $t (1 < t \leq n)$ there is an s $(1 \leq s < t)$ such that $\varphi_t(a_1, \ldots, a_n, x)$ is independent of any state variable a_k with $s \neq k$. If $t = 0$ then φ_t is independent of any state variable. We denote by $P_{\alpha_0 - \nu_i}$ the operator that forms all $\alpha_0 - \nu_i$–products of automata from a given class. We present two results. The first is due to Gécseg and Jürgensen and is taken from [17]. It provides a common extension of Theorem 4.3 and the second part of Theorem 3.4. The second result establishes a hierarchy within the α_0–product.

5.1 Theorem [17] Let K be a class not satisfying the Letičevskiĭ criteria. Then $HSP_g(K) = HSP_{\alpha_0 - \nu_1}(K)$.

5.2 Theorem There is a class K satisfying the semi-Letičevskiĭ criterion and such that $HSP_{\alpha_0 - \nu_i}(K) \subset HSP_{\alpha_0 - \nu_{i+1}}(K) \subset HSP_{\alpha_0}(K)$, for all $i \geq 1$.

We give a complete proof of Theorem 5.2. The construction is a refinement of that used in [2] and, besides providing also a new proof of Theorem 4.2, it gives a simpler class.

Given an integer $n \geq 1$, let $C_n = (C_n, \{x\}, \delta_n)$ with $C_n = \{0, \ldots, n-1\}$ and $\delta_n(i, x) = i + 1 \bmod n$, for all $i \in C_n$. Thus C_n is a counter with length n. Let $\mathcal{E} =$

$(E,\{x,y\},\delta_0)$ be the elevator, i.e., $E=\{0,1\}$, $\delta_0(0,x)=0$ and $\delta_0(0,y)=\delta_0(1,x)=\delta_0(1,y)=1$. We set

$$\mathcal{K}=\{\mathcal{E}\}\cup\{\mathcal{C}_p|p \text{ is a prime}\}$$

and show the existence of an automaton $\mathcal{M}\in \mathcal{HSP}_{\alpha_0-\nu_{i+1}}(\mathcal{K})$ which does not belong to $\mathcal{HSP}_{\nu_i}(\mathcal{K})$, where $i\geq 1$ is any fixed integer.

Let m be the product of the first $i+1$ prime numbers. We set $\mathcal{M}=(M,\{x,y\},\delta')$ with $M=\{0,\ldots,m\}$ and

$$\delta'(i,x)=\begin{cases} i+1 \bmod m & \text{if } i=0,\ldots,m-1\\ m & \text{if } i=m,\end{cases}$$

$$\delta'(i,y)=\begin{cases} i+1 \bmod m & \text{if } i=1.\ldots,m-1\\ m & \text{if } i=0 \text{ or } i=m\end{cases}$$

Claim: $\mathcal{M}\notin \mathcal{HSP}_{\nu_i}(\mathcal{K})$. Assume to the contrary that a ν_i-product $\mathcal{A}'=\mathcal{A}_1\times\ldots\times\mathcal{A}_n(\{x,y\},\varphi)$ of automata from $\mathcal{K}$ contains a subautomaton that can be mapped homomorphically onto $\mathcal{M}$. By the permutability of the ν_i-product we may assume that the counters preceed the elevators, so that $\mathcal{A}_1=\mathcal{C}_{p_1},\ldots,\mathcal{A}_k=\mathcal{C}_{p_k}$ and $\mathcal{A}_{k+1}=\ldots=\mathcal{A}_n=\mathcal{E}$ for some k and prime numbers $p_1,\ldots,p_k$. It is clear that $n>1$ and $1\leq k<n$.

Let $\mathcal{A}=(A,\{x,y\},\delta)$ be a subautomaton of $\mathcal{A}'$ with a minimum number of states that can be mapped homomorphically onto $\mathcal{M}$, and let h denote an onto homomorphism. Set $M_0=\{0,\ldots,m-1\}$, $A_0=h^{-1}(M_0)$ and $A_1=h^{-1}(\{m\})=A-A_0$. The minimality of $\mathcal{A}$ and the special structure of the component automata readily imply the following facts, where pr_t $(1\leq t\leq n)$ denotes the $t-$th projection.

(1) If $a,b\in A_0$ then $pr_t(a)=pr_t(b)$ for all $k+1\leq t\leq n$.

(2) If $a\in A_0$ and $b\in A_1$ then there exists t $(k+1\leq t\leq n)$ with $pr_t(a)=0$ and $pr_t(b)=1$.

Now let $a_0\in A_0$ be any state with $h(a_0)=0$. Then $\delta(a_0,y)=b\in A_1$ because h is a homomorphism. By (2) above, there exists t $(k+1\leq t\leq n)$ with $pr_t(a_0)=0$ and $pr_t(b)=1$, say $t=n$, for easy of notation. Let γ be a neighbourhood function of $\mathcal{A}'$ with $|\gamma(t)|\leq i$, for all t $(1\leq t\leq n)$. Set $r=l.c.m.\{p_j|j\in\gamma(n),\ j\leq k\}$, $a_1=\delta(a_0,x^r)\in A_0$ and $a_2=\delta(a_1,y)$. We derive a contradiction by proving that both $pr_n(a_2)=0$ and $pr_n(a_2)=1$ must hold.

Since m is the product of $i+1$ primes and r is the $l.c.m.$ of at most i primes we obtain $m\nmid r$, so that $a_1\notin h^{-1}(0)$ and $a_2\in A_0$ because h is a homomorphism. Since $a_0,a_2\in A_0$ and $pr_n(a_0)=0$, we have $pr_n(a_2)=0$ by (1). On the other hand, because the first k components are counters and $r=l.c.m.\ \{p_j|j\in\gamma(n),\ j\leq k\}$, it follows that $pr_j(a_1)=pr_j(a_0)$, for all $j\in\gamma(n)$, $j\leq k$. Since $a_0,a_1\in A_0$, from (1) we have $pr_{k+1}(a_1)=pr_{k+1}(a_0),\ldots,pr_n(a_1)=pr_n(a_0)$. Thus $pr_t(a_1)=pr_t(a_0)$ for all $t\in\gamma(n)$ and $\varphi_n(a_1,y)=\varphi_n(a_0,y)$. But we must have $\varphi_n(a_0,y)=y$ because $pr_n(a_0)=0$ and $pr_n(b)=1$. Therefore $\varphi_n(a_1,y)=y$ and $pr_n(a_2)=1$.

Claim: $\mathcal{M}\in \mathcal{HSP}_{\alpha_0-\nu_{i+1}}(\mathcal{K})$. We give a direct construction. Let $\mathcal{A}=\mathcal{C}_{p_1}\times\ldots\times\mathcal{C}_{p_{i+1}}\times\mathcal{E}(\{x,y\},\varphi)$ with

$$\varphi_t(a_1,\ldots,a_{i+2},z)=\begin{cases} y & \text{if } a_1=\ldots=a_{i+1}=0,\ z=y \text{ and } t=i+2,\\ x & \text{otherwise,}\end{cases}$$

where $p_1, \ldots, p_{i+1}$ are the first $i+1$ primes, $t = 1, \ldots, i+2$, $a_1 \in C_{p_1}, \ldots, a_{i+1} \in C_{p_{i+1}}$ and $a_{i+2} \in E$. We see that each feedback function φ_t, $t = 1, \ldots, i+1$, is independent of any state variable and that φ_{i+2} does not depend on a_{i+2}. Therefore $\mathcal{A}$ is an $\alpha_0 - \nu_{i+1}$ product. Now let $A_0 = \{a \in A | pr_{i+2}(a) = 0\}$, where A denotes the state set of $\mathcal{A}$. If $a = (a_0, \ldots, a_{i+2}) \in A_0$ then let $h(a) = j$ be the (unique) integer $0 \leq j < m$ with $j \equiv a_t \bmod p_t$, for all $t = 1, \ldots, i+1$. If $a \in A - A_0$ then let $h(a) = m$. It is clear that h is a homomorphism of $\mathcal{A}$ onto $\mathcal{M}$, so that $\mathcal{M} \in \mathcal{H}\mathcal{S}\mathcal{P}_{\alpha_0 - \nu_{i+1}}(\mathcal{K})$.

The fact that $\mathcal{H}\mathcal{S}\mathcal{P}_{\alpha_0 - \nu_i}(\mathcal{K}) \subset \mathcal{H}\mathcal{S}\mathcal{P}_{\alpha_0}(\mathcal{K})$ is now obvious.

6. Conclusion

We have seen that, with respect to homomorphic realization, the ν_i-products behave in a way similar to the α_i-products on classes satisfying the Letičevskiĭ criterion or not satisfying the Letičevskiĭ criteria. In particular, a class $\mathcal{K}$ is homomorphically ν_3-complete if and only if it satisfies the Letičevskiĭ criterion. As opposed to the α_i-products, the ν_i-hierarchy is proper on classes with the semi-Letičevskiĭ criterion. This holds also for homomorphic realization.

To give a summary of the rest of our comparison results, take any two of our product notions, the "β-product" and the "γ-product", say. Define $\beta \leq_{\mathcal{H}} \gamma$ if $\mathcal{H}\mathcal{S}\mathcal{P}_\beta(\mathcal{K}) \subseteq \mathcal{H}\mathcal{S}\mathcal{P}_\gamma(\mathcal{K})$ holds for all $\mathcal{K}$. Similarly let $\beta \leq_{\mathcal{I}} \gamma$ if we have $\mathcal{I}\mathcal{S}\mathcal{P}_\beta(\mathcal{K}) \subseteq \mathcal{I}\mathcal{S}\mathcal{P}_\gamma(\mathcal{K})$ for all $\mathcal{K}$. We obtain two poset structures whose exact diagrams are given in the figures below. The bottom is the quasi-direct product in both cases, for it is obvious that $q <_{\mathcal{H}} \alpha_0$ and $q <_{\mathcal{H}} \nu_1$, henceforth also $q <_{\mathcal{I}} \alpha_0$ and $q <_{\mathcal{I}} \nu_1$. (We write $\beta < \gamma$ if $\beta \leq \gamma$ but $\gamma \not\leq \beta$.)

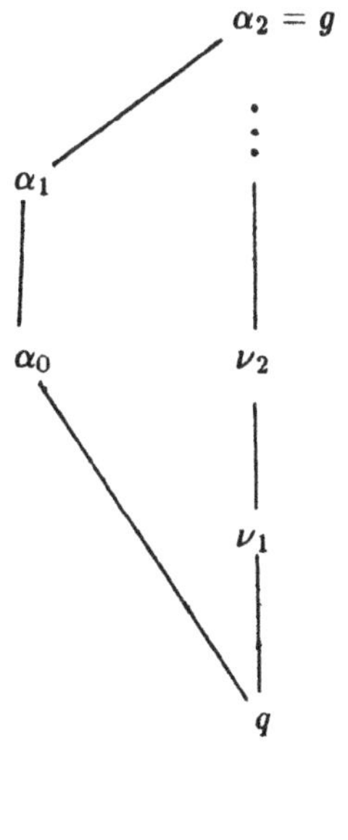

The poset for $\leq_{\mathcal{H}}$

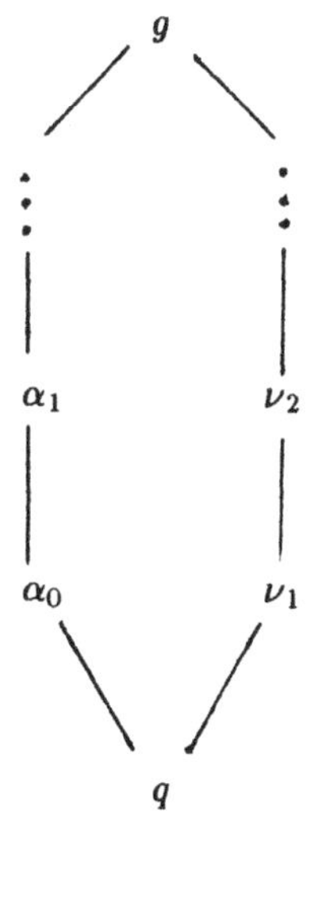

The poset for $\leq_{\mathcal{I}}$

Recently it has been shown by the first two authors that there is a class K satisfying the Letičevskiĭ criterion but which is not homomorphically ν_2−complete.

References

[1] Arbib, M.A. (Ed.), *Machines, Languages and Semigroups,* with a major contribution by K. Krohn and J.L. Rhodes, Academic Press, 1968.

[2] Dömösi, P., Ésik, Z., *On the hierarchy of ν_i−products of automata,* Acta Cybernetica, **8**(1988), 253-257.

[3] Dömösi, P., Ésik, Z., *On homomorphic simulation of automata by ν_1−products,* Papers on Automata and Languages, **IX**(1987), 91-112.

[4] Dömösi, P., Ésik, Z., *Homomorphically complete classes of automata for the ν_3−product,* in preparation.

[5] Dömösi, P., Imreh, B., *On ν_i−products of automata,* Acta Cybernetica, **6**((1983), 149-162.

[6] Eilenberg, S., *Automata, Languages and Machines,* Vol. B, Academic Press, London, 1976.

[7] Ésik, Z., *Homomorphically complete classes of automata with respect to the α_2−product,* Acta Sci. Math., **48**(1985), 135-141.

[8] Ésik, Z., *Loop products and loop-free products,* Acta Cybernetica, **8**(1987), 45-58.

[9] Ésik, Z., Gécseg, F., *On α_0−products and α_2−products,* Theoret. Comput. Sci., **48**(1986), 1-8.

[10] Ésik, Z., Horváth, Gy., *The α_2−product is homomorphically general,* Papers on Automata Theory, **V**(1983), 49-62.

[11] Gécseg, F., *Composition of automata,* 2nd Colloq. Automata, Languages and Programming, 1974, LNCS, **14**(1974), 351-363.

[12] Gécseg, F., *On products of abstract automata,* Acta Sci. Math., **38**(1976), 21-43.

[13] Gécseg, F., *On ν_1−products of commutative automata,* Acta Cybernetica, **7**(1984), 55-59.

[14] Gécseg, F., *Products of automata,* Springer Verlag, 1986.

[15] Gécseg, F., Imreh, B., *On metric equivalence of ν_i−products,* Acta Cybernetica, **8**(1987), 129-134.

[16] Gécseg, F., Imreh, B., *A comparison of α_i−products and ν_i−products,* Foundations of Control Engineering, **12**(1987), 1-9.

[17] Gécseg, F., Jürgensen, H., *On $\alpha_0 - \nu_1$−products of automata,* Univ. of Western Ontario, Report No **162**(1987), 1-14.

[18] Gluškov, V.M., *Abstract theory of automata* (Russian), Uspehi Matematiceskih Nauk, **16:5**(101), (1961), 3-62.

[19] Hartmanis, J., Stearns, R.E., *Algebraic structure theory of sequential machines,* Prentice-Hall, 1966.

[20] Imreh, B., *On α_i−products of automata,* Acta Cybernetica, **3**(1978), 301-307.

[21] Imreh, B., *A note on the generalized ν_1−product,* Acta Cybernetica, **8**(1988), 247-252.

[22] Kim, K.H., Roush, F.W., *Generating all linear transformations,* Linear Algebra and its Applications, **37**(1981), 97-101.

[23] Letičevskiĭ, A.A., *Conditions of completeness for finite automata* (Russian), Žurnal Vič. Mat. i Mat. Fiz., **1**(1961), 702-710.

[24] Tchuente, M., *Computation on binary tree-networks,* Discrete Appl. Math., **14**(1986), 295-310.

[25] Tchuente, M., *Computation on finite networks on automata,* in: Automata Networks, C. Choffrut (Ed.), LNCS, **316**, 1986, 53-67.

[26] Zeiger, H.P., *Cascade synthesis of finite state machines,* Inform. and Control, **10**(1967), 419-433.

On the communication complexity of planarity

P.ĎURIŠ AND P.PUDLÁK

ABSTRACT. *We prove $\Theta(n \log n)$ bound for the deterministic communication complexity of the graph property planarity.*

INTRODUCTION

The communication complexity measures the number of bits which must be exchanged by two computers computing cooperatively a given boolean function.We consider here the model where the input bits are split equally between the two computes and where the splitting is optimal for a given boolean function (i.e. we minimaze over all partitions into two equal parts). For this kind of a measure, the complexity of many graph properties are still open problems. For instance, Hajnal, Maass and Turán [1] ask to determine the communication complexity of planarity, bipartite matching and Hamiltonian circuits. We shall show that the communication complexity of planarity is $\Theta(n \log n)$ where n is the number of vertices. We also show such bounds for one-way communication complexity of the graph property "being a forest". We assume that the reader is familiar with the definition of communication complexity (as presented e.g. in [2]).

RESULTS

Let V be the set of vertices on wchich the graphs will be recognized by the two computers. The pairs of vertices which are accessible to the first (second) computer will be called *red* (*blue*) edges. The number of red edges is equal to the number of blue ones, possibly +1. Let $n = |V|$. Our lower bounds will be based on the following lemma.

LEMMA 1. *For some $k = \Omega(n)$ there exist vertices $p_1, \ldots, p_k, q_1, \ldots, q_k$ in V and sets of vertices $R_1, \ldots R_k$ with the following properties:*

(1) $p_1, \ldots, p_k, q_1, \ldots, q_k$ are mutually distinct and do not belong to any of R_i for $i = 1, \ldots, k$;

(2) $|R_i| = k$ for $i = 1, \ldots, k$;

(3) for every $i = 1, \ldots, k$ and every $r \in R_i, (p_i, r)$ is red edge and (q_i, r) is a blue edge.

We have proved this lemma independently, however the proof of this lemma is implicit also in [2]. For completeness we give a proof of this lemma. We shall follow the proof of [2] since it is simpler than ours.

PROOF: Call a vertex *blue* (resp. *red*) if at least $0.9n$ of its incident vertices are blue (resp. red). If a vertex is neither blue nor red then it is *mixed*. First we shall show that there are at least $0.1n$ of mixed vertices. An easy calculation shows that there are at most $0.6n$ blue vertices and at most $0.6n$ red ones. Now assume that there are less than $0.1n$ mixed vertices. Then there must be at least $0.3n$ blue vertices and at least as much red ones. But each red vertex must be connected by a red edge with more than one half of the blue vertices. Hence more than one half of the edges between the set of red vertices and the set of blue vertices is red. A symmetric argument gives the opposite, thus we have shown that there are at least $0.1n$ mixed verices.

As there are $\Omega(n)$ mixed vertices, there are $\Omega(n^3)$ triples (p,r,q), where (p,r) is red and (r,q) is blue. Hence there exist $\Omega(n^2)$ of pairs (p,q) which form such a triple with $\Omega(n)$ vertices r. Now take a maximal set of such pairs (p,q), where the pairs are not incident. It remains to show that the size of this set is $\Omega(n)$. Suppose the size is only $o(n)$. Since the set is maximal, every (p,q) above must be incident with an edge in this set. But then we would have only $n.o(n)$ such pairs.

LEMMA 2. *There exist distinct vertices $r_1,\ldots,r_k$ and a set Π of permutations of indices $1,\ldots,k$ of cardinality $2^{\Omega(n.\log n)}$ such that, for every permutation $\pi \in \Pi$ and every $i=1,\ldots,k,(p_i,r_{\pi(i)})$ is a red edge and $(q_i,r_{\pi(i)})$ is a blue edge.*

PROOF: Let $R_i, i=1,\ldots,k$ be from the lemma above. Then there are at least $k!$ sequences $r_1,\ldots,r_k$ of distinct elements such that $r_i \in R_i$ for $i=1,\ldots,k$. By the assumption $k=\Omega(n)$, hence $k!=2^{\Omega(n.\log n)}$. The number of different sets $r_1,\ldots,r_k$ is at most 2^n. As $2^{\Omega(n.\log n)}/2^n = 2^{\Omega(n.\log n)}$ the lemma follows.

THEOREM 3. *Deterministic 2-way communication complexity of the graph property planarity is $\Theta(n.\log n)$, where n is the number of vertices.*

PROOF: The upper bound follows easily from the fact that the number of edges in a planar graph is $O(n)$. For the lower bound take vertices $p_1,q_1,r_1,\ldots,p_k,q_k,r_k$ form Lemma 2. We can assume that there are two more vertices $a,b \in V$. Consider the inputs of the form shown in Figure 1, where $\pi \in \Pi$ and where the vertices which are not shown are isolated. All these graphs are accepted. Suppose the number of the communication protocols is smaller than the number of these graphs. Then we have two such different graphs with the same protocol. Combining red edges of one with blue edges of the second one we would obtain an accepted input. But it is easily seen that such a graph contains a subgraph of the form shown on Figure 2 which is not planar.

THEOREM 4. *Deterministic one-way communication complexity of the graph property "being a forest" is $\Theta(n.\log n)$.*

PROOF: The upper bound is again easy, since a forest has at most $n-1$ edges. For the lower bound we shall use Lemma 1 again. Let $\mathcal{M}$ be the set of all matchings $M=\{(x_1,y_1),\ldots,(x_{k/2},y_{k/2})\}$, where $x_i,y_i \in R_i$ for $i=1,\ldots,k/2$. Clearly $\mathcal{M}$ contains at least

$$\frac{k!}{2^{k/2}(k/2)!} = 2^{\Omega(n.\log n)}$$

such matchings, since there are at least $k!$ ordered sequences $(x_1,y_1,\ldots,x_{k/2},y_{k/2})$ which determine such matchings. Supose the *red* computer starts the computation. For each M let G_M be the forest with edges $(x_i,p_i),(p_i,y_i), i=1,\ldots,k$. Suppose that the red computer sends the same message for two different matchings $M,M' \in \mathcal{M}$. Take some $(x_j,y_j) \in M \setminus M'$ and let $\widetilde{G}_M$ ($\widetilde{G}_{M'}$ resp.) be G_M ($G_{M'}$ resp.) augmented with two blue edges (x_j,q_j) and (y_j,q_j). The computations on $\widetilde{G}_M$ and $\widetilde{G}_{M'}$ must be the same since the red computer sends the same message and the blue computer sees the same edges. This is a contradiction, since $\widetilde{G}_M$ is not a forest but $\widetilde{G}_{M'}$ is a forest. Thus there must be at least as many different messages as there are elements in $\mathcal{M}$. Hence there must be messages of length at least $\Omega(n.\log n)$.

REFERENCES

[1] A.Hajnal, W.Maass, G.Turán, *On the communication complexity of graph properties,* Proc. 20-th STOC (1988), pp.186-191.

[2] C.M.Papadimitriu, M.Sipser, *Communication complexity,* JCSS 28 (1984), pp. 260-269.

[3] A.Yao, *Some complexity questions related to distributed computing,* Proc. 11-th STOC (1979), pp. 209-213.

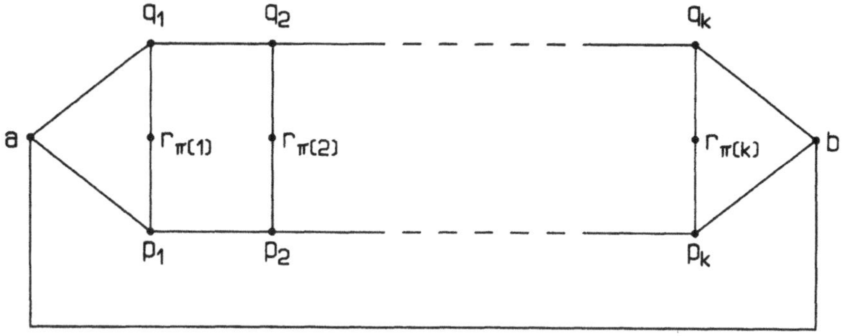

Fig. 1

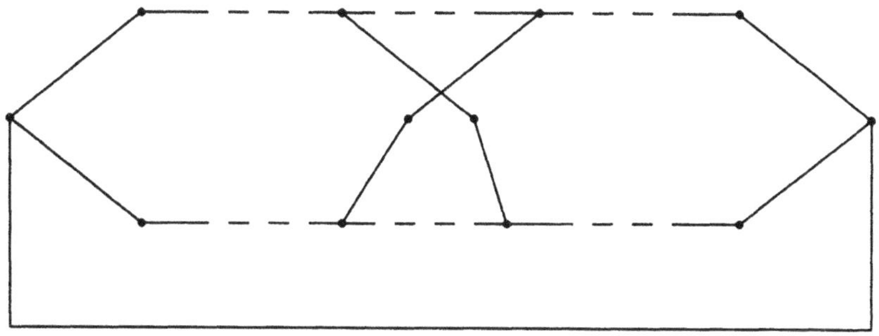

Fig. 2

CONTEXT-FREE NCE GRAPH GRAMMARS

Joost Engelfriet

Dept. of Computer Science, Leiden University,
P.O.Box 9512, 2300 RA Leiden, The Netherlands

The aim of this paper is to show some close relationships between context-free graph grammars and concepts from string language theory and tree language theory. There are many kinds of context-free graph grammars (see, e.g., [EhrNagRozRos]). Some are node rewriting and others are edge rewriting. In both cases a production of the grammar is of the form $X \to (D,B)$. Application of such a production to a labeled graph H consists of removing a node (or edge) labeled X from H, replacing it by the graph D, and connecting D to the remainder of H according to the embedding procedure B. Since these grammars are context-free in the sense that one node (or edge) is replaced, their derivations can be modeled by derivation trees, as in the case of strings. However, the grammar may still be context-sensitive in the sense that the (edges of the) graph generated according to the derivation tree may depend on the order in which the productions are applied. A graph grammar that does not suffer from this context-sensitivity is said to be confluent (or to have the finite Church-Rosser property), see [Cou] for a uniform treatment. Thus, for a confluent graph grammar G, each derivation tree of G yields a unique graph in the graph language generated by G. Due to this close relationship to (derivation) trees, the generated graph languages can be described in terms of regular tree languages and regular string languages. We will show this for the particular case of the (node rewriting) edNCE graph grammars, studied in [Kau, Bra1/2, EngLeiRoz1/2, Sch, EngLeiWel, EngLei1/2, EngRoz] (and called DNELC grammars in [Bra2]).

Graph grammars

Let Σ be an alphabet of node labels, and Γ an alphabet of edge labels. A graph over Σ and Γ is a tuple $H = (V,E,\phi)$, where V is the finite set of nodes, $\phi: V \to \Sigma$ is the node labeling function, and $E \subseteq \{(x,\gamma,y) \in V\times\Gamma\times V \mid x \neq y\}$ is the set of (labeled) edges. Thus, we consider directed graphs with labeled nodes and edges. There are no loops; there may be multiple edges, but not with the same label. The set of all graphs over Σ and Γ is denoted $GR(\Sigma,\Gamma)$. The components of a given graph H will be indicated by V_H, E_H, and ϕ_H.

The edNCE graph grammars are a special case of Nagl's grammars [Nag]. They belong to the family of NLC-like graph grammars [Roz]. The main advantage of these grammars over other NLC-like grammars is that the edge labels can be changed dynamically. In "edNCE", e stands for "edge- (and node-) labeled", d for "directed", and NCE for "neighbourhood controlled embedding".

1. Definition. An edNCE graph grammar is a system $G = (\Sigma,\Delta,\Gamma,\Omega,P,S)$, where Σ is the alphabet of node labels, $\Delta \subseteq \Sigma$ is the alphabet of terminal node labels (the elements of $\Sigma-\Delta$ are nonterminal node labels), Γ is the alphabet of edge labels, $\Omega \subseteq \Gamma$ is the alphabet of terminal edge labels (the elements of $\Gamma-\Omega$ are nonterminal edge labels), P is the finite set of productions, and $S \in \Sigma-\Delta$ is the initial nonterminal. A production $\pi \in P$ is of the form π: $X \rightarrow (D,B)$ where $X \in \Sigma-\Delta$ is the left-hand side of π (denoted $lhs(\pi)$), $D \in GR(\Sigma,\Gamma)$ is the right-hand side of π (denoted $rhs(\pi)$), and $B \subseteq V_D \times \Gamma \times \Gamma \times \Sigma \times \{in,out\}$ is the embedding relation of π (denoted $emb(\pi)$), satisfying the following condition:
if $(x,\lambda,\mu,a,d) \in B$, $\phi_D(x) \in \Delta$, and $a \in \Delta$, then $\mu \in \Omega$. □

For convenience, by the above condition, we consider "non-blocking" edNCE grammars only (cf. [EngLeiWel]).

The productions of G are applied to elements of $GR(\Sigma,\Gamma)$. For a graph $H \in GR(\Sigma,\Gamma)$, the production π: $X \rightarrow (D,B)$ is applicable to a nonterminal node $v \in V_H$ if $\phi_H(v) = X$. Application of π to v consists of the following steps. First v is removed from H, together with all edges incident with v. Then D (or, more precisely, a fresh isomorphic copy of D) is added to the rest of H. Finally, edges are established between V_D and $V_H-\{v\}$ according to the embedding relation B: for $x \in V_D$ and $y \in V_H-\{v\}$, an edge (x,μ,y) is added (an edge (y,μ,x) is added) if and only if there was an edge (v,λ,y) (an edge (y,λ,v)) in E_H and the tuple $(x,\lambda,\mu,\phi_H(y),out)$ is in B (the tuple $(x,\lambda,\mu,\phi_H(y),in)$ is in B). This results in a graph $H' \in GR(\Sigma,\Gamma)$; notation: $H \Rightarrow_{(v,\pi)} H'$ or just $H \Rightarrow H'$. As usual, the language generated by G is $L(G) = \{H \in GR(\Delta,\Omega) \mid S \Rightarrow^* H\}$, where S denotes a graph consisting of one node, labeled S. As suggested above, in any derivation step we will use "fresh" nodes for the right-hand side D, i.e., nodes that were not used previously in the derivation. A graph $H \in GR(\Sigma,\Gamma)$ such that $S \Rightarrow^* H$ will be called a sentential form of G. The class of all graph languages generated by edNCE grammars will be denoted by N-edNCE (for Non-blocking edNCE).

In other words, a tuple (x,λ,μ,a,d) in emb(π) means that if the rewritten nonterminal node v is connected to an a-labeled node by a λ-labeled edge in direction d, then, after application of the production, x will be connected to the same a-labeled node by a μ-labeled edge, in the same direction d. Thus, x takes over some edges from v, possibly changing their labels.

From now on we assume that for every production π of an edNCE grammar G the nonterminal nodes of rhs(π) are given a fixed, but arbitrary, linear order (cf. [Cou]). This allows us to define <u>left-most derivations</u> in the obvious way, similar to context-free string grammars. By $L_{lm}(G)$ we denote the subset of L(G) consisting of all terminal graphs that can be generated by a left-most derivation, and by L-edNCE we denote the class of all $L_{lm}(G)$ for edNCE grammars G. Unlike for ordinary context-free grammars, $L_{lm}(G)$ may be a proper subset of L(G), as shown in one of the next examples. Thus, it is not even clear whether L-edNCE is a subclass of N-edNCE (but we will show that it is).

<u>2. Examples</u>. To draw a production $X \to (D,B)$ of an edNCE grammar, we add B to D as follows: a tuple (x,λ,μ,a,in) of B is represented by a broken line $a \overset{\lambda/\mu}{- - \!\!>\!\! -} \bullet$ where the dot represents node x of D. For the tuple (x,λ,μ,a,out) the arrow is reversed.

As a first example consider the edNCE grammar $G_1 = (\Sigma,\Delta,\Gamma,\Omega,P,S)$ with $\Sigma = \{S,X,i,a,f\}$, $\Delta = \{i,a,f\}$, $\Gamma = \{\gamma,\lambda,\mu,\nu,\varepsilon\}$, $\Omega = \{\lambda,\mu,\nu,\varepsilon\}$, and P consists of the productions drawn in Fig.1. Thus, production π_3 is $X \to (D,B)$ with $V_D = \{x,y\}$, $E_D = \{(x,\mu,y)\}$, $\Phi_D(x) = a$, $\Phi_D(y) = f$, and $B = \{(x,\gamma,\lambda,a,in),(y,\gamma,\nu,a,out)\}$. $L(G_1)$ consists of all "ladders" of the form given in Fig.2.

As a second example consider the edNCE grammar $G_2 = (\{S,X,a\},\{a\},\{\lambda\},\{\lambda\},P,S)$, where P consists of the four productions π_1,π_3,π_4,π_5 of Fig.3. In Fig.3, an undirected line represents two directed lines (with the same label), one in each direction. $L(G_2)$ is the set of all undirected graphs in $GR(\{a\},\{\lambda\})$. In fact, repeated application of π_1 results in a complete graph with S-labeled nodes only. Any undirected edge between two nodes can then be removed by applying first π_4 to both nodes, and then π_5. Finally π_3 is used to generate a terminal graph. $L_{lm}(G_2)$ is the set of all complete graphs in $GR(\{a\},\{\lambda\})$: after application of production π_4, production π_5 has to be applied immediately (in a left-most derivation).

As a last example consider the edNCE grammar G_3 with productions π_1,π_2,π_3 of Fig.3. $L(G_3)$ is the set of all (undirected) cographs, see [CorLerSte]. □

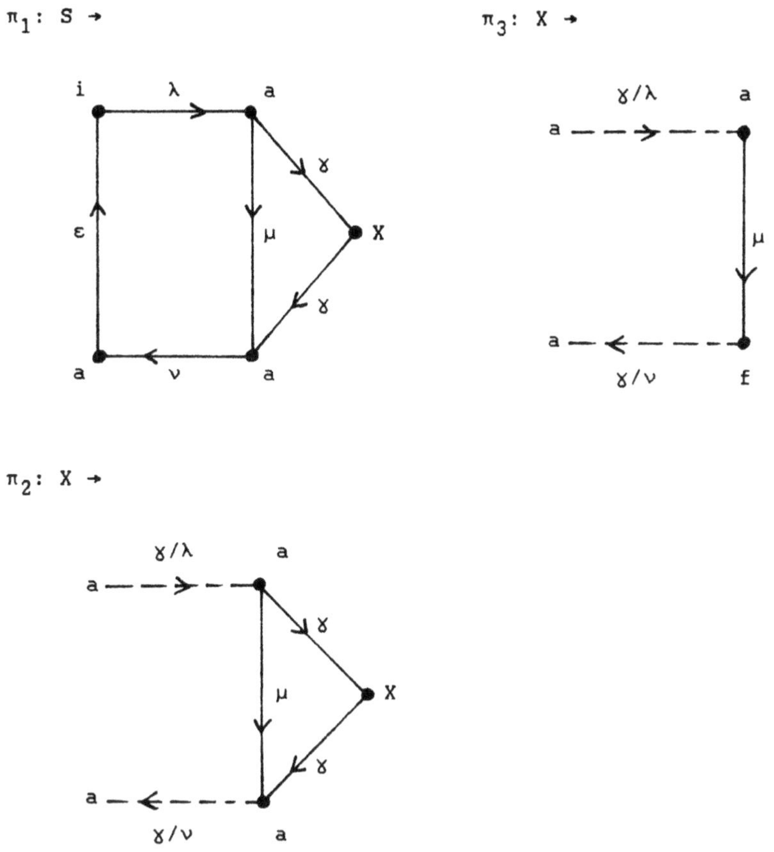

Figure 1. Grammar G_1.

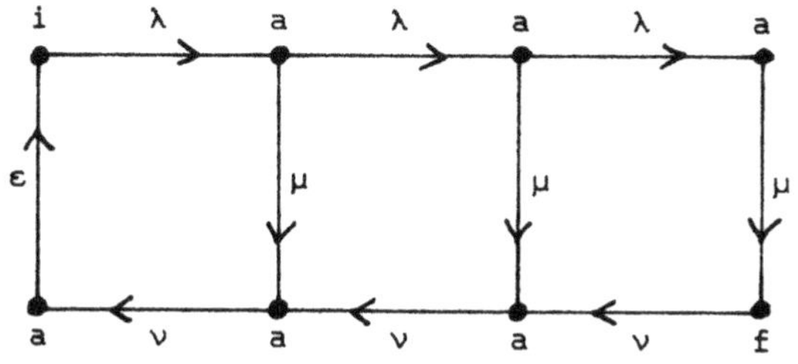

Figure 2. A ladder.

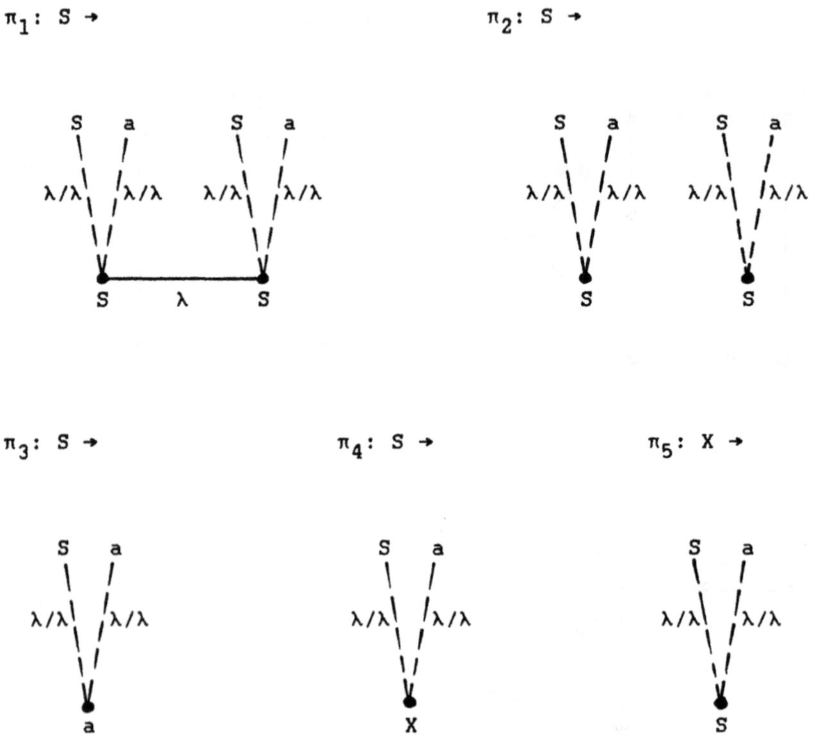

Figure 3. Grammars G_2 and G_3.

Thus, the order of application of productions in a derivation is important for G_2. We now define confluent edNCE grammars for which this order is irrelevant, cf. [Kau, Sch, Bra2, Cou].

3. Definition. Let G be an edNCE grammar. G is confluent (or C-edNCE) if the following holds for every sentential form H of G. Let x_1 and x_2 be distinct nonterminal nodes of H, and let π_1 and π_2 be productions applicable to x_1 and x_2, respectively. If $H \Rightarrow_{(x_1,\pi_1)} H_1 \Rightarrow_{(x_2,\pi_2)} H_{12}$ and $H \Rightarrow_{(x_2,\pi_2)} H_2 \Rightarrow_{(x_1,\pi_1)} H_{21}$, then $H_{12} = H_{21}$ (where we assume that in both cases the same "fresh" copies of the right-hand sides of π_1 and π_2 are used). □

It is easy to see that G_3 is confluent, but G_2 is not. Confluence is decidable ([Kau]).

The class of all graph languages generated by C-edNCE grammars is denoted C-edNCE. It is straightforward to show that for every C-edNCE grammar G, $L_{1m}(G) = L(G)$. Thus, C-edNCE $\subseteq$ L-edNCE. It is shown in [Sch, Bra2] that C-edNCE $\subsetneq$ N-edNCE, with counter-example $L(G_2)$. It is also not difficult to show that C-edNCE contains all context-free NLC languages of [Cou].

Two types of C-edNCE grammars are of special interest (see, e.g., [RozWel, EngLeiWel, EngLei1]). An edNCE grammar G is <u>boundary</u> (or B-edNCE) if there are no edges between nonterminal nodes in any rhs(π) of G. G is <u>linear</u> (or LIN-edNCE) if there is at most one nonterminal node in each rhs(π) of G (e.g., G_1). The corresponding classes of graph languages are also denoted B-edNCE and LIN-edNCE. It is easy to see that LIN-edNCE $\subseteq$ B-edNCE $\subseteq$ C-edNCE. In fact, LIN-edNCE $\subsetneq$ B-edNCE, as shown in [EngLei1], with the set of all binary trees as counter-example. B-edNCE also contains both the B-NLC languages of [RozWel], and, as shown in [EngRoz], the context-free hypergraph languages of [BauCou, HabKre].

Another special case of B-edNCE grammars is also of interest: the regular tree grammars, well known from tree language theory (see, e.g., [GecSte]). As usual, a ranked alphabet is an alphabet Δ together with a mapping rank: $\Delta \to \mathbb{N}$.

<u>4. Definition</u>. Let Δ be a ranked alphabet. An edNCE grammar $G = (\Sigma,\Delta,\Gamma,\Omega,P,S)$ is a <u>regular tree grammar</u> (or REGT grammar) if $\Gamma = \Omega = \{i \mid 1 \leq i \leq n\}$, where n is the maximal integer in the range of rank, and every production $X \to (D,B)$ in P satisfies:

- $V_D = \{v_0,v_1,\dots,v_m\}$ with $\phi_D(v_0) \in \Delta$, $m = \mathrm{rank}(\phi_D(v_0))$, and $\phi_D(v_i) \in \Sigma-\Delta$ for $1 \leq i \leq m$,
- $E_D = \{(v_0,i,v_i) \mid 1 \leq i \leq m\}$, and
- $B = \{(v_0,i,i,a,in) \mid i \in \Gamma,\ a \in \Delta\}$. □

Since Γ and Ω are uniquely determined by Δ, we will specify a REGT grammar as (Σ,Δ,P,S).

Thus, rhs(π) of a REGT grammar consists of a (terminal) father v_0, and m (nonterminal) sons $v_1,\dots,v_m$. Edges lead from the father to each son, and the sons are ordered by numbering their edges from left to right. As usual, such a production π: $X \to (D,B)$ will also be denoted $X \to aX_1\cdots X_m$, where $a = \phi_D(v_0)$ and $X_i = \phi_D(v_i)$ for $1 \leq i \leq m$. A graph language generated by a REGT grammar is called a <u>regular tree language</u>. For a ranked alphabet Δ, T_Δ denotes the set of all trees over Δ, i.e., the regular tree language generated by the REGT grammar with one nonterminal S and productions $S \to aS^m$ for every $a \in \Delta$, where $m = \mathrm{rank}(a)$. Δ is <u>monadic</u> if its elements have rank 1 or 0; in this case the trees in T_Δ

correspond to non-empty strings, as usual.

Since the derivations of an edNCE grammar $G = (\Sigma,\Delta,\Gamma,\Omega,P,S)$ are "context-free", derivation trees can easily be defined for G. The nodes of such a derivation tree are labeled by productions. As for context-free string grammars, the set D(G) of derivation trees of G is a regular tree language. It is generated by the REGT grammar (Σ',Δ',P',S') with $\Delta' = P$ (and, for every $\pi \in P$, rank(π) is the number of nonterminal nodes in rhs(π)), $\Sigma'-\Delta' = \Sigma-\Delta$, $S' = S$, and P' consists of all productions $X \to \pi X_1\cdots X_m$ with $\pi \in P$, $X = \text{lhs}(\pi)$, and $X_1,\ldots,X_m$ are the labels of the nonterminal nodes of rhs(π), according to their fixed order. As an example, for grammar G_3, $P = \{\pi_1,\pi_2,\pi_3\}$ with $\text{rank}(\pi_1) = \text{rank}(\pi_2) = 2$ and $\text{rank}(\pi_3) = 0$. $D(G_3)$ is generated by the regular tree grammar G_4 with productions $S \to \pi_1 SS$, $S \to \pi_2 SS$, and $S \to \pi_3$. Note that, in fact, $L(G_4) = D(G_3) = T_P$.

As usual, a derivation tree corresponds to a set of derivations (containing one left-most derivation). It is easy to show that for a C-edNCE grammar all these derivations yield the same result.

Regular description of graph languages

An alternative, grammar independent, way of describing a graph H is by taking a tree t, defining the nodes of H as a subset of the nodes of t (through their labels), defining the node labels in H by a node relabeling of t, and defining an edge between nodes x and y in H if the string of node labels on the shortest (undirected) path in t between x and y belongs to a certain regular string language. This means that H can be embedded in t "in a regular way". Defining such a graph for every tree in a given regular tree language, one obtains a regular description of a graph language. This idea was introduced in [Wel], for linear graph grammars, and worked out in [EngLeiWel] for LIN-edNCE and B-edNCE grammars (in the undirected case).

5. Definition. Let Σ be a ranked alphabet, let $t \in T_\Sigma$, and let $x,y \in V_t$. Then $\text{path}_t(x,y)$ is a string in $\Sigma^*\bar{\Sigma}\Sigma^*$, with $\bar{\Sigma} = \{\bar{\sigma} \mid \sigma \in \Sigma\}$, defined as follows. Let $z \in V_t$ be the least common ancestor of x and y in t. Let $u_1,\ldots,u_m$ ($m \geq 1$) and $v_1,\ldots,v_n$ ($n \geq 1$) be the paths in t from z to x and from z to y, respectively (thus, $z = u_1 = v_1$, $x = u_m$, and $y = v_n$). Then $\text{path}_t(x,y) = \phi(u_m)\cdots\phi(u_2)\bar{\sigma}\phi(v_2)\cdots\phi(v_n)$, where $\sigma = \phi(z)$ and $\phi = \phi_t$. □

Intuitively, $\text{path}_t(x,y)$ is the string of node labels on the shortest undirected path from x to y in t, with a bar indicating the change of direction of the edges on the path.

6. Definition. A regular tree embedding is a tuple $R = (\Sigma,\Delta,\Omega,T,W,h)$, where Σ is a ranked alphabet, Δ and Ω are alphabets (of node and edge labels, respectively), T is a regular tree language, $T \subseteq T_\Sigma$, W is a mapping from Ω to the class of regular string languages over $\Sigma\cup\bar{\Sigma}$, such that, for every $\lambda \in \Omega$, $W(\lambda) \subseteq \Sigma^*\bar{\Sigma}\Sigma^*$, and h is a partial function from Σ to Δ. The graph language defined by R, denoted L(R), is
$L(R) = \{gr(t) \mid t \in T\}$, where gr(t) is the graph $H \in GR(\Delta,\Omega)$ with
$V_H = \{x \in V_t \mid \phi_t(x) \text{ is in the domain of } h\}$,
$E_H = \{(x,\lambda,y) \mid x,y \in V_H,\ x \neq y,\ \lambda \in \Omega,\ path_t(x,y) \in W(\lambda)\}$, and
for every $x \in V_H$, $\phi_H(x) = h(\phi_t(x))$. □

Note that $L(R) \subseteq GR(\Delta,\Omega)$. Note that h is used both to determine which nodes of t belong to the graph gr(t) and to relabel the remaining nodes. The class of all graph languages that can be defined by a regular tree embedding is denoted RTE.

7. Examples. The language $L(G_1)$ of "ladders" can be defined by the regular tree embedding $R_1 = (\Sigma,\Delta,\Omega,T,W,h)$ with $\Sigma = \{i,a,b,f\}$, rank(i) = rank(a) = rank(b) = 1, rank(f) = 0 (thus Σ is monadic), $\Delta = \{i,a,f\}$, h(b) = a and h(x) = x for $x \in \Delta$, $T = ibab(ab)^*af$, $\Omega = \{\lambda,\mu,\nu,\varepsilon\}$, and $W(\lambda) = \{\bar{i}ba,\bar{a}ba\}$, $W(\mu) = \{\bar{a}b,\bar{a}f\}$, $W(\nu) = \{fa\bar{b},ba\bar{b}\}$, and $W(\varepsilon) = \{b\bar{i}\}$. See Fig.4 for a pictorial explanation of R_1; relabeling b by a, the graph of Fig.2 is obtained; the monadic tree ibababaf is not drawn, but suggested (horizontally).

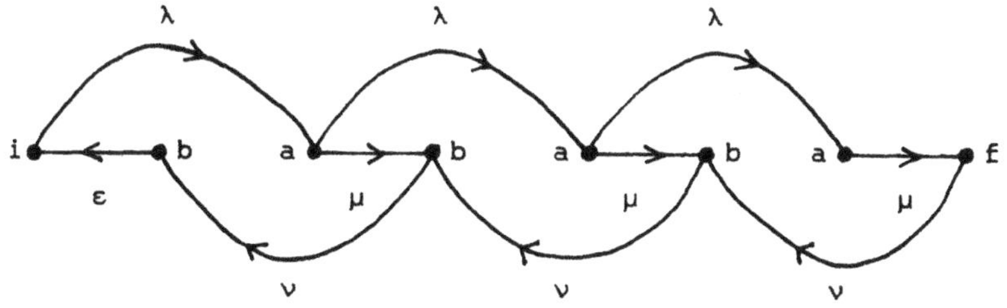

Figure 4. Regular tree embedding R_1.

As another example, the language $L(G_3)$ of cographs is defined by $R_3 = (P,\Delta,\Omega,T,W,h)$ with $P = \{\pi_1,\pi_2,\pi_3\}$, $\Delta = \{a\}$, $h(\pi_3) = a$ and h is undefined for π_1 and π_2, $\Omega = \{\lambda\}$, $T = L(G_4) = D(G_3)$, and $W(\lambda) = \pi_3\{\pi_1,\pi_2\}^*\bar{\pi}_1\{\pi_1,\pi_2\}^*\pi_3$. R_3 expresses the representation of cographs by cotrees, see [CorLerSte]. □

We now show that regular tree embeddings can be used as a grammar independent description method for confluent edNCE grammars, i.e., RTE = C-edNCE. Moreover, the proof will show that L-edNCE = C-edNCE: restricting an arbitrary edNCE grammar to left-most derivations gives a graph language that can also be generated by a confluent edNCE grammar (in contrast to the case of NLC grammars, see [JanKreRozEhr]). The latter result will be proved in a different way in [CouEngRoz]. A full proof of the next theorem is given in [Oos].

<u>8. Theorem</u>. RTE = C-edNCE = L-edNCE.

<u>Proof</u>. RTE ⊆ C-edNCE. Let $R = (\Sigma,\Delta,\Omega,T,W,h)$ be a regular tree embedding, let $G = (\Theta,\Sigma,P,S)$ be a REGT grammar generating T, and let, for every $\lambda \in \Omega$, A_λ be a finite automaton recognizing $W(\lambda)$. Let Q be the union of the state sets of all A_λ (assumed to be disjoint), let δ_λ: $Q\times(\Sigma\cup\bar{\Sigma}) \rightarrow Q$ be the transition function of A_λ, let $q_{in,\lambda}$ be the initial state of A_λ, and let $Q_{fin,\lambda}$ be the set of final states of A_λ. We now combine G and all A_λ into a C-edNCE grammar $G' = (\Sigma',\Delta,\Gamma,\Omega,P',S')$ generating L(R). First, G' has the same nonterminals as G: $\Sigma'-\Delta = \Theta-\Sigma$ and $S' = S$. Second, $\Gamma-\Omega = Q\times Q\times\Omega$. G' has one production π' for every production π of G, with the same nonterminal nodes. Thus, the derivations of G' simulate those of G. During such a simulation, an edge labeled $\langle q_1,q_2,\lambda\rangle$ from nonterminal node u to nonterminal node v (in a sentential form of G') means that $\delta_\lambda(q_1,\mathrm{path}_t(x,y)) = q_2$, where t is the tree generated sofar by G and x and y are the (terminal) fathers of u and v, respectively, in t. Similarly for edges between a nonterminal and a terminal node (where the terminal node itself should be considered, rather than its father). In this way, the A_λ are simulated in the edge labels. Formally, let $\pi = X \rightarrow \sigma X_1\cdots X_m$ be a production of G, and first assume that $h(\sigma)$ is defined. Then $\pi' = X \rightarrow (D,B)$, where

- $V_D = \{v_0,v_1,\ldots,v_m\}$ with $\phi_D(v_0) = h(\sigma)$ and $\phi_D(v_i) = X_i$ for $1 \leq i \leq m$,
- $E_D = \{(v_i,\langle q_1,q_2,\lambda\rangle,v_j) \mid i \neq j,\ \delta_\lambda(q_1,\bar{\sigma}) = q_2$, and if $i = 0$ then $q_1 = q_{in,\lambda}$, and if $j = 0$ then $q_2 \in Q_{fin,\lambda}\}$,
- B contains the following tuples (for all $q_1,q_2 \in Q$, $\lambda \in \Omega$, $a \in \Delta$, $Y \in \Sigma'-\Delta$, $\alpha \in \Sigma'$, $1 \leq i \leq m$):

if $\delta_\lambda(q_{in,\lambda},\sigma) = q_1$ then $(v_0,<q_1,q_2,\lambda>,\lambda,a,out)$ and
$(v_0,<q_1,q_2,\lambda>,<q_{in,\lambda},q_2,\lambda>,\gamma,out)$,
if $\delta_\lambda(q_2,\sigma)= q'_2 \in Q_{fin,\lambda}$ then $(v_0,<q_1,q_2,\lambda>,\lambda,a,in>)$ and
$(v_0,<q_1,q_2,\lambda>,<q_1,q'_2,\lambda>,\gamma,in)$,
if $\delta_\lambda(q'_1,\sigma) = q_1$ then $(v_i,<q_1,q_2,\lambda>,<q'_1,q_2,\lambda>,\alpha,out>$, and
if $\delta_\lambda(q_2,\sigma) = q'_2$ then $(v_i,<q_1,q_2,\lambda>,<q_1,q'_2,\lambda>,\alpha,in>$.
In case $h(\sigma)$ is undefined, just drop v_0 and everything related to v_0.
This ends the description of G'. G' is confluent because it simulates the confluent grammar G, and because the edges in G' depend only on the tree generated sofar by G.

L-edNCE ⊆ RTE. Let $G = (\Sigma,\Delta,\Gamma,\Omega,P,S)$ be an edNCE grammar. Through the use of standard constructions we may assume that each rhs(π) contains at most one terminal node, and that all nonterminal nodes of rhs(π) have distinct labels. We have to construct a regular tree embedding $R = (\Sigma',\Delta,\Omega,T,W,h)$ that defines $L_{lm}(G)$. For T we take D(G), the regular set of derivation trees of G. Thus, $\Sigma' = P$. For $\pi \in P$, $h(\pi)$ is defined iff rhs(π) contains a terminal node v, and if so, $h(\pi) = \phi_{rhs(\pi)}(v)$. It remains to show that the edges in the graph that is left-most generated according to a derivation tree t, are determined by regular languages $W(\lambda)$ over $P\cup\bar{P}$ (cf. the example of R_3). In fact, intuitively, it should be clear that whether or not there is a λ-labeled edge from node x to node y, depends on $path_t(x,y)$ only (note in particular that all nonterminal nodes in a rhs(π) have distinct labels). Moreover, this can be decided by a finite automaton that simulates G along $path_t(x,y)$. To see this, we describe W(λ) more formally.

Let $\pi_1\cdots\pi_n$, $n \geq 1$, be a sequence of productions in P, such that, for every $1 \leq i \leq n-1$, there is a (unique) node y_i in $rhs(\pi_i)$ with label $lhs(\pi_{i+1})$, and such that $rhs(\pi_n)$ has a (unique) terminal node y_n. Then we define the set $embed(\pi_1\cdots\pi_n) = \{(\lambda,\mu,a,d) \mid$ there exist $\gamma_0,\gamma_1,\ldots,\gamma_n \in \Gamma$ such that $\gamma_0 = \lambda$, $\gamma_n = \mu$, and $(y_i,\gamma_{i-1},\gamma_i,a,d) \in emb(\pi_i)$ for $1 \leq i \leq n\}$. We also define $term(\pi_1\cdots\pi_n)$ to be the label of y_n, and $lhs(\pi_1\cdots\pi_n) = lhs(\pi_1)$.

Now $W(\lambda) = W_{left}(\lambda) \cup W_{right}(\lambda)$, where $W_{left}(\lambda)$ determines the edges that run from left to right in the tree, and $W_{right}(\lambda)$ those that run from right to left. Using rev(w) to stand for the reverse of w, $W_{left}(\lambda) = \{rev(w_1)\bar{\pi}w_2 \mid w_1,w_2 \in P^*$, $\pi \in P$, there exist nonterminal nodes x_1 and x_2 in rhs(π) such that x_1 is to the left of x_2 (in the fixed order), $\phi(x_i) = lhs(w_i)$ for $i = 1,2$, and there exist $\mu,\nu \in \Gamma$ such that $(x_1,\mu,x_2) \in E_{rhs(\pi)}$, $(\mu,\nu,\phi(x_2),out) \in embed(w_1)$, and $(\nu,\lambda,term(w_1),in) \in embed(w_2)\}$. $W_{right}(\lambda)$ contains all $rev(w_2)\bar{\pi}w_1$ such that $(x_2,\mu,x_1) \in E_{rhs(\pi)}$, $(\mu,\nu,\phi(x_2),in) \in embed(w_1)$, and $(\nu,\lambda,term(w_1),out) \in embed(w_2)$. From this description of W(λ) it should not be difficult to see that it is regular. □

RTE is closed under the operation of taking the (edge) complement of all graphs (just take the complement of each $W(\lambda)$ with respect to $\Sigma^*\Sigma\Sigma^*$, see [Wel]). Since B-edNCE is not closed under this operation ([EngLeiWel]), Theorem 8 shows that B-edNCE $\subsetneq$ C-edNCE (a particular counter-example is the set of all edge complements of undirected binary trees). Thus, LIN-edNCE $\subsetneq$ B-edNCE $\subsetneq$ C-edNCE $\subsetneq$ N-edNCE.

Let RegPath denote the class of all graph languages defined by regular tree embeddings $R = (\Sigma,\Delta,\Omega,T,W,h)$ such that, for every $\lambda \in \Omega$, $W(\lambda) \subseteq \Sigma\Sigma^* \cup \Sigma^*\Sigma$. This means that edges of the graph can only be established along directed paths of the tree. RegString is the subclass where Σ is a monadic ranked alphabet. As shown in [EngLeiWel], similar to the proof of Theorem 8, RegPath = B-edNCE and RegString = LIN-edNCE.

String languages defined by graph languages

It is well known that an edge labeled graph H can be used to define a regular string language, consisting of the strings of edge labels along all directed paths in H, from certain initial nodes of H to certain final nodes of H. To define nonregular languages one might use a set of graphs rather than just one graph. Clearly, allowing arbitrary sets of graphs would give arbitrary string languages. Thus, it would be more natural to use only graph languages that can be generated by certain graph grammars. We will show that, using LIN-edNCE graph languages in this way, one obtains all one-way checking stack languages. Using all C-edNCE graph languages the output languages of tree-walking transducers are obtained.

<u>9. Definition</u>. Let $H \in GR(\Delta,\Omega)$, and let $i,f \in \Delta$. Then $path_{i,f}(H) = \{\lambda_1\cdots\lambda_n \in \Omega^* \mid$ there is a directed path $v_0,v_1,\ldots,v_n$, $n \geq 1$, in H such that $\Phi_H(v_0) = i$, $\Phi_H(v_n) = f$, and $(v_{j-1},\lambda_j,v_j) \in E_H$ for $1 \leq j \leq n\}$. For a graph language $L \subseteq GR(\Delta,\Omega)$, $path_{i,f}(L) = \cup\{path_{i,f}(H) \mid H \in L\}$. For a class of graph languages K, $Path(K) = \{path_{i,f}(L) \mid L \in K$, i and f are node labels$\}$. □

As an example, for the language $L(G_1)$ of "ladders", $path_{i,f}(L(G_1)) = \{\lambda^{n(1)}\mu\nu^{n(1)}\varepsilon\lambda^{n(2)}\mu\nu^{n(2)}\varepsilon \ldots \lambda^{n(k)}\mu\nu^{n(k)}\varepsilon\lambda^m\mu \mid k \geq 0,\ m \geq 2,\ 1 \leq n(j) \leq m \text{ for } 1 \leq j \leq k\}$. Clearly this language is not regular (not even context-free), but it is a one-way checking stack language.

A <u>tree-walking transducer</u> (abbreviated twt) is a nondeterministic automaton with a finite control, an input tree, and an output tape. The input trees are taken from a given regular tree language. At any moment of time the automaton is at a certain node of the input tree. Depending on the state of its finite control and the label of the node, it changes state, outputs a string to the output tape, and either moves to the father or to a specific son of the node. The automaton starts in its initial state at the root of the input tree, and halts whenever it reaches a final state. In this way it translates the input tree into an output string. The output language of the automaton is the set of all output strings obtained in this way. OUT(TWT) denotes the class of all output languages of twt's. In case all input trees are monadic, we may view the twt as a two-way gsm. Let OUT(2GSM) denote the class of all output languages of 2gsm's. It is well known that OUT(2GSM) equals the class of one-way checking stack languages. For more details see, e.g., [EngRozSlu] (where the twt is called ct-transducer, i.e., checking tree transducer).

10. Theorem. Path(C-edNCE) = OUT(TWT) and Path(LIN-edNCE) = OUT(2GSM).

Proof. Path(RTE) $\subseteq$ OUT(TWT). Let $R = (\Sigma,\Delta,\Omega,T,W,h)$ be a regular tree embedding, and let $i,f \in \Delta$. It is not difficult to construct a twt M with input language T and with output language $path_{i,f}(L(R))$. For a given input tree t, M first nondeterministically walks to a node x of t such that $h(\phi_t(x)) = i$. Then, repeatedly, M chooses a symbol $\lambda \in \Omega$, outputs λ, and nondeterministically walks to another node y for which $h(\phi_t(y))$ is defined, walking along the shortest path from x to y, and using its finite control to check that $path_t(x,y)$ is in $W(\lambda)$. Finally, M halts and checks that $h(\phi_t(x)) = f$ for the current node x. This also shows that Path(RegString) $\subseteq$ OUT(2GSM).

OUT(TWT) $\subseteq$ Path(RTE). Let M be a twt with input language $T \subseteq T_\Sigma$ and output alphabet Ω. We may clearly assume that there is a mapping num: $\Sigma \to \mathbb{N}$ such that, for every node x of a tree t in T, $num(\phi_t(x))$ is the label of the incoming edge of x. We may also assume that M outputs at most one symbol of Ω at each of its moves. Since it can be shown (by a nontrivial proof) that Path(RTE) is closed under erasing, we may even assume that M outputs precisely one symbol at each move. Finally, we may assume that M does not re-enter its initial state, and has a unique final state. Let Q be the set of states of M, and let $i,f \in Q$ be its initial and final state, respectively. We will construct a regular tree embedding $R = (\Sigma',\Delta,\Omega,T',W,h)$, such that $path_{i,f}(L(R))$ is the output language of M. The trees of T' are obtained from those of T as follows: to every node x of tree t in T we add #Q new sons, labeled distinctly with the elements of Q; such a node, labeled q, represents the fact that M is at node x in state q. Thus, $\Sigma' = \Sigma \cup Q$, where each element of Q has rank 0. Obviously

T' is regular. We take $\Delta = \Sigma'$, and h as the identity on Σ'. For every $\lambda \in \Omega$, $W(\lambda)$ is a finite language obtained from the finite control of M as follows. If M, in state q and reading node label σ, may go into state p, output λ, and move to the father, then $W(\lambda)$ contains $q\sigma\bar{\tau}p$ for every $\tau \in \Sigma$. If M moves to the j-th son, then $W(\lambda)$ contains $q\bar{\sigma}\tau p$ for every $\tau \in \Sigma$ with $num(\tau) = j$.

A slightly more complicated construction also works in the linear case: the additional nodes representing the states of M have to be added inbetween the nodes of the monadic input tree, to keep the tree monadic. □

If the graph corresponding to the input tree is acyclic, then it can be shown that the twt M constructed above crosses each edge of the input tree at most $O(d^2)$ times, where d is the maximal degree of the nodes of the graph (cf. [EngLei2], Theorem 15 of [EngLei1], Theorem 1.5.2 of [Sch], and Lemma 2 of [Bra2]). Thus, if $L \in$ C-edNCE is of bounded degree and contains acyclic graphs only, then $path_{i,f}(L)$ is the output language of a "finite crossing" twt, i.e., one that crosses each edge of the input tree a bounded number of times. It is shown in [EngRozSlu] that, in that case, L is the output language of a deterministic twt. Vice versa, it can be shown (cf. [EngHey]) that every output language of a deterministic twt can be obtained in this way. Analogous results hold for deterministic 2gsm's. Note that the language $\{(a^m)^n \mid m,n \geq 2\} \in$ OUT(2GSM) cannot be the output language of any deterministic twt, because it is not Parikh (cf. [EngRozSlu]).

Acknowledgement. I thank George Leih for finding (all?) my mistakes.

References

[BauCou] M.Bauderon, B.Courcelle; Graph expressions and graph rewritings, Math.Syst.Theory 20 (1987), 83-127.

[Bra1] F.J.Brandenburg; On partially ordered graph grammars, in [EhrNagRozRos], 99-111.

[Bra2] F.J.Brandenburg; On polynomial time graph grammars, Proc. STACS 88, LNCS 294, Springer-Verlag, 1988, pp.227-236.

[CorLerSte] D.G.Corneil, H.Lerchs, L.Stewart Burlingham; Complement reducible graphs, Discr.Appl.Math. 3 (1981), 163-174.

[Cou] B.Courcelle; An axiomatic definition of context-free rewriting and its application to NLC graph grammars, Theor.Comp.Sci. 55 (1988), 141-182.

[CouEngRoz] B.Courcelle, J.Engelfriet, G.Rozenberg; Handle hypergraph grammars, in preparation.

[EhrNagRozRos] H.Ehrig, M.Nagl, G.Rozenberg, A.Rosenfeld (eds.); Graph-grammars and their application to Computer Science, LNCS 291, Springer-Verlag, 1987.

[EngHey] J.Engelfriet, L.M.Heyker; The string generating power of context-free hypergraph grammars, Report 89-05, Leiden University, 1989.

[EngLei1] J.Engelfriet, G.Leih; Linear graph grammars: power and complexity, Inform. Comput. 81 (1989), 88-121.

[EngLei2] J.Engelfriet, G.Leih; Complexity of boundary graph languages, Report 88-07, Leiden University, 1988, to appear in RAIRO.

[EngLeiRoz1] J.Engelfriet, G.Leih, G.Rozenberg; Apex graph grammars, in [EhrNagRozRos], 167-185.

[EngLeiRoz2] J.Engelfriet, G.Leih, G.Rozenberg; Nonterminal separation in graph grammars, Report 88-29, Leiden University, 1988.

[EngLeiWel] J.Engelfriet, G.Leih, E.Welzl; Boundary graph grammars with dynamic edge relabeling, Report 87-30, Leiden University, 1987, to appear in JCSS.

[EngRoz] J.Engelfriet, G.Rozenberg; A comparison of boundary graph grammars and context-free hypergraph grammars, Report 88-06, Leiden University, 1988, to appear in Inform. Comput.

[EngRozSlu] J.Engelfriet, G.Rozenberg, G.Slutzki; Tree transducers, L systems, and two-way machines, J. Comput.System Sci. 20 (1980), 150-202.

[GecSte] F.Gécseg, M.Steinby; "Tree automata", Akadémiai Kiadó, Budapest, 1984.

[HabKre] A.Habel, H.-J.Kreowski; May we introduce to you: hyperedge replacement, in [EhrNagRozRos], 15-26.

[JanKreRozEhr] D.Janssens, H.-J.Kreowski, G.Rozenberg, H.Ehrig; Concurrency of node-label-controlled graph transformations, University of Antwerp, Report 82-38, 1982.

[Kau] M.Kaul; Syntaxanalyse von Graphen bei Präzedenz-Graph-Grammatiken, Ph.D.Thesis, University of Osnabrück, 1985.

[Nag] M.Nagl; "Graph-grammatiken", Vieweg, Braunschweig, 1979.

[Oos] V.van Oostrom; Graph grammars and 2nd order logic (in Dutch), M.Sc.Thesis, 1989.

[Roz] G.Rozenberg; An introduction to the NLC way of rewriting graphs, in [EhrNagRozRos], 55-66.

[RozWel] G.Rozenberg, E.Welzl; Boundary NLC graph grammars - basic definitions, normal forms, and complexity, Inform. Contr. 69 (1986), 136-167.

[Sch] R.Schuster; Graphgrammatiken und Grapheinbettungen: Algorithmen und Komplexität, Ph.D.Thesis, Report MIP-8711, University of Passau, 1987.

[Wel] E.Welzl; Boundary NLC and partition controlled graph grammars, in [EhrNagRozRos], 593-609.

DYNAMIC DATA STRUCTURES WITH FINITE POPULATION : a combinatorial analysis

J.Françon
Département informatique
Université Louis Pasteur
7, rue René Descartes
67000 - Strasbourg FRANCE

B.Randrianarimanana , R.Schott
C.R.I.N.
Université Nancy I
54506 - Vandoeuvre - lès - Nancy
FRANCE

Abstract.

This paper analyzes the average behaviour of algorithms that operate on dynamically varying data structures subject to insertions I, deletions D, positive (resp. negative) queries Q^+ (resp.Q^-) under the following assumptions :

i) the universe of keys is finite: **U** $[N] = \{1, 2, 3, ..., N\}$

ii) if the size of the data structure is k ($k \leq N$), then the number of possibilities for the operations D and Q^+ is k, whereas the number of possibilities for the i-th insertion or negative query is equal to N-i+1 for $i \leq N$.

Integrated costs for these dynamic structures are defined as averages of costs taken over the set of all their possible histories (i.e. evolutions considered up to order isomorphism) of length n. We show that the costs can be explicitely calculated for the data structures serving as implementations of linear lists, priority queues and dictionaries. Letting $N \rightarrow \infty$ we recover the results proved in [9], [10] for an infinite universe of keys (this is not obvious) and we prove also that Knuth's model can be defined as limit model of the model considered here.The method uses continued fractions and orthogonal polynomials techniques like in [10] .

1.Introduction

The difficulty of analyzing dynamic data structures even if the universe of keys is finite has been explained by A.Jonassen and D.E.Knuth in [11] where random insertions and deletions are performed on trees whose size never exceeds three.

In this paper we continue the analysis of sequences of operations on dynamic data structures.The calculation of such sequence average cost can be done under different assumptions about the choice of keys manipulated in the data structure . This work differes from the model used in previous analysis ([1] - [10], [15],[16]) by the choice of the basic assumptions, but the algebraic techniques are the same.

Consider a data structure submitted to the following operations:

- **insertion** I of a new key
- **deletion** D of a key in the structure
- **query** Q of a key in the structure (we may distinguish between successful (resp.unsuccessful) queries Q^+ (resp.Q^-)).

Assuming the data structure initially empty, *a sequence of operations* is a sequence of the form : $O_1(x_1)O_2(x_2)...O_n(x_n)$ where, for $1 \leq i \leq n$, $O_i \in \{I,D,Q^+,Q^-\}$ and x_i is a key, the operation $O_i(x_i)$ is performed on the data structure resulting from the operation O_{i-1} (O_1 is performed on the empty structure).

Our basic assumptions are :

i) the keys belong to a finite universe, called the *initial set* of keys: $[N] = \{1,2,...N\}$;

ii) if at step i the set of keys is $F_i \subset [N]$ (with $F_0=\emptyset$), then *the universe of keys at step i* is $[N]-(F_i \cup C_i)$ where C_i is the set of keys used above, called *the garbage collector at step i* ($C_0=\emptyset$) and the operation :

- $I(x_{i+1})$ is defined iff $x_{i+1} \in [N] - (F_i \cup C_i)$ and we have $F_{i+1} = F_i \cup \{x_{i+1}\}$, $C_{i+1} = C_i$;
- $D(x_{i+1})$ is defined iff $x_{i+1} \in F_i$ and we have $F_{i+1} = F_i - \{x_{i+1}\}$, $C_{i+1}=C_i \cup \{x_{i+1}\}$;
- $Q^+(x_{i+1})$ is defined iff $x_{i+1} \in F_i$ and we have $F_{i+1} = F_i$, $C_{i+1} = C_i$;

$*$ $Q^-(x_{i+1})$ is defined iff $x_{i+1} \in [N] - (F_i \cup C_i)$ and we have $F_{i+1} = F_i$, $C_{i+1} = C_i \cup \{x_{i+1}\}$

In other words, an insertion I at step i consists to carry off a key from the universe (at step i) and to add it to the structure; operation D consists to carry off a key from the structure and to throw it in the garbage collector ; operation Q^+(positive query) doesn't modify neither the universe nor the structure and the garbage collector ; whereas operation Q^- (negative query) consists to take a key from the universe and to throw it in the garbage collector, the structure isn't modified (see fig.1).

We call *history of length n* a sequence of n operations satisfying the above assumptions .

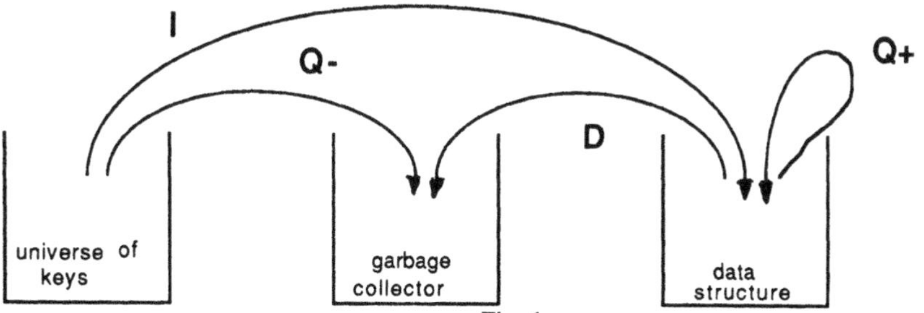

Fig. 1

Example.
For N = 10, an history of length 6 is: $I(2)I(7)Q^-(1)D(7)Q^+(2)I(9)$; the sets of keys in the structure are: $F_0=\emptyset$, $F_1=\{2\}$, $F_2=\{2,7\}$, $F_3=\{2,7\}$, $F_4=\{2\}$, $F_5=\{2\}$, $F_6=\{2,9\}$;
the garbage collectors are : $C_0=\emptyset$, $C_1=\emptyset$, $C_2=\emptyset$, $C_3=\{1\}$, $C_4=\{1,7\}$, $C_5=\{1,7\}$, $C_6=\{1,7\}$;
and the universe of keys are :
$U_0 = [10]$, $U_1 = [10] - \{2\}$, $U_2 = [10] - \{2,7\}$, $U_3 = [10] - \{1,2,7\}$, $U_4 = [10] - \{1,2,7\}$,
$U_5 = [10] - \{1,2,7\}$, $U_6 = [10] - \{1,2,7,9\} = \{3,4,5,6,8,10\}$.

Generally, to each operation O(x) of an history is associated a *cost* depending on O and x , which represents the algorithm cost realizing the operation O(x) in the data structure (at its state just before O(x) is performed).We assume here that such cost depends only on O and the size k of the structure (card F_i) ; let CO_k denote this quantity, we call it the *individual* (or *unitary*)*cost* of O for the size k (see[4],[10]).We define the cost of an history as the sum of its operations costs, and the cost of a set of histories as the sum of its elements costs.

The purpose of the present work is to calculate the cost of the set of histories of length n starting and finishing with an empty structure, called integrated cost , as function of N and the individual costs, for the following three data types :

a) **dictionaries:** all the four basic operations I, D, Q^+, Q^- are allowed without any restriction
b) **linear lists :** only I and D are allowed
c) **priority queues :** D is performed on the key of minimal value, and $Q^+$and $Q^-$are not allowed.

The plan of the paper is as follows: in section 2 we remember some basic notions and we give the integrated cost formula. In section 3 we point out the theorem of continued fractions associated to our model, and we apply it to number the histories in section 4. Section 5 is devoted to the calculation of the integrated costs and some concluding remarks are contained in Section 6.

2.Integrated cost formula

2.1.Number of possibilities of an operation

Definition 1. *For* $O \in \{I,D,Q^+,Q^-\}$, ***the number of possibilities of O is the number of keys*** $x \in [N]$ *for which* O(x) *is defined.*

Thus, at step t+1, for I or Q^-, this number is card(U_t) (i.e. the number of keys in the universe of keys at step t) because I and Q^- have respectively the following profiles $I : U_t \rightarrow F_t$ and $Q^- : U_t \rightarrow C_t$. For D and Q^+ the number of possibilities is card(F_t) because they have as profiles $D : F_t \rightarrow C_t$ and $Q^+ : F_t \rightarrow F_t$. In order to compute these numbers, we can remark an invariant :

$$\text{card}(U_0) = \text{card}(U_t) + \text{card}(C_t) + \text{card}(F_t) \quad \forall\, t \geq 1$$

$$\text{card}(C_t) = \text{card}(C_0) + \sum_{1 \leq j \leq t-1} \chi(O_j = Q^-) + \sum_{1 \leq j \leq t-1} \chi(O_j = D)$$

where χ is the characteristic function.

$$\text{card}(F_t) = \text{card}(F_0) + \sum_{1 \leq j \leq t-1} \chi(O_j = I) - \sum_{1 \leq j \leq t-1} \chi(O_j = D)$$

Hence,

$$\text{card}(U_t) = \text{card}(U_0) - \text{card}(C_0) - \text{card}(F_0) - \sum_{1 \leq j \leq t-1} \chi(O_j = I) - \sum_{1 \leq j \leq t-1} \chi(O_j = Q^-)$$

As card $(U_0) = N$, card (C_0) = card $(F_0) = 0$, if p is the number of operations I and Q^- performed above (i.e.at time j = 1,2,...,t-1) then card $(U_t) = N - p$.

It depends on previous insertions and negative queries numbers and on N.We say that ***the number of possibilities of the i-th insertion or negative query is N-i+1***; we denote it by

$\text{Npos}(i^{th}\ I \text{ or } Q^-) = N-i+1$.

At step t+1, for deletion and positive query, the number of possibilities is card F_t i.e.the size of the structure at this moment. It depends only on the size of the structure just before the deletion or the positive query considered (and not on t). If we denote by k the size of the structure we set the number of possibilities as

$\text{Npos}(D,k) = \text{Npos}(Q^+,k) = k$ called ***number of possibilities of O*** $\in \{D,Q^+\}$ ***for the size k.*** We can remark that, for priority queues, $\text{Npos}(D, k) = 1 \quad \forall\, k > 0\ (k \leq N)$.

So we can state :

i) for **dictionaries:** $\text{Npos}(i^{th}\ I \text{ or } Q^-) = N-i+1$ if $0 < i \leq N$ and 0 if not
$\text{Npos}(D, k) = d_k = k$ if $0 \leq k \leq N$ and 0 if not
$\text{Npos}(Q^+, k) = q_k = k$ if $0 \leq k \leq N$ and 0 if not

ii) for **linear lists :** $\text{Npos}(i^{th}\ I) = N-i+1$ if $0 < i \leq N$ and 0 if not
$\text{Npos}(D, k) = d_k = k$ if $0 \leq k \leq N$ and 0 if not

iii) for **priority queues :** $\text{Npos}(i^{th}\ I) = N-i+1$ if $0 < i \leq N$ and 0 if not.
$\text{Npos}(D, k) = d_k = 1$ for $0 < k \leq N$ and 0 if not .

Under these basic assumptions, we can remark that a data type is totally determined by its possibility functions .

2.2.The integrated cost formula

Definition 2. ***The schema of an history*** $h = O_1(x_1)O_2(x_2)...O_n(x_n)$ ***is the sequence*** $\Omega = O_1O_2...O_n$

The schema of an history (as well as the initial size of the structure, assumed here to be 0) suffices to define the sequence of the size of the structure, called *level sequence of h..*

Thus for an history with schema Ω = I I Q^- D Q^+ I D D the level sequence is (0,1,2,2,1,1,2,1,0). It' s dictionary type history.The number of such histories, i.e.with the same schema is : N(N-1)(N-2).2.1.(N-3).2.1 = 4 N (N - 1) (N - 2) (N - 3)

More generally, for a schema $O_1O_2...O_n$ *whose level sequence is* $(k_0,k_1,...,k_n)$,*the number of histories with he same schema is*

$$\frac{N!}{(N-i)!} \prod_{j \in J} \text{Npos}(D, k_j) \prod_{p \in P} \text{Npos}(Q^+, k_p)$$

where J (resp.P) is the set of indices j such that $O_j = D$ (resp.Q^+) and i the number of insertions I and negative queries Q^- in the schema.

For linear lists and priority queues, it's always possible to determine i, but not for dictionaries (see [9],[16]).

Let $\mathbf{H}_{k,p,n}/N/$ be the set of histories (of given type) of length n, starting at level k and finishing at level p. We set $\mathbf{H}_n/N/ = \mathbf{H}_{0,0,n}/N/$, $H_{k,p,n}/N/ = \text{card } \mathbf{H}_{k,p,n}/N/$, $H_n/N/ = \text{card } \mathbf{H}_n/N/$ and

$H_{k,p,0}/N/ = \delta_{k,p}$ (Kronecker's symbol).

For a fixed data type, we are interested by the cost of histories belonging to $\mathbf{H}_n/N/$.

With the hypothesis made on the individual costs, this is given by the following formula, called ***integrated cost formula***,

$$K_n = \sum_k (CI_k \, NI_{k,n} + CD_k \, ND_{k,n} + CQ^+_k \, NQ^+_{k,n} + CQ^-_k \, NQ^-_{k,n})$$

where the integers $NO_{k,n}$, for $O \in \{I, D, Q^+, Q^-\}$, are the number of operations O at level k in the course of all histories belonging to $\mathbf{H}_n/N/$, called *the level crossing number of O at level k*.

Thus for an history whose schema is IIQ^-DQ^+IDD, there are one insertion at level 0, two at level 1; two deletions at level 2 and one at level 1 ; one negative query at level 2; and one positive query at level 1.

If we consider the histories of length 2 of dictionary type, there are two possible schemas: Q^-Q^- and ID ; there are N(N-1) histories having the first schema and N having the second, then we have $NQ^-_{0,2} = 2N(N-1)$; $NI_{0,2} = N$, $ND_{1,2} = N$ and the other level crossing are zero.

We are also interested in the average cost of an history

$$\frac{K_n}{H_n^{/N/}}$$

3.Histories and continued fractions

3.1.Continued fractions theorem for a finite universe of keys and garbage collector

If we let H^{p+q}_{p+q+r} the number of histories of length n = p+q+r with p insertions, q negative queries,
$r = r_1 + r_2$ where r_1 (resp.r_2) is the number of deletions (resp. positive queries) and if we consider the following generating function :

$$H(t, x, z) = \sum_{p,q,r} H^{p+q}_{p+q+r} \frac{(N-(p+q))!}{N!} x^p t^q z^r$$

then we can prove that :

Theorem 3. *H(t, x, z) has the following continued fraction expansion :*

$$H(t, x, z) = \cfrac{1}{1 - t - q_0 z - \cfrac{d_1 x z}{1 - t - q_1 z - \cfrac{d_2 x z}{\cdots}}}$$

where $q_k = Npos(Q^+,k)$, $d_k = Npos(D,k)$

Sketch of the proof.
Define the alphabet $X = \{I, Q^-, Q_0^+, Q_1^+, Q_2^+, ..., D_1, D_2, ...\}$ where O_j, $O \in \{D, Q^+\}$ denotes

the operation O performed on a file of size j. Let $S^{[h]}$ denote the set of schemas represented by words over X having height $\leq h$, initial and final levels 0.

The $S^{[h]}$ have the following regular descriptions: $S^{[0]} = (Q^- + Q_0^+)^*$;

$S^{[1]} = (Q^- + Q_0^+ + I(Q^- + Q_1^+)^* D_1)^*$; $S^{[2]} = (Q^- + Q_0^+ + I(Q^- + Q_1^+ + I(Q^- + Q_2^+)^* D_2)^* D_1)^*$

and in general $S^{[h]}$ is obtained by substituting $Q^- + Q_h^+ + I(Q^- + Q_{h+1}^+)^* D_{h+1}$

for $Q^- + Q_h^+$ in the expression of $S^{[h]}$. If we let

$$H^{[h,\, p+q]}_{p+q+r}$$

denote the number of histories of height $\leq h$, length p+q+r, with p insertions, q negative queries and r deletions and positive queries, and then using the morphism :

$$I \to x \ ; \ Q^- \to t \ ; \ Q_k^+ \to q_k z \ ; \ D_k \to d_k z$$

we have :

$$H^{[0]}(t, x, z) = \frac{1}{1 - t - q_0 z}$$

$$H^{[1]}(t, x, z) = \cfrac{1}{1 - t - q_0 z - \cfrac{d_1 xz}{1 - t - q_1 z}}$$

etc.; in general $H^{[h+1]}(t,x,z)$ is obtained by substituting

$$t + q_h z + \frac{d_{h+1} xz}{1 - t - q_{h+1} z}$$

for $t + q_h z$ in $H^{[h]}(t,x,z)$. The theorem follows by letting h go to infinity.□

As N is a fixed nonnegative integer, for each given data type H(t,x,z) is a rational fraction. Thus if we apply Theorem 3 to our data stuctures and we obtain :

a) dictionary

$$H^{/N/}_{DICT}(t, x, z) = H^{Kn,[N]}_{DICT}(t, x, z) = \cfrac{1}{1 - t - 0.z - \cfrac{1.\, x\, z}{1 - t - 1.z - \cfrac{2.\, x\, z}{\cfrac{\cdots}{1 - t - N.z}}}}$$

where

$$H^{Kn,[N]}_{DICT}$$

is the rational generating function associated to histories of dictionary type in Knuth's model having height $\leq N$ [10].

b) linear list

$$H^{/N/}_{LL}(0, x, z) = H^{Kn,[N]}_{LL}(x, z) = \cfrac{1}{1 - \cfrac{1.x\, z}{1 - \cfrac{2.x\, z}{\cfrac{\cdots}{1 - N.x\, z}}}}$$

where

$$H^{Kn,[N]}_{LL}$$

is the rational generating function associated to histories of linear list type in Knuth's model having height $\leq N$ [10].

c) priority queue

$$H_{PQ}^{/N/}(0, x, z) = H_{PQ}^{Kn,[N]}(x, z) = \cfrac{1}{1 - \cfrac{x\,z}{1 - \cfrac{x\,z}{\cfrac{\cdots}{1 - x\,z}}}}$$ N symbols of fraction

In other words we can obtain the continued fractions in Knuth's model by considering an infinite universe of keys. Consequently, all the proprieties of continued fractions in Knuth's model are valid here. So we state merely the results: for the proofs one can see [10].

3.2. Histories of bounded height

As in Knuth's model we have :

Proposition 4. *If*

$$H^{[h]}(t, x, z) = \sum_{p,q,r} H_{p+q+r}^{[h,p+q]} \frac{(N-(p+q))!}{N!} t^p x^q z^r$$

then $H^{[h]}(t,x,z)$ *has a rational generating function*

$$H^{[h]}(t, x, z) = \frac{P_h(t, x, z)}{Q_h(t, x, z)}$$

where P_h *and* Q_h *are polynomials that satisfy the relations :*

$P_{-1} = 0;\ P_0 = 1;\ P_h(t,x,z) = (1 - t - q_h z)\, P_{h-1}(t,x,z) - d_h\, x\, z\, P_{h-2}(t,x,z)$

$Q_{-1} = 0;\ Q_0(t,x,z) = 1 - t - q_0 z;\ Q_h(t,x,z) = (1 - t - q_h z)\, Q_{h-1}(t,x,z) - d_h\, x\, z\, Q_{h-2}(t,x,z)$

$\deg(P_h) \le \deg(Q_{h-1}) = h$ *with* $\deg(t^i x^j z^p) = i+j+p$

From now we put : $t = x = y\,z$ (i.e.) we'll use principal variable z (to mark the contribution of operations to the number of histories) and auxillary variable y (to mark the number of insertions and negative queries).

Proposition 5. *Let*

$$H_{k,p}^{/N/}(y, z) = \sum_n \sum_i H_{k,p,n}^{/N/,\,i} \frac{(N-i)!}{N!} y^i z^n = \sum_n \sum_i \frac{H_{k,p,n}^{/N/,\,i}}{\binom{N}{i}} \frac{y^i}{i!} z^n$$

where

$$H_{k,p,n}^{/N/,\,i}$$

is the number of histories going from level k to level p in n steps with i insertions and negative queries, we have :

$$H_{k,p}^{/N/}(y, z) = \frac{1}{d_1 d_2 \ldots d_p\, y^k z^{k+p}} [Q_{k-1}(y, z)\, Q_{p-1}(y, z)\, H(y, z) - P_{\lambda-1}(y, z)\, Q_{\mu-1}(y, z)]$$

where $\lambda = \max(k,p)$ *and* $\mu = \min(k,p)$*. In particular :*

$$H_{0,k}^{/N/}(y, z) = \frac{1}{d_1 d_2 \ldots d_p\, z^k} [Q_{k-1}(y, z)\ H(y, z) - P_{k-1}(y, z)\ ]$$

$$H_{k,0}^{/N/}(y, z) = \frac{1}{y^k z^k} [Q_{k-1}(y, z)\ H(y, z) - P_{k-1}(y, z)\ ]$$

We recall that in this case

$$H(y, z) = \sum_n \sum_i \frac{H_n^i}{\binom{N}{i}} \frac{y^i}{i!} z^n$$

4.Counting of histories

4.1.Scalar product considered

An alternative way of looking at the relations between H(y, z) and $Q_h(y, z)$ is by means of orthogonality relations. Starting from the numbers (for fixed y)

$$(a_n^{/N/}(y))_{n\geq 0}$$

with

$$a_n^{/N/}(y) = \sum_i \frac{H_n^i}{\binom{N}{i}} \frac{y^i}{i!}$$

we introduce the linear form $< \ >_{N,y}$ over polynomials

$$P_y(z) = \sum_{j=0}^{k} p_j(y)\, z^j$$

defined by

$$< P_y(z) >_{N,y} = a_n^{/N/}(y)$$

This induces a scalar product $< P \mid Q>_{N,y} = < P.Q>_{N,y}$ and we have

Proposition 6. *Let* $\bar{Q}_k(y, z) = z^{k+1}Q_k(y, 1/z)$ *be the reciprocal polynomial of* $Q_k(y, z)$.*Then we have :*

$$< z^i \mid \bar{Q}_{k-1}(y, z) >_{N,y} = \begin{cases} 0 & \text{if } 0 \leq i < k \\ d_1 d_2 \dots d_k\, y^k & \text{if } i = k \end{cases}$$

Proposition 5 gives us using the scalar product $< \mid >_{N,y}$:

$$\sum_i \frac{H_{k,p,n}^i}{\binom{N}{i}} \frac{y^i}{i!} = \frac{1}{d_1 d_2 \dots d_p\, y^k} < z^n \mid \bar{Q}_{k-1}(y, z)\, \bar{Q}_{p-1}(y, z) >_{N,y}$$

4.2.Data structures and orthogonal polynomials

4.2.1.Dictionaries and Charlier polynomials

For this data type, we have :
$\bar{Q}_{-1}(y, z) = 1,\ \bar{Q}_0(y, z) = z - y,\ \bar{Q}_k(y, z) = (z - y - k)\,\bar{Q}_{k-1}(y, z) - k\,y\,\bar{Q}_{k-2}(y, z) \quad k \geq 1$
This recurrence relation translates over the generating function

$$\bar{Q}(y, z, t) = \sum_{k\geq 0} \bar{Q}_{k-1}(y, z) \frac{t^k}{k!}$$

into the differential equation

$$(1+t)\frac{\partial}{\partial t}\bar{Q}(y, z, t) = (z - (1+t)y)\,\bar{Q}(y, z, t)$$

whose solution is

$$\bar{Q}(y, z, t) = (1+t)^z e^{-yt}$$

Taking y = 1 we obtain the exponential generating function of Charlier polynomials.

Theorem 7. *The Charlier polynomials associated with dictionaries admit*

$$\sum_{k\geq 0} \bar{Q}_{k-1}(y, z) \frac{t^k}{k!} = (1+t)^z e^{-yt}$$

for exponential generating function.
As for dictionary histories :

$$h^{/N/}(y, z) = \sum_{n\geq 0} \sum_{i} H_n^{/N/, i} \frac{(N-i)!}{N!} y^i \frac{z^n}{n!} = e^{y(e^z - 1)}$$

thus

$$\sum_{n\geq 0} H_n^{/N/} \frac{z^n}{n!} = e^{Nz}$$

with

$$H_n^{/N/} = \sum_i H_n^{/N/,i} = \sum_i \binom{N}{i} H_n^{DICT\ Kn,\, i} = \sum_i \binom{N}{i} i!\, S_{n,i}$$
$$= N^n$$

where $S_{n,i}$ *are the Stirling numbers (of the second kind) and* $H_n^{KnDICT,i}$ *is the number of histories in Knuth's model for dictionaries of length n and having i insertions and negative queries.*
We have also

$$h^{/N/}(y, u, v, z) = \sum_{n,k,p} \left(\sum_i H_{k,p,n}^{i} \frac{(N-i)!}{N!} y^{i+k} \right) \frac{u^k}{k!} v^p \frac{z^n}{n!} = \exp y\,[(1+u)(1+v)\,e^z - u - v - 1]$$

and

$$\sum_{n,k,p} H_{k,p,n} \frac{u^k}{k!} v^p \frac{z^n}{n!} = [(1+u)(1+v)\,e^z - u - v]^N$$

4.2.2. Linear lists and Hermite polynomials

For this type, we have :

$$\bar{Q}_{-1} = 1;\ \bar{Q}_0(y, z) = z;\ \bar{Q}_k(y, z) = z\bar{Q}_{k-1}(y, z) - k\, y\, \bar{Q}_{k-2}(y, z)$$

Theorem 8. *The Hermite polynomials associated with linear lists admit for exponential generating function*

$$\sum_{k\geq 0} \bar{Q}_{k-1}(y, z) \frac{t^k}{k!} = e^{zt - y\frac{t^2}{2}}$$

and for y = 1 we have

$$\bar{Q}_{k-1}(z) = \sum_{0\leq i\leq \frac{k}{2}} \frac{(-1)^i\, k!}{2^i\, i!\, (k-2i)!} z^{k-2i}$$

As for histories

$$h^{/N/}(y, z) = \sum_{n\geq 0} H_{2n}^{/N/} \frac{(N-n)!}{N!} y^n \frac{z^{2n}}{2n!} = e^{y\frac{z^2}{2}}$$

Hence

$$\sum_{n\geq 0} H_{2n}^{/N/} \frac{z^{2n}}{2n!} = (1 + \frac{z^2}{2})^N$$

and

$$H_{2n}^{/N/} = \binom{N}{n} H_{2n}^{LLK} = \binom{N}{n} \frac{(2n)!}{2^n}$$

where H_{2n}^{KnLL} *is the number of histories of length* $2n$ *in Knuth's model for linear lists.*
We have also :

$$h^{/N/}(y, u, v, z) = \sum_{n,k,p} H_{k,p,n}^{/N/} \frac{\left(N - \frac{n+p-k}{2}\right)!}{N!} y^{\frac{n+p+k}{2}} \frac{u^k}{k!} v^p \frac{z^n}{n!} = \exp y\,[u\,v + (u+v)\,z + \frac{z^2}{2}]$$

and

$$\sum_{n,k,p} H_{k,p,n}^{/N/} \frac{u^k}{k!} v^p \frac{z^n}{n!} = [\,1 + u\,v + (u + v)\,z + \frac{z^2}{2}\,]^N$$

Remarks :

$$H_n^i = \delta_{i,\frac{n}{2}} H_n^{\frac{n}{2}} \text{ so } H_n = H_n^{\frac{n}{2}}, \qquad H_{k,l,n}^i = \delta_{i,n+l-k} H_{k,l,n}^{\frac{n+l-k}{2}} \quad \text{hence} \quad H_{k,l,n} = H_{k,l,n}^{\frac{n+l-k}{2}}$$

where δ is the Kronecker symbol.

4.2.3.Priority queues and Tchebycheff polynomials

For this data type : $\bar{Q}_{-1} = 1$, $\bar{Q}_0(y, z) = z$, $\bar{Q}_k(y, z) = z\,\bar{Q}_{k-1}(y, z) - y\,\bar{Q}_{k-2}(y, z)$, $k \geq 1$

thus :

$$\sum_{k\geq 0} \bar{Q}_{k-1}(y, z)\; t^k = \frac{1}{1 - z\,t + y\,t^2}$$

For $y = 1$, we obtain the generating function of Tchebycheff polynomials.

Theorem 9. *The Tchebycheff polynomials associated with priority queues have the following generating function :*

$$\bar{Q}(z,y) = \sum_{k\geq 0} \bar{Q}_{k-1}(z)\, y^k = \frac{1}{(1+y^2-zy)}$$

As for histories

$$h^{/N/}(y, z) = \sum_{n\geq 0} H_{2n}^{/N/} \frac{(N-n)!}{N!} y^n z^{2n} = \frac{1 - \sqrt{1 - 4\,y\,z^2}}{2\,y\,z}$$

hence

$$H_{2n}^{/N/} = \binom{N}{n} H_{2n}^{FPK} = \binom{N}{n} \frac{(2n)!}{(n+1)!}$$

and

$$\sum_{n,k} H_{0,k,n}^{/N/} \frac{\left(N - \frac{n+k}{2}\right)!}{N!} y^{\frac{n+k}{2}} v^k z^n = \sum_{n,k} H_{k,0,n}^{/N/} \frac{\left(N - \frac{n-k}{2}\right)!}{N!} y^{\frac{n+k}{2}} v^k z^n$$

$$= \frac{1 - \sqrt{1 - 4\,y\,z^2}}{2\,y\,z^2 - y\,v\,z\,(1 - \sqrt{1 - 4\,y\,z^2})}$$

5.The integrated cost theorem

The preceding section provides expressions for the number $H_{k,p,n}$ of histories of length n, starting at level k and finishing at level p.

Let $NO^{/N/,i}_{k,n}$ be the level crossing number of operation $O \in \{I, D, Q^+, Q^-\}$ at level k for all histories with i insertions and negative queries, initial and final level 0. Setting

$$NO_k^{/N/}(y,z) = \sum_{n\geq 0} \sum_i NO_{k,n}^{/N/,i} \frac{(N-i)!}{N!} y^i z^n$$

we have

Proposition 10.

$$NI_k^{/N/}(y, z) = y\,z\; H^{/N/}_{0,k}(y, z)\; H^{/N/}_{k+1,0}(y, z)$$
$$NQ^-_k{}^{/N/}(y, z) = y\,z\, H^{/N/}_{0,k}(y, z)\; H^{/N/}_{k,0}(y, z)$$
$$ND_k^{/N/}(y, z) = d_k\,z\; H^{/N/}_{0,k}(y, z)\; H^{/N/}_{k-1,0}(y, z)$$
$$NQ^+_k{}^{/N/}(y, z) = q_k\,z\, H^{/N/}_{0,k}(y, z)\; H^{/N/}_{k,0}(y, z)$$

where

$$H_{a,b}^{/N/}(y, z) = \sum_{n\geq 0} \sum_i H_{a,b,n}^{/N/,i} \frac{(N-i)!}{N!} y^i z^n$$

We are thus in possession of all quantities needed in order to apply the integrated cost formula. If

$$KO_n^i = \sum_{k\geq 0} CO_k NO_{k,n}^i$$

denotes the integrated cost of the operation $O \in \{I, D, Q^+, Q^-\}$ in the course of all histories of length n with i insertions and negative queries, initial and final level 0, then :

$$K_n = \sum_i (KI_n^i + KD_n^i + KQ_n^{-\,i} + KQ_n^{+\,i})$$

represents the integrated cost for the histories in $\mathbf{H}_{0,0,n}$.

5.1.Integrated cost for priority queues.

We have obtain for this data type the same expression as for priority queues in Knuth's model [9], [16], taking as costs generating function

$$K_{\mathbf{PQ}}^{/N/}(y, z) = \sum_{n\geq 0} K_{2n}^{/N/} \frac{(N-n)!}{N!} y^n z^{2n}$$

Theorem 11. *The generating function of the unitary costs*

$$C_{I\,D}(x) = \sum_k (CI_k + CD_{k+1}) x^k$$

and integrated costs

$$K_{\mathbf{PQ}}^{/N/}(y, z) = \sum_{n\geq 0} K_{2n}^{/N/} \frac{(N-n)!}{N!} y^n z^{2n}$$

for priority queues are related by the linear transformation

$$K_{PQ}^{/N/}(y, z) = y\, z^2\, B(y, z)^3\; C_{ID}[\, y\, z^2\, B(y, z)^2\,]$$

with

$$B(y, z) = \frac{1 - \sqrt{1 - 4\, y\, z^2}}{2\, y\, z^2}$$

Instead of computing explicitly the integrated cost , we can remark that

$$K_{2n}^{/N/} = \binom{N}{n} K_{2n}^{Kn,PQ}$$

where K_{2n}^{KnPQ} is the integrated cost for priority queues in Knuth's model .

This result is valid for any representation of priority queues, but we give only as example lists representations.
As

$$H_{2n}^{/N/} = \binom{N}{n} H_{2n}^{Kn,PQ}$$

if $N \rightarrow \infty$ we obtain exactly the average costs for priority queues in Knuth's model [10].

5.2.Integrated cost for linear lists

For the same reason as in priority queues case, we obtain for this data type the same result as for linear lists in Knuth's model, taking the exponential generating function :

$$K_{LL}^{/N/}(y, z) = \sum_{n \geq 0} K_{2n}^{/N/} \frac{(N-n)!}{N!}\, y^n\, \frac{z^n}{n!}$$

Theorem 12.*The generating function*

$$K_{LL}^{/N/}(y, z) = \sum_{n \geq 0} K_{2n}^{/N/} \frac{(N-n)!}{N!}\, y^n\, \frac{z^n}{n!}$$

for linear lists is related to

$$C_{ID}(x) = \sum_k (\, CI_k + CD_{k+1}\,) x^{k+1}$$

by

$$K_{LL}^{/N/}(y, z) = \frac{1}{\sqrt{1 - 2\, y\, z}}\; C_{ID}\left(\frac{y\, z}{1 - y\, z}\right)$$

Here also we obtain exactly the average costs for linear lists in Knuth's model if $N \rightarrow \infty$.

5.3.Integrated cost for dictionaries

Theorem 13. *Setting* $K_n^{/N/,i} = KI_n^{/N/,i} + KD_n^{/N/,i} + KQ^-{}_n^{/N/,i} + KQ^+{}_n^{/N/,i}$
the generating function for dictionary

$$K_{DICT}^{/N/}(y, z) = \sum_{n \geq 0} \sum_i K_n^{/N/,i} \frac{(N-i)!}{N!}\, y^i\, \frac{z^n}{n!}$$

is related to the exponential generating functions of unitary costs by the relation

$$K^{/N/}_{DICT}(y,z) = y\, e^{y(e^z-1)} \int_0^{(e^{z/2}-1)^2} [(e^z - u - 1)\, \mathcal{C}_{AS}(y\,u) + 2(\mathcal{C}_{I^-}(yu) + u\, \mathcal{C}_{I^+}(yu))] \frac{e^{-yu}\, du}{\sqrt{(e^z+1-u)^2 - 4e^z}}$$

with

$$C_{ID}(x) = \sum_k (CI_k + CD_{k+1}) \frac{x^k}{k!}$$

$$C_{Q^-}(x) = \sum_k CQ^-_k \frac{x^k}{k!} \quad \text{and} \quad C_{Q^+}(x) = \sum_k CQ^+_{k+1} \frac{x^k}{k!}$$

The right hand of Theorem 13 (i.e.the integral transform) is the expression for dictionary in Knuth's model.
That gives us

$$K^{/N/}_n = \sum_i \binom{N}{i} K_n^{DICT\ Kn,\ i}$$

The table 1 summarizes the average costs of different data representations for dictionaries, priority queues and linear lists in the model considered here.

Data type	data represent.	cost ($H^{/N/}_n$), for fixed N	cost (H^{∞}_n), $n \to \infty$
DICT	sorted list	$-N^2 + \frac{n}{2} + 1)N + \frac{n}{2} + (1 - \frac{1}{N})^n (N^2 + (\frac{n}{2} - 1)N)$	$\left(\frac{3-e}{2e}\right) n^2 + O(n)$
	unsorted list	$-\frac{5}{4} N^2 + (\frac{n}{2} + \frac{1}{4})N + \frac{n}{2} + (1 - \frac{1}{N})^n (2N(n+1) - N^2)$ $+ (1 - \frac{1}{N})^{n-1} (\frac{9}{4} N^2 + (n - \frac{27}{4}) N - n + \frac{9}{2})$	$\left(\frac{13 + 4e - 3e^2}{4e^2}\right) n^2 + O(n)$
LL	sorted list	$n(n+5)/3$ if $n \leq N$	$n(n+5)/3$
	unsorted list	$n(n+5)/6$ if $n \leq N$	$n(n+5)/6$
PQ	sorted list	$\frac{\sqrt{\pi}}{4} n^{3/2} + O(n)$ if $n \leq N$	$\frac{\sqrt{\pi}}{4} n^{3/2} + O(n)$
	unsorted list	$\frac{\sqrt{\pi}}{2} n^{3/2} + O(n)$ if $n \leq N$	$\frac{\sqrt{\pi}}{2} n^{3/2} + O(n)$
	binary tourn.	$(3\, n \,\mathrm{Log}\, n)/2 + O(n)$ if $n \leq N$	$(3\, n\, \mathrm{Log}\, n)/2 + O(n)$
	pagodas	$n\, \mathrm{Log}\, n + O(n)$ if $n \leq N$	$n\, \mathrm{Log}\, n + O(n)$

Table 1.

6.Conclusion

The integrated costs of dynamic data structures with a finite number of keys have been expliciteIy calculated in this paper for dictionaries, priority queues and linear lists. In addition, we have proved that with a 'garbage collector' and letting the number of keys going to infinity, we recover the results concerning Knuth's model which appears as a "limit model" of the model considered here.

Acknowledgments
The authors are grateful to P.Flajolet, P. Lescanne, J.L. Rémy and J.M. Steyaert for several useful discussions on this topic.*This work has been done with the help of the " P.R.C. Mathématiques et Informatique "*

References

[1] L.Chéno
Profils limites d'histoires sur les dictionnaires et les files de priorité.Application aux files binomiales.Thèse de 3ème cycle. Université Paris Sud, 1981.

[2] L.Chéno, P.Flajolet, J.Françon, C.Puech, J.Vuillemin
Finite files, limiting profiles and variance analysis. Proceedings 18th Allerton Conf. on Com. Control and Computing (Illinois 1980).

[3] P.Flajolet,J.Françon
Structures de données dynamiques en réservoir borné.Actes des journées algorithmiques de Nice, Université de Nice (1980).

[4] P.Flajolet, J.Françon, J.Vuillemin
Sequence of operations analysis for dynamic data structures.J.of algorithms 1, 111-141,1980.

[5] P.Flajolet, C.Puech, J.Vuillemin
The analysis of simple lists structures. Inf. Sc. 38, 121- 146, 1986.

[6] P.Flajolet
Analyse d'algorithmes de manipulation d'arbres et de fichiers. B.U.R.O. cahier 34-35, 1981.

[7] J.Françon
Combinatoire des structures de données.Thèse de doc. d'Etat . Université de Strasbourg, 1979

[8] J.Françon
Histoires de fichiers. RAIRO Inf.Th.12, 49-62, 1978.

[9] J.Françon, B.Randrianarimanana, R.Schott
Analysis of dynamic data structures in D.E.Knuth's model. Rapport C.R.I.N. 1986.(submitted)

[10] J.Françon, B.Randrianarimanana, R.Schott
Analysis of dynamic algorithms in D.E.Knuth's model. Proceedings C.A.A.P.'88, L.N.C.S. 299 72-88 .Springer Verlag.

[11] A.Jonassen, D.E.Knuth
A trivial algorithm whose analysis isn't. J.of Comp.and System Sc.16, 301-332, 1978.

[12] G.D.Knott
Deletion in binary storage trees.Report Stan-CS, 75-491, may 1975.

[13] D.E.Knuth
Deletions that preserve randomness. Trans.Software Eng, 351-359, 1977.

[14] D.E.Knuth
The art of computer programming : Sorting and Searching, vol.3, second printing, 1975.

[15] G.Louchard
Random walks, Gaussian processes and list structures .Th. Comp. Sc. 53, 99 - 124, 1987.

[16] B.Randrianarimanana
Analyse des structures de données dynamiques dans le modèle de D.E.Knuth.Thèse de 3ème cycle, Université Nancy 1, 1986.

ITERATED DETERMINISTIC TOP-DOWN LOOK-AHEAD

Z. FÜLÖP and S. VÁGVÖLGYI

Research Group on Theory of Automata
Hungarian Academy of Sciences
Somogyi u.7., H-6720 Szeged, Hungary

1. INTRODUCTION

It is a usual matter of theoretical study to iterate computation methods in order to obtain more powerful computing devices. For example, it was studied iterated linear control over linear context-free languages in [Khal,2] and [Vog2], iterated one-turn pushdown machines in [Vogl,2] and iterated pushdown machines in [EngVog]. In this paper we consider iterated deterministic top-down look-ahead for deterministic top-down tree transducers in the following sense.

In [FulVag], we took over the concept of look-ahead from [Eng] and created the deterministic top-down tree automaton with deterministic top-down look-ahead. We proved that these automata recognize a class of tree languages which is strictly between the class of recognizable tree languages and the class of tree languages recognizable by deterministic top-down tree automata. We iterate the look-ahead tree languages as follows. Let $DREC_0$ be the class of tree languages recognizable by deterministic top-down tree automata and let, for $n \geq 1$, $DREC_n$ be the class of tree languages recognizable by deterministic top-down tree automata with $DREC_{n-1}$ look-ahead. (So, in [FulVag] we showed that $DREC_0 \subset DREC_1 \subset REC$, where REC denotes the class of all recognizable tree languages.)

The aim of this paper is to show that $DREC_0 \subset DREC_1 \subset \ldots \subseteq REC$. Hence, by iterating the look-ahead tree languages for deterministic top-down tree automata, we obtain more and more powerful recognizing devices. Further, we show that the hierarchy does not exhaust REC, i.e., that $\bigcup_{n=0}^{\infty} DREC_n \subset REC$.

The paper is organized as follows.

In Section 2, we recall some general notations concerning trees and define the deterministic top-down tree automaton with C look-ahead, where C is a subclass of REC.

In Section 3, a few special tree languages are defined which will be used in showing the properness of the mentioned inclusions.

In Section 4, we define the concept of the iterated deterministic top-down look-ahead for deterministic top-down tree transducers and prove that the iteration hierarchy is strict with respect to recognizing capacity. It is also shown that the closure of the hierarchy of the recognized tree language classes is properly contained in REC.

2. PRELIMINARIES

For a finite set Y, we denote by $\#(Y)$ the cardinality of Y.

A ranked alphabet Σ is a finite set in which every symbol has a unique rank in the set of nonnegative integers. For every $m \geq 0$, Σ_m denotes the set of symbols of Σ which have rank m.

The set of trees or terms over Σ is denoted by T_Σ. A tree t in T_Σ is denoted by $\sigma(t_1, \ldots, t_m)$ where $\sigma \in \Sigma_m$ is the root and $t_1, \ldots, t_m \in T_\Sigma$ are the direct subtrees of t; if $m = 0$, then we write σ rather than $\sigma(\,)$.

For any set Y, $T_\Sigma(Y)$ denotes $T_{\Sigma \cup Y}$ where the elements of Y are viewed as symbols of rank 0.

For a tree $t = \sigma(t_1, \ldots, t_m) \in T_\Sigma$, the set of subtrees of t is the set consisting of t and the subtrees of $t_1, \ldots, t_m$.

We specify and keep fixed throughout the paper a countable set of symbols $X = \{x_1, x_2, \ldots\}$ called the set of variable symbols. We put $X_m = \{x_1, \ldots, x_m\}$, for every $m \geq 0$.

Any subset L of T_Σ is called a tree language over Σ.

For a tree $t \in T_\Sigma(X_m)$ and other trees $t_1, \ldots, t_m$, we denote by $t(t_1, \ldots, t_m)$ the tree which can be obtained from t by substituting each occurrence of x_i in t by t_i, for every $1 \leq i \leq m$. Moreover, for tree languages $L_1, \ldots, L_m$, we put $t(L_1, \ldots, L_m) = \{t(t_1, \ldots, t_m) \mid t_i \in L_i,\ 1 \leq i \leq m\}$.

A top-down tree automaton is a system $T = \langle \Sigma, Q, q_0, R \rangle$, where

(a) Σ is a ranked alphabet,

(b) Q, the state set of T, is a ranked alphabet with $Q = Q_1$ such that $Q \cap \Sigma = \emptyset$,

(c) q_0 is a distinguished element of Q, called the initial state of T,

(d) R is a finite set of rules of the form $q(\sigma(x_1, \ldots, x_m)) \to \sigma(q_1(x_1), \ldots, q_m(x_m))$.

T is called a deterministic top-down tree automaton if there are no different rules in R with the same left-hand side. The class of all (deterministic) top-down tree automata is denoted by $(D)FTA$.

R induces a binary relation $\Rightarrow_T$ over $T_\Sigma(Q(T_\Sigma))$ (where $Q(T_\Sigma) = \{q(t) \mid q \in Q$ and $t \in T_\Sigma\}$) as follows: for any $t, r \in T_\Sigma(Q(T_\Sigma))$, $t \Rightarrow_T r$ if and only if there is a rule $q(\sigma(x_1, \ldots, x_m)) \to \sigma(q_1(x_1), \ldots, q_m(x_m))$ in R and r can be obtained from t by substituting an occurrence of a subtree $q(\sigma(t_1, \ldots, t_m))$ of t by $\sigma(q_1(t_1), \ldots, q_m(t_m))$. The reflexive, transitive closure of $\Rightarrow_T$ is denoted by $\Rightarrow_T^*$. The tree language recognized by T is

$$L(T) = \{t \in T_\Sigma \mid q_0(t) \overset{*}{\underset{T}{\Rightarrow}} t\}.$$

The class of all tree languages recognizable by (deterministic) top-down tree automata is denoted by $(D)REC$.

We recall the well known fact that $DREC \subset REC$, a proof can be found in [GecSte].

Now we introduce the recognizing device which will be the centre of our investigations in this paper.

Let $C \subseteq REC$ be a class of tree languages. A deterministic top-down tree automaton with C look-ahead is again a system $T = \langle \Sigma, Q, q_0, R \rangle$ with the same Σ, Q and q_0 as before, however R is now a finite set of rules of the form $\langle q(\sigma(x_1, \ldots, x_m)) \to$

$\sigma(q_1(x_1),\ldots,q_m(x_m)); L_1,\ldots,L_m\rangle$, where $L_1,\ldots,L_m \in C$. Moreover, to ensure T a deterministic behaviour, it is required that for any two different rules $\langle q(\sigma(x_1,\ldots,x_m)) \rightarrow \sigma(q_1(x_1),\ldots,q_m(x_m)); L_1,\ldots,L_m\rangle$ and $\langle q(\sigma(x_1,\ldots,x_m)) \rightarrow \sigma(q'_1(x_1),\ldots,q'_m(x_m)); L'_1, \ldots, L'_m\rangle$ in R, $L_i \cap L'_i = \emptyset$ holds for some $1 \leq i \leq m$.

T recognizes a tree language over Σ as follows. Let $\Rightarrow_T^*$ be the reflexive, transitive closure of the relation $\Rightarrow_T$ over $T_\Sigma(Q(T_\Sigma))$, which is defined in the following way: for $t, r \in T_\Sigma(Q(T_\Sigma))$, $t \Rightarrow_T r$ if and only if there is a rule $\langle q(\sigma(x_1,\ldots,x_m)) \rightarrow \sigma(q_1(x_1),\ldots,q_m(x_m)); L_1,\ldots,L_m\rangle$ in R and r can be obtained from t by substituting an occurrence of a subtree $q(\sigma(t_1,\ldots,t_m))$ of t, for which $t_1 \in L_1,\ldots,t_m \in L_m$ hold, by $\sigma(q_1(t_1),\ldots,q_m(t_m))$. Then the tree language recognizable by T is again

$$L(T) = \{t \in T_\Sigma \mid q_0(t) \overset{*}{\underset{T}{\Rightarrow}} t\}.$$

It can be seen from the definition of $\Rightarrow_T$ what the term look-ahead means: a rule can be applied at a node of a tree only if the direct subtrees of that node are in the tree languages given in the rule. It should also be clear that T can apply at most one rule at a node of a tree since for any two different rules in R with the same left hand side there exist look-ahead sets corresponding to the same variable such that their intersection is empty. This is why T is called deterministic.

The class of all deterministic top-down tree automata with C look-ahead is denoted by $DFTA^C$. (We note that in [FulVag] the recognizing capacity of $DFTA^{DREC}$ was examined.)

Let $T = \langle\Sigma, Q, q_0, R\rangle$ be a deterministic top-down tree automaton with or without look-ahead. Then, for any $q \in Q$, we denote by $T(q)$ the deterministic top-down tree automaton $\langle\Sigma, Q, q, R\rangle$.

In the proof of our main result we shall consider sequences. We denote the limit of a convergent sequence $\{a_n\}$ of real numbers, as n tends to infinity, by $\lim_{n\to\infty} a_n$. Moreover,for any real number x, $|x|$ is the absolute value of x.

3. SOME SPECIAL TREE LANGUAGES

Here we define a ranked alphabet and some tree languages that will be used in the next section.

In the rest of the paper, by Σ we mean the ranked alphabet $\Sigma_0 \cup \Sigma_2$, where $\Sigma_0 = \{a, b\}$ and $\Sigma_2 = \{\sigma\}$.

For each $m \geq 1$, define the tree $e_m \in T_\Sigma(X_{m+1})$ as follows: $e_1 = \sigma(x_1, x_2)$ and, for $m \geq 2$, $e_m = \sigma(x_1, e_{m-1}(x_2,\ldots,x_{m+1}))$, i.e., e_m is the tree $\sigma(x_1,\ldots,\sigma(x_m, x_{m+1})\ldots)$.

We say that a tree in T_Σ is even (odd) if it contains even (odd) number of a's. We denote by L_e (L_o) the set of all even (odd) trees. Note that $b \in L_e$ and $a \in L_o$.

We shall use the following observation.

Observation 3.1. Let $m \geq 1$, $u \geq 1$ and let, for every $1 \leq i \leq u$, $L_{i1},\ldots,L_{i,m+1}$ be finite tree languages over Σ such that for any $1 \leq i \neq j \leq u$, there exists at least one $1 \leq l \leq m+1$ with $L_{il} \cap L_{jl} = \emptyset$. Let $L = \bigcup_{i=1}^{u} e_m(L_{i1},\ldots,L_{i,m+1})$.

Then we have

$$\#(L \cap L_e) - \#(L \cap L_o) = \sum_{i=1}^{u} d_{i1} d_{i2} \cdots d_{i,m+1}$$

where $d_{ij} = \#(L_{ij} \cap L_e) - \#(L_{ij} \cap L_o)$ for each $1 \le i \le u$ and $1 \le j \le m+1$.◇

We define, for every integer $n \ge 0$, the tree language $C_n \subseteq T_\Sigma$ as follows:

(a) $C_0 = \{a, b\}$,

(b) for $n \ge 1$, C_n is the smallest set satisfying

(i) $a, b \in C_n$ and

(ii) $\sigma(t, r) \in C_n$ whenever $t \in C_{n-1}$ and $r \in C_n$.

The elements of C_n are called n-nested combs.

Note that $C_n = \{e_m(t_1, \ldots, t_m, y) \mid m \ge 0, y \in \{a, b\}$ and $t_1, \ldots, t_m \in C_{n-1}\}$. (For $m = 0$, with $e_m(t_1, \ldots, t_m, y)$ is meant y.) Obviously, we have $C_0 \subset C_1 \subset \ldots$.

We put $C_n^e = C_n \cap L_e$ and $C_n^o = C_n \cap L_o$.

Finally, for every pair (n, k) of nonnegative integers, we define the tree language $C_{n,k}$ as follows:

(a) $C_{0,k} = C_{n,0} = \{a, b\}$, for $k = 0, 1, \ldots$ and $n = 0, 1, \ldots$.

(b) $C_{n,k} = \{\sigma(t, r) \mid t \in C_{n-1,k-1}$ and $r \in C_{n,k-1}\}$ for $k, n \ge 1$.

We observe that, for each k and n, $C_{n,k}$ is a finite tree language and that $C_{n,k} \subset C_n$, which is infinite if $n \ge 1$. Moreover, it can be seen by induction that a tree t in C_n belongs to $C_{n,k}$ if and only if the following conditions hold:

(a) the length of each path in t is at most k,

(b) each path in t chooses the left son at most n times,

(c) if a path in t chooses the left son less than n times then its length is exactly k.

(By a path we mean a path leading from the root to a terminal node of t.) For example, each tree in $C_{2,5}$ has the following form:

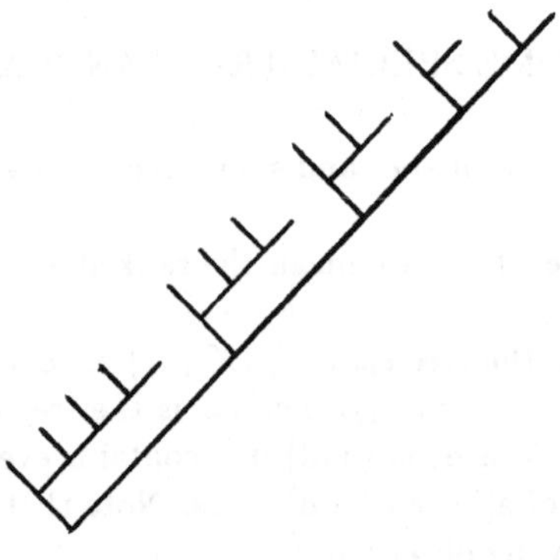

where the terminal nodes are labelled by a or b and the other nodes are labelled by σ. By the above characterization of $C_{n,k}$, we also observe that

(a) $\bigcup_{k=0}^{\infty} C_{n,k} \subset C_n$ and that

(b) if $n \ge k$, then $C_{n,k}$ is the set of all balanced trees of height k over T_Σ.

We put $C^e_{n,k} = C_{n,k} \cap L_e$ and $C^o_{n,k} = C_{n,k} \cap L_o$. Obviously, we have $\#(C^e_{n,k}) = \#(C^o_{n,k})$.

The following fact will also be used in a proof in the next section.

Observation 3.2. Let $1 \leq k, m, n$ be such that $k \geq m$. Then we have

$$\#(C_{n,k}) = \#(C_{n,k-m}) \prod_{i=1}^{m} \#(C_{n-1,k-i}).$$

Proof. Obvious, since $C_{n,k} = e_m(C_{n-1,k-1}, \ldots, C_{n-1,k-m}, C_{n,k-m})$.$\diamond$

4. THE PROBLEMS AND THE SOLUTIONS

In this section we define the concept of the iterated deterministic top-down look-ahead for deterministic top-down tree automata and prove that, by iteration, we obtain more and more powerful recognizing devices.

Let $DREC_0 = DREC$ and, for $n \geq 1$, let $DREC_n$ be the class of all tree languages which can be recognized by deterministic top-down tree automata with $DREC_{n-1}$ look-ahead. Hence, $L \in DREC_n$ if and only if $L = L(T)$ for some $T \in DFTA^{DREC_{n-1}}$. Our task is to show that $DREC_0 \subset DREC_1 \subset \ldots \subseteq REC$ and that $\bigcup_{n=0}^{\infty} DREC_n \subset REC$.

First we prove that $DREC_{n-1} \subseteq DREC_n$ for every $n \geq 1$. For $n = 1$, the inclusion is trivial since each deterministic top-down tree automaton can be thought of as a deterministic top-down tree automaton with deterministic top-down look-ahead, in which each look-ahead set is T_Δ, where Δ is the ranked alphabet of the tree automaton. Moreover, if for some n, $DREC_{n-1} \subseteq DREC_n$, then we have $DFTA^{DREC_{n-1}} \subseteq DFTA^{DREC_n}$ and hence, by definition, $DREC_n \subseteq DREC_{n+1}$.

Next we show that $DREC_n \subseteq REC$, for each $n \geq 0$. For $n = 0$, the inclusion holds by definition. Now, if $DREC_{n-1} \subseteq REC$, then each tree automaton in $DFTA^{DREC_{n-1}}$ is a special top-down tree transducer with regular look-ahead (see [Eng]) and thus its domain is in REC, for a proof see also [Eng]. Hence $DREC_n \subseteq REC$, too.

So we have $DREC_0 \subseteq DREC_1 \subseteq \ldots \subseteq REC$.

To prove that strict inclusion holds in every place we shall show that $C^e_n \in DREC_n$ but $C^e_n \notin DREC_{n-1}$ for each $n \geq 1$. This latter non-containment will follow from Lemma 4.4, which states that, for each tree language L over Σ, if $L \in DREC_{n-1}$, then

$$\lim_{k \to \infty} \frac{\#(L \cap C^e_{n,k}) - \#(L \cap C^o_{n,k})}{\#(C_{n,k})} = 0. \qquad (*)$$

This proves that $C^e_n \notin DREC_{n-1}$ because, for every $k \geq 0$ we have

$$\frac{\#(C^e_n \cap C^e_{n,k}) - \#(C^e_n \cap C^o_{n,k})}{\#(C_{n,k})} = \frac{\#(C^e_{n,k})}{\#(C_{n,k})} = \frac{1}{2}.$$

For similar reasons, it follows from $(*)$ that $L_e \notin DREC_{n-1}$ for every $n \geq 1$ and since L_e is recognizable we obtain that $\bigcup_{n=0}^{\infty} DREC_n \subset REC$.

We begin with the next lemma.

Lemma 4.1. For each $n \geq 0$, $C^e_n \in DREC_n$ and $C^o_n \in DREC_n$.

Proof. Induction on n.

For $n = 0$, take the deterministic top-down tree automaton $T = \langle \Sigma, \{q_e, q_o\}, q_e, R\rangle$, where $R = \{q_e(b) \to b,\ q_o(a) \to a\}$. Then $L(T) = C_0^e = \{b\}$ and $L(T(q_o)) = C_0^o = \{a\}$.

Let us suppose that $C_n^e \in DREC_n$ and $C_n^o \in DREC_n$. Construct the deterministic top-down tree automaton $T = \langle \Sigma, \{q_e, q_o q\}, q_e, R\rangle$ with $DREC_n$ look-ahead, where R is the set of the following rules:

$\langle q_e(\sigma(x_1, x_2)) \to \sigma(q(x_1), q_e(x_2)); C_n^e, T_\Sigma\rangle$, $\langle q_e(\sigma(x_1, x_2)) \to \sigma(q(x_1), q_o(x_2)); C_n^o, T_\Sigma\rangle$, $\langle q_o(\sigma(x_1, x_2)) \to \sigma(q(x_1), q_o(x_2)); C_n^e, T_\Sigma\rangle$, $\langle q_o(\sigma(x_1, x_2)) \to \sigma(q(x_1), q_e(x_2)); C_n^o, T_\Sigma\rangle$, $\langle q(\sigma(x_1, x_2)) \to \sigma(q(x_1), q(x_2)); T_\Sigma, T_\Sigma\rangle$, $q(a) \to a$, $q(b) \to b$, $q_e(b) \to b$, $q_o(a) \to a$.

Then, by induction hypothesis, R is a correct set of rules for a $DFTA^{DREC_n}$ automaton. On the other hand T accepts a tree of the form $e_m(t_1, \ldots, t_m, y)$, where $t_1, \ldots, t_m \in C_n$ and $y \in \{a, b\}$, if and only if odd number of t_i's are in C_n^o and $y = a$ or even number of t_i's are in C_n^o and $y = b$. Hence $L(T) = C_{n+1}^e$. For similar reasons, $L(T(q_o)) = C_{n+1}^o$.◇

Next we prepare the proof of (*) by two lemmata.

Lemma 4.2. For every $n \geq 0$, $DREC_n$ is closed under intersection.

Proof. Again induction on n.

It is a folkloric result that $DREC_0 (= DREC)$ is closed under intersection.

The proof of the induction step is not hard either. Let $L_1, L_2 \in DREC_{n+1}$, i.e., $L_1 = L(T_1)$ and $L_2 = L(T_2)$ for some $T_1 = \langle \Delta, Q_1, q_1, R_1\rangle$ and $T_2 = \langle \Delta, Q_2, q_2, R_2\rangle$ in $DFTA^{DREC_n}$. Construct the top-down tree automaton $T = \langle \Delta, Q_1 \times Q_2, (q_1, q_2), R\rangle$ such that R is the set of all rules of the form $\langle (q, r)(\delta(x_1, \ldots, x_m)) \to \delta((q_1, r_1)(x_1), \ldots, (q_m, r_m)(x_m)); L_1 \cap L'_1, \ldots, L_m \cap L'_m\rangle$ where $\langle q(\delta(x_1, \ldots, x_m)) \to \delta(q_1(x_1), \ldots, q_m(x_m)); L_1, \ldots, L_m\rangle \in R_1$ and $\langle r(\delta(x_1, \ldots, x_m)) \to \delta(r_1(x_1), \ldots, r_m(x_m)); L'_1, \ldots, L'_m\rangle \in R_2$. Since, by induction hypothesis, $DREC_n$ is closed under intersection, R is a correct set of rules for a deterministic top-down tree automaton with $DREC_n$ look-ahead. On the other hand, it is an easy exercise to show that $L(T) = L_1 \cap L_2$.◇

Now we give the intuitive meaning of the next lemma.

Let $T = \langle \Sigma, Q, q_0, R\rangle$ be an arbitrary tree automaton in $DFTA^{DREC_n}$, where $n \geq 0$ and let $q \in Q$. Then we have the following fact: there exists an integer $u \geq 0$ and, for each $1 \leq i \leq u$, there exist tree languages $L_{i1}, L_{i2} \in DREC_n$ and states $q_{i1}, q_{i2} \in Q$ with the following properties:

(a) for any $1 \leq i \neq j \leq n$ it holds that $L_{i1} \cap L_{j1} = \emptyset$ or $L_{i2} \cap L_{j2} = \emptyset$,

(b) $L(T(q)) \cap \sigma(T_\Sigma, T_\Sigma) = \bigcup_{i=1}^{u} \sigma(L_{i1} \cap L(T(q_{i1})), L_{i2} \cap L(T(q_{i2})))$.

In fact, let $\langle q(\sigma(x_1, x_2)) \to \sigma(q_{i1}(x_1), q_{i2}(x_2)); L_{i1}, L_{i2}\rangle$, $1 \leq i \leq u$ be all rules in R with left-hand side $q(\sigma(x_1, x_2))$. Then the conditions (a) and (b) are satisfied. Note that $u = 0$ implies $L(T(q)) \cap \sigma(T_\Sigma, T_\Sigma) = \emptyset$.

The next lemma generalizes the above fact from σ (which is e_1) to any e_m.

Lemma 4.3. Let $n \geq 0$ and let $T = \langle \Sigma, Q, q_0, R\rangle$ be any tree automaton in $DFTA^{DREC_n}$. Then, for every integer $m \geq 1$ and state $q \in Q$, there exists an integer $u \geq 0$ and, for $1 \leq i \leq u$, there exist tree languages $L_{i1}, \ldots, L_{i,m+1} \in DREC_n$ and states $q_{i1}, \ldots, q_{i,m+1} \in Q$ such that the following conditions are satisfied:

(a) for any $1 \leq i \neq j \leq u$, there exists at least one $1 \leq l \leq m+1$ with $L_{il} \cap L_{jl} = \emptyset$,

(b) $L(T(q)) \cap e_m(T_\Sigma, \ldots, T_\Sigma) = \bigcup_{i=1}^{u} e_m(L_{i1} \cap L(T(q_{i1})), \ldots, L_{i,m+1} \cap L(T(q_{i,m+1})))$.

Proof. Since the exact proof is rather long and technical we present only an outline here. We induct on m.

The case $m = 1$ was considered in the discussion preceding the lemma.

Let $m > 1$ and compute $L(T(q)) \cap e_m(T_\Sigma, \ldots, T_\Sigma)$.Therefore, choose a rule

$$\langle q(\sigma(x_1, x_2)) \rightarrow \sigma(q_1(x_1), r(x_2)); L_1, L\rangle \qquad (**)$$

in R. Then, for a tree $e_m(t_1, \ldots, t_{m+1}) \in T_\Sigma$, $q(e_m(t_1, \ldots, t_{m+1})) \Rightarrow_T^* e_m(t_1, \ldots, t_{m+1})$ holds such that the rule $(**)$ is applied in the first step of the derivation if and only if $t_1 \in L_1 \cap L(T(q_1))$ and $e_{m-1}(t_2, \ldots, t_{m+1}) \in L \cap L(T(r))$. We note that $L = L(T')$ for some $T' = \langle \Sigma, Q', q_0', R'\rangle \in DFTA^{DREC_{n-1}}$.

Now, we can apply the induction hypothesis for $m-1$ and r. We obtain that there exists an integer $v \geq 0$ and, for $1 \leq i \leq v$, there exist tree languages $M_{i2}, \ldots, M_{i,m+1} \in DREC_n$ and states $q_{i2}, \ldots, q_{i,m+1} \in Q$ such that

$$L(T(r)) \cap e_{m-1}(T_\Sigma, \ldots, T_\Sigma) = \bigcup_{i=1}^{v} e_{m-1}(M_{i2} \cap L(T(q_{i2})), \ldots, M_{i,m+1} \cap L(T(q_{i,m+1}))).$$

On the other hand, applying again the induction hypothesis for $m-1$ and q_0', we have that there exists an integer $z \geq 0$ and, for $1 \leq j \leq z$, there exist tree languages $N_{j2}, \ldots, N_{j,m+1} \in DREC_{n-1}$ and states $s_{j2}, \ldots, s_{j,m+1} \in Q'$ such that

$$L \cap e_{m-1}(T_\Sigma, \ldots, T_\Sigma) = \bigcup_{j=1}^{z} e_{m-1}(N_{j2} \cap L(T'(s_{j2})), \ldots, N_{j,m+1} \cap L(T'(s_{j,m+1}))).$$

Thus, $e_{m-1}(t_2, \ldots, t_{m+1}) \in L \cap L(T(r))$ if and only if there exist $1 \leq i \leq v$ and $1 \leq j \leq z$ such that $t_2 \in N_{j2} \cap L(T'(s_{j2})) \cap M_{i2} \cap L(T(q_{i2})), \ldots, t_{m+1} \in N_{j,m+1} \cap L(T'(s_{j,m+1})) \cap M_{i,m+1} \cap L(T(q_{i,m+1}))$. Hence we obtain the integer u and the desired sequence of tree languages and states by choosing the rule $(**)$ in all possible ways and forming the sequences

$L_1, L_{ij2}, \ldots, L_{ij,m+1} \in DREC_n$ and

$q_1, q_{i2}, \ldots, q_{i,m+1} \in Q$

for every choice of $(**)$ and for every $1 \leq i \leq v$ and $1 \leq j \leq z$, where

$$L_{ij2} = N_{j2} \cap L(T'(s_{j2})) \cap M_{i2}, \ldots, L_{ij,m+1} = N_{j,m+1} \cap L(T'(s_{j,m+1})) \cap M_{i,m+1}.$$

It is important that, by Lemma 4.2, $L_{ij2}, \ldots, L_{ij,m+1} \in DREC_n$.◊

Now we can prove the following lemma which is the heart of our results.

Lemma 4.4. For every $n \geq 1$ and tree language L over Σ, if $L \in DREC_{n-1}$, then

$$\lim_{k\to\infty} \frac{\#(L \cap C_{n,k}^e) - \#(L \cap C_{n,k}^o)}{\#(C_{n,k})} = 0.$$

Proof. We prove by induction on n.

(a) The case $n = 1$. Let $L \subseteq T_\Sigma$ be such that $L \in DREC_0$, i.e., $L = L(T)$ for some deterministic top-down tree automaton $T = \langle \Sigma, Q, q_0, R\rangle$. Let $k \geq 1$ be an arbitrary integer. Then, by definition, $C_{1,k} = e_k(C_{0,k-1}, \ldots, C_{0,0}, C_{1,0})$. Hence, we see that $L \cap C_{1,k} = \emptyset$ or $L \cap C_{1,k} = e_k(L(T(q_1)) \cap C_{0,k-1}, \ldots, L(T(q_k)) \cap C_{0,0}, L(T(q_{k+1})) \cap C_{1,0})$ for some states $q_1, \ldots, q_{k+1} \in Q$. Then $\#(L \cap C_{1,k}^e) - \#(L \cap C_{1,k}^o) = 0$ or, by Observation 3.1, we have that $\#(L \cap C_{1,k}^e) - \#(L \cap C_{1,k}^o) = d_1 d_2 \cdots d_{k+1}$ where, for $1 \leq j \leq k$, $d_j = \#(L(T(q_j)) \cap C_{0,k-j}^e) - \#(L(T(q_j)) \cap C_{0,k-j}^o)$ and $d_{k+1} = \#(L(T(q_{k+1})) \cap C_{1,0}^e) -$

$\#(L(T(q_{k+1})) \cap C^o_{1,0})$. Since for each $1 \leq j \leq k$, $C^e_{0,k-j} = \{b\}$ and $C^o_{0,k-j} = \{a\}$ and similarly $C^e_{1,0} = \{b\}$ and $C^o_{1,0} = \{a\}$ we have that $|d_j| \leq 1$ for every $1 \leq j \leq k+1$. On the other hand, $\#(C_{1,k}) = 2^{k+1}$, so in both cases we obtain that

$$\frac{|\#(L \cap C^e_{1,k}) - \#(L \cap C^o_{1,k})|}{\#(C_{1,k})} \leq \frac{1}{2^{k+1}}$$

from where our statement follows.

(b) The induction step. Suppose that our statement is proved for $n \geq 1$. Let $L \in DREC_n$, i.e., $L = L(T)$ for some $T = \langle \Sigma, Q, q_0, R \rangle \in DFTA^{DREC_{n-1}}$. We shall show that, for any $\varepsilon > 0$,

$$\frac{|\#(L \cap C^e_{n+1,k}) - \#(L \cap C^o_{n+1,k})|}{\#(C_{n+1,k})} < \varepsilon$$

for almost every k.

Let, to this end, $\varepsilon > 0$ be given arbitrarily and let the integer m be fixed so that $1/2^m < \varepsilon/2$. Let $k > m$. Then we have $C_{n+1,k} = e_m(C_{n,k-1}, \ldots, C_{n,k-m}, C_{n+1,k-m})$ and hence, by invoking Lemma 4.3,

$$\begin{aligned} L \cap C_{n+1,k} = \bigcup_{i=1}^{u} e_m(&L_{i1} \cap L(T(q_{i1})) \cap C_{n,k-1}, \ldots, \\ &L_{im} \cap L(T(q_{im})) \cap C_{n,k-m}, L_{i,m+1} \cap L(T(q_{i,m+1})) \cap C_{n+1,k-m}) \end{aligned}$$

where $u \geq 0$, $L_{i1}, \ldots, L_{i,m+1} \in DREC_{n-1}$ and $q_{i1}, \ldots, q_{i,m+1} \in Q$, for $1 \leq i \leq u$. Moreover, it holds that for any $1 \leq i \neq j \leq u$, $L_{il} \cap L_{jl} = \emptyset$ for some $1 \leq l \leq m+1$. Thus, by Observation 3.1,

$$\#(L \cap C^e_{n+1,k}) - \#(L \cap C^o_{n+1,k}) = \sum_{i=1}^{u} d^{(k)}_{i1} d^{(k)}_{i2} \cdots d^{(k)}_{i,m+1}$$

where, for $1 \leq j \leq m$,

$$d^{(k)}_{ij} = \#(L_{ij} \cap L(T(q_{ij})) \cap C^e_{n,k-j}) - \#(L_{ij} \cap L(T(q_{ij})) \cap C^o_{n,k-j})$$

and

$$d^{(k)}_{i,m+1} = \#(L_{i,m+1} \cap L(T(q_{i,m+1})) \cap C^e_{n+1,k-m}) - \#(L_{i,m+1} \cap L(T(q_{i,m+1})) \cap C^o_{n+1,k-m}).$$

Next we estimate $|d^{(k)}_{ij}|$, for $1 \leq j \leq m+1$.

Obviously, we have $|d^{(k)}_{i,m+1}| \leq \#(L_{i,m+1} \cap C_{n+1,k-m})$.

Moreover, for $1 \leq j \leq m$, we have

$$\begin{aligned} |d^{(k)}_{ij}| &\leq \max\{\#(L_{ij} \cap L(T(q_{ij})) \cap C^e_{n,k-j}), \#(L_{ij} \cap L(T(q_{ij})) \cap C^o_{n,k-j})\} \\ &\leq \max\{\#(L_{ij} \cap C^e_{n,k-j}), \#(L_{ij} \cap C^o_{n,k-j})\}. \end{aligned}$$

We take the equality $\#(L_{ij} \cap C_{n,k-j}) = \#(L_{ij} \cap C^e_{n,k-j}) + \#(L_{ij} \cap C^o_{n,k-j})$ and we define $a^{(k)}_{ij} = (\#(L_{ij} \cap C^e_{n,k_j}) - \#(L_{ij} \cap C^o_{n,k-j}))/2$ if $\#(L_{ij} \cap C^e_{n,k-j}) \geq \#(L_{ij} \cap C^o_{n,k-j})$ and $a^{(k)}_{ij} = (\#(L_{ij} \cap C^o_{n,k-j}) - \#(L_{ij} \cap C^e_{n,k-j}))/2$ in the opposite case. Then we obtain in both cases that

$$\max\{\#(L_{ij} \cap C^e_{n,k-j}), \#(L_{ij} \cap C^o_{n,k-j})\} \leq \#(L_{ij} \cap C_{n,k-j})/2 + a^{(k)}_{ij}$$

where, by induction hypothesis on n, it holds that $\lim_{k\to\infty} a^{(k)}_{ij}/\#(C_{n,k-j}) = 0$. Thus we can compute as follows:

$$\frac{|\#(L \cap C^e_{n+1,k}) - \#(L \cap C^o_{n+1,k})|}{\#(C_{n+1}, k)} \leq \frac{\sum_{i=1}^{u} |d^{(k)}_{i1}||d^{(k)}_{i2}| \cdots |d^{(k)}_{i,m+1}|}{\#(C_{n+1,k})} \leq$$

$$\sum_{i=1}^{u} \frac{\#(L_{i,m+1} \cap C_{n+1,k-m})}{\#(C_{n+1,k})} \prod_{j=1}^{m} \left(\frac{\#(L_{ij} \cap C_{n,k-j})}{2} + a^{(k)}_{ij} \right) =$$

$$\sum_{i=1}^{u} \left(\frac{\#(L_{i,m+1} \cap C_{n+1,k-m})}{\#(C_{n+1,k})} \prod_{j=1}^{m} \frac{\#(L_{ij} \cap C_{n,k-j})}{2} \right) +$$

$$\sum_{i=1}^{u} \left(\frac{\#(L_{i,m+1} \cap C_{n+1,k-m})}{\#(C_{n+1,k})} (\sum y^{(k)}_{i1} y^{(k)}_{i2} \cdots y^{(k)}_{im}) \right).$$

The internal sum in brackets relates to all sequences $y^{(k)}_{i1}, \ldots, y^{(k)}_{im}$ for which

(a) any $y^{(k)}_{ij}$ is either $\#(L_{ij} \cap C_{n,k-j})/2$ or $a^{(k)}_{ij}$ $(1 \leq j \leq m)$ and

(b) there exists at least one $1 \leq j \leq m$ such that $y^{(k)}_{ij} = a^{(k)}_{ij}$,

and hence it denotes the sum of $2^m - 1$ members.

We obtained the sum of two sums and we show that both sums are less than $\varepsilon/2$.

In fact, the first sum can be written as

$$\frac{1}{2^m \#(C_{n+1,k})} \sum_{i=1}^{u} \#(L_{i,m+1} \cap C_{n+1,k-m}) \prod_{j=1}^{m} \#(L_{ij} \cap C_{n,k-j}).$$

We know that $\bigcup_{i=1}^{u} e_m(L_{i1} \cap C_{n.k-1}, \ldots, L_{im} \cap C_{n,k-m}, L_{i,m+1} \cap C_{n+1,k-m}) \subseteq C_{n+1,k}$ and that, by condition (a) of Lemma 4.3, the members of this union are pairwise disjoint. Hence $\sum_{i=1}^{u} \#(L_{i,m+1} \cap C_{n+1,k-m}) \prod_{j=1}^{m} \#(L_{ij} \cap C_{n,k-j}) \leq \#(C_{n+1,k})$, showing that the first sum is at most $1/2^m$ and thus less than $\varepsilon/2$, independently of the value of k.

The second sum, using the equality $C_{n+1,k} = e_m(C_{n,k-1}, \ldots, C_{n,k-m}, C_{n+1,k-m})$, can be written as

$$\sum_{i=1}^{u} \left(\frac{\#(L_{i,m+1} \cap C_{n+1,k-m})}{\#(C_{n+1,k-m})} \sum \frac{y^{(k)}_{i1} y^{(k)}_{i2} \cdots y^{(k)}_{im}}{\#(C_{n,k-1})\#(C_{n,k-2}) \cdots \#(C_{n,k-m})} \right) \leq$$

$$\sum_{i=1}^{u} \left(\sum \frac{y^{(k)}_{i1} y^{(k)}_{i2} \cdots y^{(k)}_{im}}{\#(C_{n,k-1})\#(C_{n,k-2}) \cdots \#(C_{n,k-m})} \right).$$

We note that this double sum denotes the sum of $u(2^m - 1)$ members. Now, we examine the quotients $y_{ij}^{(k)}/\#(C_{n,k-j})$. If for some $1 \leq j \leq m$, $y_{ij}^{(k)} = \#(L_{ij} \cap C_{n,k-j})/2$, then obviously $y_{ij}^{(k)}/\#(C_{n,k-j}) \leq 1/2$. On the other hand, if $y_{ij}^{(k)} = a_{ij}^{(k)}$ then, as mentioned before, $\lim_{k\to\infty} y_{ij}^{(k)}/\#(C_{n,k-j}) = 0$. Since to any member of the above sum there is at least one $1 \leq j \leq m$ such that $y_{ij}^{(k)} = a_{ij}^{(k)}$ we get that if k is sufficiently large then any member is less than $\varepsilon/2u(2^m - 1)$, hence the second sum is less than $\varepsilon/2$, too. This ends the proof of the lemma.◇

Now we state the main results.

Theorem 4.5. For every $n \geq 1$, $DREC_{n-1} \subset DREC_n$.

Proof. It was shown in Lemma 4.1 that $C_n^e \in DREC_n$. Moreover, by Lemma 4.4, $C_n^e \notin DREC_{n-1}$ because for each $k \geq 0$

$$\frac{\#(C_n^e \cap C_{n,k}^e) - \#(C_n^e \cap C_{n,k}^o)}{\#(C_{n,k})} = \frac{\#(C_{n,k}^e)}{\#(C_{n,k})} = \frac{1}{2}.\diamond$$

Theorem 4.6. $\bigcup_{n=0}^{\infty} DREC_n \subset REC$.

Proof. It is obvious that $L_e \in REC$. Moreover, by Lemma 4.4, $L_e \notin DREC_{n-1}$ since for each $k \geq 0$

$$\frac{\#(L_e \cap C_{n,k}^e) - \#(L_e \cap C_{n,k}^o)}{\#(C_{n,k})} = \frac{\#(C_{n,k}^e)}{\#(C_{n,k})} = \frac{1}{2}.\diamond$$

REFERENCES

[Eng] Engelfriet, J., Top-down tree transducers with regular look-ahead, Math. Systems Theory 10 (1977), 289-303.

[EngVog] Engelfriet, J. and Vogler, H., Pushdown machines for the macro tree transducer, Theoret. Comput. Sci. 42 (1986), 251-368

[FulVag] Fülöp, Z. and Vágvölgyi, S., Variants of top-down tree transducers with look-ahead, Math. Systems Theory, to appear

[GecSte] Gécseg, F. and Steinby, M., Tree Automata, Akadémiai Kiadó, Budapest, 1984.

[Kha1] Khabbaz, N.A., A geometric hierarchy of languages, J. Comput. System. Sci. 8 (1974), 142-157.

[Kha2] Khabbaz, N.A., Control sets on linear grammars, Inform. Control 25 (1974), 206-221.

[Vog1] Vogler, H., Basic Tree Transducers, J. Comput. System Sci. 34 (1987), 87-128.

[Vog2] Vogler, H., Iterated linear control and iterated one-turn pushdowns, Math. Systems Theory 19 (1986), 117-133.

Using generating functions to compute concurrency

Dominique Geniet & Loÿs Thimonier

L.R.I. Orsay, Bât. 490, Université Paris XI, 91405 ORSAY CEDEX, FRANCE
Université de Picardie, 33 rue Saint Leu, 80039 AMIENS CEDEX, FRANCE

Abstract

The aim of this work is the improvement of the computation of the concurrency measure defined by ***Beauquier***, ***Bérard*** and ***Thimonier*** in **[BT87]**. First, we present the ***Arnold-Nivat***'s model for the synchronization of sequential processes, and we recall the formal definition of the concurrency measure. Then, we present a new technique for the computation of this measure often avoiding a very expensive part of the computation. The *mutual exclusion* example illustrates this method.

Introduction

The aim of this work is to improve the computing of the concurrency measure for systems of concurrent processes presented in **[BT87]**.The present technique, based on ***Darboux***'s theorem, is very expensive in a software frame. By means of structural properties of generating functions of synchronized languages, we show that a large part of the computing can be omitted...

The evaluation of the concurrency must be independent of software data, because the operating system is one of the parameters of the computing. Many authors **[F86][BT87]** proposed some evaluations, linked to different models. We'll use here this introduced in **1986** by ***Beauquier***, ***Bérard*** and ***Thimonier***. The main idea consists in the evaluation of the average waiting time for each process: the less processes wait, the more concurrent are the computations.

The modeling of concurrent processes used here is introduced in **[AN82]**. It is based on rational languages: every process is represented by a finite automaton, its behaviors correspond to the words accepted by the automaton. So, a collection of programs $(p_i)_{i\in<1,n>}$ is modeled by the homogeneous product of the automata associated to the processes. Each transition of the homogeneous product is labelled by a **n**-tuple of letters $(x_i)_{i\in<1,n>}$. For each **j**, the jth component of $(x_i)_{i\in<1,n>}$ is the action executed by the p_j. Nevertheless, the accesses to the critical sections of the system must be controlled (writing on a memory block, access to a resource, precedence constraints, etc...). It's the aim of synchronization. In our model, it is performed by eliminating some transitions of the homogeneous product automaton. The language accepted by the result automaton will be call *synchronized language*.

An average value of the waitings in the words of the synchronized language is determined by the computing of the concurrency measure. This involves the knowledge of their distribution. The multivariate generating functions **[GJ83]** constitute a very powerful tool for the combinatorial study of a given sequence according to a given criteria. A language can be interpreted as a sequence of words, and its generating function $f(z) = \sum_{n\in N} a_n . z^n$ is a way of counting them. When an alphabet is partitioned in

A and **B**, we can consider a bivariate generating function, the coefficient $a_{p,q}$ of the monomial $x^p y^q$ being for instance the number of words of the language containing **p** letters of **A** and **q** letters of **B**. In a general way, if we call π the mapping which associate to each word **w** of the language the coefficient which take it into account, then the family $(\pi^{-1}(i))_{i \in \mathbf{N}}$ must be a partition of the language, so that each word is counted only once. Thanks to ***Chomsky-Schützenberger***'s theorem **[CS63]**, the determination of generating functions of non-ambiguous rational languages is simple as well theoretically as practically. This enables to limit the complexity of the computations.

However, this algorithm gives the generating functions of languages as the ratio of two polynomials. In order to determine the general term a_n of the associated series, the most efficient method is this of ***Darboux*** **[Ben74]**: it uses the asymptotic approximation of a_n, by means of the poles of the considered function (a pole is a singularity with maximal multiplicity on the convergence circle). The complexity (as well in time as in memory space) is no trivial. To solve this problem, we introduce a new algorithm, based on the structural properties of the generating function, which cancels the computing of poles. Some experience on computing concurrency measure allows us to assume that the singularities used by *Darboux*'s formula are very often of multiplicity **1** (all the examples implemented until now verify this property). In this last case, the concurrency measure is expressed with the only help of the convergence radius, which can be easily computed by ***Newton***'s method.

As an example, we present the access to a critical section by many processes. Many possible configurations (many processes, many resources) are considered. This points out to a discussion on the best configuration.

I) Representation of concurrent processes

1) The processes and their behaviors [AN82]

We define a process by the set of its finite behaviors. We are interesting in rational processes, who are represented by the concept of non deterministic finite automaton: each word accepted by the automaton is one of the possible behaviors of the process. So, the alphabet of the language accepted by the automaton is the set of the possible atomic actions for the process. Physically, the most detailed alphabet is the set of binary actions, but it is possible to consider alphabets in which many elementary actions are grouped in a single letter. This last case corresponds to high-level languages. For a process **p**, we denote **L(p)** the rational language associated to **p** and **A(p)** the alphabet on which **L(p)** is built. The main elements of the automaton correspond to the running elements of the process:

- The ***current state*** of the automaton represents the state of the process according to its context (the trace of the path from the initial state to the current state), and its issued transitions represent its possibilities.
- During the initialization, the process is obligatorily at the ***initial state*** of the automaton.
- When the process terminates (correctly) it session, it is obligatorily in a ***final state*** of the automaton.
- The actions performed by the process are modeled by the labels of the transitions of the automaton.

This model of representation is synchronous, because it imposes the performing of one atomic action at each time unit. Each action endures exactly one time unit. Thus, the real systems can be approached by considering the **gcd** of the duration of the elementary actions of the alphabet as the time unit. For instance, if **a** endures **3.5ms**, **b** endures **4.2ms** and **c** endures **8.4ms**, the **gcd** is **0.7ms**, each occurrence of **a** may be replaced by $\mathbf{a}^5$, each occurrence of **b** by $\mathbf{b}^6$ and each occurrence of **c** by $\mathbf{c}^{12}$. To simplify, we consider here that every atomic action endures one time unit.

2) Representation of multi-processes systems [AN82]

Let **n** be an integer and $(\Sigma_i)_{i\in<1,n>}$ be a **n**-uple of alphabets. The cartesian product of the $(\Sigma_i)_{i\in<1,n>}$ is denoted $\prod_{i=1}^{i=n} \Sigma_i$ and called *product alphabet.* We call *projector on the ith component* and we denote p_i the mapping $\prod_{j=1}^{j=n} \Sigma_j \rightarrow \Sigma_i$, $(l_j)_{j\in<1,n>} \rightarrow l_i$. This definition can easily be extended to the words: $p_i(w)=p_i(w^{(1)}).p_i(w^{(2)})...p_i(w^{(|w|)})$ (we denote $w^{(i)}$ the ith letter of w). Moreover, we extend this definition to the multi-indexes: let $b\in<1,n>$ and $i=(i_j)_{j\in<1,b>}$, the *projector on the ith component* is the mapping $\prod_{j=1}^{j=n} \Sigma_j \rightarrow \prod_{j=1}^{j=b} \Sigma_{i_j}$, $(l_j)_{j\in<1,n>} \rightarrow (l_{i_j})_{j\in<1,b>}$, which is extended to the words in the same way than above.

The present problem is the modeling of the simultaneous running of a system of processes $(\pi_i)_{i\in<1,n>}$. First, the $(\pi_i)_{i\in<1,n>}$ are considered to be independent: they don't communicate and there is no synchronization. So, at each time unit, the system performs a **n**-uple of actions: the ith process executes the action modeled by the ith component of the **n**-uple. This model is synchronous: at each time unit, every process must perform an atomic action. For each process, every behavior is possible. So, the processes $(\pi_i)_{i\in<1,n>}$ don't begin and terminate obligatorily their execution at the same time. So, we add in the alphabet a special letter, @, which models the inactivity of the processes before their beginning or after their termination. This letter is called ***inactive waiting***. We denote $\mathbb{A}_@(\pi_i)$ the automaton which accepts the language $L_@(\pi_i) = @^*L(\pi_i)@^*$.

Using the notion of projection given in **I.1**, we define the notion of ***concurrent behavior***:

Definition: We call *concurrent behavior issued from the sequential behaviors* $(w_i)_{i\in<1,n>}$ every word $w \in \prod_{i=1}^{i=n} L_@(\pi_i)$ such that $\forall i\in<1,n>$, $p_i(w)\in @^*w_i@^*$.

The automaton $\mathbb{A} = \prod_{i=1}^{i=n} \mathbb{A}_@(\pi_i)$ accepts the set of concurrent behaviors of

$(\pi_i)_{i\in<1,n>}$. This automaton can effectively be constructed **[G89]**.

The following problem is the modeling of the concurrent behaviors of systems of processes which have to communicate or to synchronize themselves (we call *synchronized concurrent behavior* such a behavior). Some configurations are forbidden. So, the set of the synchronized concurrent behaviors is a subset of the set of the concurrent behaviors of the system. The synchronization constraints are modeled by a centralized control, that is an automaton which restrains the possible **n**-uple of actions at each time unit. So, we consider a set $S \subset \prod_{i=1}^{i=n} \Sigma_i$. Let A_S be an automaton accepting $\left(\prod_{i=1}^{i=n} \Sigma_i \setminus S\right)^*$. The intersection $A_S \cap \prod_{i=1}^{i=n} \mathbb{A}_{@}(\pi_i)$ accepts the set of the synchronized concurrent behaviors.

So, we insert in each $A(\pi_i)$ a new letter, #, which models the waiting of the processes. This action is called ***active waiting***. For each process, we consider the language $L_\#(\pi_i)$ which words are the sequentials behaviors of π_i shuffled with $\#^*$. This language belongs to $\mathbf{Rec}(\Sigma \cup \{\#\})$. It is shown in **[G89]** that $\forall i\in<1,n>, \forall u\in L(\pi_i), \exists w\in L((\pi_i)_{i\in<1,n>}) \mid p_i(w)\in @^* u @^*$.

3) Power of the model

Most of the programming languages use simple statements, loops and recursivity (which can always be removed using memory management **[PMS88]**). Then, the most complex programs can be represented in *Arnold-Nivat*'s model. So, this model allows the representation of any usual program.

II) The concurrency measure

The classical problems of *mutual exclusion* or *dining philosophers* may be solved in many different ways. So, a given problem may be represented by many synchronized automata. The aim of this work is to compare the efficiency of such different solutions. To perform that, we compare the average waiting time per time unit and per process. Our evaluation mode will be a ***measure of the average of the waiting time***.

1) Definition of the measure

Let **L** be the language accepted by a synchronized finite automaton associated to the system $(p_i)_{i\in<1,N>}$. For each **n**, we denote $\lambda_n(L)$ the number of words of **L** of length **n**: $\lambda_n(L)=|L\cap\Sigma^n|$ (Σ is the alphabet of **L**). $n.\lambda_n(L)$ is the summation of the lengths of such words. In the same way, we denote $\mu_n(L)$ the number of # in the words of **L** of length **n**. One can see that $\mu_n(L) = \sum_{w\in L\cap\Sigma^n} |w|_\#$. The expression

$u_L(n) = \dfrac{\mu_n(L)}{n\,.\,\lambda_n(L)}$ represents exactly the frequency of # in the words of **L** of length **n**. We remark that the $u_L(n)$ is not defined if none of the words of **L** is of length **n**. **[BT87]** profs that it exists many subsequence of the form $(u_L(n.p+k))_{n\in\mathbb{N}}$ (**p** is a characteristic integer of the sequence) who admit limits t_k. The concurrency measure **t//** is the average of the t_k: it represents the average number of # per time unit for the words of **L**.

2) Using generating functions to compute t//

Let **L** be the language associated to a multi-processes synchronized system and $F(x,y) = \sum_{p\geq 0\ ,\ q\geq 0} a_{p,q}x^p y^q$ be the generating function defined by $\forall (p,q)\in\mathbb{N},\ a_{p,q} = \left|\left\{ w\in L \,\middle|\, \begin{cases} |w| = p+q \\ |w|_{\#} = q \end{cases} \right\}\right|$. The sequences $(\lambda_n)_{n\in\mathbb{N}}$ and $(\mu_n)_{n\in\mathbb{N}}$ can be easily determined from **F**: we denote $f(z) = F(z,z)$ and $g(z) = z.\left(\dfrac{\partial F}{\partial y}\right)_{(x=z,y=z)}$. It is shown in **[BT87]** that the general term of **f** (resp. **g**) is λ_n (resp. μ_n). So, **t//** may be computed from the synchronized automaton by means of generating functions:

- Computation of **F(x,y)** from the synchronized automaton (algorithm described above).
- Computation of the series **f(z)** and **g(z)**.
- For each **n**, we compute the period **p**, and then $\left(u_L(n) = \dfrac{\mu_n}{n.\lambda_n}\right)_{n\in<0,p-1>}$
- The average of the defined values of the limits t_k of $(u_L(n.p+k))_{n\in\mathbb{N}}$ gives **t//**.

3) Determination of the generating function

The determination of the generating function presented here is given by ***Chomsky*** and ***Schützenberger*** [CS63]. Let $A=<\Sigma,Q,S,T,\delta>$ be a synchronized finite automaton. It is shown in **[G89]** that the synchronized automaton may be translated to a linear system (we denote **TS(i)** the set of the transitions issued from the state **i**, **e(j)** the label of the transition **j** and **a(j)** the state pointed by **j**)

$$\left(\sum_{j\in TS(i)} \left(x^{|e(j)| - |e(j)|_{\#}} . y^{|e(j)|_{\#}} . F_{a(j)}(x,y) \right) - F_i(x,y) = \begin{cases} -1 \text{ if } i\in T \\ 0 \text{ if } i \notin T \end{cases} \right)_{i\in<1,n>}$$

which solution is a **n**-uple $(F_i(x,y))_{i\in<1,n>}$, $F_i(x,y)$ being the generating function associated to the automaton $<\Sigma,Q,\{i\},S,T,\delta>$. So the generating function of the synchronized automaton is obtained by $F(x,y)=F_1(x,y)$, since its unique initial state is the number one.

However, the considered language must be non-ambiguous (a given word only can be accepted by a single path), because every word of the language must be only counted once. If not, the generating function is falsified. It is knew (*Kleene* theorem) that a rational language can be recognized by a deterministic finite automaton. Every non deterministic automaton being non-ambiguous, a rational language is non-ambiguous.

4) Computing of the measure using the technique of Darboux

The bivariate generating function **F** of the considered synchronized language is given by the previous algorithm of the form $\frac{P(x,y)}{Q(x,y)}$, where $(P,Q) \in \mathbb{Z}[X,Y]^2$. *Beauquier*, *Bérard* and *Thimonier* give in **[BT87]** a method for computing $t_{//}$ based on the asymptotic expansion of the generating function. This technique needs important and expensive computing **[G89]**: there are two main problems

- the computing of the integer **p** such that some sub-sequences $\left((u_L(n.p+k))_{n \in \mathbb{N}}\right)_{k \in <0,p-1>}$ converges to $t_k \in \mathbb{R}$.
- the exact computing of the t_k often needs a large evaluation **[G89]**.

It is shown in **[G89]** that the first problem needs a very large digit precision in the computing: each pole of **F** of maximal multiplicity on its convergence circle is a root of the polynomial **Q(z)** of the form $z = R.e^{2.i.\pi.\frac{p}{q}}$, and the period **p** is obtained as the **gcd** of the integers **q** for all such poles. So, for each pole, it's necessary to compute the **exact value** of **q**. **z** is always given of the form **x+i.y**, with $(x,y) \in \mathbb{R}^2$.

So, we must compute $\dfrac{\mathrm{Arctan}\left(\frac{y}{x}\right)}{2.\pi}$ (given of a float form) and detect the rational form of this real.

III) Fast computing of the measure

So, we use here a property of the generating function of synchronized languages to improve in a very significant way the computing of the concurrency measure: we present here a method which gives the **exact** value of the measure (the value obtained by ***Darboux***'s method is only an asymptotic one) avoiding the expensive computing of the integer **p**.

The algorithm of *Mac Naughton* and *Yamada* **[MY60]** gives the rational expression of the language associated to a given automaton. Such languages can be put of the form $w_1.L_1^*.w_2L_2^*...w_n.L_n^*.w_{n+1}$, where $(w_i)_{i \in <1,n+1>}$ is a family of finite words and $(L_i)_{i \in <1,n>}$ a family of rational languages of the form $w_{i,1}.L_{i,1}^*.w_{i,2}L_{i,2}^*...w_{i,n_i}.L_{i,n_i}^*.w_{i,n_i+1}$, where $(w_{i,j})_{j \in <1,n_i+1>}$ is a family of finite words and $(L_{i,j})_{j \in <1,n_i+1>}$ defined in the same way, etc... The univariate generating function of such a language is of the form $\mathbf{f(z)} = \sum_{i\in I}\left(z^{s_i}.\prod_{j\in J_i}\left(f_j^*(z)\right)\right)$, **I** and $\mathbf{J_i}$ being finite sets, $(s_i)_{i\in I}\in\mathbb{N}^I$ and $\mathbf{f_i^*}$ the univariate generating function of $\mathbf{L_i^*}$ (that is $\frac{1}{1-f_i(z)}$, if $\mathbf{f_i(z)}$ is $\mathbf{L_i}$'s).

Let us denote $\mathbf{R^{(i)}}$ the convergence radius of $\mathbf{f_i^*}$. All the singularities of $\mathbf{f_i^*}$ on its convergence circle **[E76]** are simple poles located on a regular polygonal which admits $\mathbf{R^{(i)}}$ as a vertex. So, the convergence radius of **f** is $\mathbf{Min}\ \{R^{(k)}, k\in\bigcup_{i\in I}(J_i)\}$. Let us consider two generating functions $\mathbf{f_i^*}$ and $\mathbf{f_j^*}$ used in the expression of the generating function **f** given above (their convergence radius are $\mathbf{R^{(i)}}$ and $\mathbf{R^{(j)}}$). Their convergence circle are positioned in the following way:

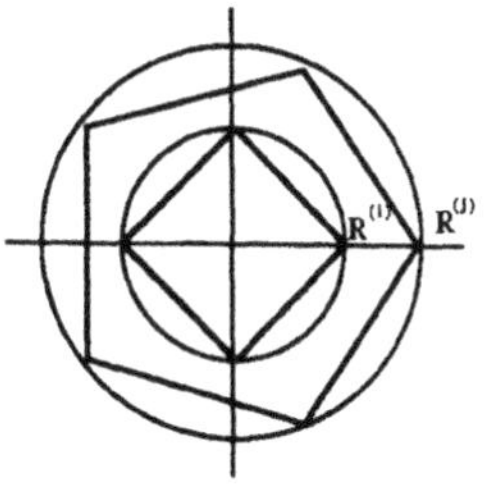

(the polygonals give the respective positions of the poles of $\mathbf{f_i^*}$ and $\mathbf{f_j^*}$). Let us study now the functions **f** with at least one non-simple pole: there are three integers **i, j** and **j'** ($j\neq j'$) such that $(j,j')\in J_i^2$ and $R^{(j)}=R^{(j')}$ (their convergence circle are the same) and $\mathbf{R^{(j)}}$ and $\mathbf{R^{(j')}}$ are ***simultaneously*** equal to the convergence radius **R** of the generating function **f**. The intuition lead us to consider that this property is often false. Instead, the poles of the generating functions are always simple in all the examples studied until now . So, we consider now the generating functions **f** whose poles are simple. Moreover, many examples model the functionment of systems associated to a language of the form **I.R*.T**, **I** corresponding to the initialization of the system, **T** its termination and **R** its running section. For such a system, the generating function **f** is a product $\mathbf{N(z).f_1^*(z)}$ ($\mathbf{f_1^*}$ being a positive generating function).

The period of a generating function $\mathbf{f_1(z)} = \sum_{n\in\mathbb{N}} a_n z^n$ is the greatest common factor of the set $\{n\in\mathbb{N} | a_n\neq 0\}$. So, if $\mathbf{f_1}$ is of period **p**, we have $f_1(z) = \sum_{n\in\mathbb{N}} a_{np} z^{np}$. Let **w** be a $\mathbf{p^{th}}$ of the unity and $\mathbf{f_1}$ a generating function of period **p**. We have $\forall z\in\mathbb{C}$, $\mathbf{f_1(z)=f_1(z.w)}$. So, we consider the following elementary results **[E76]**:

(i) Let **f** be a positive rational function of period **p** such that $\mathbf{f(0)=0}$. So, $f^*(z) = \frac{1}{1-f(z)} = \sum_{n\in\mathbb{N}} c_n z^n$ is of period **p** too. We denote $\mathbf{R^*}$ the convergence radius of $\mathbf{f^*}$. We have $\lim_{n\to+\infty}\left(c_{np}.\left(R^*\right)^{np}\right) = \frac{p}{R^*.f'\left(R^*\right)}$.

(ii) $\mathbf{R^*}$ is the unique solution of $\mathbf{f(z)=1}$, and we have $\mathbf{0<R^*<R}$.

The notations considered here are the previous ones: **F** is the bivariate generating function of a given synchronized language, **f** the associated univariate generating function and $\mathbf{g}(\mathbf{z}) = \mathbf{z}.\left(\frac{\partial \mathbf{F}}{\partial \mathbf{y}}\right)_{(x=z,y=z)}$. **f** and **g** are of the form $\frac{\mathbf{N}(\mathbf{z})}{1 - \mathbf{f}_1(\mathbf{z})}$ and $\frac{\mathbf{N}_1(\mathbf{z})}{(1 - \mathbf{f}_1(\mathbf{z}))^2}$ (**N**, $\mathbf{N_1}$ and $\mathbf{f_1}$ are polynomials of $\mathbb{Z}[X]$). We denote $\mathbf{R_1}^*$ the convergence radius of **f**. So, we obtain the following fundamental result:

Theorem: $t// = \frac{\mathbf{N_1(R_1^*)}}{\mathbf{N(R_1^*).R_1^*.f_1'(R_1^*)}}$

Proof

We denote $\sum_{n\in\mathbb{N}} \lambda_{np} z^{np}$ the expansion of **f** (with period **p**) and $\sum_{n\in\mathbb{N}} \mu_{np} z^{np}$ this of **g**.
The aim of this proof is to show that the given formula products the same result that ***Darboux***'s method result. First, we must compute the asymptotic values $\boldsymbol{\lambda}_{np}$ and $\boldsymbol{\mu}_{np}$ of λ_{np} and μ_{np}, and then the ratio $\frac{\boldsymbol{\mu}_{np}}{np.\boldsymbol{\lambda}_{np}}$.

We have $\frac{1}{1 - \mathbf{f}_1(\mathbf{z})} = \sum_{n\in\mathbb{N}} \mathbf{c}_n \mathbf{z}^n$ and $\mathbf{N}(\mathbf{z}) = \sum_{i=0}^{i=k} \mathbf{a}_i \mathbf{z}^{ip}$. So, $\forall n \geq k, \lambda_{np} = \sum_{i=0}^{i=k} \mathbf{a}_i \mathbf{c}_{(n-i)p}$.

Moreover **[Ben74]**, we have $\mathbf{c}_{np} \underset{n\to+\infty}{\sim} \mathbb{c}_{np} = \frac{1}{(\mathbf{R}_1^*)^{np}} \times \frac{\mathbf{p}}{\mathbf{R}_1^*.\mathbf{f}_1'(\mathbf{R}_1^*)}$. These two results lead us to $\lambda_{np} \underset{n\to+\infty}{\sim} \mathbb{c}_{np}.\left(\sum_{i=0}^{i=k} \mathbf{a}_i (\mathbf{R}_1^*)^{p.i}\right)$. So, we conclude that $\lambda_{np} \underset{n\to+\infty}{\sim} \boldsymbol{\lambda}_{np} = \mathbb{c}_{np}.\mathbf{N}(\mathbf{R}_1^*)$.

Each pole $\mathbf{z_j}$ of the generating function is of the form $\mathbf{R_1}^*.\mathbf{w_j}$ ($\mathbf{w_j}$ is the $\mathbf{j^{th}}$ $\mathbf{p^{th}}$ root of the unity). So, we have $\mathbf{f}_1(\mathbf{z}) \underset{z\to z_j}{\sim} \frac{\mathbf{A_j}}{1 - \frac{\mathbf{z}}{\mathbf{z_j}}} = -\frac{\mathbf{A_j}.\mathbf{z_j}}{\mathbf{z} - \mathbf{z_j}}$. The decomposition into simple elements give $\mathbf{A}_j.\mathbf{z}_j = \frac{1}{\mathbf{f}'(\mathbf{z}_j)}$ and then $\mathbf{A}_j = \frac{1}{\mathbf{R}_1^*.\mathbf{w}_j.\mathbf{f}_1'(\mathbf{R}_1^*.\mathbf{w}_j)}$. The immediate consequence of $\mathbf{f_1(z)} = \mathbf{f_1(z.w_j)} \Rightarrow \mathbf{f_1'(z)} = \mathbf{w_j.f_1'(z.w_j)}$ is the following: $\exists \mathbf{A} \in \mathbb{R} | \forall \mathbf{j} \in <1,p>, \mathbf{A}=\mathbf{A_j}$. So, we have $\mathbf{A} = \frac{1}{\mathbf{R}_1^*.\mathbf{w}_j.\mathbf{f}_1'(\mathbf{R}_1^*.\mathbf{w}_j)}$ and then $\mathbb{c}_{np} = \frac{\mathbf{p.A}}{(\mathbf{R}_1^*)^{np}}$, hence $\lambda_{np} \underset{n \to +\infty}{\sim} \frac{p.A.N(R_1^*)}{(R_1^*)^{np}}$ (I).

Let us study now the series $g(z) = \frac{N_1(z)}{(1 - f_1(z))^2}$. We know that $f_1^*(z) \underset{z \to z_j}{\sim} \frac{A}{1 - \frac{z}{z_j}}$.

Squared, this equivalence becomes $\frac{1}{(1 - f_1(z))^2} \underset{z \to z_j}{\sim} \frac{A^2}{\left(1 - \frac{z}{z_j}\right)^2}$. Let $\frac{1}{(1 - f_1(z))^2} = \sum_{n \in \mathbb{N}} b_{np} z^{np}$. By the same technique, one can obtain $\mu_{np} np \underset{n \to +\infty}{\sim} \mathbb{\mu}_{np} = \mathbb{b}_{np}.N_1(R_1^*)$.

Darboux's theorem **[Ben74]** proves that $b_{np} \underset{n \to +\infty}{\sim} np \sum_j \frac{A^2}{z_j^{np}}$, which lead us to $b_{np} \underset{n \to +\infty}{\sim} \frac{A^2.n.p}{(R_1^*)^{np}} \sum_j \frac{1}{w_j^{np}}$. $w^p=1 \Rightarrow w^{-p}=1$. So, $\sum_j w_j^{n.p} = \sum_j \frac{1}{w_j^{n.p}} = p$. These two results lead us to $b_{np} \underset{n \to +\infty}{\sim} \mathbb{b}_{np} = \frac{A^2.n.p^2}{(R_1^*)^{np}} = \frac{n.p^2}{(R_1^*)^{np+2}\left(f_1^{*\prime}(R_1^*)\right)^2}$. So, the value of μ_{np} is obtained by means of the formula $\mathbb{\mu}_{np} = N_1(R_1^*).\mathbb{b}_{np} = \frac{N_1(R_1^*).A^2.n.p^2}{(R_1^*)^{np}}$ (II).

From (I) and (II), we get $t_{//} = \lim_{n \to +\infty} \left(\frac{\mu_{np}}{n.p.\lambda_{np}}\right) = \frac{\mathbb{\mu}_{np}}{n.p.\lambda_{np}} = \frac{N_1(R_1^*)}{N(R_1^*).R_1^*.f_1'(R_1^*)}$.

qed

In the following examples, one can see how easy are the polynomials **N**, $\mathbf{N_1}$ and $\mathbf{f_1}$ to obtain from the expressions of **f** and **g**. Let us consider first the example introduced in **[BT87]**. The synchronized automaton is the following:

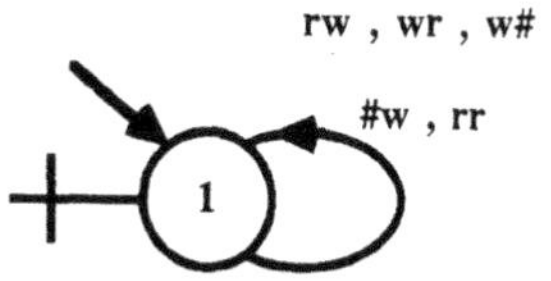

The accepted language is $L = \{rr, rw, wr, w\#, \#w\}^*$. We have

$F(x,y) = \dfrac{1}{1 - 3.x^2 - 2.x.y}$ and then $f(x) = \dfrac{1}{1 - 5.x^2}$ and $g(x) = \dfrac{2.x^2}{\left(1 - 5.x^2\right)^2}$. So, we get $f_1(x)=5x^2$, $R_1^* = \dfrac{\sqrt{5}}{5}$, $N(x)=1$ and $N_1(x)=2x^2$. The concurrency measure is obtained by the formula we get $t_{//} = \dfrac{N_1\left(R_1^*\right)}{N\left(R_1^*\right).R_1^*.f_1'\left(R_1^*\right)} = \dfrac{\frac{2}{5}}{\frac{\sqrt{5}}{5}.2\sqrt{5}} = \dfrac{1}{5}$, what is the expected value.

IV) Application to the mutual exclusion example

The formalization of the problem is simple: there is ***mutual exclusion*** between a family of processes if each one executes an infinite loop divided in two parts: the critical section (***crit***) and the rest of the program (***rest***) and if at a given time, at most one process is executing its critical section. This kind of configuration happens very often in computer science: access to a printer, to a file in a multi-users system, cpu, etc...

1) Sequential automata and synchronization

We'll consider here a family of processes that execute a program, with the following hypothesis: a process may only enter the critical section if it isn't owned by any other process. Therefore, if a process want to enter the critical section, it must make a request. If the access is impossible, the process is put in *active waiting*. Consequently, we get the representation language $\{d\#^*c^+sr^*\}^*$, with the following conventions: **d** models the request to the critical section, **#** the active waiting, **c** the critical section, **s** the termination of the critical section and **r** the remaining section. In order to simplify the present study, we assume the processes to begin and terminate their execution at the same time.

The associated finite automaton can be represented by the given graph. Once the modeling of the processes given, we consider the synchronization technique that enables to get the best optimization of the parallelism. In the case of **n** processes running concurrently, we get the following constraints:

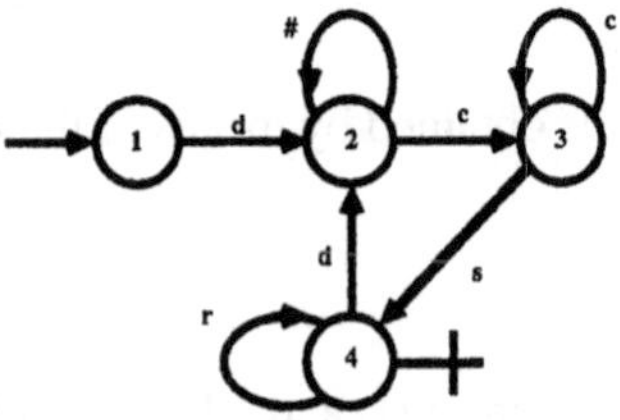

$$\text{w is forbidden} \Leftrightarrow \textit{or}\begin{cases} \exists\, i \in \langle 1,n\rangle \mid \textit{and}\begin{cases} w[i] = \text{"\#"} \\ \forall\, j \in \langle 1,n\rangle,\ w[j] \neq \text{"c"} \end{cases} \\ \exists\, (i,j) \in \langle 1,n\rangle \times \langle 1,n\rangle \backslash \{i\} \mid w[i] = w[j] = \text{"c"} \end{cases}$$

The first one forbids the waiting action for the processes if the critical section is free, and the second one verifies the mutual exclusion on the critical section. One can notice that these constraints don't manage the famine.

2) Computing the measure for a 2-processes system

The synchronized automaton that models the concurrent running of two processes which access to a critical section contains nine states (five for the compacted automaton). It can be graphically represented by the superposition of two tetrahedrals (in three dimensions). The vertices correspond to states where one of the processes is running while the other is waiting or in remaining section. The states located at the centers of the sides represent intermediate states, where one on the processes is leaving its critical section while the other enters its. The generating function given by this automaton is the ratio of two polynomials of degree **14**. The measure is **0.05676...**

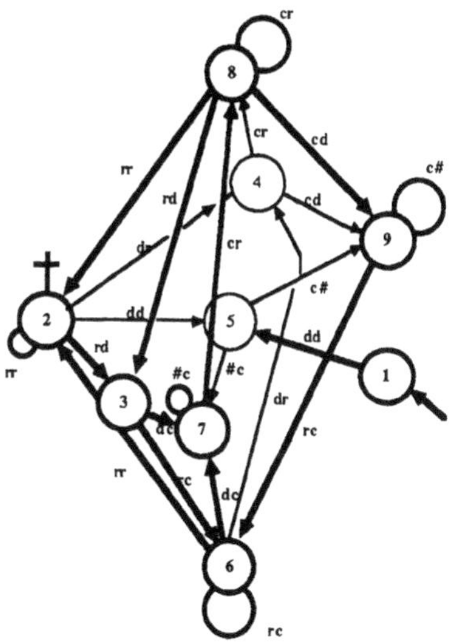

3) Computing the measure for a 3-processes system

We are now interested in three processes requesting the access to a critical section. With the help of the soft ***Automaf***, we get a synchronized automaton which contains thirteen states, but that can't be graphically represented in an easy way. The generating function is the ratio of two polynomials of degree **32**. The obtained measure is **0.13709...**

In order to be able to evaluate the possible improvement of a system containing two resources, we study a system of three processes acceding to a system with two resources. We start with the same sequential automata, but the synchronization set is slightly modified: the allowed number of **c** in an instantaneous configuration is now **2** instead of **1**. The automaton we get contains twelves states, and the concurrency measure is **0.046875..**

4) Interpretation of the results

For a system of two concurrent processes which request access to a given resource, there is in average one waiting time every twelves time units. For two processes, we have one waiting time for three time units. This mean that the adjunction of a process in the system considerably decreases the efficiency of each process. The phenomenon is confirmed by a system of four processes: two waiting time for three time units. The analysis of the results shows that the improvement of the efficiency of a multi-process system requires the access to many resources.

Conclusion

This paper is motivated by the results exposed in **[BT87]**, which shows the possibilities of computing the given measure, without presenting any implementation.

Within the frame of the software implementation of the computing of the measure for synchronized automata, the problem set by *Darboux*'s method gave rise to other theoretical approaches.The study of the general form of the obtained generating functions lead to the new method for computing the measure, which is extremely fast.

However, the generating function computing algorithm remains expensive, and can't be use in real time for realistic examples. Consequently, the conception of a cheaper technique constitutes one of the main investigation fields open by this work.

References

[AN82] **A.Arnold - M.Nivat**, "Comportements de processus", L.I.T.P. internal report n° 82-12, 1982

[Ben74] **E.A.Bender**, "Asymptotic methods in enumeration", SIAM review vol.16 n°4, p. 485-515, October 1974

[BT87] **B.Berard - L.Thimonier**, "On a concurrency measure", 2nd I.S.C.I.S. proceeding, Istanbul 1987, pp 211-225 & Nova Science Publishers, New York 1988

[CS63] **N. Chomsky - M.P.Schützenberger**, "The algebraic theory of context-free languages", Computer Programming and Formal Systems, pp 118-161, North Holland 1963

[E76] **S.Eilenberg**, "Automata Languages and machines", vol. A, Academic press 1976

[F86] **J.Françon**, "A quantitative approach of mutual exclusion", R.A.I.R.O. theoretical Informatics and Applications n° 20, 1986, pp 275-289

[G89] **D.Geniet**, "Automaf : un systemes de construction d'automates synchronises et de calcul de mesure du parallelisme", Thesis, Univ. Paris XI, to appear in 1989

[GJ83] **I.Goulden - D.Jackson**, "Combinatorial Enumerations", J. Wiley, New york 1983

[MY60] **R.Mac Naughton - Yamada**, "Regular expressions and state graphs for automata", I.E.E.E. transactions on electronic computers, EC-9, pp 39-47, 1960

[PMS88] **C.Pair - R.Mohr - R.Schott**, "Construire les algorithmes", Dunod informatique 1988

A LOGIC FOR NONDETERMINISTIC FUNCTIONAL PROGRAMS

EXTENDED ABSTRACT

ANA GIL-LUEZAS

Departamento de Informatica y Automatica.
Universidad Complutense. Madrid. Spain.

ABSTRACT.

We present a schematic functional programming language coupled with a logic of programs. Our language allows for μ-recursion, λ-abstraction, nondeterminism and calls to predefined functions. We define a denotational semantics, show that Kozen's propositional μ-calculus and Harel's first order dynamic logic for regular programs can be embedded in our logic LRF, and establish the soundness and completeness of two different proof systems: one using infinitary rules and an arithmetically complete one.

1- INTRODUCTION.

A rich variety of logics of programs has arosen in computer science, in an attempt to understand the fundamental principles behind formal verification methods. Most existing logics of programs are directed towards imperative programming languages [10], [11]. At the present time, functional programming is gaining popularity for a couple of reasons [7]. However, there are comparatively few publications devoted to investigate logical systems tailored to the behaviour of functional programs.

Cartwright and MacCarthy [2],[3],[4] have considered systems of mutually recursive definitions of functions in the framework of first order logic. Pazstor [17] investigates a dynamic logic for recursive programs in a nonstandard framework borrowed from Andréka, Németi and Sain [1]. Her programs are specified as unary recursive functions; composit'ion and mutual recursion are not expressable in the formalism. Goerdt [8] has obtained sound and relatively complete Hoare-like calculi for a typed $\lambda\mu$-calculus with higher order functions. However, his work is restricted to the investigation of partial correctness. The well known Logic for Computable Functions LCF [9] belongs to a rather different approach, more oriented towards interactive proof development, and will not be considered here. Neither Cartwright and MacCarthy nor Pazstor supply syntactic facilities to build structured functional programs and to reason about their behaviour by means of compositional, syntax directed rules.

In this communication we present a schematic functional programming language coupled with a logic of programs (LRF). The language allows for nondeterministic functions and is able to express conditionals, λ-abstractions, mutually recursive definitions ans calls to predefined functions. We define a denotational semantics (cfr. [20]) based on least fixpoint for functional expressions.

We show the expressive power of our system by embedding Kozen's propositional μ-calculus PMC (cfr. [14]) and Harel's dynamic logics for regular programs QDL (cfr. [10]) into LRF. Other logics such as dynamic logic for context-free programs CFDL (cfr. [10]), Meyer and Mitchell's logic for imperative programs with recursive procedures (cfr. [15]), etc. can be embedded similarly.

We also establish the soundness and completeness of two different proof systems: a "classical" one, which uses an infinitary rule, and arithmetically complete one which relies on Harel's notion of arithmetical structures (cfr. [10], [11]).

As an easy consequence of our results, we obtain a Löwenheim-Skolem theorem for LRF, as well as the Π^1_1-completeness of LRF's validity problem.

2- AN EXTENDED FIRST ORDER LOGIC.

As we have already said, our logic will allow for nondeterministic and predefined functions. So, using techniques inspired by Meyer and Mitchell [15] we extend first order logic with these features. We present a simple extension of first order logic obtained by adding variables for nondeterministic functions, that cannot be quantified, and a new logical symbol $\ni$, to express that an individual is a possible value of a nondeterministic function. Syntax and semantics are very similar to classical first order logic. In fact, this formalism could be reduced to ordinary first order logic by regarding function variables as predicate symbols. To view it as a different logic is merely a matter of convenience.

Definition 1. (Syntax)

Given a first order signature $\Sigma=\Sigma_{fn}\cup\Sigma_{pd}$, where Σ_{fn} is a set of ranked function symbols $\mathbf{f},\mathbf{g}$ etc. and Σ_{pd} is a set of ranked predicate symbols $\mathbf{p},\mathbf{q}$ etc., and a set of variables V=VAR$\cup$FVAR, where VAR includes individual (or data) variables $\mathbf{x},\mathbf{y},\mathbf{z}$ and FVAR includes ranked function variables $\mathbf{X},\mathbf{Y},\mathbf{Z}$. Σ-terms ($t\in TER_\Sigma$) and extended first-order Σ-formulas ($E\in EFOR_\Sigma$), are defined as follows: $t::= \mathbf{x} \mid \mathbf{f}(t_1,..,t_n)$

$$E::= \mathbf{p}(t_1,..,t_n) \mid (t_1=t_2) \mid \mathbf{X}(t_1,..,t_n)\ni t \mid \neg E_1 \mid (E_1\vee E_2) \mid \exists \mathbf{x} E_1$$

The three first kinds of formulas are called atomic. ∎

In the definition we have implicity assumed the proper arities $n\geq 0$ for $\mathbf{f},\mathbf{p}$ and $\mathbf{X}$. In the sequel, we always leave implicit such assumptions. We also accept the usual abbreviations for nonprimitive logical symbols.

Definition 2. (Valuations, environments)

Given a first order Σ-structure $\mathfrak{D}=\langle D,(\mathbf{f}^{\mathfrak{D}})_{\mathbf{f}\in\Sigma_{fn}},(\mathbf{p}^{\mathfrak{D}})_{\mathbf{p}\in\Sigma_{pd}}\rangle$:

(a) A valuation over $\mathfrak{D}$ is any mapping δ: VAR$\rightarrow$D. For $x\in D$ and $\mathbf{x}\in$VAR, we note $\delta[x/\mathbf{x}]$ the valuation δ' such that $\delta'(\mathbf{x})=x$ and $\delta'(\mathbf{y})=\delta(\mathbf{y})$ for all variables $\mathbf{y}\neq\mathbf{x}$.

(b) An enviroment over $\mathfrak{D}$ is any rank preserving mapping Δ: FVAR$\rightarrow \bigcup_{n\in N}\mathfrak{NF}_D^{(n)}$, where $\mathfrak{NF}_D^{(n)}=\{ X \mid X: D^n \rightarrow \mathfrak{P}(D) \}$ (n-ary nondeterministic functions over D). For $\mathbf{X}\in$FVAR and $X\in\mathfrak{NF}_D^{(n)}$, $\Delta[X/\mathbf{X}]$ is defined similarly as $\delta[x/\mathbf{x}]$ above.

Definition 3. (Semantical denotations)

Let a Σ-structure $\mathfrak{D}$, a valuation δ over $\mathfrak{D}$ and an enviroment Δ over $\mathfrak{D}$ be given. The denotation of terms (in symbols $t^{\mathfrak{D}}(\delta)\in D$) and extended first order formulas (in symbols $E^{\mathfrak{D}}(\Delta,\delta)\in\{true,false\}$) are defined in the standard way for the first-order logical symbols, and:

$(\mathbf{X}(t_1,..,t_n)\ni t)^{\mathfrak{D}}(\Delta,\delta)=true$ iff $t^{\mathfrak{D}}(\delta)\in\Delta(\mathbf{X})(t_1^{\mathfrak{D}}(\delta),..,t_n^{\mathfrak{D}}(\delta))$. ∎

Definition 4. (Satisfactibility, validity, logical consequence)

Let $\mathfrak{E}\cup\{E\}$ be any set of formulas. Then we define: $\mathfrak{D}\models\mathfrak{E}(\Delta,\delta)$, $\mathfrak{D}\models\mathfrak{E}$ and Sat($\mathfrak{E}$) in the usual way, $\mathfrak{D}\models\mathfrak{E}(\Delta)$ ($\mathfrak{E}$ is valid in $\mathfrak{D}$ under Δ) iff $\mathfrak{D}\models\mathfrak{E}(\Delta,\delta)$ for all valuations δ over $\mathfrak{D}$, and $\mathfrak{E}\models E$ (E is a logical consequence of $\mathfrak{E}$) iff $\mathfrak{D}\models\mathfrak{E}(\Delta)$ => $\mathfrak{D}\models E(\Delta)$ for all possible $\mathfrak{D}$, Δ. ■

Theorem 1. (Sound and complete calculus)

Any sound and complete calculus for first-order logic is a sound and complete calculus for this extension. ■

For a concrete choice, cfr. Section 5 below, where we present a calculus for the logic LRF, which includes extended first order logic as a fragment.

3- SYNTAX AND SEMANTICS OF LRF.

We now define our Logic for Recursive Functions (in sort LRF). LRF's language allows to build functional expressions, expressions and formulas. Functional expressions and expressions play the role of functional programs, while formulas provide the ability to reason on the behaviour of programs.

Definition 1. (Syntax)

Fix any signature Σ. LRF's functional Σ-expressions of arity n ($F,G\in FEXP^{(n)}_{\Sigma}$), Σ-expressions ($M,N\in EXP_{\Sigma}$) and Σ-formulas ($\varphi,\psi\in FFOR_{\Sigma}$) have the following syntax:

$$F^{(n)} ::= f \mid X \mid \lambda x_1 \ldots x_n . M \mid \mu X.F$$

$$M ::= t \mid F(M_1,\ldots,M_n) \mid (M\cup N) \mid (\varphi\rightarrow M) \mid \varepsilon x\varphi$$

$$\varphi ::= (t_1=t_2) \mid p(M_1,\ldots,M_m) \mid M\ni t \mid \neg\varphi \mid \varphi\vee\psi \mid \exists x\varphi$$ ■

We extend the notion of free variable to elements of FVAR in the natural way. Note that in a functional expression $\mu X.F$ all the occurrences of X in F are bounded.

The following technical definition is needed for the semantical treatment of the μ-operator.

Definition 2.

(a) $F\in FEXP^{(n)}$ (resp. $M\in EXP$, $\varphi\in FFOR$) is positive (resp. negative) in the functional variable X iff any free occurrence of X in F (resp. M, φ) is within the scope of an even (resp. odd) number of negation symbols.

(b) $F\in FEXP^{(n)}$ (resp. $M\in EXP$, $\varphi\in FFOR$) is existential (resp. universal) in the functional variable X iff F (resp. M, φ) is positive (resp. negative) in X and any occurrence of a formula $\exists x\psi$ in F (resp. M, φ) which lies within the scope of an odd (resp. even) number of negation symbols is such that X does not occur free in ψ. ■

The standard meaning of any recursive definition $\mu X.F$ is the least fixpoint of the equation X=F(X). As we will see in Lemma 2, least fixpoints always exist and can be constructed as the limit of a chain of approximations, provided that F is existential in X. For this reason, we restrict the syntactical especification of LRF by requiring that *a functional expression $\mu X.F$ may be formed only if F is existential in X*. In the sequel, LRF's syntax will be always understood in this restricted sense. On the other side, the meaning of any expression M is a subset of $\mathfrak{D}$'s domain, representing the possible values of M under nondeterministic evaluation.

Definition 3. (Semantical denotations)

Let a Σ-structure $\mathfrak{D}$, a valuation δ and an enviroment Δ be given. The denotations of functional expressions, expressions and formulas are defined as follows:

(1) $F^{\mathfrak{D}}(\Delta,\delta)\in\mathfrak{M}\mathfrak{F}_D^{(n)}$:

$f^{\mathfrak{D}}(\Delta,\delta)$ and $X^{\mathfrak{D}}(\Delta,\delta)$ are already defined.

$(\lambda \mathbf{x}_1\ldots\mathbf{x}_n.M)^{\mathfrak{D}}(\Delta,\delta)(x_1,\ldots,x_n)=M^{\mathfrak{D}}(\Delta,\delta[x_1/\mathbf{x}_1,\ldots,x_n/\mathbf{x}_n])$ for all $x_1,\ldots,x_n\in D$

$(\mu X.F)^{\mathfrak{D}}(\Delta,\delta)$ is the least function $X\in\mathfrak{M}\mathfrak{F}_D^{(n)}$ such that $X=F^{\mathfrak{D}}(\Delta[X/X],\delta)$ i.e., the least fixpoint of the operator $T=T(\mathfrak{D},\Delta,\delta,X.F):\mathfrak{M}\mathfrak{F}_D^{(n)}\to\mathfrak{M}\mathfrak{F}_D^{(n)}$ defined by $T(X)=F^{\mathfrak{D}}(\Delta[X/X],\delta)$. We denote $(\mu X.F)^{\mathfrak{D}}(\Delta,\delta)=fix(T)=fix(T(\mathfrak{D},\Delta,\delta,X.F))$.

(2) $M^{\mathfrak{D}}(\Delta,\delta)\subseteq D$:

$t^{\mathfrak{D}}(\Delta,\delta)=\{t^{\mathfrak{D}}(\Delta,\delta)\}$

$F(M_1,\ldots,M_n)^{\mathfrak{D}}(\Delta,\delta)=\{F^{\mathfrak{D}}(\Delta,\delta[x_1/\mathbf{x}_1,\ldots,x_n/\mathbf{x}_n])\ /\ x_i\in M_i^{\mathfrak{D}}(\Delta,\delta),1\le i\le n\}$

$(M\cup N)^{\mathfrak{D}}(\Delta,\delta)=M^{\mathfrak{D}}(\Delta,\delta)\cup N^{\mathfrak{D}}(\Delta,\delta)$

$$(\varphi\to M)^{\mathfrak{D}}(\Delta,\delta)=\begin{cases}M^{\mathfrak{D}}(\Delta,\delta) & \text{if } \varphi^{\mathfrak{D}}(\Delta,\delta)=\text{true}\\ \emptyset & \text{otherwise}\end{cases}$$

$(\varepsilon \mathbf{x}\varphi)^{\mathfrak{D}}(\Delta,\delta)=\{x\in D\ /\ \varphi^{\mathfrak{D}}(\Delta,\delta[x/\mathbf{x}])=\text{true}\}$

(3) $\varphi^{\mathfrak{D}}(\Delta,\delta)\in\{\text{true},\text{false}\}$:

$t_1=t_2$, $\neg\varphi$, $\varphi\vee\psi$ and $\exists x\varphi$ are defined as in first-order logic

$p(M_1,\ldots,M_n)^{\mathfrak{D}}(\Delta,\delta)=\text{true}$ iff there are $x_i\in M_i(\Delta,\delta)$ $1\le i\le n$ such that $p^{\mathfrak{D}}(x_1,\ldots,x_n)=\text{true}$

$(M\ni t)^{\mathfrak{D}}(\Delta,\delta)=\text{true}$ iff $t^{\mathfrak{D}}(\Delta,\delta)\in M^{\mathfrak{D}}(\Delta,\delta)$ ∎

From the above definition we have:

$(\exists x\ x=x)^{\mathfrak{D}}(\Delta,\delta)=\text{true}$, $(\exists x\ \neg\ x=x)^{\mathfrak{D}}(\Delta,\delta)=\text{false}$ for every $\mathfrak{D},\Delta,\delta$. We will denote these formulas by **tt** and **ff** respectively. We also let $\Omega^{(n)}$ stand for the functional expression $\lambda \mathbf{x}_1\ldots\mathbf{x}_n.(\mathbf{ff}\to\mathbf{x}_1)$, whose value is the totally undefined n-ary function for every $\mathfrak{D},\Delta,\delta$.

Definition 4 (Defined symbols)

We now introduce defined symbols < >, [ ] and $\downarrow$ as follows:

$\langle M\rangle\mathbf{y}\varphi$ stands for $\exists\mathbf{z}(M\ni\mathbf{z}\wedge\varphi[\mathbf{z}/\mathbf{y}])$ (where **z** is different from **y** and occurs neither in M nor in φ).

$[M]\mathbf{y}\varphi$ stands for $\neg\langle M\rangle\mathbf{y}\neg\varphi$.

$M{\downarrow}N$ stands for $\exists\mathbf{z}(M\ni\mathbf{z}\wedge N\ni\mathbf{z})$ (where **z** occurs neither in M nor in N).

< > and [ ] play here the same role as in dynamic logic [10], while $\downarrow$ acts as a kind of "existential equality" between nondeterministic expressions. ∎

We are able to prove a Coincidence lemma (cfr.[6]) which states that the meaning of formulas does depend only on the interpretation of free variables and actually occurring symbols. We also extend the concept of substitution by allowing the substitution of functional expressions for functional variables with the same arity. We claim that substitution can be defined in such a way that the following result holds

Lemma 1 (Substitution lemma)

For any $\mathfrak{D}$, $\mathbf{x}\in VAR$, $t\in TER$, $X\in FVAR$, $F\in FEXP$ and $\Theta\in FFOR$ resp. $\Theta\in EXP$ or $\Theta\in FEXP$:

(a) $(\Theta[t/\mathbf{x}])^{\mathfrak{D}}(\Delta,\delta)=\Theta^{\mathfrak{D}}(\Delta,\delta[t^{\mathfrak{D}}(\delta)/\mathbf{x}])$

(b) $(\Theta[F/X])^{\mathfrak{D}}(\Delta,\delta)=\Theta^{\mathfrak{D}}(\Delta[F^{\mathfrak{D}}(\Delta,\delta)/X],\delta)$ ∎

We are now in a position to justify the existence of the least fixed points needed for Definition 3(1). For this purpose we use the theorem of Knaster and Tarski [21], which may be applied because of the following

Lemma 2. (Monotonicity and continuity)

Given a Σ-structure $\mathfrak{D}$, a valuation δ over $\mathfrak{D}$, an environment Δ, $X \in FVAR$ and $F \in FEXP$:

(a) If F is positive in X, the operator $T(\mathfrak{D},\Delta,\delta,X,F)$ is monotonic.

(b) If F is existential in X, the operator $T(\mathfrak{D},\Delta,\delta,X,F)$ is continuous. ■

Theorem 1 (Syntactical approximations)

Let $\mathfrak{D}$ be any Σ-structure, δ a valuation over $\mathfrak{D}$, Δ an environment over $\mathfrak{D}$ and $\mu X.F \in FEXP_{\Sigma}^{(n)}$ (this forces F to be existential in X):

$$fix(T(\mathfrak{D},\Delta,\delta,X,F)) = \bigsqcup_{i\in\mathbb{N}} (\mu X.F)_i^{\mathfrak{D}}(\Delta,\delta)$$

where the ith syntactical approximation $(\mu X.F)_i$ of $\mu X.F$ is defined as:

$$(\mu X.F)_0 = \Omega^{(n)}, \quad (\mu X.F)_{i+1} = F[(\mu X.F)_i/X]$$ ■

The least fixpoint of a continuous operator $T:\mathfrak{F}_D^{(n)} \longrightarrow \mathfrak{F}_D^{(n)}$ does still admit another characterization that will play a key role for us later.

Lemma 3

Let $T:\mathfrak{F}_D^{(n)} \longrightarrow \mathfrak{F}_D^{(n)}$ be continuous. The graph of T's least fixpoint is then the set of all $\langle\vec{x},y\rangle = \langle x_1,\ldots,x_n,y\rangle \in D^{n+1}$ which have some T-derivation, where a T-derivation for $\langle\vec{x},y\rangle$ is understood as a (n+1)-tuple $\langle\vec{s},s\rangle = \langle s_1,\ldots,s_n\rangle$ of finite sequences: $s_i = \langle s_i(0),\ldots,s_i(k)\rangle \in D^{k+1}$ $(1\le i\le n)$ $\quad s = \langle s(0),\ldots,s(k)\rangle \in D^{k+1}$ all them of the same length k+1 (k≥0), satisfying the following conditions:

(1) $\langle\vec{x},y\rangle = \langle\vec{s}(k),s(k)\rangle$

(2) For all $0\le j\le k$: $s(j) \in T(fun_j(\langle\vec{s},s\rangle))(s_1(j),\ldots,s_n(j))$, where $fun_j(\langle\vec{s},s\rangle)$ is the function $x \in \mathfrak{F}_D^{(n)}$ whose graph is the finite set $\{\langle s_1(l),\ldots s_n(l),s(l)\rangle \ / \ 0\le l<j\}$ ■

4- EXPRESSIVE POWER OF LRF.

LRF is quite expressive. Other logics of programs can be embedded into it. Here, we take Kozen's propositional μ-calculus PMC (cfr. [14]) and Harel's dynamic logics for regular programs QDL [10] as typical representatives and sketch the corresponding embeddings.

Embedding PMC into LRF.

Given a propositional μ-calculus signature $\Sigma = \Sigma_p \cup \Sigma_a$ (where Σ_p is a set of proposition symbols noted **p**, **q** and Σ_a is a set of atomic action symbols **a**, **b**) and a set of propositional variables PVAR={P,Q,...}, the Σ-formulas (A,B) of PMC are defined as follows:

$B ::= \mathbf{p} \mid P \mid \langle\mathbf{a}\rangle B \mid \neg B \mid B_1 \vee B_2 \mid \mu P.B$

where B must be positive in P in $\mu P.B$.

Given a PMC's Σ-structure $\mathfrak{D} = \langle D, (\mathbf{p}^{\mathfrak{D}})_{\mathbf{p}\in\Sigma_p}, (\mathbf{a}^{\mathfrak{D}})_{\mathbf{a}\in\Sigma_a}\rangle$ and an environment Δ over $\mathfrak{D}$ $(\Delta(P)\subseteq D)$, the semantics of Σ-formulas of PMC $(B^{\mathfrak{D}}(\Delta))$ is well-known (cfr. Kozen [14]).

We associate a signature $\hat{\Sigma}=\{p_p,p_q,\ldots q_a,q_b,\ldots\}$ for LRF to any given signature $\Sigma=\{p,q,\ldots\}\cup\{a,b,\ldots\}$ for PMC. We also associate a nullary functional variable X_p to any propositional variable P. Then, we arbitrary fix an individual variable $\mathbf{x}$ and translate any PMC Σ-formula B to a LRF $\hat{\Sigma}$-formula $\varphi_B(\mathbf{x})$ having $\mathbf{x}$ as its only free variable, as follows:

B=p: $\varphi_B(\mathbf{x})= p_p(\mathbf{x})$

B=P: $\varphi_B(\mathbf{x})= X_p \ni \mathbf{x}$

B=<a>A: $\varphi_B(\mathbf{x})= \exists y(q_a(\mathbf{x},y) \wedge \varphi_A(y))$

B=¬A: $\varphi_B(\mathbf{x})= \neg\varphi_A(\mathbf{x})$

$B=A_1 \vee A_2$: $\varphi_B(\mathbf{x})= \varphi_{A2}(\mathbf{x}) \vee \varphi_{A2}(\mathbf{x})$

B=μP.A: $\varphi_B(\mathbf{x})= \mu X_p.(\lambda.\varepsilon\mathbf{x}\varphi_A(\mathbf{x}))\ni\mathbf{x}$

Remark that $\lambda.\varepsilon\mathbf{x}\varphi_A(\mathbf{x})$ is existential in X_p because A is positive in P.

<u>Theorem 1</u>.

Given a Σ-structure 𝔇, an environment Δ over 𝔇 and a Σ-formula B for PMC, we have: $B^{\mathfrak{D}}(\Delta)=\{x\in D \;/\; \hat{\mathfrak{D}}\models\varphi_B(\mathbf{x})(\hat{\Delta},\hat{\delta}[x/\mathbf{x}])\}$

provided that $\hat{\mathfrak{D}}$, $\hat{\Delta}$ and $\hat{\delta}$ are the $\hat{\Sigma}$-structure, environment over 𝔇 and valuation over 𝔇 respectively defined by: $\hat{\mathfrak{D}}$'s domain is the domain of 𝔇, $p_p^{\hat{\mathfrak{D}}}(x)$=true iff $x\in p^{\mathfrak{D}}$, $q_a^{\hat{\mathfrak{D}}}(x,y)$=true iff $x a^{\mathfrak{D}} y$, $\hat{\Delta}(X_p)=\Delta(P)$ and $\hat{\Delta}$ and $\hat{\delta}$ can be defined arbitrary otherwise. ■

<u>Embedding QDL into LRF</u>.

Let $\Sigma=\Sigma_{fn}\cup\Sigma_{pd}$ be a first-order signature and VAR a set of individual variables. The sets of QDL's Σ-formulas (A,B) and Σ-programs (α,β) are defined as follows:

$B::= p(t_1,\ldots,t_n) \mid \neg B \mid B_1\vee B_2 \mid \exists xB \mid \langle\alpha\rangle B$

$\alpha::= \mathbf{x}:=t \mid (\alpha_1;\alpha_2) \mid (\alpha_1\cup\alpha_2) \mid \alpha^{*} \mid B?$ where $t_1,\ldots,t_n,t$ are Σ-terms.

The semantics of QDL's Σ-formulas and Σ-programs is well-known [10],[11].

Following Goerdt [8], we consider any arbitrary but fixed number k and work out a translation of the fragment QDL_k of QDL involving programs and formulas which use at most k free variables $\mathbf{x}_1,\ldots,\mathbf{x}_k$ into LRF. To this purpose, we extend Σ to a LRF signature $\hat{\Sigma}$ in the following way:

$\hat{\Sigma}=\Sigma\cup\{\mathbf{asg}_j \;/\; 1\le j\le k\}\cup\{\mathbf{pr}_j \;/\; 1\le j\le k\}\cup\{\mathbf{ind},\ \mathbf{state}\}$.

For any given Σ-structure 𝔇, we consider the $\hat{\Sigma}$-structure $\hat{\mathfrak{D}}$ with domain $\hat{D}=D\cup D^k$ and interpretations:

(1) $\mathbf{asg}_j : \hat{D}\times\hat{D} \longrightarrow \hat{D}$

$\mathbf{asg}_j\ (\langle x_1,\ldots,x_k\rangle,x)=\langle x_1,\ldots,x_{j-1},x,x_{j+1},\ldots,x_k\rangle\in D^k$

$\mathbf{asg}_j$ is arbitrary defined in other cases.

(2) $\mathbf{pr}_j : \hat{D} \longrightarrow D$
$\mathbf{pr}_j\ (<x_1,..,x_k>)=x_j \in D$ and arbitrary in other cases.

(3) $\mathbf{ind} \subseteq \hat{D}$
$x \in \mathbf{ind}$ iff $x \in D$

(4) $\mathbf{state} : \hat{D} \longrightarrow \hat{D}$
$\mathbf{state}\ (x_1,..,x_k)=<x_1,...,x_k> \in D^k$ and arbitrary in other cases.

(5) Symbols in Σ are interpreted in such a way that they behave as their interpretations in $\mathfrak{D}$ for arguments belonging to D.

Now, we arbitrary fix an individual variable $\mathbf{x}$ of LRF (different from $\mathbf{x}_1,..,\mathbf{x}_k$) and define a syntactical translation assigning a monadic functional $\hat{\Sigma}$-expression F_α of LRF to each regular Σ-program α of QDL_k and a $\hat{\Sigma}$-formula φ_B of LRF to each Σ-formula B of QDL_k, as follows:

$\alpha = \mathbf{x}_j := t$: $F_\alpha = \lambda \mathbf{x}.(\neg \mathbf{ind}(\mathbf{x}) \rightarrow \mathbf{asg}_j(\mathbf{x}, t[\mathbf{pr}_1(\mathbf{x})..\mathbf{pr}_k(\mathbf{x})/\mathbf{x}_1..\mathbf{x}_k]))$

$\alpha = \alpha_1 \cup \alpha_2$: $F_\alpha = \lambda \mathbf{x}.(F_{\alpha 1}(\mathbf{x}) \cup F_{\alpha 2}(\mathbf{x}))$

$\alpha = \alpha_1 ; \alpha_2$: $F_\alpha = \lambda \mathbf{x}.F_{\alpha 2}(F_{\alpha 1}(\mathbf{x}))$

$\alpha = B?$: $F_\alpha = \lambda \mathbf{x}.(\neg \mathbf{ind}(\mathbf{x}) \rightarrow (\varphi_B[\mathbf{pr}_1(\mathbf{x})...\mathbf{pr}_k(\mathbf{x})/\mathbf{x}_1..\mathbf{x}_k] \rightarrow \mathbf{x}))$

$\alpha = \beta^*$: $F_\alpha = \mu X.(\lambda \mathbf{x}.\mathbf{x} \cup X(F_\beta(\mathbf{x})))$

$B = p(t_1,...,t_n)$: $\varphi_B = p(t_1,...,t_n)$

$B = \neg A$: $\varphi_B = \neg \varphi_A$

$B = A_1 \vee A_2$: $\varphi_B = \varphi_{A1} \vee \varphi_{A2}$

$B = \exists y A$: $\varphi_B = \exists y(\mathbf{ind}(y) \wedge \varphi_A)$

$B = <\alpha>A$: $\varphi_B = \exists z(\neg \mathbf{ind}(z) \wedge F_\alpha(\mathbf{state}(\mathbf{x}_1,...,\mathbf{x}_k)) \ni z \wedge \varphi_A[\mathbf{pr}_1(z)..\mathbf{pr}_k(z)/\mathbf{x}_1..\mathbf{x}_k])$

(where z is a variable not occurring in B)

We can prove that $\lambda \mathbf{x}.\mathbf{x} \cup X(F_\beta(\mathbf{x}))$ is existential in X by induction on β.

Theorem 2.

Assume that $\hat{\mathfrak{D}}$ has been associated to $\mathfrak{D}$ as explained above. Then:

(a) $t^{\mathfrak{D}}(\delta) = t^{\hat{\mathfrak{D}}}(\hat{\delta})$ for every $t \in TER_\Sigma$.

(b) $<y_1,...,y_k> \in F_\alpha^{\hat{\mathfrak{D}}}(\hat{\delta})(<x_1,...,x_k>)$ iff
$<\delta[x_1/\mathbf{x}_1,..,x_k/\mathbf{x}_k], \delta[y_1/\mathbf{x}_1,...,y_k/\mathbf{x}_k]> \in \alpha^{\mathfrak{D}}$
for every $x_1,..,x_k,y_1,..,y_k \in D$ and every QDL_k Σ-program α.

(c) $\varphi_B^{\hat{\mathfrak{D}}}(\hat{\delta})$=true iff $B^{\mathfrak{D}}(\delta)$=true for every QDL_k Σ-formula B. ∎

5- AN INFINITARY CALCULUS FOR LRF.

We present a sound and complete infinitary calculus for LRF. This, together with results from Section 4, will allow us to conclude that the validity problem of LRF is Π_1^1-complete.

The calculus is a Hilbert one composed by the following axioms and inference rules:

<u>Logical Axioms</u>

(EQ) Equality axioms (cfr. [13])

(PL) Kleene's axioms for propositional connectives (cfr [13])

($\exists_1$) $\varphi[t/\mathbf{x}] \to \exists\mathbf{x}\varphi$

($\exists_2$) $\exists\mathbf{x}\varphi \leftrightarrow \varphi$ if $\mathbf{x}$ does occur free in φ.

($\ni$) $t_1 \ni t_2 \leftrightarrow t_1 = t_2$

(PD) $\mathbf{p}(M_1,..,M_n) \leftrightarrow \exists y_1..\exists y_n(M_1 \ni y_1 \wedge \ldots \wedge M_n \ni y_n \wedge \mathbf{p}(y_1,..,y_n))$

where the variables y_j occur neither in $M_1,..,M_{n-1}$ nor in M_n

($\cup$) $(M \cup N) \ni t \leftrightarrow M \ni t \vee N \ni t$

($\longrightarrow$) $(\varphi \to M) \ni t \leftrightarrow \varphi \wedge M \ni t$

(ε) $(\varepsilon\mathbf{x}\varphi) \ni t \leftrightarrow \varphi[t/\mathbf{x}]$

(AP) $F(M_1,..,M_n) \ni t \leftrightarrow \exists y_1..\exists y_n(M_1 \ni y_1 \wedge \ldots \wedge M_n \ni y_n \wedge F(y_1,..,y_n) \ni t)$

where the variables y_j occur neither in $M_1,..,M_n$ nor in t.

(λ) $(\lambda\mathbf{x}_1 \ldots \mathbf{x}_n.M)(t_1,..,t_n) \ni t \leftrightarrow M[t_1/\mathbf{x}_1..t_n/\mathbf{x}_n] \ni t$

$(\mu)_i$ $(\mu X.F)_i(t_1,..,t_n) \ni t \to (\mu X.F)(t_1,..,t_n) \ni t$ for all $i \in \mathbb{N}$

<u>Inference Rules</u>

(MP) $\dfrac{\varphi,\ \varphi \to \psi}{\psi}$ ($\exists$I) $\dfrac{\varphi \to \psi}{\exists\mathbf{x}\varphi \to \exists\mathbf{x}\psi}$

(I) $\dfrac{\neg\ (\mu X.F)_i\ (t_1,..,t_n) \ni t \quad \text{for every } i \in \mathbb{N}}{\neg\ (\mu X.F)(t_1,..,t_n) \ni t}$ ■

We denote by $\vdash$ the deducibility relation of our calculus.

<u>Theorem 1</u>. (Soundness and completeness)

For any $\Phi \cup \{\varphi\} \subseteq FFOR_\Sigma$: $\Phi \models \varphi \iff \Phi \vdash \varphi$.

<u>Sketch of proof</u>. The soundness of the calculus is easy to check. The completeness proof can be obtained by tableaux method and Hintikka sets (for the use of these techniques in classical first-order logic, see Smullyan [19]) Some ideas can be borrowed from classical completeness proof for infinitary logic $L\omega_1\omega$ (cfr. Keisler [12]) ■

As an easy consequence of the previous theorem, we get

<u>Theorem 2</u>. (Löwenheim-Skolem theorem)

Any satisfiable LRF-formula has a countable model. ■

Another consequence, using the Theorem 4.2, is

<u>Theorem 3</u>.

The validity problem of LRF is Π^1_1-complete. ■

6- AN ARITHMETICAL CALCULUS FOR LRF.

In this section we present an arithmetically sound and complete finitary calculus for LRF. First we recall the notion of arithmetical structure in Harel's sense (cfr. [10]).

Definition 1. (Arithmetical structures)

A Σ-structure $\mathcal{D}$ is arithmetical if Σ contains the set of function and predicate symbols {**nat**,0,**1**,**+**,**comp**}, the domain D includes the set of natural numbers and $\mathcal{D}$ interprets the symbols **nat**,0,**1** and **+** in the expected way, and **comp**(i,x)=y meaning that "y is the ith component of the finite sequence encoded by x". ■

The completeness of a finitary calculus for LRF is based on the expressiveness of the extended first-order logic in arithmetical structures, according to Cook's notion of relative completeness [5].

Theorem 1. (Expressiveness)

Given any arithmetical Σ-structure $\mathcal{D}$, and Δ, δ an environment and a valuation over $\mathcal{D}$, respectively:

(a) For any $\varphi \in FFOR_\Sigma$ there exists an extended first-order formula E_φ such that $\varphi^{\mathcal{D}}(\Delta,\delta)=E_\varphi^{\mathcal{D}}(\Delta,\delta)$.

(b) For any $M \in EXP_\Sigma$ there exists an extended first-order formula $E_M(\mathbf{z})$ such that $M^{\mathcal{D}}(\Delta,\delta)=\{z\in D \,/\, E_M(z)^{\mathcal{D}}(\Delta,\delta[z/z])=true\}$, where $\mathbf{z}$ is a new variable not occurring in M.

(c) For any $F \in FEXP_\Sigma$ and $\mathbf{x}_1,\ldots,\mathbf{x}_n,\mathbf{z} \in VAR$ there exists an extended first-order formula $E_F(\vec{x},z)$ such that $(F^{\mathcal{D}}(\Delta,\delta))(\vec{x})=\{z\in D/ \, E_F(\vec{x},z)^{\mathcal{D}}(\Delta,\delta[\vec{x}/\vec{x},z/z])=true\}$ for all $\vec{x}\in D^n$, where $\mathbf{x}_1,\ldots,\mathbf{x}_n,\mathbf{z}$ are new variables not occurring in F.

Proof. The formulas E_φ, E_M and E_F are defined by simultaneous induction on φ, M and F:

(a) $\varphi \in EFOR_\Sigma$: $E_\varphi=\varphi$

$\varphi=p(M_1,\ldots,M_n)$: $E_\varphi=(\exists \mathbf{x}_1,\ldots,\exists \mathbf{x}_n \; E_{M_1}(\mathbf{x}_1)\wedge\ldots\wedge E_{M_n}(\mathbf{x}_n) \wedge P(\vec{\mathbf{x}}))$

$\varphi=M\ni t$: $E_\varphi=E_M(\mathbf{x})[t/\mathbf{x}]$

$\varphi=\neg\psi$: $E_\varphi=(\neg E_\psi)$

$\varphi=\psi_1\vee\psi_2$: $E_\varphi=(E_{\psi_1}\vee E_{\psi_2})$

$\varphi=\exists \mathbf{x}\psi$: $E_\varphi=(\exists \mathbf{x} E_\psi)$

(b) M=t: $E_M(\mathbf{z})=(t=\mathbf{z})$

$M=F(N_1,\ldots,N_n)$: $E_M(\mathbf{z})=(\exists \mathbf{x}_1,\ldots,\exists \mathbf{x}_n \; E_{N_1}(\mathbf{x}_1)\wedge\ldots\wedge E_{N_n}(\mathbf{x}_n) \wedge E_F(\vec{\mathbf{x}},\mathbf{z}))$

$M=N_1\cup N_2$: $E_M(\mathbf{z})=(E_{N_1}(\mathbf{z}) \vee E_{N_2}(\mathbf{z}))$

$M=\varphi\rightarrow N$: $E_M(\mathbf{z})=(\varphi \wedge E_N(\mathbf{z}))$

$M=\varepsilon \mathbf{y}\varphi$: $E_M(\mathbf{z})=\varphi[\mathbf{z}/\mathbf{y}]$

(c) $F=\mathbf{f}$: $E_F=(\mathbf{f}(\vec{\mathbf{x}})=\mathbf{z})$

$F=\mathbf{X}$: $E_F=(\mathbf{X}(\vec{\mathbf{x}})\ni\mathbf{z})$

$F=\lambda\vec{\mathbf{y}}.M$: $E_F=(E_M(\mathbf{z}))[\vec{\mathbf{x}}/\vec{\mathbf{y}}]$

$F=\mu X.G$: $E_F=(\exists \mathbf{k}\; \mathbf{nat}(\mathbf{k}) \wedge BE_F(\mathbf{k},\vec{\mathbf{x}},\mathbf{z}))$ where:

$BE_F(\mathbf{k},\vec{\mathbf{x}},\mathbf{z})=\exists \mathbf{s}_1,\ldots,\exists \mathbf{s}_n,\exists \mathbf{s}(CD_{G,X}(\mathbf{k},\vec{\mathbf{s}},\mathbf{s}) \wedge CP(\mathbf{k},\vec{\mathbf{s}},\mathbf{s},\mathbf{x},\mathbf{z})$

$CP(\mathbf{k},\vec{\mathbf{s}},\mathbf{s},\mathbf{x},\mathbf{z})=(\mathbf{s}_1(\mathbf{k})=\mathbf{x}_1\wedge\ldots\wedge\mathbf{s}_n(\mathbf{k})=\mathbf{x}_n \wedge \mathbf{s}(\mathbf{k})=\mathbf{z})$

$$CD_{G,X}(k,\vec{s},s)=\forall l(nat(l) \wedge 0\leq l\leq k \longrightarrow E_G(\vec{x},z)[\lambda\vec{x}.\varepsilon z GR(l,\vec{s},s,\vec{x},z)/X,\vec{s}(l)/\vec{x},s(l)/z])$$

$$GR(l,\vec{s},s,\vec{x},z)=\exists m(nat(l) \wedge m<l \wedge s_1(m)=x_1\wedge..\wedge s_n(m)=x_n \wedge s(m)=z)$$

where $s(l)$ stands for $comp(l,s)$ and $\vec{s}(l)$ stands for $s_1(l),..,s_n(l)$.

These formulas are inspired in Lemma 3.3 and have the following intuitive meanings:

$BE_F(k,\vec{x},z)$: "$\vec{s},s$ code a derivation sequence of lenght k+1 w.r.t. the operator given by G and X, and kth members of the coding sequence show that this derivation yields the tuple $<\vec{x},z>$"

$CP(k,\vec{s},s,x,z)$: "The kth components of coding sequences $\vec{s},s$ supply the tuple $<\vec{x},z>$"

$CD_{G,X}(k,\vec{s},s)$, $GR(l,\vec{s},s,\vec{x},z)$: "For $0\leq l\leq k$, the lth tuple $<\vec{s}(l),s(l)>$ given by the coding sequences satisfies that $s(l)$ is a possible result of the evaluation of G when the data variables $\vec{x}$ are interpreted as $\vec{s}(l)$ and the function variable X is interpreted as the function whose finite graph is given by the coding sequence through all tuples $<\vec{s}(m),s(m)>$, $m<l$" ■

The calculus consists of the following axioms and inference rules:

Logical Axioms

Axioms (EQ)-(λ) of the infinitary calculus (section 5)

Inference Rules

(MP) and (∃I) of the infinitary calculus (Section 5)

$$(C^*)\quad \frac{nat(k)\longrightarrow(E'(k,\vec{x},z) \to G[\lambda\vec{x}.\varepsilon z\ \exists l\ (nat(l) \wedge l<k \wedge E'(l,\vec{x},z))/X](\vec{x})\ni z) \qquad \varphi \to \exists k(nat(k) \wedge E'[\vec{t},t/\vec{x},z])}{\varphi \to (\mu X.G(\vec{t}))\ni t}$$

$$(I^*)\quad \frac{G[\lambda\vec{x}.\varepsilon z E(\vec{x},z)/X](\vec{x})\ni z \to E(\vec{x},z) \qquad \varphi \to \neg E[\vec{t},t/\vec{x},z]}{\varphi \to \neg(\mu X.G(\vec{t}))\ni t}$$

where E and E' are extended first-order formulas without free occurrences of X and the variable k occurring neither in $\vec{t}$ nor in t.

Notice that the axiom scheme $(\mu)_1$ and the infinitary rule (I) of section 5 have been replaced by two finitary inference rules, (C*) and (I*), which are motivated by the Lemma 3.3.

Theorem 2. (Arithmetical soundness and completeness)

For any arithmetical structure $\mathfrak{D}$ and $\varphi\in FFOR_\Sigma$: $\mathfrak{D}\models\varphi \iff Th(\mathfrak{D})\vdash_a\varphi$

where $\vdash_a$ stands for formal deducibility in the arithmetical calculus, and $Th(\mathfrak{D})$ is the extended first-order theory of $\mathfrak{D}$: $Th(\mathfrak{D})=\{E\in EFOR_\Sigma \ /\ \mathfrak{D}\models E\}$.

Sketch of proof.

Soundness: To prove that (I*) and (C*) are sound, let us fix any arithmetical Σ-structure $\mathfrak{D}$ and any environment Δ over $\mathfrak{D}$. Consider now given instances of (I*), (C*). Assume that the premises are valid in $\mathfrak{D}$ under Δ (cfr. Definition 2.4) and fix any valuation δ, we must show that the conclusions are satisfied by $\mathfrak{D}$ under Δ,δ. To this purpose, we consider:

$T:\mathfrak{N}\mathfrak{F}_D^{(n)} \longrightarrow \mathfrak{N}\mathfrak{F}_D^{(n)}$, $T(x)=G^{\mathfrak{D}}(\Delta[X/X],\delta)$ that is monotonic and continuous.

$Y \in \mathfrak{R}\mathfrak{J}_D^{(n)}$, $Y = fix(T) = (\mu X.G)^{\mathfrak{D}}(\Delta, \delta)$.

$Z \in \mathfrak{R}\mathfrak{J}_D^{(n)}$, $z \in Z(\vec{x})$ iff $E(\vec{x}, z)^{\mathfrak{D}}(\Delta, \delta[\vec{x}/\vec{x}, z/z])$.

$Z_k \in \mathfrak{R}\mathfrak{J}_D^{(n)}$, $z \in Z_k(\vec{x})$ iff $E'(k, \vec{x}, z)^{\mathfrak{D}}(\Delta, \delta[k/k, \vec{x}/\vec{x}, z/z])$, $k \in \mathbb{N}$.

$d_i, d \in D$, $d = t^{\mathfrak{D}}(\Delta, \delta)$ and $d_i = t_i^{\mathfrak{D}}(\Delta, \delta)$, $1 \le i \le n$.

Next, we consider $\varphi^{\mathfrak{D}}(\Delta, \delta) = true$ (φ the formula occurring in the rules). For (I*): The fact that the first premise is valid in $\mathfrak{D}$ under Δ means that $T(Z) \subseteq Z$ and so, $Y \subseteq Z$. The fact that the second premise is valid in $\mathfrak{D}$ under Δ implies that $d \notin Z(\vec{d})$. Hence, $d \notin Y(\vec{d})$, and the conclusion is satisfied by $\mathfrak{D}$ under Δ, δ.

For (C*): Because the first premise is valid in $\mathfrak{D}$ under Δ, we know that $Z_k \subseteq T(\bigsqcup_{l<k} Z_l)$ holds for every $k \in \mathbb{N}$ and so $Z_k \subseteq Y$ for every $k \in \mathbb{N}$. Now the fact that the second premise is valid in $\mathfrak{D}$ under Δ implies that $d \in Z_k(\vec{d})$ for some $k \in \mathbb{N}$. Hence, $d \in Y(\vec{d})$ and the conclusion is satisfied by $\mathfrak{D}$ under Δ, δ.

<u>Completeness</u>: Obviously, it is enough to prove the three following facts:

(1) For every $\varphi \in FFOR_\Sigma$: $Th(\mathfrak{D}) \vdash_a \varphi \leftrightarrow E_\varphi$

(2) For every $M \in EXP_\Sigma$: $Th(\mathfrak{D}) \vdash_a M \ni z \leftrightarrow E_M(z)$

(3) For every $F \in FEXP_\Sigma$: $Th(\mathfrak{D}) \vdash_a F(\vec{x}) \ni z \leftrightarrow E_F(\vec{x}, z)$

This can be done by simultaneous induction on φ, M and F. For instance, when F is $\mu X.G$: we must prove (a) $Th(\mathfrak{D}) \vdash_a F(\vec{x}) \ni z \rightarrow E_F(\vec{x}, z)$ and (b) $Th(\mathfrak{D}) \vdash_a E_F(\vec{x}, z) \rightarrow F(\vec{x}) \ni z$.

For (a) we take $E'(k, \vec{x}, z)$ as the formula $BE_F(k, \vec{x}, z)$ and use the rule (I*).

For (b) we take $E(\vec{x}, z)$ as the formula E_F and use the rule (C*). ∎

8- CONCLUSIONS AND FUTURE RESEARCH.

We have defined a logic for recursive nondeterministic functions with calls to global predefined functions, and illustrated its expressiveness by means of embedding results.

We have studied two sound and complete proof systems, an infinitary and an arithmetical one. The completeness of the infinitary calculus yields some nice consequences (Löwenheim-Skolem Th. and Π_1^1-completeness of the validity problem). The design of the arithmetical calculus and its completeness proof rely on a technique for expressing finite approximations of least fixpoints.

We believe that this technique can be used to develop a nonstandard view of LRF, along the lines of Andréka, Németi and Sain [1] (see also Pasztor [16]). This will be the subject of further research. In a different direction, we plan to investigate types and infinite data objects in the framework of our system.

ACKNOWLEDGEMENTS.

I wish to thank Teresa Hortala-Gonzalez and Mario Rodriguez-Artalejo for their help and hints, and Antonio Gavilanes-Franco for his comments.

REFERENCES.

[1] Andréka, H., Németi, I. and Sain, I., A Complete Logic for Reasoning about Programs via Nonstandard Model Theory, Parts I,II, T.C.S. 17 (1988), 193-212, 259-278.

[2] Cartwright, R. and McCarthy, J., Representation of Recursive Programs in First Order Logic, Stanford Art. Int. Memo AIM-324 (1979).

[3] Cartwright, R., Non-standard Fixed Points in First Order Logic, L.N.C.S. 164 (1984), 86-100.

[4] Cartwright, R., Recursive Programs as Definitions in First Order Logic. SIAM J. Comput, 13 (1984), 374-408.
[5] Cook, S.A., Soundness and Completeness of an Axiom System for Program Verification, SIAM J.Comput. 7 (1978), 70-90.
[6] Ebbinghaus, H.D., Flum, J. and Thomas, W., Mathematical Logic, Springer-Verlag (1984).84).
[7] Einsenbach,S. (ed). Functional Programming: Languages, Tools and Architectures, Ellis Horwood (1987).
[8] Goerdt, A., Ein Hoare Kalkül für getypte λ-terme. Korrektheit, Vollständigkeit, Anwendungen, Dissertation, RWTH Aachen (1985).
[9] Gordon, M.J., Milner, A.J. and Wadsworth, C.P., Edinburgh LCF, L.N.C.S. 78 Springer Verlag (1979).
[10] Harel, D., First Order Dynamic Logic, L.N.C.S. 68 (1979) Springer Verlag.
[11] Harel, D., Dynamic Logic, D.Gabbay and F. Guenthner (ed.) Handbook of Philosophical Logic 2, Reidel P.C. (1984), 479-604.
[12] Keisler, H.J., Model Theory for Infinitary Logic, North-Holland (1971).
[13] Kleene, S.C., Mathematical Logic, John Wiley and Sons (1967).
[14] Kozen,D., Results on the propositional μ-calculus, T.C.S. 27 (1983), 333-354.
[15] Meyer, A.R. and Mitchell, J.C., Termination Assertions for Recursive Programs: Completeness and Axiomatic Definiability, Inf. and Control 56 (1983), 112-138.
[16] Pasztor, A., Nonstandard Algorithmic and Dynamic Logic, J. Symbolic Comput. 2 (1986), 59-81.
[17] Pasztor, A., Recursive Programs and Denotational Semantics in Absolute Logics of Programs, Tech. Rep. FIU-SCS-87-1, Florida Int. Univ. (1987).
[18] Rogers, H., Theory of Recursive Functions and Effective Computability, McGraw-Hill (1967).
[19] Smullyan, R.M., First-order Logic, Springer-Verlag (1986).
[20] Stoy, J., Denotational Semantics: The Scott-Strachey Approach to Programming Language Theory, MIT Press (1977).
[21] Tarski, A., A Lattice Theoretical Fixpoint Theorem and its Applications, Pacific J. of Math. 5 (1955), 285-309.

Decision Problems and Coxeter Groups

Bernd Graw
Akademie der Wissenschaften der DDR
Karl–Weierstraß–Institut für Mathematik
PF 1304
DDR–1086 Berlin

1 Introduction

In this paper we will explore some remarkable connections between two apparently unrelated branches of mathematics and/or theoretical computer science, Coxeter groups and nonuniform complexity theory.

Coxeter groups originally arose as groups of symmetries of geometrical objects. H. S. M. Coxeter [15,16] described all irreducible finite groups generated by reflections and gave a presentation of such groups. This was the starting point of the theory of abstract Coxeter groups. 50 years later the Bourbaki group described Coxeter groups in terms of the exchange property [9]. Coxeter groups are important in the theories of regular polytopes, buildings, classical groups. With Coxeter groups are associated Coxeter complexes. The theory of Coxeter complexes was mainly developed by J.Tits [29,9]. The characterization of Coxeter groups by the exchange property was the starting point of A.Björner [6] to relate Coxeter groups to matroids and greedoids. Matroids are in their origin generalisations of facts of linear algebra as linear dependence, basis, linear hull. In fact matroids are now one of the basic structures of combinatorics, a concept which unifies aspects of optimisation, matchings, finite geometries, graphs, see [22]. Greedoids are generalisations of matroids as well as of abstract convexity structures. In another way matroids and greedoids appear in connection with Coxeter groups in the work of I.M.Gelfand, R.M.Goresky, R.D.MacPherson and V.V.Serganova [19], where they are special cases of so called flag-matroids.

The study of the Bruhat order of a Coxeter group and the notion of shellability gave further inside into the structure of Coxeter complexes (see [6]). Any Coxeter complex of a finite Coxeter group is homoeomorphic with a sphere.

One of the main goals of complexity theory is to bound the complexity of a given problem from below. Once this goal accepted one asks not for *ad hoc* methods but for criteria which imply lower bounds. We would like to find out (algebraic) invariants of computational problems which inherit informations on the complexity of the given problem. This approach was successfully applied f.i. by A. A. Rasborow [24,25], R. Smolensky [27], D. A. Barrington and D. Thérien [2,3,4] and J. Kahn, M. Saks and D. Sturtevant [20]. All these considerations are made for nonuniform complexity measures. We will concentrate on decision graphs and the structures related to this. An information system (see [10]) is a finite set of objects together with a set of attributes which describe properties of the objects. This structure is the unifying concept for computations with decision graphs or as they are called nowadays with branching programs (see [10,11,12,13]).

For any Coxeter group W we define faithful information system with object set W. If $f : W \to \mathbf{2}$ is a mapping then there is a decision graph computing f. How difficult are such functions? The standard counting techniques show that almost all such functions are exponential in the cardinality of the set of reflections of W. In other words, if the maximal faces of a Coxeter complex are colored with two colors $0, 1$ and we have to describe the set of all maximal faces of color 0, then this is in almost all cases a hard question.

The condition complex *Cond*(I) of an information system I is defined. If T is the set of reflections of W, then *Cond*(I) is a triangulation of a $\#T-1$–dimensional sphere and the Coxeter complex of W is embedded into *Cond*(I). Any function $f : W \rightarrow \mathbf{2}$ can in this way considered as the restriction of a language recognition problem. The complexity of f is a lower bound for the complexity of an arbitrary extension of f. This gives an algebraic frame for problems in nonuniform complexity theory. It is a generalisation since the case of the Abelian Coxeter group S_2^n yields exactly Boolean functions $f : \mathbf{2}^n \rightarrow \mathbf{2}$ and it is a specialisation since $g : W \rightarrow \mathbf{2}$ may be considered as a subfunction of $h : \mathbf{2}^T \rightarrow \mathbf{2}$.
Monotone functions are defined on W and it is shown that functions on irreducible finite Coxeter groups of type A_n, C_n, D_n are in fact functions on graphs, bipartite graphs and directed graphs, respectively. For the groups of type A_n the decision tree considerations lead to a new approach to sorting algorithms. All the considerations are made to have a better understanding of what problems have natural deciding procedures. Coxeter groups have in fact this feature. This corresponds to the relation between Coxeter groups and certain greedoids (see [8]), a fact that relies on the possible definition of this structures by exchange properties.

In the last section we summarise some techniques which work not only for Boolean functions, but for functions on Coxeter groups. These are the usual counting argument that shows that almost all functions on Coxeter groups are hard to compute, the *Cut–and–Paste* technique for one–time–only programs. [1] is used as standard reference for combinatorics. For further informations on the connections between finite posets and complexity theory see [14].

2 Coxeter Groups

2.1 Definitions

2.1.1 Let W be a finite group and S a set of generators of order 2. (W, S) is said to be a *Coxeter system* if the following condition is fulfilled: If $s, s' \in S$ and $m(s, s')$ is the order of ss' in W then

$$< s \in S; (ss')^{m(s,s')} = 1 >$$

is a presentation of W. In this case W is called a *Coxeter group*.
This might be written in the following way: Let G be a group and

$$f : S \rightarrow G$$

a mapping with

$$(f(s)f(s'))^{m(s,s')} = 1.$$

Then there is a homomorphic extension g of f from W to G.
From now on let W be a Coxeter group and S a set of generators.

2.1.2 The *length* of an element $w \in W$ is defined as the smallest natural number r such that w is the product of r elements of S. This length will be denoted by $l(w)$. Every sequence $\mathbf{s} = (s_1, s_2, ..., s_r)$ with $w = s_1 s_2 ... s_r$ and $r = l(w)$ is called a *reduced representation* of w.

2.2 The Exchange Property

Let $w \in W$, $s \in S$ and $l(sw) \leq l(w)$. Then for every reduced representation $\mathbf{s} = (s_1, s_2, ..., s_r)$ of w holds: There is an j, $1 \leq j \leq r$, such that $ss_1...s_{j-1} = s_1...s_j$.

2.2.1 Proposition. *Let $s \in S$, $w \in W$ and let $\mathbf{s} = (s_1, ..., s_r)$ be a reduced representation of w. Then either*

1. *$l(sw) = l(w) + 1$ and $(s, s_1, ..., s_r)$ is a reduced representation of sw. or*
2. *$l(sw) = l(w) - 1$ and there is an j, $1 \leq j \leq r$, such that*

$$(s_1, ..., s_{j-1}, s_{j+1}, ..., s_r) \text{ and } (s, s_1, ..., s_{j-1}, s_{j+1}, ..., s_r)$$

are reduced representations of sw and w respectively.

2.2.2 We assume now that (W, S) is a group W together with a set of quadratic generators S. In this situation one can ask if the exchange property for (W, S) is fulfilled.

2.2.3 Theorem. *(W, S) is a Coxeter system if and only if (W, S) fulfills the exchange property.*

2.3 The Coxeter Graph

2.3.1 We associate to every Coxeter system a so called *Coxeter graph* Γ_W. Γ_W is an undirected edge-labelled graph, $\Gamma_W = (S, E, f)$ defined by:

1. $E := \{\{s, s'\} | m(s, s') \geq 3\}$, where $m(s, s')$ denotes the order of ss' in W.
2. $f : E \to \mathbf{N}$, $f(\{s, s'\}) := m(s, s')$.

2.3.2 Example.

1. *Let $W = S_n$ be the symmetric group of n letters and let $S = \{s_1, ..., s_{n-1}\}$ be the set of transpositions $s_i := (i, i+1)$. Then (S_n, S) is a Coxeter system and Γ_{S_n} is the $(n-1)$-point line:*

2. *Let $W := (S_2)^n$ be the $n-th$ power of S_2 and $S := \{s_1, ..., s_n\}$ a set of generators. Then $\Gamma_{(S_2)^n}$ is the n-vertex graph without any edge.*
3. *Let $W := S_2^n \rhd S_n$ be the semidirect product defined by*

$$(\lambda_1, ..., \lambda_n; \pi)(\sigma_1, ..., \sigma_n; \psi) = (\lambda_1\sigma_{\pi^{-1}(1)}, ..., \lambda_n\sigma_{\pi^{-1}(n)}; \pi\psi).$$

Let $s_i := (1, ..., 1; (i, i+1))$, for $i = 1, ..., n-1$ and $s_n := (1, ..., 1, (0, 1); 1)$. Then $S := \{s_1, ..., s_n\}$ is a generating set for W and the Coxeter graph Γ_W is

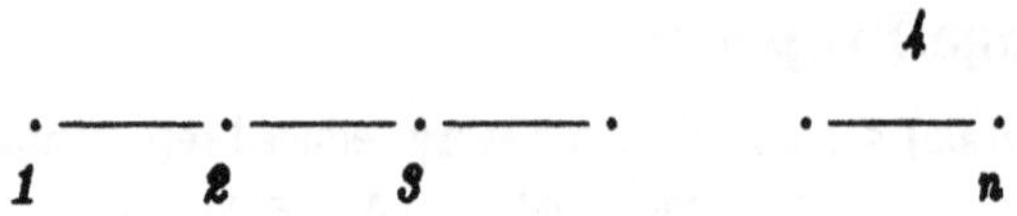

3 Coxeter Complexes

3.1 Definition and First Properties

3.1.1 Let $\mathbf{S} = (S_i)_{i \in I}$ be a family of nonempty subsets of I. The *nerve* of $\mathbf{S}$ is the following simplicial complex over I denoted by $N(\mathbf{S})$: Let $\sigma \subseteq I$. Then

$$\sigma \in N(\mathbf{S}) \text{ iff } \bigcap_{i \in \sigma} S_i \neq \emptyset.$$

3.1.2 If $I = \{(w_j, s_j) \mid 1 \leq j \leq k\}$ is a system of representatives of the left cosets $wW^{(s)}$, then

$$\mathbf{S} := (wW^{(s)})_{(w,s) \in I}$$

is a family of nonempty subsets of the set $\coprod_{J \subset S} W/W_J$
Let us consider $N(\mathbf{S})$:

$$\sigma \in N(\mathbf{S}) \iff \bigcap_{(w,s) \in \sigma} wW^{(s)} \neq \emptyset.$$

Since cosets of one and the same subgroup are disjoint or coincide holds:

$$\sigma \in N(\mathbf{S}) \iff \sigma = \{(w, s) \mid s \in J \subseteq S\}.$$

$N(\mathbf{S})$ is the *Coxeter complex* of W denoted by $\Delta(W, S)$. Hence $\Delta(W, S)$ is a pure simplicial complex with maximal faces $C_w = \{wW^{(s)} | s \in S\}$ and dimension $\#S - 1$. The maximal faces are called *chambers*.
Let Δ be a simplicial complex. Then we denote by $ch(\Delta)$ the set of its chambers.

3.1.3 Proposition.

1. *W operates transitive on $ch(\Delta(W, S))$.*
2. *W is a normal subgroup of the automorphism group of $\Delta(W, S)$ denoted by $Aut(\Delta(W, S))$.*
3. *If e denotes the unit of W then $Aut(\Delta(W, S))$ is the semidirect product of W by the isotropy group of the chamber C_e. This isotropy group in turn is isomorphic to the automorphism group of the Coxeter graph Γ_W. Hence $Aut(\Delta(W, S))$ may be considered as the semidirect product of W by a group of permutations of S.*

3.2 Irreducibility

A Coxeter group is called *irreducible* if Γ_W is connected. In this case W is not the direct product of subgroups.

3.2.1 Proposition. *Let $W = W_I \times W_J$. Then $\Delta(W,S) = \Delta(W_I) * \Delta(W_J)$, where $*$ denotes the join of simplicial complexes.*

3.2.2 Example. *Let $W = S_2^n$. Then*

$$\Delta(W,S) = *_{i=1}^{n} \Delta(S_2)$$

i.e. the n-dimensional sphere and $Aut(\Delta(W,S)) = S_2^n \rhd S_n$.

3.2.3 Fact. *Let $W = S_2^n \rhd S_n$. Then $\Delta(W,S)$ is isomorphic with the barycentric subdivision of $\Delta(S_2^n, R)$.*

3.3 Reflections

3.3.1 Let $T := \{wsw^{-1} \mid w \in W, s \in S\}$ be the set of conjugates of S. The set T is called the set of *reflections* of W. Every $t \in T$ defines a so called *wall*

$$L_t := \{w \cdot W^{(s)} \mid tw \cdot W^{(s)} = w \cdot W^{(s)}\}$$

in $\Delta(W,S)$. Every L_t is the union of faces of codimension 1 and every such face is in exactly one wall L_t.
If $w \in W$, then any representation $\mathbf{s} = (s_1, s_2, \ldots, s_p)$ of w (not necessary reduced) defines a sequence $\mathbf{t} = (t_1, t_2, \ldots, t_p)$ of elements of T by: $t_i := s_1 s_2 \ldots s_i (s_1 \ldots s_{i-1})$.
Let $t \in T$ and $n(\mathbf{s}, t) := \#\{i \mid t_i = t\}$.

3.3.2 Lemma.

1. *$\eta(\mathbf{s},t) := (-1)^{n(\mathbf{s},t)}$ is independent of the choice of the representing sequence $\mathbf{s}$ of w.*
2. *Let $T_w := \{t \in T \mid n(\mathbf{s},t) \equiv 1(2)\}$ and $\mathbf{s} = (s_1, \ldots, s_r)$ a reduced representation of w. Then $T_w = \{t_1, \ldots, t_r\}$.*
3. *Let $\mathrm{C}^0(t) := \{w \in W \mid t \notin T_w\}$, $\mathrm{C}^1(t) := W \setminus \mathrm{C}^0(t)$ and moreover $A^i(t) := \{w \cdot W^{(s)} \mid w \in \mathrm{C}^i(t)\}$, $i = 0, 1$. Then*
$$L_t = A^0(t) \cap A^1(t).$$

In geometric terms, $\mathrm{C}^0(t)$ is the set of all chambers of $\Delta(W,S)$ which are on the same side of L_t as $e = C_e$ (the unit of W). The length of an element w is the minimal number of walls to cross from e to w. The lemma shows moreover that the set T_w determines the element w uniquely. We have to remark that not all subsets of T are determined by elements of W.

4 Orderings of Coxeter Groups

4.1 The Bruhat Order

4.1.1 The definitions and facts of this section are mainly from [6]. Let (W,S) be a Coxeter system. Then we define a partial order on W:

$w \leq w'$ iff $w' = s_1...s_r$ is a reduced representation of w' then exists a reduced representation $w = s_{i_1}...s_{i_p}$ which is a subword of that of w'.
This partial order is called the *Bruhat order* on (W, S) (see [17]).
Let Δ be a simplicial complex. A linear order of $ch(\Delta)$ is said to be a shelling if for every pair (F_i, F_j) of maximal faces of Δ holds: If $i \leq j$ then there is an $k \leq j$ such that

$$F_i \cap F_j \subseteq F_k \cap F_j = F_j - \{x\}.$$

If $F_1, ..., F_q$ is a shelling of Δ, then we define the so called *restriction map* of the shelling $R : ch(\Delta) \to \Delta$

$$R(F_i) := \{x \in F_i \mid F_i - \{x\} \subseteq F_k, k \leq i-1\} \text{ for } 2 \leq i \text{ and } R(F_1) := \emptyset.$$

The *shelling characteristic* h is defined as

$$h := \#\{F_i \mid R(F_i) = F_i\}.$$

4.1.2 Theorem. *A shellable simplicial complex Δ of dimension d and shelling characteristic h has the homotopy type of a wedge of h d–dimensional spheres.*

4.1.3 Theorem. *Every linear extension of the Bruhat order defines a shelling of $\Delta(W, S)$ of characteristic 1.*

4.1.4 Corollary. *$\mid \Delta(W, S) \mid$ is piecewise linear homeomorphic with the $(\#S-1)$–dimensional sphere.*

4.1.5 Example. *$W = S_2^n$, $w \leq w'$ iff a reduced representation of w is a subword of a reduced representation of w', i.e. via the 1-1–correspondence to words over $\{0,1\}^n$ holds:*

$$w \leq w' \text{ iff } w^{-1}(1) \subseteq w'^{-1}(1).$$

A possible linear extension of this order is the lexicographic order on $\{0,1\}^n$.

4.2 The Weak Order

4.2.1 The *weak ordering* of a Coxeter group (W, S) is defined as $w \preceq w'$ iff there exist $s_1, ..., s_p \in S$ such that $w' = ws_1...s_p$ and $l(ws_1...s_i) = l(w) + i$ for $i = 1, ..., p$. (See [7] and the literature cited there.)

4.2.2 Proposition. *For $u, w \in W$ holds: $u \preceq w$ iff $T_u \subseteq T_w$.*

4.2.3 A.Björner [6] characterises those subsets $A \subseteq T$ that occur as T_w for some $w \in W$. For this let $A \subseteq T$ and $B \subseteq W$ and define

$$\varphi(A) := \{w \in W \mid T_w \subseteq T \setminus A\} \text{ and } \psi(B) := T \setminus \bigcup_{w \in B} T_w.$$

The tupel (φ, ψ) defines a Galois connection (see [5]) and closures

$$\psi\varphi(A) \supseteq A, \varphi\psi(B) \supseteq B.$$

4.2.4 A *gallery* between two chambers F_1, F_2 of a Coxeter complex is a sequence

$$F_1 = C_0, C_1, ..., C_k = F_2 \text{ of chambers such that } codim_{C_i} C_i \cap C_{i+1} = 1$$

for $i = 0, .., k-1$. F_1 and F_2 are the endpoints of the gallery and the C_i are the inner points. A gallery is called *minimal* if it has minimal length in the set of all galleries joining two fixed chambers.
Let Δ be a Coxeter complex and $L \subseteq ch(\Delta)$. Then L is said to be *convex* if it contains with the endpoints of minimal galleries also all inner points.

4.2.5 Lemma.

1. *Let $t \in T$ and let H be a halfspace defined by L_t. Then H is convex.*
2. *An arbitrary intersection of convex sets is convex.*

If we consider $B \subseteq W$ as a set of chambers of $\Delta(W, S)$, then the *convex hull* of B, $\mathbf{H}(B)$, is defined as the smallest convex set containing B.

4.2.6 Fact. $\varphi\psi(B) = \mathbf{H}(B \cup \{1\})$.

Following A.Björner we call a set $A \subseteq T$ *convex* if $\psi\varphi(A) = A$.
A is *biconvex* if A and $T \setminus A$ are both convex.

4.2.7 Proposition. *(Björner, [6]) Let (W, S) be a finite Coxeter group and let $A \subseteq T$. $A = T_w$ for some $w \in W$ iff A is biconvex.*

5 Decision Problems on Coxeter Groups

In this section we introduce an information system and classification problems for Coxeter groups and give first examples of such structures. The importance of such structures in computation theory is out of discussion and it is a challenging task to do this in a broader algebraic setting.

5.1 The Information System $I(W, T)$

5.1.1 We introduce an information system $I(W, T)$ for any Coxeter group W and the set T of reflections of W by:

- $I(W; T) := (W, T, \{0, 1\})$
- $T = \{wsw^{-1} \mid w \in W, s \in S\}$
- For $t \in T$ we define a mapping $W \to \{0, 1\}$ which we denote by abuse of notation again by t; $t(w) := i$ iff $w \in \mathbf{C}^i(t)$.

5.1.2 Fact. *The identity map on W and hence any map $f : W \to f(W)$ is dependent on $I(W, T)$.*

5.1.3 Hence we get for any Coxeter group W and any function f a classification problem $(I(W, T), f)$ and decision graphs over $I(W, T)$ which may be considered as questionnaires over $\Delta(W, S)$. If we consider decision graphs over $I(W, T)$, then we try to classify a chamber w by subsequent questions on the halfspaces defined by walls L_t including w. More precisely a decision graph over $I(W, T)$ is a quintuple $F = (Q, q_0, Y, \alpha, \delta)$:

- Q a finite set of nodes.
- Y a finite set disjoint with T.
- $\alpha : Q \to T \cup Y$ a map defining active and terminal nodes as the full inverse images of T and Y, respectively. q_0 is an active node.
- $\delta = \{\delta_q \mid q \in act\ F\}$, $\delta_q : \{0,1\} \to Q$.

5.1.4 Some complexity measures for decision graphs are defined.

1. $Size(F) := \#(actF)$
2. $Depth(F) :=$ length of a longest computation path from q_0 to a terminal node of F.

5.1.5 If $f : W \to Y$ is a mapping then we define:

1. $Size(f) := min\{Size(F) \mid \varsigma_F = f\}$
2. $Depth(f) := min\{Depth(F) \mid \varsigma_F = f\}$
3. $Treesize(f) := min\{Size(F) \mid \varsigma_F = f$ and F is a tree $\}$

5.1.6 Lemma. *If $f : W \to M$ is a map, then $\Omega(log_2(Treesize(f))) = Depth(f) \leq \#T$.*

5.1.7 The automorphism group $Aut\ \Delta(W,S) = W \triangleright G$ operates on decision graphs F over $I(W,T)$. We consider the following operation of W on $\{1,-1\} \times T$

$$w \cdot (\varepsilon, t) := (\varepsilon \cdot \eta(w^{-1}, t), wtw^{-1}), w \in W, t \in T (\text{see } 2).$$

This defines a faithful representation of W. Moreover any automorphism φ of $\Delta(W,S)$ permutes the chambers of $\Delta(W,S)$.
In fact, if C_w and C_v are neighbors (i.e. $C_w \cap C_v$ is a face of codimension 1), then $\varphi(C_w)$ and $\varphi(C_v)$ are also neighbors. Since any wall L_t is uniquely determined by a face of codimension 1 ([9]) the claim follows.
Let $(w,g) \in W \triangleright G =$ Aut $(\Delta(W,S))$. Then

$$(w,g)(\varepsilon,t) := (\varepsilon \cdot \eta(w^{-1}, g(t)), wg(t)w^{-1})$$

is this well defined operation.
Let $F = (Q, q_0, Y, \alpha, \delta)$ be a decision graph over $I(W,T)$. Then $(w,g) \cdot F := (Q, q_0, Y, \alpha', \delta')$ with

- $\alpha' : Q \to T \cup Y$,
$$q \in term(F) \Rightarrow \alpha'(q) := \alpha(q)$$
$$q \in act(F) \Rightarrow \alpha'(q) := wg(\alpha(q))w^{-1}$$
- $\delta' = \{\delta'_q \mid q \in act(F)\}$, $\delta'_q : \{0,1\} \to Q, \delta'_q(\varepsilon \cdot \eta(w^{-1}, g(t))) := \delta_q(\varepsilon)$

Hence we have to classify the orbits of functions $f : W \to M$ according to the action of $W \triangleright G$ on W and that of $Sym(M)$ on M.

6 Colorings of Coxeter Complexes

6.1 Colorings of Posets

6.1.1 Let P be a finite, pure poset and let $g : P \overset{\supseteq}{\to} C$ be a partially defined function. An element $a \in P$ is said to be *colored* if it is in *dom* g. The *color* of a is $g(a)$. We extend the partial function in such a way that all that elements defining monochromatic ideals of the partially ordered set are colored.

6.1.2 Every coloring $g : P \overset{\supseteq}{\to} C$ defines the following subposets of P:

1. $Pure(g, c) := g^{-1}(c),\ c \in C$
2. $Pure(g) := \bigcup_{c \in C} Pure(g, c)$
3. $Mix(g) := P \setminus dom\ g$

See [12,13,14] for further informations on colored posets.

6.1.3 If (W, S) is a Coxeter group and $f : W \to Y$ is a mapping, then this can be considered as a partially defined function on $\Delta(W, S)$. Thus $\Delta(W, S)$ is a colored simplicial complex and $Pure(f)$, $Mix(f)$, $Pure(f, y)$ are defined. If W is finite, then $\Delta(W, S)$ is a triangulation of the $(\#S - 1)$ - dimensional sphere and f colors the chambers and $Pure(f, f(w))$ is the set of all simplices that are only faces of chambers of color $f(w)$.

6.1.4 Example. $W = S_3$, $f : S_3 \to \{0, 1\} = Z/(2)$, $f(w) = 1$ *if* $w \in A_3$. *Then* $Pure(f, 1) = A_3$, $Pure(f, 0) = S_3 \setminus A_3$, *i.e. only chambers are colored and* $Mix(f) = \Delta(W, S) \setminus W$.

6.2 Treesize

6.2.1 If one considers information systems for sets and mappings of sets, then there is – at least – a relation between the decision tree complexity and the topological structure of some simplicial complexes, see [10] and [11]. For this let $h_0(\Delta)$ be the number of connected components of the simplicial complex Δ and define $h_0(P) := h_0(\Delta(P))$.

6.2.2 Theorem. *For any finite Coxeter group* W *and any function*

$$f : W \to [k]$$

holds: A classifying decision tree F *over* $I(W, T)$ *for* f *has size at least*

$$h_0(Pure(f)) - 1.$$

6.2.3 Example. *Let* $W = S_n$, *let* A_n *be the alternating group and let* $f : S_n \to S_n/A_n$ *be the canonical projection. This projection can be defined by* $f(s) := -1$ *for all* $s \in S$ *and homomorphic extension. Then*

$$Pure(f) = W, h_0(Pure(f)) = n! \text{ and } Treesize(f) \geq n! - 1.$$

(Any finite Coxeter group has a sign homomorphism η, *see 5.1, and* $Treesize(\eta) \geq \#W - 1.$))

6.3 The Condition Complex

6.3.1 For an arbitrary faithful information system I one can define a simplicial complex $Cond(I)$ called the condition complex of the information system I. Let $I = (X, A, V)$ be a faithful information system (see [12,13,14]), then the set of objects X is isomorphic to $Map(A, V)$ and

$$Cond(I) := \{c : A \xrightarrow{\supseteq} V | \text{dom } c \neq \emptyset\}.$$

If (W, S) is a Coxeter system, then $I(W, T)$ is a faithful information system and $Cond(I(W,T)) = \{g : T \xrightarrow{\supseteq} \{0,1\} | dom\ g \neq \emptyset\}$, i.e. $\Delta(W, S)$ is embedded into $Cond(I(W,T))$. Plainly $Cond(I(W,T))$ is isomorphic to $\{c : [\#T] \xrightarrow{\supseteq} \{0,1\} | dom\ c \neq \emptyset\}$ and this in turn is isomorphic to $\Delta(S_2^{\#T}, \{r_1, ..., r_{\#T}\})$.
Furthermore is $Cond(A, V)$ uniquely determined by $\#A$ and $\#V$ and we denote therefore this type by $Cond(\#A, \#V)$. Topologically the embedding of $\Delta(W, S)$ into $Cond(\#T, 2)$ is an embedding of a $(\#S - 1)$–dimensional sphere into a $(\#T - 1)$–dimensional one.
Hence every classification problem for a finite Coxeter group W can be considered as a language recognition problem over $\{0, 1\}$.

6.3.2 Theorem. *Let (W, S) be a finite Coxeter group and let $f : W \to Y$ be a mapping with $\#f(W) \geq 2$, then $Mix(f)$ is a pure $(\#S - 2)$-dimensional complex.*

6.3.3 The weak ordering is the tool to define *monotone functions* on a Coxeter group W. We consider on $[k]$ the natural order $1 \leq 2 \leq ... \leq k$, then a mapping $f : W \to [k]$ is said to be monotone if $u \preceq w$ implies $f(u) \leq f(w)$.

6.3.4 Example.

1. *$W = S_2^n$, then we discuss Boolean functions. In this case holds:*

$$u, w \in S_2^n,\ u \preceq w \text{ in the weak ordering of } S_2^n \text{ iff } T_u = u^{-1}(1) \subseteq T_w = w^{-1}(1),$$

where u, w are considered as elements of $\mathbf{2}^n$. Therefore the notions of monotony coincide.

2. *Let $f : \mathbf{2}^T \to \mathbf{2}$ be a monotone Boolean function, where T is the set of reflections of the Coxeter group (W, S). W is embedded into $\mathbf{2}^T$ via*

$$W \hookrightarrow \Delta(W, S) \hookrightarrow Cond(I(W,T)).$$

If g denotes the restriction of f to W, then g is a monotone function on W.

6.4 The Symmetric Group

6.4.1 We consider the case $W = S_n$ in greater detail. Remember that $S = \{s_1, ..., s_{n-1}\}$, $s_i = (i, i+1)$ is the set of Coxeter generators and that $T = \{(i,j) | 1 \leq i \leq j \leq n\}$ is the set of reflections. A decision graph over $I(W, T)$ asks questions $(i, j) \in T$ and the automorphism group of $\Delta(S_n, S)$ is $S_n \triangleright H$, where $H = \{1, h\}$, $h(s_i) = s_{n-i}$.

6.4.2 $Cond(I(S_n, T))$ may be considered as the set of partially defined graphs over the vertex set $[n]$. More exactly:
Let g be a chamber of $Cond(I(S_n, T))$, then $\Gamma(g) := ([n], g^{-1}(1))$ is a n–vertex graph. Every partially defined function $f : T \to \mathbf{2}$ is a partially defined graph on n vertices, edges in $f^{-1}(1)$ exist, edges in $f^{-1}(0)$ are forbidden and all other edges are possible.

An element $w \in S_n$ is a chamber of $Cond(I(S_n, T))$ and the graph defined is $\Gamma(w) = ([n], T_w)$.

6.4.3 Example. $W = S_4$, $w = s_1 s_2 s_1 s_3$.
$T_w = \{s_1, s_1s_2s_1, s_2, s_1s_2s_3s_2s_1\} = \{(1,2),(1,3),(2,3),(1,4)\}$

$\Gamma(w) =$ 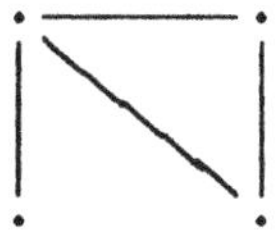

6.4.4 The group S_n acts on n–vertex graphs,

$$\pi \in S_n, \pi(\{i,j\}) := \{\pi(i), \pi(j)\}.$$

A mapping $f : \mathbf{2}^T \to \mathbf{2}$ is said to be a *graph property* , if f is constant on orbits of this action of S_n. In other words: f is constant on isomorphism classes of not numbered graphs.

6.4.5 The algebraic structure of $\Delta(W,S)$ forces some differences for decision graphs on $Cond(I(W,T))$ and $\Delta(W,S)$ respectively. If $f : \mathbf{2}^T \to \mathbf{2}$ is a mapping, then denote by g the restriction of f to W. We may ask for the relations between the complexities for f and g. Obviously lower bounds for g are also lower bounds for f and an upper bound for f implies this upper bound for g. More exactly, let $R \in \{Size, Depth, Treesize\}$ and let $g : W \to \mathbf{2}$. Then

$$R(g) = \min \{R(f) \mid f : \mathbf{2}^T \to \mathbf{2} \text{ and } f\downarrow_W = g\}.$$

Let g be as above and let $\mathcal{F}$ be an optimal decision graph for g with respect to R. Then $\zeta_{\mathcal{F}} = g$. Let $\overline{\zeta_{\mathcal{F}}} : \mathbf{2}^T \longrightarrow \mathbf{2}$ be the extension of $\zeta_{\mathcal{F}}$. (For a set $A \subset T$ we are looking for the terminal vertex of $\mathcal{F}$ which will be arrived with input A.) Then $\overline{\zeta_{\mathcal{F}}}$ is an extension of g and obviously holds:

$$R(g) = R(\overline{\zeta_{\mathcal{F}}}).$$

6.4.6 Example. *A function $f : \mathbf{2}^T \to \mathbf{2}$ is called* evasive *if $Depth(f) = \#T$.*
Let f be defined by $f(0,...,0) = 0$ and $f(a) = 1$ if $a \neq (0,...,0)$. f is evasive, since we have to ask for every coordinate.
Let g be the restriction of f to W. If $T \neq S$, then g is not evasive. For an input $w \in W$ we ask successively the questions $s \in S$.
$w = 1$ in W and w represents $(0,...,0)$ in $\mathbf{2}^T$ iff $s(w) = 0$ for all $s \in S$. Hence $depth(g) = \#S$ and g is not evasive if $S \neq T$.

6.4.7 We have considered the symmetric group on n symbols, which is a Coxeter group of type A_{n-1}. There are (see f.i. [29]) three infinite families of finite, irreducible Coxeter groups. These are A_n, $n \geq 1$,

$C_n, n \geq 2$

and

$D_n, n \geq 4$

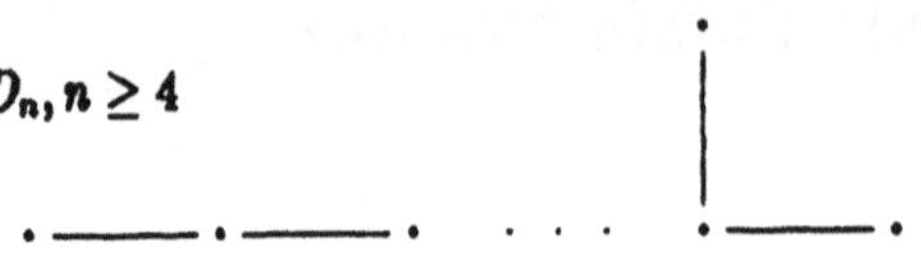

6.4.8 The Coxeter group with Coxeter diagram C_n is $W = S_2^n \triangleright S_n$ with

$$S = \{s_1, ..., s_{n-1}, s_n\}, s_i := (1, ..., 1; (i, i+1)) \text{ for } i = 1, ..., n-1$$

$$\text{and } s_n := (1, ..., 1, r; 1), \text{ where } r := (0,1) \in S_2 = Sym(\{0,1\}).$$

Let $J \subseteq [n]$ and let $r_J := (\sigma_1, ..., \sigma_n) \in S_2^n$ be defined by $\sigma_i := r$ iff $i \in J$. One easily deduces that the set of reflections is

$$T = \{(1;(k,l)) | 1 \leq k \leq l \leq n\} \cup \{(r_{\{i\}}; 1) | 1 \leq i \leq n\}$$
$$\cup \{(r_{\{k,l\}}; (k,l)) | 1 \leq k \leq l \leq n\}.$$

If $w \in W$ we may interpret T_w as the set of edges of a bipartite graph $b\Gamma_w := ([2n], E_w)$ by

- $(i, n+i) \in E_w$ iff $(r_{\{i\}}; 1) \in T_w$.
- If $i < j$, then $(i, n+j) \in E_w$ iff $(1; (i,j)) \in T_w$.
- If $i > j$, then $(i, n+j) \in E_w$ iff $(r_{\{i,j\}}; (j,i)) \in T_w$.

6.4.9 Let $H := \{r_J \in S_2^n | \#J \equiv 0(2)\}$ (see 6.4.8), then the Coxeter group $W = H \triangleright S_n$ has Coxeter diagram D_n and any element of W may be interpreted as a directed graph over $[n]$. The graph $d\Gamma = ([n], E_w)$ is defined by:

$$(i,j) \in E_w \text{ iff } \begin{array}{ll} \text{a)} & i < j \text{ and } (1;(i,j)) \in T_w \\ \text{or b)} & i > j \text{ and } (r_{\{i,j\}}; (j,i)) \in T_w. \end{array}$$

6.5 Some Observations on Complexity

6.5.1 A standard counting argument shows that the size of almost all functions $f : W \longrightarrow \mathbf{2}$ is $\Omega(\#W/\#T)$. Hence on S_n almost all functions have size at least $\Omega((n-2)!)$.

6.5.2 The *Cut-and-Paste Technique* yields exponential lower bounds for one-time-only branching programs for certain Boolean functions, see [23,30]. This technique can be transfered to functions on Coxeter groups. F.i. let $f : S_n \longrightarrow \mathbf{2}$ be the following function:

$$f(w) = 1 \text{ iff the graph } \Gamma_w \text{ has a } \lceil n/2 \rceil\text{-clique and } \lfloor n/2 \rfloor \text{ isolated points.}$$

6.5.3 Proposition. *The size of an one-time-only branching program for f is at least $\Omega(2^{n/4})$.*

6.5.4 We have considered the depth of functions $f : W \longrightarrow \mathbf{2}$ in 6.4.It was already shown that such functions are in general not evasive. A less trivial example is the following. Let $W = S_n$, let G be any graph on n vertices and let f_g be the graph property which decides wether a given graph Γ has a subgraph isomorphic to G.
Let $St(1)$ be the full star with apex 1 and let $g = f_{St(1)} \downarrow_W$.

6.5.5 Fact. *$Depth(g) = 2n - 3$, i.e. g is not evasive.*

6.5.6 We have to show that $f = f_{St(1)}$ is in fact evasive. Let
$P_f = \{w \in S_2^m \mid f(w) = 0\}, m = n(n-1)/2$. If we assume that f is not evasive, then P_f is acyclic over $\mathbf{Q}$ (see [20]). From theorems on the Burnside algebra of a finite group see [18,28] then follows that for any cyclic subgroup $H \subset S_n$ the fixed point complex P_f^H has Euler–Poincaré–characteristic 1.

Let now $H := \langle(1, \ldots, n)\rangle$. Then H is a cyclic subgroup of S_n and one easily deduces that the Euler– Poincaré–characteristic of P_f^H is $1 + (-1)^m$, where

$$m = \begin{cases} n/2 & \text{if n is even} \\ (n-1)/2 & \text{otherwise} \end{cases}$$

Hence f is an evasive graph property.

6.5.7 Now decision trees over $I(S_n, T)$ are in fact sorting trees (sorting by comparison see [21]). Hence all functions $f : S_n \longrightarrow \mathbf{2}$ have depth at most $O(n \cdot logn)$. For example *quicksort* is nothing else than decision by coset–searching.
The relation between sorting algorithms and structural considerations on $I(S_n, T)$ will be considered elsewhere.

7 Acknowledgements

I would like to thank Burkhard Molsan and Uwe Schäfer for extremely useful discussions and for first readings of the manuscript. Many of the ideas were clarified in these discussions. I am grateful to Lothar Budach for his encouragements and many hours of common thinking on decision graphs and algebraic and topological structures which may inherit informations on computational complexity.

References

[1] M. AIGNER, *Combinatorial Theory* Springer–Verlag, Berlin, 1979.

[2] D. A. BARRINGTON, *Bounded–Width Polynomial–Size Branching Programs Recognize Exactly those Languages in NC^1*. Proceedings 18–th ACM Symposium on Theory of Computing (1986),1–5.

[3] D. A. BARRINGTON, D. THÉRIEN, *Finite Monoids and the Fine Structure of NC^1*. Proceedings 19–th ACM Symposium on Theory of Computing (1987), 101-109.

[4] D. A. BARRINGTON, D. THÉRIEN, *Nonuniform Automata over Groups.* Proceedings 14–th International Colloquium on Automata, Languages and Programming (1987), 264–279.

[5] G. BIRKHOFF, *Lattice Theory,* AMS Colloquium Publications 25, American Mathematical Society, Providence, Rhode Island, 1967.

[6] A. BJÖRNER, *Some Combinatorial and Algebraic Properties of Coxeter Complexes and Tits Buildings.* Advances in Mathematics 52(1984), 173-212.

[7] A. BJÖRNER, *Orderings of Coxeter Groups.* Contemporary Mathematics 34(1984), 175-195.

[8] A. BJÖRNER, *On Matroids, Groups and Exchange Languages.* Colloquia Mathematica Societatis Janos Bolyai, 40, Matroid Theory (ed. A. Recski, L. Lovasz), North Holland, Amsterdam, 1985.

[9] N. BOURBAKI, *Groupes et algébres de Lie, Chapitre 4,5,6* Hermann, Paris 1968.

[10] L. BUDACH, *Klassifizierungsprobleme und das Verhältnis von deterministischer und nichtdeterministischer Raumkomplexität,* Berlin, Humboldt-Universität, Sektion Mathematik, Seminarbericht Nr.68, 1985.

[11] L. BUDACH, *A Lower Bound for the Number of Nodes in Decision Trees,* EIK 21(1985), 221-228.

[12] L. BUDACH, *Arsenals and Lower Bounds,* Lecture Notes in Computer Science 278(1987), 55-64.

[13] L. BUDACH, B. GRAW, *Nonuniform Complexity Classes, Decision Graphs and Homological Properties of Posets,* Lecture Notes in Computer Science 208(1985),7-13.

[14] L. BUDACH, B. GRAW, C. MEINEL, S. WAACK, *Algebraic and Topological Properties of Finite Partially Ordered Sets,* BSB Teubner Leipzig,1988.

[15] H. S. M. COXETER, *Discrete Groups Generated by Reflections.* Annals of Mathematics 35(1934), 588–621.

[16] H. S. M. COXETER, *The Complete Enumeration of Finite Groups of the Form $R_i^2 = (R_iR_j)^{k_{ij}}$.* Journal of the London Mathematical Society 10(1935), 21–25.

[17] V. V. DEODHAR, *Some Characterizations of Bruhat Ordering on a Coxeter Group and Determination of the Relative Möbius Function,* Inventiones Mathematicae 39(1977), 187-198.

[18] T. TOM DIECK, *Transformation Groups and Representation Theory.* Lecture Notes in Mathematics, vol. 766, Springer–Verlag, Berlin 1979.

[19] I.M. GELFAND, R.M. GORESKY, R.D. MACPHERSON, V.V. SERGANOVA, *Combinatorial Geometries, Convex Polyhedra, and Schubert Cells.* Advances in Mathematics 63(1987), 301-316.

[20] J. KAHN, M. SAKS, D. STURTEVANT, *A Topological Approach to Evasiveness,* Combinatorica 4(1984), 297-306.

[21] D. KNUTH, *The Art of Computer Programming, vol. 3* Addison–Wesley, Reading 1974.

[22] B. KORTE, L. LOVASZ, *Posets, Matroids, and Greedoids.* Colloquia Mathematica Societatis Janos Bolyai, 40, Matroid Theory (ed. A. Recski, L. Lovasz), North Holland, Amsterdam, 1985.

[23] K. KRIEGEL, S. WAACK, *Lower Bounds on the Complexity of Real–Time Branching Programs.* Informatique théorique et Applications 22(1988), 447–459.

[24] A. A. RAZBOROW, *Lower Bounds for the Monotone Complexity of Logical Permanent.* Matematičeskije Zametki 37(1985), 887–900.

[25] A. A. RAZBOROW, *Lower Bounds on the Size of Bounded-Depth Networks over the Basis* $\{\wedge, \oplus\}$. Preprint (in Russian), Moscow State University, 1986.

[26] A. ROSENBERG, *On the Time Required to Recognize Properties of Graphs: A Problem,* SIGACT News 5(4) (1973), 15-16.

[27] R. SMOLENSKY, *Algebraic Methods in the Theory of Lower Bounds for Boolean Circuit Complexity.* Proceedings 19–th ACM Symposium on Theory of Computing (1987), 77–82.

[28] J. THÉVENAZ, *Permutation Representations Arising from Simplicial Complexes.* Journal of Combinatorial Theory, Series A, 46(1987),121-155.

[29] J. TITS, *Buildings of Spherical Type and Finite BN–Pairs,* Lecture Notes in Mathematics 386, Springer, Berlin 1974.

[30] I. WEGENER, *Optimal Decision Trees and One–Time–Only Branching Programs for Symmetric Boolean Functions,* Information and Computation 62 (1984), 129-143.

COMPLEXITY OF FORMULA CLASSES IN FIRST ORDER LOGIC WITH FUNCTIONS

Erich Grädel*
Dipartimento di Informatica
Università di Pisa
Corso Italia 40
56100 Pisa

Abstract

We consider the complexity of deciding satisfiability of formulas in full first order logic (including function symbols and equality) which obey restrictions on their quantifier prefix and their relation and function symbols (prefix vocabulary classes). This extends results of H. Lewis and M.Fürer on the complexity of the classical solvable cases of the decision problem.

We obtain complexity results for the maximal solvable cases and some of their subcases. In particular we give a complete classification of the prefix vocabulary classes in P and in NP.

1 Solvable and unsolvable cases of the decision problem

Hilberts 'Entscheidungsproblem' — part of the formalist programme to codify all branches of mathematics by first order axioms and to reduce the proving of any mathematical theorem to mechanical derivation from these axioms — asked for an algorithm which would decide the satisfiability of any first order sentence. Before Church and Turing showed the impossibility of such an algorithm, attempts to find a positive solution had come up with decision procedures for particular subclasses of first order logic. These classes are all in the pure predicate calculus (i.e. not containing function symbols nor equality) and are defined in a purely syntactical way. They are now called the *classical solvable cases* of the decision problem:

- The monadic predicate calculus (Löwenheim 1915);
- The $\exists^*\forall\exists^*$ prefix class (Ackermann 1928);
- The $\exists^*\forall^*$ prefix class (Bernays, Schönfinkel 1928);
- The $\exists^*\forall^2\exists^*$ prefix class (Gödel 1932).

In view of these partial positive results on one side, of the negative solution due to Church and Turing on the other side, several research programmes were proposed for determining which classes of first order sentences are decidable for satisfiability and/or finite satifiability. Later, with the advent of complexity theory, it became also interesting to investigate the complexity of the decidable cases.

*Address after fall 1989: Mathematisches Institut der Universität Basel, Rheinsprung 21, CH-4051 Basel, Switzerland; e-mail: graedel@urz.unibas.ch. This work was supported by the Swiss National Science Foundation.

There is of course the question which of the uncountably many formula classes should be considered: Different answers to this question gave different research programmes. We mention some of them: Following the classical solvable cases central attention was given to *prefix classes* or *prefix vocabulary classes* in the pure predicate calculus. The classification of these classes was completed in the Sixties; Kahr, Moore and Wang [14] for prefix classes and Gurevich [10] for prefix vocabulary classes obtained the final results.

A more general programme was defined by adding equality and/or function symbols: The classification of prefix vocabulary classes in full first order logic with or without equality. This problem is also completely solved, final results were obtained by Gurevich [11,12,13], Shelah [20] and Goldfarb [8].

Other programmes are classifications of prefix vocabulary classes of Horn and Krom formulas (almost completely solved) and of decidable and undecidable mathematical theories. A survey over the history of the decision problem is presented in [3].

In this paper we consider the complexity of decidable prefix vocabulary classes in full first order logic.

Definition. A *prefix vocabulary class* is identified by a triple Π, s, t where Π is a word over the alphabet $\{\exists, \forall, \exists^*, \forall^*\}$ and s, t are either the word *all* or functions from $\mathbb{N}$ to $\mathbb{N} \cup \{\omega\}$. The class $[\Pi; s; t]$ contains all formulas ψ in prenex normal form, without equality, such that the quantifier prefix of ψ is a subword of Π (as usual every $\exists^k$ is considered as a subword of $\exists^*$) and such that the number of relation and function symbols of arity k in ψ are bounded by $s(k)$ and $t(k)$. If s and/or t is *all* then no restrictions on the relation and/or function symbols are imposed. The class $[\Pi; s; t]_=$ is defined in the same way, but in addition its formulas may contain equality.

To simplify notation we will denote a function s with $s(i) = 0$ for all $i > 2$ simply by the pair $(s(1), s(2))$. Thus e.g. $[\forall^2; (0,1); (1,0)]$ denotes the class of formulas with two universal but no existential quantifiers, with one binary relation and one unary function. A prefix Π is called finite if it contains no $\exists^*$ and no $\forall^*$; s is called finite if $s(i) \neq \omega$ for all i and only finitely many $s(i)$ are different from 0.

A formula class is called *solvable* if satisfiability is decidable for formulas in this class. A class is *conservative* if there is an algorithm which maps any first order formula to a formula of this class and preserves both satisfiability and finite satisfiability. A conservative class is of course unsolvable. For every formula class K, $sat(K)$ denotes the set of satisfiable formulas in K.

Theorem 1 (Classification without equality) *A prefix vocabulary class without equality is solvable if and only if it is included in at least one of the classes*

$[\exists^*\forall^*; all; (0)]$; $\quad [all; (\omega, 0); (\omega, 0)]$; $\quad [\exists^*\forall^2\exists^*; all; (0)]$

$[\exists^*\forall\exists^*; all; all]$; $\quad [\Pi; s; 0]$ *for finite* Π *and* s

All other prefix vocabulary classes are conservative.

Theorem 2 (Classification with equality) *A prefix vocabulary class with equality is solvable if and only if it is included in at least one of the classes*

$[\exists^*\forall^*; all; (0)]_=$; $\quad [all; (\omega, 0); (1, 0)]_=$; $\quad [\exists^*; all; all]_=$

$[\exists^*\forall\exists^*; all; (1, 0)]_=$; $\quad [\Pi; s; 0]_=$ *for finite* Π *and* s

All other prefix vocabulary classes are conservative.

The proofs of the Classification Theorems are scattered over the literature. A fairly complete account can be put together form the books of Dreben and Goldfarb [5] and Lewis [16] and the papers of Goldfarb [8], Gurevich [11,12,13] and Shelah [20].

2 The complexity of the classical solvable cases

Up to now complexity results are known only for the classical solvable cases (and some of their subcases) which were treated by H. Lewis [17] and M. Fürer [6]:

Theorem 3 (Upper bounds)

(i) $sat[\exists^*\forall\exists^*; all; (0)] \in \bigcup_{c>0} \mathrm{DTIME}(2^{cn^2/\log n})$;

(ii) $sat[\exists^*\forall^2\exists^*; all; (0)] \in \bigcup_{c>0} \mathrm{NTIME}(2^{cn/\log n})$;

(iii) $sat[all; (\omega, 0); (0)] \in \bigcup_{c>0} \mathrm{NTIME}(2^{cn/\log n})$;

(iv) $sat[\exists^*\forall^*; all; (0)] \in \bigcup_{c>0} \mathrm{NTIME}(2^{cn})$.

These results are optimal with the exception of the Ackermann class where it might be possible to get rid of the square in the exponent. In fact corresponding lower bounds do even apply to quite restricted subcases of the classical formula classes:

Theorem 4 (Lower bounds) *There exist constants $c, d, e > 0$ such that*

(i) $sat[\forall\exists^2; (\omega, 0); (0)] \notin \mathrm{DTIME}(2^{cn/\log n})$;

(ii) $sat[\forall^2\exists; (\omega, 0); (0)] \notin \mathrm{NTIME}(2^{dn/\log n})$;

(iii) $sat[\exists^*\forall^*; all; (0)] \notin \mathrm{NTIME}(2^{en})$.

For proofs we refer to the papers of Lewis and Fürer.

3 The maximal solvable classes

If we leave aside the trivial case (finite prefix, finite number of relations and no functions) then the maximal solvable classes are

- The Gödel class: $[\exists^*\forall^2\exists^*; all; (0)]$;
- The monadic class: $[all; (\omega, 0), (\omega, 0)]$;
- Formulas with one universal quantifier: $[\exists^*\forall\exists^*; all; all]$;
- The Bernays-Schönfinkel class with equality: $[\exists^*\forall^*; all; (0)]_=$;
- The existential class with equality: $[\exists^*; all, all]_=$;
- The (monadic existential second order) theory of one unary function: $[all; (\omega, 0); (1, 0)]_=$;
- The Shelah class: $[\exists^*\forall\exists^*; all; (1, 0)]_=$.

The Gödel class was fully treated by Lewis (see Theorems 3 and 4 above). We consider here some of the remaining cases.

The monadic class. Recall that the monadic class without functions (i.e. the Löwenheim class) is complete in $\bigcup_{c>0} \mathrm{NTIME}(2^{cn/\log n})$. For the full monadic class we can establish an $\mathrm{NTIME}(2^{cn^2})$-upper bound.

First we prove a bound on the size of the models that have to be checked.

Proposition 5 *Every satisfiable monadic formula of length n has a model of size $2^{O(n^2)}$.*

PROOF. The proof is a modification of Gurevichs decidability proof for the monadic class [10]. Let ψ be a monadic formula containing predicates $P_1, \ldots, P_s$ and functions $f_1, \ldots, f_t$. First we reduce the number of predicates to one: Replace every atomic formula $P_i(t)$ in ψ by $Q(g_i t)$ where Q is a new unary predicate and the g_i are new unary function symbols. This defines a new formula ψ' of length $O(n)$ in the class $[all; (1,0); (\omega, 0)]$. A model M for ψ defines a model $N = M \times \{1, \ldots, s\}$ for ψ' via

$$Q^N(a,i) :\leftrightarrow P_i^M(a); \quad f_i^N(a,j) := (f_i^M a, j); \quad g_i^N(a,j) := (a,i).$$

The increase of the model size is bounded by the factor $s = O(n/\log n)$. Conversely a model N for ψ' can be made a model for ψ with the same interpretations for the functions f_i by

$$P_i^M(a) :\leftrightarrow Q^N(g_i a).$$

Assume now that ψ has one predicate Q and functions $f_1, \ldots, f_m$. Let $\mathcal{F}$ be the set of words over the alphabet $\{f_1, \ldots, f_m\}$; in the sequel, f, g, h always denote elements of $\mathcal{F}$. We say that $h \le f$ if $f = gh$ for some g. Finally $\mathcal{G}(\psi)$ is the set of those g such that, for some $h \in \mathcal{F}$, an atom $Q(ghx)$ occurs in ψ. Obviously $|\mathcal{G}(\psi)| = n$.

Let M be a model for ψ. For every $f \in \mathcal{F}$ we define an equivalence relation E_f on M by

$$E_f(a,b) \quad \text{iff} \quad M \models \bigwedge_{h \le f} Q(ha) \leftrightarrow Q(hb).$$

If $f = gh$ then $E_f(a,b)$ implies $E_h(a,b)$ and $E_g(ha,hb)$. At last we define the equivalence relation

$$E := \bigcap_{f \in \mathcal{G}(\psi)} E_f$$

and let N be the set of equivalence classes of E.

We show that N can be made a model for ψ by appropriate definitions of the functions f_i and the predicate Q on N. This will imply the Proposition because the set of h such that $h \le f$ for some $f \in \mathcal{G}(\psi)$ has $O(n^2)$ elements and therefore $|N| = 2^{O(n^2)}$.

For $a \in M$ we denote by $[a]$ its equivalence class in N; in every class $[a]$ we choose an arbitrary member $s[a]$ and set:

$$f_i^N[a] := [f_i^M s[a]]; \qquad Q^N[a] :\leftrightarrow Q^M(s[a]).$$

We have to prove that for every $a \in M$ and every atomic formula $Q(fx)$ occurring in ψ

$$M \models Q(fa) \quad \text{iff} \quad N \models Q(f[a]).$$

Lemma. *Suppose that ψ contains an atom $Q(fx)$. Then, for all g, h with $gh = f$ and for all $a, b \in M$*

$$h[a] = [b] \implies E_g(ha, b).$$

PROOF. For $h = \lambda$ this follows immediately from the definition of E. So assume that $h = f_i h'$ and that $h[a] = [b]$; by the definition of f_i on N, $h[a] = f_i h'[a] = [f_i c]$ for some element c of the equivalence class $h'[a]$; by induction hypothesis, $h'[a] = [c]$ implies $E_{gf_i}(h'a, c)$ and thus also $E_g(f_i h'a, f_i c)$, i.e. $E_g(ha, f_i c)$.

But $[f_i c] = h[a] = [b]$ implies also $E(f_i c, b)$ and therefore $E_g(f_i c, b)$. It follows that $E_g(ha, b)$.

Setting $h = f$ the Lemma states that $f[a] = [b]$ implies $E_\lambda(fa, b)$, i.e. $M \models Q(fa) \leftrightarrow Q(b)$. Since $N \models Q([b])$ is equivalent to $M \models Q(b)$ the Proposition follows. ∎

Proposition 6 *It can be decided nondeterministically in time $2^{O(n/\log n + \log s)}$ whether a monadic formula of length n has a model of size s.*

For lack of space we omit the proof; the essential arguments can be found in [17].

Corollary 7 *There exists a constant c such that*

$$sat[all; (\omega,0); (\omega,0)] \in \text{NTIME}(2^{cn^2}).$$

Remark. The arguments in the proof of Proposition 5, together with Proposition 6, give us for free the $\text{NTIME}(2^{cn/\log n})$-upper bound for the Löwenheim class: For purely relational formulas the elimination of all but one relation symbols leads to a formula whose atoms are of the form $Q(g_i x)$ where the g_i are new function symbols; no compositions of functions occur. Therefore the set $\mathcal{G}(\psi)$ has only $O(n/\log n)$ elements and the size of the model N is bounded by $2^{O(n/\log n)}$.

It is not known whether the upper bound for the full monadic class is optimal; the best lower bound that that is known is $\text{NTIME}(2^{cn/\log n})$ (Theorem 4).

The Bernays-Schönfinkel class with equality. Using the standard technique to eliminate equality the satisfiability problem for the Bernays-Schönfinkel class with equality can be reduced to the same class without equality:

Given any first order formula ψ containing equality, let E be a binary relation symbol which does not occur in ψ, let T be the list of atoms $P(z_1,\ldots,z_j)$ where P is either E or a relation symbol of ψ; finally let r be the maximal arity of relation symbols in T. Let $P(z_i/u)$ denote the atom that is obtained by replacing z_i in P by u.

Replace every atom $(x = y)$ in ψ by Exy and take the conjunction with

$$\forall x \forall y \forall z_1 \cdots \forall z_r \Big(Exx \land (Exy \longrightarrow \bigwedge_{\substack{P\in T\\ i\leq r}} P(z_i/x) \leftrightarrow P(z_i/y))\Big).$$

Call the resulting formula ψ'. A model for ψ defines a model for ψ' over the same universe: just interprete E as equality. Conversely if ψ' has a model M, then the interpretation of E defines an equivalence relation on M and M/E is a model for ψ.

Note that the $\exists^*\forall^*$ prefix class is preserved by the elimination: If ψ has a prefix $\exists^p\forall^q$ then ψ' is equivalent to a formula with prefix $\exists^p\forall^{\max(r,q)}$. If a formula of this form is satisfiable then it has a Herbrand model of size $\max(1,p)$; hence the same bound applies for models of the original formula.

The obvious nondeterministic procedure to decide whether a formula of length n with q quantifiers holds in a structure of size s — guess a model and verify — requires time $2^{O(q\log s+\log n)}$. Since in our case $s,q = O(n/\log n)$ it follows that the Bernays-Schönfinkel class with equality is decidable in $\text{NTIME}(2^{cn})$ for some constant c, i.e. it has the same complexity as the Bernays-Schönfinkel without equality.

The existential class. Given an existential formula $\exists x_1 \cdots \exists x_k \psi$, let T be the set of terms t of height ≥ 1 that occur in ψ, possibly as a subterm of another term. Thus if, e.g. a term $t = f(t_1,\ldots,t_r)$ appears in ψ then T is meant to contain not only t, but also $t_1,\ldots,t_r$ and all their subterms which are not variables. If $T = \{t_1,\ldots,t_m\}$ then transform ψ to the formula

$$\exists x_1 \cdots \exists x_k \exists t_1 \cdots \exists t_m \Big(\psi' \land \bigwedge_{\substack{t=f(t_1,\ldots,t_r)\\ t\in T}} t = f(t_1,\ldots,t_r)\Big)$$

where ψ' is the same formula as ψ except that the terms in ψ are considered as variables in ψ'. In the new formulas all atoms are of the form $P(z_1,\ldots,z_r)$, $z_1 = z_2$ and $f(z_1,\ldots,z_r) = z_s$ where all the z_i are variables. A satisfiable formula of this form has a model of size at most $k+m$. Thus there is a straightforward procedure for deciding satisfiability of existential formulas in nondeterministic polynomial time.

Moreover this problem is certainly at least as hard as satisfiability of propositional formulas. Therefore we conclude:

Theorem 8 *The satisfiability problem for existential formulas in first order logic with equality is NP-complete.*

The theory of one unary function. The class $[all; (\omega, 0); (1,0)]_=$ properly contains the first order theory of one unary function, in our notation $[all; (0); (1,0)]_=$. This theory is not elementary recursive, a result which was first announced by Meyer [19]. A proof, even of the following somewhat stronger statement, appears in [4].

Theorem 9 *Let* $\exp_\infty(n)$ *be the 'tower of twos' function which is defined inductively by* $\exp_\infty(0) = 1$, $\exp_\infty(n+1) = 2^{\exp_\infty(n)}$. *Then there exists a constant* $c > 0$ *such that the first order theory of one unary function has a lower bound of the form* $\text{NTIME}(\exp_\infty(cn))$.

Formulas with one $\forall$. The class $[\exists^*\forall\exists^*; all; all]$ is an extension of the Ackermann class; therefore it has a $\text{DTIME}(2^{cn/\log n})$ lower bound. In [9] I could prove the following upper bound:

Theorem 10 *There exists a constant* $k \in \mathbb{N}$ *such that*

$$sat[\exists^*\forall\exists^*; all; all] \in \text{DTIME}(2^{n^k}).$$

For the Shelah class $[\exists^*\forall\exists^*; all; (1,0)]_=$, no upper complexity bound is known. A lower bound is implied by the recent result of Kolaitis and Vardi [15] that the Ackermann class with equality is NEXPTIME-complete. I conjecture that this also holds for the Shelah class.

4 The classes in P and NP

In this section we completely classify those prefix vocabulary classes whose satisfiability problems are in P and NP.

Theorem 11 *Let K be a formula class which is contained in one of the classes*

(i) $[\Pi; s; (0)]_=$ *for* Π, s *finite;*

(ii) $[\exists\forall^*; s; (0)]_=$ *for* s *finite;*

(iii) $[\Pi; (0); (1,0)]_=$ *for* Π *finite.*

Then satisfiability of formulas in K is decidable in polynomial time.

PROOF. *(i)* Formulas in $[\Pi; s; (0)]_=$ are composed from a fixed number of atomic formulas. Therefore there exist only finitely many formulas in this class in conjunctive normal form. Given an arbitrary formula of this class we can transform it in polynomial time into conjunctive normal form and then look up in a table whether it is satisfiable.

(ii) Every satisfiable formula in this class has a model with one element. Since the number of relation symbols is limited there are only finitely many structures to be checked.

The proof of *(iii)* depends on the fact that in every structure with one unary function, the number of equivalence classes of configurations of $O(1)$ elements which are distinguishable by formulas of quantifier depth $O(1)$ and length n is polynomially bounded. Note that due to Theorem 9 the complexity of $sat[\Pi; (0); (1,0)]_=$ depends in a non elementary recursive way on the length of the quantifier prefix. The details are rather complicated. ∎

We will show that no other prefix vocabulary class is in P, unless P = NP. First we determine the classes that are in NP.

Theorem 12 *Let K be a formula class contained in one of the classes* *(i)* $[\exists^*; all; all]_=$;

(ii) $[\exists\forall^*; all; (0)]_=$;

(iii) $[\exists^*\forall^q; all; (0)]_=$ *for some fixed* $q \in \mathbb{N}$.

(iv) $[\Pi_p; (q,0); (0)]_=$ *for* $p, q \in \mathbb{N}$ *and* Π_p *containing* p *universal quantifiers.*

Then $sat(K)$ *is in* NP.

PROOF. For the class $[\exists^*; all; all]_=$ this is Theorem 8. For *(ii)* and *(iii)* we use the analysis in section 3.2: A formula with prefix $\exists^p\forall^q$ without functions is satisfiable iff it has a model of size smaller or equal to $\max(p,1)$. If $p = 1$ or if the number of universal quantifiers is bounded then the obvious "guess and verify" algorithm requires nondeterministic polynomial time. For *(iv)* we prove the following

Lemma. *Every satisfiable formula* ψ *of the class* $[all; (q,0); (0)]_=$ *has a model of size at most* $2^q|\psi|$.

PROOF. Let ψ be a formula of length n with monadic predicates $P_1, \dots, P_q$ and with a model M. Define the equivalence relation

$$a \sim b \qquad \text{iff} \qquad M \models \bigwedge_{i=1}^{q} P_i a \leftrightarrow P_i b$$

and let $\tilde{M}$ be the structure that is obtained from M by choosing n arbitrary elements in every equivalence class of size $> n$ and all elements of the smaller equivalence classes. We claim that $\tilde{M}$ is a model for ψ. This is a consequence of the following statement: Let $\varphi(x_1, \dots, x_k)$ be an arbitrary subformula of ψ. Then for all $a_1, \dots, a_k \in \tilde{M}$

$$M \models \varphi(a_1, \dots, a_k) \qquad \text{iff} \qquad \tilde{M} \models \varphi(a_1, \dots, a_k).$$

This is trivial if φ is quantifier-free or a Boolean combination of formulas for which the statement is already proved. So let $\varphi(a_1, \dots, a_k) \equiv \exists x \varphi'(x, a_1, \dots, a_k)$ and suppose that the statement is true for φ'. If $M \models \varphi(a_1, \dots, a_k)$ then there exists a $b \in M$ such that $\varphi'(b, a_1, \dots, a_k)$ holds in M. If $b \in \tilde{M}$ we are done. Otherwise the equivalence class of b has more than n elements so there exist $c_1, \dots, c_n \in \tilde{M}$ with $c_i \sim b$. Because $k < n$ at least one of them, say c_1, is different from $a_1, \dots, a_k$; the same holds for b since $b \notin \tilde{M}$. Thus b is indistinguishable from c_1. It follows that $\varphi'(c_1, a_1, \dots, a_k)$ holds in M and therefore by induction hypothesis also in $\tilde{M}$. It follows that $\tilde{M} \models \varphi(a_1, \dots, a_k)$ and the Lemma is proved.

Now the Theorem follows from the following Lemma which was proved by Lewis [17]:

Lemma. *For some polynomial* g *it can be decided nondeterministically in time* $g(ns^p)$ *whether a formula of length* n *with* p *universal quantifiers has a model of size* s. ∎

The next theorem lists the minimal NP-complete classes.

Theorem 13 *Let* K *be one of the prefix vocabulary classes*

$[\exists^*; (0); (0)]_=$, $\quad [\exists^*; (1,0); (0)]$, $\quad [\exists; (\omega,0); (0)]$,

$[\exists; (1,0); (1,0)]$, $\quad [\forall; (\omega,0); (0)]$.

Then $sat(K)$ *is* NP*-complete.*

PROOF. SAT can be reduced to any of these classes. Let $\psi(X_1, \dots, X_n)$ be a propositional formula. First we introduce the following notation: If $\alpha_1, \dots, \alpha_n$ are arbitrary formula then $\psi[X_i \equiv \alpha_i]$ is the formula that is obtained by replacing every occurrence of X_i by α_i. We have the reductions

- $\psi \longrightarrow \exists x \exists y_1 \cdots \exists y_n \; \psi[X_i \equiv (y_i = x)]$;
- $\psi \longrightarrow \exists x_1 \cdots \exists x_n \; \psi[X_i \equiv Px_i]$;
- $\psi \longrightarrow \exists x \; \psi[X_i \equiv P_i x]$;
- $\psi \longrightarrow \exists x \; \psi[X_i \equiv Pf^i x]$;
- $\psi \longrightarrow \forall x \; \psi[X_i \equiv P_i x]$.

■

Let K be a prefix vocabulary class that is not contained in any of the classes which are in P due to Theorem 11 and that does not include any of the NP-complete classes of Theorem 13. Then K contains at least one of the classes

(i) $[\exists^2\forall^*; (0); (0)]_=$	(iv) $[\forall; (1,0); (1,0)]$
(ii) $[\exists^2\forall^*; (1,0); (0)]$	(v) $[\forall; (0); (2,0)]_=$
(iii) $[\forall^*; (0); (1,0)]_=$	(vi) $[\forall; (0); (0,1)]_=$.

Gurevich proved that the last two of these classes are unsolvable (even conservative) [13]. The next two theorems show that neither of the other four classes is likely to be in P; in fact they are not even in NP unless the polynomial time hierarchy collapses to NP.

Theorem 14 *The satisfiability problem for the classes*

$$[\exists^2\forall^*; (0); (0)]_= \qquad [\exists^2\forall^*; (1,0); (0)] \qquad [\forall^*; (0); (1,0)]_=$$

is Co-NP-hard.

PROOF. Let $\psi(X_1, \ldots, X_n)$ be a propositional formula. It is valid iff the formulas

$$\exists x \exists y \forall z_1 \cdots \forall z_n (x \neq y \wedge \psi[X_i \equiv (z_i = x)])$$

$$\exists x \exists y \forall z_1 \cdots \forall z_n ((Px \oplus Py) \wedge \psi[X_i \equiv Pz_i])$$

$$\forall x \forall y_1 \cdots \forall y_n (fx \neq x \wedge \psi[X_i \equiv (y_i = x)])$$

are satisfiable. ■

Theorem 15 *The class sat*$[\forall; (1,0); (1,0)]$ *is* PSPACE*-hard.*

PROOF. Let A be a problem in PSPACE. Without loss of generality we make the following assumptions: A is decided by the Turing machine M in space n^k and time 2^{n^k}; after acceptance M remains in the accepting configuration. Let Σ be the alphabet and Q the set of states of the Turing machine; then a configuration of M is encoded by a word of length n^k over the alphabet $\Gamma = \Sigma \cup (Q \times \Sigma)$ and the transition function of M is described by function $\delta : \Gamma^3 \longrightarrow \Gamma$ (the jth symbol of the successor configuration of c is determined by applying δ on the $(j-1)^{\text{th}}$, j^{th} and $(j+1)^{\text{th}}$ symbol of c). If we use instead of Γ the alphabet $\{0,1\}$ then, for some constant d, the encoding of a configuration has length dn^k and the computational behaviour of M is decribed by a function δ from $\{0,1\}^{3d}$ to $\{0,1\}^d$.

Set $m := dn^k$. We want to describe computations by M on inputs of length n by structures $(M; f, P)$ (with unary function f and monadic predicate P) in the following way: Given any element a of M, the set $\{f^i a \mid i \in \mathbb{N}\}$ is a homomorphic image of the natural numbers with successor, and P defines an infinite binary word on this set. We want this word to be divided into segments of length $4m + 3$ which have the form

$110b_0 0b_1 0 \cdots b_{m-1} 0 c_0 0 c_1 0 \cdots c_m 0.$

This is asserted by the formula $\forall x \alpha$ where

$$\alpha \equiv \Big((Px \wedge Pfx) \longrightarrow \bigwedge_{i=1}^{2m+1} \neg P(f^{2i}x)\Big) \wedge \bigvee_{i=0}^{4m+2} (Px \wedge Pfx).$$

Given a structure $(M; f, P)$ which satisfies $\forall x\alpha$, every $a \in M$ with $M \models (Pa \wedge Pfa)$ is the leading element of such a segment; it defines a natural number $b(a) < 2^m$ with binary notation $b_0 \cdots b_{m-1}$ and a word $c(a) = c_0, \ldots, c_{m-1} \in \{0,1\}^*$ via

$$b_i = 1 \quad \text{iff} \quad M \models P(f^{2i+3}a)$$
$$c_i = 1 \quad \text{iff} \quad M \models P(f^{2i+2m+3}a).$$

A computation is encoded by a sequence of 2^m such segments; if a_i is the leading element of the ith segment then $b(a_i) = i$ and $c(a_i)$ encodes the ith configuration of the computation.

It is not difficult to construct from the transition function δ quantifier free formulas of length polynomial in m which express the following properties:

$\text{Next}(x, y)$:	$c(y)$ encodes the successor configuration of $c(x)$.
$\text{Acc}(x)$:	$c(x)$ encodes an accepting configuration.
$\text{Inp}_w(x)$:	$c(x)$ is the initial configuration on input w.
$\text{Succ}(x, y)$:	$b(y) = b(x) + 1 \pmod{2^m}$.
$\text{First}(x)$:	$b(x) = 0$.
$\text{Last}(x)$:	$b(x) = 2^m - 1$.

Given an input w of length n construct the formula

$$\begin{aligned}\psi \equiv (Px \wedge Pfx) \quad \longrightarrow \quad & \big(\text{Succ}(x, f^{4m+3}x) \wedge \\ & (\text{First}(x) \longrightarrow \text{Inp}_w(x)) \wedge \\ & (\neg\text{Last}(x) \longrightarrow \text{Next}(x, f^{4m+3}x)) \wedge \\ & (\text{Last}(x) \longrightarrow \text{Acc}(x))\big).\end{aligned}$$

M accepts w if and only if the formula $\forall x(\alpha \wedge \psi)$ is satisfiable. This proves the Theorem. ∎

Remark. In fact the satisfiability problem for $[\forall; (1,0); (1,0)]$ is PSPACE-complete.

Thus Theorem 11 gives a complete classification of the prefix vocabulary classes in P. Next we consider the case where K is not contained in any of the classes that are in NP due to Theorem 12. Then K includes one of the classes *(i)* - *(vi)* of the list above or

(vii) $[\forall\exists; (\omega, 0); (0)]$	*(ix)* $[\forall^*\exists; (1,0), (0)]$
(viii) $[\forall\exists^*; (0,1); (0)]$	*(x)* $[\forall^*\exists; (0); (0)]_=$.

Using essentially the same ideas as in the proof of Theorem 15 it can shown that also the classes $[\forall\exists; (\omega, 0); (0)]$ and $[\forall\exists^*; (0,1); (0)]$ are PSPACE-hard. The classes $[\forall^*\exists; (1,0), (0)]$ and $[\forall^*\exists; (0); (0)]_=$ are Co-NP-hard: Reduce the propositional formula $\psi(X_1, \ldots, X_n)$ to

$$\forall x \forall y_1 \cdots \forall y_n \exists z((Px \oplus Pz) \wedge \psi[X_i \equiv (Px \oplus Py_i)])$$

$$\forall x \forall y_1 \cdots \forall y_n \exists z((x \neq z) \wedge \psi[X_i \equiv (x \neq y_i)]).$$

These formulas are satisfiable (in a model with two elements) if and only if ψ is valid.

Thus also the list of classes in NP as given by Theorem 12 is exhaustive and every prefix vocabulary class in NP$-$ P is actually NP-complete.

References

[1] W. Ackermann, *Über die Erfüllbarkeit gewisser Zählausdrücke*, Math. Annalen **100** (1928), 638–649.

[2] P. Bernays and M. Schönfinkel, *Zum Entscheidungsproblem der mathematischen Logik*, Math. Annalen **99** (1928), 342–372.

[3] E. Börger, *Decision problems in predicate logic*, in: "Logic Colloquium 82", Elsevier (North Holland) 1984, 263–301.

[4] K. Compton and C. W. Henson, *A uniform method for proving lower bounds on the computational complexity of logical theories*, to appear in Annals of Pure and Applied Logic.

[5] B. Dreben and W. Goldfarb, *The decision problem*, Addison-Wesley, Reading (MA) 1979.

[6] M. Fürer, *Alternation and the Ackermann case of the decision problem*, in: "Logic and Algorithmic", Monographie Nr. 30 de L'Enseignement Mathématique, Genève 1982, 161–186.

[7] K. Gödel, *Ein Spezialfall des Entscheidungsproblems der theoretischen Logik*, Ergebnisse eines mathematischen Kolloquiums **2** (1932), 27–28.

[8] W. Goldfarb, *The unsolvability of the Gödel class with identity*, J. Symbolic Logic **49** (1984), 1237–1252.

[9] E. Grädel, *Satisfiability of formulas with one ∀ is decidable in exponential time*, submitted for publication.

[10] Y. Gurevich, *Ob effektivnom raspoznavanii vipolnimosti formul UIP*, Algebra i Logika **5** (1966), 25–55 (in Russian).

[11] Y. Gurevich, *The decision problem for the logic of predicates and operations*, Algebra and Logic **8** (1969), 294–308.

[12] Y. Gurevich, *Formuly s odnim ∀* (Formulas with one ∀), in: "Izbrannye voprosy algebry i logiki", (Selected Questions in Algebra and Logic; in memory of A. Malćev), Nauka, Novosibirsk 1973, 97–110 (in Russian).

[13] Y. Gurevich, *The decision problem for standard classes*, J. Symbolic Logic **41** (1976), 460–464.

[14] A. Kahr, E. Moore and H.Wang, *Entscheidungsproblem reduced to the ∀∃∀ case*, Proc. Nat. Acad. Sci. U.S.A. **48** (1962), 365–377.

[15] P. Kolaitis and M. Vardi, *0-1 Laws and Decision Problems for Fragments of Second-Order Logic*, Proceedings of 3rd IEEE Symposium on Logic in Computer Science 1988, 2–11.

[16] H. Lewis, *Unsolvable Classes of Quantificational Formulas*, Addison Wesley, Reading (MA) 1979.

[17] H. Lewis, *Complexity Results for Classes of Quantificational Formulas*, J. of Computer and System Sciences **21** (1980), 317–353.

[18] L. Löwenheim, *Über Möglichkeiten im Relativkalkül*, Math. Annalen **76** (1915), 447–470.

[19] A. Meyer, *The inherent computational complexity of theories of ordered sets*, in: "Proc. 1974 Int. Cong. of Mathematicians", Vancouver (1974), 477–482.

[20] S. Shelah, *Decidability of a portion of the predicate calculus*, Israel J. Math **28** (1977), 32–44.

Normal and Sinkless Petri Nets*

Rodney R. Howell
Dept. of Computing and Information Sciences
Kansas State University
Manhattan, KS 66506

Louis E. Rosier
Dept. of Computer Sciences
The University of Texas at Austin
Austin, TX 78712

Hsu-Chun Yen
Dept. of Computer Science
Iowa State University
Ames, IA 50011

Abstract

We examine both the modeling power of normal and sinkless Petri nets and the computational complexities of various classical decision problems with respect to these two classes. We argue that although neither normal nor sinkless Petri nets are strictly more powerful than persistent Petri nets, they nonetheless are both capable of modeling a more interesting class of problems. On the other hand, we give strong evidence that normal and sinkless Petri nets are easier to analyze than persistent Petri nets. In so doing, we apply techniques originally developed for conflict-free Petri nets — a class defined solely in terms of the structure of the net — to sinkless Petri nets — a class defined in terms of the behavior of the net. As a result, we give the first comprehensive complexity analysis of a class of potentially unbounded Petri nets defined in terms of their behavior.

1 Introduction

Many aspects of the fundamental nature of computation are often studied via formal models, such as Turing machines, finite-state machines, and push-down automata (see, e.g., [HU79]). One formalism that has been used to model parallel computations is the Petri net (PN) [Pet81, Rei85]. As a means of gaining a better understanding of the PN model, the decidability and computational complexity of typical automata theoretic problems concerning PNs have been examined. These problems include boundedness, reachability, containment, and equivalence. Lipton [Lip76] and Rackoff [Rac78] have shown exponential space lower and upper bounds, respectively, for the boundedness problem. Also, Rabin [Bak73] and Hack [Hac76] have shown the containment and equivalence problems, respectively, to be undecidable. No tight bounds, however, have yet been established for the complexity of the reachability problem. The best lower bound known for this problem is exponential space [Lip76], but the only known algorithm is nonprimitive recursive [May84]. (See also [Kos82, Lam87].) Even the decidability of this problem was an open question for many years.

Early efforts to show the reachability problem to be decidable included the study of various restricted subclasses of PNs [CLM76, CM75, GY80, Gra80, HP79, LR78, May81, MM81, MM82, Mul81, VV81]. The only classes for which completeness results have been given concerning all four of the problems mentioned above are 1-conservative PNs [JLL77], 1-conservative free choice PNs [JLL77], symmetric PNs [CLM76, MM82, Huy85], and conflict-free PNs [JLL77, HRY87, HR88a] (the proofs in [JLL77] also apply to 1-bounded PNs and elementary nets). Of these classes only symmetric and conflict-free PNs are potentially unbounded, and both symmetric and conflict-free PNs are defined only in terms of their structure, not in terms of their behavior. Thus, until now there has been no class of potentially unbounded PNs defined in terms of their behavior for which a comprehensive complexity analysis has been given.

Of all of the PN classes for which completeness results have been shown concerning all four of the problems mentioned above, the class for which the decision procedures are most efficient is that of conflict-free PNs. In

*This work was supported in part by National Science Foundation Grant No. CCR-8711579.

particular, the boundedness problem is complete for polynomial time [HRY87], the reachability problem is NP-complete [HR88a, JLL77], and the containment and equivalence problems are Π_2^P-complete [HR88a], where Π_2^P is the set of all languages whose complements are in the second level of the polynomial-time hierarchy [Sto77]. However, since conflict-free PNs comprise such a simple class, their modeling power is very limited. In particular, we can show that conflict-free PNs cannot model the producer/consumer problem if more than one consumer is involved and the actions of each consumer are to be modeled by separate transitions. Furthermore, the obvious generalization of conflict-free PNs to persistent PNs (see [LR78]) is not very helpful: not only is it impossible to model the above problem with persistent PNs, but the complexities of the various problems regarding persistent PNs appear to be worse than for conflict-free PNs. In particular, all four problems are PSPACE-hard [JLL77]. The known upper bounds are much worse: the best upper bound known for boundedness is exponential space [Rac78], and no primitive recursive algorithms are known for the other three problems, though they are known to be decidable [Gra80, May81, Mul81]. Even the problem of recognizing a persistent PN, though known to be decidable [Gra80, May81, Mul81], is not known to be primitive recursive, and is PSPACE-hard [JLL77].

More recently, Yamasaki [Yam84] has defined two other generalizations of conflict-free PNs, normal PNs and sinkless PNs. The relationship of normal PNs to sinkless PNs is analogous to the relationship of conflict-free PNs to persistent PNs; i.e., normal PNs are those PNs that are sinkless for every initial marking [Yam84], just as conflict-free PNs are those PNs that are persistent for every initial marking [LR78]. In addition, normal PNs, like conflict-free PNs, are defined in terms of the structure of the net, whereas sinkless PNs, like persistent PNs, are defined in terms of the behavior of the net. However, both normal and sinkless PNs are incomparable to the class of persistent PNs; i.e., there are persistent PNs that are not sinkless, and normal PNs that are not persistent. In this paper, we examine both the modeling power of normal and sinkless PNs and the computational complexities of the four problems mentioned above with respect to these two classes. We show that both in terms of modeling power and ease of analysis, normal and sinkless PNs compare very favorably to conflict-free and persistent PNs.

Concerning the modeling power, we can show that the producer/consumer problem mentioned above, which cannot be modeled by persistent PNs, can be modeled by normal PNs. Another problem we consider is the mutual exclusion problem. We can show that although this problem cannot be modeled by sinkless PNs, a version in which a bounded number of exclusions take place can be modeled by normal PNs (persistent PNs cannot even model one exclusion). We therefore conclude that although not all persistent PNs are sinkless, the class of problems that can be modeled by sinkless (or even normal) PNs is somewhat more interesting than that class modeled by persistent PNs.

We then examine whether the more "useful" nature of sinkless PNs causes a corresponding increase in the complexities of the classical problems (as compared to persistent PNs). We show that this is not the case. To the contrary, we give strong evidence that these problems are actually more efficiently decidable for sinkless PNs than for persistent PNs. In particular, we show that for both normal and sinkless PNs, the boundedness problem is co-NP-complete, the reachability problem is NP-complete, and the containment and equivalence problems are Π_2^P-complete. Note that with the exception of the boundedness problem, the complexities of these problems are identical to those for conflict-free PNs — an extremely simple class. In fact, the techniques used in deriving these results for normal and sinkless PNs are simply more sophisticated applications of the techniques developed in [HR88a] for conflict-free PNs. Thus, techniques originally developed for analyzing problems involving conflict-free PNs — a class defined solely in terms of the structure of the net — have been generalized to apply not only to normal PNs, but also to sinkless PNs — a class defined in terms of the behavior of the net. Furthermore, these results represent the first comprehensive complexity analysis of the classical problems concerning a class of potentially unbounded PNs defined in terms of their behavior.

We also examine the question of how much easier it is to recognize a normal PN than to recognize a sinkless PN. Recall that the problem of recognizing a persistent PN is not known to be primitive recursive, and is at least PSPACE-hard [JLL77], whereas a conflict-free PN can easily be recognized in polynomial time. The main problem in deciding persistence is that some sort of reachability analysis is necessary due to the fact that persistence is a behavioral property. Since sinkless PNs are likewise defined in terms of their behavior, whereas normal PNs are defined solely in terms of their structure, one might suppose that normal PNs would be easier to recognize than sinkless PNs. However, in order to determine whether a PN is normal, a rather complex property of the graphical representation of the PN must be tested. The end result is that both the problem of determining whether a PN is normal and the problem of determining whether a PN is sinkless are co-NP-complete. We therefore conclude that in most applications, one might as well consider the entire class of sinkless PNs rather than the more restricted class of normal PNs.

The remainder of the paper is organized as follows. In Section 2, we formally define the concepts used throughout the paper. In Section 3, we compare the modeling power of normal and sinkless PNs with that of persistent PNs. Finally, in Section 4, we examine the complexities of the various problems regarding normal and sinkless PNs. Due to space limitations, we are only able to give general proof strategies; detailed proofs may be found in [HRY88].

2 Definitions

Let N denote the set of nonnegative integers, R the set of rational numbers, N^k (R^k) the set of vectors of k nonnegative integers (rational numbers, respectively), and $N^{k\times m}$ ($R^{k\times m}$) the set of k × m matrices of nonnegative integers (rational numbers, respectively). For a k-dimensional vector v, let v(i), $1 \leq i \leq k$, denote the *i*th component of v. For a k × m matrix A, let A(i,j), $1 \leq i \leq k$, $1 \leq j \leq m$, denote the element in the *i*th row and the *j*th column of A, and let a_j denote the *j*th column of A. For a given value of k, let **0** denote the vector of k zeros (i.e., **0**(i) = 0 for i = 1,...,k). Given k-dimensional vectors u, v, and w, we say:

- v = w iff v(i) = w(i) for i = 1,...,k;
- $v \geq w$ iff $v(i) \geq w(i)$ for i = 1,...,k;
- $v > w$ iff $v \geq w$ and $v \neq w$; and
- u = v + w iff u(i) = v(i) + w(i) for i = 1,...,k.

A *Petri Net* (PN, for short) is a tuple (P,T,φ,μ_0), where P is a finite set of *places*, T is a finite set of *transitions*, φ is a *flow function* $\varphi : (P \times T) \cup (T \times P) \rightarrow N$, and μ_0 is the *initial marking* $\mu_0 : P \rightarrow N$ (in this paper, we only consider PNs for which the range of φ is {0,1}). A *marking* is a mapping $\mu : P \rightarrow N$. A transition $t \in T$ is *enabled* at a marking μ iff for every $p \in P$, $\varphi(p,t) \leq \mu(p)$. A transition t may *fire* at a marking μ if t is enabled at μ. We then write $\mu \xrightarrow{t} \mu'$, where $\mu'(p) = \mu(p) - \varphi(p,t) + \varphi(t,p)$ for all $p \in P$. A sequence of transitions $\sigma = t_1...t_n$ is a *firing sequence* from μ_0 (or a firing sequence of (P,T,φ,μ_0)) iff $\mu_0 \xrightarrow{t_1} \mu_1 \xrightarrow{t_2} \cdots \xrightarrow{t_n} \mu_n$ for some sequence of markings $\mu_1,...,\mu_n$. We also write $\mu_0 \xrightarrow{\sigma} \mu_n$. For $\sigma,\sigma' \in T^*$, $\sigma' = t_1...t_n$, let $\sigma \dot{-} \sigma'$ be inductively defined as follows. Let σ_0 be σ. If $t_i \in \sigma_{i-1}$, let σ_i be σ_{i-1} with the last occurrence of t_i deleted; otherwise, let $\sigma_i = \sigma_{i-1}$. Finally, let $\sigma \dot{-} \sigma' = \sigma_n$.

Since we only consider Petri nets (P,T,φ,μ_0) such that the range of φ is {0,1}, we may view them from two different perspectives. The first perspective is graph-theoretical: $P \cup T$ forms the set of vertices of a directed graph, and (u,v) is an edge iff $\varphi(u,v) = 1$. Since φ is a function on $(P \times T) \cup (T \times P)$, the graph representation of the PN is bipartite. The other perspective is algebraic. Suppose P contains k elements and T contains m elements. By establishing an ordering on the elements of P and T (i.e., $P = \{p_1,...,p_k\}$ and $T = \{t_1,...,t_m\}$), we define the k × m *addition matrix* $\bar{T}$ of (P,T,φ,μ_0) so that $\bar{T}(i,j) = \varphi(t_j,p_i) - \varphi(p_i,t_j)$. Thus, if we view a marking μ as a k-dimensional column vector in which the *i*th component is $\mu(p_i)$, each column $\bar{t}_j$ of $\bar{T}$ is then a k-dimensional vector such that if $\mu \xrightarrow{t_j} \mu'$, then $\mu' = \mu + \bar{t}_j$. (Note that by this convention, the notations $\mu(p_i)$ and $\mu(i)$ are interchangeable.) For a given alphabet Σ, let $\Psi : \Sigma^* \rightarrow (\Sigma \rightarrow N)$ be the *Parikh mapping* so that for $\sigma \in \Sigma^*$, $a \in \Sigma$, $\Psi(\sigma)(a)$ is the number of occurrences of a in σ. For $\Sigma = T$, we can view $\Psi(\sigma)$ as an m-dimensional column vector in which the *j*th component is $\Psi(\sigma)(t_j)$. Then if $\mu_0 \xrightarrow{\sigma} \mu$, $\mu_0 + \bar{T}\cdot\Psi(\sigma) = \mu$ (note that the converse does not necessarily hold).

Let $\mathcal{P} = (P,T,\varphi,\mu_0)$ be a PN. The *reachability set* of $\mathcal{P}$ is the set $R(\mathcal{P}) = \{\mu \mid \mu_0 \xrightarrow{\sigma} \mu \text{ for some } \sigma\}$. Let c = $u_1,u_2,...,u_n,u_1$ be a circuit in the graph of $\mathcal{P}$, and let μ be a marking of $\mathcal{P}$. For convenience, we will always assume

u_1 is a place. Let pl(c) = $\{u_1,u_3,...,u_{n-1}\}$ denote the set of places in c, and let tr(c) = $u_2u_4...u_n$ denote the sequence of transitions in c. We define $\mu(c) = \sum_{p_i \in pl(c)} \mu(i)$. We say c is *token-free* in μ iff $\mu(c) = 0$. c is said to be *minimal* iff pl(c) does not properly include the set of places in any other circuit. (Note that the transitions in c are ignored in this definition.) c is said to have a *sink* iff for some $\mu \in R(\mathcal{P})$ and some σ and μ' such that $\mu \xrightarrow{\sigma} \mu'$, $\mu(c) > 0$, but $\mu'(c) = 0$. c is said to be *sinkless* iff it does not have a sink. $\mathcal{P}$ is said to be *sinkless* iff each minimal circuit of $\mathcal{P}$ is sinkless. $\mathcal{P}$ is said to be *normal* iff for every minimal circuit c and each transition t_j, $\sum_{p_i \in pl(c)} \bar{T}(i,j) \geq 0$; i.e., no transition can decrease the token count of a minimal circuit by firing at any marking. We say $\mathcal{P}$ is *persistent* if for every $\mu \in R(\mathcal{P})$, when any pair of distinct transitions t_1 and t_2 are both enabled at μ, t_1t_2 is a firing sequence from μ; i.e., no enabled transition can ever be disabled by firing some other transition. $\mathcal{P}$ is said to be *conflict-free* iff for every place p for which there are two or more transitions t such that $\varphi(p,t) = 1$, for each such transition, $\varphi(t,p) = 1$. (This definition of conflict-freedom was given in [LR78] for PNs whose flow function has a range of {0,1}; see [CM75, HRY87, HR88a, HR88b] for somewhat more general definitions.) As was shown in [Yam84], the relationship of normal PNs to sinkless PNs is analogous to the relationship of conflict-free PNs to persistent PNs: normal PNs are those PNs that are sinkless for every initial marking [Yam84], while conflict-free PNs are those PNs that are persistent for every initial marking [LR78].

Given a marking μ of a given PN $\mathcal{P}$, the *reachability problem* (RP) is to determine whether $\mu \in R(\mathcal{P})$. The *boundedness problem* (BP) is to determine whether $R(\mathcal{P})$ is finite. The *sink-detection problem* is the problem of determining whether there is a minimal circuit of $\mathcal{P}$ with a sink. Given two PNs $\mathcal{P}$ and $\mathcal{P}'$, the *containment* and *equivalence problems* (CP and EP, respectively) are to determine whether $R(\mathcal{P}) \subseteq R(\mathcal{P}')$ and whether $R(\mathcal{P}) = R(\mathcal{P}')$, respectively. In examining the latter two problems, we use concepts from linear algebra and the theory of semilinear sets. For any vector $v_0 \in N^k$ and any finite set $V = \{v_1,...,v_m\} \subseteq N^k$, the set $\mathcal{L}(v_0,V) = \{x \mid \exists c_1,...,c_m \in N$ such that $x = v_0 + \sum_{i=1}^{m} c_i \cdot v_i\}$ is called the *linear set* with *base* v_0 over the set of *periods* V. A finite union of linear sets is called a *semilinear set* (SLS for short). If $x = \sum_{i=1}^{m} a_i \cdot v_i$ for some $a_1,...,a_m \in R$, then x is a *linear combination* of the vectors in V. If $a_i \geq 0$ for all i, then x is a *nonnegative linear combination* of the vectors in V. If in addition for some i, $a_i > 0$, then x is a *positive linear combination* of the vectors in V.

3 Modeling Power

In this section, we examine the modeling power of normal and sinkless PNs, comparing it with that of conflict-free and persistent PNs. It has already been shown that the class of conflict-free PNs is properly contained in both the classes of persistent PNs [LR78] and sinkless PNs [Yam84]. Furthermore, it is clear from the definitions that the class of sinkless PNs properly includes the class of normal PNs (see also [Yam84]). On the other hand, it is not hard to see that the class of persistent PNs is incomparable to both the class of normal PNs and the class of sinkless PNs; i.e., there is a persistent PN that is not sinkless, and there is a normal PN that is not persistent (see also [Yam84]).

One of the shortcomings of conflict-free and persistent PNs is that their modeling power is severely limited. By the definition of persistence, only a very limited type of nondeterminism is allowed: if more than one transition is enabled, the next transition to fire may be nondeterministically chosen, but the firing of this transition cannot disable any others. Although there are modeling problems, such as a simple version the producer/consumer problem, that can be modeled by conflict-free PNs, the severely restricted version of nondeterminism prohibits even persistent PNs from modeling many "interesting" problems. In particular, we can show that neither a more general producer/consumer problem nor the mutual exclusion problem can be modeled by persistent PNs. Given these limitations of persistent PNs and the fact that there are normal PNs that are not persistent, it is natural to ask whether normal PNs can model some of the classical modeling problems that persistent PNs cannot. In this section, we argue that although the mutual exclusion problem cannot be modeled even by sinkless PNs, the generalized producer/consumer problem and a version of the mutual exclusion problem in which the total number of exclusions in any computation is bounded by a fixed constant both can be modeled by normal PNs (and, hence, by sinkless PNs). On the other hand, we can show that persistent PNs cannot even model one exclusion. In the next section, we will present evidence that sinkless

PNs may be significantly easier to analyze than persistent PNs, but not significantly more difficult to analyze than normal PNs. These results suggest that sinkless PNs may be a more useful class of PNs than many of the classes that have been studied in the past.

Due to space limitations, we give here only a brief summary of the results of this section; formal proofs may be found in [HRY88]. The proof strategies used there involve a language-theoretic approach similar to the one employed in [Hac75].

We consider three general modeling problems:

1. $PC_{m,n}$ – the producer/consumer problem with m producers, n consumers, and one buffer;

2. ME_n – the mutual exclusion problem for n processes; and

3. ME_n^k – the mutual exclusion problem for n processes such that the total number of exclusions is bounded by k.

Although we will not formally define here what it means to model each of the above problems, we do point out the following requirement: different actions in the above problems must be modeled by different transitions of a PN. For example, the "consume" action of two different consumers must be modeled by separate transitions.

Our results are summarized in Table 3-1. Intuitively, it is not hard to see that persistent PNs cannot model even such simple problems as $PC_{1,2}$ or ME_2^1 (though $PC_{1,1}$ can be modeled by even conflict-free PNs). Regarding $PC_{1,2}$, there must arise a situation in which the buffer contains one item and both consumers are ready to receive; i.e., two actions are simultaneously enabled. However, when one of these actions occurs, the other becomes disabled — a conflict. A similar situation occurs for ME_2^1. On the other hand, it is not hard to construct normal PNs to model $PC_{m,n}$ or ME_n^k for any positive k, m, and n; in fact, the normal PNs given in [HRY88] to model these problems are perhaps the most obvious PNs (normal or otherwise) for this purpose.

Problem	Modeled by persistent?	Modeled by normal/sinkless?
$PC_{1,2}$	no	yes
$PC_{m,n}$	no	yes
ME_2	no	no
ME_2^1	no	yes
ME_n^k	no	yes

Table 3-1

Normal and sinkless PNs do have significant limitations in their modeling power, however. In particular, by using results from [Yam84], we can show that sinkless PNs cannot model ME_2. See [HRY88] for details.

4 Complexity Results

In this section, we examine the complexities of various problems involving normal and sinkless PNs. These problems include the boundedness, reachability, containment, and equivalence problems, as well as the sink detection problem and the problem of determining whether a PN is normal. We give strong evidence to suggest that, even though sinkless PNs possess some important modeling capabilities that are absent in persistent PNs, sinkless PNs may be easier to analyze[1] than persistent PNs. In particular, it was shown in [JLL77] that the problem of deciding persistence of an arbitrary PN is PSPACE-hard; it is a straightforward matter to modify this proof to show that the BP, RP, CP, and EP are all PSPACE-hard for persistent PNs. Furthermore, even though all of these problems are known to be decidable [Gra80, May81, Mul81], no primitive recursive upper bound has been shown for any of them. On the other hand, we will show that for normal and sinkless PNs, the RP is NP-complete, the BP is co-NP-complete, and the CP and EP are Π_2^P-complete. These results not only imply that the problems for persistent PNs are at least as hard as the corresponding problems for sinkless PNs, they also imply that if there exist polynomial-time reductions from any of the persistent PN problems to one of the sinkless PN problems, then the polynomial-time hierarchy

[1] All of our complexity measures are with respect to polynomial-time many-one reductions.

must collapse [Sto77]. Furthermore, such a reduction would give a dramatic improvement over any previous upper bounds known for the problem in question (regarding persistent PNs). Our results also imply that sinkless PNs are no harder to analyze than normal PNs (at least with respect to the problems mentioned above), and that with the exception of the BP (see [HRY87]), sinkless PNs are no harder to analyze than conflict-free PNs (see [HR88a]). Since conflict-free PNs can be efficiently recognized, at least in this sense they are easier to work with than sinkless PNs. The question then arises as to whether normal PNs can be more efficiently recognized than sinkless PNs. That such might be the case is suggested by the fact that no type of reachability analysis must be done to decide whether a PN is normal; only the structure of the graphical representation of the PN need be analyzed. However, the graph property in question is far from trivial. In fact, we are able to show that both the problem of deciding whether a PN is normal and the question of deciding whether a PN is sinkless are co-NP-complete. Thus, we see in sinkless PNs a class of PNs having modeling capabilities that are absent in persistent PNs, but which can be recognized and analyzed as efficiently as normal PNs, and for many problems can be analyzed as efficiently as conflict-free PNs.

In [HR88a], we developed several techniques for analyzing conflict-free PNs. These techniques relied heavily upon the structural property of conflict-freedom. In this section, we show that these same techniques also apply, albeit in a more sophisticated manner, to sinkless PNs, a class defined not in terms of structure, but in terms of behavior. Thus, we are able to use techniques developed for analyzing a class defined in terms of structure to give the first comprehensive complexity analysis for a class of potentially unbounded PNs defined in terms of behavior.

We begin by developing an important lemma (Lemma 4.4) which will be used in deriving most of the upper bounds in this section. In [HR88a], we showed that for any conflict-free PN $\mathcal{P}$ and any marking μ of $\mathcal{P}$, there is an instance of integer linear programming that has a solution iff μ is reachable in $\mathcal{P}$. Furthermore, we showed that this instance of integer linear programming can be guessed in polynomial time. It therefore followed that the reachability problem for conflict-free PNs is in NP. We will show in Lemma 4.4 that a similar fact holds for sinkless PNs. In so doing, we make use of the following lemmas from [Yam84].

Lemma 4.1 (from [Yam84]): Let $\mathcal{P} = (P,T,\varphi,\mu_0)$ be a PN with k places and m transitions. For each w $\in$ N^m, there is some firing sequence σ of $\mathcal{P}$ with $\Psi(\sigma) = w$ if

1. $\mu_0 + \bar{T}\cdot w \geq 0$, and
2. for each firing sequence σ' of $\mathcal{P}$ and each circuit c, if $\Psi(\sigma' tr(c)) \leq w$, then $\mu(c) > 0$, where $\mu_0 \xrightarrow{\sigma'} \mu$.

Lemma 4.2 (from [Yam84]): If a PN $\mathcal{P} = (P,T,\varphi,\mu_0)$ has no token-free circuits in every reachable marking, then $R(\mathcal{P}) = \{\mu \mid \mu = \mu_0 + \bar{T}\cdot x \geq 0$ for some $x \in N^m\}$, where m is the number of transitions in T.

The above lemma gives an instance of integer linear programming whose solution set gives the reachability set of a PN. The only requirement is that no reachable marking can have a token-free circuit. In order to enable us to work with PNs whose reachability sets contain no markings with token-free circuits, we give the following easily shown lemma.

Lemma 4.3: Let $\mathcal{P} = (P,T,\varphi,\mu_0)$ be a sinkless PN, and let $\mathcal{P}' = (P,T',\varphi',\mu)$ be such that $\mu_0 \xrightarrow{\sigma} \mu$ in $\mathcal{P}$ for some σ, $T' \subseteq T$ such that each $t \in T'$ is enabled at some point in the firing of σ from μ_0, and φ' is the restriction of φ to $(P \times T') \cup (T' \times P)$. Then $\mathcal{P}'$ has no token-free circuits in every reachable marking.

Given the above lemmas, we can now outline our strategy for showing the RP to be in NP, and the BP to be in co-NP. This strategy will then be the basis for most of the other upper bounds shown in this paper. Let $\mathcal{P} = (P,T,\varphi,\mu_0)$ be a sinkless PN, and consider the sequence $\mathcal{P}_1,\ldots,\mathcal{P}_n$, where each $\mathcal{P}_h = (P,T_h,\varphi_h,\mu_{h-1})$ such that $T_0 = \emptyset$, and for $1 \leq h \leq n$,

- $T_h = T_{h-1} \cup \{t_{j_h}\}$ for some $t_{j_h} \notin T_{h-1}$ enabled at μ_{h-1}, for $1 \leq h \leq n$;
- φ_h is the restriction of φ to $(P \times T_h) \cup (T_h \times P)$, for $0 \leq h \leq n$; and
- $\mu_{h-1} \xrightarrow{\sigma_{h-1}} \mu_h$ for some $\sigma_{h-1} \in T_h^*$, $1 \leq h \leq n$.

From Lemma 4.3, no $\mathcal{P}_h$ contains a token-free circuit for any marking in $R(\mathcal{P}_h)$. Thus, from Lemma 4.2, there is an instance of integer linear programming S_h whose solution set describes the reachability set of $\mathcal{P}_h$. Furthermore, S_h can clearly be constructed from $\mathcal{P}_h$ in polynomial time. Also note that the initial marking of $\mathcal{P}_h$ is given by a solution to S_{h-1}. Thus, a portion of $R(\mathcal{P})$ can be given by the solution set of a system S of linear inequalities over the integers, where S is constructed in polynomial time from P and a sequence of distinct transitions of $\mathcal{P}$. To formalize these ideas, let $P = \{p_1,...,p_k\}$, $T = \{t_1,...,t_m\}$, and let $\tau = t_{j_1}...t_{j_n}$ be any sequence of distinct transitions from T. We define the *characteristic system of inequalities* for $\mathcal{P}$ and τ as $S(\mathcal{P},\tau) = S_0 \cup ... \cup S_n$, where $S_0 = \{x_0 = \mu_0\}$, $S_h = \{x_{h-1}(i) \geq \varphi(p_i,t_{j_h}),\ x_h = x_{h-1} + A_h \cdot y_h \mid 1 \leq i \leq k\}$, and A_h is the k × h matrix whose columns are $\bar{t}_{j_1},...,\bar{t}_{j_h}$ for $1 \leq h \leq n$. The variables in S are the components of the k-dimensional column vectors $x_0,...,x_n$ and the h-dimensional column vectors y_h, $1 \leq h \leq n$. We now claim that for any marking μ, $\mu \in R(\mathcal{P})$ iff there is a sequence of distinct transitions $\tau = t_{j_1}...t_{j_n}$ such that $S(\mathcal{P},\tau)$ has a nonnegative integer solution in which $x_n = \mu$. It is precisely this fact that allows us to apply the techniques of [HR88a], developed specifically for the structurally-defined conflict-free PNs, to the behaviorally-defined sinkless PNs.

Lemma 4.4: Let $\mathcal{P} = (P,T,\varphi,\mu_0)$ be a sinkless PN, and let μ be any marking of $\mathcal{P}$. Then there is some $\sigma \in T^*$ such that $\mu_0 \xrightarrow{\sigma} \mu$ iff there is some sequence $\tau = t_{j_1}...t_{j_n}$ of distinct transitions in T such that $S(\mathcal{P},\tau)$ has a nonnegative integer solution in which $x_n = \mu$. Furthermore, σ and τ can be chosen such that $\sigma = \sigma_1...\sigma_n$, where $\mu_0 = x_0 \xrightarrow{\sigma_1} x_1 \xrightarrow{\sigma_2} \cdots \xrightarrow{\sigma_n} x_n = \mu$, t_{j_h} is enabled at x_{h-1}, $\sigma_h \in \{t_{j_1},...,t_{j_h}\}^*$, and $y_h(h')$ gives the number of times $t_{j_{h'}}$ occurs in σ_h, for $1 \leq h' \leq h \leq n$.

The following corollary follows from Lemmas 4.2, 4.3 and 4.4.

Corollary 4.1: Let $\mathcal{P} = (P,T,\varphi,\mu_0)$ be a sinkless PN, τ be a sequence of n distinct transitions from T, T′ be the set of transitions in τ, φ' be the restriction of φ to $(P \times T') \cup (T' \times P)$, and μ be any marking of $\mathcal{P}$ such that for some nonnegative integer solution of $S(\mathcal{P},\tau)$, $x_n = \mu$. Then $R(P,T',\varphi',\mu) = \{\mu' \mid \mu' = \mu + \bar{T}' \cdot x \text{ for some } x \in N^n\}$.

Lemma 4.4 can now be coupled with the fact that integer linear programming is in NP [BT76] to show that the RP for sinkless PNs is in NP. Since the RP is NP-hard for conflict-free PNs [JLL77], it will then follow that the RP for sinkless (normal) PNs is NP-complete. We therefore have the following theorem.

Theorem 4.1: The RP for sinkless (normal) PNs is NP-complete.

In [HRY87], we showed the BP to be PTIME-complete for conflict-free PNs. However, we can show the problem to be co-NP-complete for both normal and sinkless PNs. The upper bound is shown in a manner similar to the proof of Theorem 4.1; the lower bound is shown using a standard reduction from 3SAT.

Theorem 4.2: The BP for sinkless (normal) PNs is co-NP-complete.

We can also use Lemma 4.4 to show the sink detection problem for PNs to be NP-complete in a manner similar to Theorems 4.1 and 4.2.

Theorem 4.3: The sink detection problem for PNs is NP-complete.

We have seen that with respect to the BP and the RP, sinkless PNs are no harder to analyze than normal PNs. Since Theorem 4.3 shows that the problem of recognizing a sinkless PN is co-NP-complete, we now address the problem of recognizing a normal PN. Recall that the property of being normal is defined solely in terms of the structure of the net. Thus, recognizing a normal PN may be thought of as a problem concerning directed graphs. Using standard techniques, we can show this problem to be co-NP-complete.

Theorem 4.4: The problem of determining whether a PN is normal is co-NP-complete.

We conclude this section by showing the CP and EP for normal and sinkless PNs to be Π_2^P-complete, where Π_2^P is the class of languages whose complements are in the second level of the polynomial-time hierarchy (see, e.g., [Sto77]). The strategy we use is again similar to that developed in [HR88a] (see also [HRHY86, Huy86]). Recall that sinkless PNs have effectively computable semilinear sets [Yam84]. We use Lemma 4.4 to give an upper bound on the size of the SLS representation of the reachability set of a given sinkless PN. Although the CP and EP for SLSs are known to be Π_2^P-complete [Huy82], the bound on the size of the SLS representation of the reachability set must be at least exponential in the sizes of the PNs even for conflict-free PNs (see [HR88a]). However, as is the case with conflict-free PNs [HR88a], we can show that for sinkless PNs the SLS representation can be chosen to have a high degree of symmetry. Proceeding as in [HR88a], we use this symmetry to show the CP (and, hence, the EP) to be in Π_2^P. Since the EP is known to be Π_2^P-hard for conflict-free PNs [HR88a], the CP and EP will have then been shown to be Π_2^P-complete for sinkless and normal PNs. The following lemma gives an SLS representation of the reachability set of a sinkless PN. The strategy follows that developed in [HR88a].

Lemma 4.5: Let $\mathcal{P} = (P,T,\varphi,\mu_0)$ be a sinkless PN with k places and m transitions such that no component of μ_0 is larger than $n \geq 1$. Then there exist constants c_1, c_2, d_1, and d_2, independent of k, m, and n, such that $R(\mathcal{P}) = \bigcup_{\mu \in \beta} \mathcal{L}(\mu, \rho_\mu)$, where β is the set of all reachable markings with no component larger than $(c_1 \cdot k \cdot m \cdot n)^{c_2 \cdot k \cdot m}$, and ρ_μ is the set of all $\delta \in N^k$ such that:

1. for some $\sigma \in T^*$, $\mu \xrightarrow{\sigma} \mu + \delta$;
2. δ has no component larger than $(d_1 \cdot k \cdot m \cdot n)^{d_2 \cdot k \cdot m}$;
3. if $\mu(i) = 0$, then $\delta(i) = 0$, for $1 \leq i \leq k$; and
4. $\delta \neq 0$.

The property of conflict-free PN reachability sets that allowed the CP to be shown to be in Π_2^P in [HR88a] is that their SLS representations have a certain symmetry. In particular, for any two markings μ and μ' of a conflict-free PN $\mathcal{P} = (P,T,\varphi,\mu_0)$ such that $\mu(i) = 0$ iff $\mu'(i) = 0$, if $\mu \xrightarrow{\sigma} \mu + v$ for some $\sigma \in T^*$ and some $v \geq 0$, then there is some $\sigma' \in T^*$ such that $\mu' \xrightarrow{\sigma'} \mu' + v$. The following lemma shows that a similar symmetry extends to sinkless PNs. The lemma follows in a straightforward manner from Lemma 4.1.

Lemma 4.6: Let μ and μ' be reachable markings of a sinkless PN $\mathcal{P} = (P,T,\varphi,\mu_0)$ with k places such that $\mu(i) = 0$ iff $\mu'(i) = 0$. For any vector $v \in N^k$ such that $v(i) = 0$ if $\mu(i) = 0$, if there is a $\sigma \in T^*$ such that $\mu \xrightarrow{\sigma} \mu + v$, then there is a $\sigma' \in T^*$ such that $\mu' \xrightarrow{\sigma'} \mu' + v$.

Now that we have given a convenient SLS representation of the reachability set of a sinkless PN and shown the symmetry therein, we can outline our strategy for showing the CP and EP to be in Π_2^P. Again, this strategy was first developed for the structurally-defined conflict-free PNs in [HR88a], borrowing some techniques from [Huy86]; we use Lemmas 4.5 and 4.6 to show that this strategy also applies to the behaviorally defined sinkless PNs. Let $SL_1 = \bigcup_{\mu \in \beta_1} \mathcal{L}(\mu, \rho_\mu^1)$ and $SL_2 = \bigcup_{\mu \in \beta_2} \mathcal{L}(\mu, \rho_\mu^2)$ be the SLS representations given by Lemma 4.5 for the sinkless PNs $\mathcal{P}_1$ and $\mathcal{P}_2$, respectively. In order to show that $R(\mathcal{P}_1) \not\subseteq R(\mathcal{P}_2)$, our algorithm will prove the existence of a $\mu \in SL_1 - SL_2$. (Note that since the SLS representations are exponential in the sizes of the PN representations, the SLS representations cannot be generated by the algorithm.) Let $\mu \in SL_1$. Then $\mu \in \mathcal{L}(\mu_1, \rho_{\mu_1}^1)$ for some $\mu_1 \in \beta_1$. If, in addition, $\mu \in SL_2$, then $\mu \in \mathcal{L}(\mu_2, \rho_{\mu_2}^2)$ for some $\mu_2 \in \beta_2$. Note from the definition of the SLS representations in Lemma 4.5 that for any place p_i, $\mu_1(i) = 0$ iff $\mu(i) = 0$ iff $\mu_2(i) = 0$. Furthermore, we may assume without loss of generality that $\mu_1 \in R(\mathcal{P}_2)$; otherwise, we will have found a witness to the fact that $R(\mathcal{P}_1) \not\subseteq R(\mathcal{P}_2)$. Thus, from Lemma 4.6, $\rho_{\mu_2}^2 = \rho_{\mu_1}^2$, where $\rho_{\mu_1}^2$ is as defined in Lemma 4.5. So to show the existence of a $\mu \in SL_1 - SL_2$, it is sufficient to show the existence of a $\mu \in \mathcal{L}(\mu_1, \rho_{\mu_1}^1) - \bigcup_{\mu_2 \in \beta_2'} \mathcal{L}(\mu_2, \rho_{\mu_1}^2)$ for some $\mu_1 \in \beta_1$, where $\beta_2' = \{\mu' \in \beta_2 \mid \mu'(i) = 0 \text{ iff } \mu_1(i) = 0\}$. Note that once μ_1 is chosen, we are only concerned with two period sets, $\rho_{\mu_1}^1$ and $\rho_{\mu_1}^2$.

Consider two sets $\mathcal{L}(\mu_1, \rho_1)$ and $\bigcup_{\mu_2 \in \beta} \mathcal{L}(\mu_2, \rho_2)$. In order to show the existence of a $\mu \in \mathcal{L}(\mu_1, \rho_1) - \bigcup_{\mu_2 \in \beta} \mathcal{L}(\mu_2, \rho_2)$, we consider two cases. First, suppose that every vector in ρ_1 is a positive linear combination of the vectors in ρ_2. In [HR88a], we showed that in this case there must be a $\mu \in \mathcal{L}(\mu_1, \rho_1) - \bigcup_{\mu_2 \in \beta} \mathcal{L}(\mu_2, \rho_2)$ whose size is polynomial in the sizes of the elements of ρ_1, ρ_2, β, and μ_1 and exponential in the dimension of these vectors; i.e., the witness is small enough to be written down in space polynomial in the sizes of the representations of the PNs from which the SLSs are derived. On the other hand, suppose some vector in ρ_1 is not a linear combination of the vectors in ρ_2. We also showed in [HR88a] that in this case, $\mathcal{L}(\mu_1, \rho_1)$ cannot be contained in $\bigcup_{\mu_2 \in \beta} \mathcal{L}(\mu_2, \rho_2)$. For the sake of completeness, we now reproduce the relevant lemmas from [HR88a].

Lemma 4.7 (from [HR88a]): Let ρ_1, ρ_2, and β be finite subsets of N^k, $\mu_1 \in N^k$, and $n \in N$ such that no integer in ρ_1, ρ_2, β, or μ_1 exceeds n. If every vector in ρ_1 is a positive linear combination of vectors in ρ_2 and

$\mu \in \mathcal{L}(\mu_1, \rho_1) - \bigcup_{\mu_2 \in \beta} \mathcal{L}(\mu_2, \rho_2)$, then there is a μ' with no component larger than $k(n+1)^{2k+1} + n$ such that $\mu' \in \mathcal{L}(\mu_1, \rho_1) - \bigcup_{\mu_2 \in \beta} \mathcal{L}(\mu_2, \rho_2)$.

Lemma 4.8 (from [HR88a]): Let $\delta, \mu_1 \in \mathrm{N}^k$ such that $\delta \neq \mathbf{0}$, and let ρ and β be finite subsets of N^k. If δ is not a positive linear combination of the vectors in ρ, then there is an $n \in \mathrm{N}$ such that $\mu_1 + n\delta \notin \bigcup_{\mu \in \beta} \mathcal{L}(\mu, \rho)$.

Finally, using Lemmas 4.5–4.8 and techniques from [HR88a] (see also [Huy86]) we are able to show the following theorem.

Theorem 4.5: The CP and EP for sinkless (normal) PNs are Π_2^P-complete.

References

[Bak73] H. Baker. *Rabin's Proof of the Undecidability of the Reachability Set Inclusion Problem of Vector Addition Systems.* Memo 79, MIT Project MAC, Computer Structure Group, 1973.

[BT76] I. Borosh and L. Treybig. Bounds on positive integral solutions of linear Diophantine equations. *Proc. AMS*, 55:299–304, March 1976.

[CLM76] E. Cardoza, R. Lipton, and A. Meyer. Exponential space complete problems for Petri nets and commutative semigroups. In *Proceedings of the 8th Annual ACM Symposium on Theory of Computing*, pages 50–54, 1976.

[CM75] S. Crespi-Reghizzi and D. Mandrioli. A decidability theorem for a class of vector addition systems. *Information Processing Letters*, 3:78–80, 1975.

[Gra80] J. Grabowski. The decidability of persistence for vector addition systems. *Information Processing Letters*, 11:20–23, 1980.

[GY80] A. Ginzburg and M. Yoeli. Vector addition systems and regular languages. *J. of Computer and System Sciences*, 20:277–284, 1980.

[Hac75] M. Hack. *Petri Net Languages.* Memo 124, MIT Project MAC, Computer Structure Group, 1975.

[Hac76] M. Hack. The equality problem for vector addition systems is undecidable. *Theoret. Comp. Sci.*, 2:77–95, 1976.

[HP79] J. Hopcroft and J. Pansiot. On the reachability problem for 5-dimensional vector addition systems. *Theoret. Comp. Sci.*, 8:135–159, 1979.

[HR88a] R. Howell and L. Rosier. Completeness results for conflict-free vector replacement systems. *J. of Computer and System Sciences*, 37:349–366, 1988.

[HR88b] R. Howell and L. Rosier. On questions of fairness and temporal logic for conflict-free Petri nets. In G. Rozenberg, editor, *Advances in Petri Nets 1988*, pages 200–226, Springer-Verlag, Berlin, 1988. LNCS 340.

[HRHY86] R. Howell, L. Rosier, D. Huynh, and H. Yen. Some complexity bounds for problems concerning finite and 2-dimensional vector addition systems with states. *Theoret. Comp. Sci.*, 46:107–140, 1986.

[HRY87] R. Howell, L. Rosier, and H. Yen. An $O(n^{1.5})$ algorithm to decide boundedness for conflict-free vector replacement systems. *Information Processing Letters*, 25:27–33, 1987.

[HRY88] R. Howell, L. Rosier, and H. Yen. *Normal and Sinkless Petri Nets.* Technical Report TR-CS-88-14, Kansas State University, Manhattan, Kansas 66506, 1988.

[HU79] J. Hopcroft and J. Ullman. *Introduction to Automata Theory, Languages, and Computation.* Addison-Wesley, Reading, Mass., 1979.

[Huy82] D. Huynh. The complexity of semilinear sets. *Elektronische Informationsverarbeitung und Kybernetik*, 18:291–338, 1982.

[Huy85] D. Huynh. The complexity of the equivalence problem for commutative semigroups and symmetric vector addition systems. In *Proceedings of the 17th Annual ACM Symposium on Theory of Computing*, pages 405–412, 1985.

[Huy86] D. Huynh. A simple proof for the Σ_2^P upper bound of the inequivalence problem for semilinear sets. *Elektronische Informationsverarbeitung und Kybernetik*, 22:147–156, 1986.

[JLL77] N. Jones, L. Landweber, and Y. Lien. Complexity of some problems in Petri nets. *Theoret. Comp. Sci.*, 4:277–299, 1977.

[Kos82] R. Kosaraju. Decidability of reachability in vector addition systems. In *Proceedings of the 14th Annual ACM Symposium on Theory of Computing*, pages 267–280, 1982.

[Lam87] J. Lambert. Consequences of the decidability of the reachability problem for Petri nets. In *Proceedings of the Eighth European Workshop on Application and Theory of Petri Nets*, pages 451–470, 1987. To appear in *Theoret. Comp. Sci.*

[Lip76] R. Lipton. *The Reachability Problem Requires Exponential Space.* Technical Report 62, Yale University, Dept. of CS., Jan. 1976.

[LR78] L. Landweber and E. Robertson. Properties of conflict-free and persistent Petri nets. *JACM*, 25:352–364, 1978.

[May81] E. Mayr. Persistence of vector replacement systems is decidable. *Acta Informatica*, 15:309–318, 1981.

[May84] E. Mayr. An algorithm for the general Petri net reachability problem. *SIAM J. Comput.*, 13:441–460, 1984. A preliminary version of this paper was presented at the *13th Annual Symposium on Theory of Computing*, 1981.

[MM81] E. Mayr and A. Meyer. The complexity of the finite containment problem for Petri nets. *JACM*, 28:561–576, 1981.

[MM82] E. Mayr and A. Meyer. The complexity of the word problems for commutative semigroups and polynomial ideals. *Advances in Mathematics*, 46:305–329, 1982.

[Mul81] H. Müller. On the reachability problem for persistent vector replacement systems. *Computing, Suppl.*, 3:89–104, 1981.

[Pet81] J. Peterson. *Petri Net Theory and the Modeling of Systems.* Prentice Hall, Englewood Cliffs, NJ, 1981.

[Rac78] C. Rackoff. The covering and boundedness problems for vector addition systems. *Theoret. Comp. Sci.*, 6:223–231, 1978.

[Rei85] W. Reisig. *Petri Nets: An Introduction.* Springer-Verlag, Heidelberg, 1985.

[Sto77] L. Stockmeyer. The polynomial-time hierarchy. *Theoret. Comp. Sci.*, 3:1–22, 1977.

[VV81] R. Valk and G. Vidal-Naquet. Petri nets and regular languages. *J. of Computer and System Sciences*, 23:299–325, 1981.

[Yam84] H. Yamasaki. Normal Petri nets. *Theoret. Comp. Sci.*, 31:307–315, 1984.

Descriptive and Computational Complexity

Neil Immerman

Dept. of Computer and Information Science
University of Massachusetts
Amherst, MA 01003
U.S.A.

Abstract

Computational complexity began with the natural physical notions of time and space. Given a property, S, an important issue is the complexity of checking whether or not an input satisfies S. For a long time, complexity referred to the time or space used in the computation. A mathematician might ask, "What is the complexity of expressing the property S" It should not be surprising that these two questions – that of checking and that of expressing – are related. It is startling how closely tied they are when the second question refers to expressing the property in first-order logic. Many complexity classes originally defined in terms of time or space resources have precise definitions as classes in first-order logic.

In 1974 Fagin gave a characterization of nondeterministic polynomial time as the set of properties expressible in second-order existential logic. We will begin with this result and then survey some more recent work relating first-order expressibility to computational complexity. Some of the results arising from this approach include characterizing polynomial time as the set of properties expressible in first-order logic plus a least fixed point operator, and showing that the set of first-order inductive definitions for finite structures is closed under complementation. We will end with an unexpected result that was derived using this approach: For all s(n) greater than or equal to log n, nondeterministic space s(n) is closed under complementation.

The following figures give some of the relationships among computational and descriptive complexity classes.

$$LH = AC^0 \subset Th^0 \subseteq NC^1 \subseteq L \subseteq NL \subseteq AC^1 \subseteq Th^1 \subseteq NC$$

$$LH \subset NC^1 \subseteq FO[\log n/ \log\log n] \subseteq AC^1 \subseteq P \subseteq NP \subseteq PH \subseteq PSPACE$$

Known relationships between complexity classes

Class	FO	SO	Parallel
LH	FO		CRAM[1]
Th^0	FOM		
L	$(FO + DTC)$		
NL	$(FO + TC)$		
AC^1	$FO[\log n]$		$CRAM[\log n]$
NC	$FO[(\log n)^{O[1]}]$		$CRAM[(\log n)^{O[1]}]$
P	$(FO + LFP) = FO[n^{O[1]}]$		$CRAM[n^{O[1]}]$
NP		$(SO\exists)$	
PH		SO	$CRAM[1]\text{-}PROC[2^{n^{O[1]}}]$
PSPACE	$FO[2^{n^{O[1]}}]$	$SO[n^{O[1]}]$	$CRAM[2^{n^{O[1]}}]$

Computational versus Descriptive Complexity

THE EFFECT OF NULL-CHAINS ON THE COMPLEXITY OF CONTACT SCHEMES

Stasys P. Jukna

Institute of Mathematics and Cybernetics
Lithuanian Academy of Sciences
232600 Vilnius, MTP-1
Akademijos str., 4
Lithuania, USSR

Abstract

The contact scheme complexity of Boolean functions has been studied for a long time but its main problem remains unsolved: we have no example of a simple function (say in NP) that requires $\Omega(n^3)$ contact scheme size. The reason is, perhaps, that although the contact scheme model is elegantly simple, our understanding of the way it computes is vague.

On the other hand, it is known (see, e.g. [2,3]) that the main tool to reduce the size of schemes is to use "null-chains", i.e. chains with zero conductivity.(These chains enable one to merge non-isomorphic subschemes). So, in order to better understand the power of this tool, it is desirable to have lower bound arguments for schemes with various restrictions on null-chains.

In this report such an arguments are described for schemes without null-chains (Theorems 1-2), for schemes with restricted topology of null-chains (Theorem 3), and for schemes with restricted number and/or restricted length of null-chains (Theorem 4). In all these cases nearly-exponential lower bounds are established. Finally, we prove that null-chains do not help at all if schemes are required to realize sufficiently many prime implicants (Theorem 5).

1. Preliminaries

We deal with the standard model of contact schemes but we need some notations. Fix some set of Boolean variables $\mathcal{X}^+ = \{x_1,...,x_n\}$ and their negations $\mathcal{X}^- = \{\neg x_1,...,\neg x_n\}$. The elements of $\mathcal{X} = \mathcal{X}^+ \cup \mathcal{X}^-$ are called <u>contacts</u>. A contact scheme S is a labelled digraph with two distinguished nodes (the source and the output), and edges labelled by contacts. The <u>size</u> of S, size(S), is the number of edges in S. A chain is (a sequence of edges in)a path from the source to output. A <u>subchain</u> is a subsequence of (not necessarily consecutive) edges in a chain. A <u>cut</u> is a minimal set of edges which contains an edge from each chain. We will often identify a chain [cut] A with the set $A \subseteq \mathcal{X}$ of contacts it consists of; the current meaning will be clear from the context. A chain [cut] $A = \{y_1,...,y_m\} \subseteq \mathcal{X}$ (m $\le$

2n) defines the monomial $K_A = \&_{i=1}^{m} y_i$ [the clause $D_A = \bigvee_{i=1}^{m} y_i$]. A chain [cut] A is <u>redundant</u> if $K_A \equiv 0$ [$D_A \equiv 1$]. Thus a chain (as well as a cut) is redundant iff it contains some pair of contrary contacts x_i and $\neg x_i$. Redundant chains [cuts] are also called <u>null-chains</u> [<u>one-cuts</u>]. A contact scheme <u>computes</u> a Boolean function f_S iff

$$f_S = \bigvee\{ K_A : A \text{ is a chain of } S \},$$

or equivalently, iff

$$f_S = \& \{ D_A : A \text{ is a cut of } S \}.$$

We will also need the following notions from extremal set theory. Let $\mathcal{F}$ be a family of subsets of a finite set N . For an integer i ($0 \le i \le |N|$), put

$$\#_i\mathcal{F} = \max \{ |\mathcal{G}| : \mathcal{G} \subseteq \mathcal{F} \text{ and } |\bigcap_{A\in\mathcal{G}} A| \ge i \}$$

i.e. $\#_i\mathcal{F}$ is the maximum number of sets in $\mathcal{F}$ that have at least i elements in common. Thus

$$|\mathcal{F}| = \#_0\mathcal{F} \ge \#_1\mathcal{F} \ge \dots \ge \#_{|N|}\mathcal{F} = 1 .$$

The rate to which $\#_i\mathcal{F} \longrightarrow 1$ as $i \longrightarrow |N|$ characterizes the "dispersion" of elements from N over the subsets of $\mathcal{F}$.

A family $\mathcal{F}$ is (t,r)-<u>dispersed</u> if

$$\#_i\mathcal{F} \,/\, \#_{i+1}\mathcal{F} \ge t \quad \text{for all } i = 0,1,\dots,r-1.$$

A family $\mathcal{F}$ is (k,r)-<u>disjoint</u> ($k\ge 2, r\ge 0$) if $\#_r\mathcal{F} \le k - 1$. Notice that any (t,r)-dispersed family is also (k,r)-disjoint with $k = |\mathcal{F}|\cdot t^{-r}$.

In this report we show that for any sufficiently dispersed family $\mathcal{F}_o \subseteq 2^N$, the characteristic function $f_{\mathcal{F}} : 2^N \longrightarrow \{0,1\}$ of any family $\mathcal{F} \subseteq 2^N$, given by

$$A \in \mathcal{F} \quad \Longleftrightarrow \quad \exists\, B \in \mathcal{F}_o : \; A \supseteq B ,$$

requires super-polynomial size to be computed by contact schemes with various restrictions on null-chains and one-cuts. The consequence is that, under these restrictions, almost all NP-complete functions require super-polynomial contact scheme size.

2. Schemes Without Null-Chains

For a Boolean function f, let $L^*(f)$ denote the minimum size of a contact scheme without null-chains computing f .

The first non-trivial lower bound for π-schemes without null-chains has been proved by A.K. Pulatov in [8] and improved to contact schemes by S.E. Kuznetsov in [6]. Somewhat later similar results have been obtained for one-time-only branching programs - a special type of contact sheme without null-chains - by now a long list of authors (see, e.g. references in [3] or [12]).

Associate with a Boolean vector $\alpha = (\alpha_1, \dots, \alpha_n)$ the set of contacts $N_\alpha = \{x_1^{\alpha_1}, \dots, x_n^{\alpha_n}\} \subset \mathfrak{X}$ where $x^1 = x$ and $x^0 = \neg x$, and put $\mathbb{N}_f = \{ N_\alpha : \alpha \in f^{-1}(1) \}$. Let

$$d(f) = 1 + \min \{ r : \mathbb{N}_f \text{ is } (2,n-r)\text{-disjoint} \} .$$

Notice that $d(f)$ is actually the minimal Hamming distance between any two vectors in $f^{-1}(1)$.

Theorem 1 (Pulatov [8], Kuznetsov [6]): For any Boolean function f

$$L^*(f) \geq |\mathbb{N}_f|^{d(f)/n} .$$

The theorem enables to obtain non-trivial lower bounds for functions f with $d(f)$ large enough with respect to $|\mathbb{N}_f|$. Recently, S.V. Zdobnov has announced in [13] the following improvement of this result.

Theorem 2 (Zdobnov [13]): If $d(f) \geq 3$ then

$$L^*(f) \geq |\mathbb{N}_f| . n^{(1/2-\varepsilon)\log n} . 2^{-n} .$$

The theorem already yields super-polynomial lower bounds for some functions f with small $d(f)$, including the characteristic function of the Hamming code. Unfortunately, this argument (as well as Theorem 1) does not work for functions f with small $|\mathbb{N}_f|$.

Example 1 : Let $m \geq 2$ be a prime power and let $1 \leq s \leq m/2$. The Galois function is the following function $g_{m,s}(X)$ of $n=m^2$ Boolean variables $X = \{ x_{i,j} : i,j \in GF(m) \}$:

$g_{m,s}(X) = 1$ iff there exists a polynomial σ of degree at most $s-1$ over the Galois field $GF(m)$ such that $\forall i,j \in GF(m)$ $x_{i,j} = 1$ iff $j = \sigma(i)$.

Since $d(g) \leq 2m$, we have that

$$|\mathbb{N}_g|^{d(g)/n} \leq m^2 = n \qquad \text{and} \qquad \log |\mathbb{N}_g| = s\log m = o(n),$$

and therefore, both arguments fail for g, whereas it is known (see [2]) that $L^*(g_{m,s}) \geq m^s$.

So, even for schemes without null-chains new arguments are desirable. General technique for schemes with restrictions on the topology of null-chains have been proposed in [2,3]. Let us briefly describe a modification of this argument .

3. Schemes With Free Subchains

Let $\mathcal{R}(S)$ be the set of all subchains in a contact scheme S . For $A\in\mathcal{R}(S)$, let $ext(A) = \{ C\in\mathcal{R}(S) : A\cup C$ is a chain in S $\}$ be the set of all extentions of A in S, and let $sp(S) = \{ B\in\mathcal{R}(S) : ext(B) = ext(A) \}$ be the "span" of A in S. For families of sets $\mathcal{F}$ and $\mathcal{G}$, set $\mathcal{F}\otimes\mathcal{G} = \{A\cup B : A\in\mathcal{F}$ and $B\in\mathcal{G} \}$. A subchain $A \in \mathcal{R}(S)$ is called to be <u>free</u> in S if it produces no new null-chain, i.e. if

$$\forall\, C \in ext(A) : \qquad K_C \neq 0 \;\Rightarrow\; K_{A\cup C} \neq 0 .$$

A collection of subchains $\mathcal{A} \subseteq \mathcal{R}(S)$ is a <u>separator</u> of S if

$$S_\emptyset = \bigcup_{A\in\mathcal{A}} S_A \qquad \text{and} \qquad |\mathcal{A}| \leq size(S).$$

where $S_A = sp(A) \otimes ext(A)$ (and hence, $S_\emptyset$ is the set of all chains in S). A separator $\mathcal{A}$ is an [a,b]-separator if $a \leq |A^+| \leq b$ for all $A\in\mathcal{A}$. (Throughout, A^+ stands for the set of all unnegated variables (not edges !) in a subchain A). Thus, any cut defines an obvious [0,1]-separator. Moreover, any scheme has at least one [a,b]-separator for any $0 \leq a \leq b \leq \min\{|A^+| : A\in S_\emptyset \}$.

We call a contact scheme S to be [a,b]-<u>separable</u> if there exists an [a,b]-separator $\mathcal{A}$ of S such that all $A\in\mathcal{A}$ are free in S . Let $L_{a,b}(f)$ denote the minimum size of an [a,b]-separable contact scheme computing f . It is clear that for all $a \leq b$

$$L^*(f) \geq L_{a,b}(f).$$

Let $\mathbb{N}_f(m) \subseteq \mathbb{N}_f$ be the m-th slice of f , i.e $\mathbb{N}_f(m) = \{ A\in\mathbb{N}_f : |A^+| = m \}$. A function f is called $(k,r)_m$-<u>disjoint</u> if the following two conditions are fulfilled :

(i) $\quad \#_r\{ A^+ : A \in \mathbb{N}_f(m) \} \leq k-1$,

(ii) $\quad$ if $A \in \mathbb{N}_f$ but $A \notin \mathbb{N}_f(m)$ then $|A^+| \geq 2m$.

Theorem 3 : If f is $(k,r)_m$-disjoint for some $k\geq 2$ and $m\geq 2r\geq 0$ then

$$L_{r,m-r}(f) \geq |\mathbb{N}_f(m)|\cdot(k-1)^{-2}$$

Proof : Let S be an [r,m-r]-separable scheme computing f, and let $\mathcal{A} \subseteq \mathcal{R}(S)$ be the corresponding free separator of S . Notation: for a set of chains $\mathscr{C}$ we will write $||\mathscr{C}||$ instead of $|\mathscr{C} \cap \mathbb{N}_f(m)|$. Then

$$|\mathbb{N}_f(m)| = ||S_\emptyset|| \le \sum_{A\in\mathscr{A}} ||S_A|| \le \delta\ |\mathscr{A}| \le \delta \text{ size}(S) ,$$

where

$$\delta = \max\left\{ ||S_A|| \ :\ A\in\mathscr{A} \right\}.$$

So it remains to prove that $\delta \le (k-1)^2$. Take $A\in\mathscr{A}$. Then $r\le|A^+|\le m-r$ and A is free in S. Consider $\text{Ext} = \{ C \in \text{ext}(A) : ||\text{sp}(A)\otimes\{C\}|| \ge 1 \}$. Ext is the set of all the extentions of A that are used to compute the m-th slice of f . Other extentions of A are of no interest for us since

$$||S_A|| = ||\text{sp}(A)\otimes\text{Ext}|| .$$

Let $\mathscr{D} := (\{A\}\otimes\text{Ext}) \cap \mathbb{N}_f(m)$. Then $|\mathscr{D}| \le k-1$ since $|\cap\{D^+ : D\in\mathscr{D}\}| \ge |A^+| \ge r$. The crucial observation is that $\mathscr{D} = \{A\}\otimes\text{Ext}$. This follows from (ii) because if $B = A\cup C$ with $C\in\text{Ext}$, then $K_B \ne 0$ and $|B^+| \le |A^+| + |C^+| \le (m-r)+m \le 2m$. Hence, Ext may be partitioned into $|\mathscr{D}| \le k-1$ pairwise disjoint subsets $\text{Ext}_D = \{ C\in\text{Ext} : A\cup C = D \}$, $D\in\mathscr{D}$. By (i) we have, for each $D\in\mathscr{D}$, that $||\text{sp}(A)\otimes\text{Ext}_D|| \le k-1$ because

$$|\cap \{C^+ : C\in\text{Ext}_D \}| \ge |D^+\setminus A^+| \ge m-(m-r) = r .$$

Therefore, $\delta \le |\mathscr{D}|(k-1) \le (k-1)^2$ and the theorem follows. ∎

The class of schemes without null-chains is not closed under the negation in a sense that $L^*(\neg f) \ll L^*(f)$ for some f . Let, for example, p_n be the function of $n = m^2$ Boolean variables representing the elements of an mxm-matrix M, whose value is 1 iff each row and each column of M has exactly one 1. Then $|\mathbb{N}_p^+| = m!$ and, therefore, p_n is $(k,r)_m$-disjoint for $r = m/2$ and $k = r!$. By Theorem 3, $L^*(p_n) \ge \exp(\Omega(\sqrt{n}))$, whereas one may easily verify that

$$L^*(\neg p_n) = O(n^{3/2}).$$

On the other hand, Theorem 3 enables one to construct an explicit functions f such that both f and ¬f are hard to compute by schemes without null-chains.(Notice that Theorems 1 and 2 both fail in this situation, because $d(\neg f) = 1$ for any function f with $d(f) \ge 3$).

Example 2 : Define the function $f_{m,s}$ of $n = m^2$ variables by :

$$f_{m,s}(\alpha)=\begin{cases} g_{m,s}(\alpha) & \text{if} \quad 0 \le |N_\alpha| < n/2, \\ g^*_{m,s}(\alpha) & \text{otherwise,} \end{cases}$$

where f^* stands for the dual of f, i.e. $f^* = \neg f(\neg x_1,,,.\neg x_n)$.

Since f is $(2,s)_m$-disjoint and self-dual (i.e. $f = f^*$), Theorem 3 immediately yields the following lower bound.

Corollary 1 : $\min\left\{ L_{s,m-s}(f),\ L_{s,m-s}(\neg f) \right\} \ge m^s$.

Specifically, both f and $\neg f$ are hard to compute if null-chains are forbidden

4. Schemes With Long Null-Chains

As we have seen above, there is an exponential gap between the complexity of schemes with and without null-chains. This means that although the usedge of null-chains and one-cuts has no influence on the function computed, such chains and cuts may lead to great reduction of size.

In this section we will show that null-chains and one-cuts do not help in both of the following situations:

(i) if we restrict the number of null-chains and one-cuts in a scheme, or

(ii) if we do not use "very short" null-chains or one-cuts.

Given a contact scheme S, let $\mathfrak{m}(S)$ $[\mathfrak{m}^\perp(S)]$ denote the number of all minimal subsets $A^+ \subseteq \mathscr{X}^+$ where A ranges over all null-chains [one-cuts] in S . (Recall that A^+ is the set of unnegated variables in A). Let $\mathfrak{l}(S)$ $[\mathfrak{l}^\perp(S)]$ stand for $\min|A^+|$ where A ranges over all null-chains [one-cuts] in S . Thus for any contact scheme S , we have that

$$0 \le \mathfrak{l}(S) \le n \quad \text{and} \quad 0 \le \mathfrak{m}(S) \le \binom{n}{\mathfrak{l}}.$$

Define

$$L_{\mu,\lambda}(f) = \min\left\{ \text{size}(S) : S \text{ computes } f \text{ and } \mathfrak{m}(S) \le \mu \text{ and } \mathfrak{l}(S) \le \lambda \right\}.$$

In case of one-cuts we will write $L^\perp$ instead of L .Notice that $L_{\mu,\lambda}(f) = L(f)$, the unrestricted contact scheme complexity of f, if either $\lambda = n$ or $\lambda < n$ but $\mu = \binom{n}{\mathfrak{l}}$.

We will estimate these complexity functionals in terms of the dispersion of minterms and maxterms. A <u>minterm</u> [<u>maxterm</u>] of a Boolean function f is a minimal set, of contacts $A \subseteq \mathscr{X}$ such that

$$f \geq \underset{y \in A}{\&} y \neq 0 \qquad [\, f \leq \bigvee_{y \in A} y \neq 1 \,].$$

Define $\mathit{min}(f)$, $\mathit{Max}(f)$ as the set of minterms, respectively maxterms of f . Let $\mathfrak{r}(f)$, $\mathfrak{R}(f)$ denote the minimum cardinality of a set in $\mathit{min}(f)$, respectively in $\mathit{Max}(f)$.

For integers $t, r \geq 1$ and real numbers $p, \aleph \in [0,1]$, let $H_f(t,r,p,\aleph)$ denote the following number:

$$H_f = t^{-r/2} \min \left\{ \Delta_f(r/2) \,,\, \left[1 - \aleph - \#_0 \mathit{min}(f) p^{\mathfrak{r}(f)}\right] 2^{tp^r - r\log\sqrt{t}} \right\},$$

where

$$\Delta_f(i) = \max_{\mathscr{F}} \left[\#_0 \mathscr{F} \,/\, \#_i \mathscr{F} \right]$$

and $\mathscr{F}$ ranges over all (t,r)-dispersed subfamilies $\mathscr{F} \subseteq \mathit{min}(f)$.

Theorem 4 : For any monotone Boolean function f , the following bound holds:

$$L_{\mu,\lambda}(f) \geq \max_{p \in [0,1)} H_f(t,r,p,\aleph)$$

where

$$\aleph = \min \left\{ \mu p^{\lambda} \,,\, np/\lambda \right\}.$$

The same bound holds also for $L^{\perp}_{\mu,\lambda}(f)$ with $\mathit{min}(f)$ and $\mathfrak{r}(f)$ replaced by $\mathit{Max}(f)$ and $\mathfrak{R}(f)$.

Proof (sketch): Let S be a minimal contact scheme computing f with $\mathfrak{m}(S) \leq \mu$ and $\mathfrak{l}(S) \leq \lambda$. Replace in S all the negated contacts by constant 1 (or by 0 in case of one-cuts). Let f^+ be the monotone function computed by the resulting scheme S^+. Then $\mathrm{size}(S) \geq \mathrm{size}(S^+)$ and $f^+ \geq f$. From ([4], Theorem 4) it follows that

$$\mathrm{size}(S^+) \geq \max_{p \in [0,1)} H_{f^+}(t,r,p,\aleph_+),$$

where

$$\aleph_+ = \mathrm{Prob}\left[K_A \leq f^+ \& \neg f \right] \leq \mathrm{Prob}\left[K_A \leq f^+ \right]$$

and $A \subseteq \{x_1, \ldots, x_n\}$ is a random monomial in which each variable x_i appears independently and with equal probability $p \in [0,1)$.

Let g be the disjunction of all the monomials in $min(f^+)\setminus min(f)$. Then $f^+ = f \vee g$, $\mathfrak{r}(g) \geq \mathfrak{l}(S)$ and $\#_0 min(g) \leq \mathfrak{m}(S)$. So,

$$\aleph_+ \leq \text{Prob}\left[K_A \leq f \right] + \text{Prob}\left[K_A \leq g \right].$$

It remains to notice that for any monotone f, we have that

$$\text{Prob}\left[K_A \leq f \right] \leq \#_0 min(f)\ p^{\mathfrak{r}(f)}$$

and, by Chebyshev's inequality,

$$\text{Prob}\left[K_A \leq f \right] \leq \text{Prob}\left[|A| \geq \mathfrak{r}(f) \right] \leq np/\mathfrak{r}(f).$$ ▮

Example 3: Let f_n be the monotone function of $n = \binom{m}{2}$ Boolean variables representing the edges of an undirected graph G, which is 1 iff G contains an s-clique where $s = \lceil (m/\ln m)^{2/3} \rceil$. Then $\#_i min(f_n) = \binom{m-i}{s-i}$, and hence $min(f_n)$ is (t,r)-dispersed for any $t \leq \lceil m/3 \rceil$ and $r \leq s$.

Corollary 2 : If $\lambda = \Omega(n^{1-1/s})$ or $\mu \leq (1-\varepsilon)n^{\lambda/s}$, $\varepsilon > 0$, then

$$L_{\mu,\lambda}(f_n) \geq \exp(\Omega(n^{1/6-o(1)})).$$

Proof: Take $r = \lceil \sqrt{s} \rceil$, $t = \lceil 4r \ln m \rceil$ and $p = m^{-2/s}$. Then $\#_0 min(f) p^{\mathfrak{r}(f)} \leq \binom{m}{s} p^{s^2} < m^{-s}$, and by Theorem 4, the bound holds for any μ, λ such that $\min\{ \mu p^{\lambda}, np/\lambda \} \leq const < 1$. ▮

Example 4 : Define

$$g_n^+ = \mathop{\&}_{\sigma \in \Pi} \bigvee_{i \in GF(m)} x_{i,\sigma(i)}$$

where Π is the set of all polynomials over GF(m) of degree at most s-1, and $s = \lceil \ln m \rceil$. As $\#_i Max(g_n^+) = m^{s-i}$, the family $\#_i Max(g_n^+)$ is (t,r)-dispersed for any $t \leq m/3$ and $r \leq s$.

Corollary 3: If $\lambda = \Omega(n)$ or $\log_2 \mu \leq O(\lambda)$ then

$$L^{\perp}_{\mu,\lambda}(g_n^+) \geq n^{\Omega(\log n)}.$$

Proof: Take $t = \lceil \sqrt{m} \rceil$, $r = \lceil s/2 \rceil$ and $p = (t^{-1} \ln^2 t)^{1/s}$, and apply Theorem 4. ∎

This bound is almost tight because g_n^+ is computable by a trivial contact scheme S with $\mathfrak{m}^{\perp}(S) = 0$ and $size(S) \leq n^{\log n}$.

Theorem 4 yieds also the following criterion for the monotone scheme complexity $L^+(f)$. For a random monomial $A \subseteq X^+$, put

$$P_A(r) = \max\left\{ \text{Prob}\,[\, A \supseteq B \,] : \; B \subseteq X^+ \text{ and } |B| = r \right\}.$$

We say A islocally independent if for any two monomials $B_1, B_2 \subseteq X^+$, the events $\{ A \supseteq B_i \mid A \supseteq B_1 \cap B_2 \}$ are independent. We say f is (t,r)-good if there exists a locally independent monomial A such that

$$\text{Prob}\,[\, K_A \leq f \,] \leq \text{const} < 1 \quad \text{and} \quad P_A(r) \gg t^{-1} \ln \Delta_f(r).$$

Criterion: If f is (t,r)-good and $min(f)$ is (t,r)-dispersed for some t and r such that $\ln t \ll r^{-1} \ln \Delta_f(r)$, then

$$L^+(f) \geq \Delta_f(r) t^{-r}.$$

5. Shemes With Necessary Minterms

For a contact scheme S, let f_S denote the Boolean function it computes, and let $\mathcal{D}_S$ denote the set of all monomials corresponding to non-null chains of S. A minterm $A \in min(f)$ is necessary if there exists a vector $\alpha \in \{0,1\}^n$ with $K_A(\alpha) = 1$ but $K_B(\alpha) = 0$ for all other minterms $B \in min(f) - \{A\}$. (These minterms belong necessarily to each shortest DNF of f). Define $nec(f)$ as the set of all necessary minterms of f .

A contact scheme S is called to be a δ-scheme ($\delta \in [0,1]$) if

$$|\mathcal{D}_S \cap nec(f_S)| \geq \delta \, |nec(f_S)|,$$

i.e. if S realizes at least δ fraction of all the necessary minterms of f_S . A scheme S is ω-scheme if

$$nec(f_S) \subseteq \mathcal{D}_S \subseteq min(f_S).$$

Note that any scheme is δ-scheme for some $\delta \in [0,1]$. An ω-scheme is a special type of δ-scheme for $\delta = 1$.

For $\delta \in [0,1] \cup \{\omega\}$, let $L_\delta(f)$ denote the minimum size of a δ-scheme computing f . Thus, $L_0(f)$ is the unrestricted contact scheme complexity of f .

The functional $L_\delta(f)$ has been studied for a long time. The first non-trivial result in this direction has been obtained by E. A. Okol'nishnikova in [7], where a sequence of functions $f_n(x_1,\ldots,x_n)$ is given such that

$$L_1(f_n) \le 2n \quad \text{but} \quad L_\omega(f_n) \ge \exp(\Omega(n^{1/4})) .$$

The next major development was made by A. A. Razborov [9,10] and A. E. Andreev [1] where super-polynomial lower bounds for $L^+(f)$, the monotone scheme complexity of f, have been proved . One may transfere these bounds also to $L_1(f)$, because any minimal 1-scheme for monotone f has no null-chains, and therefore $L_1(f) = L^+(f)$.

However, we have seen before that the presence of null-chains may substantially reduce the size of schemes (see also [2,3,6,8-11]). Thus we need a technique to prove lower bounds for $L_\delta(f)$ with $\delta < 1$, as well as for $L_1(f)$ and non-monotone f (in these cases null-chains may be used in a non-trivial manner to reduce the size of schemes).

We say f is $(t,r)_\delta$-dispersed if each sub-family $\mathcal{A} \subseteq nec(f)$ with $|\mathcal{A}| \ge \delta\, |nec(f)|$ is a (t,r)-dispersed family.

Using an extention of Andreev-Razborov argument [1,9] to non-monotone circuits, given in [4,5], one may prove the following lower bound. Let H^*_f stand for H_f with $min(f)$ replaced by $nec(f)$.

Theorem 5: For any $\delta \in [0,1]$ and any $(t,r)_\delta$-dispersed Boolean function f , we have that

$$L_\delta(f) \ \ge\ \delta \min_{p\in[0,1)} H^*_f(t,r,p,0).$$

Example 5: Let us consider the following non-monotone version of $\mathfrak{g}_n^+$ (see Example 4):

$$\mathfrak{h}_n = \bigvee_{\sigma\in\Pi} K_\sigma \qquad \text{where} \qquad K_\sigma = \mathop{\&}_{i\in GF(m)} (x_{i,\sigma(i)} \ \&\ \neg x_{i,\sigma(i)\oplus 1}) .$$

Then $nec(\mathfrak{h}_n) = \{ K_\sigma : \sigma \in \Pi \}$ and $\mathfrak{h}_n$ is $(t,r)_\delta$-dispersed for any $t \le \delta m/3$ and $r < s$ ([5]). Taking t,r and $p\in[0,1)$ as in Corollary 3, we obtain from Theorem 5 the following lower bound.

Corollary 4 : For any $\delta \ge n^{-o(\log n)}$, and hence, for any constant $\delta \in (0,1]$, we have that

$$L_{\delta}(\mathfrak{h}_n) \;\ge\; n^{\Omega(\log n)} .$$

Thus, for an arbitrary small constant $\delta \in (0,1]$, the δ-scheme size of $\mathfrak{h}_n$ is almost the same as the size $|\mathit{nec}(\mathfrak{h}_n)| = \mathcal{O}(n^{\log n})$ of its shortest DNF $\mathit{nec}(\mathfrak{h}_n)$, and so, if $\delta \ge$ const>0 then, for some Boolean functions, null-chains do not help at all.

REFERENCES

[1] A. E. Andreev, On one method of obtaining lower bounds of individual monotone function complexity, Dokl. Akad. Nauk SSSR, 282 (1985).

[2] S. P. Jukna, Lower bounds on the complexity of local circuits, Springer Lecture Notes in Comput. Sci., 233 (1986).

[3] S. P. Jukna, Entropy of contact circuits and lower bounds on their complexity, Theoretical Computer Science, 57, n.1 (1988).

[4] S. P. Jukna, Two lower bounds for circuits over the basis (&,V,¬), Springer Lecture Notes in Comput. Sci., 324 (1988).

[5] S. P. Jukna, Method of approximations for obtaining circuit size lower bounds, Preprint n.6 (Vilnius, 1988).

[6] S. E. Kuznetsov, The influence of null-chains on the complexity of contact circuits, in : Probabilistic Methods in Cybernetics, 20 (Kazan, 1984).

[7] E. A. Okol'nishnikova, On the influence of one type of restrictions to the complexity of combinational circuits, in : Discrete Analysis, 36 (Novosibirsk, 1981).

[8] A. K. Pulatov, Lower bounds on the complexity of implementation of characteristic functionds of group codes by π-schemes, in : Combinatorial-Algebraic Methods in Applied Mathematics (Gorki, 1979).

[9] A. A. Razborov, Lower bounds on the monotone complexity of some Boolean functions, Dokl. Akad. Nauk SSSR, 281 (1985).

[10] A. A. Razborov, A lower bound on the monotone network complexity of the logical permanent, Mat. Zametki,37,n.5 (1985).

[11] E. Tardos, The gap between monotone and non-monotone circuit complexity is exponential, Combinatorica 7 (1987).

[12] I. Wegener, The complexity of Boolean functions, Wiley-Teubner, 1987.

[13] S. V. Zdobnov, Lower bounds on the complexity of schemes without null-chains, in : Proc. 9th All-Union Conf. on Math. Logic (Leningrad, 1988).

Monte-Carlo Inference and its Relations to Reliable Frequency Identification

by

Efim Kinber
Computing Center
Latvian State University
29 Rainis Boulevard
226250 Riga, U.S.S.R.

Thomas Zeugmann
Department of Mathematics
Humboldt University at Berlin
P.O. **Box 1297**
Berlin, 1086, G.D.R.

Abstract. For EX and BC-type identification Monte-Carlo inference as well as reliable frequency identification on sets of functions are introduced. In particular, we relate the one to the other and characterize Monte-Carlo inference to exactly coincide with reliable frequency identification, on any set $\mathfrak{M}$. Moreover, it is shown that reliable EX and BC-frequency inference forms a new discrete hierarchy having the breakpoints 1, 1/2, 1/3,... .

1. Introduction

Inductive inference has its historical origins in the philosophy of science. Within the last two decades it has attracted much attention of computer scientists. The theory of inductive inference can be considered as a form of machine learning with potential applications to artificial intelligence (cf. Osherson et al. (1986)). Nowadays inductive inference is a well-developed mathematical theory which has been the subject of collections of papers (cf. Barzdin, ed. (1974), (1975), (1977) and of excellent survey papers (cf. Angluin/Smith (1983), (1986), Daley (1986), and Klette/Wiehagen (1980)). Part of the following work was suggested by an open problem presented in Daley (1986). As in previous studies, we deal with the synthesis of programs for recursive functions. An inductive inference machine (abbr. IIM) is a recursive device (deterministic, probabilistic, or pluralistic) which, when fed more and more ordered pairs from the graph of some function, outputs more and more hypotheses, i.e. programs. There are many possible requirements on the sequence of all actually created programs. when fed more and more ordered pairs from the graph of some function, outputs more and more hypotheses, i.e. programs. There are many possible requirements on the sequence of all actually created programs. Considering deterministic and probabilistic IIM we shall study ex-

---

† The results were obtained during the author's visit of the computing center of the Latvian State University.

planatory (EX, EX_{prob}) inference, where the sequence of programs has to converge to a single program correctly computing the function to be identified, as well as behaviorally correct (BC, BC_{prob}) inference, i.e., all but finitely many programs have to satisfy the relevant correctness criterion. Furthermore the correctness criteria range from absolutely correct to finite error tolerance. On the other hand, investigating pluralistic IIMs we shall distinguish between the following two cases: First, the sequence of programs is required to contain a *particular* correct hypothesis with a certain *frequency* (EX_{freq}). In the second case the sequence of programs is required to contain with a certain *frequency* correct but possibly *distinct* programs (BC_{freq}). The reader is encouraged to consult Case/Smith (1983), Pitt (1985), and Podnieks (1974), (1975) for further information.

Moreover, we generalize the reliability notion originally introduced by Blum/Blum (1975) and Minicozzi (1976) to all these modes of identification. For EX-type inference, an IIM works reliably on a certain set $\mathfrak{M}$ of functions (e.g. the total functions, or the recursive functions) if for every function from $\mathfrak{M}$ the sequence of created programs does not converge to an incorrect solution. Hence the IIM itself recognizes whether its last hypothesis may be or may not be correct. In the latter case it performs a mind change, i.e., it outputs a program being different from the previous one. Thereby the IIM M implicitly transmits an error message to the outside word. If M identifies some function from $\mathfrak{M}$ then its sequence of error messages is finite, otherwise it is infinite. Thus our generalization works as follows: Instead of outputting programs, now the IIM M is required to output ordered pairs (i_n,b), where the i_n are the programs and $b \in \{0,1\}$. If $b = 0$, we interprete $(i_n,0)$ as an error message. In other words, $b = 1$ indicates that M trusts in its current hypothesis. If M does not identify some function from $\mathfrak{M}$ then it again produces an infinite sequence of error messages. Otherwise, the output sequence contains only finitely many error messages and among all created programs there are, with a certain frequency, correct ones.

Transmitting this approach to probabilistic IIMs we get that on some function from $\mathfrak{M}$ all possible computations either yield an infinite sequence of error messages or a finite one, independently of the sequence of coin-flips. Furthermore, in the latter case, with a certain *probability*, there must occur sequences of programs satisfying the particular identification criterion. Hence, all uncertainty lies in the domain of identification. Consequently we can interprete this type of probabilistic identification as *Monte-Carlo inference*.

In the present paper, first we extent Pitt's (1984), (1985) unification results in characterizing *Monte-Carlo inference* to coincide with reliable frequency identification. Second, it is proved that the introduced reliability notion ensures the useful properties known for the or dinary case, i.e., closure under union and finite invariance (cf. Minicozzi (1976)).
Third we investigate the power of reliable EX-type and BC-type frequency inference in comparing them with ordinary frequency identification. Thereby we obtain the strongest possible result, e.g. we show that there are classes being reliably EX-identifiable on the set of all total functions with frequency 1/(k+1) not contained in $BC_{freq}(1/k)$, for all numbers k. This directly yields *four infinite hierarchies*.

2. Basic Definitions and Notations

Unspecified notations follow Rogers (1967). $\mathbb{N} = \{0,1,2,...\}$ denotes the set of all natural numbers. The class of all partial recursive and recursive functions of one variable over $\mathbb{N}$ is denoted by $\mathbb{P}$ and $\mathbb{R}$, respectively. For n=1 we omit the upper index. The class of all partial and total functions over $\mathbb{N}$ is denoted by $\mathbb{PF}$ and $\mathbb{TF}$, respectively. Let $f \in \mathbb{PF}$, then we set Arg f = {x/f(x) is defined}. For $n \in \mathbb{N}$, we denote by $\mathbb{TF}_n$ and $\mathbb{R}_n$ the class of all functions $f \in \mathbb{PF}$ and $f \in \mathbb{P}$, respectively, for which $card(\mathbb{N}\text{-Arg } f) \leq n$. The class of all functions $f \in \mathbb{P}$ and $f \in \mathbb{PF}$ with cofinite domain is denoted by $\mathbb{R}_*$ and $\mathbb{TF}_*$, respectively. Let f, $g \in \mathbb{TF}_*$, and $n \in \mathbb{N}$, we write $f(n) \leq g(n)$ if both f(n) and g(n) are defined and f(n) is not greater than g(n). Furthermore, for f, $g \in \mathbb{TF}_*$ and $n \in \mathbb{N}$ we write $f =_n g$ and $f =_* g$ iff $card(\{x/f(x) \neq g(x)\}) \leq n$ and $card(\{x/f(x) \neq g(x)\}) < \infty$, respectively. By φ_0, φ_1, φ_2,... we denote a fixed acceptable programming system of all (and only all) the partial recursive functions (cf. Machtey/Young (1978)). If $f \in \mathbb{P}$ and $i \in \mathbb{N}$ is such that $\varphi_i = f$ then i is called a program for f.
Using a fixed effective encoding <...> of all finite sequences of natural numbers onto $\mathbb{N}$ we write f^n instead of <(f(0),...,f(n)>, for any $n \in \mathbb{N}$, $f \in \mathbb{PF}$, where f(x) is defined for all $x \leq n$.
Proper set inclusion is denoted by $\subset$ in difference to $\subseteq$; and by # we denote incomparability of sets.
A sequence $(j_n)_{n \in \mathbb{N}}$ of natural numbers is said to be convergent to a number j iff $j_n = j$ for almost all n.
Now we define several concepts of identification. In the sequel we deal only with the inference of everywhere defined functions, since this suffices to get the desired results. An IIM M is just a recursive

function. Suppose an IIM M is given the graph of some function $f \in \mathbb{TF}$ as input. We may suppose without loss of generality that f is given in its natural order (f(0),f(1),...) to M (cf. Blum/Blum (1975)).

Definition 1: Let $a \in \mathbb{N} \cup \{*\}$, and let $f \in \mathbb{TF}$. An IIM M EX^a-identifies f iff the sequence $(M(f^n))_{n\in\mathbb{N}}$ converges to a number i such that $\varphi_i =_a f$

If M does EX^a-identify f, we write $f \in EX^a(M)$. The collection of EX^a-inferrible sets is denoted by EX^a, formally, $EX^a = \{U/\exists M\ [U \subseteq EX^a(M)]\}$ For a = 0 we omit the upper index.

Definition 2: Let $\mathfrak{M} \subseteq \mathbb{TF}$, and let $a \in \mathbb{N} \cup \{*\}$. An IIM $M \in \mathbb{R}$ works EX^a-reliably on the set $\mathfrak{M}$ iff for every function $f \in \mathfrak{M}$ *either* the sequence $(M(f^n))_{n\in\mathbb{N}}$ converges to a number i such that $\varphi_i = f$ *or* it diverges.

Definition 3: Let $a \in \mathbb{N} \cup \{*\}$, and let $\mathfrak{M} \subseteq \mathbb{TF}$. An IIM M reliably EX^a-identifies $f \in \mathbb{TF}$ on the set $\mathfrak{M}$ iff M works EX^a- reliably on the set $\mathfrak{M}$ and M EX^a-identifies f.

If M does reliably EX^a-identify f on the set $\mathfrak{M}$ we write $f \in \mathfrak{M}\text{-}REX^a(M)$. The collection of reliably on $\mathfrak{M}$ EX^a-identifiable sets is denoted by $\mathfrak{M}\text{-}REX^a$. Again, for a = 0 we omit the upper index.

Reliably working IIMs were originally introduced and studied by Minicozzi (1976) and Blum/Blum (1975), in case a = 0, * . In Kinber/Zeugmann (1985) the collections $\mathbb{TF}\text{-}REX^a$ and $\mathbb{R}\text{-}REX^a$ have been considered.

Next we consider behaviorally correct inference which has been introduced by Barzdin (1974) and which has been studied intensively in Case/Smith (1983).

Definition 4: Let $a \in \mathbb{N} \cup \{*\}$, and let $f \in \mathbb{TF}$. An IIM M BC^a-identifies f iff $\varphi_{M(f^n)} =_a f$, for almost all n.

We write $f \in BC^a(M)$, if M does BC^a-identify f and set $BC^a = \{U/\exists M[U \subseteq BC^a(M)]\}$.

Moreover, Leo Harrington discovered that $\mathbb{R} \in BC^*$ (cf. Case/ Smith (1983)). Now we introduce reliable BC^a-identification. As already explained in the introduction, we now require the IIM to output ordered pairs (i_n,b), where $b \in \{0,1\}$, instead of only outputting programs. If b = 0, then (i_n,b) is said to be an error message.

Definition 5: Let $\mathfrak{M} \subset \mathbb{TF}$, and let $a \in \mathbb{N} \cup \{*\}$. An IIM $M \in \mathbb{R}$ works BC^a-reliably on the set $\mathfrak{M}$ iff for every function $f \in \mathfrak{M}$ the output sequence (i_n,b_n) *either* satisfies $(i_n,b_n) = (i_n,1)$ and $\varphi_{i_n} =_a f$, for almost all n, *or* there are infinitely many n such that $(i_n,b_n) = (i_n,0)$. In other words, either M does BC^a-identify f or it produces infinitely many error messages.

Definition 6: Let $a \in \mathbb{N} \cup \{*\}$, and let $\mathfrak{M} \subseteq \mathbb{TF}$. An IIM M reliably

BC^a-identifies $f \in \mathbb{TF}$ on the set $\mathfrak{M}$ iff M works BC^a-reliably on $\mathfrak{M}$ and M does BC^a-identify f.

If M does reliably BC^a-identify f on $\mathfrak{M}$ we write $f \in \mathfrak{M}\text{-}RBC^a(M)$ and we denote by $\mathfrak{M}\text{-}RBC^a$ the collection of on $\mathfrak{M}$ reliably BC^a- inferrible sets.

Next we deal with frequency identification due to Podnieks (1974).

Definition 7: Let $0 < p \leq 1$, and let $f \in \mathbb{TF}$. An IIM M BC-identifies f with frequency p iff

$$\liminf_{k \to \infty} \frac{card(\{n/\varphi_{M(f^n)} = f,\ 0 \leq n \leq k\})}{k} \geq p$$

We set $BC_{freq}(p) = \{U/\exists M$ [M identifies every $f \in U$ with frequency p]}.

Pitt (1985) has introduced what is essentially the EX version of Podnieks' BC-frequency identification.

Definition 8: Let $0 < p \leq 1$, and let $f \in \mathbb{TF}$. An IIM M EX- identifies f with frequency p iff there is a *particular* program i such that

$$\liminf_{k \to \infty} \frac{card(\{n/M(f^n) = i,\ 0 \leq n \leq k\})}{k} \geq p \text{ and } \varphi_i = f.$$

By $EX_{freq}(p)$ we denote the collection of all function classes U being EX-identifiable with frequency p.

In order to introduce reliable frequency identification, the IIMs are again required to output pairs (i_n, b_n). Furthermore, looking at EX-frequency inference we require the particular program i computing the considered function to occur at least with frequency p. Moreover, the created output sequence is only allowed to contain finitely many error messages, i.e., the program i is at almost all places accompanied by b = 1.

Definition 9: Let $\mathfrak{M} \subseteq \mathbb{TF}$, and let $0 < p \leq 1$. An IIM M reliably BC-identifies (EX-identifies) $f \in \mathbb{R}$ on the set $\mathfrak{M}$ with frequency p iff

(1) for all $g \in \mathfrak{M}$ the output sequence $(M(g^n))_{n \in \mathbb{N}}$ *either* contains only finitely many error messages and M BC-identifies (EX-identifies) g with frequency p, *or* in the sequence $(M(g^n))_{n \in \mathbb{N}}$ error messages occur infinitely often, and

(2) M BC-identifies (EX-identifies) f with frequency p.

If M does reliably BC-identify (EX-identify) f on the set $\mathfrak{M}$ with frequency p we write $f \in \mathfrak{M}\text{-}RBC_{freq}(p)(M)$ $(f \in \mathfrak{M}\text{-}REX_{freq}(p)(M))$.

Furthermore we set $\mathfrak{M}\text{-}RBC_{freq}(p) = \{U/\exists M[U \subseteq \mathfrak{M}\text{-}RBC_{freq}(p)(M)]\}$, and analogously define $\mathfrak{M}\text{-}REX_{freq}(p)$.

Next we define team inference which was originally introduced by Smith (1982). A team is a finite collection of IIMs. A team $(M_1,\ldots,M_n)$ successfully BC-infers (EX-infers) a set $U \subseteq \mathbb{R}$ if, for each $f \in U$, some team member M_i successfully BC-identifies (EX-identifies) f.

Furthermore we set $BC_{team}(n) = \{U/\exists(M_1,\ldots,M_n)[U \subseteq BC(M_1,\ldots,M_n)]\}$, and declare analogously $EX_{team}(n)$.

Finally in this section we define Monte-Carlo inference.
A probabilistic IIM P is simply a deterministic IIM which is allowed to flip a "t-sided coin". For any fixed sequence S of coin-flips P behaves like a deterministic IIM, which we denote by P^S. We require P^S, for any sequence S of coin-flips, to output ordered pairs (i_n, b_n).

Definition 10: Let $\mathfrak{M} \subseteq \mathbb{TF}$, and let $0 < p \leq 1$. A Monte-Carlo IIM P BC-identifies (EX-identifies) $f \in \mathbb{R}$ on the set $\mathfrak{M}$ with probability p iff

(1) for all $g \in \mathfrak{M}$ and all sequences S of coin-flips the output sequence $(P^S(g^n))_{n\in\mathbb{N}}$ *either* satisfies (α) *or* (β).

(α) $(P^S(g^n))_{n\in\mathbb{N}}$ contains only finitely many error messages (independently of S) and the probability (taken over all sequences S) that P^S BC-identifies (EX-identifies) g is greater than or equal to p.

(β) in the sequence $(P^S(g^n))_{n\in\mathbb{N}}$ error messages occur infinitely often (again independently of S).

(2) The sequence $(P^S(f^n))_{n\in\mathbb{N}}$ fulfills (α).

If the Monte-Carlo IIM P does BC-identify f on $\mathfrak{M}$ with probability p, we write $f \in \mathfrak{M}\text{-RBC}_{prob}(p)(P)$ and set $\mathfrak{M}\text{-RBC}_{prob}(p) = \{U / \exists P\ [U \subseteq \mathfrak{M}\text{-RBC}_{prob}(p)(P)]\}$. In an analogous way we define $\mathfrak{M}\text{-REX}_{prob}(p)(P)$ and $\mathfrak{M}\text{-REX}_{prob}(p)$.
In the sequel we shall study the unknown relationships between the above defined modes of inference.

3. Results

3.1. Monte-Carlo Inference and Reliable Frequency Inference

In this section we extend Pitt's (1985) unification result in showing that Monte-Carlo inference is exactly the same as reliable frequency identification. Moreover, as an immediate consequence from the theorems below one gets that, if there is any hierarchy for reliable frequency identification, then it must be a discrete one.

Theorem 1: Let $n \in \mathbb{N}$, and let $\mathfrak{M} \subseteq \mathbb{TF}$. Furthermore, let $U \in \mathfrak{M}\text{-RBC}_{prob}(p)$ with $1 \geq p > 1/(n+1)$. Then

(1) there is an IIM M reliably BC-identifying U on $\mathfrak{M}$ with frequency 1/n.

(2) for any $f \in U$ the sequence $((i_k, b_k))_{k\in\mathbb{N}}$ of M's hypotheses on f has the property that there is some $r \in \{0,\dots,n-1\}$ such that $\varphi_{i_k} = f$, for almost all k with $k \equiv r \bmod n$.

Theorem 2: Let $\mathfrak{M} \subseteq \mathbb{TF}$, $n \in \mathbb{N}$ and let $U \in \mathfrak{M}\text{-RBC}_{freq}(p)$ with $1 \geq p > 1/(n+1)$. Then there is an IIM M reliably BC-identifying U with frequency 1/n. Moreover, the output sequence $((i_k,b_k))_{k\in\mathbb{N}}$ of M's hypotheses has for any $f \in U$ the property that there is an $r \in \{0,...,n-1\}$ such that $\varphi_{i_k} = f$ for al- most all $k \equiv r \bmod n$.

Corollary 3: Let $\mathfrak{M} \subseteq \mathbb{T}\mathbb{F}$. Then $\mathfrak{M}\text{-RBC}_{prob}(1/n) = \mathfrak{M}\text{-RBC}_{freq}(1/n)$, for any number $n \geq 1$.

Finally in this section we show that the Theorems 1 and 2 as well as Corollary 3 remain valid if reliable BC-frequency identification is replaced by reliable EX-frequency inference.

Theorem 4: Let $n \in \mathbb{N}$, $\mathfrak{M} \subseteq \mathbb{T}\mathbb{F}$, and let $U \in \mathfrak{M}\text{-REX}_{prob}(p)$ with $1 \geq p > 1/(n+1)$. Then

(1) there is an IIM M reliably EX-identifying U on $\mathfrak{M}$ with frequency 1/n.

(2) for any $f \in U$ the sequence $((i_k,b_k))_{k\in\mathbb{N}}$ of M's hypotheses on f is such that there are some $r \in \{0,...,n-1\}$ and $j \in \mathbb{N}$ satisfying $i_k = j$ and $\varphi_j = f$ for almost all k with $k \equiv r \bmod n$.

Theorem 5: Let $\mathfrak{M} \subseteq \mathbb{T}\mathbb{F}$, $n \in \mathbb{N}$ and let $U \in \mathfrak{M}\text{-REX}_{freq}(p)$ with $1 \geq p > 1(n+1)$. Then there is an IIM M reliably EX-inferring U with frequency 1/n. Moreover, the output sequence $((i_k,b_k))_{k\in\mathbb{N}}$ of M's hypotheses on any $f \in U$ behaves such that there is an $r \in \{0,...,n-1\}$ and a $j \in \mathbb{N}$ satisfying $i_k = j$ and $\varphi_j = f$ for almost all $k \equiv r \bmod n$.

Corollary 6: Let $\mathfrak{M} \subseteq \mathbb{T}\mathbb{F}$. Then $\mathfrak{M}\text{-REX}_{prob}(1/n) = \mathfrak{M}\text{-REX}_{freq}(1/n)$ for any number $n \geq 1$.

Monte-Carlo inference is thus completely characterized. Consequently, in the sequel it suffices to deal with frequency identification.

3.2. Closure Properties

In this section we show that the reliability notions introduced above preserve the closure properties originally pointed out by Minicozzi (1976) for reliable EX-identification. For all $a \in \mathbb{N} \cup \{*\}$ the $\mathfrak{M}\text{-REX}^a$ case has already been handled in Kinber/Zeugmann (1985). Thus it remains to deal with $\mathfrak{M}\text{-RBC}_{freq}(1/n)$, $\mathfrak{M}\text{-RBC}^a$, and $\mathfrak{M}\text{-REX}_{freq}(1/n)$.

The following proposition states that any reliable inference is closed under recursively enumerable unions.

Proposition 1: Let $\mathfrak{M} \subseteq \mathbb{T}\mathbb{F}$, and let $ID \in \{RBC_{freq}(1/n), RBC^a, REX_{freq}(1/n) / n \geq 1, n \in \mathbb{N}, a \in \mathbb{N} \cup \{*\}\}$. Furthermore, let $(M_i)_{i\in\mathbb{N}}$ be a recursive enumeration of IIMs working in the sense of $\mathfrak{M}$-ID. Then there exists an IIM such that $\mathfrak{M}\text{-ID}(M) = \bigcup_{i\in\mathbb{N}} \mathfrak{M}\text{-ID}(M_i)$.

The next proposition states that reliable identification is closed under finite variance. For any class $U \subseteq \mathbb{R}$ let $[U] = \{g \,/\, g \in \mathbb{R}, \exists f[f \in U, f =_* g]\}$.

Proposition 2: Let $\mathfrak{M} \subseteq \mathbb{T}\mathbb{F}$, and let $ID \in \{RBC_{freq}(1/n),\ RBC^a,\ REX_{freq}(1/n) \,/\, n \geq 1,\ n \in \mathbb{N},\ a \in \mathbb{N} \cup \{*\}\}$. Then $U \in \mathfrak{M}\text{-}ID$ implies $[U] \in \mathfrak{M}\text{-}ID$.

3.3. Hierarchy Results

The main goal of this section consists in clarifying the basic identification power of reliable frequency inference on sets $\mathfrak{M}$. We confine ourselves to consider exclusively the cases $\mathfrak{M} = \mathbb{T}\mathbb{F}$ and $\mathfrak{M} = \mathbb{R}$, since they are of basic interest. We start with several fundamental observations that give a first answer to the question of what actually can and cannot be reliably inferred with a certain frequency. Above all, in accordance with our theorems in section 3.1., it generally suffices to deal with the discrete frequencies 1, 1/2, 1/3,.... Note that by definition $\mathbb{T}\mathbb{F}\text{-}REX_{freq}(1/n) \subseteq \mathbb{T}\mathbb{F}\text{-}RBC_{freq}(1/n) \subseteq \mathbb{R}\text{-}RBC_{freq}(1/n)$, for any number $n \geq 1$. In order to obtain as sharp results as possible we proceed in the sequel almost always as follows: Studying the power of reliable frequency inference we deal, whenever appropriate, with with $\mathbb{T}\mathbb{F}\text{-}REX_{freq}(1/n)$ or $\mathbb{T}\mathbb{F}\text{-}RBC_{freq}(1/n)$; whereas its limitations are shown in dealing with $\mathbb{R}\text{-}RBC_{freq}(1/n)$.
First of all, we point out that reliable frequency identification is generally less powerful than the ordinary frequency inference.

Theorem 7: $EX \setminus \mathbb{R}\text{-}RBC_{freq}(1/n) \neq \emptyset$ for every number $n \geq 1$.

As an immediate consequence one gets:
Corollary 8: For any number $n \geq 1$:
$\mathbb{R}\text{-}REX_{freq}(1/n) \subset EX_{freq}(1/n)$ and $\mathbb{R}\text{-}RBC_{freq}(1/n) \subset BC_{freq}(1/n)$

Next we ask whether $\mathfrak{M}\text{-}REX_{freq}(1/n)$ is always properly contained in $\mathfrak{M}\text{-}RBC_{freq}(1/n)$. The affirmative answer is given by the following theorems which, beyond that, lead to a much deeper insight into the capabilities of reliable frequency inference.
Theorem 9: $\mathbb{T}\mathbb{F}\text{-}RBC \setminus EX^* \neq \emptyset$
Theorem 10: Let $\mathfrak{M} \in \{\mathbb{T}\mathbb{F}, \mathbb{R}\}$, and let $a \in \mathbb{N} \cup \{*\}$. Then $\mathfrak{M}\text{-}REX^a \subset \mathfrak{M}\text{-}RBC$.
The following corollary summerizes the results concerning reliable BC-inference.
Corollary 11: $EX^* \;\#\; \mathbb{T}\mathbb{F}\text{-}RBC$ and $EX^* \;\#\; \mathbb{R}\text{-}RBC$
By the next theorem we heighten the known results dealing with the

capabilities of reliable EX^a-identification.

Theorem 12: $\mathbb{UF}\text{-}REX^a \setminus EX_{team}(a) \neq \emptyset$ for any $a \in \mathbb{N}$.

Exploring the above results the announced relation between the two reliable frequency identification modes (i.e., EX and BC) can be obtained

Theorem 13: Let $\mathfrak{M} \in \{\mathbb{UF}, \mathbb{R}\}$. Then $\mathfrak{M}\text{-}REX_{freq}(1/n) \subset \mathfrak{M}\text{-}RBC_{freq}(1/n)$

Smith (1982) completely relates the EX and BC team hierarchy, thereby in particular showing that $EX_{team}(n+1) \setminus BC_{team}(n+1) \neq \emptyset$. Unfortunately, none of his proofs can be applied in our setting. This is caused by the fact that the capabilities of pluralism are mainly based on non-union problems. Therefore, at first glance it may seem to be hopeless to transfer the power of team inference, at least partially, to reliable frequency identification. However, this is a misleading impression. Reliable frequency inference is closed under union since all the inferrible classes U share the *common property* that functions functions not contained in U lead to infinite sequences of error messages. Nevertheless, even in the limit it may be undecidable into which subclass of U the considered function falls. Yet we were surprised to find the next theorem.

Theorem 14: $\mathbb{UF}\text{-}REX_{freq}(1/(n+1)) \setminus BC_{team}(n) \neq \emptyset$ for any number $n \geq 1$.

The latter Theorem directly yields infinite hierarchies of reliable frequency identification starting from $\mathfrak{M}$-REX and $\mathfrak{M}$-RBC, respectively, where $\mathfrak{M} \in \{\mathbb{UF}, \mathbb{R}\}$. However, until now we knew only a little about uniform upper limitations concerning the power of reliable frequency inference. On the other hand, the BC- frequency hierarchy is properly contained in BC^* since $\mathbb{R} \in BC^*$. Hence it would be interesting to know whether $\mathfrak{M}\text{-}RBC_{freq}(1/n) \subset \mathfrak{M}\text{-}RBC^*$. For $\mathfrak{M} = \mathbb{R}$ no new insight can be expected since any IIM M which BC^*-identifies $\mathbb{R}$ obviously works BC^*-reliably on $\mathbb{R}$, i.e., $\mathbb{R} \in \mathbb{R}\text{-}RBC^*$. Nevertheless, the case $\mathfrak{M} = \mathbb{UF}$ seems to be promising if one looks to the next theorem.

Theorem 15: $EX \# \mathbb{UF}\text{-}RBC^*$

The next theorem states that reliable frequency identification is generally less powerful than reliable BC^*-inference.

Theorem 16: $\mathbb{UF}\text{-}RBC_{freq}(1/n) \subset \mathbb{UF}\text{-}RBC^*$ for all $n \geq 1$.

Summarizing the results pointed out above, we obtain the following figure:

$$
\begin{array}{ccccccc}
EX_{freq}(1) & \subset & EX_{freq}(1/2) & \subset\ldots\subset & EX_{freq}(1/n) & \subset & BC^{*} \\
\cup & & \cup & & \cup & & \cup \\
\mathfrak{R}\text{-}REX_{freq}(1) & \subset & \mathfrak{R}\text{-}REX_{freq}(1/2) & \subset\ldots\subset & \mathfrak{R}\text{-}REX_{freq}(1/n) & \subset & \mathfrak{R}\text{-}RBC^{*} \\
\cap & & \cap & & \cap & & \cap \\
\mathfrak{R}\text{-}RBC_{freq}(1) & \subset & \mathfrak{R}\text{-}RBC_{freq}(1/2) & \subset\ldots\subset & \mathfrak{R}\text{-}RBC_{freq}(1/n) & \subset & \mathfrak{R}\text{-}RBC^{*} \\
\cap & & \cap & & \cap & & \cap \\
BC_{freq}(1) & \subset & BC_{freq}(1/2) & \subset\ldots\subset & BC_{freq}(1/n) & \subset & BC^{*}
\end{array}
$$

References

ANGLUIN, D., AND SMITH, C. H. (1983), Inductive inference: theory and methods. Computing Surveys 15, 237 - 269

ANGLUIN, D., AND SMITH, C. H. (1986), Formal inductive inference, in Encyclopedia of Artificial Intelligence, (S. Shapiro, Ed.), to app.

BARZDIN, YA. M. (1974), Two theorems on the limiting synthesis of functions, in Theory of Algorithms and Programs I, (Ya. M. Barzdin, Ed.), pp. 82 - 88, Latvian State University.

Theory of Algorithms and Programs I, II, III, (1974), (1975), (1977) (Ya. M. Barzdin, Ed.). Latvian State University.

BLUM, L. AND BLUM, M. (1975), Toward a mathematical theory of inductive inference. Inform. and Control 28, 122 - 155.

CASE. J., AND SMITH, C. H. (1983), Comparison of identification criteria for machine inductive inference. Theo. Comp. Sci. 25, 193-220

DALEY, R. (1986), Towards the development of an analysis of learning algorithms, in Proc. Internat. Workshop of Analogical and Inductive Inference, Wendisch-Rietz, 1986, (K. P. Jantke, Ed.), Lect. Notes Comp. Sci. 265, pp. 1 - 18.

GOLD, E. M. (1965), Limiting recursion. J. Symbolic Logic 30, 28 - 48.

KINBER, E. B., AND ZEUGMANN, T. (1985), Inductive inference of almost everywhere correct programs by reliably working strategies. J. Inf. Processing and Cybernetics (EIK) 21, 91 - 100.

KLETTE; R., AND WIEHAGEN, R. (1980), Research in the theory of inductive inference by GDR mathematicians - a survey. Inf. Sci. 22, 149 - 169.

MACHTEY, M., AND YOUNG, P. (1978) "An Introduction to the General Theory of Algorithms", North-Holland, New York

MINICOZZI, E. (1976), Some natural properties of strong-identification in inductive inference. Theo. Comp. Sci. 2, 345 - 360.

OSHERSON, D., STOB, M., AND WEINSTEIN, S. (1986), "Systems that Learn" MIT Press, Cambridge.

PITT, L. B. (1984), A characterization of probabilistic inference. In Proc. 25th Annual Symp. Foundations of Comp. Sci.,pp. 485 - 494.

PITT, L. B. (1985), Probabilistic inductive inference. Yale University, YALEU/DCS/TR-400, Ph.D. Thesis.

PODNIEKS, K.M. (1974), Comparing various concepts of function prediction and program synthesis I, in Theory of Algorithms and Programs, (Ya. M. Barzdin, Ed.), pp. 68 - 81, Latvian State University.

PODNIEKS, K.M. (1975), Comparing various concepts of function prediction and program synthesis II, Theory of Algorithms and Programs, (Ya. M. Barzdin, Ed.), pp. 35 - 44 Latvian State University.

ROGERS, H. JR. (1967) "Theory of Recursive Functions and Effective Computability", Mc-Graw Hill, New York

SMITH, C. H. (1982), The power of pluralism for automatic program synthesis. Journal of the ACM 29, 1144 - 1165.

SEMILINEAR REAL-TIME SYSTOLIC TRELLIS AUTOMATA

Ivan Korec,
Mathematical Institute of Slovak Academy of Sciences,
Obrancov mieru 49, 81473 Bratislava, Czechoslovakia

ABSTRACT. A real-time systolic trellis automaton/RTSTA/ will be called semilinear of degree (at most) k if its underlying trellis T is semilinear of degree (at most) k, i. e. the set of rows of T can be expressed as a finite union of languages of the form

$$\left\{u_0 v_1^i u_1 v_2^i u_2 \ldots u_{k-1} v_k^i u_k ;\ \ i \geqq 1\right\} .$$

It is proved that every semilinear RTSTA of arbitrary degree k is equivalent with a semilinear regular RTSTA of degree 3. If regularity is not requested the degree can be diminished to 2, but not to 1.

1. Introduction.

Systolic trellis automata were defined and studied in [CGS1], [CGS2] as a (rather abstract) model of VLSI computations and pipelining. Later in [Gru] they were called real-time systolic trellis automata; we shall use the abbreviation RTSTA for them. The simplest RTSTA are the homogeneous ones, in which all processors are identical. Regular RTSTA and modular RTSTA are more general; in both cases several types of processors are used and the processors are laid out by simple rules. The kind of a RTSTA will be usually defined by the kind of the underlying trellis. (Definitions will follow later.) Both regular and modular RTSTA were shown to be stronger than homogeneous RTSTA in the sense that they can accept more languages.

Semilinear RTSTA which will be studied here are also a generalization of the homogeneous RTSTA. They seem to be the simplest RTSTA which are stronger than the homogeneous ones (of course, the words "the simplest" is understood only informally). The semilinear RTSTA can be naturally classified into infinitely many classes by their degrees (and in more details by their types). It seems rather surprising that every semilinear RTSTA of arbitrary degree k is equivalent with a semilinear regular RTSTA of degree 3. Without the regularity the degree can be diminished to 2. Moreover, in both cases the latest RTSTA can be simultaneously modular.

2. Trellises

In this paragraph we introduce some general notions and notations and also the notions concerning trellises. We introduce them for the later study of RTSTA.

The set of all nonnegative integers will be denoted by N and the set $\{0, 1, \dots, k-1\}$, $k \in N$, by N_k. A word of length q over an alphabet A is a mapping of N_q into A. The length of a word w will be denoted $|w|$. We generalize the notion of word to two-dimensional case. A (p, q)-word over A is a mapping of $N_p \times N_q$ into A. If p, q are not specified we speak about array words (over A), if $p = q$ about square words. We imagine that the symbols of a (p, q)-word w are written into the unit squares of a $p \times q$ rectangular with sides parallel with coordinate axis (as they will be chosen below: $w(0, 0)$ is written into the top unit square). The set of all (p, q)-words over A will be denoted $A^{p,q}$.

If A is an alphabet then a trellis over A, or simply a trellis, is a mapping of $N \times N$ into A. The set of all trellises over A will be denoted A^{NN}. Trellises can be imagined in the plane similarly as the classical Pascal triangle. To do that, introduce the coordinate system in which the axis x is oriented right-down and the axis y left-down. In the Pascal triangle the unit square with the top vertex (x, y) contains the binomial coefficient $\binom{x+y}{x}$. If we draw a trellis T then this square will contain the value $T(x, y)$. We shall speak about columns (infinite sequences) and rows (finite sequences, i. e. words) in the obvious way. The i-th row of a trellis T consists of

$$T(0, i),\ T(1, i-1), \dots,\ T(i-1, 1),\ T(i, 0).$$

For every $i \in N$, the top symbol of the i-th column is $T(i, 0)$, the top symbol of the $(-i)$-th column is $T(0, i)$. The 0-th column of T will be called the axis of T. It consists of

$$T(0, 0),\ T(1, 1),\ T(2, 2),\ T(3, 3), \dots$$

The element $T(0, 0)$ will be called the root of T, the elements $T(i, j+1)$, $T(i+1, j)$ are sons of $T(i, j)$ and $T(i, j)$ is their father. A subtrellis of T consists of an element $T(x, y)$ and all its descendants.

We shall deal with array subwords (abbreviation: ASW) of trellises. The symbol $ASW(T; m, n; x, y)$ will denote the (m, n)-subword of the trellis T with the top symbol at (x, y); hence its lowest element is $T(x+m-1, y+n-1)$. If m, n, x, y are not specified we shall speak simply about array subwords, if $m = n$ we shall speak about square subwords.

A mapping F of A^{NN} into B^{NN} will be called $((k, l), (m, n))$-substitution if there is a mapping f of $A^{k,l}$ into $B^{m,n}$ such that for every $T \in A^{NN}$ and $S \in B^{NN}$ it holds $S = F(T)$ if and only if for all $i, j \in N$

$$ASW(S; m, n; mi, nj) = f(ASW(T; k, l; ki, lj)).$$

In this case we shall say that F is induced by f. If k, l, m, n are not specified we shall speak about array substitution. We shall say that $T \in A^{NN}$ is a fixed point of the array substitution F if it holds $T = F(T)$.

Definition 2.1. A trellis T over A is said to be:

a) strictly (p, q)-modular $(p, q \geqq 2)$ if T is a fixed point of a $((1, 1), (p, q))$-substitution;

b) (p, q)-modular $(p, q \geqq 2)$ if there is a strictly (p, q)-modular trellis S and an array substitution F such that $T = F(S)$;

c) /strictly/ modular if it is /strictly/ (p, q)-modular for some integers p, q;

d) strictly regular if there is an algebra $\mathbf{A} = (A; K, l, r, .)$ of signature $(0, 1, 1, 2)$ such that for all x, y N

$$T(x, y) = \begin{cases} K & \text{if } x = y = 0, \\ l(T(0, y - 1)) & \text{if } x = 0, y > 0, \\ r(T(x - 1, 0)) & \text{if } x > 0, y = 0, \\ T(x - 1, y).T(x, y - 1) & \text{if } x > 0, y > 0; \end{cases}$$

in this case we also write $T = GPT(\mathbf{A})$;

e) regular if there is a strictly regular trellis S and a $((1, 1), (1,1))$-substitution h such that $T = h(S)$;

f) semihomogeneous if it has only finitely many different subtrellises.

Here we shall study trellises with the property that the sets of their rows are semilinear languages (i. e., they correspond to semilinear sets in the sense of [Gin]). However, if these languages are semilinear then they are also simple semilinear (abbreviation: SSL) in the following sense.

Definition 2.2. 1) A language L is called a simple linear language of degree at most k if there are words

$$u_0, v_1, u_1, v_2, \ldots, v_k, u_k \tag{2.2.1}$$

such that

$$L = \{u_0 v_1^i u_1 v_2^i u_2 \ldots v_k^i u_k; \quad i \geqq 1\} \tag{2.2.2}$$

2) A language L is said to be a simple semilinear language (SSL

language) of degree at most k if L is the union of finitely many pairwise disjoint simple linear languages of degree at most k.

3) A language L is a simple /semi-/linear language of degree k if it is such language of degree at most k, but it is not a simple /semi-/linear language of degree at most $k-1$. If the degree is not important we shall delete the words "of degree (at most) k".

4) The finite sequence (2.2.1) will be called a generating sequence for L if (2.2.2) holds. A generating system for an SSL language L is a set which consists of generating sequences for a finite set M of pairwise disjoint simple linear languages such that L is the union of M. We say that an SSL language is given if a generating system for it is given.

Degrees of semilinear languages allows us to characterize their properties. For example:

Theorem 2.3. The set of rows of a trellis is a regular (resp. context-free) language if and only if it is an SSL language of degree 1 (resp. at most 2).

Degrees are very suitable for the classification of strictly regular semilinear trellises (they will be defined; in essential, they coincide with semilinear generalized Pascal triangles, see [Kor2]). However, they will be less suitable to classify (general) semilinear trellises. For that purpose, the types introduced below will be more appropriate. To define them, call the empty set and finite sets of positive rationals with the greatest element 1 briefly R-sets.

Definition 2.4. a) A R-set X is a type of the simple linear language L from (2.2.2) if it contains all the positive numbers among

$$\frac{|v_1|}{|v_1v_2\cdots v_k|},\quad \frac{|v_1v_2|}{|v_1v_2\cdots v_k|},\quad \ldots,\quad \frac{|v_1v_2\cdots v_k|}{|v_1v_2\cdots v_k|}.$$

b) A R-set X is a type of a simple semilinear language L if L can be written as a disjoint union of finitely many simple linear languages of type X.

We can immediately see that a SSL language L has (infinitely) many types. However, to every generating system for L we can find a type of L in the obvious canonical way. The empty set is the least type of every finite language. We state the hypothesis that every SSL language has the least type. Every R-set is the least type of a SSL language.

Definition 2.5. Let T be a trellis, L be the set of their rows and X be a R-set. Then T will be called:

a) semilinear trellis if L is a SSL language;

b) semilinear trellis of degree (at most) k or of type X if L is a SSL language of degree (at most) k or of type X, respectively;

c) strictly regular semilinear trellis if T is simultaneously strictly regular and semilinear; analogously for other conjuction of properties.

The degree of any strictly regular semilinear trellis is equal to the cardinality of its minimal type. For regular semilinear trellises the degree can be smaller. Consider for example the trellis with the set of rows

$$L = \{a^{2n-2}b^n;\ n \geq 1\} \cup \{a^{n-1}b^{2n};\ n \geq 1\} \cup \{a^{3n};\ n \geq 1\}.$$

Its minimal type is $\{1/3,\ 2/3,\ 1\}$ and its degree is 2.

3. Real-time systolic trellis automata

Since the notion of RTSTA has been often considered (see e. g. [CGS1], [Gru], [Vol]) we may explain it here briefly. We begin with a rather formal definition:

Definition 3.1. a) A (A_i, A_o)-processor (where A_i, A_o are alphabets) is an ordered pair (f, g), where f is its input function which maps A_i into A_o and g is its transition function which maps $A_o \times A_o$ into A_o.

b) An RTSTA is an ordered sixtuple

$$\mathbf{A} = (A_l,\ A_i,\ A_o,\ T,\ P,\ A_a) \tag{3.1.1}$$

where A_l is a labelling alphabet, A_i is an input alphabet, A_o is an operating alphabet, T is a trellis over A_l, P is a mapping of A_l into the set of (A_i, A_o)-processors and $A_a \subseteq A_o$ is the set of accepting symbols.

c) $L(\mathbf{A})$, the language accepted by RTSTA $\mathbf{A}$, is the set of all $w \in A_i^+$ such that the computation of $\mathbf{A}$ which begins with the input w at the $(|w| - 1)$-th row gives an output $x \in A_a$ in the root of $\mathbf{A}$.

d) Two RTSTA will be called equivalent if they accept the same language.

Remember that the computation of $\mathbf{A}$ with the input word w

starts at the row of $\mathbb{A}$ which contains exactly $|w|$ processors; each of them obtains one symbol of w. Each of these processors uses its input function and sells the output symbol to both its fathers (marginal processors to their unique fathers). Each of them obtains two symbols from A_0. In the next step the row of these fathers is active; they use their transition functions and again send the outputs upwards. So the computation continues until the root processor of $\mathbb{A}$ is active; if its output belongs to A_a the word w is accepted.

The function $G(x, y) = P(T(x, y))$ also can be considered as a trellis, and this trellis is only important for the computations of $\mathbb{A}$; P, T need not be known if G is given directly. Roughly speaking, T determines the (possible) structure of $\mathbb{A}$ and G determines the computations of $\mathbb{A}$. Different kinds of RTSTA can be defined by the appropriate properties of T, as we shall see from the next definition. Notice that although Definition 3.1 allows arbitrary trellisses T, only finitely presented trellises have a computational sense. For various kinds of RTSTA T can be determined by various ways (hence these kinds of RTSTA have formally different definitions; they need not be ordered sixtuples). For example, T is determined by a finite algebra for regular RTSTA and is completely unnecessary for homogeneous RTSTA. For semilinear RTSTA T can be determined by a generating system for the language of their rows.

Definition 3.2. The RTSTA $\mathbb{A}$ from (3.1.1) is:

a) regular if the trellis T is strictly regular;
b) modular if the trellis T is strictly modular;
c) semilinear if the trellis T is semilinear;
d) semilinear of type X /or of degree k/ if T is semilinear of type X /or of degree k, respectively/;
e) homogeneous if the trellis T is a constant function;
f) semihomogeneous if the trellis T is semihomogeneous;
g) internally homogeneous if the transition functions of all processors $P(T(x, y))$, $x, y \in N$, coincide.

If $\mathbb{A}$ is a regular or modular trellis then the trellis $G(x, y) = P(T(x, y))$ is also regular or modular, respectively. The converse is not true because $\mathbb{A}$ need not effectively use the whole complexity of T.

4. Theorems.

Proofs of theorems will be only sketched or will be deleted at all. The proofs uses various programming techniques from [CGS1], [CGS2], [Vol] and [Kor3], e. g. simulation, parallel guessing, sending signals upwards by different side speeds and collecting information during a bounded number of steps.

Theorem 4.1. Every semilinear RTSTA **A** is equivalent with an internally homogeneous semilinear RTSTA (of the same type).

The proof uses the idea that the underlying trellis T of **A** can be expressed in the form T = F(S), where F is a ((1, 1), (1, 1))-substitution and S is a semilinear trellis in which almost all elements are uniquely determined by their sons. Notice that S, T can have the same type(s) but the degree of S can be greater than the degree of T.

Theorem 4.2. Every semilinear RTSTA **A** is equivalent with a regular semilinear RTSTA **B** of the type $\{1/3, 2/3, 1\}$.

Proof. A possible underlying trellis of **B** is shown in Figure 1; notice that it is strictly regular (and also strictly (2, 2)-modular).

<u>Figure 1:</u>
The underlying trellis of **B**.

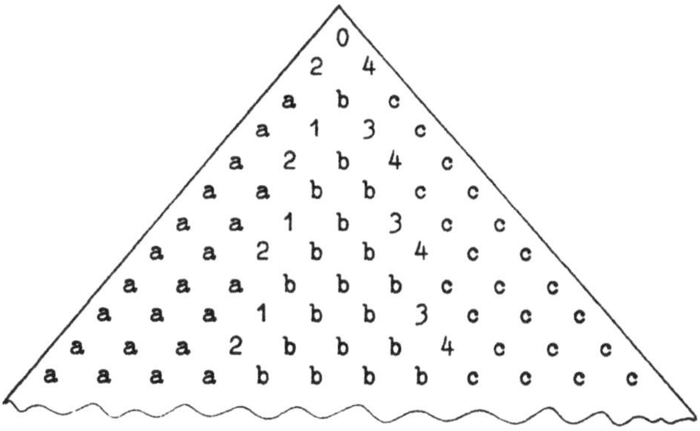

The RTSTA **B** will compute as follows (see Fig. 2). Let the computation of **A** consists of 3n steps (the cases of $3n \pm 1$ steps are similar). Then in the first 2n steps **B** will simultaneously:

a) Solve firing squad problem with "generals" G_1, G_2 which divides the input word into three parts with equal lengths; their position are determined by the underlying trellis. To each of the generals belong the soldiers between him and the axis.

b) Concentrate the input word; after 2n steps every of n active processor will know three input symbols; here two auxiliary signals from the general G_1 are used.

c) Sell signals from one of the general upwards so that the concentrated input word will be divided into pieces in such a way that the simulation of **A** can begin.

Figure 2:
Computation of **B**.

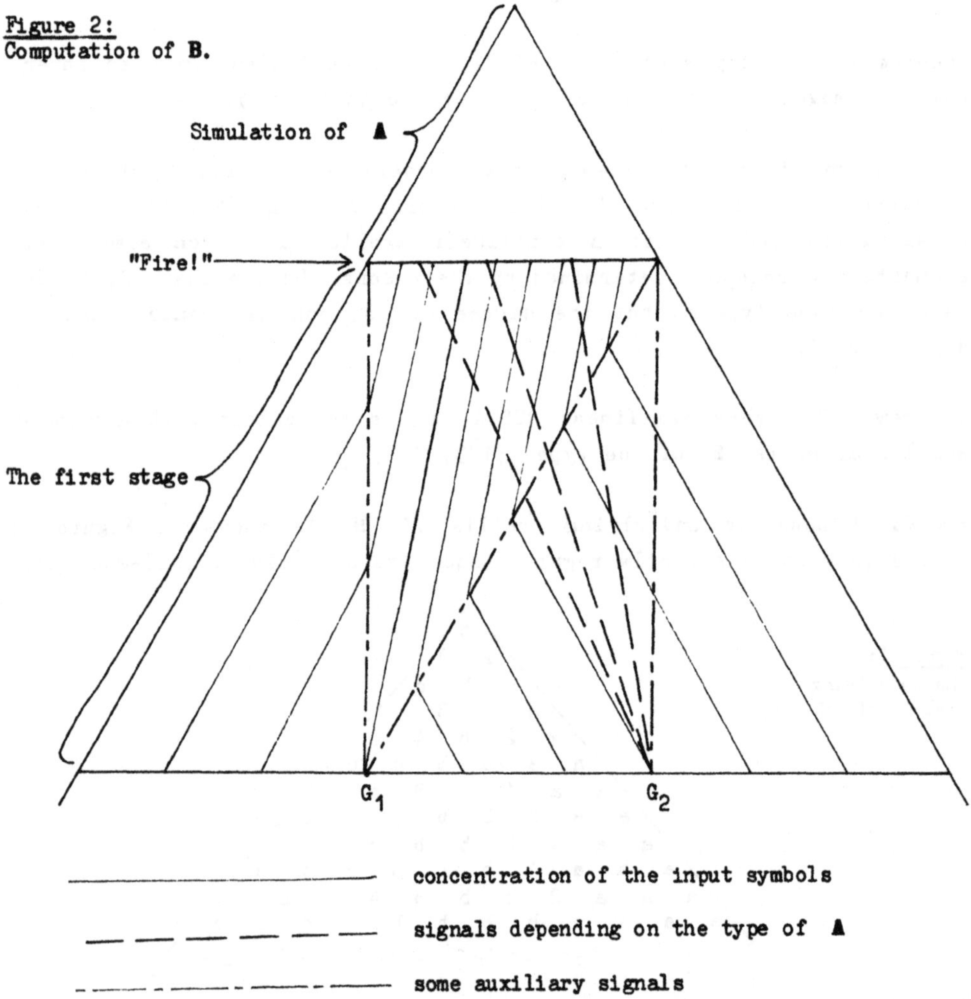

In the last n steps **B** simulates the computation of an internally homogeneous RTSTA equivalent with **A**; the simulation is started by the "Fire" and is three times faster than the original computation. In the part a) a time optimal solution of firing squad problem (mentioned in [Min]) is assumed. However, with a small modification of the proof every solution in linear time is sufficient.

Theorem 4.2 can be generalized as follows.

Theorem 4.3. Let X be a R-set of cardinality at least 3. Then for every semilinear RTSTA **A** there is an equivalent RTSTA **B** which is semilinear of type X and (simultaneously) regular and modular.

Hence the semilinear RTSTA of type $\{1/3, 2/3, 1\}$ or of arbitrary type X, card(X) $\geqq 3$, are as strong as the general semilinear RTSTA. For languages of a special form we can prove a stronger result:

Theorem 4.4. Let **A** be a semilinear RTSTA such that the length of every word $w \in L(\mathbf{A})$ is a power of 2. Then there is a semilinear RTSTA **B** of type $\{1/2, 1\}$ which is equivalent to **A**.

Proof. **B** works similarly as in Theorem 4.2. Solution of firing squad problem is replaced by computation of Pascal triangle modulo 2 (with the root near the center of the input word and in the direction upwards). Simulation of **A** (twice faster than the original computation) begins when a row consisting of 1's is computed. However, since any processor obtains only local information, it starts the simulation whenever it obtains "1". Too early simulations are cancelled whenever a new simulation begins. The concentrated input symbols are continually sent upwards; they are lost at the margins of **B**.

Theorem 4.4 can be generalized analogously as Theorem 4.2; in this case the cardinality of X will be at least two. Further, powers of 2 can be replaced by powers of another prime.

Theorem 4.5. The language $\{a^n;\ n$ a power of $2\}$ is accepted by a semilinear RTSTA of type $\{1/2, 1\}$.

The set $\{1/2, 1\}$ can be replaced by any R-set of cardinality 2. By [CGS1] this language is not accepted by any homogeneous RTSTA. (However, the language $\{a^n b^n;\ n$ a power of $2\}$ is accepted by a homogeneous RTSTA.) Therefore semilinear RTSTA are stronger than homogeneous ones.

The degree 3 of RTSTA **B** in Theorem 4.2 can be diminished (without changing of the type of **B**) if regularity is not requested. Simplify the underlying trellis of **B** displayed in Figure 1 so that the elements 2, 3 remain unchanged and all other elements are replaced by 0. The new trellis is semilinear of degree 2, and can be used similarly as the original one. So we obtain:

Theorem 4.6. Every semilinear RTSTA is equivalent with a semilinear RTSTA of degree 2.

Open problem. Is every semilinear RTSTA equivalent with a semilinear RTSTA of type $\{1/2, 1\}$?

References:

[CeG] Černý, A., Gruska, J., "Modular trellises", in Rozenberg, G., Salomaa, A., "The book of L", Springer (1986), 45-61.

[CGS1] Culik, K. II., Gruska, J. and Salomaa, A., "Systolic trellis automata", Intern. J. Computer Math., 15 (1984), 195-212 and 16 (1984), 3-22.

[CGS2] Culik, K. II., Gruska, J. and Salomaa A., "Systolic trellis automata: Stability, decidability and complexity", Research report CS-82-04, Department of Computer Science, University of Waterloo (1982).

[Gin] Ginsburg, S., "The mathematical theory of context-free languages", Mc Graw - Hill (1966).

[Gru] Gruska, J., "Systolic automata - Power, characterizations, nonhomogeneity", Proceedings Math. Foundations of Computer Science, Praha 1984. Springer Verlag, Lect. Notes in Computer Science 176 (1984), 32-49.

[HU] Hopcroft, J. E. and Ullman, J. D., "Formal languages and their relation to automata". Addison-Wesley (1969).

[Kor1] Korec, I., "Generalized Pascal triangles" (in Slovak), Doctor thesis, UK Bratislava (1984).

[Kor2] Korec, I., "Generalized Pascal triangles - Decidability results", Acta Math. Univ. Comen. 46-47 (1985), 93-130.

[Kor3] Korec, I., "Two kinds of processors are sufficient and large operating alphabets are necessary for regular trellis automata languages", Bull. EATCS 23 (June 1984), 35-42.

[Kor4] Korec, I., "Undecidable problems concerning generalized Pascal triangles of commutative algebras", Proceedings Math. Foundations of Computer Science Bratislava 1986, Springer Verlag, Lect. Notes in Comp. Science 233 (1986), 458-466.

[Min] Minsky, M.: "Computation: Finite and infinite machines", Prentice Hall (1967).

[Vol] Vollmar, R.: "Some remarks on pipeline processing by cellular automata", Computers and Artif. Intelligence, Vol. 6 (1987), No. 3, 263-278.

Inducibility of the Composition of Frontier-to-Root Tree Transformations

Tibor Kovács
Kalmár Laboratory of Cybernetics, University of Szeged
Árpád tér 2., H-6720 Szeged, Hungary

1 Introduction

The concept of tree automaton is based on the realization that the finite automata can be regarded as finite unary algebras which, if viewed this way, have polynomial symbols as input. If we drop the requirement of unarity when defining finite automata, we get the notion of tree automata. This generalization can be done in two different ways: by deciding on the direction of processing of the trees we get the notions of frontier-to-root or root-to-frontier tree automata respectively. The same way of generalization leads us from the notion of generalized sequential machines to the notions of frontier-to-root and root-to-frontier tree transducers (see [11], [1], [8], [9]).

The importance of the notions gained this way are demonstrated by the results according to which the class of forests recognized by tree automata is basically identical with the class of forests consisting of the production trees of context-free grammars; or by those which concern the composition of tree transformations, the hierarchy of the n-surface forests and the classes of the n-transformational languages (see [1], [3],[4], [5], [6], [8], [9]). The importance of the issue is again shown by the fact that it is obviously related to the issue of syntax-directed compilation (see [11]).

The way of generalization by which we can get the notion of tree transducers, naturally, raises the problem of whether the classes of tree transformations (like the class of transformations induced by generalized sequential machines) are closed under composition. It is well-known that the property of being closed is not transmitted during the process of generalization. However, the question is still open whether it is possible to give the exact dividing lines between the classes of tree transformations.

In this paper we give the first definite and positive result concerning the issues outlined above, namely that it is possible to give an algorithmically decidable condition which tells if the composition of transformations induced by two frontier-to-root tree transducers is inducible with a frontier-to-root tree transducer. As a consequence, this result will also answer the problem of which root-to-frontier tree transformations are inducible with frontier-to-root tree transducers.

We wish to emphasize that although the condition is, in both cases, based on the inspection of the transducer(s) inducing the transformation, it provides the same result starting with any transducer(s) inducing the same transformation, i.e. it is characteristic of the transformation itself and not the specific way of inducing.

2 Notions and Notations

We shall use the notions and notations in the same sense and form as [8] and [9]. So by the set of *operational symbols* we mean the $F = \bigcup(F_m \mid m \geq 0)$ union of $F_0, F_1, \ldots$ set of symbols, which are assumed to be pairwise disjoint. The symbols in F_m will be called *m-ary operational symbols.* If F is finite, it will be called a *ranked alphabet.*

The $\mathbf{A} = (A, \{(f)_A \mid f \in F\})$ (or shortly $\mathbf{A} = (A, F)$) system is an *F-algebra,* where for every nonnegative integer m and $f \in F_m$ $(f)_A : A^m \mapsto A$ is the realization of the m-ary operational symbol f on the nonempty set A.

We shall define trees as *polynomial symbols.* Let Z be an arbitrary set disjoint with F. The set of *F-trees over* Z ($T_F(Z)$) is the smallest set satisfying

(i) $F_0 \cup Z \subseteq T_F(Z)$

(ii) if $m \geq 1; f \in F_m; p_1, \ldots, p_m \in T_F(Z)$ then $f(p_1, \ldots, p_m) \in T_F(Z)$.

It is well-known that $(T_F(Z), F)$ is the free F-algebra generated by Z, and that elements of $T_F(Z)$, called *trees,* can be represented by tree shape graphs where the nodes are labelled with the appropriate symbols from $F \cup Z$.

For an arbitrary set of operational symbols F and a set Z the sets of F-trees over Z are called *forests.* The forests recognizable by *(n-ary deterministic) frontier-to-root tree automata* are called *regular forests.* (For the notion of tree automaton see [1], [8], [9].)

The important notions of the *root*,the *set of subtrees* and the *height of trees* are defined as follows:

(i) $p \in F_0 \cup Z : root(p) = p, sub(p) = \{p\}, h(p) = 0$.

(ii) $p = f(p_1, \ldots, p_m) : root(p) = f, sub(p) = \{p\} \bigcup(sub(p_i) \mid i = 1, \ldots, m)$, $h(p) = (max\{h(p_i) \mid i = 1, \ldots, m\}) + 1$.

It is obvious that trees can be represented by tree shape graphs, too. Based on this representation the *set of occurences of subtrees* $sub^*(p)$ can naturally be defined in the same way as $sub(p)$ and it contains those tree shape subgraphs of the graph representing p which represent *occurences* of subtrees (in sense $sub(p)$) of p. Likewise, we can define the notion of path leading from the root to a node and its length.

We can define a partial ordering relation on $sub(p)$ and one on $sub^*(p)$ $(p \in T_F(Z))$:

(i) $p_1, p_2 \in sub(p) : p_1 \leq p_2 \Longleftrightarrow p_1 \in sub(p_2)$

(ii) $p_1, p_2 \in sub^*(p) : p_1 \preceq p_2 \Longleftrightarrow p_1 \in sub^*(p_2)$

If the graphs representing p_1 and p_2 are identical, too, we will write $p_1 \equiv p_2$.

From now on capitals will be used to denote sets whereas the elements will be denoted by the appropriate lower case letters unless stated otherwise.

If F, G are ranked alphabets, X, Y are finite sets of variables then the sets $\tau \subseteq T_F(X) \times T_G(Y)$ are called *tree transformations.*

Let $\Xi = \{\xi_1, \xi_2, \ldots\}, \Xi_l = \{\xi_1, \ldots, \xi_l\}$ be sets of auxiliary symbols.

The system $\underline{A}=(T_F(X),A,T_G(Y),A',\Sigma_A)$ is called a *frontier-to-root (or bottom-up) tree transducer* if F, G are ranked alphabets, X, Y are finite sets of variables, A is a finite set disjunkt with the others (*set of states*), $A' \subseteq A$ (*set of final states*), and Σ_A is a finite set of *(rewriting) rules* of form:

(i) $x_i \longrightarrow aq; a \in A, q \in T_G(Y)$.

(ii) $f(a_1\xi_1,\dots,a_l\xi_l) \longrightarrow a\bar{q}(\xi_1,\dots,\xi_l)$; $f \in F_l$, $l \geq 0$, $a,a_1,\dots,a_l \in A$, $\bar{q} \in T_G(Y \cup \Xi_l)$.

A frontier-to-root tree transducer is *linear* if on the right hand side of all rewriting rules in Σ_A all auxiliary symbols occur at most once, and it is *deterministic* if different rules have different left hand sides.

If we wish to emphasize how many times each auxiliary symbol occurs in a tree $\bar{q} \in T_G(Y \cup \Xi_l)$, we use the notation $\bar{q}(\xi_{11},\dots,\xi_{1s_1},\dots,\xi_{l1},\dots,\xi_{ls_l})$ or, in more complicated cases, the notation $\bar{q}(\xi_{111},\dots,\xi_{11t_1 1},\dots,\xi_{ls_l 1},\dots,\xi_{ls_l t_{ls_l}})$ for the identification of their occurences. For example, $\xi_{l1},\dots,\xi_{ls_l}$ denote occurences of ξ_l, and $\xi_{ls_l 1},\dots,\xi_{ls_l t_{ls_l}}$ also denote occurences of ξ_l if we have already differentiated occurences of ξ_l.

Let p and q be trees in $T_F(X \cup A \times T_G(\Xi_l \cup Y))$. Then q can directly be derived from p by $\underline{A}$, if q is derived from p in the following way:

(i) one occurence of $x_i \in X$ in p is substituted by the right hand side (aq) of a rule in Σ_A; or

(ii) one occurence of a subtree $f(a_1p_1,\dots,a_kp_k)$ in p is substituted by tree $a\bar{q}(p_1,\dots,p_k)$, where $f(a_1\xi_1,\dots,a_k\xi_k) \longrightarrow a\bar{q}(\xi_1,\dots,\xi_k)$ is a rule from Σ_A.

Direct derivation is denoted by $p \Rightarrow_A q$. The reflexive-transitive closure of this relation is denoted by $\Rightarrow_A^*$, and is called *derivation*. Naturally, each derivation can be given by a sequence of direct derivations. We shall always use the notion of derivation in this sense (i. e. a sequence of direct derivations).

The relation $\tau_A = \{(p,q) \mid p \in T_F(X), q \in T_G(Y),\ p \Rightarrow_A^* aq\ , a \in A'\}$ is called the *tree transformation induced by frontier-to-root tree transducer* $\underline{A}$, or shortly frontier-to-root tree transformation.

The system $\underline{A}=(T_F(X),A,T_G(Y),A',\Sigma_A)$ is called *root-to-frontier (or top-down) tree transducer with regular look-ahead* if F, G are ranked alphabets; X, Y are finite sets of variables; A is a finite set (*set of states*); $A' \subseteq A$ (*set of starting states*); and Σ_A is a set of rewriting rules of form:

(i) $ax_i \longrightarrow q; a \in A, q \in T_G(Y)$.

(ii) $(af(\xi_1,\dots,\xi_k) \longrightarrow \bar{q}(a_{11}\xi_{11},\dots,a_{ks_k}\xi_{ks_k}); R_1,\dots,R_k)$; $f \in F_k$, $k \geq 0$, $a,a_{11},\dots,a_{ks_k} \in A$, $\bar{q} \in T_G(Y \cup \Xi)$, $R_1,\dots,R_k$ are regular subsets of $T_F(X)$.

The definition of direct derivation parallels with direct derivation by frontier-to-root tree transducers with the restriction that in (ii) p_i must be a tree in R_i ($i=1,\dots,k$). The relation $\Rightarrow_A^*$ is again the reflexive-transitive closure of the relation $\Rightarrow_A$, and the transformation induced by the transducer $\underline{A}$ is $\tau_A = \{(p,q) \mid p \in T_F(X), q \in T_G(Y),$ $ap \Rightarrow_A^* q\ , a \in A'\}$.

The root-to-frontier tree transducer $\underline{A}$ with regular look-ahead is called *root-to-frontier tree transducer* if in the rules of form (ii) in Σ_A every forest R_i is identical with $T_F(X) (i=1,\dots,k)$.

Let $\mathcal{L}$, $\mathcal{F}$, $\mathcal{F}_R$ denote the class of frontier-to-root tree transformations, the root-to-frontier tree transformations and the root-to-frontier tree transformations with regular look-ahead, respectively. The composition of transformations, denoted by $\circ$, means the product of relations. The classes of transformations resulting from the composition of n transformations of the same class are denoted by $\mathcal{L}^n$, $\mathcal{F}^n$ and $\mathcal{F}_{\mathcal{R}}{}^n$.

The following well-known facts motivated us: (i) the classes $\mathcal{L}$ and $\mathcal{F}$ are uncomparable, (ii) $\mathcal{L}^n \subset \mathcal{L}^{n+1}$, $\mathcal{L}^n \subset \mathcal{F}^{n+1}$, (iii) $\mathcal{F}^n \subset \mathcal{L}^{n+1}$, $\mathcal{F}^n \subset \mathcal{F}^{n+1}$, (iv) $\mathcal{L} = \mathcal{LL} \circ \mathcal{L} = \mathcal{L} \circ \mathcal{DL}$, where $\mathcal{LL}, \mathcal{DL}$ denote the classes of tree transformations induced by linear or deterministic frontier-to-root tree transducers, respectively. (See, e.g. [1], [8], [9].)

3 The Inducibility of Transformations in $\mathcal{L}^2$ by Frontier-to-Root Tree Transducers

The properties of the operation of frontier-to-root and root-to-frontier tree transducers (and so of the tree transformations induced by them) follow from the forms of the rewriting rules , from the definition of direct derivation and from the finiteness of the sets of rules and states. *Engelfriet*'s remarks about the capabilities of transducers in [1],[2] are an important starting point for the present issue. These characteristic features are:

(T) "Copying of an input tree and processing the copies differently."

(B1) "Copying of an output tree after nondeterministic processing of the input tree."

(B2) "Deciding whether to delete a tree or not after processing it."

These are, however, not sufficient for the exact determination of the difference between transformations from $\mathcal{L}$ and $\mathcal{L}^2$. Beyond the relation defined on $T_F(X) \times T_G(Y)$, by the inner structure of the derivations, tree transducers also realize a more subtle relation, namely a relation between subtrees of the trees in the relation.

Here and from now on, by $\underline{A}, \underline{B}$ and $\underline{C}$ we mean frontier-to-root tree transducers $\underline{A} = (T_F(X), A, T_G(Y), A', \Sigma_A)$, $\underline{B} = (T_G(Y), B, T_H(Z), B', \Sigma_B)$ and $\underline{C} = (T_F(X), C, T_H(Z), C', \Sigma_C)$.

If we say "derivation" we mean a sequence of direct derivations, so if we say "subderivation" we mean a subsequence of this derivation, such that it is a derivation of a subtree (in sense $sub^*(p)$) of the starting tree of original derivation. If we take the derivation $d = (p \Rightarrow_A^* ar, r \Rightarrow_B^* bq)$ then we mean the concatenation of this two sequences of direct derivations. In this case, its subderivation $(p' \Rightarrow_A^* a'r', r' \Rightarrow_B^* b'q')$ means the concatenation of these subderivations. The starting (sub)trees of (sub)derivations by $\underline{B}$ must be identical with the result of (sub)derivations by $\underline{A}$ in sense $sub^*(p)$.

Let $d_1 = (p \Rightarrow_A^* ar)$ and , and $p' \in sub^*(p)$.

Then $W_A(p')/d_1 = \{r' \mid r' \in sub^*(r)$ and $p' \Rightarrow_A^* a'r'$ is a subderivation in $d_1\}$, $W_{AB}(p')/d_2 = \{q' \mid q' \in sub^*(q)$ and $p' \Rightarrow_A^* a'r'$, $r' \Rightarrow_B^* b'q'$ is a subderivation in $d_2\}$.

If $r' \in W_A(p')/d_1$ $(q' \in W_{AB}(p')/d_2)$ then r' (q') are called the "images" of p'.

The finite set of states and rules, however, creates an important and necessary connection also between different derivations.

Let $\mathcal{N}$ denote the set of natural numbers. The set $I \subseteq \mathcal{N}^n$ is called *complete n-ary derivation heap index set* if there exist $N_1, \ldots, N_n \subseteq \mathcal{N}$, $N_1, \ldots, N_n$ infinite and $I = N_1 \times \ldots \times N_n$.

Let $I \subseteq \mathcal{N}^n$, $1 \le j \le n$ and $\mathrm{i} = (i_1, \ldots, i_j, \ldots, i_n)$. We put $P_j(\mathrm{i}) = i_j$ (the j' th projection), $first_j = min\{P_j(\mathrm{i}) \mid \mathrm{i} \in I\}$, $next_j(l) = min\{P_j(\mathrm{i}) \mid \mathrm{i} \in I, P_j(\mathrm{i}) > l\}$. I is called an *n-ary derivation heap index set*, provided that it only differs from a complete n-ary derivation heap index set J in that, if the elements of I are represented by the reticular nodes of the n-dimensional space, then on all lines including i ($\mathrm{i} \in I$) and parallel with any co-ordinate axis there is only a finite number of reticular nodes j for which $\mathrm{j} \in J \setminus I$.

Definition 1. Let $\underline{A}$ and $\underline{B}$ be frontier-to-root tree transducers, and let D be a set of pairs constructed from derivations by $\underline{A}$ and $\underline{B}$. D is called an *n-ary derivation heap* (in $\tau_A \circ \tau_B$) if

(i) I is an n-ary derivation heap index set

(ii) $D = \{d_\mathrm{i} \mid d_\mathrm{i} = (\, p_\mathrm{i} \Rightarrow^*_A a r_\mathrm{i} \,,\, r_\mathrm{i} \Rightarrow^*_B b q_\mathrm{i} \,), a \in A', b \in B', \mathrm{i} \in I\}$

(iii) $(\exists s_1, \ldots s_n > 0)\,(\exists t_{11}, \ldots, t_{1s_1}, \ldots, t_{n1}, \ldots, t_{ns_n} \ge 0)\quad (\forall 1 \le j \le s_j)$,
$\sum_{k=1}^{s_j} t_{jk} > 0$
and $(\forall \mathrm{i} \in I)$

$$p_\mathrm{i} = \bar{p}(p^1_\mathrm{i}, \ldots, p^n_\mathrm{i}),$$
$$r_\mathrm{i} = \bar{r}(r^1_\mathrm{i}, \ldots, r^1_\mathrm{i}, \ldots, r^n_\mathrm{i}, \ldots, r^n_\mathrm{i}),$$
$$q_\mathrm{i} = \bar{q}(q^{11}_\mathrm{i}, \ldots, q^{11}_\mathrm{i}, \ldots, q^{ns_n}_\mathrm{i}, \ldots, q^{ns_n}_\mathrm{i}),$$

where r^j_i occurs in r_i s_j times and q^{jk}_i occurs in q_i t_{jk} times.

(iv) for $(\forall \mathrm{i} \in I)$ $\bar{p}(a_1\xi_1, \ldots, a_n\xi_n) \Rightarrow^*_A a\bar{r}(\xi_{11}, \ldots, \xi_{1s_1}, \ldots, \xi_{n1}, \ldots, \xi_{ns_n})$ and $\bar{r}(b_{11}\xi_{11}, \ldots, b_{ns_n}\xi_{ns_n}) \Rightarrow^*_B b\bar{q}(\xi_{111}, \ldots, \xi_{11t_{11}}, \ldots, \xi_{ns_n1}, \ldots, \xi_{ns_nt_{ns_n}})$ are subderivations in d_i.

(v) for $(\forall \mathrm{i} \in I)(\forall 1 \le j \le n)(\forall 1 \le k \le s_j)$
$(\, p^j_\mathrm{i} \Rightarrow^*_A a_j r^j_\mathrm{i} \,, r^j_\mathrm{i} \Rightarrow^*_B b_{jk} q^{jk}_\mathrm{i} \,)$ is subderivation in d_i.

(vi) $(\forall \mathrm{i}_1, \mathrm{i}_2 \in I)(\forall 1 \le j \le n)(\forall 1 \le k \le s_j)$
if $P_j(\mathrm{i}_1) = P_j(\mathrm{i}_2)$ then the derivations
$d_1 = (\, p^j_{\mathrm{i}_1} \Rightarrow^*_A a_j r^j_{\mathrm{i}_1} \,, r^j_{\mathrm{i}_1} \Rightarrow^*_B b_{jk} q^{jk}_{\mathrm{i}_1} \,)$ and
$d_2 = (\, p^j_{\mathrm{i}_2} \Rightarrow^*_A a_j r^j_{\mathrm{i}_2} \,, r^j_{\mathrm{i}_2} \Rightarrow^*_B b_{jk} q^{jk}_{\mathrm{i}_2} \,)$ are identical.

(vii) for $(\forall 1 \le j \le n)$ let

(a) $(\forall \mathrm{i} \in I) p^j_\mathrm{i} = \bar{p}^j_l(p^j_l)$, where $l = P_j(\mathrm{i})$.

(b) $(\forall \mathrm{i} \in I) \bar{p}^j_{first_j}(\xi) = \bar{p}^j_{first_j}(\xi)$,
$\bar{p}^j_{next_j(l)}(\xi) = \bar{p}^j_l(\tilde{p}_{next_j(l)}(\xi))$,
where $0 < h(\tilde{p}^j_l(\xi))$ and $l = P_j(\mathrm{i})$.

(viii) Let $Q_{jk} = \{q^{jk}_\mathrm{i} \mid \mathrm{i} \in I\}, 1 \le j \le n, 1 \le k \le s_j, t_{jk} > 0$. Then Q_{jk} must be either singleton or infinite.

(ix) for $(\forall 1 \le j \le n)(\exists 1 \le k \le s_j), t_{jk} > 0$ and Q_{jk} is infinite.

(x) for $(\forall 1 \le j \le n)(\forall 1 \le k \le s_j)$, if $t_{jk} > 0$ and Q_{jk} is infinite then $h(q^{jk}_\mathrm{i}) \to \infty$ whenever $(\, P_j(\mathrm{i}) \to \infty, \mathrm{i} \in I\,)$.

The n-ary derivation heap is called n-ary complete derivation heap if I/D is a complete derivation heap index set.

From now the notations of form α/β will mean "the α in the definition of β". For example $\bar{q}/D$ will denote the tree $\bar{q}(\xi_{111}, \ldots, \xi_{11t_{11}}, \ldots, \xi_{ns_n1}, \ldots, \xi_{ns_nt_{ns_n}})$ in (iv) of the definition of the n-derivation heap D.

The forest containing the trees from which the derivations start in a complete derivation heap is in fact the concatenation of the singleton forest $(\{\bar{p}/D\})$ and n infinite and not necessarily regular forests $(\{p^j_{\mathrm{i}}/D \mid \mathrm{i} \in I\}; j = 1, \ldots, n)$. The derivations in D are constructed from the derivations (fixed in definition of the transducers) of the trees. Condition (vii) imposes a similarity requirement on trees in the set $\{p^j_{\mathrm{i}}/D \mid \mathrm{i} \in I\}$ whereby if $P_j(\mathrm{i}_1) \to \infty$, $P_j(\mathrm{i}_2) \to \infty$, then a greater and greater part of $p^j_{\mathrm{i}_1}$ and $p^j_{\mathrm{i}_2}$ are identical. Conditions (viii)-(x) are made necessary by the fact that finite n-surface forests are encodable in the set of states and the set of rewriting rules of a frontier-to-root tree transducer , if we want to construct a frontier-to-root tree transducer $\underline{C}$, such that $\tau_C = \tau_A \circ \tau_B$.

Note that the definition of the n-ary derivation heap can also be given for one single frontier-to-root tree transducer $\underline{B}$, by selecting a frontier-to-root tree transducer inducing (with singleton set of states) the identical transformation as $\underline{A}$.

Let $\tau(D) = \{(p,q) \mid \exists \mathrm{i} \in I/D : d_{\mathrm{i}} = (p \Rightarrow^*_A ar, r \Rightarrow^*_B bq), d_{\mathrm{i}} \in D\}$.

The n-ary derivation heaps D_1 and D_2 are called similar if $\bar{p}/D_1 = \bar{p}/D_2$ if the maximum identical part of $\bar{q}/D_1$ and $\bar{q}/D_2$ (starting from their roots) is such that the occurences of the auxiliary symbols in it are separated (the occurences determined by the finite set Q_{jk}/D_1 and Q_{jk}/D_2 may be exceptions). The formal definition is the following:

Definition 2. The n-ary derivation heaps D_1 and D_2 are called *similar*, if

(i) $\bar{p}/D_1 = \bar{p}/D_2$

(ii) $\bar{q}/D_1 = \tilde{q}(\bar{q}^1_{11}, \ldots, \bar{q}^1_{1u_1}, \ldots, \bar{q}^1_{n1}, \ldots, \bar{q}^1_{nu_n})$,
$\bar{q}/D_2 = \tilde{q}(\bar{q}^2_{11}, \ldots, \bar{q}^2_{1u_1}, \ldots, \bar{q}^2_{n1}, \ldots, \bar{q}^2_{nu_n})$, where $u_1, \ldots, u_n \geq 0$

(iii) Let $\Xi^w_j = \{\xi_{vkl} \mid 1 \leq v \leq n, 1 \leq k \leq s_v/D_w, 1 \leq l \leq t_{vk}/D_w$ and $v = j$ or Q_{vk}/D_w is singleton $\}$ $(w = 1,2)$.
Then for $\forall j (1 \leq j \leq n) \forall u (1 \leq u \leq u_j) \forall w (1 \leq w \leq 2) : \bar{q}^w_{ju} \in T_H(Z \cup \Xi^w_j)$

(iv) $\forall j (1 \leq j \leq n) \forall u (1 \leq u \leq u_j) : h(\bar{q}^1_{ju}) \cdot h(\bar{q}^2_{ju}) = 0$.

(v) If $\bar{q}^w_{ju} \in T_H(Z) (1 \leq j \leq n, 1 \leq u \leq u_j, 1 \leq w \leq 2)$
then $\bar{q}^{3-w}_{ju} \notin T_H(Z)$, but if ξ_{vkl} occurs in $\bar{q}^{3-w}_{ju}$, then Q_{vu}/D_{3-w} is singleton.

The following theorem gives an important description of the tree transformations induced by frontier-to-root tree transducers:

Theorem 3. Let $\underline{A}, \underline{B}$ and $\underline{C}$ be frontier-to-root tree transducers, $\tau_C = \tau_A \circ \tau_B$ and D an n-ary derivation heap in $\tau_A \circ \tau_B$. Then there exists an n-ary derivation heap D' in $\tau_A \circ \tau_B$ and an n-ary derivation heap D^* in τ_C (i.e. in *identical* $\circ\, \tau_C$) such that $D \supseteq D', \tau(D) \supseteq \tau(D') = \tau(D^*)$ and D, D', D^* are similar.

The proof is based on the fact that any tree can only be processed in a finite number of ways by one tree transducer , and so there is at least one derivation of $\bar{p}/D$ by $\underline{C}$ such that it occurs in an infinite number of times among the derivations equivalent to those in D. It can be proved that among these there exists at least one derivation which determines (by the identity of the result of derivations) such a subset of I/D which is itself an n-ary derivation heap index set. This way of selecting determines a derivation of $\bar{p}/D$ by $\underline{C}$. By using this derivation we can construct the derivation heap D^* (in τ_C) for which we can prove that D, D', D^* are similar.

We give the exact dividing lines between $\mathcal{L}$ and $\mathcal{L}^2$ on the basis of Theorem 3. and *Engelfriet*'s remarks (T),(B1),(B2). We notice that these together mean that only those $\mathcal{L}^2$ transformations can be in $\mathcal{L}$ for which in any derivation $d = (p \Rightarrow^*_A ar, r \Rightarrow^*_B bq)$

any two different images of any $p' \in sub^*(p)$ subtrees in d can only differ from each other in a finite number of ways. Furthermore,the frontier-to-root tree transducer inducing the composition must operate so that for every $p' \in sub^*(p)$ it derives the smallest common subgraph of the images in d as the result of subderivation, and encodes the derivations from this in its own states.

Let us consider the following examples for the possible forms of derivations. Let $\underline{A} = (T_F(X_1), \{a_1, a_2\}, T_F(X_1), \{a_1\}, \Sigma_A)$, $F = F_1 \cup F_2$, $F_1 = \{f, h\}$, $F_2 = \{g\}$, $\Sigma_A = \{ \quad x_1 \to a_2 x_1 , \quad f(a_2 \xi_1) \to a_2 f(\xi_1), \quad f(a_2 \xi_1) \to a_1 g(\xi_1, \xi_1) \quad \}$, $\underline{B} = (T_F(X_1), \{b_1, b_2, b_3\}, T_F(X_1), \{b_1\}, \Sigma_B)$ and the set of rules are

(I) $\Sigma_B = \{ x_1 \to b_2 x_1 , x_1 \to b_3 x_1 , f(b_2 \xi_1) \to b_2 \xi_1 , f(b_3 \xi_1) \to b_3 f(\xi_1) ,$
$g(b_2 \xi_1, b_3 \xi_2) \to b_1 g(\xi_1, \xi_2)\}$.
$\tau_A \circ \tau_B = \{(f^k(x_1), g(x_1, f^{k-1}(x_1))) \mid k \geq 1\}$. (Here $f^k(x_1) = f(\ldots(f(x_1))\ldots)$ i.e. the operational symbol f occurs k-times.)

(II) $\Sigma_B = \{ x_1 \to b_2 f(x_1) , x_1 \to b_3 x_1 , f(b_2 \xi_1) \to b_2 f(\xi_1) , f(b_3 \xi_1) \to b_3 f(\xi_1) ,$
$g(b_2 \xi_1, b_3 \xi_2) \to b_1 g(\xi_1, \xi_2)\}$.
$\tau_A \circ \tau_B = \{(f^k(x_1), g(f^k(x_1), f^{k-1}(x_1))) \mid k \geq 1\}$.

(III) $\Sigma_B = \{ x_1 \to b_2 f(x_1) , x_1 \to b_3 h(x_1) ,$
$f(b_1 \xi_1) \to b_1 f(\xi_1) , f(b_2 \xi_1) \to b_1 g(\xi_1, h(x_1)) ,$
$f(b_3 \xi_1) \to b_3 g(f(x_1), \xi_1) , g(b_1 \xi_1, b_1 \xi_2) \to b_1 g(\xi_1, \xi_2)\}$.
$\tau_A \circ \tau_B = \{(f^k(x_1), g(f^{k-2}(t), f^{k-2}(t))) \mid k \geq 2 , t = g(f(x_1), h(x_1))\}$.

(IV) $\Sigma_B = \{ x_1 \to b_2 x_1 , x_1 \to b_3 x_1 , f(b_2 \xi_1) \to b_2 h(\xi_1) , f(b_3 \xi_1) \to b_3 f(\xi_1) ,$
$g(b_2 \xi_1, b_3 \xi_2) \to b_1 g(\xi_1, \xi_2)\}$.
$\tau_A \circ \tau_B = \{(f^k(x_1), g(h^{k-1}(x_1), f^{k-1}(x_1))) \mid k \geq 1\}$.

(I)-(III) of the examples above are such that the deviations between the different images of the same (sub)tree are possible to encode with the finite number of states of the frontier-to-root tree transducer inducing the composition, but in (IV) this encoding is not possible and so $\tau_A \circ \tau_B \notin \mathcal{L}$.

For example, in case (I), all subtrees of form $f^l(x_1)(l < k)$ have two images x_1 and $f^l(x_1)$, so the frontier-to-root tree transducer $\underline{C}$ (for which $\tau_C = \tau_A \circ \tau_B$) can operate on this subtrees as identical tree transducer and encode the image x_1 in its states and rewriting rules so as to end the derivations. In case (II), the images of subtrees of form $f^l(x_1)(l < k)$ are $f^l(x_1)$ and $f^{l-1}(x_1)$, so the deviation can be encoded by $f(\xi)$. This deviation must be encoded in the set of states of $\underline{C}$ and to end the derivations it must be encoded in the set of rewriting rules of $\underline{C}$ by $g(f(\xi), \xi)$. In case (III), only the subtree x_1 has different images. Thus to end the subderivations of $f(x_1)$ this deviation must be encoded in the set of rewriting rules of $\underline{C}$ by $g(f(x_1), \xi)$ or $g(\xi, h(x_1))$. In case (IV), an infinite number of deviations $h^l(\xi)$ or $f^l(\xi)$ should be encoded in the set of states and in the set of rewriting rules of $\underline{C}$.

Although the examples do not cover all the possible cases, they indicate how the deviations between the images of a subtree in a derivation can be classified according to the degrees of the deviations and the number of their occurences; and it is possible to give the derivations to be encoded, which are here

(I): $\{x_1\}$
(II)/a: $\{f(\xi)\}$ based on difference of the images of the subtree $f^l(x_1)$ $(l < k)$,
(II)/b: $\{g(f(\xi), \xi)\}$ based on the image of the root of $f^k(x_1)$ and on (II)/a,
(III): $\{g(f(x_1), \xi), g(\xi, h(x_1))\}$

(IV): $\{h^{k-1}(\xi), f^{k-1}(\xi)\}$, respectively.

On the basis of these remarks we can define the notion of T-property and its variations, and also the n-ary derivation heaps having the deviations (i.e. all derivations in the derivation heap have the same special T-property). In the next proofs we need the the definition of T-property and its variations in this form:

Definition 4. Let $\underline{A}$ and $\underline{B}$ be frontier-to-root tree transducers, and let d be a derivation, where $d=(p \Rightarrow_A^* ar, r \Rightarrow_B^* bq)$, $a \in A'$, $b \in B'$, $p = \bar{p}(p_T)$, $r_T^1, r_T^2 \in W_A(p_T)$, $r_T^1 \not\equiv r_T^2$, $q_T^1 \in W_B(r_T^1), q_T^2 \in W_B(r_T^2)$. Let q_T be the smallest element of $sub^*(q)$ containing q_T^1 and q_T^2 as subgraphs.

(i) $q_T^1 \underline{T}\, q_T^2 \iff q_T^1 \neq q_T^2$

(ii) $q_T^1 \underline{weak - T}\, q_T^2 \iff$
$(q_T^1 \underline{T}\, q_T^2)$ and $\exists q_T^*$
for which $q_T^* \in sub^*(q_T)$ and
$((q_T^1 = q_T^*$ and $q_T^2 \in sub^*(q_T^*))$ or $(q_T^2 = q_T^*$ and $q_T^1 \in sub^*(q_T^*)))$

(iii) $q_T^1 \underline{strong - T}\, q_T^2 \iff ((q_T^1 \underline{T}\, q_T^2)$ and not $(q_T^1 \underline{weak - T}\, q_T^2))$

(iv) $q_T^1 \underline{hard - T}\, q_T^2 \iff ((q_T^1 \underline{strong - T}\, q_T^2)$ and not $(q_T^1 \leq q_T^2$ or $q_T^1 \geq q_T^2))$

(v) $q_T^1 \underline{not - hard - T}\, q_T^2 \iff ((q_T^1 \underline{strong - T}\, q_T^2)$ and not $(q_T^1 \underline{hard - T}\, q_T^2))$

For every derivation having a special T-property and for derivation heaps having a T-property indicated in (I),(II)/a and (IV) we can define a set of augmented trees (denoted by $VAL(D)$) for the description of the deviations of the different images of subtrees. Here D is either an n-ary derivation heap indicated in (I),(II)/a or (IV), or a derivation having a special T-property. By $h(VAL(D))$ we mean the maximum height of trees in $VAL(D)$). The unary derivation heaps defined on the basis of (I) will be called *finite derivation heaps having T-property* (*finite T-heaps*), and the n-ary derivation heaps defined on the basis of (II)/a and (IV) will be called *n-ary derivation heaps having strong T-property* (*strong T-heaps*).

Based on Theorem 3. and the fact that all frontier-to-root tree transducers inducing $\tau_A \circ \tau_B$ can only possess a finite number of states and rules, the following can be proved:

Lemma 5. If $\underline{A}, \underline{B}$ and $\underline{C}$ are frontier-to-root tree transducers with $\tau_C = \tau_A \circ \tau_B$ and D is a strong T-heap in $\tau_A \circ \tau_B$, then there exists a rule in Σ_C which contains all the elements of $VAL(D)$ as a subgraph in its right hand side.

Lemma 6. If $\underline{A}, \underline{B}$ and $\underline{C}$ are frontier-to-root tree transducers such that $\tau_C = \tau_A \circ \tau_B$ and D is a binary strong T-heap in $\tau_A \circ \tau_B$ and u is that node of $\bar{p}/D$ graph where the paths leading from root to letters ξ_1 and ξ_2 split, then the rule for D in Σ_C existing according to Lemma 5. is to be applied in node u in the derivations in D^* existing according to Theorem 3.

This means that if there exists a ternary strong T-heap D in $\tau_A \circ \tau_B$ where, according to Lemma 6., the node u_1 determined by ξ_1 and ξ_2 and the node u_2 determined by ξ_2 and ξ_3 are not identical, then $\tau_A \circ \tau_B \notin \mathcal{L}$. Such strong T-heaps will be called *uninducible (strong) T-heaps.*

Lemma 7. Let $\underline{A}$ and $\underline{B}$ be frontier-to-root tree transducers. If there is an infinite number of derivations having a special T-property or an infinite number of the finite T-heaps, different in $VAL(D)$, then in $\tau_A \circ \tau_B$ there exist an infinite number of strong T-heaps different in $VAL(D)$.

Theorem 8. Let $\underline{A}$ and $\underline{B}$ be frontier-to-root tree transducers. There exists a $\underline{C}$ frontier-to-root tree transducer $\underline{C}$ such that $\tau_C = \tau_A \circ \tau_B$, if and only if there is only a finite number of strong T-heaps different in $VAL(D)$, and none of them are uninducible.

This theorem follows from Lemmas 5.,6. and 7. and from the fact that — provided that we know the maximum height of the trees in $VAL(D)$ — we can construct a frontier-to-root tree transducer which operates in the following way: processing from the letters to the root, it derives the smallest common subgraph of the images for all the subtrees, and encodes the deviations in its own states.

It is an essential question whether this condition is decidable .

Lemma 9. For every k natural number and for every $\underline{A}$ and $\underline{B}$ frontier-to-root tree transducers it is decidable whether in $\tau_A \circ \tau_B$ there exists a finite T-heap or a strong T-heap or a derivation having a special T-property (notated D) for which $h(VAL(D)) > k$.

The proof of decidability is based on a usual method: based on the derivations, the starting trees of derivations are relabelled with a label (from a finite label set) describing the examined properties, and so, by cutting the trees at the nodes having the same label, from the original derivation we construct new derivations, which also possess the examined property. Thus, for example, if we cut p into $p = p_1(p_2(p_3))$, then by leaving out p_2 (*reduction*) we can gain a lower tree and its derivation, whereas by multiplying p_2 (*iteration*) we can gain however an infinite sequence of trees and their derivations. In case of iteration we construct the derivation of $p_1(p_2(\ldots(p_2(p_3)\ldots)))$ from subderivations starting from p_1, p_2 and p_3 (these are sequences of direct derivations). The properties of the label set guarantee that an infinite number of these derivations have a special T-property if the original derivation of $p_1(p_2(p_3)))$ has had the same special T-property, too. As a label set, we can use a set constructed from states and such deviations which contain trees not higher than a given bound.

Lemma 10. For every $\underline{A}$ and $\underline{B}$ frontier-to-root tree transducers it is decidable whether in $\tau_A \circ \tau_B$ there is a finite number of finite T-heaps different in $VAL(D)$.

Lemma 11. For every natural number k and for every frontier-to-root tree transducers $\underline{A}$ and $\underline{B}$ it is decidable whether there is at least one uninducible strong T-heap in $\tau_A \circ \tau_B$ among the strong T-heaps with $h(VAL(D)) \leq k$.

Lemma 12. For every natural number k and for every frontier-to-root tree transducers $\underline{A}$ and $\underline{B}$ it is decidable whether among strong T-heaps and derivations having a special T-property for which $h(VAL(D)) \leq k$, there exists at least one which is possible to iterate (i.e. which proves the existence of an infinite number of strong T-heaps different in $VAL(D)$).

Theorem 13. The condition given in Theorem 8. is decidable for every frontier-to-root tree transducers $\underline{A}$ and $\underline{B}$.

Corollary 14. For every transformation $\tau \in \mathcal{L}^2$, if two frontier-to-root tree transducers $\underline{A}$ and $\underline{B}$ are known such that $\tau = \tau_A \circ \tau_B$, it is decidable whether $\tau \in \mathcal{L}$.

It is known that $\mathcal{F} \subseteq \mathcal{F}_\mathcal{R}$, $\mathcal{F}_\mathcal{R} \subseteq \mathcal{L}^2$. For any root-to-frontier tree transducer with regular look-ahead $\underline{A}$ it is easy to construct such $\underline{B_1}$ and $\underline{B_2}$ frontier-to-root tree transducers that $\tau_A = \tau_{B_1} \circ \tau_{B_2}$.

Thus the following are true:

Theorem 15. For every root-to-frontier tree transducer with regular look-ahead $\underline{A}$ (and, especially, for any root-to-frontier tree transducer) it is decidable if $\tau_A \in \mathcal{L}$.

Corollary 16. For every transformation $\tau \in \mathcal{F}_{\mathcal{R}}$, (and, especially, every $\tau \in \mathcal{F}$) if a root-to-frontier tree transducer with regular look-ahead $\underline{A}$ is known such that $\tau = \tau_A$, it is decidable whether $\tau \in \mathcal{L}$.

4 References

[1] Engelfriet, J. : *Tree automata and tree grammars.* DAIMI FN-10, Inst. Math., Univ. Aarhus (1975).

[2] Engelfriet, J. : *Bottom-up and top-down tree transformations — A comparison.* Math. Syst. Theory 9 (1975), 198-231 pp.

[3] Engelfriet, J. : *Top-down tree transducers with regular look-ahead.* Math. Syst. Theory 10 (1977), 289-303 pp.

[4] Engelfriet, J. : *A hierarchy of transducers.* Troisième Coll. les Arbres en Algebre et en Programmation, Lille (1978), 103-106 pp.

[5] Engelfriet, J. Rozenberg, G. and G. Slutzki :*Tree transducers, L-systems and two-way machines.* J. Comput. Syst. Sci. 20 (1980), 150-202 pp.

[6] Engelfriet, J. and S. Skyum : *Copying theorems.* Information Processing Letters 4 (1976), 157-161 pp.

[7] Ésik Z.: *Decidability results concerning tree transducers I.* Acta Cybernetica 5 (1980) 1-20 pp.

[8] Gécseg F. and M. Steinby : *Algebraic theory of tree automata II.* Matematikai Lapok XXVII. 3-4. (1979), 283-336 pp. (Hungarian).

[9] Gécseg F. and M. Steinby : *Tree automata.* Akadémiai Kiadó, Budapest (1984).

[10] Ogden, W. F. and Rounds, W. C. :*Compositions of n tree transducers.* 4. Ann. ACM STC (1972), 198-206 pp.

[11] Rounds, W. C. : *Mappings and grammars on trees.* Math. Syst. Theory 4. (1970), 257-287 pp.

[12] Zachar Z. : *The solvability of the equivalence problem for deterministic frontier-to-root tree transducers.* Acta Cybernetica 4 (1979),167-177 pp.

ON OBLIVIOUS BRANCHING PROGRAMS OF LINEAR LENGTH

(*Extended Abstract*)

Matthias Krause
Humboldt-Universität zu Berlin
Sektion Mathematik
Unter den Linden 6
DDR - Berlin
1086

Stephan Waack
Karl-Weierstrass-Institut fuer
Mathematik der Akademie der
Wissenschaften der DDR
Mohrenstr. 39
DDR - Berlin
1086

1. Introduction

One major goal of complexity theory is to seperate complexity classes such as $\mathbb{L}$, and $\mathbb{NL}$ or to prove their coincidence. As usually $\mathbb{L}$ and $\mathbb{NL}$ denote the classes of all languages A which can be accepted by a deterministic and nondeterministic logspace bounded Turing machine, respectively. The nonuniform counterparts $\mathcal{L}$ and $\mathcal{NL}$ are the languages for which there are a polynomial p(n) and an advice $\alpha_n \in \{0,1\}^*$, where $|\alpha_n| \le p(n)$, such that a deterministic or nondeterministic Turing machine, resp., accepts $w\#\alpha_n$ within logspace, $|w| = n$ (# is an additional tape symbol) if and only if w belongs to A.

A *Σ-decision graph* (DG) T_n, for Σ a finite alphabet, is a directed acyclic graph with the following properties.

- It has exactly one source, i.e. a node with indegree 0.
- Every node has outdegree 0 or $|\Sigma|$.
- Sinks, i.e. nodes with outdegree 0 are labelled by 0 or 1.
- Branching nodes, i.e. nodes with outdegree $|\Sigma|$, are labelled i, for some $1 \le i \le n$, and the $|\Sigma|$ outgoing arcs are labelled by the element of Σ, where each $\sigma \in \Sigma$ occurs exactly once.

To every word $w_1 w_2 \ldots w_n = w \in \Sigma^n$ there corresponds a unique path

p_w from the source to a sink (at a branching node labelled i, it chooses the arc labelled by w_i). The decision graph T_n decides a set $L^{(n)} \subseteq \Sigma^n$ iff for every $w \in \Sigma^n$ the sink at the end of the path p_w is labelled by $L^{(n)}(w)$. [Throughout this work we make no difference between $L^{(n)}$ and its characteristic function denoted by $L^{(n)}$, too.]
The *size* of a decision graph T_n, which we denote by $SIZE(T_n)$, is the number of branching nodes of T_n.

A {0,1} - decision graph is a *branching program* (BP). Branching programs compute Boolean functions. They have been studied more extensively than decision graphs over larger alphabets, although the latter ones are more adapted in many cases. The logarithm of the size of a smallest decision graph deciding a language is a lower bound on the space requirement for any reasonable sequential model of computation.

It is well-known that $\mathcal{P}_{BP} = \mathcal{L} \cap 2^{\{0,1\}^*}$ and $\mathcal{P}_{DG} = \mathcal{L}$, where $\mathcal{P}_{BP}$ and $\mathcal{P}_{DG}$ are the classes of languages which can be accepted by branching programs and decision graphs, resp., of polynomially bounded size. Efforts to prove lower bounds for branching programs are eventually aimed at separating $\mathbb{L}$ from other complexity classes.

Nonlinear lower bounds $(\Omega(n^2/(\log n)^2)$ have already been given by Nechiporuk [7] (in the more general framework of contact schemes). In order to obtain larger lower bounds for branching programs and decision graphs, restricted models are considered. First we turn to decision graphs the multiplicity of reading of which is restricted.

A *read - k - times - only* decision graph is allowed to encounter each input variable at most k times along any computation path. It is called *real-time*, if for every w the length of the computation path p_w is less than or equal to n. $\mathcal{P}_{DGk}$ is defined to be the class of all formal languages which can be decided by a sequence of read-k-times-only decision graphs the size of which is polynomially bounded.

Read-once-only branching programs were studied by Wegener [8], Zak [10], and Ajtai et al. [1]. Wegener and Zak gave $2^{\Omega(\sqrt{n})}$ lower bounds, whereas in [1] a $2^{c \cdot n}$ lower bound was proved, for c is approximately 10^{-13}. For example, the graph property studied by Hajnal et al. in [1] is "G has an even num- ber of triangles". Zak investigated the property "G is a halfclique". Clearly, the real-time model is more powerful than the read-once-only model. Kriegel and Waack studied in [4] the real-time decision graph complexity of the Dyck language D_m^*. It is known that the word problem for the Dyck language D_m^* is identical with the word problem of the free group of rank m. A $(2m)^{n/24}$

lower bound was obtained for real-time decision graphs, and a $2^{n/48}$ lower bound for the real-time branching program complexity of an encoding of D_m^*. No superpolynomial lower bound is known even in the case of read-twice-only branching programs.

Another approach that recently gained popularity is proving lower bounds for levelled branching programs for which several additional constraints are imposed.

A decision graph is called *levelled* iff its nodes are organized in pairwise disjoint levels so that arcs go from each level to the next level only. The *width* of a levelled decision graph is the maximum number of nodes on any level. The *length* is the number of levels.

In [1] an $\Omega(n \log n / \log\log n)$ bound was proved for the size of levelled branching programs the width of which is bounded by $(\log n)^{O(1)}$ for almost all symmetric Boolean functions.

Alon/Maass [2] studied input oblivious decision graphs of bounded width. A decision graph is called *input oblivious* (ODG) if it is levelled and the nodes of any level are labelled by one and the same input variable. In [2] among others the sequence equality function Q_{2n} is investigated. Q is defined over the 3-letter alphabet {0,1,2}. $Q_{2n}(a_1, a_2, \ldots, a_n, b_1, b_2, \ldots, b_n) = 1$ iff the sequence obtained from $a_1a_2 \ldots a_n$ by omitting all ocurrences of 2 coincides with the one obtained in the same way from $b_1b_2 \ldots b_n$. It is shown that for any $1 \le s \le 1/4 \log n$, if the width of an input oblivious decision graph computing Q_{2n} is at most $2^{n/2^s}$, then its length is $\Omega(n.s)$.

Krause considered in [3] oblivious read-k-times-only branching programs with the additional restriction that the variables occur only blockwise and in each block in the same order (k^*-programs). He gave examples of functions which do not belong to $\mathcal{P}_{BP1}$ but which can be computed by the help of polynomially bounded 2^*-branching programs. Further, an exponential lower bound for k^*-programs ($k \in \mathbb{N}$ arbitrarily fixed) was proved for the decision whether a given subset of $\mathbb{F}_n \times \mathbb{F}_n$ contains the graph of a polynomial over $\mathbb{F}_n$ of degree less than n/2, where n is assumed to be a prime number.

For $\log n \le s(n) \le n$, s nondecreasing, let $\mathbf{TISP}_{\mathcal{O}}(n, s(n))$ be the class of all formal languages over a finite alphabet which can be decided by a sequence of oblivious decision graphs of linear bounded length, and of $2^{O(s(n))}$ bounded size.

Our investigations are motivated as follows.

(i) There is no superpolynomial lower bound known for read-k-times-only decision graphs, if $k \ge 2$. It is interesting whether this is possible when imposing further constraints.

(ii) Up to now there are essentially two types of models of restricted decision graphs for which superpolynomial lower bounds can be proved. These are read-once-only programs and input oblivious decision graphs of small length. How are these models related to each other ?

We consider problems belonging to $\mathbb{L}$. The results are the following.

(i) We prove exponential lower bounds for the graph accessibility problems GAP, and GAP1 (Theorem 3.5), and for the word problem of the free group (Theorem 3.6) of finite rank.

(ii) We prove the following results.

- $\mathcal{P}_{DG1}$ is not contained in $\mathbf{TISP}_{\mathcal{O}}(n, \log n)$. The sequence equality function Q_{2n} for which an exponential lower bound on input oblivious decision graphs was proved in [2] belongs to $\mathcal{P}_{DG1}$ (Proposition 2.6).
- $\mathbf{TISP}_{\mathcal{O}}(n, \log n)$ is not contained in $\mathcal{P}_{DG1}$. $(\text{HALFCLIQUE}_n)_{n \in \mathbb{N}}$ which belongs to $\mathbf{TISP}_{\mathcal{O}}(n, \log n)$ (Proposition 2.5), is not in $\mathcal{P}_{DG1}$ [10].
- The union of $\mathcal{P}_{DG1}$ and $\mathbf{TISP}_{\mathcal{O}}(n, \log n)$ is properly contained in $\mathcal{L}$.

The word problem of the free group for which there are exponential lower bounds for both models (see [4], and Theorem 3.6) belongs to $\mathbb{L}$. This result suggests that current techniques don't suffice to separate $\mathbb{L}$ from larger complexity classes.

2. Reducibility, and Upper Bounds

It is standard in complexity theory to introduce reducibility notions in order to compare the complexity of two given problems. As in [9] we say that a mapping $\pi_n : \{y_1, y_2, \ldots, y_m\} \longrightarrow \{x_1, \bar{x}_1, \ldots x_n, \bar{x}_n, 0, 1\}$ is a projection reduction from a set $A \subseteq \{0,1\}^n$ to a set $B \subseteq \{0,1\}^m$ iff $A(x_1, x_2, \ldots, x_n) = B(\pi_n(y_1), \pi_n(y_1), \ldots, \pi_n(y_m))$.

Equivalently, this means that $A = (\pi_n^*)^{-1}(B)$, where $\pi_n^* : \{0,1\}^n \longrightarrow \{0,1\}^m$ is the canonical map resulting from π_n.

$\left\{\pi_n : \{y_1, y_2, \ldots, y_{p(n)}\} \longrightarrow \{x_1, \bar{x}_1, x_2, \bar{x}_2, \ldots, x_n, \bar{x}_n, 0, 1\} \mid n \in \mathbb{N}\right\}$ is called a p-projection reduction from $L \subseteq \{0,1\}$ to $L' \subseteq \{0,1\}$, if for each $n \in \mathbb{N}$ π_n is a projection from $L^{(n)}$ to $L'^{(p(n))}$ and $p(n) = n^{O(1)}$.

In this case we say that L is p-projection reducible to L'. A sequence $\{\pi_n \mid n \in \mathbb{N}\}$ is called an ℓ-projection reduction iff $p(n) = O(n)$.

For practical reasons we generalize the notion of a projection reduction to languages over an arbitrary alphabet. Let Σ and Γ be finite alphabets, and let $A \subseteq \Sigma^n$ and $B \subseteq \Gamma^m$ be two sets. A projection reduction π_n from A to B is defined to be $\pi_n = \{\pi_{n,0}, \pi_{n,i} | i \in \mathcal{I}\}$ so that the following conditions are fulfilled.

(i) $\pi_{n,0}$ is a map from $\{1,2,\ldots,m\} \longrightarrow \{1,2,\ldots,n\} \sqcup \Gamma$, where for any two sets the binary operation symbol "$\sqcup$" means the disjoint union.

(ii) The index set $\mathcal{I}$ is defined to be $\pi^{-1}_{n,0}(\{1,2, \ldots ,n\})$.

(iii) The local functions $\pi_{n,i}$, $i \in \mathcal{I}$, map Σ to Γ.

(iv) $A = (\pi^*_n)^{-1}(B)$, where $\pi^*_n : \Sigma^n \longrightarrow \Gamma^m$ is defined as follows.

$$(\pi^*_n(w))(i) := \begin{cases} \pi_{n,i}(w(\pi_n(i))), & \text{if } i \in \mathcal{I} \\ \pi_{n,0}(w(i)) & \text{otherwise.} \end{cases}$$

We agree that if w is a word w(i) or w_i denotes the i-th letter of w.

We remark that the reducibility notion for arbitrary alphabets coincides with the usual one for Boolean functions, if we restrict ourselves to $\Sigma = \Gamma = \{0,1\}$. Now we can define p-projection reductions and ℓ-projection reductions for languages over arbitrary alphabets in the straightforward way.

One natural way to get ℓ-projection reductions is to consider reductions via a balanced homomorphism. A homomorphism $\phi : \Sigma^* \longrightarrow \Gamma^*$, where Σ and Γ are finite alphabets, is called *balanced* iff for all $\sigma, \sigma' \in \Sigma$ we have $|\phi(\sigma)| = |\phi(\sigma')| =: |\phi|$.

A language $L \subseteq \Sigma^*$ is called *bh-reducible* to a language $L' \subseteq \Gamma^*$ if there is a balanced homomorphism $\phi : \Sigma^* \longrightarrow \Gamma^*$ such that $L = \phi^{-1}(L')$. We observe that $L^{(n)} = \phi^{-1}(L'^{(|\phi|.n)})$. It is not hard to prove

Lemma 2.1. If $L \subseteq \Sigma^*$ is bh-reducible to $L' \subseteq \Gamma^*$, then L is ℓ-projection reducible to L'. ∎

We consider the *graph accessibility problems* $GAP=(GAP_{N(N-1)})_{N \in \mathbb{N}}$, $GAP1 =(GAP1_{N(N-1)})_{N \in \mathbb{N}}$ and $GAPMON1=(GAPMON1_{N(N-1)})_{n \in \mathbb{N}}$ for directed graphs. Hereby, we identify directed graphs $G=(V,E)$ with adjacency matrices.

For each directed graph $G=(V,E)$, where $V=\{v_1,\ldots,v_N\}$, let

$GAP_{N(N-1)}(G) = 1$ iff there is a directed path from v_1 to v_N in G,

$GAP1_{N(N-1)}(G) = 1$ iff $GAP_{N(N-1)}(G) = 1$ and $\text{outdegree}(G) = 1$, and

$GAPMON1_{N(N-1)}(G) = 1$ iff $GAP1_{N(N-1)}(G)=1$ and $(v_i,v_j)\in E$ implies $i<j$.

In [6] the following theorem is proved.

Theorem 2.2. GAP1 as well as GAPMON1 are complete for $\mathcal{L}$, and GAP is complete for $\mathcal{NL}$ with respect to p-projection reductions. ∎

The proof of the following proposition is easy and that's why omitted.

Proposition 2.3. Let $s:\mathbb{N} \longrightarrow \mathbb{N}$ be a nondecreasing function such that $\log n \le s(n) \le n$, and for each $\varepsilon > 0$ there is a $\delta > 0$ such that $s(\varepsilon.n) \le \delta.s(n)$. Then the class $\mathbf{TISP}_{\mathcal{O}}(n,s(n))$ is closed with respect to ℓ-projection reductions. ∎

By giving an explicite algorithm it is possible to show the following.

Proposition 2.4. GAPMON1 belongs to $\mathbf{TISP}_{\mathcal{O}}(n, \log n)$. ∎

In order to separate $\mathbf{TISP}_{o}(n,\log n)$ from $\mathcal{P}_{BP1}$ let us define the language $HALFCLIQUE=(HALFCLIQUE_n)_{n\in\mathbb{N}}$, where $HALFCLIQUE_n:\{0,1\}^n \longrightarrow \{0,1\}$ is defined if $n = \binom{N}{2}$ for some $N\in\mathbb{N}$.

For each undirected graph G over N nodes let $HALFCLIQUE_n(G)=1$ iff G consists of a $(N/2)$-clique and $N/2$ isolated vertices.
In [10] it is shown that HALFCLIQUE does not belong to $\mathcal{P}_{BP1}$. But it is possible to define an algorithm computing HALFCLIQUE which provides

Proposition 2.5. $HALFCLIQUE_n$ belongs to $\mathbf{TISP}_{\mathcal{O}}(n,\log n)$.

However, the following proposition yields that polynomial-size read-once-only decision graphs cannot be simulated by oblivious decision graphs of linear length and polynomial width.

Proposition 2.6. The sequence equality function Q_{2n} belongs to $\mathcal{P}_{DG1}$ but not to $\mathbf{TISP}_{\mathcal{O}}(n,o(n))$.

The **Proof** of "Q_{2n} does not belong to $\mathbf{TISP}_{o}(n,\log n)$" is similar to that of lemma 3.7 ∎

Now we turn to word problems of free groups. For $A=\{a_1,a_2,\ldots,a_m\}$, $m\ge2$, let $\langle A\rangle$ denote the *free group on* A. The integer m is called *the rank* of the group. Then each element of $\langle A\rangle$ can be represented as a word over the alphabet $\underline{A} := A \cup \{a_1^{-1}, a_2^{-1}, \dots, a_m^{-1}\}$.

In general, a group G is called *finitely presented* iff there are a finite set $A = \{a_1, a_2, \dots, a_m\}$ and a finite set of reduced words $R = \{r_1, r_2, \dots, r_s\}$ from $\underline{A}^*$ so that $G \cong \langle A\rangle/cl(R)$, where cl(R) denotes the smallest normal subgroup containing R. Then we write $G = \langle A; R\rangle$.

The *word problem* of $\langle a_1, a_2, \dots, a_m; R\rangle$ is the following language. $W(\langle a_1, a_2, \dots, a_m; R\rangle) := \{w \in \underline{A}^* \mid w = 1 \text{ in } G\}$. Let us denote $W^{(n)}(\langle a_1, a_2, \dots, a_m; R\rangle) := W(\langle a_1, a_2, \dots, a_m; R\rangle) \cap \underline{A}^n$.

Let us have a look at the presentation $\langle a_1, a_2, \dots, a_m, \nabla; \nabla\rangle$. It is trivial that this is a presentation of the free group of rank m.

Lemma 2.7. The word problem $W(\langle a_1, a_2, \dots, a_m, \nabla; \nabla\rangle)$ is bh-reducible to the word problem $W(\langle a_1, a_2, \dots, a_m\rangle)$.

Proof. We define $\phi\colon \{a_1, a_2, \dots, a_m, \nabla\}^* \longrightarrow \{a_1, a_2, \dots, a_m\}^*$ by $\phi(a_i) = a_i a_i$, $\phi(a_i^{-1}) = a_i^{-1} a_i^{-1}$, $i = 1, 2, \dots, m$, and $\phi(\nabla) = \phi(\nabla^{-1}) = a_1 a_1^{-1}$. This ϕ defines a group monomorphism. Hence claim (i) is proved. ∎

3. Lower Bounds

Put $[n] = \{1, 2, \dots, n\}$ and let $\mathfrak{y} = (y_1, y_2, \dots, y_r)$ be a sequence of elements of [n]. Let Z_1 and Z_2 be two disjoint subsets of [n].
We say that a $\{Z_1, Z_2\}$-alternation occurs at index i in the sequence $\mathfrak{y}$ iff the following conditions are satisfied.

(i) y_i belongs to $Z_1 \cup Z_2$. (ii) There is a $k > i$ such that $y_k \in Z_1 \cup Z_2$.

(iii) $y_i \in Z_1$ iff $y_{\kappa(i)} \in Z_2$, where $\kappa(i) = \min\{k \mid k > i,\ y_k \in Z_1 \cup Z_2\}$.

The number of indices i at which there occurs a $\{Z_1, Z_2\}$-alternation is called the *alternation length* of $\mathfrak{y}$ *with respect to* $\{Z_1, Z_2\}$.

The following lemma is a straightforward consequence of a Ramsey-theoretic lemma due to Alon and Maass (see [2]).

Lemma 3.1. Assume that in $\mathfrak{y}$ each $a \in [n]$ appears at most k times. Then for any partition (X_1, X_2) of [n] into two disjoint sets there are subsets $Y_i \subseteq X_i$, $i = 1, 2$, so that $|Y_i| \geq |X_i| \cdot 2^{-(2k-1)}$, $i = 1, 2$, and, the alternation length of $\mathfrak{y}$ with respect to $\{Y_1, Y_2\}$ is at most 2k. ∎

We associate with each input oblivious decision graph DG of length λ a sequence $\mathfrak{y} = (y_1, y_2, \dots, y_\lambda)$ of indices, where y_i is that number the nodes of level i are labelled with. $\mathfrak{y}$ is called *index sequence* of DG.

We need one technical notion. Let c_1,c_2: $[n]\xrightarrow{\supseteq}\Sigma$ be partial assignments so that c_1 and c_2 coincide on $dom(c_1)\cap dom(c_2)$. Define the union $c_1 \vee c_2$ of domain $dom(c_1) \cup dom(c_2)$ as $(c_1 \vee c_2)(i) = c_1(i)$ if $i \in dom(c_1)$ and $(c_1 \vee c_2)(i) = c_2(i)$ if $i \in dom(c_2)$.

Definition. Let Z_1 and Z_2 be two disjoint subsets of $[n]$, let s_0 be a partial assignment, $dom(s_0)=[n]-Z_1\cup Z_2$, and let $S_i \subseteq \{c \mid dom(c)=Z_i\}$, $i=1,2$. Assume φ: $S_1\longrightarrow S_2$ to be a bijection. Further let $L^{(n)}\subseteq \Sigma^n$. The set $S=\{s_0 \vee s_1 \vee \varphi(s_1) \mid s_1\in S_1\}$ is defined to be a *sheaf* in $L^{(n)}$, if for all $s_1\in S_1$ and $s_2\in S_2$ it holds that $s_0 \vee s_1 \vee s_2$ belongs to $L^{(n)}$ if and only if $s_2 = \varphi(s_1)$.

$\{Z_1,Z_2\}$ is called the *support* of the sheaf S . The number $\log_2|S|$ is called the *thickness* of the sheaf S.

In this work we use sheafs in the context of the following lemma. Informally speaking it claims that if palindroms are reducible to a language, then that language contains a sheaf.

Lemma 3.2. Let Z_1 and Z_2 be disjoint subsets of $[n]$, and let $L^{(n)}\subseteq \Sigma^n$. Assume that we are given a projection reduction $\pi_n=\{\pi_{n,0},\pi_{n,i}\ (i\in\mathcal{I})\}$, where $\pi_{n,0}$: $\{1,2,..,n\}\longrightarrow \{1,2,...,2\tau\} \cup \Sigma$, $\mathcal{I}=\pi^{-1}_{n,0}(\{1,2,...,2\tau\})$, and $\pi_{n,i}$:$\{0,1\}\longrightarrow\Sigma$, from $PAL^{(2\tau)}=\left\{ww^R \mid w\in\{0,1\}^\tau\right\}$ to $L^{(n)}$ so that $\pi^{-1}(\{1,2,...,\tau\})=Z_1$ and $\pi^{-1}(\{\tau+1,..,2\tau\})=Z_2$, or vice versa. Then $\pi^*_n(PAL^{(2\tau)})$ is a sheaf in $L^{(n)}$ of thickness τ and support $\{Z_1,Z_2\}$. ■

The next lemma supplies a lower bound for oblivious decision graphs in terms of sheafs. Similar methods were developed in [1], [2], and [3].

Lemma 3.3. Let T_n be an input oblivious decision graph of width ω and length λ deciding a set $L^{(n)}$. Let α be the alternation length of T_n with respect to $\{Z_1,Z_2\}$, where Z_1, and Z_2 are disjoint subsets of $[n]$.

If S is sheaf in $L^{(n)}$ of thickness τ and support $\{Z_1,Z_2\}$, then

$$\omega \geq 2^{\tau/\alpha}. \qquad ■$$

The following theorem claims that the complexity of a language is high if it contains a sheaf in a rather general position.

Theorem 3.4. Let $s : \mathbb{N} \longrightarrow \mathbb{N}$ be a nondecreasing function, $\log n \leq s(n) \leq n$, and let $L \subseteq \Sigma^*$. Assume that for all ε, $0<\varepsilon< 1/2$, there is a $\delta > 0$ so that for all $n \in \mathbb{N}$ the following condition is fulfilled.

There is a partition $I_1 \cup I_2 = [n]$ such that $|I_j| \geq \lfloor n/2 \rfloor$, $j=1,2$, and for any two subsets $Y_1 \subseteq I_1$, $Y_2 \subseteq I_2$, $|Y_i| \geq \varepsilon \cdot n$, there is a sheaf with support $\{Y_1,Y_2\}$ in $L^{(n)}$ of thickness not less than $\delta \cdot s(n)$.

Then $L \notin \mathbf{TISP}_{\mathcal{O}}(n,o(s(n)))$.

Proof. Let $(T_n)_{n \in \mathbb{N}}$ be a sequence of input oblivious decision graphs of length $c \cdot n$ computing $L^{(n)}$. Using Lemma 3.1 and lemma 3.3 combinatoric arguments show that $\log_2(\mathrm{SIZE}(T_n)) \geq (\delta/8c) \cdot s(n)$. ∎

This technique enables us to prove the following two theorems.

Theorem 3.5. Both GAP1 and GAP do not belong to $\mathbf{TISP}_{\mathcal{O}}(n,o(\sqrt{n}))$. ∎

Theorem 3.6. For each $m \geq 2$ the word problem of the free group over $A=\{a_1,..,a_m\}$ does not belong to $\mathbf{TISP}_{\mathcal{O}}(n.o(n))$.

Theorem 3.6 is proved by lemma 2.7 and the following.

Lemma 3.7. The word problem of the presentation $\langle a_1,a_2,..,a_m,\nabla;\ \nabla \rangle$ does not belong to $\mathrm{TISP}_{\mathcal{O}}(n,o(n))$.

Proof. We have to apply Theorem 3.4. Let Y_1, Y_2 be two disjoint subsets of $[n]$, $|Y_i| \geq \varepsilon.n$, $i=1,2$. Let $Z_i \subseteq Y_i$ be maximal subsets with $i \in Z_1$ and $j \in Z_2$ implies, w.l.o.g., $i<j$. We know that $|Z_i| \geq (\varepsilon/2).n$ for $i=1,2$. Assume that $|Z_1|=|Z_2|$. We show that there is a sheaf in $W^n(\langle a_1, a_2,..,a_m,\nabla;\nabla\rangle)$ of thickness $|Z_1|=|Z_2| \geq (\varepsilon/2)\cdot n = n'$ and support $\{Z_1,Z_2\}$.

We put all input variables x_j, $j \notin Z_1 \cup Z_2$, to be ∇. Since ∇ equals 1 in the group we have again a word problem of shorter length n'. We consider words of length n' over the alphabet $\{a_1,a_2, \ldots ,a_m,a \ ,a_1^{-1}, \ldots a_m^{-1}\}$ of the type $uv^{-1} =: w(u,v)$, where $u,v \in \{a_1,a_2, \ldots ,a_m\}^{n'/2}$. Obviously, $w(u,v)=1$ iff $u \equiv v$. Now it is easy to define a projection reduction from the palindroms to the word problem in the required way. ∎

Let us finish the paper with the following observation.

Corollary 3.8. $\mathcal{P}_{DG1} \cup \mathbf{TISP}_{\mathcal{O}}(n,\log n) \subset \mathcal{L}$.

Proof. The result follows from Theorem 3.8 and the well-known theorem due to Lipton and Zalcstein [5] which states that the word problem of the free group is solvable in logspace. ∎

References.

[1] Ajtai, M., L. Babai, P. Hajnal, J. Komolos,.P. Pudlak, V. Roedel, E.Semeredi, and G.Turan , Two lower bounds for branching programs. Proc. 18-th ACM STOC, 1986, 30 - 39.

[2] Alon, N., W. Maass, Meanders, Ramsey theory and lower bounds for branching programs, Proc. 27-th IEEE - FOCS, 1986, 410 -417.

[3] Krause, M., Lower bounds for depth-restricted branching programs, to appear in Information and Computation.

[4] Kriegel, K., S. Waack, Lower bounds on the complexity of real-time branching programs, Informatique theorique et Applications/ Theoretical Informatics & Applications, Vol.22,No.4,1988, pp. 447-459.

[5] Lipton R. J., Y. Zalcstein, Word problems solvable in logspace, J. of the ACM, Vol. 24, No. 3, 1977, pp. 522 - 526.

[6] Meinel, Ch., The nonuniform complexity classes $\mathcal{NC}^1$, $\mathcal{L}$, and $\mathcal{NL}$, J.Inf.Process.Cybern. EIK, Vol. 23, No. 10/11, 1987, pp. 545 -558.

[7] Nechiporuk, E. I., On a Boolean function, Dokl. Akad. Nauk USSR, Vol. 169, No. 4, 1966, pp. 765 - 766.

[8] Wegener, I., On the complexity of branching programs and decision trees for clique functions, J.of the ACM, Vol.35,2,1988, 461-471.

[9] Skyum, L.G. Valiant, A complexity theory based on Boolean algebra, Proc. Proc. 22-th FOCS, 1981, pp. 244 - 253.

[10] Zak, S., An exponential lower bound for one-time-only branching programs, Proc.MFCS'84, LNCS, Vol. 176, 1984, pp. 562 - 566.

SOME TIME-SPACE BOUNDS FOR ONE-TAPE DETERMINISTIC TURING MACHINES

Maciej Liśkiewicz and Krzysztof Loryś

Institute of Computer Science, Wrocław Uniwersity,
Przesmyckiego 20, 51-151 Wrocław, Poland

Abstract.

Every single-tape deterministic Turing machine (DTM) of time complexity $T(n) \geqslant n^2$ can be simulated by a single-tape DTM in space $T^{1/2}(n)$. It is shown that the time of the simulation can be bounded by $T^{3/2}(n)$. Similar results are shown for offline machines and for machines with multidimensional tape.

1. Introduction.

One of the most significant results relating time and space complexity says that space $S(n)=T(n)/\log T(n)$ is sufficient for simulation of multitape deterministic Turing machines (DTMs) of time complexity $T(n)$ ([2],[1]). In case of one-tape Turing machines the function $S(n)$ can be considerably reduced. Namely, in [7] Paterson proved that the space $S(n)=T^{1/2}(n)$ is sufficient when single-tape (even nondeterministic) machines are simulated. For offline machines the adequate space limit equals to $(T(n)\cdot \log n)^{1/2}$ when the machines have 1-dimensional work tape ([7]) and to $(T(n)\cdot \log T(n))^{r/(r+1)}$ when their work tape is r-dimensional ([6]).

Recall that single-tape TM has a single two-way read-write tape which initially contains the input word. (We assume that the tape is infinite to the right only.) In an offline TM the input word is placed on an additional read-only input tape.

Of course the space-bounded simulating machines may work much longer than the simulated ones. For example, it is not known if $T(n)/\log T(n)$--space bounded multitape DTMs simulating $T(n)$-time bounded ones can operate in a time shorter than exponential in $T(n)/\log T(n)$. The situation is quite different in the case of one-tape machines. Let us summarize the best known results.

(1) Any language accepted by a single-tape DTM in time $T(n)\geq n^2$ can be accepted by a single-tape DTM in space $T^{1/2}(n)$ and time $T^2(n)$ ([3], [4]).

(2) Any language accepted by an offline DTM in time $T(n)\geq n$ can be accepted by an offline DTM in space $(T(n)\cdot log\ n)^{1/2}$ and time $T^{3/2}(n)\cdot(T^{1/2}(n)+n/(log\ n)^{1/2})$ ([3], [4]).

(3) Any language accepted by an offline DTM with an r-dimensional storage tape in time $T(n)\geq n$ can be accepted by an offline DTM with an r-dimensional storage tape in space $(T(n)\cdot log\ T(n))^{r/(r+1)}$ and time $T^2(n)\cdot G_r(n)$, where $G_r=n/(T(n)\cdot log\ T(n))^{1/(r+1)}+log\ T(n)$ ([3]).

(4) Any language accepted by a single-tape NTM in time $T(n)\geq n^2$ can be accepted strongly by a single-tape NTM in space $T^{1/2}(n)$ and time $T(n)$ ([5]).

(5) Any language accepted by an offline NTM in time $T(n)\geq n$ can be accepted strongly by an offline NTM in space $(T(n)\cdot log\ n)^{1/2}$ and time $T(n)+n\cdot(T(n)/log\ n)^{1/2}$ ([5]).

In this paper we sharpen the time bounds in (1),(2) and (3). Our new time limits on the simulations are as follows:

(1) for single-tape DTM: $T^{3/2}(n)$,

(2) for offline DTM: $T(n)\cdot((T(n)\cdot log\ n)^{1/2}+n)$,

(3) for offline DTM with an r-dimensional storage tape: $T^{2-r'}(n)\cdot G_r(n)/log^{r'}T(n)$, where $r'=1/(r+1)$.

2. One-tape deterministic TMs.

Ibarra and Moran [3] constructed an algorithm for simulation of $T(n)$-time bounded single-tape DTMs by $T^{1/2}(n)$-space bounded ones working in time $T^2(n)$. We modify this algorithm to reduce the time of simulation to $T^{3/2}(n)$. Then we adapt the algorithm to offline DTMs.

Definition 1. A sequence $B=(B_0,B_1,B_2,\ldots)$ is a *partition* of a semiinfinite tape F if for each i, B_i consists of a finite number of consecutive cells of F, B_{i+1} is directly to the right of B_i, and each cell of F belongs to a unique B_i. The B_i's are called *tape segments*. If B_0 consists of p cells, and for each $i\geq 1$, B_i consists of s cells, then B is called *(p,s) partition*.

Definition 2. Let M be a single-tape TM, let w be a word in the input alphabet of M and let $B=(B_0,B_1,B_2,\dots)$ be a partition of M's tape. *Crossings* of M are pairs (d,q) where $d\in\{-1,1\}$ and q is a state of M. For each i, the *crossing sequence* of M on input w between B_{i-1} and B_i after t steps, denoted by $CS(M,w,i,t)$, is a finite sequence of crossings defined as follows:

(a) $CS(M,w,i,0)$ is the empty sequence,

(b) $CS(M,w,i,t)=CS(M,w,i,t-1)$ if M's head does not cross the boundary between B_{i-1} and B_i during its tth move on input w,

(c) $CS(M,w,i,t)=CS(M,w,i,t-1)(d,q)$ if M in the state q moves its head across the boundary between B_{i-1} and B_i in the tth move; d is -1 if M's head moved left and is 1 otherwise.

Definition 3. Let M, w, B be as in Definition 2. The *history* of M on input w with respect to B after t steps, denoted by $HIST(M,w,B,t)$, is a finite sequence of crossings and the symbol # defined as follows:

$$HIST(M,w,B,t)=CS(M,w,1,t)\#CS(M,w,2,t)\#\dots\#CS(M,w,j_t,t)$$

where j_t is the number of the last nonempty crossing sequence. By $|HIST(M,w,B,t)|$ we denote the number of crossings in $HIST(M,w,B,t)$.

Lemma 1. Let M be a single-tape TM of time complexity $T(n)$, let $w\in\Sigma^*$ be a word of length n and let s be a natural number. Then there is $p\leqslant s$ such that for the $\langle p,s\rangle$ partition B we have

$$|HIST(m,w,B,T(n))|\leqslant T(n)/s.$$

Proof. Omitted (see [3]).

Definition 4. $S(n)$ is *fully space constructible* in time $T(n)$ by a single-tape DTM if there is a single-tape DTM, which on any input of length n halts within time $T(n)$ and uses exactly $S(n)$ cells.

Theorem 1. Let A be accepted by a single-tape DTM M in time $T(n)$, where $T(n)\geqslant n^2$. Assume that $S(n)=T^{1/2}(n)$ is fully space constructible in time $T^{3/2}(n)$ by a single-tape DTM. Then A can be accepted by a single-tape DTM M_1 in space $S(n)$ and time $T^{3/2}(n)$.

Proof. We construct a modification of the algorithm given in [3].

Let B_p be the $\langle p,S(n)\rangle$ partition of M's tape, for $p=1,2,\dots,S(n)$. Let w be a word from Σ^*. By Lemma 1, for at least one p_o, $|HIST(M,w,B_{p_o},T(n))|\leqslant S(n)$. We use the fact that knowing $HIST(M,w,B_{p_o},t)$, M_1 can reconstruct the contents of any segment after t steps of M. During the entire simulation M_1 records a contents of a single block of the tape, consisting of three segments, and simulates M on this block. Every time when M crosses a boundary between segments, M_1 actualizes the history stored on a separate track of the tape. When M attempts to move its head out of the block, M_1 reconstructs the contents of the new block and resumes simulation on this block.

But M_1 does not know the value of p_o. Therefore it simulates M for a fixed p, starting from $p=1$, in hope that this p is a good one. However, when it turns out that $HIST(M,w,B_p,t)$ already contains $S(n)+1$ elements after simulation of t steps, then M_1 seeks the least $p_1 > p$ such that $|HIST(M,w,B_{p_1},t)| \leq S(n)$ and resumes simulation from the step t. This is the main difference between our algorithm and that from [3], since the algorithm of [3] for each new p simulates M from the very beginning.

In the algorithm $HIST(p,w)$ denotes $HIST(M,w,B_p,t)$, where t is the current number of M's steps.

Algorithm.

Step 1. Construct $S(n)=T^{1/2}(n)$.

Step 2. $p:=1$; $HIST(p,w)$:= the empty sequence.

Step 3. (* reconstruction of a block *)
Using $HIST(p,w)$ recompute the contents of the M's tape block consisting of three segments of B_p: the segment in which the M's head was in step t and the segments adjacent to it. (In the case when the head is in B_0, the block consists of two segments only: B_0 and B_1).

Step 4. (* simulation *)
Simulate M on this block from the step t until M reaches a final state or crosses a boundary of segments. If M has reached an accepting (rejecting) state then accept (reject, resp.), otherwise go to Step 5.

Step 5. If $|HIST(p,w)|=S(n)$, then go to Step 6, otherwise insert a new element to $HIST(p,w)$. If M has crossed one of the boundaries of the middle segment of the block then go to Step 4, otherwise (i.e. when M has crossed a block's boundary) go to Step 3.

Step 6. $p_o:=p$.

Step 7. $p:=p+1$.

Step 8. (* creation of a new history *)
Using $HIST(p_o,w)$ simulate M in succession on all segments which were visited by M before step t and create $HIST(p,w)$. If $HIST(p,w)$ contains more than $S(n)$ elements to the time t, then go to Step 7, otherwise go to Step 3.

End of algorithm.

It is easy to see that M_1 uses $O(S(n))$ cells and recognizes the same language as M, so we focus on the time analysis of the algorithm.

Let us note that a single insertion of an element into $HIST(p,w)$ may be done in time $O(T^{1/2}(n))$. This time is also sufficient for finding a single crossing across a boundary of a given segment (in steps 3 and 8).

We show that the total time of performing each separate step does not exceed $O(T^{3/2}(n))$.

For steps 1,2,6 and 7 this is obvious.

Step 3 seeks $O(T^{1/2}(n))$ elements in $HIST(p,w)$, which costs $O(T(n))$, and needs $O(T(n))$ operations of simulation of M. This step is performed once after Step 2, $O(T^{1/2}(n))$ times after Step 8 and after Step 5 whenever M's head crosses a block boundary. Note that whenever M_1 begins the simulation of M on a new block, M's head is at least $T^{1/2}(n)$ cells from the boundaries of the block. So after Step 5, Step 3 may be performed only $O(T^{1/2}(n))$ times. Thus the total cost of Step 3 is $O(T^{3/2}(n))$.

In Step 4 M_1 simulates M step by step, so the total cost of this step is $O(T(n))$.

The total cost of Step 5 is $O(T^{3/2}(n))$. This follows from the fact that $\sum_{1\leq p\leq T^{1/2}(n)} |HIST(p,w)| \leq T(n)$.

The cost of Step 8 consists of three components bounded by $O(T(n))$, namely: the cost of simulation, the cost of seeking elements in $HIST(p_o,w)$ and the cost of insertions into $HIST(p,w)$. Since this step can be performed $O(T^{1/2}(n))$ times only, its total cost is $O(T^{3/2}(n))$.

Using the well-known linear speed-up technic we obtain the time complexity $T^{3/2}(n)$.

□

The algorithm constructed in the proof of Theorem 1 can also be used to simulate offline DTMs. In this case crossings must contain, apart from a state and a move direction, information about the position of the input head, what causes that the algorithm needs more than $T^{1/2}(n)$ space to store a history. On the other hand, $S(n)=(T(n)\cdot \log n)^{1/2}$ space is sufficient, since by Lemma 1 for each input w of length n there is a $\langle p,S(n)\rangle$ partition B such that $|HIST(m,w,B,T(n))| \leq (T(n)/\log n)^{1/2}$. Let us note also that a single insertion of one element into history, organized as previously, can cost as much as $T^{1/2}(n)\cdot \log^{3/2} n$, which together with the fact that even $T(n)$ insertions may be needed implies that the algorithm would run for at least $(T(n)\cdot \log n)^{3/2}$ steps. However we can reduce this cost to $T(n)\cdot[(T(n)\cdot \log n)^{1/2}+n]$.

<u>Theorem 2.</u> Let A be accepted by an offline DTM M in time $T(n)$, where $T(n) \geq n$. Let $S(n)=(T(n)\cdot \log n)^{1/2}$ be fully space constructible by an offline DTM in time $T_1(n)=T(n)\cdot[(T(n)\cdot \log n)^{1/2}+n]$. Then A can be accepted by an offline DTM M_1 in space $S(n)$ and time $O(T_1(n))$.

<u>Proof.</u> The machine M_1 performs the algorithm given in the proof of Theorem 1. However now $HIST(p,w)$ has the following form after the simulation of t steps of M :

(*) $CS(p,w,1,t_o),\ldots,CS(p,w,j,t_o),k,(d_1,q_1,l_1),\ldots,(d_r,q_r,l_r)$,

where

- t_o is the number of M's steps simulated before the last performance of Step 8 by M_1,
- k is the number of the work tape segment, visited by M's head in the step t_o,
- (d_i,q_i,l_i) are consecutive crossings between the segments, recorded between the t_o^{th} and t^{th} step, (d_i-direction of move, q_i-state of M, l_i-input head position).

Such an organization of $HIST(p,w)$ enables M_1 to insert one crossing into it in $O((T(n)\cdot \log n)^{1/2}+n)$ steps, so the total cost of all insertions is $O(T_1(n))$.

In order to evaluate the cost of Step 3 we describe it in more detail. Let $HIST(p,w)$ be stored in the form (*). M_1 does the following:

(i) evaluates $m=k+\sum_{i=1}^{r} d_i$ (i.e. m is the number of the segment currently visited by M's head),

(ii) marks in $HIST(p,w)$ all crossings through the left boundary of the $(m-1)^{st}$ segment and the right boundary of the $(m+1)^{st}$ one,

(iii) simulates M; whenever M attempts to cross the block boundaries, M_1 finds a succeeding crossing marked at (ii), places the input head at the position stored in this crossing and resumes the simulation.

It is easy to check, that both (i) and (ii) may be performed in time $O(S^{1/2}(n)\cdot \log T(n))$, and (iii) in time $O(T(n)+n\cdot(T(n)/\log n)^{1/2})$. Since Step 3 is executed at most $O((T(n)\cdot \log n)^{1/2})$ times, its total cost is $O(T_1(n))$.

In order to achieve the stated time complexity, we have to improve the way $HIST(p,w)$ is used in Step 8, in particular the method of marking the crossings at the beginning of each segment simulation.

At the beginning of Step 8, M_1 evaluates m as in Step 3, but while doing this divides the part of the tape occupied by $(d_1,q_1,l_1),\ldots,(d_r,q_r,l_r)$ into sectors $t_1,t_2,\ldots,t_s$ of length not greater than $2\lceil log\ T(n)\rceil$. Within each sector t_j $(j=1,\ldots,s)$, M_1 writes down in binary the number $m_j=k+\sum_{i=1}^{u_j} d_i$, where u_j is the number of the last crossing stored in t_j (see Fig.1).

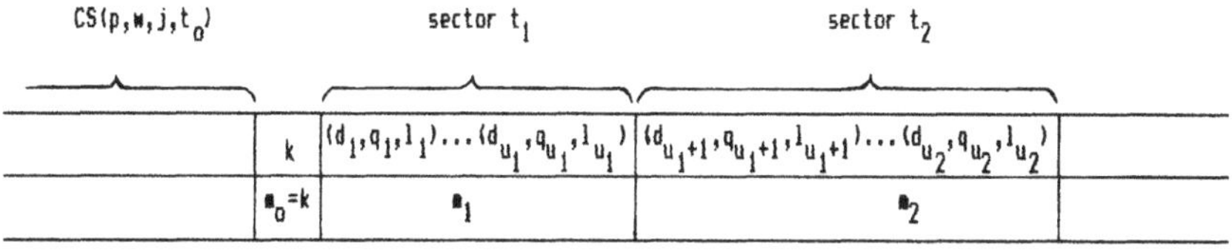

Fig. 1

It is easy to see that this can be done in time $O(S(n)\cdot log\ T(n))$.

Let us note that only crossings across the boundaries of the segments of the numbers from $\langle m_{j-1}-v_j, m_{j-1}+v_j\rangle$ may be stored in t_j, where v_j is the number of crossings in t_j. M_1 avails itself of this fact during the marking of crossings in $HIST(p_o,w)$. M_1 simulates M segment by segment and decreases the numbers $m_o,m_1,\ldots,m_{s-1}$ by one at the end of the simulation of each segment so that each m_j describes the distance of the currently simulated segment from the segment reached by M after the last crossing written in t_j. In order to mark the crossings across the succeeding boundary, M_1 looks through its tape searching for the sectors t_j such that m_{j-1} is from the interval $\langle -v_j,v_j\rangle$. This may be easily done since we can assume that the values v_j $(j=1,\ldots,s)$ are stored on the third track below m_{j-1}. The values v_j may be evaluated in time $O((T(n)\cdot log\ n)^{1/2} log\ log\ T(n))$. If m_{j-1} is from $\langle -v_j,v_j\rangle$, then M_1 marks in t_j all crossings (d_x,q_x,l_x), for which $z_x=m_{j-1}+\sum_{i=1+u_{j-1}}^{x} d_i=0$. Note that for each t_j, the values z_x are evaluated at most $log\ T(n)$ times and clearly they can be computed in such a way that each cell of t_j is visited at most $loglog\ T(n)$ times. So the total cost of computing the z_x's is $O((T(n)log\ n)^{1/2} logT(n)\cdot loglogT(n))=O(T(n))$. To the cost of the crossings marking one must add the cost of looking for "good" m_j's, which is $O(T(n))$. Thus $O(T(n))$ is the total cost of crossing marking.

It is easy to check that the cost of simulation of M on consecutive segments is $O(T(n)+n\cdot(T(n)/log\ n)^{1/2})$ and it is as well the cost of the whole Step 8, because the cost of all insertions into $HIST(p,w)$ has been already evaluated separately. Since Step 8 may be performed at most $O((T(n)\cdot \log n)^{1/2})$ times, its total cost is

$$O((T(n)\cdot \log n)^{1/2}[T(n)+n(T(n)/\log n)^{1/2}])=O(T_1(n)). \qquad \square$$

The same algorithm can be used if $T(n)$ is not fully space constructible. However in this case the time of simulation is a little longer.

<u>Corollary 1.</u>

(a) Let A be accepted by a single-tape DTM M in time $T(n)$, where $T(n)\geq n^2$. Then A can be accepted by a single-tape DTM M_1 in space $S(n)=T^{1/2}(n)$ and time $T^{3/2}(n)\cdot \log T(n)$.

(b) Let A be accepted by an offline DTM M in time $T(n)$, where $T(n)\geq n$. Then A can be accepted by an offline DTM M_1 in space $S(n)=(T(n)\cdot \log n)^{1/2}$ and time $T(n)[(T(n)\cdot \log n)^{1/2}+n]\cdot \log T(n)$.

<u>Proof.</u> M_1 does the simulation for $S(n)=1,2,4,8,\ldots,2^i,\ldots$. Note that the simulation succeeds for the smallest i such that $2^i\geq S(n)$.

$\square$

<u>3. Multidimensional TMs.</u>

The algorithm used in the proof of Theorem 1 can also be adapted for simulation of offline DTMs with multidimensional tape. We describe the modifications required in the case of 2-dimensional machines. We show that a 2-dimensional $T(n)$-time bounded DTM M can be simulated by a 2-dimensional DTM M_1 in space $S(n)=(T(n)\cdot \log T(n))^{2/3}$ and time $O(T_1(n))$, where

$$T_1(n)=T^{5/3}(n)\cdot [\log T(n)+n/(T(n)\cdot \log T(n))^{1/3}]/\log^{1/3}T(n).$$

As in [3] by the $\langle p,s\rangle$ partition ($1\leq p\leq s$) of the 2-dimensional tape we mean the partition in which each segment is a square which for some u and v contains all cells (x,y) satisfying $us-p\leq x<(u+1)s-p$, and $vs-p\leq y<(v+1)s-p$. The numbers u,v are called <u>*segment coordinates.*</u> Lemma 1 has the following analogue (see [3]):

<u>Lemma 2.</u> Let a 2-dimensional offline DTM M be of time complexity $T(n)$. Let w be in Σ^*, $|w|=n$ and let $s\leq T(n)$. Then for some $\langle p,s\rangle$ partition B the machine M crosses the boundaries between the segments of B at most $T(n)/s$ times working on w.

Let us note that the information indicating a position of heads, a state of M, and a direction of move when M crossed a boundary between two segments, is not sufficient to reconstruct the contents of segments. This is due to the fact that M_1 cannot set the crossings of history created in Step 8 in chronological order. Such an order was not nesessary in the case of one-dimensional tape, since then the segments are bounded by two boundaries and the head can return to a segment across this boundary only, across which it retired this segment last time. Now the situation is quite different, the head can return across any of four boundaries.

In order to secure the correctness of the simulation, each crossing additionally contains the number of the step in which this crossing was made. Thus the history is now a sequence of quintuples $\langle d_i, q_i, l_i, h_i, t_i\rangle$ ordered according t_i's, where q_i, l_i are as in the definition of the history for one-dimensional offline TM, $d_i \in \{\langle -1,0\rangle, \langle 1,0\rangle, \langle 0,-1\rangle, \langle 0,1\rangle\}$ indicates one of four directions, h_i indicates the exact position of the work head and t_i is the number of the step in which this crossing was made. Since a single crossing can be stored in $O(\log T(n))$ cells, by Lemma 2 for some $\langle p, (T(n)\cdot \log T(n))^{1/3}\rangle$ partition the whole history can be stored in a single segment.

In Step 3 M_1 reconstructs the contents of the block consisting of the segment currently visited by the head and the eight segments adjacent to it. At the beginning of this step M_1 marks each crossing in $HIST(p,w)$, after which M's head entered this block. It is easy to see that this marking can be made in $O(S(n)\cdot \log T(n))$ steps. Since the crossings are chronologically ordered (i.e. according to t_i) the simulation can be done in time

$$O(T(n)\cdot [\log T(n)+n/(T(n)\cdot \log T(n))^{1/3}]).$$

This results in $O(T_1(n))$ as the total cost of Step 3, because this step can be performed at most $T^{2/3}(n)/\log^{1/3}T(n)$ times.

In Step 4, M_1 counts the simulated steps. Since M_1 keeps the counter near the head, the total cost of Step 4 is $O(T(n)\cdot \log T(n))$.

More radical modifications are needed in Step 8. The main problems M_1 must solve are as follows:

(i) before the simulation on each segment, M_1 has to mark the proper crossings in a short time,

(ii) M_1 has to be able to designate all segments visited till then by M's head,

(iii) after the simulation on all these segments M_1 must restore the chronological order of the history.

Thus Step 8 assumes now the following shape:

8.1 *HIST(p,w)* := "the empty sequence".

8.2 Label all crossings in $HIST(p_o,w)$ as "unmarked".

8.3 For each crossing *e* in $HIST(p_o,w)$ compute the coordinates of the segment that *M*'s head entered when making *e*. Store these coordinates on the second track below e.

8.4 Simulate *M* on the segment of the coordinates (0,0). Whenever *M* crosses any boundary between segments in the new $\langle p,S^{1/2}(n)\rangle$ partition, insert a new element into *HIST(p,w)*. Note that the parameter t_i in the new crossings can be easily computed. If *HIST(p,w)* contains more than $T^{2/3}(n)/log^{1/3}T(n)$ elements (i.e. *S(n)* cells do not suffice to store *HIST(p,w)*), then go to Step 7. Label all crossings with the coordinates (0,0) as "marked". If all crossings in $HIST(p_o,w)$ are already "marked" then go to 8.6.

8.5 **For each** crossing $\langle d,q,l,h,t\rangle$ from $HIST(p_o,w)$ **do** the following:
   a) move the coordinate system in the direction *d*; to this end subtract *d* from the coordinates stored below each crossing,
   b) if crossings with the coordinates (0,0) are "unmarked" then **perform** 8.4.

8.6 Sort *HIST(p,w)* chronologically.
   HIST(p,w) can be treated as a table with $(T(n)\cdot log\ T(n))^{1/3}$ rows and $T^{1/3}(n)/log^{2/3}T(n)$ columns. Each element of this table occupies *log T(n)* cells. Note that the cost of a transposition of two adjacent elements is less if they are in the same column than if they are in the same row. This justifies the choice of the following algorithm of table sorting:
   (i) select the smallest element in each column and place them in the first row,
   (ii) select the smallest element in the first row and remove it from the table,
   (iii) if the table is nonempty then go to (i).

Now we evaluate the cost of Step 8. Of course 8.1 and 8.2 have no effect on this cost. Step 8.3 can be in an obvious manner performed in $O(S(n)\cdot log\ T(n))$ steps. The most expensive part of Step 8 are steps 8.4 and 8.5. We divide the cost of these steps into the cost of simulation and the cost of the remaining operations. The first component can be easily bounded by $O(T(n)\cdot log\ T(n))$, the second one can be bounded by $O(T^{4/3}(n)\cdot log^{1/3}T(n)+n\cdot T^{2/3}(n)/log^{1/3}T(n))$. Since Step 8.6 can be implemented for M_1 to consume no more than $O(T(n)\cdot log\ T(n))$ time, the cost of the whole Step 8 is $O(T^{4/3}(n)\cdot log^{1/3}T(n)+n\cdot T^{2/3}(n)/log^{1/3}T(n))$.

Since this step can be performed at most $(T(n)\cdot log\ T(n))^{1/3}$ times, its total cost is $O(T^{5/3}(n)\cdot log^{2/3}T(n)+n\cdot T(n))=O(T_1(n))$.

Clearly, the above considerations generalize to offline DTMs of any dimension.

Theorem 3. Let A be accepted by an offline DTM M with an r-dimensional storage tape in time $T(n)\geq n$. Let $r'=1/(r+1)$ and $T_1(n) = (T(n))^{2-r'}(log\ T(n)+n/(T(n)\cdot log\ T(n))^{r'})/log^{r'}T(n)$. Let $S(n)=(T(n)\cdot log\ T(n))^{r/(r+1)}$ be fully space constructible by a 1-dimensional offline DTM in time $T_1(n)$. Then A can be accepted by an offline DTM M_1 with an r-dimensional storage tape in space $S(n)$ and time $O(T_1(n))$.

Acknowledgments. The authors would like to thank Prof. Leszek Pacholski for his guidance and illuminating discussions.

References.

[1] L.M. Adleman and M.C. Loui, *Space-bounded simulation of multitape Turing machines*, Math. Systems Theory, 14 (1980), 215-222.

[2] J.E. Hopcroft, W.J. Paul and L.G. Valiant, *On time versus space*, J. Assoc. Comput. Mach., 24 (1977), 332-337.

[3] O.H. Ibarra and S. Moran, *Some time-space tradeoff results concerning single-tape and offline TM's*, SIAM J. Comput., 12 (1983), 388-394.

[4] S.Kurtz and W. Maass, *Some time, space and reversal tradeoffs*, Conf. Comput. Complexity Theory, March 21-25, 1983, Santa Barbara, pp. 30-40.

[5] K. Loryś and M. Liśkiewicz, *Two applications of Fürer's counter to one-tape nondeterministic TMs*, in Mathematical Foundations of Computer Science, Ed. by M.P. Chytil, L. Janiga and V. Koubek, LNCS 324, Springer-Verlag, Berlin 1988, pp.445-453.

[6] M.C. Loui, *A space bound for one-tape multidimensional Turing machines*, Theoret. Comput. Sci., 15 (1981), 311-320.

[7] M.S. Paterson, *Tape bounds for time-bounded Turing machines*, J. Comput. Systems Science., 6 (1972), 116-124.

RANK OF RATIONAL FINITELY GENERATED W-LANGUAGES

Igor Litovsky
Laboratoire LABRI, ENSERB, Université de Bordeaux I
351, cours de la libération, F-33405 Talence Cedex

Abstract. Let R be a rational language of (finite) words. We propose a way to decide whether R^w is equal to F^w for some finite set F. Next, in this case, we prove that one can compute the least integer, n, that can be the cardinal of some finite set F satisfying $R^w = F^w$. Furthermore one can construct all finite sets F satisfying both $R^w = F^w$ and card(F) = n.

INTRODUCTION.

Given a language of (finite) words, R, some ways to associate a language of infinite words with it are known, e.g. the infinite power [2], the limit [4] or the adherence [1]. Here we are interested in the infinite power of R (denoted by R^w) that is the set of infinite words such as $u_1 \ldots u_n \ldots$ where each u_n is a word of R. Then any language R defines one w-language R^w, but many other languages have the same infinite power as R, such languages are called generators of R^w. So (in accordance with the law of least effort) it is natural to seek the minimal generators of R^w with respect to the inclusion. In particular the case of finite generators (if any) is worth solving : which is the least integer that can be the cardinal of a generator of R^w ?

Our starting point is a paper by M. Latteux and E. Timmerman [4] where it is proved that, for any rational language R included in A^+, one can decide whether R^w is finitely generated, i.e. whether R^w is equal to F^w for some finite set F. First we propose a new way (theorem 8) to decide this question : we prove that there exists a constructible language, denoted by R^t, such that R^w is finitely generated if and only if R^t is a finite generator of R^w.

Then we assume that R^w is finitely generated, that is to say that we assume that R is a finite set (since otherwise we can remove R in R^t which is a finite set). Two questions come naturally :

- which is the least integer that can be the cardinal of a generator of R^w ?
- how to obtain one (indeed all) generator(s) of least cardinal, i.e. having the above integer for cardinal ?

As R^t is the generator containing all shorter "useful" words of generators of R^w (see definition of R^t), it seemed that it should have contained the generators of least cardinal : a non-elementary example (example 5) explodes that conjecture. Thus R^t does not allow us to solve the above questions. To answer them we use another way (theorem 10) : we prove that the length of any word of any generator of least cardinal is less than a calculable integer N. Hence any generator of least cardinal is included in $A \cup \ldots \cup A^N$, it follows that it is possible to construct all generators of least cardinal.

A. DEFINITIONS AND NOTATIONS.

Let A be a finite alphabet, we denote by A^*, resp. A^w, the set of all finite words, resp. infinite words (or w-words) over A. Subsets of A^*, resp. A^w, are called languages, resp. w-languages. We denote by ϵ the empty word and by A^+ the set $A^*-\epsilon$. The length of any word u of A^* is denoted by lg(u).

Let X be a language of A^* and Y be either a language or a w-language, XY stands for the set $\{xy \ / \ x \in X \text{ and } y \in Y\}$ and $X^{-1}Y$ stands for the set $\{v \in A^* \cup A^w \ / \ xv \in Y \text{ for some } x \in X\}$. Let X be a language, we denote by Root(X) the following set $(X-\epsilon)-(X-\epsilon)(X-\epsilon)$.

Let u be a word and v be either a word or a w-word, u is a prefix of v (denoted by $u < v$) if and only if $v \in u(A^* \cup A^w)$. We denote by Pref(v) the set of all prefixes of v and by Ppref(v) the set of all proper prefixes of v that is Pref(v)-v. For any subset of X, included in $A^* \cup A^w$, Pref(X) is the union of Pref(v) for each v in X. Let u,v be two words, u&v denotes the greatest common prefix to u and v.

Let $M = (A,Q,I,T,\delta)$ be an automaton where A is the alphabet, Q is a finite set, the set of states, $I \subset Q$ is the set of initial states, $T \subset Q$ is the set of terminal states, and δ is the transition relation (i.e. a mapping from AxQ into 2^Q). When δ is a mapping from QxA into Q and I is equal to $\{q_0\}$, M is a deterministic automaton and $\delta(a,q)$ is denoted by $q{*}a$. We denote by T(M) the set of accepted words. For the w-words we consider the "Büchi acceptance" [2] defined as follows. Let

w be a w-word, a run of M on w is a sequence of Q, $q_0 \dots q_n \dots$ such that $q_0 \in I$ and for $i \geq 0$, $q_{i+1} \in \delta(w(i), q_i)$; the run is called successful if some state of T occurs infinitely often in it; M accepts w if there is a successful run of M on w. We denote by $T^w(M)$ the set of w-words accepted by M.

Let L be a language of A^*, the limit of L denoted by lim(L) is equal to the w-language $\{w \in A^w \ / \ Pref(w) \cap L$ is an infinite set$\}$.

Let L be a language included in $A^* \cup A^w$, the adherence of L is denoted by Adh(L) and is the w-language $\{w \in A^w \ / \ Pref(w) \subset Pref(L)\}$.

Let R be a language of A^+, R^w stands for the w-power of R that is the w-language $\{u_1 u_2 \dots u_n \dots \ / \ u_n \in R\}$. We denote by $[R]_w$ the family $\{ G \subset A^+ \ / \ R^w = G^w \}$ and then G is called a generator of R^w. G is said to be a minimal generator of R^w if and only if : $G \in [R]_w$ and $\forall G' \in [R]_w \quad G' \subset G \Rightarrow G' = G$.

B. A NEW WAY TO DECIDE WHETHER R^w IS FINITELY GENERATED.

Let us recall the definition :

Definition 1 [4]. Given a language R, R^w is finitely generated if $R^w = F^w$ for some finite set F.

The notion of adherence is introduced in [1], its interest for our problem is contained in the following.

Lemma 1 [4]. Let R be a language, then R^w is an adherence if and only if $R^w = Adh(R^*)$. Furthermore, if R is a rational language, then one can decide whether R^w is an adherence.

Lemma 2 [4]. Let R be a language. If R^w is finitely generated then R^w is an adherence.

Let us note that the converse of the above statement does not hold as shown by the following example.

Example 1. $R = a + b + ab^*c$. As $Adh(R) = ab^w \subset R^w$, R^w is an adherence. Assume that R^w is finitely generated : $R^w = F^w$ for some finite set F. Let n be the greatest integer of the set $\{lg(u) \ / \ u \in F\}$. Since $ab^nca^w \in R^w$, we have $b^*ca^w \cap R^w \neq \emptyset$: that is a contradiction.■

We now define a language which plays a main role in the family $[R]_w$, specially when R^w is an adherence, as shown afterwards.

Definition 2. Let R be a language, $R^c = \{u \in A^+ / uR^w \subset R^w\}$.

Let us note that R^c only depends on R^w (and not on R), in other words : $\forall\, G \in [R]_w$, $G^c = R^c$.

Lemma 3. Let R be a language. R^c is a semi-group and is greater (with respect to the inclusion) than any generator of R^w. Furthermore whenever R is a rational language, R^c is a rational and constructible language.

Proof.

It is easy to see that : $\forall\, G \in [R]_w$, $G \subset R^c$ and R^c is a semi-group. Then we consider an automaton M such that $T^w(M) = R^w$. It is well known that the relation denoted by $\sim$ and defined by : $u \sim v$ iff $\delta(I,u) = \delta(I,v)$, is a right congruence of finite and calculable index. As it saturates R^c, R^c is rational and constructible.■

R^w may not be an adherence and then R^c is generally not a generator of R^w, as shown below :

Example 2. $R = a^*b$. $R^c = a^*b + a^+$, hence $a \in R^c$ but $a^w \notin R^w$, thus $R^{cw} \neq R^w$.■

Now we assume that R^w is an adherence, and we have :

Lemma 4. Let R be a language. If R^w is an adherence then R^c is a generator of R^w.

Proof.

As $R \subset R^c$, we have $R^w \subset R^{cw}$. For the converse inclusion, as $R^cR^w \subset R^w$, we have : $Adh(R^cR^w) \subset Adh(R^w)$. Now, as R^w is an adherence, $Adh(R^w) = R^w$ and as R^c is a semi-group, $R^{cw} \subset Adh(R^c)$. On the other hand : $Adh(R^c) \subset Adh(R^cR^w)$. Hence $R^{cw} \subset R^w$.■

Consequently, according to lemma 3 :

Corollary 5. Let R be a language. If R^w is an adherence then R^c is the greatest generator of R^w.

Let us remark that the converse of lemma 4 does not hold, as shown by the following example.

Example 3. $R = ba^*$. $R^c = b(a + b)^*$, we have $R^c \in [R]_w$, though R^w is not an adherence, indeed for example $ba^w \in Adh(R)$ but $ba^w \notin R^w$.■

Definition 3. Let R be a language. $R^t = \{u \in R^c \ / \ uR^w$ is not included in $[R^c \cap Ppref(u)]R^w\}$.

In other words, R^t is the sub-language of R^c, in which "useless" words are taken away : indeed a word u of R^c such that uR^w is included in $[R^c \cap Ppref(u)]R^w$, satisfies a fortiori $(R^c-u)^w = R^w$.

Proposition 6. Let R be a rational language such that R^w is an adherence. R^t is a generator of R^w, furthermore it is a rational and constructible language.

Proof .

First as $R^t \subset R^c$, we have $R^{tw} \subset R^{cw} = R^w$. On the other hand, we are going to prove that $R^w \subset R^tR^w$, which implies $R^w \subset R^{tw}$.
Let $w \in R^w$ and u the least word of R^c (w.r.t. the prefix order) such that : $w \in u(R^c)^w$. By construction u is a word of R^t.
We now prove that R^t is a rational and constructible language.
Let us consider a deterministic automaton $M = (A,Q,q_0,T,*)$ which recognizes R^c, i.e. $T(M) = R^c$. As M is a deterministic automaton, we have : $T^w(M) = \lim(T(M))$ that is to say $T^w(M) = \lim(R^c)$.
As $R^w = (R^c)^w$ is an adherence : $\lim(R^c) \subset (R^c)^w$, the converse inclusion being always true for a semi-group we deduce :
$T^w(M) = (R^c)^w$.
Let $\approx$ be the relation on A^+ defined by : $u \approx v$ if and only if

a) $q_0 * u = q_0 * v$

and b) $q_0 * (R^c \cap Ppref(u))^{-1}u = q_0 * (R^c \cap Ppref(v))^{-1}v$.

It is easy to verify that $\approx$ is an equivalence relation of finite and calculable index.
Furthermore $u \approx v$ implies : $\forall\, m \in A^+ \quad um \approx vm$, indeed :
on the one hand, according to a) : $q_0 * um = q_0 * vm$
on the other hand :
let $r \in R^c \cap Ppref(um)$, two cases are to be considered :

. $r \in Ppref(u)$: according to b), $\exists\, r' \in Ppref(v)$ such that : $q_0 * r^{-1}u = q_0 * r'^{-1}v$, hence $q_0 * r^{-1}(um) = q_0 * r'^{-1}(vm)$

. $r = um'$ where $m' \in A^*$: according to a), $q_0 * um' = q_0 * vm'$, hence $vm' \in R^c \cap Ppref(vm)$ and $q_0 * (um')^{-1}um = q_0 * (vm')^{-1}vm$.

Finally : $q_0 * (R^c \cap Ppref(um))^{-1}um = q_0 * (R^c \cap Ppref(vm))^{-1}vm$.
That is to say $\approx$ is a right congruence on A^+.
Furthermore we can see that, by construction, $\approx$ saturates R^t.
It follows that R^t is a rational and constructible language.■

Remark : The previous proposition holds even if R^w is not an adherence [6], but the proof is longer and the result is here useless.

Lemma 7. Let R be a language such that R^w is an adherence. $\forall G \in [R]_w$, R^t is included in Pref(G).

Proof.

$\forall G \in [R]_w$, $R^{tw} = G^w$. As for each $u \in R^t$ there exists $w \in R^w$ with $uw \notin (G \cap \mathrm{Ppref}(u))R^w$, there exists $g \in G$ such that $u < g$ and $uw \in gR^w$. That is to say : $u \in \mathrm{Pref}(G)$.■

We proceed now to the main result concerning R^t :

Theorem 8. Let R be a rational language. The following conditions are equivalent :

(i) R^w is finitely generated

(ii) R^t is a finite generator of R^w

(iii) R^t is a finite language and R^w is an adherence.

Proof.

(i) => (iii). If R^w is finitely generated then R^w is an adherence and according to lemma 7 : $R^t \subset \mathrm{Pref}(F)$ for some finite set $F \in [R]_w$. That is to say R^t is a finite language.

(iii) => (ii). It follows immediatly from proposition 1.

(ii) => (i). This part is obvious.■

Let us remark that the condition "R^t is a finite set" is not sufficient, we must add "R^t is a generator of R^w" or "R^w is an adherence", as shown by the following example :

Example 2 (continued). $R^t = a + b$ is a finite set but R^w is not finitely generated. R^t is not in $[R]_w$ and R^w is not an adherence.■

According to theorem 8 we can propose an algorithm to decide, for any rational language R, whether R^w is finitely generated :

- decide whether R^w is an adherence.
- if R^w is an adherence then construct R^t
decide whether R^t is a finite language.

Let us note that, if R^w is finitely generated, there exist infinitely many finite generators. Indeed, F being a finite generator, for each $i > 0$, F^i also is a finite generator of R^w.

C. THE GENERATORS OF LEAST CARDINAL.

Now we suppose that R is a finite language, such that $R = R^t$ (since R^t is a constructible language, that is not a restriction in view of section B).

We are going to solve the two natural following questions :
. Which is the least possible cardinal for a finite generator of R^w ? (This number will be called the rank of R^w)
. How to construct one (or better all) generator(s) having for cardinal the rank of R^w ?

Definition 4. Let R be a finite language. The rank of R^w, denoted by $rk(R^w)$, is the least integer of the set $\{card(G) \mid G \in [R]_w\}$.
A generator of R^w such that $card(G) = rk(R^w)$ is called a generator of least cardinal for R^w .

Obviously there is always at least one generator of least cardinal, but it is generally not unique.

Example 4. $R = aa+aaa+b$. $R^c = R^*$ and $R^t = R$. Here R and $G = aa+aaab+b$ are two generators of least cardinal for R^w.■

First we have seen that R^t is necessarily a finite language when R^w is finitely generated, however it is not necessarily a generator of least cardinal. Furthermore it is even possible that not any generator of least cardinal is included in $Pref(R^t)$ as shown by the following example. Thus R^t does not allow us to solve both previous questions.

Example 5. (This example, though non-trivial, seems to be the shorter one, indeed when it is reduced R^t get a generator of least cardinal for R^w!).
Let $F = c+ca+ca^2+ca^3+a^8+a^2b+ca^5$. Let $D = \epsilon+a+a^2+b$. Let $R = DF$.
Without detailing the calculations, we give the different steps of the proof.
1) $R^c = R^+$ and $Root(R^c) = R$, where $Root(R^c) = R^c-R^cR^c$.
2) $\forall\ u \in R-D(ca^2+ca^3+ca^5)$, uR^w is not included in $(R-u)R^w$, i.e. only the words of $D(ca^2+ca^3+ca^5)$ can be taken away from R, to keep the same w-power.
3) $[R-D(ca^2+ca^3)]^w = [R-D(ca^5)]^w = R^w$ but $[R-D(ca^2+ca^3+ca^5)]^w \neq R^w$
4) According to 1),2) and 3) $R^t = R-Dca^5$.

5) $\forall$ $G \subset Pref(R^t)$, $G \in [R]_w \Rightarrow G = R^t$, i.e. the unique generator of R^w included in $Pref(R^t)$ is R^t .
6) Finally, as $card(R-D(ca^2+ca^3)) = 20$ and $card(R^t) = 24$, we obtain that R^t is not a generator of least cardinal and (by 5) that $Pref(R^t)$ does not contain any one of them.■

Of course the generators of least cardinal are minimal generators of R^w, and for these last ones we have the useful following :

Lemma 9. Let G be a minimal generator of R^w. For each $g \in G$, there exists $w \in G^w$ such that $gw \notin (G-g)G^w$.

Proof.
Assume that : $\exists\, g \in G$ / $gG^w \subset (G-g)G^w$. We deduce : $GG^w \subset (G-g)G^w$ and then $G^w \subset (G-g)^w$. Hence $(G-g)$ is a generator of R^w, that is a contradiction with "G is a minimal generator of R^w ".■

Now we state and prove the main result of this paper.

Theorem 10. Let R be a finite language (such that $card(R^w) > 1$). One can compute the rank of R^w. Furthermore the generators of least cardinal for R^w are finite in number and constructible languages.

Let us remark that if $card(R^w) = 1$, $R^w = \{u^w\}$. Thus $rk(R^w) = 1$ and for each $i > 0$, $\{u^i\}$ is a generator of least cardinal for R^w. That is the single case where there exist infinitely many generators of least cardinal in $[R]_w$.

Proof .
Let $M = (A,Q,q_0,T,*)$ be a deterministic automaton, which recognizes R^c, i.e. $T(M) = R^c$.
It has been shown in the proof of proposition 6 that :

$$T^w(M) = R^w = (R^c)^w$$

$\forall$ $n > 0$, $\forall$ $G \in [R]_w$, $\approx$ denotes the following relation over A^nA^* :
$\forall$ $u,v \in A^nA^*$: $u \approx v$ iff $q_0 * u = q_0 * v$ and

$$q_0 * (G_{<n})^{-1}u = q_0 * (G_{<n})^{-1}v,$$

where $G_{<n} = G \cap (A \cup \ldots \cup A^{n-1})$.
(the notation $\approx$ conceals the dependance of $\approx$ from n and G but a precise notation would be very heavy)
$\approx$ satisfies the four following properties :
(P1) $\approx$ is an equivalence relation such that for each u and $u' \in A^nA^*$, and each $v \in A^*$: $u \approx u' \Rightarrow uv \approx u'v$

(without defining $\approx$ over A^nA^*, the later implication does not hold)

(P2) $\approx$ is of finite index and furthermore :

$\exists$ C > 0 such that $\forall$ n > 0, $\forall$ G $\in [R]_w$ index($\approx$) < C

(we can take C = card(2^Q).card(Q) + 1)

Note a main point that C is dependant on neither n nor G.

(P3) $\forall$ u,u' $\in A^nA^*$: u $\approx$ u' => (u $\in R^c$ iff u' $\in R^c$)

(P4) $\forall$ u,u' $\in A^nA^*$: u $\approx$ u' => (uw $\in G_{<n}R^w$ iff u'w $\in G_{<n}R^w$)

(without defining $\approx$ over A^nA^*, the later equivalence does not hold)

(P1), (P2), (P3) are immediate consequences of the definition of $\approx$. For (P4), let w $\in A^w$, if uw $\in G_{<n}R^w$ then there exists $g_1 \in G_{<n}$ with u = g_1u' and u'w $\in R^w$. Hence the set {i / q_0 * u'w[i] $\in$ T} is an infinite set. As there exists $g_2 \in G_{<n}$ such that v = g_2v' and q_0 * u' = q_0 * v', we obtain that {i / q_0 * v'w[i] $\in$ T} is an infinite set, hence v'w $\in R^w$ and vw $\in G_{<n}R^w$. That proves (P4).

State now the conclusive lemma for our proof:

Lemma 11. Let G be a generator of least cardinal for R^w. For each word ff' of Pref(G) where lg(f') = C, there exists f" $\in A^+$ such that ff" $\in$ G and lg(f'&f") < C.

In view of this lemma, we deduce that, if there exists g in G with lg(g) > k.C, then card(G) > k. Consequently any generator of least cardinal is included in $(A^*)_{<M}$, where M = C.card(R)+1. That ends the proof of theorem 10.∎

Proof of lemma 11.

Assume that there exists a word ff' $\in$ Pref(G) satisfying both lg(f') = C and (H) : $\forall$ f" $\in A^+$ / ff" $\in$ G => ff' < ff".

Let us denote by $\approx$ the relation associated with G and n = lg(f).

As lg(f') = C, there exist u,v $\in A^+$ with f < u < uv < ff' and u $\approx$ uv.

Fact. uv^w is the unique w-word such that $uv^w \in (G^w \cap fA^w)-(G_{<n}G^w)$.

Proof of fact.

Let ff" $\in$ G (f" exists since ff' $\in$ Pref(G)). First we prove that $uv^w \in G^w$. According to (H) we can write ff" = uvf"'. As for each i > 0, uvf"' $\approx$ uv^if"' we have uv^if"' $\in R^c$, hence $uv^w \in Adh(R^c)$. $(R^c)^w$ being an adherence, $(R^c)^w = Adh(R^c)$ and $R^c \in [R]_w$, hence $uv^w \in G^w$.

Prove now that uv^w is the unique w-word being in $G^w \cap fA^+$ and not in $G_{<n}G^w$. Let w $\in G^w$ such that ff"w $\notin G_{<n}G^w$ (w exists according to lemma 9). Assume that ff"w $\neq uv^w$, then there exists k > 0 such that $ff''w = uv^k w'$ with $v \not< w'$. As $uv^k \approx u$, $uw' \in G^w$, however $uw' \notin G_{<n}G^w$. Hence, according to (H), $uw' \in uvA^w$, that is to say that $v < w'$ (contradiction with the definition of w').■

We conclude so the proof of lemma 11 : as $card(R^w) > 1$, there exist two w-words w, w' in R^w with $w \neq w'$. As $uv^w = gg'^w$ for some $g \in G^+ \cap fA^+$ and $g' \in G^+$, and as furthermore for each $i > 0$, $gg'^i w \neq uv^w$ or $gg'^i w' \neq uv^w$, we deduce $gg'^i \in G_{<n}Pref(G^*)$. Thus $gg'^w \in Adh(G_{<n}Pref(G^*))$ which is equal to $G_{<n}G^w$ (since $G_{<n}$ is a finite language and G^w is an adherence). That yields $uv^w \in G_{<n}G^w$: contradiction.■

REFERENCES.

[1] L. Boasson and M. Nivat, Adherences of languages; Journal of Computer and System Sciences, 20 (1980) 285-309.
[2] J.R. Büchi, On decision method in restricted second-order arithmetics; Proc. Congr. Logic, Method. and Philos. Sci. (Stanford Univ. Press, Stanford, 1962) 1-11.
[3] S. Eilenberg, Automata, Languages and Machines; Vol. A (Academic Press, New York, 1974).
[4] L.H. Landweber, Decision problems for w-automata, Math. Syst. Theory 3 (1969) 376-384.
[5] M. Latteux and E. Timmerman, Finitely generated w-languages; Information Processing Letters 23 (1986) 171-175.
[6] I. Litovsky, Générateurs des langages rationnels de mots infinis; Thèse Univ. Lille I, 1988.
[7] I. Litovsky and E. Timmerman, On generators of rational w-power languages; Theoretical Computer Science 53 (1987) 187-200.
[8] R. MacNaughton, Testing and generating infinite sequences by a finite automaton; Information and Control 9 (1966) 521-530.
[9] L. Staiger, A Note on Connected w-languages; EIK 16 (1980) 5/6, 245-251.
[10] L. Staiger, Finite-state w-languages; Journal of Computer and System Sciences, 27 (1983) 434-448.

Extensional properties of sets of time bounded complexity (extended abstract)

WOLFGANG MAASS* AND THEODORE A. SLAMAN**

Abstract. We analyze the fine structure of time complexity classes for RAM's, in particular the equivalence relation $A =_C B$ ("A and B have the same time complexity") $\Leftrightarrow$ (for all time constructible f : $A \in DTIME_{RAM}(f) \Leftrightarrow B \in DTIME_{RAM}(f)$). The $=_C$-equivalence class of A is called its ***complexity type***. We prove that every set X can be partitioned into two sets A and B such that $X =_C A =_C B$, and that the partial order of sets in an arbitrary complexity type under $\subseteq^*$ (inclusion modulo finite sets) is dense. The proofs employ a new strategy for finite injury priority arguments.

We consider the following set of time bounds:

$$T := \{f : \mathbf{N} \to \mathbf{N} \mid f(n) \geq n \text{ and } f \text{ is time constructible on a RAM}\},$$

where f is called time constructible on a RAM if some RAM can compute the function $1^n \mapsto 1^{f(n)}$ in $O(f(n))$ steps. We do not allow arbitrary recursive functions as time bounds in our approach in order to avoid pathological phenomena (e.g. gap theorems [HU], [HH]). In this way we can focus on those aspects of complexity classes that are relevant for concrete complexity (note that all functions that are actually used as time bounds in the analysis of algorithms are time constructible). We use the random access machine (RAM) with uniform cost criterion as machine model (e.g. as defined in [CR]; see also [AHU], [MY]) because this is the most frequently considered model in algorithm design, and because a RAM allows more sophisticated diagonalization - constructions than a Turing machine. One defines $DTIME_{RAM}(f) := \{A \subseteq \{0,1\}^* \mid$ there is a *RAM*

---

*Department of Mathematics, Statistics, and Computer Science, University of Illinois at Chicago, Chicago, IL. 60680. Written under partial support by NSF-Grant CCR 8703889.

**Department of Mathematics, University of Chicago, Chicago, IL. 60637. Written under partial support by Presidential Young Investigator Award DMS-8451748 and NSF-Grant DMS-8601856.

of time complexity $O(f)$ that computes $A\}$. We write $DTIME(f)$ for $DTIME_{RAM}(f)$ in the following.

For sets $A, B \subseteq \{0,1\}^*$ we define

$$A =_C B(\text{"}A \text{ has the same det. time complexity as } B\text{"})$$
$$:\Leftrightarrow \forall f \in T(A \in DTIME(f) \Leftrightarrow B \in DTIME(f)).$$

A *complexity type* is an equivalence class of this equivalence relation $=_C$.

We write **0** for $DTIME(n)$, which is the "minimal" complexity type. Note that for every complexity type $\mathcal{C}$ and every $f \in T$ one has either $\mathcal{C} \subseteq DTIME(f)$ or $\mathcal{C} \cap DTIME(f) = \emptyset$.

In this paper we investigate some basic properties of the partial order

$$PO(\mathcal{C}) := \langle \{X | X \in \mathcal{C}\}, \subseteq^* \rangle,$$

where C is an arbitrary complexity type and $\subseteq^*$ denotes inclusion modulo finite sets (i.e. $X \subseteq^* Y :\Leftrightarrow X - Y$ is finite).

This work is part of the long range project to study the relationship between extensional properties of a set and its computational complexity. Among other work in this direction we would like to mention in particular the study of the complexity of sparse sets (see e.g. [Ma]), and the investigation of the relationship between properties of recursively enumerable sets under $\subseteq^*$ and their degree of computability (see e.g. Chapter XI in [So]). Our approach differs from this preceding work insofar as it also applies to "actually computable" sets (i.e. sets in P). Therefore it provides an opportunity to develop finer construction tools that can be used to examine also the structure of sets of small complexity. In this paper we introduce a new strategy for a finite injury priority argument that allows us to prove splitting and density theorems for sets of arbitrarily given complexity type $\mathcal{C}$ (for example sets of time complexity $\Theta(n^2)$). Further results about the structure of complexity types can be found in [MS].

Theorem 1. Every set X can be split into two sets A, B of the same complexity type as X (i.e. $X = A \cup B$, $A \cap B = \emptyset$, $X =_C A =_C B$).

In order to ***prove*** this result one needs a technique for controlling the complexity type of the constructed sets A, B. This is less difficult if X has an "optimal" time bound $f_X \in T$ for which $\{f \in T | X \in DTIME(f)\} = \{f \in T | f = \Omega(f_X)\}$ (in this case we say that X is of ***principal*** complexity type). However Blum's speed-up theorem [B] asserts that there are for example sets $X \in P$ such that

$$\{f \in T \mid X \in DTIME(f)\} = \left\{f \in T \mid \exists i \in \mathbf{N}\left(f(n) = \Omega\left(\frac{n^2}{(\log n)^i}\right)\right)\right\}.$$

Note that this effect occurs even if one is only interested in time constructible time bounds (and sets X of "low" complexity).

In order to prove Theorem 1 also for sets X whose complexity type is non-principal, we show that in some sense the situation of Blum's speed up theorem (where we can characterize the functions f with $X \in DTIME(f)$ with the help of a "cofinal" sequence of functions) is already the worst that can happen (unfortunately this is not quite true, since we cannot always get a cofinal sequence of functions f_i where $f_{i+1}(n) = O\left(\frac{f_i(n)}{g(n)}\right)$ for a fixed function g with $g(n) \to \infty$ for $n \to \infty$, as required for the proof of the speed-up theorem).

Definition. $(t_i)_{i \in \mathbf{N}} \subseteq \mathbf{N}$ is called a characteristic T-sequence if $t : i \mapsto t_i$ is recursive and

a) $\forall i \in \mathbf{N}(\{t_i\} \in T$ and program t_i is a witness for the time-constructibility of $\{t_i\})$

b) $\forall i, n \in \mathbf{N}(\{t_{i+1}\}(n) \leq \{t_i\}(n))$.

Lemma 1. ("inverse of the speed-up theorem"). For every recursive set A there exists a characteristic T-sequence $(t_i)_{i \in \mathbf{N}}$ such that $(t_i)_{i \in \mathbf{N}}$ is characteristic for A (i.e. $\forall f \in T(A \in DTIME(f) \Leftrightarrow \exists i \in \mathbf{N}(f(n) = \Omega(\{t_i\}(n)))))$.

One can also prove the *converse of Lemma 1* and construct for any given characteristic T-sequence $(t_i)_{i \in \mathbf{N}}$ a set A such that

$$\forall f \in T(A \in DTIME(f) \Leftrightarrow \exists i \in \mathbf{N}\ (f(n) = \Omega(\{t_i\}(n)))).$$

This construction, which is a refinement of the proof of Blum's speed-up theorem, is one component in the priority-constructions of Theorems 1 and 2. . It is more delicate because in our more general situation it is not guaranteed that there is a function g with $g(n) \to \infty$ (or at least $g(n) \geq 2$) such that $\forall i(\{t_{i+1}\} \leq \{t_i\}/g)$ (the existence of such uniform "gap" g is used in the customary proof of Blum's speed-up theorem). Some further details of the proof are described in [MS].

Remark. The idea of characterizing the complexity of an arbitrary recursive set by a sequence of "cofinal" complexity bounds is rather old (see e.g. [MF], [L], [LY], [SS], [MW]). However none of these results provide the here needed characterization of the time complexity of an arbitrary recursive set in terms of a uniform cofinal sequence of time constructible time bounds. [L] and [MW] give corresponding results for space complexity of Turing-machines. These results exploit the linear speed-up theorem for space complexity on Turing-machines, which is not available for time complexity on RAM's. Time complexity on RAM's has been considered in [SS], but only sufficient conditions are given for the cofinal sequence of time bounds (these conditions are stronger than ours, and they are probably not necessary). The more general results on complexity sequences in axiomatic complexity theory ([MF], [SS]) involve "overhead functions", or deal with nonuniform sequences, which makes the specialization to the notions considered here impossible. It is an open problem whether one can characterize the time complexity of any recursive set on Turing-machines in the same way as it is done here for RAM's (because of the lack of a fine time hierarchy theorem for multi-tape Turing-machines).

Idea of the Proof of Theorem 1. Associate with the given set X a characteristic T-sequence $(t_i)_{i \in N}$ as in Lemma 1. For every e, $n \in \mathbf{N}$ and $x \in \{0,1\}^*$ define

$$TIME(e, x) := (\text{number of steps in the computation of } \{e\} \text{ on input } x)$$

and

$$MAXTIME(e, n) := \max\{TIME(e, x) \mid |x| = n\}.$$

It is sufficient to partition X into sets A and B in such a way that for every $e \in \mathbf{N}$ the following requirements R_e^A, R_e^B, S_e^A, S_e^B are satisfied:

$$R_e^A :\Leftrightarrow (A = \{e\} \Rightarrow \forall f \in T(\forall n(MAXTIME(e,n) \leq f(n))$$
$$\Rightarrow \exists j \in \mathbf{N}(f(n) = \Omega(\{t_j\}(n))))))$$
$$S_e^A :\Leftrightarrow A \in DTIME(\{t_e\}(n)).$$

R_e^B, S_e^B are defined analogously.

Note that it is not possible to satisfy R_e^A by simply setting $A(x) := 1 - \{e\}(x)$ for some x: in order to achieve that $A \subseteq X$ we can only place x into A if $x \in X$.

Instead, we adopt the following strategy to satisfy R_e^A (the strategy for R_e^B is analogous): For input $x \in \{0,1\}^*$ compute $\{e\}(x)$.

Case I. If $\{e\}(x) = 0$, then this strategy issues the constraint "$x \in A \Leftrightarrow x \in X$",

Case II. If $\{e\}(x) = 1$, then this strategy issues the constraint "$x \notin A$" (which forces x into B if $x \in X$).

In the case of a conflict for some input x between strategies for different requirements one lets the requirement with the highest priority (i.e. the smallest index e) succeed (this causes in general an "injury" to the other competing requirements).

The interaction between the described strategies is further complicated by the fact that in the case where R_e^A is never satisfied via Case II, or via Case I for some $x \in X$, we have to be sure that Case I issues a constraint *for almost every input* x (provided that the simulation of $\{e\}(x)$ is not prematurely halted by some requirement S_i^A with $i \leq e$, see below). Consequently the number of requirements whose strategies act on the same input x grows with $|x|$ (only those R_i^A, R_i^B with $i < |x|$ can be ignored where one can see by "looking back" for $|x|$ steps that they are already satisfied).

The strategy for requirement $S_e^A(S_e^B)$ is as follows: it issues the constraint that for all inputs x with $|x| \geq e$ the *sum* of all steps that are spent on simulations for the sake of requirements $R_i^A, R_i^B, S_i^A, S_i^B$ with $i \geq e$ has to be bounded by $O(\{t_e\}(|x|))$. One can prove that in this way $S_e^A(S_e^B)$ becomes satisfied (because only finitely many inputs are placed into A or B for the sake of requirements of higher priority). One also has to prove that the constraint of S_e^A does not hamper the requirements of lower priority in a serious manner.

This part of the construction is more difficult than its counterpart in Blum's speed-up-theorem [B], because it need not be the case that $\{t_{i+1}\} = o(\{t_i\})$. A further complication is caused by the fact that although there are constants K_i, K_{i+1} such that $\{t_i\}(n)$ converges in $\leq K_i \cdot \{t_i\}(n)$ steps and $\{t_{i+1}\}(n)$ converges in $\leq K_{i+1} \cdot \{t_{i+1}\}(n)$ steps, we may have that $K_i \ll K_{i+1}$ (and therefore $K_i \cdot \{t_i\}(n) \ll K_{i+1} \cdot \{t_{i+1}\}(n)$). Therefore the requirements S_j^A with $j > i$ are not able to "take over" the job of S_i^A, and *all* computations $\{t_i\}(|x|)$, $i \leq |x|$, have to be simulated simultaneously for each input x.

In order to show that a single RAM R can carry out simultaneously all of the described strategies, one exploits in particular that a RAM can dovetail an unbounded number of simulations in such a way that the number n_e of steps that it has to spend in order to simulate a single step of a simulated program $\{e\}$ does not grow with the number of simulated programs (the precise construction of R is rather complex).

In order to verify that this construction succeeds, one has to show that each requirement R_e^A, R_e^B is "injured" at most finitely often. This is not obvious, because we may have for example that R_{e-1}^B (which has higher priority) issues overriding constraints for infinitely many arguments x according to Case I. However in this case we know that only finitely many of these x are elements of X (otherwise R_{e-1}^B would have been seen to be satisfied from some point of the construction on), and all of its other constraints are "compatible" with the strategies of lower priority (since we make $A, B \subseteq X$).

Finally we verify that each requirement $R_e^A (R_e^B)$ is satisfied. This is obvious if Case II occurs in the strategy for R_e^A for some input x where R_e^A is no longer injured; or if Case I occurs for such input x with $x \in X$ (in both cases we can make $A \neq \{e\}$). However it is also possible that $x \notin X$ for each such x (and that $\{e\} = A$), in which case R_e^A becomes satisfied for a different reason. In this case we have $\{e\}(x) = 0 = X(x)$ for each such x. Therefore we can use $\{e\}$ to design a new algorithm for X that is (for every input) at least as fast as the algorithm $\{e\}$ for A (it uses $\{e\}$ for those inputs where $\{e\}$ is faster than the "old" algorithm for X of time complexity $\{t_e\}$). Therefore one can prove that $X \in DTIME(f)$ for every $f \in T$ that bounds the running time of algorithm $\{e\}$ for A. This implies that $f(n) = \Omega(\{t_j\}(n))$ for some $j \in \mathbf{N}$ (by construction of the characteristic T-sequence $(t_i)_{i \in \mathbf{N}}$). □

Corollary 1. For every complexity type $\mathcal{C} \neq \mathbf{0}$ the partial order $PO(\mathcal{C})$ of sets in $\mathcal{C}$ has neither minimal nor maximal elements.

Let $PO_{0,1}(\mathcal{C})$ be the partial order

$$\langle \{X | X \in \mathcal{C} \vee X = \{0,1\}^* \vee X = \phi\}, \subseteq^* \rangle$$

(thus $PO_{0,1}(\mathcal{C})$ results from $PO(\mathcal{C})$ by adding the smallest set ϕ and the largest set $\{0,1\}^*$).

Corollary 2. For every complexity type $\mathcal{C}$ there is an embedding E of the partial order of the countable atomless Boolean algebra AB into $PO_{0,1}(\mathcal{C})$ with $E(1) = \{0,1\}^*$, $E(0) = \phi$, and $E(a \vee b) = E(a) \cup E(b)$ as well as $E(a \wedge b) = E(a) \cap E(b)$ for all elements $a, b \in AB$.

Remark. In order to define this embedding E one starts with any two sets X, $\{0,1\}^* - X$ in $\mathcal{C}$, and applies the splitting theorem iteratively. The idea is to represent the elements of AB by arbitrary finite unions of those sets in $\mathcal{C}$ that are constructed in this way. In order to guarantee that these finite unions U are also in $\mathcal{C}$ (unless $U =^* \{0,1\}^*$) one has to prove a slightly stronger version of the splitting theorem. The following additional property of A, B is needed: For every $f \in T$ for which there exists some $U \in DTIME(f)$ with $U \cap X = A$ or $U \cap X = B$ one has $X \in DTIME(f)$. One can prove this stronger version with a small variation in the strategy for requirement R_e^A (R_e^A can now be satisfied via Case II at a single input x only if $\underline{x \in X}$; if R_e^A never gets satisfied at any input x via Case I or Case II one can argue that $x \notin X$ whenever the simulation of $\{e\}(x)$ can be finished in an attempt for R_e^A, independently of the output of $\{e\}(x)$).

In the following we write $Y \underset{\infty}{\subset} X$ if $Y \subseteq X$ and $X - Y$ is infinite.

Theorem 2. (density theorem)
Assume that $Y \underset{\infty}{\subset} X$ and $Y \leq_C X$ (i.e. $\forall f \in T(X \in DTIME(f) \Rightarrow Y \in DTIME(f))$). Then there is a set A such that $Y \underset{\infty}{\subset} A \underset{\infty}{\subset} X$ and $A =_C X$.

Idea of the proof of Theorem 2. Let $(t_i)_{i\in\mathbf{N}}$ be a characteristic T-sequence for X. It is sufficient to construct A such that $Y \subseteq A \subseteq X$ and for all $e \in \mathbf{N}$ the requirements R_e, S_e, T_e, U_e are satisfied, where R_e, S_e are identical with the requirements R_e^A, S_e^A in the proof of Theorem 1 (together they ensure that $A =_C X$) and

$$T_e: \quad |A - Y| \geq e$$

$$U_e: \quad |X - A| \geq e.$$

The strategy to satisfy R_e is similar to the strategy for satisfying R_e^A. However in Case II (where $\{e\}(x) = 1$), unlike in the splitting theorem, R_e does not have the power to keep x out of A (even if R_e has the highest priority) because x may later enter Y. Instead, R_e issues in Case II the constraint "$x \notin A \Leftrightarrow x \notin Y$" (i.e. R_e wants to keep x out of A if it turns out that $x \notin Y$).

It is easy to see that R_e becomes satisfied if Case I occurs for some $x \in X$, or if Case II occurs for some $x \notin Y$ (provided that R_e is not "injured" at x by requirements of higher priority). If neither of these events occurs, then we can conclude that $\{e\}(x) = X(x)$ whenever the simulation of $\{e\}(x)$ can be finished before it is halted for the sake of some requirement S_i with $i \leq e$. This information can be used (as in the proof of Theorem 1) to design an algorithm for X that converges for every input x "at least as fast" as the computation $\{e\}(x)$. □

Corollary 3. For every complexity type $\mathcal{C}$ the partial order $PO(\mathcal{C})$ is dense.

It is easy to see that $PO(\mathcal{C})$ is isomorphic to the countable atomless Boolean algebra AB if $\mathcal{C} = \mathbf{0}$. Furthermore it was shown that AB can be embedded into $PO_{0,1}(\mathcal{C})$ for every complexity type $\mathcal{C}$. However the following corollary suggests that the structure of the partial order $PO(\mathcal{C})$ is substantially more complicated than that of AB if $\mathcal{C} \neq \mathbf{0}$. Obviously any complexity type $\mathcal{C} \neq \mathbf{0}$ is closed under complementation, but not under union or intersection. However, it could still be the case that any two sets $A, B \in \mathcal{C}$ have a least upper bound in the partial order $PO(\mathcal{C})$. This is ruled out by the following result.

Corollary 4. Consider an arbitrary complexity type $\mathcal{C} \neq \mathbf{0}$. Then any two sets $A, B \in \mathcal{C}$ have a least upper bound in the partial order $PO(\mathcal{C})$ if and only if $A \cup B \in \mathcal{C}$. In particular one can define with a first order formula over $PO(\mathcal{C})$ whether $A \cup B \in \mathcal{C}$ (respectively $A \cap B \in \mathcal{C}$) for $A, B \in \mathcal{C}$.

Proof. Assume that $A, B \in \mathcal{C}$, $A \cup B \notin \mathcal{C}$, $A \cup B \subseteq D$ and $D \in \mathcal{C}$. Then $(A \cup B) \leq_{\mathcal{C}} D$ and $(A \cup B) \underset{\infty}{\subset} D$. Thus there exists by Theorem 2 a set $D' \in \mathcal{C}$ with $(A \cup B) \underset{\infty}{\subset} D' \underset{\infty}{\subset} D$. Therefore D is not a least upper bound for A and B in $PO(\mathcal{C})$. □

Remark. This result suggests that the first order theory of the partial order $PO(\mathcal{C})$ is nontrivial for $\mathcal{C} \neq \mathbf{0}$.

Acknowledgement. We would like to thank Joel Berman for helpful comments.

REFERENCES

[AHU] A.V. Aho, J.E. Hopcroft, J.D. Ullman, The Design and Analysis of Computer Algorithms, Addison-Wesley (Reading, 1974).

[B] M. Blum, A machine-independent theory of the complexity of recursive functions, *J. ACM,* **14**(1967), 322-336.

[CR] S.A. Cook, R.A. Reckhow, Time-bounded random access machines, *J. Comp. Syst. Sc.,* **7**(1973), 354-375.

[HH] J. Hartmanis, J.E. Hopcroft, An overview of the theory of computational complexity, *J. ACM,* **18**(1971), 444-475.

[L] L.A. Levin, On storage capacity for algorithms, *Soviet Math. Dokl.,* **14**(1973), 1464-1466.

[Ly] N. Lynch, Helping: several formalizations, *J. of Symbolic Logic,* **40**(1975), 555-566.

[MY] M. Machtey, P. Young, An Introduction to the General Theory of Algorithms, North-Holland (Amsterdam, 1978).

[Ma] S. Mahaney, Sparse complete sets for NP: solution of a conjecture of Berman and Hartmanis, *J. Comp. Syst. Sc.,* **25**(1982), 130-143.

[MF] A.R. Meyer, P.C. Fischer, Computational speed-up by effective operators, *J. of Symbolic Logic,* **37**(1972), 55-68.

[MW] A.R. Meyer, K. Winklmann, The fundamental theorem of complexity theory, *Math. Centre Tracts,* **108**(1979), 97-112.

[MS] W. Maass, T.A. Slaman, On the complexity types of computable sets (extended abstract), to appear in : *Proc. of the Structure in Complexity Theory Conference* 1989.

[SS] C.P. Schnorr, G. Stumpf, A characterization of complexity sequences, *Zeitschr. f. math. Logik u. Grundlagen d. Math.,* **21**(1975), 47-56.

[So] R.I. Soare Recursively Enumerable Sets and Degrees, *Springer* (Berlin, 1987).

Learning under uniform distribution

A.Marchetti-Spaccamela
Dept. of Mathematics
University of L'Aquila
L'Aquila, Italy

M.Protasi
Dept. of Mathematics
University of Roma "Tor Vergata"
Roma, Italy

Abstract We study the learnability from examples of boolean formulae assuming that the examples satisfy a uniform distribution assumption. We analyze the requirements of known algorithms (upper and lower bounds) under uniform distribution and we propose a new combinatorial measure in order to characterize the complexity of boolean formulae.

1 Introduction

Algorithms for learning boolean formulae have been widely studied in the last years since Valiant's paper [V84] that showed how it is possible to study learning algorithms in the framework of computational complexity theory. In this approach we assume that the learner has access to data that exemplify the formula to be learned in a positive or in a negative way and that he has performed its task when he can find a good approximation of the formula with sufficiently large probability.

One of the most important classes from the learning point of view is the class DNF, that is the class of formulae in disjunctive normal form; however, since the class DNF is not efficiently learnable, several subclasses have been investigated. Let k-DNF (k-CNF) be the class of boolean formulae in disjuntive (conjunctive) normal form where each term (clause) is the

---

Work partially supported by Italian project MPI 'Algoritmi e Strutture di Calcolo'.

product (sum) of at most k literals. Let d-term-DNF (d-clause-CNF) be the class of boolean formulae in disjuntive (conjunctive) normal form with at most d terms. Furthermore, let k-d-term-DNF (k-d-clause-CNF) be the subclass of k-DNF (k-CNF) formulae with at most d terms (clause). A formula f is monotone if it has not negated variables.

The majority of the results have been given assuming that the examples are drawn according to a probability distribution which is fixed but unknown (distribution free model). However, because of the generality of the approach, few classes can be learned and many negative results occur. For example in [KLPV87] it has been shown that the problem of learning monotone d-term DNF by d-term DNF formulae ($d \geq 2$) is NP-hard. There is also a stronger result; namely it is hard to learn a formula even if we allow the formula to have about twice as many disjuncts as the formula to be learned. Formally, if $d > 5$ then monotone d-term DNF are not learnable by (2 d - 5)-term-DNF formulae.

A further limitation of the distribution free model is that even if the number of examples is polynomially bounded it can be too large from a practical point of view. For these reasons it is useful to restrict the class of probability distribution used for generating the examples. The case in which the uniform distribution is considered has been previously considered by several authors [KLPV87], [BI88], [KMP88]; in this paper we continue this approach. The aim is to give stronger and more efficient results for some classes of formulae. More precisely we are mainly interested in studying the sample complexity, that is the number of examples required.

In Section 3 we analyse Valiant's algorithm for learning k-DNF formulae in the case of k-d-term-DNF formulae and we show that, for formulae with small k and d, the use of uniform distribution allows to increase the efficiency of the algorithm. On the other side we show that for a whole class of algorithms the performance obtained in the distribution free model cannot be improved significantly. In Section 4 we introduce the notion of term similarity of a formula. This notion is an attempt towards the aim of characterizing the performance of a learning algorithm from a combinatorial point of view under uniform distribution and we show how term similarity affects the complexity of learning process in the case of the k-d-term-DNF formulae. In Section 5, we concentrate on d-term-DNF formulae; we prove a lower bound on the number of examples for the class of algorithms studied in Section 3. Furthermore we study how the uniform distribution affects the sample complexity of an algorithm introduced in [KMP88] for DNF formulae.

2 Basic definitions

First of all we give some standard definitions. Let n be a natural number. A concept F is a boolean function with domain $\{0,1\}^n$. Given a vector X, X is a positive example of the concept F, if $F(X) = 1$, otherwise X is a negative example.

For any concept F there are many possible boolean formulae f such that f is consistent with the concept F. A class of representations of a concept F is a set of boolean formulae that is used to represent F. Given a formula f and an example X, f(X) denotes the value of the formula applied to the example X, while $|f|$ denotes its length.

We assume that the learning algorithm can call two procedures pos and neg that give a positive and a negative example X, respectively.

Let D^+ and D^- be the given probability distribution of the generation of positive and negative examples, respectively, such that

$$\Sigma_{f(X)=1} D^+(X) = 1 \text{ and } \Sigma_{f(X)=0} D^-(X) = 1$$

and such that the examples are independently given.

If we assume that the distributions of positive and negative examples are fixed but unknown, we say that we learn in the distribution free model. Throughout the paper we will assume that the distributions D^+ and D^- are uniform.

Definition 1 Let B be a class of representations. Then B is polynomially learnable iff there exists a polynomial q and an algorithm L such that

$(\forall n)$ $(\forall f \in B$ with n variables) $(\forall D^+ \forall D^-)$ $(\forall \varepsilon > 0)$

L, given a set of $q(n, |f|, 1/\varepsilon)$ examples of f, runs in time polynomial in n, $|f|$ and $1/\varepsilon$ and outputs a formula $g \in B$ with n variables that with probability at least $1-\varepsilon$ has the following properties:

i) $\Sigma_{\forall X\, g(X)=0} D^+(X) < \varepsilon$ and ii) $\Sigma_{\forall X\, g(X)=1} D^-(X) < \varepsilon$

If the algorithm L uses only positive (negative) examples, then B is polynomially learnable from positive (negative) examples.

3 Learning k-d-term-DNF formulae under uniform distribution

The classes k-DNF and k-d-term-DNF have already been studied in the distribution free model by several authors. The basic algorithm has been introduced by Valiant [V84]; therefore before proceeding it is useful to sketch algorithm V(k) proposed by Valiant to learn k-DNF formulae. Let T be the set of all possible terms of up to k literals. Then the algorithm calls for $2((2n)^{k+1} + \log(1/\varepsilon))/\varepsilon$ negative examples (here and in the following all logarithms are natural). At each step it eliminates from T those terms that satisfy the negative example (and hence cannot be present in the formula to be learned). Valiant has obtained the following result.

Theorem 1 [V 84] For any given positive integer k, k-DNF is polinomially learnable via the algorithm V(k) that uses at most $2((2n)^{k+1} + \log(1/\varepsilon))/\varepsilon$

negative examples and runs in time polynomial in n^{k+1} and $1/\varepsilon$.

The number of examples required by the algorithm is polynomially bounded in n^{k+1}. Hence if the value of k is not small the number of examples required can be too large. On the other hand, the number of examples cannot be significantly improved: in fact it has been proven a lower bound that shows that Valiant's algorithm is optimal in terms of the number of examples [EHKV88].

The above considerations suggest to investigate what happens to Valiant's algorithm when we restrict ourselves to the uniform distribution. A first step in this direction is given by Theorem 2.

Lemma 1 Let f be a k-d-term-DNF formula. A term C that does not belong to f is satisfied by at least 2^{n-k-d} negative examples.

Proof The lemma is proved if we show that, given a formula f with d terms, the maximum number of positive examples that satisfy C occurs when all the terms of f have k-1 variables in common with C and all remaining variables are different. We prove it by induction on d.

d=1

In order to prove this case let D be the term that constitues the formula f; we count the number of positive examples that satisfy C. Without loss of generality let $C = x_1 x_2 ... x_k$; if X is a positive example that satisfies C, it must also satisfy D. This implies that variables belonging to both C and D have the same truth value and that variables belonging to the union of C and D are fixed. The maximum number of positive examples that verify the above condition occurs when $|C \cap D| = k-1$.

Inductive step.

Assume that the claim is true for d-1 we show it for d. Suppose that f is a formula such that the maximum number of positive examples that satisfy C and d terms has maximum cardinality. Let g be the formula obtained by considering the first d-1 tems of f and let S be the d-th term. We consider two cases whether the following condition I is satisfied or not.

I: all the terms of g have k-1 variables in common with C and all remaining variables are different.

If I is true, when we add the term S some of the negative examples that satisfy C become positive examples. The maximum number of positive examples that we obtain occurs when $|C \cap S| = k-1$. It is easy to see that in this case the number of negative examples halves; this completes the case when I is true.

If I is not true let us consider a term T of g that has not k-1 variables in common with C. Now we consider the formula g' obtained from g by deleting T and adding a term T' that has k-1 variables in common with C and one variable in common with T with the same truth value. Since d is less than n-k such a term exists. Let us denote by E (E') the set of negative examples that do not satisfy g and satisfy C (g' and C). It is easy to see that E' is a subset of E. In fact if a positive example satisfies C and T' it also satisfies C and T. Now we consider the effect of adding to g and g' the new term S. Clearly some of the negative examples of E and E' now become positive examples. Let us denote by D (D') the set of negative examples that do not satisfy (g or S) and satisfy C ((g' or S) and satisfy C).Since E is a subset of E' we have that D is a subset of D'. Hence it is sufficient to consider the bound obtained when condition I is satisfied.

QED

Theorem 2 Valiant's algorithm learns k-d-term-DNF formulae, $n > k+d$, under uniform distribution using at most $O(2^{k+d}$ $(k \log n + \log (1/\varepsilon))$ examples.

Proof The lemma implies that, in the uniform distribution model, each term is eliminated by a negative example with probability greater than $1/ 2^{k+d}$.

Now we estimate the probability that after r examples there is a term $C \notin f$ that is not eliminated in Valiant's algorithm.

$$\text{Prob}\,(\exists\, C \notin f \text{ that is not eliminated}) \le E[\text{ no. of } C \notin f \text{ that are not eliminated}]$$
$$\le \binom{n}{k-1} (1 - 1/2^{k+d})^r$$
$$\le (ne/k)^k \exp(-r/2^{k+d})$$
$$\le \exp(-r/2^{k+d} + k\log n + k - k\log k)$$

If r is equal to 2^{k+d} $(k \log n + \log (1/\varepsilon))$ we obtain the desired result.

QED

Note that in the above theorem there is a significant improvement on the number of examples required in the case of a formula with a small number of terms. In fact, in Valiant's algorithm the leading term is n^{k+1}, while in the above theorem is 2^{k+d} and therefore does not depend on the number of variables. We also observe that the reduction in the number of examples required implies a corresponding reduction in the running time of the algorithm. Furthermore we note that the proof of the above theorem implies that the algorithm exactly finds the formula to be learned and not an approximation to it as required in Definition 1.

If we change the model by allowing the use of both positive and negative examples, then it has been shown that k-DNF formulae with a small number of terms can be learned more efficiently with respect to Valiant's algorithm in the distribution free model through a greedy approach [H88].

It is known that, in the distribution free model, the lower bound on the number of the examples needed to learn k-DNF formulae is matched by Valiant's algorithm [EHKV88]. In Theorem 2 we have proved that if we restrict ourselves to the uniform distribution model then it is possible to learn a meaningful subset of k-DNF formulae in a more efficient way. Now we wonder if it possible to obtain a lower bound in this restricted case.

Definition 2 Given a concept F, a set of examples E and a class of representations B, a consistent hypothesis h with respect to E, F and B is a formula belonging to B such that $F(X) = h(X)$ for all $X \in E$.

A maximally consistent hypothesis h is a consistent hypothesis such that for all consistent hypotheses k and for all examples X if $k(X)=1$ then $h(X)=1$.

Definition 3 A maximally consistent algorithm is an algorithm that, given a set E of examples of a concept F and a class of representations B, finds the maximally consistent hypothesis h with E, F and B if such an hypothesis exists.

Theorem 3 If $k < n/4$, then there exists a k-DNF formula that requires $\Omega((n/k)^k \log(1/\varepsilon))$ examples to be learned by a maximally consistent algorithm that only uses negative examples under uniform distribution .

Proof Let f be the following formula to be learned:

$f(X) = 1$ if and only if there are at least k positive variables in X.

Let us define now the following set:

$G = \{C \mid C$ is a k-term with exactly k-1 positive variables$\}$.

We partition G in disjoint subsets G(I): for any (k-1) tuple $I=(i_1,i_2,....,i_{k-1})$

$G(I) = \{C \mid C \in G$ and the positive variables of C are those indexed by I$\}$

Clearly there are $\binom{n}{k-1}$ subsets G(I).

Note that for any I, terms in G(I) are eliminated exactly by one negative example. In fact , assume without loss of generality that $I = \{1, 2, ...,k-1\}$; then the terms in G(I) have the following form $x_1 x_2 .... x_{k-1} \bar{x}_j$, for some j between k and n; clearly the definition of f implies that for all j there is only a negative example that satisfy the terms belonging to G(I).

Now for any (k-1) tuple $I =(i_1,i_2,....,i_{k-1})$ a maximally consistent hypothesis h with respect a set of negative examples E verifies the following condition:

if the negative example $(x_{i_1},x_{i_2},\ldots,\bar{x}_{i_{k-1}},\bar{x}_{i_k},\ldots,x_{i_n})$ does not belong to E then

$$h(x_{i_1},x_{i_2},\ldots,\bar{x}_{i_{k-1}},\bar{x}_{i_k},\ldots,x_{i_n}) = 1.$$

In fact if this is not true, the formula $h \vee (x_{i_1}x_{i_2}\ldots x_{i_{k-1}}\bar{x}_{i_k})$ is a consistent hypothesis that contradicts the fact that h is maximally consistent. Hence in order to obtain a formula that approximate f it is necessary to eliminate at least a fraction of (1 - ε) of all terms with k-1 positive variables. The evaluation of the number of negative examples needed to eliminate at least (1 - ε) of all terms with k-1 positive variables is an application of the well known coupon collector's problem [F71].
Since the number of negative examples that allow to eliminate terms with k-1 positive variables is

$$\binom{n}{k-1} \leq \sum_{i=1}^{k-1} \binom{n}{i-1} \leq 2 \binom{n}{k-1} \quad \text{(for } k < n/4)$$

then, in order to eliminate at least (1 - ε) of all terms with k-1 positive variables requires at least $\Omega((n/k)^{k-1}\log(1/\varepsilon))$ negative examples.

QED

This theorem shows that the performance of Valiant's algorithm under uniform distribution is comparable to the distribution free case. Furthermore, because the difference between the upper bound of Valiant's algorithm and the lower bound shown in Theorem 2 is small, we derive that, using a maximal consistent algorithm, we cannot improve significantly. This fact suggests to exploit other combinatorial conditions in order to improve the performance of the algorithm in the distribution dependent case. In the next paragraph we propose such an approach.

4 A combinatorial measure for learning

In the study of learning algorithms one of the most important technical tools is given by the so called Vapnik-Chervonenkis dimension (see for example VC[71], LMR[88], [H88]). This dimension allows to characterize meaningful classes of boolean formulae: lower and upper bounds on the number of examples required to learn a formula can be expressed in terms of this dimension.

Here we propose a different notion of combinatorial measure with the aim to pursue the study of structural properties that affect the learnability of classes of formulae in the case of uniform distribution.

Definition 4 Given a DNF formula $F = T_1 + T_2 + ... T_d$, let s be the longest common subsequence among pairs of terms.
In this case we say that the formula F has *term similarity* s.

Intuitively a formula with high term similarity is a formula in which there are terms that do not differ substantially; this increases the possibility of misclassifying the concept and makes the process of learning harder.
It is immediate to prove the following fact that will be used in the following.

Fact 1 The term similarity of a k-DNF formula is no greater than k.

Now, as we did in Section 3, we study the complexity of learning k-d-term-DNF formulae using the similarity of the formula as a parameter.

Theorem 4 Valiant's algorithm learns k-d-term-DNF formulae with similarity s, with $d < 2^{(k-s)/2-1} + 1$, under uniform distribution, using $O(2^k /(1/2 - (d-1)\, 2^{(s-k)/2}) (k \log n + \log (1/\varepsilon))$ examples.
Proof Let f be the formula to be learned. A term D that does not belong to the formula to be learned is satisfied by at least 2^{n-k} examples.
In order to prove the theorem we must evaluate the number of positive examples that satisfy D. Without loss of generality let $D = x_1 x_2 ... x_k$; if X is a positive example that satisfies D it must also satisfy a term C belonging to the formula to be learned. This implies that variables belonging to both C and D have the same truth value and that variables belonging to the union of C and D are fixed. Let $C_1, C_2, ..., C_d$, be the terms of f. Let e_i be the number of variables common to C_i and D, $i = 1,2,...d$. Without loss of generality we assume that $e_i \geq e_{i+1}$ for every i. It is easy to see that $e_i \leq k - 1$, and that the number of examples that satisfy D and C_1 is 2^{n-2k+e_1}.
If the similarity is s any other term C_j must satisfy $e_j \leq k - e_i + s$. Since $e_i \geq e_j$, we obtain $e_j \leq (s+k)/2$.
The number of examples that satisfy D and C_j is $2^{n-2k+e_j} \leq 2^{n-2k+(s+k)/2}$.
Summing over all terms we have that the number of positive examples that satisfy D is less than

$2^{n-2k+e_1} + \sum_{i \neq j} 2^{n-2k+(s+k)/2} = 2^{n-k}(2^{-1} + (d-1)\, 2^{(s-k)/2})$

Hence the number of negative examples that satisfy D is at least $2^{n-k}(1 - 2^{-1} + (d-1)\, 2^{(s-k)/2})$

Since the total number of negative examples is less than 2^n and the examples are uniformly distributed, we obtain that the probability that D satisfies a negative example is at least

$2^{-k} (1 - 2^{-1} - (d-1)\, 2^{(s-k)/2}) = 2^{-k} (2^{-1} - (d-1)\, 2^{(s-k)/2})$

Let g be equal to $(d-1)\, 2^{(s-k)/2}$. If g is smaller than 1/2, we have

Prob($\exists$ D $\notin$ f, D does not satisfy r negative examples) $\leq$

E(no. of D $\notin$ f, D does not satisfy r negative examples) $\leq$

$n^k (1 - 2^{-k} (1/2 - g))^r \leq e^{-r\, 2^{-k}(1/2 - g) + k \ln n}$

If $r = 2^k(k \ln n + \ln (1/\varepsilon)) / (1/2 - g)$ we obtain that $e^{-r\, 2^{-k}(1/2 - g) + k \ln n} \leq e^{-\ln (1/\varepsilon)} \leq \varepsilon$.

This completes the proof of the theorem.

QED

Note that the number of examples required to learn depends on the number of variables only for a logarithmic factor. Furthermore the theorem shows how similarity can affect the complexity of learning boolean formulae. In fact the similarity measures the difficulty of learning k-d-term-DNF formulae: the more similar are terms in the formula to be learned, the more difficult is the learning activity.

Finally we observe that if the value of s is unknown it is possible to modify Valiant's algorithm to obtain an algorithm for learning k-d-term-DNF formulae under uniform distribution with $O(2^k /(1/2 - (d-1)\, 2^{(s-k)/2}) (k \log n + \log (1/\varepsilon) \log (1/\varepsilon))$ examples (i.e. with an increase of a $\log (1/\varepsilon)$ factor).

5 Learning d-term-DNF formulae under uniform distribution

In this section we extend the results of Section 3 to broader classes of formulae. More precisely in Section 3 we studied k-d-term-DNF formulae; now we consider d-term-DNF formulae. To study this class of formulae we use an idea introduced in [KMP88] for the distribution free model. We try to apply Valiant's algorithm for learning k-DNF with k=1, then k=2 and so on, until the formula is learned. Since we want to learn d-term-DNF formulae, we do not know the value of k. Hence the termination condition in Valiant's algorithm necessitates to be modified.

For an integer value h, let T be the set of all terms of up to h literals; let Ex be the set of examples seen in so far (obviously at the beginning Ex is empty). Let M be the set of all terms in T that do not satisfy any negative example from Ex. Then the algorithm calls for a sequence of negative examples. At each step it eliminates from M those terms that satisfy the negative example (and hence cannot be present in the formula to be learned).

There are two possibilities. If there is a sufficiently long sequence of negative examples in which few terms are eliminated then condition ii) of definition 2.1 is satisfied (i.e. the formula is learned as far as negative examples); otherwise after a sufficiently number of examples have been called, we know that the formula to be learned has more than h literals in at least one term. Hence we increase the value h and we continue in the same way.

When condition ii) of definition 2.1 is verified we still have to check condition i) of the definition. To this aim let g be the sum of all remaining terms. We call for a suitable set of positive examples and we check for each example X whether g(X) = 1 for a sufficiently number of examples. If this is true then condition i) is satisfied, otherwise we increase h and we proceed in the same way by considering all possible terms of up to h literals and calling for a sequence of negative examples.

Algorithm A

<u>begin</u>

h := 0; M := ∅; Ex := ∅;

<u>repeat</u>

  h := h +1;

  G := { set of terms with h literals per term};

  eliminate terms from G that satisfy examples belonging to Ex;

  M := M ∪ G;

  neg := false; pos := false;

  <u>while</u> ($|Ex| < 2^{k+d}$ (k log n + log (1/ε)) log (1/ε)) and (neg = false)

  <u>do</u> <u>begin</u>

    call for a negative example X and add it to Ex; eliminate those terms from M that satisfy X;

    let g be the disjuntion of all terms in M;

    {Check negative examples}

    <u>if</u> in the most recent sequence of 3 log (1/ε) /ε negative examples

      the number of examples X such that g(X) =0 is less than log (1/ε)

    <u>then</u> neg := true

    <u>end</u> ;

  <u>if</u> neg = true

  <u>then</u> {Check positive examples}

  <u>begin</u>

    call for a sequence of log (1/ε) /ε positive examples;

    <u>if</u> for all examples X of the sequence g(X) = 1

<u>then</u> pos := true
<u>end</u>
<u>until</u> (pos and neg)
<u>end</u>.

The analysis of the above algorithm is given in the following two theorems.

<u>Theorem 5</u> [KMP88] Given a DNF formula f with at most k literals per term, algorithm A correctly learns it in polynomial time. Furthermore, the value of the variable h is, at the end of the algorithm, no greater than k.

<u>Theorem 6</u> Let f be a d-term-DNF formula such that h, the maximum number of literals in a term, verifies $h < n - d$. Algorithm A learns f under uniform distribution in polynomial time using $O(2^{k+d} (k \log n + \log (1/\varepsilon)) \log (1/\varepsilon))$ examples.
<u>Proof</u> The proof is a modification of theorem 2. Details will be given in the full paper.

The second result of this section concerns the intrinsic difficulty of learning any DNF formula applying a maximal consistent algorithm.

<u>Theorem 7</u> For each n, there exists a set of DNF formulae that requires $\Omega(2^n / \sqrt{n})$ examples to be learned under uniform distribution by a maximal consistent algorithm.
<u>Proof</u> The proof is a modification of theorem 3. Details will be given in the full paper.

6 References

[BEHW 86] A.Blumer,A.Ehrenfeucht,D.Haussler,M.Warmuth **Classifying learnable geometric concepts with the Vapnik-Chervonenkis dimension** Proc. 18th Acm Symposium on Theory of Computing (1987).
[BR 87] P.Berman,R.Roos **Learning one-counter languages in polynomial time** Proc. 28th Symposium on Foundations of Computer Science (1987).
[EHKV88] A.Ehrenfeucht, D.Haussler,M.Kearns,L.Valiant **A general lower bound on the number of examples needed for learning** Information and Computation (1988).
[F71] W.Feller **An introduction to probability theory and its applications** Vol.1 Wiley and Sons (1971).
[H88] D.Haussler **Quantifying inductive bias: AI learning and Valiant's learning framework** Artificial Intelligence, 36 (1988).

[KLPV 87] M.Kearns,M.Li,L.Pitt,L.G.Valiant **On the learnability of boolean formulae** Proc.19th Acm Symposium on Theory of Computing (1987).

[LMR88] N.Linial,Y.Mansour,R.L.Rivest **Results on learnability and Vapnik Chervonenkis dimension** Proc. 29th Symposium on Foundations of Computer Science (1988).

[L 87] N.Littlestone **Learning quickly when irrelevant attributes abound : a new linear threshold algorithm** Machine learning 2,(4) (1988).

[M82] T.M.Mitchell **Generalization as search** Artificial Intelligence, 18 (1982).

[N 87] B.K.Natarajan **On learning boolean functions** Proc. 19th Symposium on Theory of Computing (1987).

[V 84] L.G.Valiant **A theory of learnable** Comm.Acm, 27,(11) (1984).

[VC71] V.Vapnik,A.Y.Chervonenkis **On the uniform convergence of relative frequencies of events to their probabilities** Theor. Prob. and Appl., 16, (2) (1971).

An extended framework for default reasoning

M.A. Nait Abdallah
Department of Computer Science
University of Western Ontario
London, Ontario, Canada

Abstract

In this paper, we investigate the proof theory of default reasoning. We generalize Reiter's framework to a *monotonic* reasoning system, and in particular allow formulae with *nested* defaults. We give proof rules for this extended default logic, called *default ionic logic*, and give deduction theorems. We also give examples of applications of our framework to some well-known problems: weak implication, disjunctive information, default transformation, and normal versus non-normal defaults.

1 Introduction

In this paper we investigate the proof theory of default reasoning. We have as a goal a general theory for combining defaults, with the possiblity of having *nested defaults*, and a logical tool for choosing among defaults for implementation purposes.

The calculus on extensions developed by Reiter [5] has yielded a kind of reasoning that has been called *non-monotonic.* This non-monotonicity is the source of some major problems for implementers.

The applicability of *modus tollens*, a fundamental tool in resolution, is unclear. For example L. Sombé [7] p 143 indicates that the contrapositive of default $\frac{u \ : \ v}{v}$ should be $\frac{\neg v \ : \ \neg u}{\neg u}$, but that this is missing in default logic. On the other hand, Poole [4], encountering a similar problem, writes: "*Assume we have the following Theorist fragment:*

$$\textbf{default}\ birdsfly \ : \ bird(x) \rightarrow flies(x)$$

... Using the default we can also explain $\neg bird(b)$ from $\neg flies(b)$. ... Theorist users have found that they needed a way to say "this default should not be applicable in this case" [4] p 15. He then proceeds to introduce an operational device he calls *constraints*, "*a very useful mechanism in practice*" [4].

In this paper, we generalize Reiter's framework [5, 6] to a *monotonic* reasoning system, and in particular allow formulae with nested defaults. We give proof rules for this new default logic, called *default ionic logic*, and prove deduction theorems.

We aim at a *general theory for combining defaults*, with the possibility of having *nested defaults*, and a *logical tool for choosing among defaults* for implementation purposes. It should be clear from the beginning that, in this paper, *we are not using defaults* in the usual sense (e.g. for building extensions), but *we are talking about defaults.* In other words, we are using defaults the way predicate formulae are used in predicate logic, i.e. as the basic building stones upon which we are going to build our proofs.

The main ideas of our approach are:

- Defaults are not used as simply *tools* for approximating propositional (or first-order) theories. Rather, we are developing a logic on its own, where *defaults are considered as statements*, i.e. as first-class citizens, that should be studied on their own.

- We attempt to *rehabilitate the classical point of view* that *reasoning amounts to establishing logical consequences* in the sense of the deductive system we present in this paper (See Section 3.2).

- The classical properties of deduction in predicate logic should be—and are, in our system—preserved, and no additional *operational* features, such as Poole's [4] *constraints* saying when modus tollens should not be applied to a default, are needed.

- We have a logic that is *monotonic*. The idea here is that the apparent non-monotonicity of default reasoning is due to some essential parameters of the reasoning being hidden. Their hiding is the sole cause of non-monotonicity. The display of these *hidden parameters* shows the obvious monotonicity of the logic involved.

- We handle a much richer language than Poole [4] and Reiter [5, 6]. This allows us to have conclusions, requisites and justifications that contain, or are themselves defaults in the sense of Reiter.

Some advantages of default ionic logic are:

- Prerequisites, justifications and conclusions may contain (nested) defaults.

- We can easily handle *disjunctive information.*

- Our logic is trivially *compact* and we are able to show *deduction theorems.*

- The notion of *contrapositive* of a default is clarified: The contrapositive of $a \rightarrow [b]$ (cf Section 2 for our notation) is indeed $\neg[b] \rightarrow \neg a$ and not $\neg b \rightarrow \neg[a]$. *Weak implication* $a \rightarrow [b]$ (first form) is somehow transitive, since $a \rightarrow [b]$ and $b \rightarrow [c]$ implies $a \rightarrow (b, [c])_{\star}$. *Weak implication* $[a \rightarrow b]$ (second form) is "less transitive" since, using the *deduction theorem*, one can show that $[a \rightarrow b]$ and $[b \rightarrow c]$ both imply $(\{a \rightarrow b\ ,\ b \rightarrow c\}\ ,\ c)_{\star}$.

The contents of this paper are as follows. Section 2 gives the intuition behind *default ions.* Section 3 gives the syntax and axiomatics of propositional default ionic logic. Using these definitions, we are able to prove two deduction theorems, and discover some new default transformation rules. Section 4 gives some applications: we discuss weak implication and modus tollens, show how to use disjunctive information, prove the correctness of two transformation rules suggested by Lukaszewicz [1], and reexamine the notion of constraint in the sense of Poole [4].

2 Default ions

The basic idea of our approach is the same as the one used by D. Scott when he made "*absence of information/termination*" into an element of the domain (of computation) called *bottom.* It was also used in [3] to solve the local definition problem in logic programming. Essentially, we make Reiter's default

$$\frac{: n}{p}$$

into an element of a set of generalized formulae, called *ionic logic formulae*, and denote it by the formula:

$$(n, p)_\star$$

Default ion $(n,p)_\star$ should be read: If n is consistent with our beliefs (i.e. current logical scope), then infer (or assert) p. For the sake of having a shorter notation, default ion $(n,n)_\star$ will be sometimes abbreviated as $[n]$. The intuitive meaning of $[n] = (n,n)_\star$ is *'n' is true if it is consistent with the current context.* In other words, we "weakly" assert n. From now on, statements of the form $[n]$ will be called *weak statements* and they shall be read "*weakly n*". Symmetrically, statements from "usual" logic will be called *strong statements.*

The default

$$\frac{m : n}{p}$$

will be translated by the ionic formula

$$m \rightarrow (n, p)_\star$$

In the terminology of default logic, m is the prerequisite, n is the justification, and p is the conclusion.

3 Propositional default ionic logic

3.1 Syntax

We take a set P of propositional letters together with the single higher-order binary operator (we call *ionic operator*) $(\;.\;,\;.\;)_\star$, and the usual set of connectives.

Formulae of *propositional default ionic logic* are called *(default) ionic formulae*, and are recursively defined as follows:

1. Each propositional letter is a formula.
2. The set of formulas is closed under connectives $\wedge$, $\vee$, $\neg$ and $\rightarrow$.
3. The set of ionic formulae is closed under the ionic operator: For every *finite* set of formulae $\Phi = \{f_1 \;,\; \ldots \;,\; f_n\}$ and formula g, the expression $(\Phi, g)_\star$ is a formula, called *default (propositional) ion.*

The intuitive meaning of *default ion* $(\{f_1 \;,\; \ldots \;,\; f_n\}, g)_\star$ is as follows: *If, separately, each one of the f_i is consistent with the current set of beliefs (logical context), then one can assert g.* [1] In the case where $\Phi = \{f\}$ is a singleton, ion $(\Phi, g)_\star = (\{f\}, g)_\star$ will be simply denoted by $(f,g)_\star$. If, furthermore, f and g are identical, then the abbreviation $[g]$ shall be used for ion $(g,g)_\star$.

Examples of ionic formulae are : $p \wedge \neg(\Phi, p)_\star$, $p \rightarrow (\Phi, p)_\star$ and $(\Phi, p)_\star \rightarrow p$.

3.2 Axioms and proof rules for propositional default ionic logic (PDIL)

Axioms of our logic are of two kinds: *axioms inherited from propositional logic*, and specific axioms dealing with *default ions.*

[1] Notice that, as the framework presented here is purely syntactical, the interpretation where *all* of the f_i must be simultaneously consistent with the current set of beliefs is also allowed.

3.2.1 Axioms inherited from propositional logic

A sentence is a propositional logic axiom of PDIL iff it has one of the following forms (where a, b, c are arbitrary default ionic formulae.)

- $a \to (b \to a)$
- $a \to (b \to c) \to ((a \to b) \to (a \to c))$
- $(\neg a \to \neg b) \to (b \to a)$

3.2.2 Axioms that are specific to default ions

Axioms and inference rules that are specific to default ions are as follows. Our notations are as follows : a, b, f, g, h, k, j are arbitrary default ionic formulae, J, Γ and Δ are arbitrary finite sets of ionic formulae.

- justification elimination : $j \wedge (j \wedge k, g)_\star \to (k, g)_\star$
- elementary transformations :
  1. $(J, a \to b)_\star \to (a \to (J, b)_\star)$
  2. $(a \to (J, b)_\star) \to (J, a \to b)_\star$
  3. $(J, (J, a)_\star \to b)_\star \to (J, a \to b)_\star$
- $\vee$-intro in justification : $(\Gamma \cup \{j\}, g)_\star \wedge (\Gamma \cup \{k\}, g)_\star \to (\Gamma \cup \{j \vee k\}, g)_\star$
- $\wedge$-intro in conclusion : $(\Gamma, f)_\star \wedge (\Gamma, g)_\star \to (\Gamma, f \wedge g)_\star$
- Thinning : $(\Gamma, g)_\star \to (\Gamma \cup \Delta, g)_\star$
- Empty set of justifications : $(\emptyset, g)_\star \to g$
- Justification introduction : $g \to (\Gamma, g)_\star$
- Abstraction :
  1. $(\Gamma, (\Delta, g)_\star)_\star \to (\Gamma \cup \Delta, g)_\star$
  2. $(\Gamma \cup \Delta, g)_\star \to (\Gamma, (\Delta, g)_\star)_\star$

3.2.3 Proof rules

- J-modus ponens : $\dfrac{(\Gamma \cup \{j\}, g)_\star \quad k \to j}{(\Gamma \cup \{k\}, g)_\star}$
- C-modus ponens : $\dfrac{(\Gamma, a \to b)_\star \quad (\Delta, a)_\star}{(\Gamma \cup \Delta, b)_\star}$
- I-modus ponens : $\dfrac{a \quad a \to b}{b}$

The meaning of most of these axioms and proof rules is intuitively obvious.

For example, the *J-modus ponens rule* says that, if from the fact that j is consistent with our beliefs, g can be inferred, and if k implies j, then from the fact that k is consistent with our beliefs, g can also be inferred. Indeed, if k is consistent with our beliefs, then j has to be consistent with our beliefs since $k \to j$, and the first default can then be used to infer g. Also in the *elimination rule*, the truth of j makes the requirement of $j \wedge k$ being consistent redundant, the consistenty of k suffices.

We now have the following deduction theorems for propositional default ionic logic.

Theorem 3.1 (Deduction theorem - weak form) *For any set A of default ionic formulae, and any default ionic formulae φ and ψ, if $A \cup \{\psi\} \vdash \varphi$ then $A \vdash (\psi \rightarrow \varphi)$.*

This theorem implies the following.

Theorem 3.2 (Deduction theorem—strong form) *For any set A of ionic formulae, and any default ions $(J, \psi)_\star$ and $(J, \varphi)_\star$, if $A \cup \{(J, \psi)_\star\} \vdash (J, \varphi)_\star$ then $A \vdash (J, \psi \rightarrow \varphi)_\star$*

We can also deduce the following.

Theorem 3.3 *The following formulae are theorems of propositional default ionic logic.*

- *derived elimination rules :*
  1. $(j \rightarrow k) \wedge (\{j \ , \ k\} \ , \ g)_\star \rightarrow (j, g)_\star$
  2. $k \wedge (k, g)_\star \rightarrow (\emptyset, g)_\star$
  3. $j \wedge (\Gamma \cup \{j \vee k\}, g)_\star \rightarrow (\Gamma, g)_\star$
  4. $k \wedge (\Gamma \cup \{j \vee k\}, g)_\star \rightarrow (\Gamma, g)_\star$

- *$\wedge$-introduction in justification :*
  1. $(j, g)_\star \rightarrow (k \wedge j, g)_\star$
  2. $(j, g)_\star \rightarrow (j \wedge k, g)_\star$

- *default transformation :* $((J, a)_\star \rightarrow (J, b)_\star) \rightarrow (J, a \rightarrow b)_\star$

- *$\vee$-introduction in conclusion :*
  1. $(\Gamma, g)_\star \rightarrow (\Gamma, g \vee f)_\star$
  2. $(\Gamma, g)_\star \rightarrow (\Gamma, f \vee g)_\star$

- *$\wedge$-elimination in conclusion :*
  1. $(\Gamma, f \wedge g)_\star \rightarrow (\Gamma, f)_\star$
  2. $(\Gamma, f \wedge g)_\star \rightarrow (\Gamma, g)_\star$

4 Applications of default ionic logic to "non-monotonic" reasoning

4.1 Notions of weak implication

4.1.1 $a \rightarrow [b]$

As we have mentioned earlier, default ion $[a]$ may be read "*Weakly a*". Thus the statement $a \rightarrow [b]$ should be read "*a implies weakly a*". If we take some liberties with our terminology, it can also be seen as stating a relation between a and b: "*a weakly implies b*". This yields our first notion of *weak implication*. Weak implication in this sense is somewhat *transitive*, since:

1.	$a \rightarrow (b, b)_\star$	
2.	$b \rightarrow (c, c)_\star$	
3.	a	
4.	$(b, b)_\star$	1,3 I-modus ponens
5.	$(b, (c, c)_\star)_\star$	2,4 C-modus ponens
6.	$a \rightarrow (b, (c, c)_\star)_\star$	3,5 deduction theorem

Whence $\{a \rightarrow [b], b \rightarrow [c]\} \vdash a \rightarrow (b, [c])_\star$. By the deduction theorem, we can conclude $(a \rightarrow [b]) \wedge (b \rightarrow [c]) \rightarrow (a \rightarrow (b, [c])_\star)$.

Notice that the contrapositive of "weak implication" $(a \rightarrow [b])$ is $(\neg[b] \rightarrow \neg a)$ and not $(\neg b \rightarrow \neg[a])$.

4.1.2 $[a \to b]$

A second notion of weak implication is given by $[a \to b]$ i.e. "*Weakly "a implies b"*". The contrapositive of $[a \to b]$ is not clearly defined, but *modus tollens* reasoning can be performed in a straightforward manner:

1.	$(a \to b, a \to b)_\star$	alternative notation for $[a \to b]$
2.	$(a \to b) \to (\neg b \to \neg a)$	propositional calculus tautology
3.	$(a \to b, (a \to b) \to (\neg b \to \neg a))_\star$	justification introduction
4.	$(a \to b, \neg b \to \neg a)_\star$	1,3 application

Now if we add the additional hypothesis $\neg b$, the deduction may proceed as follows:

5.	$\neg b$	hypothesis
6.	$(a \to b, \neg b)_\star$	justification introduction
7.	$(a \to b, \neg a)_\star$	4,6 application

Therefore we have $\{[a \to b], \neg b\} \vdash (a \to b, \neg a)_\star$, i.e. $[a \to b]$ entails $\neg b \to (a \to b, \neg a)_\star$.

Putting this in the customary ornithological terminology, we have established from "*It is typically the case that birds fly*" and "*Tweety does not fly*" that "*If it is consistent to believe that birds fly, then Tweety is not a bird*".

4.2 Use of disjunctive information

4.2.1 Case analysis using two strong statements

Consider the following example: *Tweety is a bird or an airplane, we do not know which. Birds typically fly, and airplanes also typically fly. Does Tweety fly?* Since universal quantification is (almost) irrelevant in this case, we can reduce this problem to propositional logic[2]. Our abbreviations are: a = Tweety is an airplane, b = Tweety is a bird, c = Tweety flies. The answer is supplied by the following deduction.

1.	$a \vee b$	
2.	$a \to [c]$	
3.	$b \to [c]$	
4.	$(a \to [c]) \to ((b \to [c]) \to (a \vee b \to [c]))$	tautology
5.	$(b \to [c]) \to (a \vee b \to [c])$	4, 2 modus ponens
6.	$a \vee b \to [c]$	3, 5 modus ponens
7.	$[c]$	1, 6 modus ponens

The conclusion is the *weak statement* that Tweety flies.

4.2.2 Case analysis using two weak statements

Consider the following problem:

Nixon is a quaker and a republican. Quakers are typically pacifists. Republicans are typically non-pacifists. Both pacifists and non-pacifists are typically politically active. Is Nixon politically active?

[2]In fact, we *are* replacing the given problem by another problem where we have : "If Tweety is a bird, then if it is consistent to assume that Tweety flies, deduce that Tweety flies. If Tweety is an airplane, etc ..."

Using the same reduction to propositional logic as above, together with the following abbreviations: r = nixon is a republican, q = nixon is a quaker, p = nixon is a pacifist, a = nixon is politically active, we get the following deduction:

1.	q	
2.	r	
3.	$q \to [p]$	
4.	$r \to [\neg p]$	
5.	$p \to [a]$	
6.	$\neg p \to [a]$	
7.	$[p]$	1,3 I-modus ponens
8.	$[\neg p]$	2,4 I-modus ponens
9.	$(p, [a])_\star$	5,7 C-modus ponens
10.	$(\neg p, [a])_\star$	6,8 C-modus ponens
11.	$(p, [a])_\star \wedge (\neg p, [a])_\star$	9,10 conjunction
12.	$(p \vee \neg p, [a])_\star$	11, $\vee$-intro in justification, I-modus ponens
13.	$p \vee \neg p$	tautology
14.	$[a]$	justification elimination

Thus we conclude that the *weak* statement "*nixon is politically active*" holds.

4.3 Lukaszewicz rules as theorems

In [1] Lukaszewicz suggests, on the basis of their "reasonability", two transformations rules whose intent is to allow one to replace non-normal defaults by normal ones. The rules are :

1. $a \to (b, c)_\star$ can be replaced by $a \to (b \wedge c, c)_\star$.
2. If $(a \wedge c, b)_\star$ holds, then $a \to (b \wedge c, c)_\star$ can be replaced by $a \to (b \wedge c, b \wedge c)_\star$.

In fact the following can be proved in our framework:

1. $(a \wedge c, b)_\star$ entails $a \to (b, c)_\star$.
2. $a \to (b, c)_\star$ entails $a \to (b \wedge c, c)_\star$.
3. If $(a \wedge c, b)_\star$ holds, then $a \to (b \wedge c, c)_\star$ and $a \to [b \wedge c]$ entail each other.

We show here that Lukaszewicz' two rules above can be proved as theorems in our system in the following sense: *each formula entails the formula which replaces it.*

Proof of Lukaszwicz first transformation

1.	$a \to (b, c)_\star$	
2.	$(b, c)_\star \to (b \wedge c, c)_\star$	$\wedge$-intro in justification
3.	$(a \to (b, c)_\star) \to (((b, c)_\star \to (b \wedge c, c)_\star) \to (a \to (b \wedge c, c)_\star))$	axiom propositional logic
4.	$((b, c)_\star \to (b \wedge c, c)_\star) \to (a \to (b \wedge c, c)_\star)$	1,3 I-modus ponens
5.	$a \to (b \wedge c, c)_\star$	2,4 I-modus ponens

Proof of Lukaszewicz second transformation

1.	$(a \wedge c, b)_\star$	
2.	$a \rightarrow (b \wedge c, c)_\star$	
3.	a	hypothesis
4.	$(b \wedge c, c)_\star$	2,3 I-modus ponens
5.	$(c, b)_\star$	1,3 justification elimination
6.	$(b \wedge c, b)_\star$	5, $\wedge$-intro in justification
7.	$(b \wedge c, b \wedge c)_\star$	4,6 $\wedge$-intro in conclusion
8.	$a \rightarrow (b \wedge c, b \wedge c)_\star$	above deduction 1 thru 7, deduction theorem

Thus both of Lukaszewicz's transformations are obtained as consequences of the *deduction theorem.*

4.4 An example of Reiter, Criscuolo and Lukaszewicz revisited

In this example, we shall use some *first-order logic* language, but as we shall see, propositional logic is sufficient for treating the example.

Confusion between *weak* and *strong* statements may lead to an ambiguous understanding of what some authors write. In [1], the following example, borrowed from [6], is discussed : *Typically adults are employed, except when they are high-school dropouts.* Our abbreviations are $a(x) = x$ is an adult, $d(x) = x$ is a dropout, $e(x) = x$ is employed, and $j =$ John. This yields :

$$\forall x \ . \ a(x) \rightarrow (\neg d(x) \wedge e(x), e(x))_\star$$

In [6] an alternative representation is suggested in the case where the additional assumption h "*Typically adults are not high-school dropouts*", i.e. $\forall x \ . \ a(x) \rightarrow (\neg d(x), \neg d(x))_\star$ holds :

$$RC = \{a(x) \rightarrow [\neg d(x)] \ ; \ a(x) \wedge \neg d(x) \rightarrow [e(x)]\}$$

In [1] it is argued on the basis of the *same* additional assumption that the original statement can be replaced by :

$$L = \{a(x) \rightarrow [\neg d(x) \wedge e(x)]\}$$

We then have the following:

$$a(j) \vdash_L [\neg d(j) \wedge e(j)]$$
$$a(j) \vdash_{RC} [\neg d(j)] \ , \ (\neg d(j), [e(j)])_\star$$

If we further assume that $\neg e(j)$ also holds, then Lukaszewicz claims: "*The representation of Reiter and Criscuolo forces John not to be a high-school dropout, while ours remains agnostic on this point.*" We show here that the original statement together with hypothesis h is equivalent to Reiter and Criscuolo's formulation (RC), which is equivalent to Lukaszewicz' formulation (L) together with hypothesis h. In other words, the latter formulation is more agnostic than the former *only if* one forgets the additional assumption h that was used to justify it.

We now proceed to prove this formally in our system.

- Original statement + hypothesis imply L :

1.	$a \rightarrow (\neg d \wedge e, e)_\star$	given original
2.	$a \rightarrow [\neg d]$	hypothesis h
3.	a	our hypothesis
4.	$(\neg d, \neg d)_\star$	3,2 I-modus ponens
5.	$(\neg d \wedge e, e)_\star$	1,3 I-modus ponens
6.	$(\neg d \wedge e, \neg d)_\star$	4, justification introduction
7.	$(\neg d \wedge e, \neg d \wedge e)_\star$	5,6 $\wedge$-intro
8.	$a \rightarrow [\neg d \wedge e]$	above deduction+deduction theorem (weak form)

- Original + hypothesis h imply RC

1.	$a \rightarrow (\neg d \wedge e, e)_\star$	original
2.	$a \rightarrow [\neg d]$	h
3.	$a \wedge \neg d$	hypothesis
4.	a	3, $\wedge$-elim
5.	$(\neg d, \neg d)_\star$	4,2 I-modus ponens
6.	$(\neg d \wedge e, e)_\star$	1,4 I-modus ponens
7.	$\neg d$	3, $\wedge$-elim
8.	$(e, e)_\star$	6,7 justification elimination
9.	$a \rightarrow [e]$	above deduction +deduction theorem

- RC implies Original statement + hypothesis.

Since RC contains the hypothesis, it is enough to show that RC implies $(a \rightarrow (\neg d \wedge e, e)_\star)$

1.	$a \wedge \neg d \rightarrow [e]$	
2.	$a \rightarrow [\neg d]$	
3.	a	hypothesis
4.	$(\neg d, \neg d)_\star$	2,3 I-modus ponens
5.	$\neg d \wedge e \rightarrow \neg d$	tautology
6.	$(\neg d \wedge e, \neg d)_\star$	4,5 C-modus ponens
7.	$(\neg d \wedge e, a)_\star$	3,jsutification intro
8.	$(\neg d \wedge e, \neg d \wedge e)_\star$	6,7 $\wedge$-intro
9.	$(\neg d \wedge e, [e])_\star$	1,8 C-modus ponens
10.	$(\{\neg d \wedge e \ ; \ e\} \ , \ e)_\star$	10, abstraction
11.	$(\neg d \wedge e, e)_\star$	justification elimination
12.	$a \rightarrow (\neg d \wedge e, e)_\star$	deduction theorem

- L implies the original statement

1.	$a \to [\neg d \wedge e]$	given L
2.	a	hypothesis
3.	$(\neg d \wedge e, \neg d \wedge e)_\star$	1,2 I-modus ponens
4.	$\neg d \wedge e \to e$	tautology
5.	$(\neg d \wedge e, e)_\star$	3,4 C-modus ponens
6.	$a \to (\neg d \wedge e, e)_\star$	previous deduction +deduction theorem

Therefore, Reiter and Criscuolo's representation RC is equivalent to Lukaszewicz' representation L together with additional assumption h.

4.5 Syntactic constraints a la Poole are unnecessary

Introducing default ions allows one to avoid some syntactic problems encountered by some authors. For example, the default "*People are normally not diabetic*" [4] p 15, is translated in our formalism by:

$$\forall x \;.\; person(x) \to [\neg diabetic(x)]$$

Notice that $diabetic(john)$ does not imply $\neg person(john)$ as in Poole's formalism [4]. To avoid this, Poole introduces *constraints*, an operational device with no clear logical semantics. In our formalism, we can deduce nothing: indeed, $\neg[\neg diabetic(john)]$ implies $\neg person(john)$, but $diabetic(john)$ does not imply $\neg person(john)$, since the *strong* statement $diabetic(john)$ and the negation of the *weak* statement $\neg[\neg diabetic(john)]$ are very different formulae with different meanings.

Another way of interpreting Poole's default [4] p 15 is to read it as *Typically it is not the case that people are diabetic*; we must then use the weak implication

$$\forall x \;.\; [person(x) \to \neg diabetic(x)]$$

But this does not change the essence of our argument, since the difference will be then between *ion* $(person(john) \to \neg diabetic(john), \neg person(john))_\star$, deduced via modus tollens, and *strong statement* $\neg person(john)$.

Thus, in both cases, our more adequate notation makes the problem encountered by Poole disappear.

References

[1] Lukaszewicz W. *Two results on default logic*, Proc. IJCAI-85, (1985), pp. 459-461

[2] McDermott D. and Doyle J. *Non-monotonic logic I*, Artificial Intelligence 13, (1980), pp. 27-39

[3] Nait Abdallah M. A. *Ions and local definitions in logic programming*, Springer LNCS 210 (1986), pp. 60-72

[4] Poole D. *A logical framework for default reasoning*, U. of Waterloo report CS-87-59 (1987)

[5] Reiter R. *A logic for default reasoning* , Artificial Intelligence 13, (1980), pp. 81-132

[6] Reiter R. and Criscuolo G. *On interacting defaults* , Proc IJCAI-81, pp. 270-276

[7] Sombé L. *Inférences non-classiques en intelligence artificielle* in Actes des Journees Nationales Intelligence Artificielle, D. Pastre ed., Teknea (1988) pp. 137-230

Logic programming of some mathematical paradoxes

M.A. Nait Abdallah
Department of Computer Science
University of Western Ontario
London, Ontario, Canada

Abstract

Using the notions of logic field and ion defined in [7, 9], we give an algorithmic analysis, in terms of logic programming, of three paradoxes: Protagoras, Newcomb and the Hangman. We show that each one of these paradoxes points out a *programming* mistake to be avoided in logic programming.

1 Introduction

The question we discuss in this paper is: how does logic programming relate to fundamental problems of expression and reasoning in mathematics, i.e. how it relates to *mathematical paradoxes.*

A careful analysis of Russell paradox from the point of view of combinatory logic, leads to the fixpoint combinator $Y = \lambda f.(\lambda x.f(xx))(\lambda x.f(xx))$. The logical paradox then resolves into an infinite computation. The Y combinator turns out to be interpreted as the least fixpoint operator in Scott-Strachey denotational semantics [3, 15], thus giving a semantics for recursive procedure calls in programming.

Not all paradoxes are so easily analyzed in terms of *combinatory* logic, however, as constructs other than *abstraction* and *application* may intervene. Some of these other paradoxes are especially important from the point of view of a *formal logic of knowledge* [2]. The following question then arises: can we learn as much about *knowledge-oriented programming languages*, from a computational study ot these paradoxes, as we have learned about *imperative and functional programming languages* from a computational study of Russell paradox? In order to examine this question, we shall consider in this paper three such paradoxes, namely *Protagoras paradox*, *Newcomb's paradox* and the *Hangman paradox*. We shall not attempt to give comprehensive references about these paradoxes, as they have been discussed in innumerable books and articles.

Our method of investigation for solving the above problems is based on programming language semantics. Such a *prescriptive* use of semantics has been advocated in [1].

The general approach we adopt is that of *predicate logic*, as most of mathematics is based on predicate logic. More specifically, we shall use a *logical extension* of Horn-clause programming [8, 9]. This extension is based on the two concepts of *ions* and *logic fields*, and may be seen as an amalgamation of algebraic semantics [5] and Horn-clause programming [4]. It uses equation-solving in the set of logic programs.

Intuitively the notion of *(generic) procedure* corresponds to what is available in ordinary programming languages: the possibility to handle a λ-abstracted form of a piece of code. Here pieces of code are sets of Horn clauses, and the variables to be abstracted may be any one of the following: object variables, function variables, predicate variables, or procedure variables.

The equations on logic programs to be solved (called here, following [5], *schemes* or *rewriting systems*) correpond to the evaluation of a given program in the context of a given set of (recursive) procedure definitions, or to the evaluation of those procedure values that allow to solve a given goal (checking theories against facts, see below). Pairs of the form (logic program, system of equations) are called *logic fields* [9].

The paper is organized as follows.

We first give an outline of our theoretical framework. We define ions, procedures, rewriting systems and logic fields. We also define derivations on ions, and give an example.

We then describe an algorithmic approach to Protagoras, Newcomb's, and the Hangman paradoxes. Each one of these paradoxes is given a full formalization in the framework of logic fields, and its mechanisms are analyzed in computational terms. We argue that each one of these paradoxes stresses the importance of a feature of logic fields: Protagoras indicates the importance of *ions*, Newcomb's illustrates procedure parameter transmission, and the Hangman illustrates the *non-determinacy of rewriting systems.*

The yield of our study of mathematical paradoxes is as follows. Protagoras unfolds into a set of computations that can be represented by a *finite automaton.* Newcomb's paradox turns out to arise from an ill-defined notion of prediction, and one of its components, the *dominance argument,* is based on the incorrect assumption that some derivation is successful. The Hangman comes out both from an incorrect management of procedure variable assignments, and a set of unjustified but hidden assumptions.

In other words, each one of these paradoxes points out some programming mistake which should be avoided in logic programming.

2 Logic fields and ions

Logic fields were first introduced in [9]. A more detailed exposition can be found in [8, 9, 12], where [8] describes how to divide a classical logic program into the main program P (see below) and the rewriting system Σ (see below), and [9, 12] describe the formalism. For the sake of brevity, we shall only recall here some definitions.

2.1 Logic fields and ions

Definitions. Let Ξ be a set of *procedure variables*, R be a set of *constant relation* symbols, Ψ be a set of *relation variables*, F be a set of *constant function* symbols, Φ be a set of *function variables*, and V be a set of *variables of individuals.*

A *logic field* is a pair (Σ, P) where P is a program with no occurrence of function or relation variables, and Σ is a rewriting system. A *program* is a set of definite clauses and procedure calls. A *definite clause* is an implicitly universally quantified well-formed formula of the form:

$$A \leftarrow B_1 \wedge \ldots \wedge B_m \qquad m \geq 0$$

where each one of A, B_1, ..., B_m is an atom or an ion. *Atoms* (or *atomic formulae*) are defined in the usual way. A *procedure call* is an expression of the form $\xi(\vec{\alpha}; \vec{\beta})$, where ξ is a procedure variable, $\vec{\alpha}$ is a list of actual input parameters, and $\vec{\beta}$ is a list of actual output (name) parameters. We designate by $\Pi(\Xi, R \cup \Psi, F \cup \Phi, V)$ the set of all logic programs constructed in this way.

A *rewriting system* over $\Pi(\Xi, R \cup \Psi, F \cup \Phi, V)$, where Ξ is the set of *procedure* variables, R (resp. Ψ) is the set of constant (resp. variable) relation symbols, F (resp. Φ) is the set of constant (resp. variable) function symbols, and V is the set of variables of individuals, is a system Σ of n equations:

$$\Sigma: \;\; \xi_i = \forall \vec{\varphi_i} \exists \vec{\chi_i} . T_i \qquad i = 1, \ldots, n$$

where $\xi_i \in \Xi$ is a procedure variable, $\forall\vec{\varphi_i}\exists\vec{\chi_i}.T_i$ is a set of procedures $\forall\vec{\varphi_i}\exists\vec{\chi_i}.Q_i$, where $T_i \subseteq \Pi(\Xi, R \cup \Psi_i, F \cup \Phi_i, V)$ is a subset whose elements are logic programs such that $\Phi_i = \Phi \cap (\vec{\varphi_i} \cup \vec{\chi_i})$, $\Psi_i = \Psi \cap (\vec{\varphi_i} \cup \vec{\chi_i})$. A *procedure* is a closed expression of the form $\forall\vec{\varphi}\exists\vec{\chi}.Q$, where Q is a logic program.

An *ion* is a pair $((\Sigma, P), g)$ where (Σ, P) is a logic field and g is a conjunction of atoms or ions; the g part can be seen intuitively as the "goal component" of the ion.

If a logic program consists of a single procedure call $\{\xi(\vec{\alpha}; \vec{\beta}) \leftarrow\}$, then it will be abbreviated as $\xi(\vec{\alpha}; \vec{\beta})$. Also, if we are in the context of one single rewriting system Σ, ions $((\Sigma, P), g)$ may be abbreviated as (P, g), the rewriting system thus remaining implicit.

2.2 Derivations on ions

We define derivations steps on ions as follows.

An *SLD step* is any instance of the following ion rewriting rule:

$$\frac{g \rightarrow^P g'}{((\Sigma, P), g) \rightarrow ((\Sigma, P), g')}$$

where $g \rightarrow^P g'$ means that g yields g' in one SLD-step in the context of logic program P, more precisely, in the clausal context of the occurrence of g in P [7, 9].

A *rewriting step* (or *procedure call step*) is any instance of the following ion rewriting rule:

$$\frac{P \rightarrow^\Sigma P'}{((\Sigma, P), g) \rightarrow ((\Sigma, P'), g)}$$

where $P \rightarrow^\Sigma P'$ means that logic program P yields logic program P' in one rewriting step in the context of rewriting system Σ. For the needs of this paper, rewriting steps shall be submitted to the following *unique procedure call rewriting rule*: *Whenever in a given derivation, a given procedure variable is rewritten into some specific procedure, in all later steps of this derivation, this same procedure variables can only be rewritten into the* same *procedure value.*

A derivation *succeeds* if it yields an *empty ion*, i.e. one whose "goal component" is empty.

Since ions are pairs $((\Sigma, P), g)$, where g is itself a conjunction of atoms or ions, the question arises about what SLD-steps are permitted in the case of *nested ions*, for example $((\Sigma, P), g_1 \wedge ((\Sigma_2, P_2), g_2))$. In this case, for the needs of this paper, we shall take as a convention that only definite clauses belonging to P_2 will be applicable to g_2 for generating SLD-steps. More details on this question are given in [12].

2.3 Example

An example of a derivation with procedure calls is as follows. Let (Σ, P) be the following deterministic logic field.

$$\begin{array}{llcl}
\Sigma = \{ & \xi_1 = \exists\chi.\{\ \chi(0, x) & \leftarrow & \\
 & \qquad\qquad \chi(s(x), s(y)) & \leftarrow & \chi(x, y)\} \\
 & \xi_2 = \forall\varphi_1\exists\chi_1.\{\ \chi_1([x]) & \leftarrow & \\
 & \qquad\qquad \chi_1([xy \mid z]) & \leftarrow & \varphi_1(x, y) \wedge \chi_1([y \mid z])\}\}
\end{array}$$

with logic program:

$$\begin{array}{lll}
P = \{ & 1. & \xi_1(; le) \leftarrow \\
 & 2. & \xi_2(le; ordered) \leftarrow \\
\} & &
\end{array}$$

In rewriting system Σ, procedure ξ_1 *exports* the binary predicate χ, procedure ξ_2 *imports* binary predicate ϕ_1 end *exports* unary predicate χ_1. Logic program P acts like a pipe: it feeds the predicate exported by ξ_1 (and called here le) to ξ_2, which will return the unary predicate *ordered*.

Then we obtain the following derivation:

1.	$((\Sigma, P), ordered([0\ s(0)])$	
2.	$((\Sigma, P_1), ordered([0\ s(0)])$	rewriting step for ξ_2
3.	$((\Sigma, P_1), le(0, s(0)) \wedge ordered([s(0)])$	$2, P_1(2.2)$
4.	$((\Sigma, P_1), le(0, s(0)))$	$3, P_1(2.1)$
5.	$((\Sigma, P_2), le(0, s(0)))$	rewriting step for ξ_1
6.	$((\Sigma, P_2),\ )$	$5, P_2(1.1)$

where P_1 is obtained from P, and P_2 from P_1, by means of a rewriting step. More precisely P_1 and P_2 have values:

$$
\begin{array}{lll}
P_1 = \{ & 1. & \xi_1(;le) \leftarrow \\
 & 2.1 & ordered([x]) \leftarrow \\
 & 2.2 & ordered([x\ y|z]) \leftarrow le(x,y) \wedge ordered([y|z]) \\
\} & &
\end{array}
$$

$$
\begin{array}{lll}
P_2 = \{ & 1.1 & le(0,x) \leftarrow \\
 & 1.2 & le(s(x),s(y)) \leftarrow le(x,y) \\
 & 2.1 & ordered([x]) \leftarrow \\
 & 2.2 & ordered([x\ \ y|z]) \leftarrow le(x,y) \wedge ordered([y|z]) \\
\} & &
\end{array}
$$

The derivation yields the empty ion $((\Sigma, P_2),\ )$ and thus is a successful derivation.

3 Protagoras paradox

Protagoras was a law teacher who had a bright, but pennyless student. In order to allow this student to complete his studies, he signs a contract with him, stating that the student would have to pay him only once he (the student) would have won his first lawsuit in court. The student successfully completes his studies to become a lawyer. But he then refuses to take any client, and also refuses to pay his teacher, arguing on the basis of their contract that he has not won any lawsuit yet. The teacher finally takes his student to court, reasoning that he would certainly get his money back, whether he wins (decision of the court is in his favour), or loses (the contract then applies). The student is not worried in the least, arguing that he does not have to pay in any case, whether he loses (the contract then says he does not have to pay yet), or wins (the court decided he did not have to pay).

In our formalization, we shall assume the following about this paradox: (i) it is a fully stated and fully logical problem (e.g. no police is available). (ii) the contract itself is never challenged in court, (iii) the student is *logically honest* in the sense that he will do his best to avoid paying, but he shall pay when logically forced to do so, (iii) the teacher will do everything he can to get his money, and is ready to sue as many times as necessary.

3.1 The logic field

The statement of the paradox will be expressed in terms of a logic field (Σ, P) defined as follows. The rewriting system Σ defines the following set of procedures :

Court describes the enforcing of the court's decision, **Contract** describes the clauses of the contract, **Student** describes the student's point of view, **Teacher** describes the teacher's point of view. Finally **History** records the events so far i.e., the history of the lawsuit; as the solution of the case is likely to take some time, this procedure will be defined by a non-deterministic equation, thus permitting updates.

The main program P defines the "rules of the game".

More precisely, Σ is defined as being the following set of equations:

$$
\begin{array}{lcl}
Court & = & \{ \; doesnotpay(x,t) \leftarrow wins(x,t) \\
 & & \quad pays(x,t) \leftarrow loses(x,t) \\
 & & \} \\
Contract & = & \{ \; pays(x,t) \leftarrow haswonatrial(x,t) \\
 & & \quad doesnotpay(x,t) \leftarrow \neg haswonatrial(x,t) \\
 & & \} \\
Student & = & \{ \; doesnotpay(student,t) \leftarrow \\
 & & \qquad (Contract \cup \{loses(student,t) \leftarrow\}, doesnotpay(student,t)) \\
 & & \qquad \wedge\, loses(student,t) \\
 & & \quad doesnotpay(student,t) \leftarrow \\
 & & \qquad (Court \cup \{wins(student,t) \leftarrow\}, doesnotpay(student,t)) \wedge \\
 & & \qquad \wedge\, wins(student,t) \\
 & & \quad pays(x,t) \leftarrow \neg doesnotpay(x,t) \\
 & & \} \\
Teacher & = & \{ \; pays(student,t) \leftarrow \\
 & & \qquad (Court \cup \{loses(student,t) \leftarrow\}, pays(student,t)) \wedge \\
 & & \qquad \wedge\, loses(student,t) \\
 & & \quad pays(student,t) \leftarrow \\
 & & \qquad (Contract \cup \{wins(student,t) \leftarrow\}, pays(student,t)) \wedge \\
 & & \qquad \wedge\, wins(student,t) \\
 & & \} \\
History & = & \{ \; \{ \; goestocourt(student,k) \leftarrow \\
 & & \qquad \varphi_k(student,k) \leftarrow \; : \\
 & & \qquad k = 1,\ldots,n \text{ and } \varphi_k \in \{loses, wins\} \\
 & & \quad \} \; : \quad n = 1,2,3,\ldots \\
 & & \}
\end{array}
$$

The main program P is given by:

$$
\begin{array}{ll}
\{ & 1.\; pays(student,t) \leftarrow (Student, pays(student,t)) \\
 & 2.\; haswonatrial(x,t) \leftarrow goestocourt(x,t') \wedge wins(x,t') \wedge (t' \leq t) \\
 & 3.\; History \leftarrow \\
\} &
\end{array}
$$

The *court decision making process* itself is not formalized by a procedure, since no information is given in the statement of the paradox. We assume that it is described by an oracle giving, each time the student and the teacher go to court, a decision that is either in favour or against the student.

3.2 The computation

The sequences of facts to be solved are :

$$F_i = goestocourt(student, i) \wedge \varphi_i(student, i) \wedge \psi_i(student, i)$$

where $\varphi_i \in \{wins, loses\}$ and $\psi_i \in \{pays, doesnotpay\}$. (As we said earlier, the exact value of the actual sequence depends on the successive decisions of the court.) In other words the ion to be solved is $((\Sigma, P), \bigwedge_{i \in N} F_i)$, where $N = \{1, 2, \ldots, n\}$ is some initial subset of the natural numbers, corresponding to the number of times n the teacher, trying to get his money, takes the student to court. The resolution of this ion leads to a regular language over the alphabet $\{w, f\}$ denoted by the regular expression f^*ww^*f, which gives the set of "paying sequences", where letter w means that the *student wins*, and f that the *student loses.* Words of this language correspond to sequences of court decisions.

The behaviour of the system is given by the *prefix language* of f^*ww^*f, i.e. by the language $ww^*+ff^*w^*+f^*ww^*f$. In other words, each string from this language gives a possible behaviour of the system. Only strings belonging to f^*ww^*f, however, will be considered as satisfying by the teacher, since they represent the only computations ending with his getting paid.

The transition table of the 4-state finite automaton corresponding to these languages is as follows:

	start	*use contract*	*use court decision*
student wins	use court decision	use court decision	use court decision
student loses	use contract	use contract	pay teacher

4 Newcomb's paradox

The paradox involves a Being who has the ability to predict the choices you will make.

There are two boxes. Box 1 contains 1,000 gold coins. Box 2 contains either 1 million gold coins or no money. You have the choice between two actions: taking what is in both boxes or taking only what is in the second box. If the Being predicts you will take what is in both boxes, (or you will base your choice on some random event), he does not put the 1 million coins in the second box. If he predicts you will take only what is in the second box, he puts 1 million gold coins in the second box. You know these facts, he knows that you know them and so on. The Being makes his prediction of your choice, does (or does not) put the 1 million coins in the second box, and then you choose. It is also assumed that you are a greedy person, and that you want to get as much gold as possible. What do you do?

There are two plausible arguments for reaching two different decisions. The *expected utility argument* is as follows. Whatever I am doing, the Being has already predicted, and thus it is better to take only the second box and thus get 1 million gold coins, instead of only 1,000 coins if I took both. On the other hand the *dominance argument* says that the Being has already made his prediction, and has acted accordingly. He has, or has not put the 1 million gold coins in the second box, and whatever I choose to do cannot change this. Therefore it is better to take both boxes, and do better by 1,000 coins, whether the second box contains 1 million coins or not.

We now provide an analysis, in terms of logic fields, of this paradox. We shall see that the paradox is ill-stated, as no precise definition of *choice* and *prediction* is given. More technically, Newcomb's paradox illustrates non-determinacy and parameter-passing in procedure calls in logic fields computations.

In our formalization, we shall assume the following about this paradox: *It is a fully logical problem i.e., the Being makes no mistake*, and thus, no probabilities are involved.

4.1 The logic field

We shall use a logic field (Σ, P) in order to express the above paradox. The rewriting system Σ defines the following two procedures : β (or $History$) describes the prediction made by the Being, and ξ describes the box contents.

The main program P defines the "rules of he game", i.e. here the description of the possible choices to be made.

The ion to be solved in order to express the solution of the paradox in terms of logic fields is : $((\Sigma, P), choose(i))$, i.e., which box is to be taken ? The resolution of this ion leads to the following results.

4.2 Newcomb's paradox under the randomness assumption

To clarify ideas about Newcomb's paradox we first define some other logic field (Σ, P) which does not formalize Newcomb as stated above, but a slight variant of it. More precisely, in the following we shall make the following *randomness assumption* (formalized by the definition of β given below): *the Being has made some* guess *and acted accordingly.* That guess has nothing to do with my own acts, and whatever I choose to do cannot change that guess (*Dominance argument*). The Being's guess may be seen as a *free* prediction.

In that case, rewriting system Σ is defined as being the following set of equations:

$$\beta = \{ \; \{ \sigma(both) \leftarrow \; \}, \; \{ \sigma(second) \leftarrow \; \} \; \}$$

$$\xi = \forall\gamma\exists B_1 B_2. \; \{ \; B_1(1,000) \leftarrow \quad B_2(0) \leftarrow (\gamma, \sigma(both)) \quad B_2(1,000,000) \leftarrow (\gamma, \sigma(second)) \; \}$$

The main program P is given by:

1. $c(both, x+y) \leftarrow B_1(x) \wedge B_2(y)$
2. $c(second, y) \leftarrow B_2(y)$
3. $choose(i) \leftarrow c(i,z) \wedge c(j,w) \wedge i \neq j \wedge (w \leq z)$
4. $\xi(\beta \; ; \; B_1, \; B_2) \leftarrow$

4.2.1 The computation

The ion to be solved is : $((\Sigma, P), choose(i))$ i.e., which box is to be taken ? The resolution of this ion leads to the following results. We have two successful derivations. The first one is obtained in the case the Being has predicted that I would take both boxes. The derivation yields the empty ion $((\Sigma, P_2), \;)$, and thus is a successful derivation is reached by taking *both* boxes and obtaining a *gain* of 1,000 coins.

We have a second successful derivation in the case the Being has predicted I would take only the second box:

$$((\Sigma, P), choose(i)) \rightarrow^* \Box, \text{ where } \beta = \{\sigma(second) \leftarrow\}, i = both, \text{ and } gain = 1 \; million + 1,000$$

where $\Box$ is an abbreviation of the fact that the derivation succeeds.

We also have two finitely failed derivations, namely:

- $((\Sigma, P), choose(i)) \rightarrow^* f\!f$, where $\beta = \{\sigma(both) \leftarrow\}$, $i = second$, and $gain = 0$.
- $((\Sigma, P), choose(i)) \rightarrow^* f\!f$, where $\beta = \{\sigma(second) \leftarrow\}$, $i = second$, and $gain = 1million$.

where $f\!f$ is an abbreviation of the fact that the derivation finitely fails.

Thus we may conclude from the previous derivations that if the Being makes a *random guess* (procedure β) before putting any money in the boxes (procedure ξ), and if my goal is to get as much money as possible, then I am better off by 1,000 coins if I pick *both* boxes. Notice that the only difference between this argument and the *dominance argument* discussed earlier in the exposition of the paradox, is that here, the Being's prediction is a random guess.

In the next step of the discussion, we shall examine what happens when we establish a link between β, i.e. the *prediction*, and $choose(i)$, i.e. the *choice*. This will give a formalization of the dominance argument itself.

4.3 Newcomb's paradox under the accuracy assumption

We now modify the previous logic field as follows. We assume that the prediction made by the Being is dependent on the action I choose to perform. This may be called the *accuracy assumption*: *The Being will predict some act if I do perform that act later on.* This yields the following new definition of β ($History$).

$$\begin{array}{lll} \beta & = & \{ \; \{ \; \sigma(both) \leftarrow ((\Sigma, P), choose(both)) \; \}, \\ & & \phantom{\{} \; \{ \; \sigma(second) \leftarrow ((\Sigma, P), choose(second)) \; \} \\ & & \} \end{array}$$

The remaining of the logic field is unchanged.

4.3.1 The computation

The resolution of the ion $((\Sigma, P), choose(i))$ leads to the following results. We have two infinite, and thus unsuccessful, derivations:

$$((\Sigma, P), choose(i)) \rightarrow^* ((\Sigma, P), (\beta, ((\Sigma, P), choose(both)))) \rightarrow^* \ldots$$

where $\beta = \{\sigma(both) \leftarrow ((\Sigma, P), choose(both))\}$, $i = both$, and $gain = 1,000$, and

$$((\Sigma, P), choose(i)) \rightarrow^* ((\Sigma, P), (\beta, ((\Sigma, P), choose(second)))) \rightarrow^* \ldots$$

where $\beta = \{\sigma(second) \leftarrow ((\Sigma, P), choose(second))\}$, $i = both$, and $gain = 1million + 1,000$.

In both cases, we have a vicious circle: I shall choose, say, both boxes, if the Being predicts that '*I shall choose both boxes*', if the Being predicts that "*the Being predicts that* 'I shall choose both boxes'", etc ...

We also have two finitely failed derivations:

- $((\Sigma, P), choose(i)) \rightarrow^* ((\Sigma, P), (\beta, ((\Sigma, P), choose(both))) \wedge (0 \leq 1,000)) \rightarrow^* f\!f$ where $\beta = \{\sigma(both) \leftarrow ((\Sigma, P), choose(both))\}$, $i = second$, and $gain = 0$.
- $((\Sigma, P), choose(i)) \rightarrow^* ((\Sigma, P), (\beta, ((\Sigma, P), choose(second))) \wedge (1million + 1,000 \leq 1million)) \rightarrow^*$ $f\!f$ where $\beta = \{\sigma(second) \leftarrow ((\Sigma, P), choose(second))\}$, $i = second$, and $gain = 1million$.

where $f\!f$ means that the derivation finitely fails.

The only way to obtain successful derivations under the accuracy assumption, is to somehow try to modify the logic field in order to make the infinite derivations successful, i.e. stop the infinite recursion.

Thus, we see that Newcomb's paradox under the *accuracy assumption*, is a combination of non-determinism plus recursiveness.

The *dominance argument* assumes that the recursion present in the above two infinite derivations is already *terminated* (i.e. solved), and thus assumes that the correct formalization of the paradox is the logic field we gave earlier by using the *randomness assumption*. Thus it appears that the meaning of the notion of "*prediction*" in the framework of the dominance argument is unclear; it is obviously not expressed by our accuracy assumption. In other words the dominance argument assumes that the Being has predicted our action, but otherwise discards any link between that prediction and our action, thus making that prediction *computationally* equivalent to a random choice. *This makes the dominance argument logically unsound.*

4.4 Newcomb's paradox as an instance of Schrödinger's Cat

The *expected utility argument* uses another way to solve the aforementioned recursion. It says "Whatever I am doing, the Being has predicted." It links the explanation substitution with the possible success of the goal *choose*(i), by feeding the value of i back to the "prediction" β of the Being. The notion of *physical determinism* vanishes, and the Being's prediction β becomes a *function* of the box choice i I am making. In other words, my choice determines the Being's prediction and the box contents. The content of the boxes remains completely undefined until I open them. This is reminiscent of Schrödinger's Cat. In terms of logic fields, this can be done as follows.

$$
\begin{aligned}
\beta \;=\; & \forall i. \\
& \{ \quad \{ \ \sigma(both) \leftarrow (i = both) \quad \}, \\
& \qquad \{ \ \sigma(second) \leftarrow (i = second) \ \} \\
& \}
\end{aligned}
$$

$$
\begin{aligned}
\xi \;=\; & \forall i \forall \gamma . \exists B_1 B_2 . \\
& \{ \ B_1(1{,}000) \leftarrow \\
& \quad B_2(0) \leftarrow (\gamma(i), \sigma(both)) \\
& \quad B_2(1{,}000{,}000) \leftarrow (\gamma(i), \sigma(second)) \\
& \}
\end{aligned}
$$

The main program P is given by:

1. $c(both, x+y) \leftarrow (\xi(both, \beta; B_1, B_2), B_1(x) \wedge B_2(y))$
2. $c(second, y) \leftarrow (\xi(second, \beta; B_1, B_2), B_2(y))$
3. $choose(i) \leftarrow c(i,z) \wedge c(j,w) \wedge i \neq j \wedge (w \leq z)$
4. $\xi(\beta; B_1, B_2) \leftarrow$

4.4.1 The computation

The resolution of ion $((\Sigma, P), choose(i))$ leads to an optimal *gain* of 1 million gold coins, and it will be obtained if I pick the second box only.

5 The Hangman paradox

"A judge decrees on Sunday that a prisoner shall be hanged on noon on the following Monday, Tuesday or Wednesday, that he shall not be hanged more than once, and that he shall not know until the morning of the hanging the day on which it will occur. By familiar arguments it appears that the decree cannot be fulfilled and that it can."[6].

The analysis in terms of logic fields of this paradox turns out to be an *effective* (i.e. *computable*) formalization of those of Quine [14] and Montague [6].

5.1 A formalization in terms of logic fields

We shall assume that $h(x)$ stands for " The prisoner is hanged at time x ". The days of the week Sunday, Monday, Tuesday and Wednesday shall be abbreviated by their initials. The sequence of times is given by the sequence of days $\{\mathbf{S},\mathbf{M},\mathbf{T},\mathbf{W}\}$. If $t \in \{\mathbf{S},\mathbf{M},\mathbf{T},\mathbf{W}\}$ we denote by $[\mathbf{S},t]$ the smallest initial subsequence containing t. A *history* is defined as a sequence of events. At any given time (day), only two mutually exclusive events are possible: "the prisoner is hanged" or "the prisoner is not hanged". No event occurs on Sunday.

5.1.1 The rewriting system Σ

The rewriting system Σ of logic field (Σ, P) will contain three procedure definitions: the prisoner K, the decree δ, and the possible sequence of events $History$.

The *prisoner* will suppose he is at some time t, and assume that the sequence of events is following some course γ where t occurs. He knows that he shall be hanged at some later time x if and only if, whenever he assumes he assumes he is at time t of history γ, he is able to prove that he will actually be hanged at time x. Obviously time t occurs earlier in history γ than time x, and the prisoner will only be able to use in his proof that part of γ that took place until time t : that part of γ we will denote by $\gamma\,|_{[\mathbf{S},t]}$.

$$
\begin{aligned}
K \;=\; & \forall t \forall \gamma. \\
& \{\;\; 1.\;\; \textit{is-history}(\gamma) \leftarrow \textit{occurs-in}(t,\gamma) \\
& \quad\;\; 2.\;\; h(x) \leftarrow \textit{is-history}(\gamma) \wedge (\gamma\,|_{[\mathbf{S},t]}, h(x)) \\
& \}
\end{aligned}
$$

As far as our syntax is concerned, the *ion* $((K(y,\gamma),q)$ stands for the statement " The prisoner knows, in the case the course of events is equal to γ, at time $y \in \gamma$, that q holds". For example, $((K(\mathbf{M},Q),h(\mathbf{T}))$ means that the prisoner, on the basis of that portion of events from Q that took place so far, knows on Monday that he will be hanged on Tuesday. Notice that we omit, in our ion notation, rewriting system Σ, as no other rewriting system occurs in our problem.

The *punishment* defined in the decree consists of two components: first, the prisoner shall be hanged at time $t \in \{\mathbf{S},\mathbf{M},\mathbf{T},\mathbf{W}\}$; second at any given time he shall not know that he will be hanged on the following day. This is expressed by the following procedure, where γ is the "history" parameter, and p is the punishment predicate defined in the decree.

$$
\begin{aligned}
\delta \;=\; & \forall \gamma \exists p. \\
& \{\;\; 1.\;\; p(\mathbf{M}) \leftarrow h(\mathbf{M}) \wedge \neg(K(\mathbf{S},\gamma), h(\mathbf{M})) \\
& \quad\;\; 2.\;\; p(\mathbf{T}) \leftarrow h(\mathbf{T}) \wedge \neg(K(\mathbf{M},\gamma), h(\mathbf{T})) \\
& \quad\;\; 3.\;\; p(\mathbf{W}) \leftarrow h(\mathbf{W}) \wedge \neg(K(\mathbf{T},\gamma), h(\mathbf{W})) \\
& \}
\end{aligned}
$$

The possible *sequences of events* are defined by the following non-deterministic equation:

$$History = \{Q_0, Q_1, Q_2, Q_3, Q_4,\}$$

where the programs Q_i are defined by:

$$\begin{aligned} Q_0 &= \emptyset \\ Q_1 &= \{h(\mathbf{M})\} \\ Q_2 &= \{\neg h(\mathbf{M}), h(\mathbf{T})\} \\ Q_3 &= \{\neg h(\mathbf{M}), \neg h(\mathbf{T}), h(\mathbf{W})\} \\ Q_4 &= \{\neg h(\mathbf{M}), \neg h(\mathbf{T}), \neg h(\mathbf{W})\} \end{aligned}$$

More intuitive names for the Q_i's would be respectively: *Is-hanged on Monday*, *Is-hanged on Tuesday*, *Is-hanged on Wednesday* and *Is-not-hanged at all*. But since we want to submit the paradox to a formal treatment, we shall stick to the Q notation.

5.1.2 The main program

The main program is defined by:

$$\begin{aligned} P \; = \; \{ \;\; & 1. \;\; History \leftarrow \\ & 2. \;\; \delta(History, p) \leftarrow \\ \} \end{aligned}$$

The formal definition of predicate $occurs\text{-}in(t, \gamma)$ used in procedure K, and which says whether a given day t occurs in a given history γ, should be included in the main program P. Such a formalization is straightforward, however, and is omitted for reasons of clarity.

5.2 The computation

The intuitive argument leading to the paradox is some kind of backward "induction": one first shows that the prisoner cannot be hanged on Wednesday, then by the same argument one eliminates Tuesday and Monday. Therefore he cannot be hanged. But the hangman comes, say, on Tuesday without being expected, and the decree is fulfilled.

Using our logic programming model, we show that this intuitive argument contains in fact quite a few hidden assumptions, and is indeed flawed. Indeed, we find three successful ions:

$$\begin{aligned} ((\Sigma, P_2^{1,Q_3}), \neg p(\mathbf{W})) &= ((\Sigma, P_2^{Q_3} \cup H_1), \neg p(\mathbf{W})) \\ ((\Sigma, P_2^{2,Q_2}), \neg p(\mathbf{T})) &= ((\Sigma, P_2^{Q_2} \cup H_2), \neg p(\mathbf{T})) \\ ((\Sigma, P_2^{3,Q_1}), \neg p(\mathbf{M})) &= ((\Sigma, P_2^{Q_1} \cup H_3), \neg p(\mathbf{M})) \end{aligned}$$

where we define $P_i^{j,Q_k} = P_i^{Q_k} \cup H_j$, where $P_i^{Q_k}$ is the value obtained from P after the i-th rewriting step, with Q_k as the assumed sequence of events, and the set of additional hypotheses H_j is defined in each case. More precisely,

$$\begin{aligned} P_2^{1,Q_3} \; = \\ \{ \;\; & 1.1 \;\; \neg h(\mathbf{M}) \leftarrow \\ & 1.2 \;\; \neg h(\mathbf{T}) \leftarrow \\ & 1.3 \;\; h(\mathbf{W}) \leftarrow \\ & 2.1 \;\; p(\mathbf{M}) \leftarrow h(\mathbf{M}) \wedge \neg(K(\mathbf{S}, Q_3), h(\mathbf{M})) \\ & 2.2 \;\; p(\mathbf{T}) \leftarrow h(\mathbf{T}) \wedge \neg(K(\mathbf{M}, Q_3), h(\mathbf{T})) \\ & 2.3 \;\; p(\mathbf{W}) \leftarrow h(\mathbf{W}) \wedge \neg(K(\mathbf{T}, Q_3), h(\mathbf{W})) \\ & 3. \;\;\;\; ((K(\mathbf{T}, Q_3), h(\mathbf{W})) \leftarrow ((\Sigma, P_2^{Q_3}), \neg h(\mathbf{M})) \wedge ((\Sigma, P_2^{Q_3}), \neg h(\mathbf{T})) \\ \} \end{aligned}$$

$$P_2^{2,Q_2} = $$

$$\{ \begin{array}{ll} 1.1 & \neg h(\mathbf{M}) \leftarrow \\ 1.2 & h(\mathbf{T}) \leftarrow \\ 2.1 & p(\mathbf{M}) \leftarrow h(\mathbf{M}) \wedge \neg(K(\mathbf{S}, Q_3), h(\mathbf{M})) \\ 2.2 & p(\mathbf{T}) \leftarrow h(\mathbf{T}) \wedge \neg(K(\mathbf{M}, Q_3), h(\mathbf{T})) \\ 2.3 & p(\mathbf{W}) \leftarrow h(\mathbf{W}) \wedge \neg(K(\mathbf{T}, Q_3), h(\mathbf{W})) \\ 3. & ((K(\mathbf{T}, Q_3), h(\mathbf{W})) \leftarrow ((\Sigma, P_2^{Q_3}), \neg h(\mathbf{M})) \wedge ((\Sigma, P_2^{Q_3}), \neg h(\mathbf{T})) \\ 4. & (K(\mathbf{M}, Q_2), h(\mathbf{T})) \leftarrow (K(\mathbf{M}, Q_2), \neg h(\mathbf{M})) \wedge (K(\mathbf{M}, Q_2), \neg h(\mathbf{W})) \\ 5. & (K(\mathbf{M}, Q_2), \neg h(\mathbf{W})) \leftarrow ((\Sigma, P_2^{1,Q_3}), \neg p(\mathbf{W})) \end{array} \}$$

In other words, ion $((\Sigma, P \cup H_1), \neg p(\mathbf{W}))$ succeeds with $History = Q_3$, ion $((\Sigma, P \cup H_2), \neg p(\mathbf{T}))$ succeeds with $History = Q_2$, and ion $((\Sigma, P \cup H_3), \neg p(\mathbf{M}))$ succeeds with $History = Q_1$. Thus the only logic field where all three goals $\neg p(\mathbf{W})$, $\neg p(\mathbf{T})$, and $\neg p(\mathbf{M})$ succeed is $(\Sigma, P \cup H_3)$, where H_3 is given by:

$$\begin{array}{l} ((K(\mathbf{T}, Q_3), h(\mathbf{W})) \leftarrow ((\Sigma, P_2^{Q_3}), \neg h(\mathbf{M})) \wedge ((\Sigma, P_2^{Q_3}), \neg h(\mathbf{T})) \\ (K(\mathbf{M}, Q_2), h(\mathbf{T})) \leftarrow (K(\mathbf{M}, Q_2), \neg h(\mathbf{M})) \wedge (K(\mathbf{M}, Q_2), \neg h(\mathbf{W})) \\ (K(\mathbf{M}, Q_2), \neg h(\mathbf{W})) \leftarrow ((\Sigma, P_2^{1,Q_3}), \neg p(\mathbf{W})) \\ (K(\mathbf{S}, Q_1), \neg h(\mathbf{T})) \leftarrow ((\Sigma, P_2^{2,Q_2}), \neg p(\mathbf{T})) \\ (K(\mathbf{S}, Q_1), h(\mathbf{M})) \leftarrow (K(\mathbf{S}, Q_1), \neg h(\mathbf{M})) \wedge (K(\mathbf{S}, Q_1), \neg h(\mathbf{W})) \\ (K(\mathbf{S}, Q_1), \neg h(\mathbf{W})) \leftarrow ((\Sigma, P_2^{1,Q_3}), \neg p(\mathbf{W})) \end{array}$$

This logic field is quite different from the initial logic field (Σ, P). In other words, the intuitive argument does not address the situation (Σ, P) described by the decree, but a different situation described by $(\Sigma, P \cup H_3)$. Therefore, there is, *stricto sensu*, no paradox.

6 Conclusion

A precise computational analysis of three mathematical paradoxes has been made. Each one of these paradoxes sheds light on some important aspect of the formal logic of knowledge.

Protagoras paradox stresses the importance of the notion of *ion*. Indeed, apart from tautologies, no statement is true except in the context of some axiomatic theory. Furthermore, there may be several coexisting, or even competing, such theories [11]. Protagoras paradox occurs because the notion of an ion is missing.

Newcomb's paradox illustrates *non-determinacy* and *parameter-passing* in procedure calls in logic field computations. This paradox occurs because these mechanisms have been ignored.

The Hangman illustrates non-determinacy and the importance of the coherent management of procedure variable rewritings provided by our framework. The intuitive argument in the Hangman paradox rests on a logical patchwork made up of the three ions mentioned in the above discussion: $((\Sigma, P_2^{1,Q_3}), \neg p(\mathbf{W}))$, $((\Sigma, P_2^{2,Q_2}), \neg p(\mathbf{T}))$, and $((\Sigma, P_2^{3,Q_1}), \neg p(\mathbf{M}))$, where the values of the three different programs P_2^{1,Q_3}, P_2^{2,Q_2}, and P_2^{3,Q_1} were also given. If the procedure variable $History$ is seen as a variable which gets assigned during the computation, it appears that this variable has been badly treated here, as it is not uniformly assigned throughout the argument. For example, we have clause (5) of P^3:

$$(K(\mathbf{M}, Q_2), \neg h(\mathbf{W})) \leftarrow ((\Sigma, P_2^{1,Q_3}), \neg p(\mathbf{W}))$$

where a property that holds in the case the course of events is Q_3 is used to deduce some fact the prisoner would know in the case the course of events is Q_2. Thus we are here simply mixing "temporal lines".

In other words, each one of these paradoxes points out some programming mistake which should be avoided in logic programming.

References

[1] E. A. Ashcroft, W. W. Wadge. *Prescription for semantics*, ACM TOPLAS 4 (2), (1982) pp. 238-293

[2] N. Asher, J. Kamp. *The knower's paradox and representational theories of attitudes*, in Theoretical Aspects of Reasoning about Knowledge, J. Y. Halpern ed, Morgan Kaufmann (1986), pp. 131-147

[3] H. Barendregt. *The type free lambda calculus*, in Handbook of Mathematical Logic, J. Barwise ed, North Holland (1977), pp. 1091-1142.

[4] van Emden M.H. and Kowalski R. *The semantics of logic as a programming language*, J. ACM 23, (1976) pp. 733-742

[5] Guessarian I. *Algebraic semantics*, Springer LNCS 99, Berlin (1981)

[6] D. Kaplan, R. Montague. *A paradox regained* , Notre Dame Journal of Formal Logic 1, (1960), pp. 79-90

[7] Nait Abdallah M. A. *Ions and local definitions in logic programming*, Springer LNCS 210 (1986), pp. 60-72

[8] Nait Abdallah M. A. *Procedures in logic programming*, Springer LNCS 225 (1986), pp. 433-447

[9] Nait Abdallah M. A. *AL-KHOWARIZMI: A formal system for higher order logic programming*, Springer LNCS 233 (1986), pp. 545-553

[10] Nait Abdallah M.A. *Logic programming with ions*, Springer LNCS 267 (1987), pp. 11-20

[11] Nait Abdallah M.A. *Heuristic logic and the process of discovery*, Proc. Fifth International Conference of Logic Programmaing, Bowen and Kowalski ed., Vol 2. MIT Press (1988)

[12] Nait Abdallah M.A. *A logico-algebraic approach to the model theory of knowledge*, (Theoretical Computer Science, to appear)

[13] R. Nozick. *Newcomb's problem and two principles of choice*, in Essays in Honor of Carl G. Hempel, N. Rescher ed., Humanities Press (1969)

[14] W.V.O. Quine, *On a so-called paradox*, Mind 57, (1953), pp. 65-67

[15] J. Stoy, *The Scott-Strachey approach to programming language semantics*, MIT Press (1977)

ANALYSIS OF COMPACT 0-COMPLETE TREES: A NEW ACCESS METHOD TO LARGE DATABASES

Ratko Orlandic
Prirodno-matematicki Fakultet
Univerzitet "V. Vlahovic", Titograd, Yugoslavia

John L. Pfaltz
Department of Computer Science
University of Virginia, Charlottesville, VA 22903

1. Introduction

A new retrieval method, called **compact 0-complete trees** or **C_0-trees**, was first presented on VLDB '88 conference [OrP88]. In that paper we have described the basic algorithms (e.g. search, insertion and deletion operators) and emphasized the usefulness of the structure for searching large files. For example, compared with B-trees [BaM72] this organization reduces the size of secondary indices (50%-80%), and hugely increases branching factor of the index tree, thus providing a reduction of number of disk accesses per exact match query. In this paper we derive the expected values of those parameters which are critical for the performance of the structure.

The analysis of the expected performance of C_0-trees can be enriched and often simplified by the ability to directly translate the results obtained by examining the conceptual structure, a special kind of a binary trie, into the statements about the actual representation. In the next section we give some preliminary remarks about the binary tries, which serve as the underlying concept upon which the final structure is based. Then we review the abstract model, called a 0-complete tree, and link it to the binary trie by the invariant property 2.2. Finally, we introduce the actual representation, which is a compact image of its conceptual 0-complete tree. In section 3 we derive the expected values of some of the parameters that govern the performance of C_0-trees. The main results of that section, expressed by theorems 3.1 and 3.2, are related to binary tries. But, using the invariant property 2.2 and the relationship between the 0-complete trees and the C_0-trees, these results are easily interpreted to obtain the desired values.

2. Background

In any edge labeled trie [Fre60] keys are presumed to be strings over an alphabet $A = \{a_i\}$. In this paper we will consider the case when $A = \{0,1\}$ giving rise to a special kind of tree called a binary trie, in which every node, except the root, can be described as either a **0-node** or a **1-node**, depending upon the label of its unique incoming edge. A binary trie is said to be **complete** if every interior (non-leaf) node has precisely two descendents (see figure 1a). By a "binary trie" we always mean a complete binary trie. Leaves of a binary trie are of special importance because all data items (or pointers to actual data items) are stored in them. If the leaf capacity, c, is one (as in figure 1), then such leaves are said to be **singleton**. Otherwise, if $c > 1$, the leaves are referred to as **pages**. Thus, we distinguish **binary tries with singleton leaves** from **paginated binary tries**. An empty leaf is called a **dummy**.

Using arc labels, every node u (leaf or non-leaf) can be uniquely identified by its access **path** from the root, which we define to be a binary string obtained by concatenating the labels of edges traversed from the root down to that node, and denote with $path(u)$. The length of $path(u)$ is the **depth** of the node u. Since the path to the root node is an empty string its depth is 0. Let the subset of keys contained in a leaf be denoted by $[K_i]$ and let $prefix_{\min}([K_i])$ be the unique, shortest common prefix of all keys in the subset which differentiates $[K_i]$ from all other subsets in the trie. Then we can define a trie in terms of the following invariant property.

Invariant Property 2.1: If a leaf L_i (with capacity $c \geq 1$) in a binary trie T contains a subset of keys $[K_i]$, then $path(L_i) = prefix_{\min}([K_i])$.

In the light of this property, every node can be assigned a *key interval* consisting of all strings from the key space that begin with the $path(u)$. Let k be the depth of the node u and let M be the key length (M=6 in figure 1). Then the key space consists of 2^M distinct bit strings and the key interval corresponding to u contains precisely 2^{M-k} such strings. Consequently, the *length* of the key interval assigned to u is $2^{M-k}/2^M = 2^{-k}$. Using the key interval assignments to the nodes of a binary trie we can say that a node is interior node if its key interval of length 2^{-k} receives more than c input items, where c is the leaf capacity. Otherwise, it is a leaf.

In the subsequent discussion we consider the binary tries with singleton leaves (i.e. c = 1), and introduce a special trie called a 0-complete tree. Essentially, a 0-complete tree can be defined as follows.

Invariant Property 2.2: Every 0-complete tree, T_0, is a complete binary trie with singleton leaves, but without dummy 1-leaves.

Notice, empty 0-leaves must be retained. Figure 1b illustrates a 0-complete tree, which has been derived from figure 1a by deleting all dummy 1-leaves. An implicit consequence of this property is that 0-complete trees with singleton leaves satisfy the trie invariant 2.1. This suggests that a missing 1-leaf can appear only if no key in the tree belongs in its key interval. As soon as such key appears, the 1-leaf is created and linked at the appropriate place.

The compact computer representation of a 0-complete tree T_0 relies on the notion of a **bounding node**, defined to be the successor node of a leaf L_i of T_0 in the *pre–order traversal* of its nodes. Since the bounding nodes are defined in terms of the pre-order traversal, it follows that in a 0-complete tree each leaf, except the last one, has its own unique bounding node. It can be shown that a node in a 0-complete tree is bounding if and only if it is a 1-node.

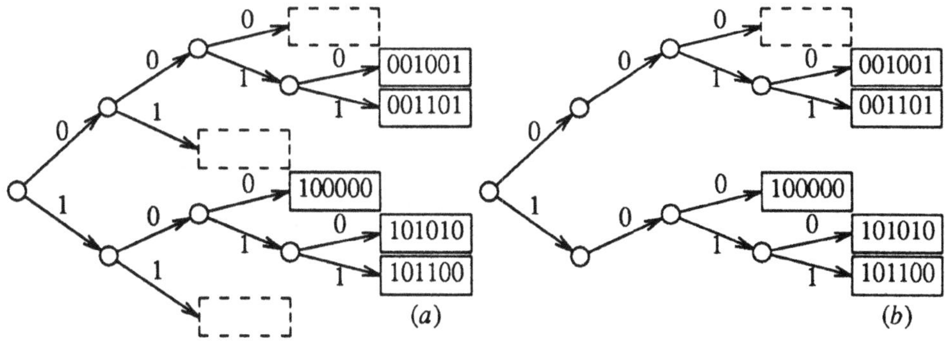

Figure 1. (a) Complete Binary Trie and (b) 0-complete Tree.

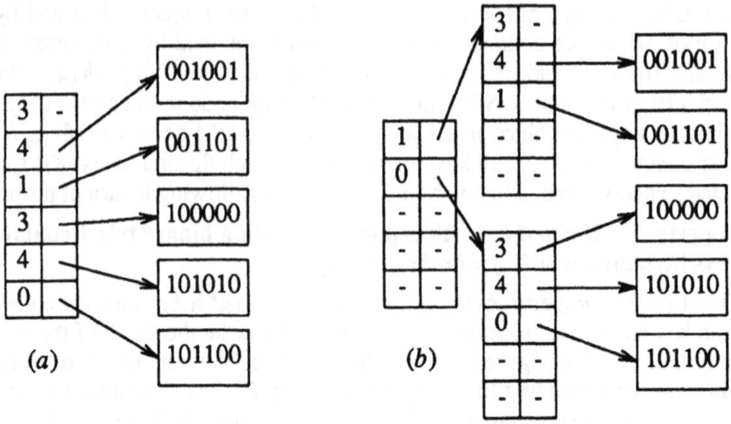

Figure 2. Compact Representation of Figure 1b (a) before and (b) after Splitting.

A compact 0-complete tree, or C_0**-tree**, is just a compact equivalent of a 0-complete tree T_0. It consists of (d_i, p_i) pairs, where d_i is the depth of the ith bounding node in the pre-order traversal of T_0 and p_i indicate the record stored in the leaf L_i of T_0 (see figure 2a). For the last entry, corresponding to the last leaf in the pre-order with no bounding node, we choose an imaginary bounding node depth 0. Dummy entries of the C_0-tree, with pointers set to *nil*, correspond to empty 0-leaves in T_0. Thus, a C_0-representation contains as many entries as there are leaves in its conceptual 0-complete tree T_0, and as many dummy entries as there are empty 0-leaves in T_0. Actual records, along with their keys, are stored in a separate data file, in arbitrary order; but they can be accessed sequentially using the C_0-tree.

Notice, only $\log_2(M+1)$ bits are needed to represent a depth value, where M is the maximal allowed key length in bits. This ensures that for all realistic key sizes (of less than 32 bytes) a 1 byte depth field will suffice, so that for large files, of up to 2^{24} data items, an index entry need not be longer than 4 bytes, yielding large storage savings. In addition, a simple and efficient retrieval algorithm [OrP88], can be used to search the index structure.

Index blocks of a compact representation are strictly bounded; they can overfill and require splitting in the fashion of B-trees. The resulting structure is a typical multiway tree hierarchy of index blocks. One splitting option, called **minimal partition splitting** [OrP88], is generated by finding the entry in the index block (other than the last entry in the block) with the minimal depth value and splitting immediately after that entry (see figure 2b). This splitting option can be analyzed since it generates a structure in which index blocks can be perceived as the leaves of a paginated 0-complete tree T'_0. For instance, the entries of the lowest level blocks of a C_0-tree are only the abstractions of the records they denote. In the paginated 0-complete tree T'_0, in which there are as many leaves as there are index blocks at the lowest level of the C_0-tree, the leaves can contain actual records instead of the representative (depth, pointer) pairs. However, paginated 0-complete trees, as oppose to 0-complete trees with singleton leaves, violate the trie invariant 2.1, implying that some items may be "out of place". The reason for this deviation is the following.

Imagine a paginated 0-complete tree T'_0 with some missing 1-leaves. If the item to be inserted in T'_0 belongs to a missing 1-leaf, then instead of creating that leaf, we rather store the item in the *first existing leaf preceding the missing leaf in the pre-order traversal.* Thus, in general, there is no correspondence between the leaves of T'_0 and the equivalent paginated binary trie generated with the same input items. But the subsequent splitting of leaves in T'_0 will tend to restore the correspondence by creating the missing leaves and storing appropriate items in them. Other splitting options can be devised to deliberately manipulate the placement of entries in the index blocks (i.e. in the leaves of a paginated 0-complete tree) to achieve higher storage utilization.

3. Analysis

This section is devoted to the derivation of analytic results concerning the performance and the expected size of a C_0-tree as a function of the number n of items indexed. In particular, we consider three critical parameters: $L_{C_0}(n)$, the number of entries at the lowest level of the tree; $D_{C_0}(n)$, the number of dummy entries at that level; and S_{C_0}, the storage utilization of its index blocks. Each of these parameters of C_0-trees has an appropriate analog in both complete and 0-complete trees, which we distinguish by a different subscript. For instance, the number of leaves $L(n)$ of complete, 0-complete and C_0-representation will be denoted by $L_T(n)$, $L_{T_0}(n)$ and $L_{C_0}(n)$, respectively. Our interest here is focused on estimation of the expected values of these parameters, for which it is traditional to assume that records are independently and uniformly distributed over the key space.

3.1. Analysis of Binary Tries

To analyze a binary trie T we assume that n items are distributed over the key space according to the Binomial (Bernoulli) law in which n is fixed and the number of keys falling in an interval of length $t = 2^{-k}$ is a Bernoulli distributed random variable Y given by:

$$P(2^{-k},\mathrm{Y}=j) = \binom{n}{j} 2^{-jk}(1-2^{-k})^{n-j} . \tag{1}$$

Unfortunately, expressions derived using this model are often cumbersome and difficult to evaluate.

More suitable expressions can be obtained using the Poisson model which is a close approximation of the Binomial distribution. In this model the number of items n in a trie T is not fixed, but randomly varies according to the Poisson law. The probability that a particular tree is built from a set of precisely n items is $P(n) = e^{-r} r^n / n!$, where r is the average number of elements. Then the number of entries, Y, which fall in an interval of length $t = 2^{-k}$ (corresponding to a node at level k) is a Poisson variable [Fae79]:

$$P(2^{-k},\mathrm{Y}=j) = \frac{(r2^{-k})^j}{j!} e^{-r2^{-k}} . \tag{2}$$

Letting the parameter r of the Poisson model be n, expression (2) becomes:

$$P(2^{-k},\mathrm{Y}=j) = \frac{(n2^{-k})^j}{j!} e^{-n2^{-k}} . \tag{3}$$

It will be shown by proposition 3.3 that this approximation causes asymptotically negligible error.

Theorem 3.1: Let T be a complete binary trie with leaf capacity c, obtained by entering n uniformly distributed items, where n is large. Let $D_T(n)$ denote the expected number of dummy 0-leaves (or dummy 1-leaves) in T. Then,
(a) $D_T(n) < ((n/c) \cdot \log_2 e) \cdot 2^{-(c+1)}$, for $c \geq 1$; and
(b) $D_T(n) \approx n \cdot (\log_2 e - 1)/2$, for $c = 1$.

Proof: Let L be a leaf at level $k \geq 1$ of T, let w be its sibling and p their common parent. Let Y be the number of items falling in the key interval of length $2^{-(k-1)}$ corresponding to p, and let X denote the number of entries falling in the interval of length 2^{-k}, corresponding to L. Then L will be a dummy leaf (0-dummy or 1-dummy) if w is an interior node and no item falls in the key interval of L. The probability that this happens is the probability that the key interval of p receives more than c items, none of which fall in the interval of L. We denote this probability with $P(Y > c,\ X = 0)$. The probability that interval of p receives j items is:

$$P(Y = j) = \frac{(n2^{-(k-1)})^j}{j!} e^{-n2^{-(k-1)}} .$$

The conditional probability that no item enters the interval of L if j items fall into its parent's interval is:

$$P(X = 0 \mid Y = j) = \binom{j}{0} 2^{-0}(1 - 2^{-1})^{j-0} = 2^{-j} .$$ Therefore, we have:

$$P(Y > c,\ X = 0) = \sum_{j=c+1}^{\infty} P(Y{=}j) \cdot P(X{=}0 \mid Y{=}j) = \sum_{j=c+1}^{\infty} \frac{(n2^{-(k-1)})^j}{j!} e^{-n2^{-(k-1)}} \cdot 2^{-j} .$$

Since there are $2^{(k-1)}$ potential 0-dummies (1-dummies) at level k of T, the expression for the total number of 0-dummies (1-dummies) is:

$$D_T(n) = \sum_{k=1}^{\infty} 2^{k-1} P(Y > c,\ X = 0) = \sum_{k=1}^{\infty} 2^{k-1} \sum_{j=c+1}^{\infty} \frac{(n2^{-(k-1)})^j}{j!} e^{-n2^{-(k-1)}} \cdot 2^{-j}$$

$$= \sum_{l=0}^{\infty} 2^l \sum_{j=c+1}^{\infty} \frac{(n2^{-l})^j}{j!} e^{-n2^{-l}} \cdot 2^{-j} . \qquad (4)$$

We introduce the following notation:

$$D_T(n) = \sum_{l=0}^{\infty} f(l), \text{ where } f(l) = 2^l \sum_{j=c+1}^{\infty} \frac{(n2^{-l})^j}{j!} e^{-n2^{-l}} \cdot 2^{-j} . \qquad (5)$$

Then the expression (5) can be evaluated using the Euler's summation formula by substituting the discrete variable l with the continuous y as follows:

$$D_T(n) = \sum_{l=0}^{\infty} f(l) = \int_0^{\infty} f(y)dy + \frac{1}{2}(f(0) - f(\infty)) + \int_0^{\infty} B_1^*(y) f'(y) dy . \qquad (6)$$

$B_1^*(y)$ is the Bernoulli polynomial $B_1(y - \lfloor y \rfloor)$. By definition, $B_1(z) = z - 1/2$, and therefore $|B_1^*(y)| = |y - \lfloor y \rfloor - 1/2| \leq 1/2$. It then follows that:

$$\left| \int_0^{\infty} B_1^*(y) f'(y) dy \right| \leq \frac{1}{2} \left| \int_0^{\infty} f'(y) dy \right| = \frac{1}{2} |f(\infty) - f(0)| .$$

Readily, expression (6) can be reduced to just the first integral term, provided we can show that $f(0)$ and $f(\infty)$ converge to some small constants.

$$f(0) = 2^0 \sum_{j=c+1}^{\infty} \frac{(n2^{-0})^j}{j!} e^{-n2^{-0}} \cdot 2^{-j} = \sum_{j=c+1}^{\infty} \frac{n^j}{j!} e^{-n} \cdot 2^{-j} < 2^{-c} \sum_{j=c+1}^{\infty} \frac{n^j}{j!} e^{-n} < 2^{-c} < 1.$$

To estimate $f(\infty)$ we substitute x for 2^{-l} obtaining:

$$f(\infty) = \lim_{x \to 0} \left[\left[\sum_{j=c+1}^{\infty} \frac{(nx)^j}{2^j j!} e^{-nx} \right] / x \right].$$

Since this is a 0/0 limit form we can use the L'Hospital's rule to obtain:

$$f(\infty) = \left[n \sum_{j=c+1}^{\infty} \frac{(nx)^{j-1}}{2^j (j-1)!} e^{-nx} - n \sum_{j=c+1}^{\infty} \frac{(nx)^j}{2^j j!} e^{-nx} \right]_{x=0} = 0.$$

Therefore, the limits $f(0)$ and $f(\infty)$ exist. Then, $D_T(n)$ can be evaluated as:

$$D_T(n) \approx \int_0^{\infty} f(y)dy = \int_0^{\infty} 2^y \sum_{j=c+1}^{\infty} \frac{(n2^{-y})^j}{j!} e^{-n2^{-y}} \cdot 2^{-j} dy. \tag{7}$$

Setting $x = n2^{-y}$ we obtain:

$$D_T(n) \approx \int_0^{n} \frac{n}{x} \sum_{j=c+1}^{\infty} \frac{x^j}{2^j j!} e^{-x} \frac{dx}{x \ln 2} = n \cdot \log_2 e \sum_{j=c+1}^{\infty} \frac{1}{2^j j!} \int_0^{n} x^{j-2} e^{-x} dx.$$

For n large, we have: $\int_0^{n} x^{j-2} e^{-x} dx \approx \int_0^{\infty} x^{j-2} e^{-x} dx = \Gamma(j-1) = (j-2)!$,

where $\Gamma(i+1)$ is the gamma function, which is $i!$ for i=0,1,2,... Then,

$$D_T(n) \approx n \cdot \log_2 e \sum_{j=c+1}^{\infty} \frac{(j-2)!}{2^j j!} = n \cdot \log_2 e \sum_{j=c+1}^{\infty} \frac{1}{2^j j(j-1)}. \tag{8}$$

$$D_T(n) < \frac{n \cdot \log_2 e}{2^{c+1}} \sum_{j=c+1}^{\infty} \frac{1}{j(j-1)} = \frac{n \cdot \log_2 e}{2^{c+1}} \left[\frac{1}{c} - \frac{1}{c+1} + \frac{1}{c+1} - \frac{1}{c+2} + \cdots \right] = \frac{n \cdot \log_2 e}{c \cdot 2^{c+1}}.$$

This completes the proof of part (a) of the theorem. Observe that the derivations above are valid for c ≥ 1. So, to prove part (b) we set c = 1 in the expression (8) to obtain:

$$D_T(n) \approx n \cdot \log_2 e \sum_{j=2}^{\infty} \frac{1}{2^j j(j-1)} = \frac{n}{\ln 2} \cdot \left[\sum_{j=2}^{\infty} \frac{1}{(j-1)2^j} - \sum_{j=2}^{\infty} \frac{1}{j2^j} \right] = \frac{n}{\ln 2} \cdot \left[\frac{1}{2} \sum_{i=1}^{\infty} \frac{1}{i2^i} - \sum_{j=1}^{\infty} \frac{1}{j2^j} + \frac{1}{2} \right]$$

By Taylor's expansion $\ln 2 = \sum_{j=1}^{\infty} \frac{1}{j2^j}$, and hence:

$$D_T(n) \approx \frac{n}{\ln 2} \cdot \left(\frac{1}{2} \ln 2 - \ln 2 + \frac{1}{2} \right) = \frac{n}{\ln 2} \cdot (1 - \ln 2)/2 = n \cdot (\log_2 e - 1)/2.$$

We have just proved part (b) of this theorem. □

The expected number of leaves, $L_T(n)$, in a binary trie with exactly n items has been already estimated in [Fae79]. Knuth [Knu73] has also obtained a comparable result, but in the context of his analysis of radix-exchange sorting. Their investigations show that the parameter $L_T(n)$ fluctuates around a mean value, but its oscillations decay in amplitude as n goes larger, eventually converging to the mean value. We will here present a different derivation of the expected value of $L_T(n)$, paying no attention to the problem of how this parameter converges to it.

Theorem 3.2: Let T be a complete binary trie with leaf capacity c, obtained by entering n uniformly distributed items, where n is a large constant. Let $L_T(n)$ denote the expected number of leaves in T. Then, $L_T(n) \approx (n/c)\cdot\log_2 e$, for $c \geq 1$.

Proof: We will use the fact that in any complete binary trie the number of leaves is one greater than the number of interior nodes. The probability that a node at level k of T is an interior node is:

$$P(Y > c) = \sum_{j=c+1}^{\infty} \frac{(n2^{-k})^j}{j!} e^{-n2^{-k}} .$$

Since there are 2^k potential interior nodes at level k, the total number of leaves in T is:

$$L_T(n) = 1 + \sum_{k=0}^{\infty} 2^k P(Y > c) = 1 + \sum_{k=0}^{\infty} 2^k \sum_{j=c+1}^{\infty} \frac{(n2^{-k})^j}{j!} e^{-n2^{-k}} . \tag{9}$$

Let $f(k) = 2^k \sum_{j=c+1}^{\infty} \frac{(n2^{-k})^j}{j!} e^{-n2^{-k}}$. Then, $f(0) = \sum_{j=c+1}^{\infty} \frac{n^j}{j!} e^{-n} < 1$.

Also, by substituting 2^{-k} with x and applying the L'Hospital's rule, as in the proof of theorem 3.1, we obtain $f(\infty) \to 0$. Thus, both $f(0)$ and $f(\infty)$ converge to some small constants, and the application of the Euler's summation formula produces:

$$L_T(n) = \sum_{k=0}^{\infty} f(k) + O(1) \approx \int_0^{\infty} f(y)dy = \int_0^{\infty} 2^y \sum_{j=c+1}^{\infty} \frac{(n2^{-y})^j}{j!} e^{-n2^{-y}} dy .$$

Substituting x for $n2^{-y}$ we obtain:

$$L_T(n) \approx \int_0^n \frac{n}{x} \sum_{j=c+1}^{\infty} \frac{x^j}{j!} e^{-x} \frac{dx}{x\ln 2} = n\cdot\log_2 e \int_0^n \sum_{j=c+1}^{\infty} \frac{x^{j-2}}{j!} e^{-x} dx = n\cdot\log_2 e \sum_{j=c+1}^{\infty} \frac{1}{j!} \int_0^n x^{j-2} e^{-x} dx$$

$$\approx n\cdot\log_2 e \sum_{j=c+1}^{\infty} \frac{\Gamma(j-1)}{j!} = n\cdot\log_2 e \sum_{j=c+1}^{\infty} \frac{(j-2)!}{j!} = n\cdot\log_2 e \sum_{j=c+1}^{\infty} \frac{1}{j(j-1)}$$

$$\approx n\cdot\log_2 e\cdot\left[\frac{1}{c} - \frac{1}{c+1} + \frac{1}{c+1} - \frac{1}{c+2} + \cdots\right] = \frac{n}{c}\cdot\log_2 e .$$

Thus, $L_T(n) \approx (n/c)\cdot\log_2 e$ for $c \geq 1$. □

We will now demonstrate that using a Bernoulli instead of a Poisson distribution yields similar result for theorem 3.2. We will show that any expression for $L_T(n)$, obtained assuming the Bernoulli distribution, can be reduced to the expression (9) with the small error introduced.

Proposition 3.3: The error term in the expression for the expected number of leaves, in a binary trie T with n items, introduced by approximating the Binomial distribution with the Poisson model, asymptotically converges to $O(c)$, where c is the leaf capacity of T.

Proof: Substituting the Bernoulli distribution (1) into expression (9) yields:

$$L_T^B(n) = 1 + \sum_{k=0}^{\infty} 2^k \sum_{j=c+1}^{n} \binom{n}{j} 2^{-jk}(1-2^{-k})^{n-j} . \tag{10}$$

We have to show that this expression differs from

$$L_T^P(n) = 1 + \sum_{k=0}^{\infty} 2^k \sum_{j=c+1}^{\infty} \frac{(n2^{-k})^j}{j!} e^{-n2^{-k}} \tag{11}$$

by an error term which converges to $O(c)$ for large number of entries n. Thus, we have to show that $|L_T^B(n) - L_T^P(n)| = \varepsilon(n) = O(c)$. Our approach will be the following. (a) We will show that $L_T^B(n) \ge L_T^P(n)$, for all n, obtaining $L_T^B(n) - L_T^P(n) = \varepsilon(n)$, thus, removing the absolute signs. (b) Then, we will find a $L'_T(n) \ge L_T^P(n)$, and compute $L'_T(n) - L_T^P(n) = \varepsilon'(n)$. (c) Finally, we will conclude that, since $L'_T(n) \ge L_T^P(n)$, $\varepsilon(n)$ must be upper bounded by $\varepsilon'(n)$. Let us first develop equation (10) as follows:

$$L_T^B(n) = 1 + \sum_{k=0}^{\infty} 2^k \left[1 - \sum_{j=0}^{c} \binom{n}{j} 2^{-jk} (1-2^{-k})^{n-j} \right]$$

$$= 1 + \sum_{k=0}^{\infty} 2^k \left[1 - \sum_{j=0}^{c} \frac{n(n-1)\cdots(n-j+1)}{n^j} \frac{n^j}{j!} \cdot 2^{-jk} (1-2^{-k})^{n-j} \right].$$

Since for large n, $\dfrac{n(n-1)\cdots(n-j+1)}{n^j} \approx 1$, we obtain:

$$L_T^B(n) \approx 1 + \sum_{k=0}^{\infty} 2^k \left[1 - \sum_{j=0}^{c} \frac{(n2^{-k})^j}{j!} (1-2^{-k})^{n-j} \right]. \tag{12}$$

To approximate $(1-2^{-k})^{n-j}$ we do the following: $\ln(1-2^{-k})^{n-j} = (n-j) \cdot \ln(1-2^{-k})$. For $n \gg c \ge j$ we have: $\ln(1-2^{-k})^{n-j} \approx n \cdot \ln(1-2^{-k})$. The first two terms in the Taylor's expansion for $ln(1-x)$ are $-x - x^2/2$. (Implicit in this proof is the fact that case $k = 0$, i.e. $x = 2^{-0} = 1$, does not contribute significantly to the expressions (10) and (11)). Hence,

$$\ln(1-2^{-k})^{n-j} \approx n \cdot (-2^{-k} - \frac{2^{-2k}}{2}) = -n2^{-k} - \frac{n2^{-2k}}{2}, \quad \text{implying}$$

$$(1-2^{-k})^{n-j} \approx e^{-n2^{-k}} \cdot e^{-n2^{-2k}/2}. \tag{13}$$

(a) We now show that $L_T^B(n) \ge L_T^P(n)$. Since $e^{-x} \le 1$ for all $x \ge 0$, then

$$(1-2^{-k})^{n-j} \le e^{-n2^{-k}}. \tag{14}$$

Substituting (14) in (12) we obtain:

$$L_T^B(n) \ge 1 + \sum_{k=0}^{\infty} 2^k \left[1 - \sum_{j=0}^{c} \frac{(n2^{-k})^j}{j!} e^{-n2^{-k}} \right] = 1 + \sum_{k=0}^{\infty} 2^k \sum_{j=c+1}^{\infty} \frac{(n2^{-k})^j}{j!} e^{-n2^{-k}} = L_T^P(n).$$

Thus, since $L_T^B(n) \ge L_T^P(n)$, we let $\varepsilon(n) = L_T^B(n) - L_T^P(n)$.

(b) Next, we estimate an upper bound of $L_T^B(n)$, and call it $L'_T(n)$. To do so we use the well known relation $e^{-x} \ge 1 - x$, for all $x \ge 0$. Then equation (13) can be written as:

$$(1-2^{-k})^{n-j} \ge e^{-n2^{-k}} \cdot (1 - \frac{n2^{-2k}}{2}). \tag{15}$$

Substituting (15) in (12) we obtain:

$$L_T^B(n) \le L'_T(n) \approx 1 + \sum_{k=0}^{\infty} 2^k \left[1 - \sum_{j=0}^{c} \frac{(n2^{-k})^j}{j!} e^{-n2^{-k}} \cdot (1 - \frac{n2^{-2k}}{2}) \right]$$

$$L'_T(n) \approx 1+\sum_{k=0}^{\infty}2^k\left[1-\sum_{j=0}^{c}\frac{(n2^{-k})^j}{j!}e^{-n2^{-k}}\right]+\sum_{k=0}^{\infty}2^k\sum_{j=0}^{c}\frac{(n2^{-k})^j}{j!}e^{-n2^{-k}}\frac{n\,2^{-2k}}{2}$$

$$\approx 1+\sum_{k=0}^{\infty}2^k\sum_{j=c+1}^{\infty}\frac{(n2^{-k})^j}{j!}e^{-n2^{-k}}+\frac{1}{2}\sum_{k=0}^{\infty}n\,2^{-k}\sum_{j=0}^{c}\frac{(n2^{-k})^j}{j!}e^{-n2^{-k}}. \tag{16}$$

Thus, by subtracting (11) from (16), we obtain the error term $\varepsilon'(n)$ to be:

$$\varepsilon'(n) \approx \frac{1}{2}\sum_{k=0}^{\infty}n\,2^{-k}\sum_{j=0}^{c}\frac{(n2^{-k})^j}{j!}e^{-n2^{-k}}. \tag{17}$$

Let $f(k) = n\,2^{-k}\sum_{j=0}^{c}\frac{(n2^{-k})^j}{j!}e^{-n2^{-k}}$. Then, $f(\infty)\to 0$. Also, $f(0) = n\sum_{j=0}^{c}\frac{n^j}{j!}e^{-n}$, which tends to 0 for large n. Subsequently, both $f(0)$ and $f(\infty)$ converge to some small constants. So, we can apply Euler's summation formula to estimate $\varepsilon'(n)$.

$$\varepsilon'(n) \approx \frac{1}{2}\int_0^{\infty}n\,2^{-y}\sum_{j=0}^{c}\frac{(n2^{-y})^j}{j!}e^{-n2^{-y}}dy. \tag{18}$$

Setting $x = n2^{-y}$ yields: $\varepsilon'(n) \approx \frac{1}{2}\int_0^{n}x\sum_{j=0}^{c}\frac{x^j}{j!}e^{-x}\frac{dx}{x\ln 2} = \frac{\log_2 e}{2}\sum_{j=0}^{c}\frac{1}{j!}\int_0^{n}x^je^{-x}dx$

$$\approx \frac{\log_2 e}{2}\sum_{j=0}^{c}\frac{\Gamma(j+1)}{j!} = \frac{\log_2 e}{2}\sum_{j=0}^{c}\frac{j!}{j!} = \frac{\log_2 e}{2}\cdot(c+1).$$

Therefore, $O(c)$ is the upper bound of the error term obtained by approximating the Binomial distribution with the Poisson model. □

This proposition justifies the result of the theorem 3.2 by showing that it is a close approximation of the Bernoulli model. One can similarly show that the error term of the result of theorem 3.1 is also small.

3.2. Basic Parameters of the Actual Structure

So far we have considered binary tries. Using the invariant property 2.2 and the relationship between the 0-complete trees and their C_0-representations, we will translate the results of theorems 3.1 and 3.2 to obtain $D_{C_0}(n)$, the number of dummy entries at the lowest level of the C_0-tree; $L_{C_0}(n)$, the total number of entries at the lowest level of the tree; and S_{C_0}, the storage utilization of its index blocks.

Corollary 3.4: The expected number of dummy entries, $D_{C_0}(n)$, at the lowest level of a C_0-tree with n records, asymptotically converges to $n(\log_2 e-1)/2 \approx 0.221\cdot n$.

Proof: A dummy entry appears in a C_0-tree for every empty 0-leaf in the equivalent 0-complete tree with singleton leaves, T_0. By the invariant property 2.2 and by theorem 3.1, part (b), the average number of empty 0-leaves in T_0, and subsequently the average number of dummies in the C_0-tree, converges to $n(\log_2 e-1)/2$ for large n. □

Corollary 3.5: The expected total number of entries, $L_{C_0}(n)$, at the lowest level of a C_0-tree with n records, asymptotically converges to $n\,(log_2 e + 1)/2 \approx 1.221 \cdot n$.

Proof: $L_{C_0}(n)$ is the same as $L_{T_0}(n)$, the average total number of leaves in the equivalent 0-complete tree T_0 with singleton leaves. By the invariant property 2.2, T_0 can be considered as a tree obtained from a complete binary trie T, by eliminating all dummy 1-leaves. Then, by theorems 3.1, part (b), and 3.2 we have: $L_{T_0}(n) = L_T(n) - D_T(n) \approx n \cdot \log_2 e - n \cdot (log_2 e - 1)/2 = n \cdot (log_2 e + 1)/2$. Hence, $L_{C_0}(n) = L_{T_0}(n) \approx n \cdot (log_2 e + 1)/2$. □

With the minimal partition splitting, a deviation of paginated 0-complete trees T'_0 from true trie behavior can happen only if, at some point in time, the corresponding paginated binary trie T' has an empty interval belonging to a missing 1-leaf. The theorem 3.1, part (a), suggests that for uniform distribution of items the probability that we find an empty interval in T' is indeed very slim. For instance, even for 100,000,000 records and the index block capacity c = 100 (in practice c is several times larger) the number of empty 1-leaves will be less than 10^{-20}, which is for all intents and purposes 0. Thus, assuming the uniform distribution of keys and minimal partition splitting we can say that a paginated 0-complete tree T'_0 is just a complete binary trie.

Consequently, if there are v entries at the lowest level of the C_0-tree then, by theorem 3.2, the number of index blocks B(v) at that level converges to $(v/c) \cdot \log_2 e$, where $\log_2 e$ is the storage expansion factor, i.e. the average utilization of index blocks is $1/log_2 e = ln\,2 \approx 0.693$. However, due to the presence of dummies at this level of the tree, the effective utilization drops. By corollary 3.5, $v \approx n \cdot (log_2 e + 1)/2$, and therefore, $B(n) = (n/c) \cdot \log_2 e \cdot (log_2 e + 1)/2$. Thus, counting only "useful" entries, the effective storage utilization, S_{C_0}, at this level of a C_0-tree, drops to about $2 \,/\, (log_2 e \cdot (log_2 e + 1)) \approx 0.567$. The upper layer blocks of a C_0-tree do not have dummies and their utilization does not deteriorate. Then, as the number of entries at the upper layers grows, they can be expected to attain utilization of $ln\,2 \approx 0.693$.

All of the expected values presented here have been observed in an extensive series of experimental runs using both uniform and non-uniform key distributions. Observed patterns of variances around the expected values are described in [Orl89].

References

[BaM72] R. Bayer and E. McCreight, "Organization and Maintenance of Large Ordered Indexes", *Acta Informatica*, 1972, 173-189.

[Fae79] R. Fagin and et.al., "Extendible Hashing---A Fast Access Method for Dynamic Files", *Trans. Database Systems 4*, 3 (Sep. 1979), 315-344.

[Fla83] P. Flajolet, "On the Performance Evaluation of Extendible Hashing and Trie Searching", *Acta Inf. 20* (1983), 345-369.

[Fre60] E. Fredkin, "Many-way Information Retrieval", *Comm. of the ACM 3*, 9 (Sep. 1960), 490-499.

[Knu73] D. Knuth, *The Art of Computer Programming: Sorting and Searching*, Addison Wesley, Reading, MA, 1973.

[OrP88] R. Orlandic and J. L. Pfaltz, "Compact 0-Complete Trees", *Proc. 14th Conf. on VLDB*, Long Beach, CA, Aug. 1988, 372-381.

[Orl89] R. Orlandic, "Design, Analysis and Applications of Compact 0-Complete Trees", Ph.D. Dissertation, University of Virginia, Apr. 1989.

REPRESENTATION OF RECURSIVELY ENUMERABLE LANGUAGES USING ALTERNATING FINITE TREE RECOGNIZERS

Kai Salomaa
University of Turku, Department of Mathematics
SF-20500 Turku, Finland

1. Introduction

Here we continue the investigation of alternating yield-languages begun in [9], that is, languages obtained as yields of forests recognized by alternating finite bottom-up tree recognizers.

Alternation increases essentially the recognition power of nondeterministic bottom-up tree automata, cf. [8, 9]. This can be compared with the fact that the computations of alternating finite automata and top-down tree automata can be simulated by corresponding nondeterministic automata, see [1, 4, 11]. Alternating pushdown tree automata recognize already all recursively enumerable forests, cf. [10], whereas alternation enables ordinary pushdown automata to recognize the family of deterministic exponential time languages, cf. [1, 4]. Alternating automata operating on infinite trees are studied in [5, 6].

Intuitively, the power of alternating bottom-up computation is due to the fact that in this case alternation is a global concept: the computations in different subtrees can affect each other. Thus for instance, if the automaton has to make different existential choices at a node n_1 corresponding to different universal choices at node n_2 in another subtree, then it is clear that the order in which the independent nodes n_1 and n_2 are read can be crucial. Hence an alternating bottom-up computation cannot be simulated by a nondeterministic one using a subset construction as is possible in the top-down case where computations in different subtrees are independent and thus alternation can be seen to be a local concept.

In [9] it is shown that every context-sensitive language can be represented as the yield of a forest recognized by an alternating bottom-up recognizer but the question of the characterization of the family of alternating yield-languages is left open. Here we show that the family of alternating yield-languages equals to the recursively enumerable (r.e.) languages (modulo the empty word). This might at first seem quite strange: How can strictly read-only machines with a finite-state memory perform the computations of arbitrary Turing machines? The catch here is that the size of the input trees can become arbitrarily large as a function of the length of the yield, and thus the input tree can be made to be a guess that codes an entire computation of the Turing machine on the yield. The alternating tree recognizer is then able to check that this guess represents a correct accepting computation.

The representation result of context-sensitive languages in [9] was much easier because there the alternating tree recognizer was able to directly check that the input tree represents a correct derivation of the context-sensitive grammar. (Note that using the same technique as in representing the context-sensitive languages, it was shown in [9] that every r.e. language is a so called extended alternating yield-language. These are extensions of ordinary yield-languages where some nullary symbol of the corresponding ranked alphabet is allowed to represent the empty word in the yield.) Also, it was seen in [9] that the membership problem for alternating yield-languages is undecidable. Here in Theorem 3.4 we use actually basically the same idea as in the proof of this undecidability result.

The size of a tree as a function of its yield can become arbitrarily large only if the tree has symbols of rank one. Thus it is seen that alternating yield-languages corresponding to forests where the ranked alphabet does not contain unary symbols (or more generally to so called yield-bounded forests) are deterministic exponential time languages. Also, an interesting observation is that the representation of r.e. languages as alternating yield-languages is effective only for languages where all words have length at least two.

2. Alternating tree recognizers

In this section we recall the definition of an alternating bottom-up tree recognizer from [8, 9]. We assume the reader to be familiar with tree automata and some basic notions from formal language theory, cf. [2, 3, 7]. For a more detailed presentation of alternating tree recognizers and examples see [8, 9]. First we briefly explain some notations.

The set of (nonempty) words over an alphabet Z is denoted by Z^* (Z^+), the length of a word $w \in Z^*$ by $|w|$, and the empty word by λ.

The symbols Σ and Ω stand always for finite **ranked alphabets**. The set of m-ary, $m \geq 0$, symbols of Σ is denoted by Σ_m and the rank of an element $\sigma \in \Sigma$ is denoted by $\mathrm{rank}_\Sigma(\sigma)$ or simply by $\mathrm{rank}(\sigma)$ if there is no danger of confusion. If A is a set disjoint with Σ and we define a ranked alphabet Ω by setting $\Omega = \Sigma \cup A$ and specifying the ranks of elements of A, this is always taken to mean that $\mathrm{rank}_\Omega(\sigma) = \mathrm{rank}_\Sigma(\sigma)$ for all $\sigma \in \Sigma$.

The set of **Σ-trees, F_Σ**, is the smallest set B such that for all $m \geq 0$ and $\sigma \in \Sigma_m$, if $t_1, \ldots, t_m \in B$ then $\sigma(t_1, \ldots, t_m) \in B$. For a finite set X, the set of **ΣX-trees, $F_\Sigma(X)$**, is defined to be F_Ω, where $\Omega = \Sigma \cup X$ and $\mathrm{rank}(x) = 0$ for all $x \in X$. Subsets of $F_\Sigma(X)$ (respectively F_Σ) are called **ΣX-forests** (resp. Σ-forests). The number of nodes of a Σ-tree t is denoted by **size(t)**.

The **yield-function** $yd_\Sigma : F_\Sigma \to (\Sigma_0)^+$ is defined inductively as follows.

(i) $yd_\Sigma(\sigma) = \sigma$ for all $\sigma \in \Sigma_0$.

(ii) If $m \geq 1$, $\sigma \in \Sigma_m$, and $t_1, ..., t_m \in F_\Sigma$, then

$$yd_\Sigma(\sigma(t_1, ..., t_m)) = yd_\Sigma(t_1) \cdots yd_\Sigma(t_m).$$

Usually yd_Σ is denoted just by yd when Σ is known. The word yd(t) is obtained from a Σ-tree t simply by concatenating the labels of the leaves of t from left to right.

2.1. Definition. An **alternating bottom-up tree recognizer** is a four-tuple $\underline{A}$ = (A, Σ, g, A'), where

(i) A is a finite set of states.

(ii) Σ is a ranked alphabet.

(iii) g is a mapping that associates with every element $\sigma \in \Sigma_m$ ($m \geq 0$) a function σ_g : $A^m \to P(P(A))$. (Here $P(B)$ denotes the power set of a set B. If above $\sigma \in \Sigma_0$, then σ_g is interpreted to be an element of $P(P(A))$.)

(iv) A' is a subset of A consisting of so called accepting final states.

The class of alternating bottom-up recognizers is denoted by **AR**.

Let $\underline{A}$ = (A, Σ, g, A') be as above. Elements of $F_\Sigma(A)$ represent intermediate stages of a computation of $\underline{A}$ and they are called **$\underline{A}$-configurations.** Let K be an $\underline{A}$-configuration. Subtrees of K of the form $\sigma(a_1, ..., a_m)$, $m \geq 0$, $\sigma \in \Sigma_m$, $a_1, ..., a_m \in A$, are called **active subtrees** of K. (Here by subtrees we actually mean "occurrences of subtrees".) The set of active subtrees of K is denoted by **act(K)**. If $f = \sigma(a_1, ..., a_m) \in act(K)$, we denote $f_g = \sigma_g(a_1, ..., a_m)$.

A **configuration tree** of $\underline{A}$ is a rooted tree graph the nodes of which are labeled by $\underline{A}$-configurations, and the set of configuration trees of $\underline{A}$ is denoted by **CT($\underline{A}$)**. If T $\in$ CT($\underline{A}$), then **conf(T)** denotes the set of all $\underline{A}$-configurations that label some node of $\underline{A}$.

Now we are ready to define the computations of an alternating tree recognizer.

2.2. Definition. The transition relation of a recognizer $\underline{A} \in$ AR, $\Rightarrow_{\underline{A}}$, is a binary relation on CT($\underline{A}$) defined as follows. For T, T' $\in$ CT($\underline{A}$) we define $T \Rightarrow_{\underline{A}} T'$ iff T' is obtained from T as follows. Choose a leaf n of T and suppose that n is labeled by a configuration K $\in F_\Sigma(A)$. Let $f = \sigma(a_1, ..., a_m)$ be an active subtree of K. Then T' is obtained from T by attaching for the node n the successors

$$K(f \leftarrow c_1), ..., K(f \leftarrow c_h),$$

where $\{ c_1, ..., c_h \}$ is some nonempty element of f_g, $h \geq 1$. (Here $K(f \leftarrow c_i)$ denotes the ΣA-tree obtained from K by replacing the specific occurrence of the subtree f with c_i.)

The transition relation defined above is the so called EU-computation mode (existential-universal) of [8, 9]. In each computation step the automaton first chooses existentially a set from f_g and then branches universally to all of its states. Here we need not concern ourselves with the other computation mode (UE-mode), because it is shown in [8, 9] that the families of yield-languages defined by the EU- and UE-mode of computation are identical.

2.3. Definition. Let $\underline{A} \in$ AR and $K \in F_\Sigma(A)$. The set of **K-computation trees of $\underline{A}$** is

$$COM(\underline{A}, K) = \{ T \in CT(\underline{A}) \mid K \Rightarrow_{\underline{A}}^{*} T \}.$$

(Above in the right-hand side K is interpreted to be the configuration tree with one node labeled by K.) A computation tree is **accepting** if all its leaves are labeled by elements of A′, and a configuration K is said to be accepting if COM($\underline{A}$, K) contains at least one accepting computation tree. The set of accepting configurations is denoted by **H($\underline{A}$)** and the **forest recognized by $\underline{A}$** is

$$L(\underline{A}) = H(\underline{A}) \cap F_\Sigma .$$

The family of forests recognized by alternating tree recognizers is denoted by L(AR).

Thus, intuitively a recognizer $\underline{A}$ accepts such input trees t that in every branch of an alternating computation $\underline{A}$ is able to reach the root of t in an accepting final state.

A forest that can be recognized by a deterministic bottom-up tree recognizer, cf. [2], is said to be **regular.** (An AR-recognizer $\underline{A}$ = (A, Σ, g, A′) is said to be deterministic if for all $m \geq 0$, $\sigma \in \Sigma_m$, and $a_1, \ldots, a_m \in A$, we can write $\sigma_g(a_1, \ldots, a_m) = \{\{b\}\}$, $b \in A$.) The family of regular forests is denoted by **REG**.

It is known, cf. [8, 9], that alternating bottom-up recognizers recognize also non-regular forests; however explicit construction of such recognizers would usually be quite difficult. When we want to show that a given nonregular forest can be recognized by an alternating recognizer the following result of Theorem 2.5 is very useful. It states that alternating recognizers can check that all configurations appearing in accepting computation trees belong to a given regular forest, i.e., that alternating computation is closed with respect to regular control. For many quite "difficult" forests one can construct a simple recognizer that recognizes the forest using a suitable regular control-forest.

2.4. Definition. Let $\underline{A}$ = (A, Σ, g, A′) ∈ AR, $K \in F_\Sigma(A)$, and let N be a ΣA-forest. The set of **N-controlled K-computation trees** of $\underline{A}$ is

$$COM(\underline{A}, K)[\, N \,] = \{ T \in COM(\underline{A}, K) \mid N \supseteq conf(T) \}.$$

The configuration K is **N-controlled accepting** if COM($\underline{A}$, K)[N] contains an accepting computation tree, and the set of N-controlled accepting configurations is denoted by **H($\underline{A}$)[N]**. The forest **N-controlled recognized** (or **recognized with the control-forest** N) **by $\underline{A}$** is

$$L(\underline{A})[\, N \,] = H(\underline{A})[\, N \,] \cap F_\Sigma .$$

The family of forests recognized by an alternating recognizer using a regular control-forest is denoted by L(AR)[REG].

The following result stating that $L(\mathrm{AR})$ is closed with respect to regular control is proved in [9].

2.5. Theorem. $L(\mathrm{AR})[\ \mathrm{REG}\ ] = L(\mathrm{AR})$.

For the sake of completeness we still recall the definition of a Turing machine. A (nondeterministic) **Turing machine** is a six-tuple $\underline{M} = (I, Z, Q, q_0, Q(acc), \delta)$, where I is a finite input alphabet, Z ($\supseteq$ I) is a finite alphabet of tape symbols containing a so called end marker ¢, (¢ $\notin$ I), Q is a finite set of states, q_0 is the initial state, Q(acc) is a subset of Q consisting of so called accepting final states and δ is a set of rewrite rules of the following forms. Here $q_1, q_2 \in Q$ and $z_1, z_2, z \in Z - \{ ¢ \}$.

(i) $q_1z_1 \to z_2q_2$ (move right and replace z_1 by z_2);

(ii) $zq_1z_1 \to q_2zz_2$ (move left and replace z_1 by z_2);

(iii) $q_1¢ \to q_2z¢$, $¢q_1 \to ¢q_2z$ (expand workspace).

Elements of Z^*QZ^* are called configurations of $\underline{M}$ and Z^*QZ^* is denoted by $C(\underline{M})$. The rewrite rules of δ determine in the usual way a binary relation $\to_{\underline{M}}$ on $C(\underline{M})$, which is called the transition relation of $\underline{M}$.

The **language accepted by** $\underline{M}$ is

$$L(\underline{M}) = \{ w \in I^* \mid ¢q_0w¢ \to_{\underline{M}}^* w_1qw_2 ,\ q \in Q(acc),\ w_1, w_2 \in Z^* \}.$$

A language L over an alphabet I is said to be **recursively enumerable (r.e.)** if there exists a Turing machine $\underline{M} = (I, Z, Q, q_0, Q(acc), \delta)$ such that $L = L(\underline{M})$.

We make the convention that in the following by an r.e. language we always mean a language not containing the empty word. This is done because we consider presentations of languages as yields of forests and the presence of the empty word would necessitate the introduction of some special "empty tree".

3. Alternating yield-languages

In this section we prove the main result that every r.e. language is an alternating yield-language and, furthermore, the presentation is effective if the language contains only words of length at least two.

3.1. Lemma. Let $\underline{M} = (I, Z, Q, q_0, Q(acc), \delta)$ be a Turing machine. Denote $M = Z \cup Q$, and $M' = \{ m' \mid m \in M \}$. Define the ranked alphabets

$\Sigma = \Sigma_0 \cup \Sigma_1$, where $\Sigma_0 = M$ and $\Sigma_1 = M'$,

$\Omega = \Sigma \cup \{ \omega \}$, where $\mathrm{rank}(\omega) = 2$.

The mapping $h : F_\Sigma \to M^+$ is defined inductively as follows:

(i) $h(m) = m$ if $m \in \Sigma_0$, and

(ii) $h(m'(t)) = h(t)m$ for all $m' \in \Sigma_1, t \in F_\Sigma$, ($m \in M$).

We define the Ω-forest L_1 by

$L_1 = \{ \omega(t_1, t_2) \mid t_1, t_2 \in F_\Sigma , h(t_1), h(t_2) \in C(\underline{M}), h(t_1) \rightarrow_{\underline{M}} h(t_2) \}$.

Then $L_1 \in \mathcal{L}(AR)$.

We prove the above lemma in two parts. First in Lemma 3.2 we construct an alternating recognizer for a forest modified from L_1, where the rewrite relation of $\underline{M}$ is "coded" so that the lengths of left- and right-hand sides of rewrite rules have length one. After that we show how this construction can be extended to prove Lemma 3.1.

3.2. Lemma. Let $\Sigma = \Sigma_1 \cup \Sigma_0$ be a ranked alphabet and $h_\Sigma : F_\Sigma \rightarrow \Sigma^+$ be defined inductively by $h_\Sigma(\sigma) = \sigma$ if $\sigma \in \Sigma_0$, and $h_\Sigma(\sigma(t)) = h_\Sigma(t)\sigma$ for $\sigma \in \Sigma_1$, $t \in F_\Sigma$. Let R be a binary relation on Σ. The relation R is extended to a relation on Σ^+ by defining

$w_1 R w_2$, $\quad w_1, w_2 \in \Sigma^+$,

iff we can write

$w_i = u a_i v$,

$u, v \in \Sigma^*$, $a_i \in \Sigma$, $i = 1,2$, and $a_1 R a_2$.

Again let $\Omega = \Sigma \cup \{ \omega \}$, where rank($\omega$) = 2. Define

$L_2 = \{ \omega(t_1, t_2) \mid t_1, t_2 \in F_\Sigma , h_\Sigma(t_1) R h_\Sigma(t_2) \}$.

Then $L_2 \in \mathcal{L}(AR)$.

Proof. We construct a recognizer $\underline{A} = (A, \Sigma, g, A') \in$ AR and a regular control-forest L_3 such that

(3.2.1) $\quad L_2 = L(\underline{A})[L_3]$.

From this it then follows by Theorem 2.5 that $L_2 \in \mathcal{L}(AR)$.

The recognizer $\underline{A}$ constructed below will in fact be "nondeterministic", that is, for all $m \geq 0$, $\sigma \in \Sigma_m$, and $a_1 ,\dots, a_m \in A$, we have

$\sigma_g(a_1 ,\dots, a_m) = \{ \{b_1\} ,\dots, \{b_n\} \}$,

$b_1 ,\dots, b_n \in A$, (i.e., computations of $\underline{A}$ do not branch universally.) To simplify the notation, in the rest of this proof we will always denote $\sigma_g(a_1 ,\dots, a_m)$ as above just by $\{ b_1 ,\dots, b_n \}$.

Denote $\Sigma' = \{ \sigma' \mid \sigma \in \Sigma \}$. The set of states of $\underline{A}$ is chosen to be

$A = \{ x[i, j] \mid x \in \Sigma, i = 0,1,2, j = 0,1 \} \cup \{ x'[i, 1] \mid x' \in \Sigma', i = 0,1,2 \} \cup \{ f \}$,

and $A' = \{ f \}$.

The rules of g are defined as follows. Here the symbol "+" indicates addition modulo 3.

Let $\sigma \in \Omega_0$. Then

(3.2.2) $\sigma_g = \{ \sigma[0, 0], \sigma'[0, 1] \}$.

Let $\sigma \in \Omega_1$ and $z \in \Sigma$, $0 \leq i \leq 2$. Then

(3.2.3) $\sigma_g(z[i, 0]) = \{ \sigma[i+1, 0], \sigma'[i+1, 1] \}$,

(3.2.4) $\sigma_g(z[i, 1]) = \{ \sigma[i+1, 1] \}$,

and

(3.2.5) $\sigma_g(z'[i, 1]) = \{ \sigma[i+1, 1] \}$.

Finally,

(3.2.6) $\omega_g(x, y) = \{ f \}$ iff $x = y = z[i, 1]$ or $x = z_1'[i, 1], y = z_2'[i, 1], z_1 R z_2, \quad z, z_1, z_2 \in \Sigma, 0 \le i \le 2$; and $\omega_g(x, y) = \emptyset$ otherwise.

Let H be a subset of $(\Omega_0 \cup A)^2 \cup A$ defined as follows:

$$H = \{ f \} \cup \Omega_0{}^2 \cup \{ x[0, 0] \mid x \in \Sigma \}\Omega_0 \cup \{ x'[0, 1] \mid x \in \Sigma \}\Omega_0$$
$$\cup \{ x[i, j]x[i, j] \mid x \in \Sigma, 0 \le i \le 2, 0 \le j \le 1 \}$$
$$\cup \{ x'[i, 1]y'[i, 1] \mid x, y \in \Sigma, 0 \le i \le 2, xRy \}$$
$$\cup \{ x[i+1, j]y[i, j] \mid x, y \in \Sigma, 0 \le i \le 2, 0 \le j \le 1 \}$$
$$\cup \{ x'[i+1, 1]y[i, 0] \mid x, y \in \Sigma, 0 \le i \le 2 \}$$
$$\cup \{ x[i+1, 1]y'[i, 1] \mid x, y \in \Sigma, 0 \le i \le 2 \}.$$

The ΩA-forest L_3 is defined by

$$L_3 = yd^{-1}(H).$$

Because H is finite, clearly L_3 is regular.

Intuitively in a state $\sigma[i, j]$, σ indicates the last unary symbol that has been read and the value of j indicates whether the automaton has passed the point where it checks for the relation R. The control-forest L_3 uses the indices i to force the automaton to read alternately unary symbols from subtrees t_1 and t_2 of an input tree $\omega(t_1, t_2)$.

From rule (3.2.6) it is seen that (L_3-controlled) $\underline{A}$ accepts only trees of the form $\omega(t_1, t_2)$, $t_1, t_2 \in F_\Sigma$. Suppose that $t = \omega(t_1, t_2)$ is as above and that $T \in COM(\underline{A}, t)[L_3]$ is accepting. Suppose that a configuration

$$K = \omega(\sigma_1 \cdots \sigma_r(x), \tau_1 \cdots \tau_s(y)),$$

$\sigma_i, \tau_j \in \Sigma_1$, $x, y \in A$, $i = 1, \ldots, r$, $j = 1, \ldots, s$, $(r, s \ge 0)$, labels a node of the computation tree T. We claim that

Claim 1. If $x = y = \sigma[i, 0]$, $\sigma \in \Sigma$, $i \in \{0,1,2\}$, then $r, s \ge 1$ and one of the following conditions holds:

(1a) $\sigma_r = \tau_s$ and T contains a configuration

$$\omega(\sigma_1 \cdots \sigma_{r-1}(\sigma_r[i+1, 0]), \tau_1 \cdots \tau_{s-1}(\sigma_r[i+1, 0])),$$

or (1b) $\sigma_r R \tau_s$ and T contains a configuration

$$\omega(\sigma_1 \cdots \sigma_{r-1}(\sigma_r'[i+1, 1]), \tau_1 \cdots \tau_{s-1}(\tau_s'[i+1, 1])).$$

Claim 2. If $x = y = \sigma[i, 1]$ or ($x = \sigma'[i, 1]$ and $y = \tau'[i, 1]$), $\sigma, \tau \in \Sigma$, $i \in \{0,1,2\}$, then either

(2a) $r = s = 0$,

or (2b) $r, s \ge 1$, $\sigma_r = \tau_s$ and T contains a configuration

$$\omega(\sigma_1 \cdots \sigma_{r-1}(\sigma_r[i+1, 1]), \tau_1 \cdots \tau_{s-1}(\sigma_r[i+1, 1])).$$

Here we prove Claim 1; the proof of Claim 2 is completely analogous. If x and y are of the form $\sigma[i, 0]$, the rule (3.2.6) is not applicable to the configuration K. Hence necessarily

$r, s \geq 1$ and the computation has to be continued by reading σ_r or τ_s using the rule (3.2.3). If A would read the symbol τ_s, the resulting configuration K_1 would have a yield

$\sigma[i, 0]\tau_s[i+1, 0]$ or $\sigma[i, 0]\tau_s'[i+1, 1]$

and hence K_1 would not belong to L_3.

(i) Suppose now that A reads σ_r making the existential choice $\sigma_r[i+1, 0]$, i.e., the successor of K is the configuration

$K_2 = \omega(\sigma_1 \cdots \sigma_{r-1}(\sigma_r[i+1, 0]), \tau_1 \cdots \tau_s(\sigma[i, 0]))$.

If A would here next read the symbol σ_{r-1}, the resulting configuration would have a yield $\sigma_{r-1}[i+2, 0]\sigma[i, 0]$ or $\sigma_{r-1}'[i+2, 1]\sigma[i, 0]$. Thus by the choice of H it follows that in K_2 A has to read τ_s making the existential choice $\tau_s[i+1, 0]$ and, furthermore, that $\tau_s = \sigma_r$. Hence the condition (1a) holds.

(ii) Suppose then that in K A reads the symbol σ_r making the existential choice $\sigma_r'[i+1, 1]$. By exactly the same argument as above in (i), in the resulting configuration A next has to read the symbol τ_s. The existential choice $\tau_s[i+1, 0]$ is prohibited by H, and thus A obtains the configuration

$\omega(\sigma_1 \cdots \sigma_{r-1}(\sigma_r'[i+1, 1]), \tau_1 \cdots \tau_{s-1}(\tau_s'[i+1, 1]))$,

which belongs to L_3 only if $\sigma_r R \tau_s$. This completes the proof of Claim 1.

Clearly in an L_3-controlled t-computation A must first read the leaves of t_1 and t_2 making an identical existential choice of the form $\sigma[0, 0]$ or choices $\sigma_1'[0, 1]$ and $\sigma_2'[0, 1]$ such that $\sigma_1 \mathbf{R} \sigma_2$. From Claims 1 and 2 it follows that in the computation T after this A must alternately read symbols from the subtrees t_1 and t_2 and check that t_1 and t_2 are identical except for one pair of symbols that belongs to the relation R. Thus A verifies that $h_\Sigma(t_1) R h_\Sigma(t_2)$ and (3.2.1) holds. q.e.d.

Proof of Lemma 3.1 (Outline). Using essentially the same idea as in the proof of Lemma 3.2 one can construct a recognizer that with a suitable control-forest recognizes the forest L_1 of Lemma 3.1. The only difference is that instead of just checking that some pair of symbols in the unary subtrees t_1 and t_2 belongs to the relation R, now the automaton has to guess when it has reached the position in t_1 and t_2 where the rewrite rule of M is applied, and to guess the rule that is used and then check that the next symbols in t_1 (respectively t_2) correspond to the left-hand (resp. right-hand) side of the rule. (Since the two sides of the rewrite rule may have different lengths, for this phase of the computation the automaton has to be freed of the restriction of having to read alternately symbols from t_1 and t_2. Also, using the control-forest it is easy to check that the encountered left- and right-hand side correspond to the same rewrite rule.)

q.e.d.

3.3. Lemma. Let $\underline{M} = (I, Z, Q, q_0, Q(acc), \delta)$ be a Turing machine and define the ranked alphabet Σ associated with M as in Lemma 3.1. Define

$$\Omega = \Sigma \cup \{ \omega \} \cup \{ \$' \},$$

where rank(ω) = 2 and rank($\$'$) = 1, ($\omega$, $\$' \notin \Sigma$). Also the mapping

$$h : F_{\Sigma \cup \{\$'\}} \to (M \cup \{ \$ \})^+$$

is defined similarly as in Lemma 3.1. (Here "\$" should not be confused with the end marker "¢" of $\underline{M}$.) Denote

$$L_4 = \{ \omega(t_1, t_2) \mid t_1, t_2 \in F_{\Sigma \cup \{\$'\}} \text{ and the condition (*) below holds } \}.$$

(*) $h(t_i) = u_1{}^i\$u_2{}^i\$ \cdots \$u_n{}^i$,

$n \geq 1$, $u_j{}^i \in C(\underline{M})$, and $u_j{}^1 \to_{\underline{M}} u_j{}^2$, $j = 1, \ldots, n$, $i = 1,2$.

Then $L_4 \in \mathcal{L}(AR)$.

Proof. This follows immediately from the construction in the proofs of Lemmas 3.1 and 3.2. After encountering boundary markers "$\$'$" in the subtrees t_1 and t_2, the tree recognizer can start to check that also the next pair of "subwords" represents a correct computation step of $\underline{M}$. q.e.d.

3.4. Theorem. Let L be a recursively enumerable language over an alphabet I. Then there exists a forest $L' \in \mathcal{L}(AR)$ such that

$$yd(L') = L - (I^2 \cup I).$$

Proof. Suppose that $\underline{M} = (I, Z, Q, q_0, Q(acc), \delta)$ is a Turing machine such that $L(\underline{M}) = L$. Without loss of generality we may assume that for all $w \in L(\underline{M})$ there exists a computation of $\underline{M}$ accepting w that has odd length. We construct a ranked alphabet Σ as follows:

$$\Sigma = \Sigma_0 \cup \Sigma_1 \cup \Sigma_2 \cup \Sigma_3,$$

where

$$\Sigma_0 = I, \quad \Sigma_1 = (Z \cup Q)' \cup \{ \$' \}, \quad \Sigma_2 = \{ \tau \} \text{ and } \Sigma_3 = \{ \omega \}.$$

Here $(Z \cup Q)' = \{ x' \mid x \in Z \cup Q \}$. Define the mapping

$$h : F_{\Sigma_1 \cup \Sigma_0} \to (Z \cup Q \cup \{ \$ \})^+$$

inductively by

(i) $h(\sigma) = \sigma$ if $\sigma \in \Sigma_0$, and

(ii) $h(\sigma'(t)) = h(t)\sigma$ if $\sigma' \in \Sigma_1$, $t \in F_{\Sigma_1 \cup \Sigma_0}$.

We say that a Σ-tree t is **well formed** if t is of the form

(3.4.1) $\omega(t_1, t_2, t_3)$,

where $t_1, t_2 \in F_{\Sigma_1 \cup \Sigma_0}$, $t_3 \in F_{\Sigma_2 \cup \Sigma_0}$, and we can write

$$h(t_1) = x\$u_1\$u_2\$ \cdots \$u_m ,$$

$m \geq 1$, $x \in I$, $u_1 \in \{¢q_0\}I^+\{¢\}$, $u_i \in C(\underline{M})$, $i = 2, \ldots, m$, and

$$h(t_2) = y\$v_1\$ \cdots \$v_n ,$$

$n \geq 1$, $y \in I$, $v_i \in C(\underline{M})$, $i = 1, \ldots, n$, $v_n = w_1 q w_2$, $q \in Q(acc)$, $w_1, w_2 \in Z^*$.

We define the Σ-forest L' to consist of all well formed trees t as in (3.4.1) such that

$m = n$, $u_1 = ¢q_0 yd(t)¢$ $(= (¢q_0 xy)yd(t_3)¢)$,

$u_i \to_{\underline{M}} v_i$, $i = 1, \ldots, m$,

$v_i \rightarrow_{\underline{M}} u_{i+1}$, $i = 1$,..., m-1.

Thus in the trees t of L′ (as in (3.4.1)) the subtrees t_1 and t_2 code an accepting computation of $\underline{M}$ on the word yd(t). By the assumption that every word of L($\underline{M}$) can be accepted by a computation of odd length, it thus follows that for every word $w \in I^+$ such that $|w| \geq 3$ we have

$$w \in yd(L') \text{ iff } w \in L(\underline{M}).$$

(Note that the yield of a well formed tree has always length at least 3 and that every word of I^+ can be represented in the form yd(t_3), where t_3 is as above.)

In the following we show how an alternating recognizer $\underline{A} = (A, \Sigma, g, A') \in$ AR can L_1-controlled recognize the forest L′ using a suitably chosen regular control-forest L_1. We do not try to give an explicit construction of $\underline{A}$ and L_1.

By choosing the ΣA-forest L_1 so that $yd(L_1) \cap \Sigma_0{}^+ A \Sigma_0{}^* = \varnothing$, we can guarantee that in all L_1-controlled computations on an input tree t the recognizer $\underline{A}$ must first read the left-most leaf of t. Furthermore, L_1 can force $\underline{A}$ to read the left-most leaf using a special existential choice { c_1, c_2, c_3, c_4 } that branches universally the computation tree into four separate computations denoted respectively C_1, C_2, C_3 and C_4.

The computation C_1 just checks that the input tree t is well formed. This is easy since the set of well formed trees is regular. In the following when explaining the computations C_2, C_3 and C_4 we can thus always assume that the input tree is of the form (3.4.1).

In the computation C_2, $\underline{A}$ reads first the leaf y of t_2 and then the first \$′-markers from t_1 and t_2. After this $\underline{A}$ can as in Lemma 3.3 check that m = n and $u_i \rightarrow_{\underline{M}} v_i$, i = 1,...,m.

In the computation C_3 on the other hand, $\underline{A}$ starts by reading symbols from t_1 until it encounters the second \$′-marker (i.e., $\underline{A}$ "passes by deterministically" the word u_1) and by reading y and the first symbol \$′ from t_2. Now again by (the proof of) Lemma 3.3, $\underline{A}$ is able to check that $v_i \rightarrow_{\underline{M}} u_{i+1}$, i = 1 ,..., m-1. (In the end $\underline{A}$ "passes by" v_m.) Note that by defining the control-forest L_1 appropriately one can guarantee that the sets of states of $\underline{A}$ used respectively in computations C_2 and C_3 (as well as C_4) are completely disjoint, and thus the control-forests obtained from Lemma 3.3 for C_2 and C_3 do not "interfere" with each other (or with the control-forest of C_4.)

For the computation C_4 it remains to check that $u_1 = ¢q_0yd(t)¢$. This is done by forcing the automaton to read alternately symbols from u_1 and the yield of t using an idea similar to that of the proof of Lemma 3.2. After the first \$′-marker in t_1 $\underline{A}$ checks that the next three symbols are ¢′, q_0' and x′ ($\underline{A}$ has kept x "in memory"). After this always when reading a symbol c′ from u_1, $\underline{A}$ enters a state c′[i], i ∈ { 0, 1, 2 }, where at each step the index i is increased by one modulo 3. All remaining leaves of t (that is, all leaves except x) are read by making an existential choice of the form d[i], i = 0,1,2, where d is the label of the leaf. Now the control-forest L_2 corresponding to the computation C_4 (L_2 is the subset of L_1 affecting C_4) is defined to consist of trees where the yield is of the form

$c'[\ i\ ]d_1[\ j_1\ ]d_2[\ j_2\ ]\cdots d_n[\ j_n\ ]w, \quad w \in I^*,$

where $j_1 \cdots j_n \in \{012\}^*(\ \{\ \lambda\ \} \cup \{\ 0\ \} \cup \{\ 01\ \}\ )$, and either

(3.4.2) $\quad i = j_n$ and $c = d_n$, or

(3.4.3) $\quad i = j_n + 1 \pmod 3$.

Clearly L_2 is regular. From (3.4.2) and (3.4.3) it follows that in L_2-controlled computations always after reading a symbol c' from u_1, A has to read the left-most remaining leaf of t and check that it is labeled by c, after which A again has to read the next symbol of u_1. Finally, A checks that the last symbol of u_1 is ¢'.

Now we have shown that $L(A)[\ L_1\ ] = L'$, and thus from Theorem 2.5 it follows that $L' \in \mathcal{L}(AR)$. q.e.d.

For an arbitrary r.e. language L over an alphabet I, an alternating tree recognizer can of course recognize a suitable finite forest that has as its yield the set $L \cap (I \cup I^2)$. On the other hand, for a recognizer $A \in$ AR a Turing machine can check whether a given word w belongs to yd(L(A)) by first guessing a tree t such that yd(t) = w and then checking whether $t \in L(A)$. Thus we have (remembering the convention made in Section 2 that r.e. languages are assumed not to contain the empty word):

3.5. Corollary. The family of alternating yield-languages is equal to the family of recursively enumerable languages.

One should note that the construction in the proof of Theorem 3.4 is effective, however the result of Corollary 3.5 is not. Actually by an easy modification of the proof of Theorem 3.4 one can show that every r.e. language containing only words of length at least two has an effective representation as an alternating yield-language. (In the proof of Theorem 3.4 one can code words of length two by allowing also "well formed" trees where the root is labeled by a binary symbol that has subtrees t_1 and t_2 as in (3.4.1).) On the other hand, it is easy to see (cf. [1, 4, 8]) that alternating tree recognizers with unary input trees can be simulated by deterministic automata. Thus it is, for instance, decidable whether a given word of length one belongs to an alternating yield-language, and the representation for words of length one cannot be effective. To sum up we have:

3.6. Corollary. Suppose that L is an arbitrary r.e. language over I.

(i) There exists an algorithm that given (a Turing machine accepting) L produces a forest $L' \in \mathcal{L}(AR)$ such that $yd(L') = L - I$.

(ii) There does not exist an algorithm that given L would produce a forest $L' \in \mathcal{L}(AR)$ such that $yd(L') = L$.

In the proof of Theorem 3.4 it is essential that input trees can have arbitrarily many consequtive unary symbols. We say that a forest L is (linearly) **yield-bounded** if there exists a constant c such that for every $t \in L$: $size(t) \leq c \cdot |yd(t)|$. (It is easy to see that L is yield-bounded iff for every $t \in L$ the number of unary symbols occurring in t is linearly bounded by $|yd(t)|$.)

If L is a Σ-forest, define the function $s_L : \Sigma_0^+ \rightarrow N \cup \{ \infty \}$ by

$s_L(w) = \inf\{ size(t) \mid t \in L, yd(t) = w \}$.

(If $w \notin yd(L)$, then $s_L(w) = \infty$.) Let $\underline{A} = (A, \Sigma, g, A') \in AR$. It is easy to see that an alternating Turing machine can recognize the language $yd(L(\underline{A}))$ with the space bound $s_{L(\underline{A})}$. Hence it follows that alternating yield-languages corresponding to yield-bounded forests are alternating linear space (that is deterministic exponential time, cf. [1]) languages. The characterization of alternating yield-languages corresponding to yield-bounded forests remains an open question. More generally, one can consider also forests where the sizes of trees are bounded by some nonlinear function of the length of the yield.

Acknowledgement. I have greatly benefited from discussions with Joost Engelfriet and Magnus Steinby.

References

[1] A. K. Chandra, D. C. Kozen and L. J. Stockmeyer, Alternation, *J. Assoc. Comput. Mach.* **28** (1981) 114-133.

[2] F. Gécseg and M. Steinby, Tree automata, Akadémiai Kiadó, Budapest, 1984.

[3] J. E. Hopcroft and J. D. Ullman, Introduction to automata theory, languages, and computation, Addison-Wesley, 1979.

[4] R. E. Ladner, R. J. Lipton and L. J. Stockmeyer, Alternating pushdown and stack automata, *SIAM J. Comput.* **13** (1984) 135-155.

[5] D. E. Muller, A. Saoudi and P. E. Schupp, Alternating automata, the weak monadic theory of the tree, and its complexity, Proc. of 13th ICALP, Lect. Notes Comput. Sci. **226** (1986) 275-283.

[6] D. E. Muller and P. E. Schupp, Alternating automata on infinite trees, *Theoret. Comput. Sci.* **54** (1987) 267-276.

[7] A. Salomaa, Formal languages, Academic Press, New York, 1973.

[8] K. Salomaa, Alternating bottom-up tree recognizers, Proc. of 11th CAAP, Lect. Notes Comput. Sci. **214** (1986) 158-171.

[9] K. Salomaa, Yield-languages recognized by alternating tree recognizers, *RAIRO Inform. Théor.* **22** (1988) 319-339.

[10] K. Salomaa, Alternating tree pushdown automata, *Ann. Univ. Turku Ser.* AI **192** (1988).

[11] G. Slutzki, Alternating tree automata, Proc. of 8th CAAP, Lect. Notes Comput. Sci. **159** (1983) 392-404.

ABOUT A FAMILY OF BINARY MORPHISMS WHICH STATIONARY WORDS ARE STURMIAN.

Patrice SÉÉBOLD

LIFL
Laboratoire d'Informatique Fondamentale de LILLE
UA 369 du CNRS

Université LILLE I - 59655 VILLENEUVE d'ASCQ Cédex
FRANCE

Abstract: In n° 32 of the EATCS Bulletin, Márton Kósa stated five conjectures concerning the study of some binary morphisms and the factors of the infinite words these morphisms generate [11].

I prove the first four conjectures, the main result being the characterization of a large class of morphisms which stationary words are Sturmian.

Résumé: Dans le n° 32 du Bulletin de l'EATCS, Márton Kósa a présenté cinq conjectures relatives à l'étude de quelques morphismes binaires et aux facteurs des mots infinis engendrés par ces morphismes [11].

Je prouve les quatre premières de ces conjectures, le principal résultat étant la mise en évidence d'une large classe de morphismes dont les mots stationnaires sont Sturmiens.

Mailing address:
Patrice SÉÉBOLD
I.U.T. du Littoral
Département Informatique
Rue David
BP. 689
62228 CALAIS Cédex - FRANCE

1) Introduction

The study of repetitions in words is an old problem since it has been initiated by Axel Thue in 1906 when he published the first of a series of papers on this subject (see [21], [22] and also [9]).

Beyond its own interest, this study is important because it has connexion with many other fields of theoretical computer science, and also number theory, ergodic theory, symbolic dynamics ... (see, in particular, [1], [2], [6], [7], [8], [10], [13], [14]).

After Thue, many authors studied this problem and, recently, a lot of papers and books appeared related to it (see [4], [12], [15], [16], [17], [19], [20]).

Among infinite words, the Fibonacci sequence (which can be defined as the fixed point of the morphism $\varphi:\{a,b\} \to \{a,b\}^+$, $\varphi(a) = ab$, $\varphi(b) = a$) is particularly interesting and has been widely studied (see [3], [7], [10], [15], [17], [18]).

In particular, recent papers appeared dealing with this sequence which is a basic example of a unification theorem of J. Shallit [20] and is connected, in Physics, with "Quasicrystals" [5].

Furthermore, it is the best known example of a Sturmian word (words which have exactly (n+1) factors of length n, for any $n \in \mathbf{N}$).

In his paper, Márton Kósa introduces the following morphisms on the two letter alphabet $\mathcal{A} = \{0,1\}$:

$$\mathcal{X} = (0 \mapsto 1,\ 1 \mapsto 0)\ ,\ \mathcal{L} = (0 \mapsto 0,\ 1 \mapsto 01)\ ,\ \mathcal{R} = (0 \mapsto 0,\ 1 \mapsto 10).$$

Calling $\mathbf{St}(\mathcal{A})$ the monoid generated by $\{\mathcal{X}, \mathcal{L}, \mathcal{R}\}$, he gives three relations obviously true in it and conjectures that these form a complete set of generating relations of $\mathbf{St}(\mathcal{A})$ (see [11] - Conjecture 1). Then, calling $\mathbf{Reg}(\mathcal{A})$ a set of morphisms on $\mathcal{A}$ which elements generate infinite words, he conjectures that if $\mathcal{G} \in \mathbf{St}(\mathcal{A}) \cap \mathbf{Reg}(\mathcal{A})$ then its stationary words (words $\underline{\underline{w}}$ for which there exists $i \in \mathbf{N}$ such that $\mathcal{G}^i(\underline{\underline{w}}) = \underline{\underline{w}}$) are Sturmian ([11] - Conjecture 2).

Finally, considering a particular element of $\mathrm{St}(\mathcal{A})$, he gives two conjectures about the factors of the infinite word generated by this morphism ([11] - Conjecture 3), a last conjecture generalizing these.

Here, I prove the first four conjectures of Márton Kósa, the main result being the proof of the second conjecture (see theorem 2).

This proof, based on the relationship between morphisms in $\mathbf{St}(\mathcal{A}) \cap \mathbf{Reg}(\mathcal{A})$ and the Fibonacci morphism, indicates that Fibonacci morphism and Sturmian words are very closely related. Moreover, this result is very important because it is, to the best of my knowledge, the first known characterization of a simple way to obtain, automatically, Sturmian words.

2) Notations

The main notations are classical (see [12]) and I just recall them briefly.

Let $\mathcal{A}$ be an *alphabet* . $\mathcal{A}^*$ denotes the free monoid generated by $\mathcal{A}$ and $\mathcal{A}^+ = \mathcal{A}^* - \{\varepsilon\}$ where ε is the *empty word.*

Let $u \in \mathcal{A}^*$, $|u|$ denotes the *length* of u and $|u|_a$ denotes the number of occurrences of the letter a in the word u.

A word $v \in \mathcal{A}^*$ is a *factor* (resp. *left factor*) of the word $u \in \mathcal{A}^*$ if there exist $u_1 \in \mathcal{A}^*$ and $u_2 \in \mathcal{A}^*$ such that $u = u_1 v u_2$ (resp. $u_1 = \varepsilon$).

An *infinite word* (or *sequence*) on $\mathcal{A}$ is a function $\underline{\underline{x}} : \mathbb{N} \to \mathcal{A}$.
The set of infinite words on $\mathcal{A}$ is denoted by $\mathcal{A}^\omega$ and $\mathring{\mathcal{A}} = \mathcal{A}^+ \cup \mathcal{A}^\omega$.

An infinite word $\underline{\underline{x}}$ is said to be *ultimately periodic* if there exist $u \in \mathcal{A}^*$ and $v \in \mathcal{A}^*$ such that $\underline{\underline{x}} = uv^\omega$.

Let f be a morphism on $\mathcal{A}$ and $\underline{\underline{x}} \in \mathcal{A}^\omega$. $\underline{\underline{x}}$ is said to be *generated by* f if there exist $i \in \mathbb{N}$ and $a \in \mathcal{A}$ such that, for any $n \in \mathbb{N}$, $(f^i)^n(a)$ is a left factor of $\underline{\underline{x}}$ and $|(f^i)^n(a)| < |(f^i)^{n+1}(a)|$.
In this case, the usual notation is : $\underline{\underline{x}} = (f^i)^\omega(a)$.

Finally, two morphisms f and g on $\mathcal{A}$ are said to be *equal* if, for any $a \in \mathcal{A}$, $f(a) = g(a)$.

3) Definitions and results

a) Presentation of a family of binary morphisms

Let $\mathcal{A} = \{0,1\}$ be an alphabet and let $\mathbf{Em}(\mathcal{A})$ be the monoid obtained by using the composition of endomorphisms on $\mathring{\mathcal{A}}$.

Let $\mathcal{X} : \mathcal{A} \to \mathcal{A}^+$, $\mathcal{L} : \mathcal{A} \to \mathcal{A}^+$, $\mathcal{R} : \mathcal{A} \to \mathcal{A}^+$ be endomorphisms

$\mathcal{X}$	$\mathcal{L}$	$\mathcal{R}$
$0 \mapsto 1$	$0 \mapsto 0$	$0 \mapsto 0$
$1 \mapsto 0$	$1 \mapsto 01$	$1 \mapsto 10$

Let $\mathbf{St}(\mathcal{A})$ be the submonoid of $\mathbf{Em}(\mathcal{A})$ generated by $\{\mathcal{X},\mathcal{L},\mathcal{R}\}$ and $\mathbf{Id}$ the identical transformation on $\mathring{\mathcal{A}}$ ($\mathbf{Id}(0) = 0$, $\mathbf{Id}(1) = 1$) which is the unit element of $\mathbf{Em}(\mathcal{A})$ and $\mathbf{St}(\mathcal{A})$.

3 equalities are obviously true in $\mathbf{St}(\mathcal{A})$:

(1) $\mathfrak{X}^2 = \mathfrak{Id}$

(2) $\mathfrak{L}\mathfrak{R} = \mathfrak{R}\mathfrak{L}$

(3) For any $k \in \mathbf{N}$, $\mathfrak{R}\mathfrak{X}\mathfrak{R}^k\mathfrak{X}\mathfrak{L} = \mathfrak{L}\mathfrak{X}\mathfrak{L}^k\mathfrak{X}\mathfrak{R}$

and I prove the following

PROPOSITION 1: (1) , (2) and (3) form a complete set of generating relations of $\mathbf{St}(\mathcal{A})$.

b) The main result

DEFINITION 1: $\mathfrak{G} \in \mathfrak{Em}(\mathcal{A})$ is *regular* if and only if the three following conditions hold:

1) $\mathfrak{G}(0)$ contains letter 1 at least once

2) $\mathfrak{G}(1)$ contains letter 0 at least once

3) At least one of the two following conditions holds

a) $\mathfrak{G}(0)$ contains letter 0 at least once

b) $\mathfrak{G}(1)$ contains letter 1 at least once.

Regularity of $\mathfrak{G} \in \mathfrak{Em}(\mathcal{A})$ assures that any stationary word w of $\mathbf{St}(\mathcal{A})$ (word such that there exists $n \in \mathbf{N}$, $\mathfrak{G}^n(w) = w$) is in $\mathcal{A}^{\omega}$.

The set of regular morphisms on $\mathcal{A}$ is denoted by $\mathfrak{Reg}(\mathcal{A})$.

DEFINITION 2: For any $n \in \mathbf{N}$, let $p(n)$ be the number of factors of length n of an infinite word $\underline{u}$.

Then $\underline{u}$ is *Sturmian* if and only if $p(n) = n+1$, for any $n \in \mathbf{N}$.

There are several different ways to obtain Sturmian words on a two letter alphabet (see some of these in [15]). However, there is not any "very simple" method to construct such words. So, it would be very helpful to know which Sturmian words can be obtained by iterating a morphism on a two letter alphabet.

The best known example of such a word is the Fibonacci sequence, fixed point of the morphism $\varphi:\{a,b\} \to \{a,b\}^+$, $\varphi(a) = ab$, $\varphi(b) = a$.

In what follows, I show that any morphism in $\mathbf{St}(\mathcal{A}) \cap \mathfrak{Reg}(\mathcal{A})$ is closely related to the Fibonacci morphism and, then, I prove the

THEOREM 2: If $\mathfrak{G} \in \mathbf{St}(\mathcal{A}) \cap \mathfrak{Reg}(\mathcal{A})$, then its stationary words are Sturmian.

c) Arithmetic properties of a particular morphism of $\mathbf{St}(\mathcal{A})$

DEFINITION 3: Let $\mathfrak{G} : \mathcal{A} \to \mathcal{A}^+$ $\quad (\mathfrak{G} = XRXRXL)$

$$0 \mapsto 101$$
$$1 \mapsto 10101$$

and $\underline{Q}$ its only fixed point ($\underline{Q}$ is an infinite word , $\underline{Q} = \mathfrak{G}^{\omega}(0) = \mathfrak{G}^{\omega}(1)$).

Let $\mathbf{B}_0 = \{1,2\}$ and $\mathbf{B}_1 = \{2,3\}$ be alphabets.

Let $\mathcal{H}_0 : \mathcal{A} \to \mathbf{B}_0{}^+$ and $\mathcal{H}_1 : \mathcal{A} \to \mathbf{B}_1{}^+$

$$0 \mapsto 2 \qquad\qquad 0 \mapsto 23$$
$$1 \mapsto 21 \qquad\qquad 1 \mapsto 233$$

Let $\mathbf{t}(j,i)$ be the i-th letter of $\mathcal{H}_j(\underline{Q})$, $j = 0,1$, $i \in \mathbf{N}$.

Finally, let $f(r,j) = -1 + \sum_{i=1}^{r} \mathbf{t}(j,i)$, $j = 0,1$, $r \in \mathbf{N}$.

($f(r,j)$ represents the sum of the first r letters of $\mathcal{H}_j(\underline{Q})$, minus 1).

(Note that, here, the first letter of $\mathcal{H}_j(\underline{Q})$ has indice "1" and not indice "0" as in the usual definition of infinite words.)

The $f(r,j)$'s give a numerical characterization of the factors of the infinite word $\mathcal{H}_j(\underline{Q})$. Consequently, it would be useful to produce a formula to compute, for any $r \in \mathbf{N}$, the value of $f(r,j)$.

But, as we saw before, any morphism in $\mathbf{St}(\mathcal{A}) \cap \mathbf{Reg}(\mathcal{A})$ is closely related to the Fibonacci morphism and, since the number $\phi = \frac{1+\sqrt{5}}{2}$ is also connected with the Fibonacci word (see[3]), it is natural for this number to arise in the definition of the $f(r,j)$'s.

Actually, I prove the

PROPOSITION 3-4: For any $r \in \mathbf{N}$, $$f(r,0) = \left\lfloor \frac{1+\sqrt{5}}{2} r + \frac{1-\sqrt{5}}{4} \right\rfloor \qquad (3)$$

and $$f(r,1) = \left\lfloor \frac{3+\sqrt{5}}{2} r - \frac{1+\sqrt{5}}{4} \right\rfloor \qquad (4)$$

(where $\lfloor x \rfloor$ denotes the integer part of x).

4) Sketches of proofs

a) I don't present here the proof of proposition 1 which is by induction and, though it is rather technical, does not require special new notion.

b) The proof of theorem 2 is more interesting and needs auxiliary results.

First, I recall that the Fibonacci word $\underline{\underline{f}}$ is the fixed point of the morphism $\varphi : \mathfrak{B} \to \mathfrak{B}^+$ ($\mathfrak{B} = \{a,b\}$), $\varphi(a) = ab$, $\varphi(b) = a$ and can also be generated by the morphism $\tilde{\varphi} : \mathfrak{B} \to \mathfrak{B}^+$, $\tilde{\varphi}(a) = ba$, $\tilde{\varphi}(b) = a$.

$\underline{\underline{f}}$ is a Sturmian word and one has the following

PROPERTY 1: The image of any Sturmian word of $\mathfrak{B}^\omega$ by φ or $\tilde{\varphi}$ is also a Sturmian word.

Now, turning again to $\mathbf{St}(\mathcal{A}) \cap \mathfrak{Reg}(\mathcal{A})$, I show the next two properties:

PROPERTY 2: Let $\mathfrak{G} \in \mathbf{St}(\mathcal{A}) \cap \mathfrak{Reg}(\mathcal{A})$ and $\underline{\underline{w}} \in \mathcal{A}^\omega$.
If $\underline{\underline{w}}$ is a stationary word of $\mathfrak{G}$, then $\underline{\underline{w}}$ is generated by $\mathfrak{G}$ or $\mathfrak{G}^2$ (this is due to the fact that, if u is a left factor of $\underline{\underline{w}}$, since $\mathfrak{G}$ is necessarily strictly growing, then $\mathfrak{G}(u)$ or $\mathfrak{G}^2(u)$ is a left factor of $\underline{\underline{w}}$ strictly lengther than u and starting with u).

PROPERTY 3: Let $\mathfrak{G} \in \mathbf{St}(\mathcal{A}) \cap \mathfrak{Reg}(\mathcal{A})$.
Then $\mathfrak{G} \in \{\mathsf{X}, \mathsf{X}\mathfrak{L}, \mathfrak{L}\mathsf{X}, \mathsf{X}\mathfrak{R}, \mathfrak{R}\mathsf{X}\}^+ \setminus \{\mathsf{X}\}^+$.

The proof is now based on the following remark:

$\mathfrak{L}\mathsf{X}$ is obtained from φ by replacing a by 0 and b by 1,

$\mathsf{X}\mathfrak{L}$ is obtained from φ by replacing a by 1 and b by 0,

$\mathfrak{R}\mathsf{X}$ is obtained from $\tilde{\varphi}$ by replacing a by 0 and b by 1,

$\mathsf{X}\mathfrak{R}$ is obtained from $\tilde{\varphi}$ by replacing a by 1 and b by 0.

Hence any factor of a stationary word of a morphism $\mathfrak{G} \in \mathbf{St}(\mathcal{A}) \cap \mathfrak{Reg}(\mathcal{A})$ is a factor of a Sturmian word, thus $p(\mathrm{n}) \leq \mathrm{n}+1$, for any $\mathrm{n} \in \mathbf{N}$.

Now, to show that $p(n) \geq n+1$, I show that $\mathfrak{G} \in \mathbf{St}(\mathcal{A}) \cap \mathfrak{Reg}(\mathcal{A})$ cannot be ultimately periodic, using the following characterization of all the binary morphisms which generate ultimately periodic words (see [19]):

PROPERTY 4: Let $\underline{x}$ be an ultimately periodic infinite word on $\mathcal{A}$ and $f : \mathcal{A} \to \mathcal{A}^+$ a morphism such that $\underline{x} = f^{\omega}(0)$.

Then f has one of the following six forms:

1) $f(0) = 0^p$, $p \geq 2$ and $f(1) \in \mathcal{A}^*$ $(\underline{x} = 0^{\omega})$
2) $f(0) = 01^p$, $f(1) = 1^q$, $p,q \geq 1$ $(\underline{x} = 01^{\omega})$
3) $f(0) \in \mathcal{A}^+$ with $|f(0)|_0 \geq 2$ and $f(1) = \varepsilon$ $(\underline{x} = (f(0))^{\omega})$
4) $f(0) = (01^p)^q0$, $f(1) = 1$, $p,q \geq 1$ $(\underline{x} = (01^p)^{\omega})$
5) $f(0) = v^p$, $f(1) = v^q$, $p,q \geq 1$ $(\underline{x} = v^{\omega})$
6) $f(0) = (01)^p0$, $f(1) = (10)^q1$, $p,q \geq 1$ $(\underline{x} = (01)^{\omega})$

c) The proof of Proposition 3 is very long and technical (needed deep results on the Fibonacci numbers and word), thus I just give here the main steps of this proof.

Some properties of the Fibonacci numbers

(For details on these properties, see in particular [3])

Fibonacci numbers can be defined as follows:

$F_0 = 1$, $F_1 = 2$

$F_{n+2} = F_{n+1} + F_n$ for any $n \in \mathbf{N}$.

THEOREM: (Zeckendorf) Any integer $n > 0$ can be represented by a sum of distinct Fibonacci numbers:

$$n = F_{k_1} + \ldots + F_{k_{r-1}} + F_{k_r} \quad (k_1 > \ldots > k_{r-1} > k_r).$$

Furthermore, this decomposition is unique under the following condition:

(z) $k_i > k_{i+1} + 2$ for any $i < r$ (which means that there is no consecutive indices).

Now, let $\phi = \dfrac{1+\sqrt{5}}{2} = 1{,}618 \ldots$ and $\hat{\phi} = \dfrac{1+\sqrt{5}}{2} = -\dfrac{1}{\phi} = -0{,}618 \ldots$

One has: a) $\hat{\phi}^2 = \hat{\phi} + 1$

b) For any $n \in \mathbf{N}$, $\phi\,\hat{\phi}^n + \hat{\phi}^{n+1} = \hat{\phi}^n$

c) For any $n \in \mathbf{N}$, $F_{n+1} = \phi F_n + \hat{\phi}^{n+2}$

Now, I give the main properties needed to prove Proposition 3 (proof of Proposition 4 could be done in the same manner).

1) $\underline{f} = (\varphi \circ \tilde{\varphi}^2)(a)\,(\varphi \circ \tilde{\varphi}^2)^2(a) \ldots (\varphi \circ \tilde{\varphi}^2)^i(a) \ldots$

2) $\mathcal{H}_0(\underline{Q}) = \mathcal{G}_0^{\omega}(2)$ with $\mathcal{G}_0 : \mathcal{B}_0 \to \mathcal{B}_0^+$

$$2 \mapsto 21221$$
$$1 \mapsto 221$$

3) $\mathcal{G}_0$ is obtained from $\varphi \circ \tilde{\varphi}^2$ by replacing a by 2 and b by 1

4) Let $\underline{f}'$ be obtained from $\underline{f}$ by replacing a by 2 and b by 1 ($\underline{f}'$ is the Fibonacci word on the alphabet $\mathcal{B}_0$).
$\underline{f}' = \mathcal{G}_0(2)\,\mathcal{G}_0^2(2) \ldots \mathcal{G}_0^i(2) \ldots$

5) Let v be a left factor of $\underline{f}$ such that

$$F_3 + F_6 + \ldots + F_{3p} + F_{3p} < |v| \leq F_3 + F_6 + \ldots + F_{3p} + F_{3(p+1)},\ p \in \mathbf{N}.$$

Then, there exists $\{l_1, \ldots, l_k\} \subset \mathbf{N}$ such that

$$|v| = F_{l_1} + \ldots + F_{l_k} \text{ and } k \leq p+3$$

6) Let u be a left factor of $\underline{f}$.
If there exists some $p \in \mathbf{N}$, $p \geq 2$, such that $|u| = F_p$ then it is well known that $|u|_a = F_{p-1}$ and $|u|_b = F_{p-2}$.

Now, let $u \in \mathcal{B}^+$ be a left factor of $\mathcal{H}_0(\underline{Q}) = \mathcal{G}_0^{\omega}(2)$ and let $p \in \mathbf{N}$ be such that $F_{3p} < |v| \leq F_{3p+1}$. There exists $v \in \mathcal{B}^+$ such that v is a left factor of $\underline{f}'$ and $v = \mathcal{G}_0(2) \ldots \mathcal{G}_0^p(2)u$

Let $\{l_1, \ldots, l_k\} \subset \mathbf{N}$ be such that $|v| = F_{l_1} + \ldots + F_{l_k}$.

Let $r = |u|$, one has: $r = F_{l_1} + \ldots + F_{l_k} - F_3 - \ldots - F_{3p}$ and $k \le p+3$.

Now, applying properties a), b) and c) above, one has:

$$f(r,0) = -1 + \phi(F_{l_1} + \ldots + F_{l_k} - F_3 - F_6 - \ldots - F_{3p})$$

$$+ \hat{\phi}^{l_1+2} + \ldots + \hat{\phi}^{l_k+2} - \hat{\phi}^{5} - \hat{\phi}^{8} - \ldots - \hat{\phi}^{3p+2}$$

and $$\frac{1+\sqrt{5}}{2}\, r + \frac{1-\sqrt{5}}{4} = \phi(F_{l_1} + \ldots + F_{l_k} - F_3 - F_6 - \ldots - F_{3p}) + \frac{\hat{\phi}}{2}$$

Thus, one has to prove the two following inequalities:

$$-1 + \hat{\phi}^{l_1+2} + \ldots + \hat{\phi}^{l_k+2} - \hat{\phi}^{5} - \hat{\phi}^{8} - \ldots - \hat{\phi}^{3p+2} \le \frac{\hat{\phi}}{2}$$

and $$\frac{\hat{\phi}}{2} < \hat{\phi}^{l_1+2} + \ldots + \hat{\phi}^{l_k+2} - \hat{\phi}^{5} - \hat{\phi}^{8} - \ldots - \hat{\phi}^{3p+2}$$

which is done using properties of geometrical progressions, and the fact that $k \le p+3$.

5) Conclusion

In this paper, we introduced a large class of binary morphisms ($\mathbf{St}(\mathcal{A}) \cap \mathbf{Reg}(\mathcal{A})$) which stationary words are Sturmian.

However, there exist other binary morphisms which stationary words are Sturmian: for example, the morphism f on $\mathcal{A}$ ($f(0) = 0101101$, $f(1) = 0101$) generates a Sturmian word, but $f = \mathsf{X R X L X} f'$ with $f'(0) = 101$, $f'(1) = 11$ and, since $f' \notin \mathbf{St}(\mathcal{A})$, $f \notin \mathbf{St}(\mathcal{A}) \cap \mathbf{Reg}(\mathcal{A})$.

Consequently, an interesting question is to know whether it is possible to give a characterization of <u>all</u> the binary morphisms which stationary words are Sturmian.

Another interesting problem is the fifth conjecture of Márton Kósa which concerns a possible generalization of the definition of the $f(r,j)$'s for any $j \ge 2$.

References

[1] S. ADJAN: *The Burnside problem and identities in groups*, Math. Grenzgeb. vol.95 - Springer (1979).

[2] J.M. AUTEBERT - J. BEAUQUIER - L. BOASSON - M. NIVAT: *Quelques problèmes ouverts en théorie des langages algébriques*, RAIRO - Inf. Th. 13 (1979), p. 363 - 379.

[3] J. BERSTEL: *Mots de Fibonacci*, L.I.T.P. Séminaire d'Informatique Théorique 1980-81, p. 57 - 78.

[4] J. BERSTEL: *Some recent results on square-free words*, STACS Paris (1984).

[5] E. BOMBIERI - J.E. TAYLOR: *Which distributions of matter diffract? An initial investigation*, J. Physique 47 (1986), p. 19 - 28.

[6] J. BRZOZOWSKI - K. CULIK II - A. GABRIELIAN: *Classification of noncounting events*, J. Comp. Syst. Science 5 (1971), p. 41 - 53.

[7] G. CHRISTOL - T. KAMAE - M. MENDÈS-FRANCE - G. RAUZY: *Suites algébriques, automates et substitutions*, Bull. Soc. Math. France 108 (1980), p. 401 - 419.

[8] A. EHRENFEUCHT - G. ROZENBERG: *Elementary homomorphisms and a solution to the D0L-sequence equivalence problem*, Theor. Comp. Sci. 7 (1978), p. 169 - 183.

[9] G. HEDLUND: *Remarks on the work of Axel Thue on sequences*, Nord. Mat. Tidskr. 16 (1967), p. 148 - 150.

[10] J. KARHUMÄKI: *On the equivalence problem for binary morphisms*, Inform. and Control 50 (1981), p. 276 - 284.

[11] M. KÓSA: in 'Problems and Solutions', EATCS Bulletin 32 (1987), p. 331 - 333.

[12] M. LOTHAIRE: 'Combinatorics on words', Addison-Wesley (1983).

[13] M. MORSE: *Recurrent geodesics on a surface of negative curvature*, Trans. Amer. Math. Soc. 22 (1921), p. 84 - 100.

[14] M. MORSE - G. HEDLUND: *Unending chess, symbolic dynamics and a problem in semi-groups*, Duke Math. J. 11 (1944), p. 1 - 7.

[15] G. RAUZY: *Mots infinis en arithmétique*, 12ième école de printemps d'Informatique Théorique - Le Mont Dore (1984) - in 'Automata on infinite words' , Lecture Notes in Computer Science 192 Springer, p. 165 - 171.

[16] A. SALOMAA: 'Jewels of formal language theory', Pitman (1981).

[17] P. SÉÉBOLD: *Sequences generated by infinitely iterated morphisms*, Discrete Applied Mathematics 11 (1985), p. 255 - 264.

[18] P. SÉÉBOLD: 'Propriétés combinatoires des mots infinis engendrés par certains morphismes', Thèse de Doctorat (1985), Rapport L.I.T.P. 85-16.

[19] P. SÉÉBOLD: *An effective solution to the DOL periodicity problem in the binary case*, EATCS Bulletin 36 (1988), p. 137 - 151.

[20] J. SHALLIT: *A generalization of automatic sequences*, Theor. Comp. Sci. 61 (1988), p. 1 - 16.

[21] A. THUE: *Uber unendliche Zeichenreihen*, Norske Vid. Selsk. Skr. I. Math. Nat. kl. Christiania (1906), p. 1 - 22.

[22] A. THUE: *Uber die gegenseitige Lage gleicher Teile gewisser Zeichenreihen*, Norske Vid. Selsk. Skr. I. Math. Nat. kl. Christiania (1912), p. 1 - 67.

On the Finite Degree of Ambiguity of Finite Tree Automata

Helmut Seidl

Fachbereich Informatik
Universitaet des Saarlandes
D-6600 Saarbruecken
West Germany

ABSTRACT

The degree of ambiguity of a finite tree automaton A, da(A), is the maximal number of different accepting computations of A for any possible input tree. We show: it can be decided in polynomial time whether or not da(A)<∞. We give two criteria characterizing an infinite degree of ambiguity and derive the following fundamental properties of an finite tree automaton A with n states and rank L>1 having a finite degree of ambiguity:
for every input tree t there is a input tree t_1 of depth less than $2^{2n}\cdot n!$ having the same number of accepting computations;
the degree of ambiguity of A is bounded by $2^{2^{2\cdot\log(L+1)\cdot n}}$

0. Introduction

Generalizing a result of [SteHu81,SteHu85,Kuich88] from finite word automata to finite tree automata we showed in [Sei89] that, for any fixed constant m it can be decided in polynomial time whether or not two m-ambiguous finite tree automata are equivalent. Since the equivalence problem of finite tree automata is logspace complete in deterministic exponential time in general, this result justifies our special interest in the class of finitely ambiguous finite tree automata. In this paper we continue the investigations of [Sei88].
In [WeSei86] it is shown that it can be decided in polynomial time whether or not the degree of ambiguity of a finite word automaton is finite. For this a criterion (IDA) is given characterizing an infinite degree of ambiguity. Moreover, this paper proves an upper bound $5^{n/2}\cdot n^n$ for the maximal degree of ambiguity of a finitely ambiguous finite word automaton A having n states. Using an estimation of Baron [Ba88] Kuich slightly improves this upper bound [Kui88]. In [WeSei88] the analysis of finitely ambiguous finite word automata is completed by proving a non-ramification lemma which allows for every word w to construct a word w' of length less than $2^{2n}\cdot n!$ having the same number of accepting computation paths.
In this paper we extend the methods of [WeSei86,WeSei88] to finite tree automata. For a finite tree automaton A we employ the branch automaton A_B. A_B is a finite word automaton canonically constructed from A which accepts the set of all branches of trees in L(A). A_B allows to formulate two reasons (T1) and (T2) for A to be infinitely ambiguous. The second one originates in an appropriate extension of the criterion (IDA) of [WeSei86] whereas the first one has no analogon in the word case. We prove a non-ramification lemma for finite tree automata. We apply this lemma to prove: if the branch automaton A_B of a finite tree automaton A with n states neither complies with (T1) nor with (T2) then for every input tree t there is a input tree t_1 of depth less than $2^{2n}\cdot n!$ having the same number of accepting computations as t. Since the number of computations for a tree of bounded depth is bounded, this proves: da(A) < ∞ iff A_B doesn't comply with (T1) or (T2). Since the criteria (T1) and (T2) are testable in polynomial time, it follows that it can be decided in polynomial time whether or not the degree of ambiguity of a finite tree automaton is finite.
Finally, we investigate the maximal number of accepting computations of a finitely ambiguous finite

tree automaton A for a given tree t. Now, it no longer suffices to analyze the set of traces of the set of accepting computations for t on a single branch. We estimate the number of nodes in t where an accepting computation of A for t "leaves" the first strong connectivity component of the state set of A. This allows to perform an induction on the number of strong connectivity components yielding $da(A)<\infty$ iff $da(A)<2^{2^{2\cdot\log(L+1)\cdot n}}$ where n is the number of states and L is the rank of A. (As usual, log denotes the logarithm with base 2). A simple example shows that this upper bound is tight up to a constant factor in the highest exponent.
A full version of this paper will appear in Acta Informatica.

1. General Notations and Concepts

In this section we give basic definitions and state some fundamental properties.
A ranked alphabet Σ is the disjoint union of alphabets $\Sigma_0,..,\Sigma_L$.The rank of $a\in\Sigma$, rk(a) , equals m iff $a\in\Sigma_m$. T_Σ denotes the free Σ-algebra of (finite ordered Σ-labeled) trees, i.e. T_Σ is the smallest set T satisfying (i) $\Sigma_0\subseteq T$, and (ii) if $a\in\Sigma_m$ and $t_0,..,t_{m-1}\in T$, then $a(t_0,..,t_{m-1})\in T$. Note: (i) can be viewed as the subcase of (ii) where m=0.
The depth of a tree $t\in T_\Sigma$, depth(t), is defined by depth(t)=0 if $t\in\Sigma_0$, and $\text{depth}(t)=1+\max\{\text{depth}(t_0),..,\text{depth}(t_{m-1})\}$ if $t=a(t_0,..,t_{m-1})$ for some $a\in\Sigma_m$, m>0.
The set of nodes of t , S(t) is the subset of $\mathbb{N}_0^*$ defined by $S(t)=\{\varepsilon\}\cup\bigcup_{j=0}^{m-1} j\cdot S(t_j)$ where $t=a(t_0,..,t_{m-1})$ for some $a\in\Sigma_m$, $m\geq 0$. t defines maps $\lambda_t(\_):S(t)\to\Sigma$ and $\sigma_t(\_):S(t)\to T_\Sigma$ mapping the nodes r of t to their labels or the subtrees of t with root r, respectively. We have

$$\lambda_t(r) = \begin{cases} a & \text{if } r=\varepsilon \\ \lambda_{t_j}(r') & \text{if } r=j\cdot r' \end{cases}$$

and

$$\sigma_t(r) = \begin{cases} t & \text{if } r=\varepsilon \\ \sigma_{t_j}(r') & \text{if } r=j\cdot r' \end{cases}$$

Let $t,t_1\in T_\Sigma$ and $r\in S(t)$. Then $t[t_1/r]$ denotes the tree obtained from t by replacing the subtree with root r with t_1.

A finite tree automaton (abbreviated: FTA) is a quadruple $A = (Q,\Sigma,Q_I,\delta)$ where:
Q is a finite set of states,
$Q_I\subseteq Q$ is the set of initial states,
$\Sigma=\Sigma_0\cup..\cup\Sigma_L$ is a ranked alphabet, and
$\delta\subseteq\bigcup_{m\geq 0} Q\times\Sigma_m\times Q^m$ is the set of transitions of A.
$rk(A)=\max\{rk(a)\mid a\in\Sigma\}$ is called the rank of A.
Let $t=a(t_0,..,t_{m-1})\in T_\Sigma$ and $q\in Q$. A q-computation of A for t consists of a transition $(q,a,q_0..q_{m-1})\in\delta$ for the root and q_j-computations of A for the subtrees t_j, $j\in\{0,..,m-1\}$. Especially, for m=0, there is a q-computation of A for t iff $(q,a,\varepsilon)\in\delta$. Formally, a q-computation ϕ of A for t can be viewed as a map $\phi:S(t)\to Q$ satisfying (i) $\phi(\varepsilon)=q$ and (ii) if $\lambda_t(r)=a\in\Sigma_m$, then $(\phi(r),a,\phi(r\cdot 0)..\phi(r\cdot(m-1)))\in\delta$. ϕ is called accepting computation of A for t, if ϕ is a q-computation of A for t with $q\in Q_I$. For $t\in T_\Sigma$ and $q\in Q$ $\Phi_{A,q}(t)$ denotes the set of all q-computations of A for t, $\Phi_{A,Q_I}(t)$ denotes the set of all accepting computations of A for t. If A is known from the context, we will omit A in the index of Φ.
For any $r\in S(t)$ and any q-computation $\phi\in\Phi_q(t)$ let ϕ_r denote the subcomputation of A for the subtree $\sigma_t(r)$ of t induced by ϕ, i.e. ϕ_r is defined by $\phi_r(r') = \phi(rr')$.
Assume $t\in T_\Sigma$, $r\in S(t)$ and $q,q_1\in Q$. A map $\phi:(S(t)\setminus r\cdot S(\sigma_t(r))\cup\{r\}\to Q$ is called partial q-computation of A for t relative to q_1 at node r, if

- $\phi(\varepsilon) = q$; $\phi(r) = q_1$; and
- $\lambda_t(r')=a\in\Sigma_m$ implies $(\phi(r'),a,\phi(r'0)..\phi(r'(m-1))\in\delta$ for all $r'\notin r\cdot S(\sigma_t(r))$.

If $q\in Q_I$, then ϕ is called accepting partial computation of A for t relative to q_1 at r. The set of all partial q-computations of A for t relative to q_1 at r is denoted by $\Phi^{P}_{A,q,q_1}(t,r)$. The set of all accepting partial computations of A for t relative to q_1 at r is denoted by $\Phi^{P}_{A,Q_I,q_1}(t,r)$. Again, if A is known from the context we omit A in the index.

Finally, we define the ambiguity of A for a tree t, $da_A(t)$, as the number of different accepting computations of A for t. Note: $da_A(t)$ is finite for every $t \in T_\Sigma$.
The (tree) language accepted by A, L(A) is defined by $L(A) = \{t \in T_\Sigma \mid da_A(t) \neq 0\}$.
The degree of ambiguity of A, da(A) is defined by $da(A) = \sup\{da_A(t) \mid t \in T_\Sigma\}$.
A is called

- unambiguous, if $da(A) \leq 1$;
- ambiguous, if $da(A) > 1$;
- finitely ambiguous, if $da(A) < \infty$; and
- infinitely ambiguous, if $da(A) = \infty$.

For measuring the computational costs of our algorithms relative to the size of an input automaton, we define the size of A, |A|, by $|A| = \sum_{(q,a,q_0..q_{m-1}) \in \delta} (m+2)$.

An FTA $A = (Q,\Sigma,Q_I,\delta)$ is called reduced, if

- $Q \times \{a\} \times Q^m \cap \delta \neq \emptyset$ for all $m \geq 0$ and $a \in \Sigma_m$, and
- $\exists t \in T_\Sigma, \phi \in \Phi_{Q_I}(t) : q \in im(\phi)$ for all $q \in Q$. [1]

The following fact is well-known:

Proposition 1.1
For every FTA $A=(Q,\Sigma,Q_I,\delta)$ there is an FTA $A_r=(Q_r,\Sigma_r,Q_{r,I},\delta_r)$ with the following properties:

(1) $Q_r \subseteq Q$, $Q_{r,I} \subseteq Q_I$, $\delta_r \subseteq \delta$;

(2) A_r is reduced;

(3) $L(A_r)=L(A)$; and

(4) $da(A_r)=da(A)$.

A_r can be constructed from A by a RAM (with the uniform cost criterion and without multiplications) in time O(|A|). □

Actually, the construction of A_r is analogous to the reduction of a contextfree grammar.
Proposition 1.1 can be used to decide in polynomial time whether or not L(A) is empty. The next proposition shows that it also can be decided in polynomial time whether or not A is unambiguous.

Proposition 1.2
Given FTA A, one can decide in time $O(|A|^2)$ whether or not $da(A)>1$. □

As usual, a finite word automaton is defined as a 5-tuple $M = (Q,\Gamma,\delta,Q_I,Q_F)$ where
Q is a finite set of states;
Γ is a finite alphabet;
$Q_I \subseteq Q$ is the set of initial states;
$Q_F \subseteq Q$ is the set of final states; and
$\delta \subseteq Q \times \Gamma \times Q$ is the transition relation of M .
A word $\pi = q_0x_1q_1..q_{m-1}x_mq_m \in Q(\Gamma Q)^*$ with $x_j \in \Gamma$ and $q_j \in Q$ is called computation path of M for $w=x_1..x_m$ from q_0 to q_m if $(q_{j-1},x_j,q_j) \in \delta$ for all $j \in \{1,..,m\}$. π is said to start in q_0 and end in q_m. The set of all computation paths of M for w from q_0 to q_m is denoted by $\Pi_{M,q_0,q_m}(w)$. A computation path π of

[1] im(ϕ) denotes the image of the map ϕ.

M for w is called accepting, if π starts in an initial state and ends in a final state. If M is known from the context, we omit M in the index of Π.
The language L(M) accepted by M is defined by $L(M) = \{w \in \Gamma^* \mid$ there is an accepting computation path of M for w$\}$.
For the ranked alphabet Σ let Σ_B be the (ordinary) alphabet $\Sigma_B = \{(a,j) \mid m>0, a\in\Sigma_m, j\in\{0,..,m-1\}\}$. The set B(t) of branches of a tree t is defined by $B(t) = \{\varepsilon\}$ if $t=a\in\Sigma_0$ and $B(t) = \bigcup_{j=0}^{m-1}(a,j)\cdot B(t_j)$ if $t=a(t_0,..,t_{m-1})$ for some $a\in\Sigma_m$, $m>0$. Note: the sequence of the second components of the symbols of a branch forms a leaf, whereas the sequence of the first components gives the labels on the path in t from the root of t to this leaf (omitting the label of the leaf itself). A prefix $w=(a_1 j_1)..(a_k j_k)$ of a branch of t is called path in t. A subtree $\sigma_t(rj)$ of t is called associated to the path w if $r=j_1..j_\kappa$ for some $\kappa<k$ and $j\neq j_{\kappa+1}$.
For a given reduced FTA $A = (Q,\Sigma,Q_I,\delta)$ we define the branch automaton A_B. A_B is the finite word automaton defined by $A_B = (Q,\Sigma_B,\delta_B,Q_I,Q_F)$ where $Q_F = \{q\in Q \mid \exists a\in\Sigma_0 : (q,a,\varepsilon)\in\delta\}$, and the transition relation δ_B is obtained from δ as follows. If $(q,a,q_0..q_{k-1})\in\delta$, then $(q,(a,j),q_j)\in\delta_B$ for all $j\in\{0,..,k-1\}$.
Since A is assumed to be reduced, every $q\in Q$ lies on an accepting computation path of A_B as well. By [Cou78, prop. 4.9] we have: $L(A_B) = \{v\in\Sigma_B^* \mid \exists t\in L(A) : v$ branch of $t\}$.
Assume $t\in T_\Sigma$, ϕ is a q-computation of A for t, and $w=(a_1,j_1)..(a_k,j_k)$ is a path in t. The trace of ϕ on w is the computation path ϕ_w of A_B for w with $\phi_w=q_0(a_1,j_1)q_1..(a_k,j_k)q_k$ where $q_\kappa=\phi(j_1..j_\kappa)$ for all $\kappa\in\{0,..,k\}$.

2. Characterizing an Infinite Degree of Ambiguity

In this section we give a complete characterization of those FTA's having an infinite degree of ambiguity. In terms of the branch automaton corresponding to a FTA A we state two reasons (T1) and (T2) for an infinite degree of ambiguity of A. Both (T1) and (T2) are decidable in polynomial time. We formulate a non-ramification lemma for FTA's. This lemma enables us to prove: if the branch automaton of a FTA neither satisfies (T1) nor (T2), then for every tree t there is a tree of depth less than $2^{2n}\cdot n!$ having the same number of accepting computations. Since the number of different accepting computations for a tree of depth at most $2^{2n}\cdot n!$ is bounded by some constant, we conclude that (T1) and (T2) precisely characterize an infinite degree of ambiguity.

For the following, $A = (Q,\Sigma,Q_I,\delta)$ is a fixed reduced FTA with n states. For an arbitrary state $q\in Q$, A_q denotes the FTA $A_q=(Q,\Sigma,\{q\},\delta)$.

Proposition 2.1
If A_B satisfies (T1), then $da(A) = \infty$:

(T1) $\exists p,q,q_j\in Q\ \exists w_1,w_2\in\Sigma_B^*, (a,j)\in\Sigma_B\ \exists \pi_1\in\Pi_{p,q}(w_1), \pi_2\in\Pi_{q_j,p}(w_2)$: (T1.1) or (T1.2) is true:

(T1.1) There exist two different transitions $(q,a,q_0^{(i)}..q_{j-1}^{(i)}q_jq_{j+1}^{(i)}..q_{k-1}^{(i)})\in\delta$, $i=1,2$, with $L(A_{q_{j'}^{(1)}})\cap L(A_{q_{j'}^{(2)}})\neq\emptyset$ for all $j'\neq j$.

(T1.2) There exists a transition $(q,a,q_0..q_j..q_{k-1})\in\delta$ with $da(A_{q_{j'}})>1$ for some $j'\neq j$.

Whether or not A_B satisfies (T1) can be decided in time $O(|A|^2)$.

Proof:
Assume A_B satisfies (T1). Since A is reduced, we can construct a tree $t\in T_\Sigma$, $r_0=r_1r_2\in S(t)$, $r_2\neq\varepsilon$ and $\phi^{(0)},\phi^{(1)}\in\Phi_{Q_I}(t)$ such that:

(1) $\phi^{(0)}(r_1) = \phi^{(0)}(r_1r_2) = \phi^{(1)}(r_1) = \phi^{(1)}(r_1r_2)$ and

(2) $\exists r'j$ prefix of $r_2\ \exists j'\neq j : \phi^{(0)}_{r_1r'j'}\neq\phi^{(1)}_{r_1r'j'}$.

Define $u_1=\sigma_t(r_1)$ and $u_k=u_1[u_{k-1}/r_2]$ for $k>1$. Let $t_k=t[u_k/r_1]$. Then $da(t_k)\geq 2^k$. Intuitively, t_k is obtained from t by iterating "u_1 minus $\sigma_{u_1}(r_2)$" k times. By (2), $\phi^{(0)}$ and $\phi^{(1)}$ differ at the iterated part. Furthermore, we can mend the corresponding subcomputations of $\phi^{(0)}$ and $\phi^{(1)}$ together to obtain accepting

computations for t_k. Since for different occurrences of the iterated part we can independently choose subcomputations either according to $\phi^{(0)}$ or according to $\phi^{(1)}$, we get at least 2^k accepting computations for t_k .
The description of the algorithm is omitted. □

A set of transitions $\{(q^{(i)},(a,j),q_j^{(i)})\in\delta_B \mid i\in I\}$ for some index set I, is said to match if there are transitions $(q^{(i)},a,q_0^{(i)}..q_j^{(i)}..q_{m-1}^{(i)})\in\delta$, $i\in I$, such that $\bigcap_{i\in I} L(A_{q_{j'}^{(i)}}) \neq \emptyset$ for all $j'\neq j$.
A set of computation paths $\{q_0^{(i)}(a_1,j_1)q_1^{(i)}..(a_k,j_k)q_k^{(i)} \mid i\in I\}$ is said to match if the sets of transitions $\{(q_{\kappa-1}^{(i)},(a_\kappa,j_\kappa),q_\kappa^{(i)}) \mid i\in I\}$ match for all $\kappa\in\{1,..,k\}$.

Proposition 2.2
If A_B satisfies (T2), then da(A)=∞ :

(T2) $\exists\, p,q\in Q,\ p\neq q,\ w\in\Sigma_B^+ : \exists\, \pi_1\in\Pi_{p,p}(w),\pi_2\in\Pi_{p,q}(w),\pi_3\in\Pi_{q,q}(w) : \pi_1,\pi_2,\pi_3$ match .

Whether or not A_B satisfies (T2) can be decided in time $O(|A|^3)$.

Proof:
If (T2) is fulfilled we can construct $t\in T_\Sigma$, $r_0=r_1r_2\in S(t)$ with $r_2\neq\varepsilon$ and $u_1=\sigma_t(r_1)$ such that there are:

$\phi_0\in\Phi_{Q_I}(t)$ with $\phi_0(r_1)=p$ and $\phi_0(r_1r_2)=q$;

$\phi_1\in\Phi_{p,p}^P(u_1,r_2)$, and

$\phi_2\in\Phi_{q,q}^P(u_1,r_2)$.

Define $t_k=t[u_k/r_1]$ where, for $k>1$, $u_k=u_1[u_{k-1}/r_2]$. Then $da_A(t_k)\geq k$.
Intuitively, one can construct accepting computations $\phi^{(\kappa)}$, $\kappa\in\{1,..,k\}$, for t_k which accept the first $\kappa-1$ occurrences of "u_1 minus $\sigma_{u_1}(r_2)$" according to ϕ_1, the next occurrence according to ϕ_0 and the remaining $k-\kappa$ occurrences according to ϕ_2.
Again, we omit the description of the algorithm. □

Thus, (T1) and (T2) give two polynomially decidable reasons for the infinite degree of ambiguity of A. (T2) is the extension of the criterion (IDA) in [WeSei86] characterizing the infinite degree of ambiguity of finite word automata (additionally we demand the three computation paths of A_B from q to q, q to p and p to p to match), whereas (T1) solely arises from the tree structure. We now formulate the non-ramification lemma for FTA's.

Assume $t\in T_\Sigma$, and $w=(a_1,j_1)..(a_K,j_K)$ is a branch of t. By $G_t(w)$ we denote the acyclic digraph which describes all traces of accepting computations of A for t on w. $G_t(w)=(V,E)$ is defined as follows.

Vertices: $V\subseteq Q\times\{0,..,K\}$ is the set of all (q,k) such that $\exists\,\phi\in\Phi_{Q_I}(t) : \phi(j_1..j_k)=q$.

Edges: $E\subseteq V\times V$ is the set of all pairs ((q,k),(q',k+1)) such that $\exists\,\phi\in\Phi_{Q_I}(t) : \phi(j_1..j_k)=q \;\&\; \phi(j_1..j_kj_{k+1})=q'$.

If $((q_0,k),(q_1,k+1))\ ((q_1,k+1),(q_2,k+2))\ ..\ ((q_{d-1},k+d-1)),(q_d,k+d))$ is a path in $G_t(w)$ where $k\in\{0,..,K-1\}$ and $d>0$, then the following holds:

(1) $q_0(a_{k+1},j_{k+1})q_1(a_{k+2},j_{k+2})..q_{d-1}(a_{k+d},j_{k+d})q_d$ is a computation path of A_B for $(a_{k+1},j_{k+1})..(a_{k+d},j_{k+d})$;

(2) there is a partial q_0-computation ϕ of A for $\sigma_t(j_1..j_k)$ relative to q_d at node $j_{k+1}..j_{k+d}$ such that $\phi(j_{k+1}..j_{k+\kappa})=q_\kappa$, $\kappa\in\{0,..,d\}$.

The proof of the following proposition can be found in the full version.

Proposition 2.3 (Non-Ramification Lemma for FTA's)
Assume A_B does not comply with (T2). Let $t\in T_\Sigma$, let w be a branch of t, and $G_t(w)=(V,E)$. For $k\in\{0,..,|w|\}$ define $D_k = \{q\in Q \mid (q,k)\in V\}$. If $D_k = D_{k+d}$ for some $d\geq1$, then

(1) for every vertex (q,k) in V there is exactly one path in $G_t(w)$ starting in (q,k), and

(2) for every vertex $(q',k+d)$ in V there is exactly one path in $G_t(w)$ ending in $(q',k+d)$. □

Theorem 2.4
Assume that A_B neither satisfies (T1) nor (T2). Then, for every tree $t\in T_\Sigma$, there is a tree t_1 with $depth(t_1) < 2^{2n}\cdot n!$ such that $da_A(t) = da_A(t_1)$.

As a consequence of theorem 2.4 we get the main theorem of this section:

Theorem 2.5
Assume A is a reduced FTA. Then

(1) $da(A) = \infty$ iff A_B satisfies (T1) or (T2).

(2) It can be decided in polynomial time whether or not $da(A) < \infty$.

(3) If $da(A) < \infty$, then there is a tree $t\in T_\Sigma$ with $depth(t) < 2^{2n}\cdot n!$ such that $da(A) = da_A(t)$. □

One can easily construct FTA's such that the corresponding branch automata satisfy any of the criteria (T1.1), (T1.2) or (T2) but none of the others. Therefore, the characterization given in (1) is irredundant. Note further: (3) implies a triple exponential upper bound on the degree of ambiguity of finitely ambiguous FTA's. However, we will prove a (tight) double exponential upper bound in section 3.

Proof of 2.4:
For every tree $t\in T_\Sigma$ and position $r\in S(t)$, define $ACC_t(r)$ as the set of all states q for which there is a accepting partial computation of A for t relative to q at r, i.e.

$$ACC_t(r) = \{q\in Q \mid \Phi^P_{Q_I,q}(t,r)\neq\emptyset\}$$

Define $DER_t(r)$ as the set of all q for which there is a q-computation of A for $\sigma_t(r)$, i.e.

$$DER_t(r) = \{q\in Q \mid \Phi_q(\sigma_t(r))\neq\emptyset\}$$

Assume A_B neither satisfies (T1) nor (T2). Let t denote an arbitrary tree of T_Σ. We show: if there is a branch $w=(a_1,j_1)..(a_K,j_K)$ of t of length $K\geq n!\cdot 2^{2n}$, then we can find a tree $t_1\in T_\Sigma$ with fewer nodes such that $da_A(t)=da_A(t_1)$. This implies the assertion of theorem 2.4.
W.l.o.g. we assume $da_A(t)>0$, i.e. $\Phi_{Q_I}(t)\neq\emptyset$. Consider the acyclic graph $G_t(w) = (V,E)$ and for $k\in\{0,..,K\}$ the sets $D_k = \{q\in Q \mid (q,k)\in V\}$. Clearly, $D_k = ACC_t(j_1..j_k)\cap DER_t(j_1..j_k)$.
Assume $K \geq 2^{2n}\cdot n!$. Then there exist $B_1,B_2\subseteq Q$ and a set $I\subseteq\{0,..,K\}$ with $\#I\geq n!+1$ such that $B_1=ACC_t(j_1..j_k)$ and $B_2=DER_t(j_1..j_k)$ for all $k\in I$. It follows that there are $k_1<k_2$ in I such that
$B_1=ACC_t(j_1..j_{k_1})=ACC_t(j_1..j_{k_2})$ and
$B_2=DER_t(j_1..j_{k_1})=DER_t(j_1..j_{k_2})$ and
for every $q\in B_1\cap B_2$ there is a unique path in $G_t(w)$ from (q,k_1) to (q,k_2).
Define $r_1=j_1..j_{k_1}$, $r_2=j_{k_1+1}..j_{k_2}$, $u=\sigma_t(r_1r_2)$, and $t_1=t[u/r_1]$. We prove: $da_A(t)=da_A(t_1)$.
$da_A(t) \leq da_A(t_1)$:
For every $\phi\in\Phi_{Q_I}(t)$, we have $\phi(r_1)=\phi(r_1r_2)$. Therefore, ϕ gives rise to an accepting computation $\bar\phi$ for t_1 where $\bar\phi$ is defined by:

$$\bar\phi(r) = \begin{cases}\phi(r_1r_2r') & \text{if } r=r_1r' \\ \phi(r) & \text{else}\end{cases}$$

We have to show that this map is injective. Assume ϕ_1,ϕ_2 are two accepting computations of A for t with $\phi_1(r_1)=\phi_2(r_1)$. By the construction of r_1 and r_2 , ϕ_1 and ϕ_2 agree at every node $r_1r'j$ where $r'j$ is a prefix of r_2. Since A_B does not satisfy (T1), we furthermore have that ϕ_1 and ϕ_2 also agree at every subtree of t with root $r_1r'j'$, $j'\neq j$. It follows: if $\bar\phi_1=\bar\phi_2$ then also $\phi_1=\phi_2$. This proves the injectivity.
$da_A(t) \geq da_A(t_1)$:
Assume $\bar\phi$ is an accepting computation of A for t_1 and $\bar\phi(r_1)=p$. Then $p\in ACC_{t_1}(r_1)\cap DER_{t_1}(r_1)$. Observe $ACC_{t_1}(r_1)=ACC_t(r_1)=B_1$ and $DER_{t_1}(r_1)=DER_t(r_1r_2)=B_2$ which by the construction of r_1 and r_2 also equals

$DER_t(r_1)$. Thus, $p \in D_{k_1}$, and there is a path in $G_t(w)$ from (p,k_1) to (p,k_2). Therefore, there is a partial p-computation of A for $\sigma_t(r_1)$ relative to p at node r_2. It follows that we can extend $\overline{\phi}$ to an accepting computation ϕ for t. Clearly, two different accepting computations $\overline{\phi}_1,\overline{\phi}_2$ for t_1 give rise to two different accepting computations for t. This proves the stated inequality. □

3. A Tight Upper Bound for the Finite Degree of Ambiguity

In this section we prove the following theorem.

Theorem 3.1
Assume A is a reduced FTA with n states and rank L>1. If A_B does not comply with (T1) or (T2), then $da(A) < 2^{2^{2\cdot\log(L+1)\cdot n}}$.

Theorem 3.1 gives an alternative proof for the correctness of our characterization of an infinite degree of ambiguity by the criteria (T1) and (T2). The following example shows that the upper bound for the maximal degree of ambiguity of a finitely ambiguous FTA given in theorem 3.1 is optimal up to a constant factor in the highest exponent.

Theorem 3.2
For every $n \geq 3$ and $L \geq 2$ there is a finitely ambiguous FTA $A_{n,L}$ with n states and rank L such that $da(A_{n,L}) = 2^{2^{\log(L)\cdot(n-2)}}$.

Proof:
Define $A_{n,L}$ by $A_{n,L} = (\{1,..,n\},\Sigma,\{1\},\delta_{n,L})$ where $\Sigma_0=\{\#\}$, $\Sigma_L=\{o\}$ and $\Sigma_m=\emptyset$ else, and $\delta_{n,L} = \{(i,o,(i+1)^L) \mid 1 \leq i \leq n-3\} \cup \{n-2\}\times\{o\}\times\{n-1,n\}^L \cup \{(n-1,\#,\varepsilon),(n,\#,\varepsilon)\}$. Then $L(A_{n,L}) = \{\Delta_{n,L}\}$ where $\Delta_{n,L}$ denotes the complete L-ary tree of depth n-2 whose inner nodes are labeled with o and whose leafs are labeled with #. Since $L(A_{n,L})$ is finite, the degree of ambiguity of $A_{n,L}$ is finite, too. There is a bijection between $\Phi_{\{1\}}(\Delta_{n,L})$ and the set of all words of length L^{n-2} over a two letter alphabet. Therefore, $da(A_{n,L}) = 2^{L^{n-2}}$. □

We now prove theorem 3.1 . Let $A=(Q,\Sigma,Q_I,\delta)$ be a fixed reduced FTA with n>0 states and rank L>1 (the case n=0 is trivial).
We partition the set Q according to accessibility. For states $p,q \in Q$, we say q is accessible from p (short: $p \rightarrow_A q$) iff there is a computation path of A_B from p to q. The equivalence relation $\leftrightarrow_A$ on Q is defined by $p \leftrightarrow_A q$ iff $p \rightarrow_A q$ and $q \rightarrow_A p$. The equivalence classes of Q w.r.t. $\leftrightarrow_A$ are denoted by $Q_1,..,Q_k$. They are also called the strong connectivity components of Q. W.l.o.g. we assume for $p \in Q_i$ and $q \in Q_j$, $p \rightarrow_A q$ implies $i \leq j$.

We first deal with FTA's having just one initial state. Define d(k) to be the maximal degree of ambiguity of a reduced FTA A with 1 initial state, rank L, at most n states and at most k strong connectivity components such that A_B does not comply with (T1) or (T2). Observe: in order to prove theorem 3.1 it suffices to compute an upper bound for d(n).
So, for our FTA A assume $Q_I=\{q_I\}$. Since A is reduced, q_I is in Q_1. Let t be a fixed tree in L(A). We classify the q_I-computations of A for t relative to Q_1. The following observation is crucial.

Fact 3.3
Assume A_B does not comply with (T1), and da(A)>1. Assume $\phi \in \Phi_{q_I}(t)$ and $r \in S(t)$. If $\phi(r) \in Q_1$, then there is at most one j such that $\phi(rj) \in Q_1$.

Proof:
For a contradiction assume there are $j_1 \neq j_2$ such that $q_1=\phi(rj_1) \in Q_1$ and $q_2=\phi(rj_2) \in Q_1$. Since Q_1 is strongly connected, we have $q_1 \rightarrow_A q_I$. Therefore, since A_B does not comply with (T1), A_{q_2} is unambiguous. Since also $q_2 \rightarrow_A q_I$, $A_{q_I}=A$ must be unambiguous as well: contradiction. □

Fact 3.3 already implies:

Fact 3.4
If $L>1$, $d(1)=1$. □

Thus, if A_B does not comply with (T1) and $da(A)>1$, then for every accepting computation ϕ of A for t, there is a unique maximal trace of ϕ such that every state on it lies in Q_1. This trace is denoted by $\pi_1(\phi)$.

The following fact is an easy consequence of propositions 2.1 and 2.2:

Fact 3.5
Assume A_B does not comply with (T1) or (T2). Assume ϕ, ϕ' are two q_I-computations for t where $\pi_1(\phi)$ is a computation path for w, $\pi_1(\phi')$ is a computation path for w' and $v=(a_1,j_1)..(a_K,j_K)$ is the maximal common prefix of w and w'. If $\phi(j_1..j_K)=\phi'(j_1..j_K)$, then the following holds:

(1) ϕ and ϕ' agree on v, i.e. $\phi_v=\phi'_v$;

(2) ϕ and ϕ' also agree on every subtree of t associated to v, i.e. if $\sigma_t(r)$ is a subtree of t associated to v then $\phi_r=\phi'_r$. □

Now assume $da(A)>1$, and Q has $k>1$ strong connectivity components. We want to perform an induction on k. Therefore, we calculate the cardinality of the set $\{\pi_1(\phi) \mid \phi\in\Phi_{q_I}(t)\}$. Let w be a branch of t and $G_t(w) = (V,E)$ be defined as in section 2. Let J(w) denote the set of all i such that $i=|w|$ or there is an edge $((q,i),(q',i+1))$ in $G_t(w)$ with $q\in Q_1$ and $q'\notin Q_1$. Applying the non-ramification lemma for FTA's we get:

Fact 3.6
Assume A_B does not comply with (T2). Then for every branch w of t, $\#J(w) < 2^n$.

Proof:
For $i\in J(w)$ define $D_i=\{q\in Q \mid (q,i)\in V\}$. Assume $\#J(w)\geq 2^n$. Then there exist $i<i'$ such that $D_i=D_{i'}$. By the non-ramification lemma 2.3 there is exactly one path in $G_t(w)$ starting in (q,i) for every q in D_i. Since $Q_1\cap D_i = Q_1\cap D_{i'}$ and every vertex (p',i') of $G_t(w)$ with $p'\in Q_1$ only can be reached from a vertex (p,i) with $p\in Q_1$, we conclude that for every edge $((q,i),(q',i+1))$ in $G_t(w)$, $q\in Q_1$ implies $q'\in Q_1$: contradiction. □

Fact 3.6 is the appropriate extension of a corresponding result in [WeSei86] for finite word automata. However, to apply fact 3.6 we need the following additional observation.

Fact 3.7
Assume A_B does not comply with (T1). Assume ϕ, ϕ' are different q_I-computations of A for t where $\pi_1(\phi)$ is a computation path for v, $\pi_1(\phi')$ is a computation path for v', u is the maximal common prefix of v and v', and v is a prefix of the branch w. Then the following holds:

(1) $|v|\in J(w)$;

(2) $|u|\in J(w)$.

Proof:
Assertion (1) is immediately clear from the definition of $\pi_1(\_)$.
Ad (2):
W.l.o.g. $v\neq u\neq v'$. Assume $u=(a_1,j_1)..(a_m,j_m)$, $v=u(a,j)u_1$ and $v'=u(a,j')u'_1$. By fact 3.3 there is at most one $\bar{j}$ such that $\phi(j_1..j_m\bar{j})\in Q_1$. Hence, $\phi(j_1..j_mj')\notin Q_1$. □

Together the facts 3.4, 3.5, 3.6 and 3.7 allow to estimate the cardinality of the set $\{\pi_1(\phi) \mid \phi\in\Phi_{q_I}(t)\}$.

Lemma 3.8

Assume A_B does not comply with (T1) or (T2). Assume da(A)>1. Then

$$\#\{\pi_1(\phi) \mid \phi \in \Phi_{q_I}(t)\} < (L+1)^{2^n} \cdot n$$

Proof:
Define $T = \{u \in \Sigma_B^* \mid \exists \phi \in \Phi_{q_I}(t)$: $\pi_1(\phi)$ is a computation path for u}. By fact 3.5, $\#\{\pi_1(\phi) \mid \phi \in \Phi_{q_I}(t)\} \leq n \cdot \#T$. Consider the smallest superset $\overline{T}$ of T which for every two elements $v,v' \in T$ contains the maximal common prefix of v and v'. The set $\overline{T}$ can be viewed as the set of nodes of a tree $s=(\overline{T},E)$ where $(v_1,v_2) \in E$ iff

(i) v_1 is a prefix of v_2 different from v_2; and

(ii) there is no v in $\overline{T}$ different from v_1 and v_2 such that v_1 is a prefix of v and v is a prefix of v_2.

By the facts 3.6 and 3.7 depth(s) $\leq 2^n$. Moreover, fact 3.6 implies that every node of s has at most L successors. Therefore, s has less than $(L+1)^{2^n}$ nodes. From this, the result follows. □

Now we are able to prove:

Lemma 3.9
For every k>1,
$\log d(k) < \log(L+1) \cdot 2^n \cdot (L+1)^{k-2} + \log n \cdot (L+1)^{k-1}$.

Proof:
W.l.o.g. assume da(A)>1. Assume $t \in L(A)$, $w=(a_1,j_1)..(a_K,j_K)$ is a path in t, $r=j_1..j_K$, $\sigma_t(r)=a(t_0,..,t_{m-1})$, and $q \in Q_1$. Let $\Phi^{(r,q)}$ denote the set of all accepting computations ϕ of A for t such that $\pi_1(\phi)$ is a computation path of A_B for w from q_I to q. By lemma 3.4 all $\phi \in \Phi^{(r,q)}$ agree on w and on every subtree of t associated to w. They possibly differ in the transition chosen at node r and in the subcomputations chosen for the subtrees t_j, $0 \leq j \leq m-1$. By the definition of $\pi_1(\_)$ we may view the set of q'-computations $\{\phi_{rj} \mid \phi \in \Phi^{(r,q)}, \phi(rj)=q'\}$, $j \in \{0,..,m-1\}$, as the set of all accepting computations for t_j of a reduced FTA $A'_q=(Q',\Sigma,\{q'\},\delta')$ where $Q' \subseteq Q \setminus Q_1$ and $\delta' \subseteq \delta$ and Q' has at most k-1 strong connectivity components. Since there are at most n^m different transitions applicable at node r, we conclude that $\Phi^{(r,q)} \leq n^m \cdot d(k-1)^m$. By lemma 3.8 we get the following inductive inequation for d(k):

$$d(k) < (L+1)^{2^n} \cdot n \cdot n^L \cdot d(k-1)^L$$

Since by fact 3.4 d(1)=1, the assertion follows. □

Proof of Theorem 3.1:
Assume $A=(Q,\Sigma,Q_I,\delta)$ is a reduced FTA with n states and rank L>1. W.l.o.g. n>1. Assume A_B does not comply with (T1) or (T2). Since Q has at most n strong connectivity components and $\#Q_I \leq n$, we have $da(A) \leq n \cdot d(n)$, and therefore by lemma 3.9,

$$\begin{aligned}
\log da(A) &< \log(L+1) \cdot 2^n \cdot (L+1)^{n-2} + \log n \cdot [(L+1)^{n-1}+1] \\
&\leq 2^n \cdot (L+1)^{n-1} \cdot [\frac{\log(L+1)}{L+1} + \frac{2 \cdot \log n}{2^n}] \\
&\leq 2^n \cdot (L+1)^{n-1} \cdot 2 \\
&\leq 2^{2 \cdot \log(L+1) \cdot n}
\end{aligned}$$
□

Acknowledgement
I thank Andreas Weber for many fruitful discussions and carefully reading of an earlier version of this paper.

References

[Aho74] A.V. Aho, J.E. Hopcroft, J.D. Ullman: The design and analysis of computer algorithms. Addison-Wesley 1974

[Ba88] G. Baron: Estimates for bounded automata. Technische Universitaet Graz und oesterreichische Computer Gesellschaft, Report 253, part 2, June 1988

[Cou78] B. Courcelle: A representation of trees by languages, part II. Theor. Comp. Sci. 7 (1978) pp. 25-55

[GeStei84] F. Gecseg, M. Steinby: Tree automata. Akademiai Kiado, Budapest, 1984

[Kui88] W. Kuich: Finite automata and ambiguity. Technische Universitaet Graz und oesterreichische Computer Gesellschaft, Report 253, part 1, June 1988

[Paul78] W. Paul: Komplexitaetstheorie. B.G. Teubner Verlag Stuttgart 1978

[Sei89] H. Seidl: Deciding equivalence of finite tree automata. Proc. STACS'89, LNCS 349, pp. 480-492

[SteHu81] R. Stearns, H. Hunt III: On the equivalence and containment problems for unambiguous regular expressions, regular grammars and finite automata. 22th FOCS (1981) pp. 74-81

[SteHu85] R. Stearns, H. Hunt III: On the equivalence and containment problems for unambiguous regular expressions, regular grammars and finite automata. SIAM J. Comp. 14 (1985) pp. 598-611

[WeSei86] A. Weber, H. Seidl: On the degree of ambiguity of finite automata. MFCS 1986, Lect. Notes in Comp. Sci. 233, pp. 620-629

[WeSei88] A. Weber, H. Seidl: On finitely generated monoids of matrices with entries in $\mathbb{N}$. Preprint 1988

[We87] A. Weber: Ueber die Mehrdeutigkeit und Wertigkeit von endlichen Automaten und Transducern. Doct. Thesis Frankfurt/Main 1987

APPROXIMATION ALGORITHMS FOR CHANNEL ASSIGNMENT IN CELLULAR RADIO NETWORKS

Hans Ulrich Simon
Fachbereich Informatik, Universität des Saarlandes
Im Stadtwald, D–6600 Saarbrücken

Abstract

A radiocommunication system has to satisfy the channel requests in its service area subject to the constraints that the number of channels is limited and co–channel interference must be excluded. This paper explores the possibility of finding optimal or nearly optimal channel assignments in polynomial time. It relates this problem to 'graph–coloring' and, if the network is assumed as cellular, to 'packing in the plane'. An approximation algorithm is designed which satisfies at least a fraction of $1-e^{-1}$ of the maximum number of satisfiable requests. Several related versions of the problem are also analysed. For some of them, approximation schemes are discovered. Finally, NP–completeness is shown for a very restricted version and lower bounds for the best performance ratios are discussed (under the $P \neq$ NP–assumption).

1 Introduction

A *radiocommunication system* has to satisfy the channel requests in its service area subject to the constraints that the number of channels is limited and co–channel interference must be excluded. In a *cellular radio network*, the service area is subdivided into a net of hexagonal cells (although our results remain valid for other regular patterns of cells, their presentation is simplified if one pattern is fixed; see figure 1a). Two cells are called *adjacent* if they share a common edge. This relation defines the dual network called the *cell–graph*. The *distance* between two cells C_i, C_j is the length of a shortest path from C_i to C_j in the cell-graph. Two cells may use the same channel simultaneously without co–channel interference if their distance is greater than $2r$ where r denotes the so–called *interference radius*. In the discrete model of radiocommunication, the system periodically records the current *load* (see figure 1b), i.e., the number r_i of currently unsatisfied requests for each cell C_i. During one period, it tries to satisfy a maximum number of requests in order to reduce the load for the next period as much as possible.

We differentiate between 3 kinds of channel assignment:

Static Channel Assignment An invariant set of channels is assigned to each cell according to a statistical long–term prediction of the submitted load. This assures nearly optimal assignments for slightly varying load distributions.

Dynamic Channel Assignment Basing only on the current load, the system decides how to assign the channels. This assures flexible response to varying load distributions.

Hybrid Channel Assignment The system applies a static preassignment in a first phase and dynamically computes an assignment for remaining unsatisfied requests in a second phase.

This paper is concerned with the dynamic decisions of the system and thus with the dynamic channel assignment and the phase 2 of hybrid channel assignment. It explores the possibility of obtaining optimal or nearly optimal assignments in polynomial time. The problem is explored empirically very well (see [5] who compare several heuristics via statistics, for instance) whereas it seems that this paper is the first in presenting mathematical performance guarantees for the approximation algorithms. Our investigations are structured as follows:

The next section provides a basis of formal definitions and notations. The following 2 sections are devoted to the design of approximation algorithms for dynamic and hybrid channel assignment, respectively. The last section contains a short description of several related results, in particular negative–ones (see the full paper [10] for a detailed description). Finally, the open problems are summarized.

2 Definitions and Notations

The problems of channel assignment belong to the class of *combinatorial optimization problems*. Each problem Π of this class consists of:

1. a set INPUT(Π) of *problem instances* I of size $|I|$,
2. for each instance I a set $S(I)$ of *legal solutions*,
3. for each solution $\sigma \in S(I)$ its *value* $|\sigma|$, which is called *profit* in the case of maximization and *cost* in the case of minimization. The values are nonnegative integers.

Example 2.1 Dynamic Channel Assignment(DCA[*]) *An instance of DCA[*] is specified by the number k of channels, the interference radius r and the labelled cell–sequence consisting of all cells with a positive load. Each cell with a positive load, called active cell in the following, is labelled by its load and its relative position*

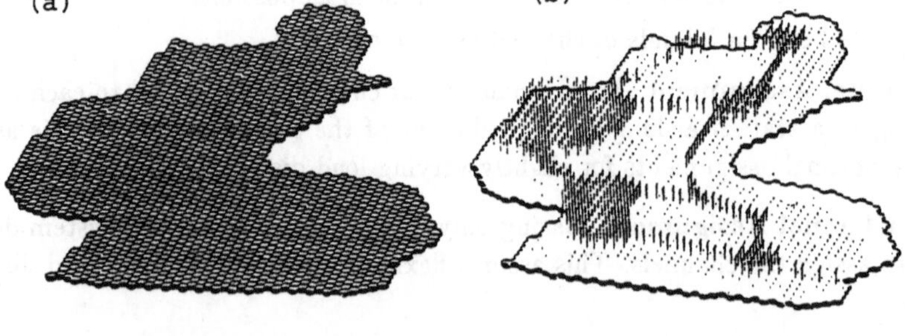

Figure 1: (a) Federal Republic of Germany, covered by a net of hexagonal cells; (b) the current load.

in the hexagonal net. The size of the instance is the number n of active cells. A legal solution is an assignment of channels to cells which excludes co-channel interference. The profit is the number of satisfied requests. The corresponding restricted problem with a fixed number k of channels is abbreviated by $DCA[k]$.

Hybrid Channel Assignment(HCA[*]) *An instance of $HCA[*]$ or $HCA[k]$ is specified like an instance of the corresponding DCA-problem, augmented with a set $f(i) \subseteq [1:k]$ of forbidden channels for each active cell C_i. The set $f(i)$ contains those channels which have already been preassigned during the static phase to a cell with a distance from C_i not exceeding $2r$. The size of an instance is $k \cdot n$. Legal solutions and profit are defined in the obvious way.*

The value of an optimal solution for I is denoted by $\text{opt}_{\Pi}(I)$ or $\text{opt}(I)$ if Π is self-evident. An *approximation algorithm* A is an algorithm which produces a legal solution $\sigma \in S(I)$ with a value $A(I) = |\sigma|$ for each given instance I and has a running-time (this notion refers to the random access machine with the uniform cost measure) which is polynomially bounded in $|I|$. If the decision problem associated with the optimization problem Π is NP-complete, an approximation algorithm for Π cannot produce optimal solutions for all instances unless $P =$ NP (see [2]). However, it may come 'close' to optimum. One common notion, formalizing what we mean by 'close', is the *performance ratio* of A on I defined as:

$$R_A(I) = \begin{cases} \text{opt}(I)/A(I) & \text{in the case of maximization} \\ A(I)/\text{opt}(I) & \text{in the case of minimization} \end{cases}$$

Its value is 1 if the algorithm A produces an optimal solution for I and gives the multiplicative deviation from the optimum in general. The function $R_A(n,m)$ is defined as follows:

$$R_A(n,m) = \sup\{R_A(I) |\; I \in \text{INPUT}(\Pi),\; |I| \le n,\; \text{opt}(I) \ge m\}.$$

Setting $R_A(n) = R_A(n,0)$, *the (absolute) performance ratio* of A is defined as

$$R_A = \lim_{n\to\infty} R_A(n).$$

The *asymptotic performance ratio* is defined as

$$R_A^{\infty} = \lim_{m\to\infty} \lim_{n\to\infty} R_A(n,m).$$

The *best (absolute) performance ratio* for Π is given by

$$R_{MIN}(\Pi) = \inf\{R_A |\; A \text{ is an approximation algorithm for } \Pi\}.$$

The *best asymptotic performance ratio* for Π is defined analogously. $R_{MIN}(\Pi)$ has value 1 iff for each ϵ there exists an approximation algorithm A_ϵ with $R_{A_\epsilon} < 1 + \epsilon$. The family A_ϵ is then called an *approximation scheme.* The scheme is called *full* if the running time of the family A_ϵ is polynomially bounded in $|I|$ and $1/\epsilon$. The problem is called *strongly NP-complete* if it remains NP-complete for polynomially bounded values of the numerical input parameters. In this case, no full approximation scheme exists unless $P =$ NP. For more details see [2].

DCA is related to graph coloring. A *legal coloring* of an undirected graph is an assignment of colors to vertices subject to the constraint that adjacent vertices obtain different colors. 'Graph Coloring' (GC) is the problem of legally coloring a given graph G with a minimum number of colors. The value $\mathrm{opt}_{GC}(G)$ is called the *chromatic number* of G. GC[*] denotes the problem of legally coloring a maximum number of vertices of a given undirected graph with a given number k of colors. GC[k] denotes the corresponding problem if k is fixed. GC[1] is also known as 'Maximum Independent Set' (MIS).

Given an instance I of DCA, we define the *cellular interference graph* $G(I)$ as follows: We replace each active cell C_i with r_i requests by a clique of size r_i. Each pair of the obtained cliques is connected completely bipartitely iff there is a co–channel interference between the corresponding cells. $G(I)$ has a k–colorable subgraph of size s iff s requests can be satisfied by k channels (although the sizes of the encodings of $G(I)$ and I may differ in one order of magnitude, some of our algorithms are easier explained for the input $G(I)$; the reader will verify during the course of this paper that they always transform by obvious methods into algorithms with the input I.)

DCA is also related to geometrical packing problems in the plane. If we draw a circle of radius r around a given cell, we obtain a hexagonal supercell (see figure 2). The assignment of a channel F to a cell C is modelled by placing a supercell, let's say of type F, around C. DCA is then equivalent to the following geometrical optimization problem: given k types of identically shaped supercells of radius r and from each type an arbitrary number, pack a maximum number of supercells around cells in the service area subject to

1. supercells of the same type must not overlap,
2. for all $i = 1, \ldots, n$, at most r_i supercells are placed around C_i.

The relations between HCA, coloring and packing problems are analogous.

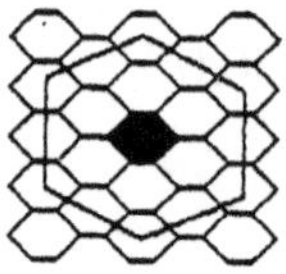

Figure 2: An active (=black) cell and its supercell of radius 2.

3 The analysis of DCA

It is well known [7] that an optimal algorithm A for MIS can be transformed into an approximation algorithm A' for GC with performance ratio $R_{A'}(n) \leq \ln n$. For this, we proceed in rounds. In the first round, A is applied to the input graph G. The resulting independent set is colored with the first color and removed from G afterwards. While the surviving subgraph G' of G, consisting of all remaining uncolored vertices, is nonempty, we start the next round by applying A to G', coloring the resulting independent set by the next color and removing it afterwards. Johnson [7] has shown that this coloring process uses at most $\text{opt}_{\text{CG}}(G) \cdot \ln n$ rounds.

Of course, this construction of A' may start with an arbitrary approximation algorithm A of MIS. In addition, if we stop the coloring process after k rounds, we obtain an approximation algorithm A^k for GC[k]. If k is an input parameter, we obtain from A an appproximation algorithm A^* for GC[*]. A generalization of the reasoning in [7] then leads to the following result.

Theorem 3.1 *With* $R = R_A(n), R' = R_{A'}(n), R^k = R_{A^k}(n), R^* = R_{A^*}(n)$*:*

$$\begin{aligned} R' &\leq R \cdot \ln n. \\ R^k &\leq \frac{1}{1-(1-1/(kR))^k}. \\ R^* &\leq \frac{1}{1-e^{-1/R}}. \end{aligned}$$

Proof If c denotes the chromatic number of G, all surviving subgraphs G', generated during the coloring process A', can be colored with c colors, too. Thus, in each round, at least a fraction of $1/c$ of remaining uncolored vertices forms an independent set. At least a fraction of $1/(cR)$ is actually found by the algorithm A. After k rounds, the number of still uncolored vertices is bounded above by

$$n \cdot (1 - 1/(cR))^k < n \cdot e^{-k/cR}.$$

After $k = c \cdot R \cdot \ln n$ rounds, all vertices are colored. This proves the first assertion of the theorem.

We have to be a little bit more careful with the analysis of $CG[k]$ and $CG[*]$. $\langle S \rangle$ denotes a maximum k–colorable subgraph of G. The size of S is denoted by s. The number of vertices in G, which are colored after i rounds of the coloring process, is denoted by s_i. Certainly:

$$s_1 \geq \frac{s}{kR}.$$

After i rounds at least $s - s_i$ of the vertices from S are still uncolored. Therefore:

$$s_{i+1} \geq s_i + \frac{s - s_i}{kR}.$$

If we define $t_i = s - s_i$ and perform an obvious induction:

$$t_i \leq (1 - 1/(kR))^i \cdot s.$$

If we set $i = k$:

$$s_k \geq s \cdot (1 - (1 - 1/(kR))^k) > (1 - e^{-1/R}) \cdot s.$$

This proves the second and the third assertion. •

When applied to a particular input graph G, the coloring process A^* might perform much better than suggested by the worstcase bound given in theorem 3.1. $G_1, \ldots, G_k$ denote the input graphs for the rounds $1,\ldots,k$ of A^* , and $R_1,\ldots,R_k$ denote upper bounds for $R_A(G_1),\ldots,R_A(G_k)$, respectively. With

$$\bar{D} = \frac{1}{k}\cdot\sum_{i=1}^{k}\frac{1}{R_i}$$

the following holds:

Corollary 3.2

$$R_{A^*}(G) \leq \frac{1}{1-e^{-\bar{D}}}.$$

The proof of corollary 3.2 follows the outline of the proof for theorem 3.1 and is omitted.

In general, approximation algorithms for MIS, performing well in the worstcase, are hard to design. No approximation algorithm with a finite performance ratio is known to date. But, 'MIS in cellular interference graphs' (by this, we mean the following problem: given an instance I of DCA[*] or DCA[k], find a maximum set of active cells such that the supercells around them are nonoverlapping) is solvable by an approximation scheme. This scheme makes use of a method due to Hochbaum and Maass, called shifting strategy, which applies to covering and packing problems in the plane. A detailed description is contained in [6]. For our purpose, the following characteristics of the method are essential: It controls the performance ratio by an adjustable parameter l such that $R_A \leq (l/(l-1))^2$. Given l and the diameter d of the geometrical objects, it appropriately partitions the whole service area into local squares of size $dl \times dl$. In [6], an approximation scheme for packing problems is obtained whenever the following assumption holds:

Assumption for the shifting strategy There is an algorithm for the local problem of finding an optimal solution within a local square of size $dl \times dl$ which has a running–time polynomially bounded in the size of the whole input.

The problem of packing nonoverlapping supercells around active cells fulfils this assumption because at most $O(l^2)$ supercells fit in the local square. An optimal solution is then found by exhaustively searching all $n^{O(l^2)}$ legal solutions. Thus:

Corollary 3.3 *There exists an approximation scheme for 'MIS in cellular interference graphs'.*

Remark 3.4 *The running time of the approximation scheme depends exponentially on the precision parameter l, i.e. it is not full. This cannot be improved unless P = NP because the packing problem (even the special–one with hexagonal cells and supercells, as we shall see later) is strongly NP–complete.*

We now obtain the following approximation algorithm for DCA:

1. In a first stage, we write an algorithm A which is close to optimal for 'MIS in cellular interference graphs'.

2. In a second stage, we transform this algorithm into the algorithms A^k for DCA[k] and A^* for DCA[$*$].

Since the performance ratio of A is arbitrarily close to 1, we obtain the following result:

Corollary 3.5

$$R_{MIN}(DCA[k]) \leq \frac{1}{1-(1-1/k)^k}.$$
$$R_{MIN}(DCA[*]) \leq \frac{1}{1-e^{-1}}.$$

Corollary 3.6 *If the interference radius is fixed, the corresponding restricted version of $DCA[*]$ is solved by an approximation scheme.*

The proof follows from a stronger result (corollary 4.3) in the next section.

4 The Analysis of HCA

As we have seen in section 3, DCA[*] is strongly related to GC[*]. Similarly, HCA[*] is related to a modified version HGC[*] of GC[*]. An input of HGC[*] is specified by a number k of colors, an undirected graph G and a mapping

$$f : V \rightarrow 2^k$$

associating with each vertex a subset of forbidden (preassigned) colors. The objective is to find a maximum k–colorable subgraph of G subject to the constraint that any vertex must not be colored with one of its forbidden colors. HCA[*], as defined in section 2, is then a subproblem of HGC[*]. HMIS denotes the problem of finding a maximum set U of vertices in G which is independent and can be colored with one color i not forbidden for any vertex in U. An approximation algorithm A for HMIS transforms into an approximation algorithm A^* for HGC[*] by a coloring process similar to that for the corresponding dynamic problems: HMIS is applied during k rounds. Each round determines a color i which is assigned to the respective independent set. Afterwards, color i is discarded.

Theorem 4.1 *With $R = R_A(n)$ and $R^* = R_{A^*}(n)$: $R^* \leq 1 + R$.*

Proof S denotes an optimal solution of HGC[*] and s its size. The number of vertices colored after i rounds of the coloring process is denoted by s_i. Certainly:

$$s_1 \geq \frac{s}{kR}.$$

$T_i \subseteq S$ denotes the set of vertices x in S such that:

1. x is not colored during the first i rounds of the coloring process.
2. The color of x in the optimal solution S has not yet been discarded during the first i rounds of the coloring process.

We claim and will show inductively that the size of T_i is not less than

$$s - (1+R)s_i.$$

In round $i+1$, the algorithm A for HMIS finds a color j and a set P, let's say of size p, which can be legally colored with j. If Q denotes the set of vertices in T_i with the color j in the optimal solution, the size of Q is at most Rp. Thus, the size of T_{i+1} is not less than

$$s - (1+R)s_i - (1+R)p = s - (1+R)s_{i+1}.$$

This completes the inductive proof of the claim.

The definition of T_i implies that this set is colored with at most $k-i$ colors by the optimal solution. Therefore, we obtain the following recursive relation:

$$s_{i+1} \geq s_i + \frac{s - (1+R)s_i}{R(k-i)}.$$

We assume w.l.o.g. that the term $s - (1+R)s_i$ does not become negative during the first $k-1$ rounds. Then:

$$s_k \geq s_{k-1} + \frac{s-(1+R)s_{k-1}}{R} = \frac{(1+R)s_{k-1}}{1+R} + \frac{s-(1+R)s_{k-1}}{R} \geq \frac{s}{1+R}.$$

This completes the proof of the theorem. •

As in the dynamic case, the methods of Hochbaum and Maass [6] can be combined with this theorem in order to show the following result:

Corollary 4.2

$$\begin{aligned} R_{MIN}(HMIS\ in\ cellular\ interference\ graphs) &= 1. \\ R_{MIN}(HCA[*]) &= 2. \end{aligned}$$

Corollary 4.3 *If the interference radius is bounded by a constant, the corresponding restricted version of HCA[*] can be solved by an approximation scheme.*

Proof We claim that the assumption (see section 3) for the shifting strategy holds. We will present the local HCA[*] problem of finding an optimal solution within a local square of size $dl \times dl$ as an integer programming problem with a fixed dimension (exhaustive search is still applicable for a constant k but leads to exponential running time for a variable k). The diameter $d = 2r$ of a hexagonal supercell is constant. Certainly, the number of active cells in the local square is bounded by the constant $p = O((dl)^2)$. We associate with each *legal set* S, i.e., a set of active cells in the local square with nonoverlapping supercells around them, the 0,1–vector $\vec{x}(S) = (x_1, \ldots, x_p)$ whose i'th component is 1 iff the i'th cell belongs to S. If the sequence $S_1, \ldots, S_q$ runs through all legal sets, the corresponding vectors $\vec{x}(S_1), \ldots, \vec{x}(S_q)$, written as column vectors, form a $(p \times q)$–matrix A. The numbers $w_j = |S_j|$ for $j = 1, \ldots, q$ form a vector $\vec{w}$. Its size $q = O(2^p)$ is a constant. The numbers r_i of requests in the i'th active cell form a vector $\vec{r}$ of size p. We say, that a color $c \in [1:k]$ is *forbidden* for the column $j \in [1:q]$ iff it is forbidden for one of the active cells in S_j. In order to define an integer programming problem with a fixed dimension equivalent to local HCA[*], we require the following equivalence relation on the set $[1:k]$ of colors: for all $c, c' \in [1:k]$: $c \sim c'$ iff

$$\text{for all } j \in [1:q] : \ (c \text{ is forbidden for } j \Leftrightarrow c' \text{ is forbidden for } j).$$

The resulting color classes are denoted by $C_1, \ldots, C_z$ and its sizes by $k_1, \ldots, k_z$. Then, $k = \sum_{h=1}^{z} k_i$ and $z = O(2^q)$ is a constant.
We associate a column vector

$$\vec{y}_h = (y_{h1}, \ldots, y_{hq})^T$$

of variables with each color class C_h. The value of y_{hj} is intended to be the number of colors from class h such that the set of active cells to which they are assigned is identical to S_j. By setting

$$B = 1 + \max\{k_h |\ h = 1, \ldots, q\},$$

the vector $\vec{b}_h = (b_{h1}, \ldots, b_{hq})$ is defined as follows: for all $h \in [1:z]$ and $j \in [1:q]$,

$$b_{hj} = \begin{cases} B & \text{if the colors of class } h \text{ are forbidden for } j \\ 1 & \text{otherwise} \end{cases}$$

With these definitions, an integer programming problem with a fixed dimension equivalent to local HCA[*] is specified as follows:
maximize $\vec{w} \cdot \sum_{h=1}^{z} \vec{y}_h$, subject to

$$A \cdot \sum_{h=1}^{z} \vec{y}_h \leq \vec{r} \text{ and for } h = 1, \ldots, z : \ (\vec{b}_h \cdot \vec{y}_h \leq k_h) \wedge \vec{y}_h \geq \vec{0}.$$

By applying the result of Lenstra [8] about integer programming with a fixed dimension, the corollary follows. •

5 Final Remarks and Open Problems

This section states several additional results which are contained in more detail in the full paper [10]. It also lists several open problems.

Final Remarks

1. An analysis of an algorithm for DCA[*] with a linear running time, such as the GLADYC–algorithm in [5] for instance, can be performed by applying theorem 3.1 and corollary 3.2. The full paper [10] shows that

$$R_{GLADYC}(DCA[*]) \leq \frac{1}{1 - e^{-\bar{D}}}$$

where $\bar{D}$ denotes the so-called average density ratio of the input instance which is conjectured to be close to 1 for practical cases.

2. The following holds for all $k \geq 1$ unless $P = NP$:

$$\text{Either } R_{MIN}GC[k]) \leq \frac{1}{1-(1-1/k)^k} \quad \text{or} \quad R_{MIN}GC[k]) = \infty.$$

$$\text{Either } R_{MIN}GC[*]) \leq \frac{1}{1-e^{-1}} \quad \text{or} \quad R_{MIN}GC[*]) = \infty.$$

   Moreover, either all best performance ratios are unbounded simultaneously or none of them is unbounded. This result follows from the corresponding gap–result for MIS [2] and theorem 3.1.

3. The full paper shows the NP–completeness of the following decision problems:
   - Given an instance of DCA[3], decide whether all requests are satisfiable.
   - Given an instance of DCA[1] and a number B, decide whether at least B requests are satisfiable.

   In addition, the problems remain NP–complete for a fixed interference radius of 2 and for a fixed number 1 of requests per active celle. Thus, the NP–completeness is strong. The source problems of the two reductions are 'Graph 3–Colorability for Planar Graphs with no Vertex–Degree Exceeding 4' and 'MIS in Cubic Planar Graphs' [3], respectively. The technique employed is related to the technique of Fowler, Paterson and Tanimoto in [1]. The corresponding hybrid versions are more general and thus also NP–complete. However, it is easily shown that, for $k = 2$, the first decision problem becomes polynomially solvable, even for the hybrid version and for arbitrary graphs ('Graph 2–Colorability' with the additional constraint of forbidden colors).

4. The minimization problem, corresponding to DCA[k] for $k \geq 3$ (minimize the number of unsatisfied requests), has unbounded best absolute and asymptotic performance ratios unless $P = NP$. This follows from the preceeding remark and standard techniques (see [9] for instance). For $k = 1$ the problem becomes a special case of 'Vertex Cover' for which approximation algorithms with a performance ratio of 2 are known [4,2].

Open Problems 1. Nothing is known about the best performance ratios of the minimization problems corresponding to GC[2] (i.e. given an undirected graph, minimize the number of vertices being removed in order to obtain a bipartite subgraph) or to DCA[2] (i.e. given an input instance of DCA[2], minimize the number of unsatisfied requests). The corresponding decision problems are NP–complete.

2. Are there approximation schemes solving DCA[*] and HCA[*] for unbounded interference radius ?

3. HMIS is easily shown to be computationally equivalent to MIS. The hybrid versions HGC[k] and HGC[*] might be computationally harder than GC[k] or GC[*]. Assuming $P \neq NP$, are there nontrivial lower bounds for their performance ratios?

References

[1] R. J. Fowler, M. S. Paterson, and S. L. Tanimoto. Optimal packing and covering in the plane are NP–complete. *Information Processing Lett.*, 12(3):133–137, June 1981.

[2] M. R. Garey and D. S. Johnson. *Computers and Intractability: A Guide to the Theory of NP–Completeness.* W.H.Freeman and Company, San Francisco, 1979.

[3] M. R. Garey, D. S. Johnson, and L. Stockmeyer. Some simplified NP–complete graph problems. *Theor. Comput. Sci.*, 1(3):237–267, 1976.

[4] F. Gavril. 1974. unpublished.

[5] M. Grevel and A. Sachs. A graph theoretical analysis of dynamic channel assignment algorithms for mobile radiocommunication systems. *Siemens Forsch.- u. Entwickl.-Ber.*, 12(5):298–305, 1983.

[6] D. S. Hochbaum and W. Maass. Approximation schemes for covering and packing problems in image processing and VLSI. *J. Assoc. Comput. Mach.*, 32(1):130–136, 1985.

[7] D. S. Johnson. Approximation algorithms for combinatorial problems. *J. Comp. System Sci.*, 9(3):256–278, 1974.

[8] H. W. Lenstra, Jr. *Integer Programming with a Fixed Number of Variables.* Report 81–03, University of Amsterdam, Apr. 1981.

[9] S. Sahni and T. Gonzales. P–complete approximation problems. *J. Assoc. Comput. Mach.*, 23(3):555–565, 1976.

[10] H. U. Simon. *The Analysis of Dynamic and Hybrid Channel Assignment.* SFB 124–B1 10/1988, Universität des Saarlandes, D–6600 Saarbrücken, FRG, May 1988.

THE BOREL HIERARCHY IS INFINITE IN THE CLASS OF REGULAR SETS OF TREES

Jerzy Skurczyński
Merchant Marine Academy
Czerwonych Kosynierów 83
81-962 Gdynia, Poland

1. Introduction.

It has been known from [BL 69] that regular sets of ω-sequences are located on low levels of the Borel hierarchy, namely they are at most Δ^0_3 sets ($F_{\sigma\delta} \cap G_{\delta\sigma}$ sets in a more traditional notation). Borel classes of regular sets of ω-sequences are completely characterized in automata structure and ω-regular expression terms. For a complete survey of the results and the bibliography see [Wa 79].

In the case of regular sets of infinite trees the full topological classification is still unknown (according to the author's knowledge). The state of affairs for today is as follows: 1)it follows indirectly from [Ra 69] and more directly from [Mos1 80] and [Mos2 80] that every regular set of trees is in Δ^1_2 (PCA $\cap$ CPCA projective class); 2)Niwiński in [Ni 85] has given an example of non-Borel regular set, which is Σ^1_1-complete and is accepted by a Büchi automaton; 3)every set accepted by an automaton with so-called weak condidions (conditions for sets of all states appearing on the paths) is weakly definable (definable by a weak SkS formula), so it is obviously a Borel set ([MSS 86],[Sk 86]); 4)in contrary to the ω-sequences case there exist sets accepted by automata with weak conditions, which are not Σ^0_2 sets ([Sk 86]).

In this situation at least two natural questions arise: 1)Is the Borel hierarchy infinite in the class of regular sets of trees or does it collapse at some level ? 2)Do any regular sets exist which are Borel, but not weakly definable ?

In the present paper we give a positive answer to the first question. Moreover, we show infinity of the Borel hierarchy in the class of tree languages acceptable by automata with weak conditions (however, there exist simple examples of weakly definable sets, which are not accepted by such automata).

2. Notations and definitions.

For simplicity we will consider only the space of infinite binary trees valued from the alphabet $\Sigma=\{0,1\}$, but it is easy to generalize our construction to trees of any fixed arity $k\geq 2$, valued from any Σ such that $\mathrm{card}(\Sigma)\geq 2$.

As usual, a (infinite, binary, Σ-valued) tree is a pair $t=(v,T)$, where $T=\{0,1\}^*$ and $v:T\rightarrow\Sigma$ is a total function. For $x\in T$ length(x) denotes the lenght of the word x (in particular length(Λ)=0, where Λ is the empty word). By t_x we denote the infinite subtree of t starting from the node x. More formally, $t_x=(v_x,T)$, where $v_x(y)=v(xy)$ for $y\in T$.

An automaton on trees is a quintuple $\mathfrak{A}=\langle\Sigma,S,\delta,s_o,\mathcal{F}\rangle$, where S is a finite set of states, $\delta:S\times\Sigma\rightarrow P(S\times S)$ is a (nondeterministic, partial) transition function, $s_o\in S$ is an initial state, $\mathcal{F}\subseteq P(S)$ is a family of subsets of accepting states. A run of $\mathfrak{A}$ on t is any total function $r:T\rightarrow S$ such that $r(\Lambda)=s_o$ and for every $x\in T$ we have $(r(x0),r(x1))\in\delta(r(x),v(x))$. $\mathfrak{A}$ accepts t in the sense of weak conditions if there exists such a run r of $\mathfrak{A}$ on t that for every path Π in T $r(\Pi)=F$ for some $F\in\mathcal{F}$ (a path is any subset Π of T such that $\Lambda\in\Pi$ and for every $x\in T$: 1) if $x\in\Pi$ then exactly one of x0, x1 belongs to Π; 2) if $x\notin\Pi$ then none of x0, x1 belongs to Π).

In the literature strong acceptance conditions are more often dealt with. We obtain the definition of the strong acceptance from the above one by substituting $r(\Pi)=F$ by $\mathrm{In}(r|\Pi)=F$, where $\mathrm{In}(r|\Pi)=\{s\in S \mid r(x)=s \text{ for infinitely many } x\in\Pi\}$.

Both defined types of acceptance conditions (either weak or strong) are called Muller-type conditions (Muller conditions). Putting $r(\Pi)\nsubseteq F$ instead of $r(\Pi)=F$ ($\mathrm{In}(r|\Pi)\nsubseteq F$ instead of $\mathrm{In}(r|\Pi)=F$) we obtain weak (strong) Büchi conditions. In the rest of this paper we will deal with weak Muller conditions only.

In the space $\mathcal{T}$ of all trees we define a metric d by taking

$$d(t_1,t_2)=\begin{cases}\dfrac{1}{1+\min\{\mathrm{length}(x)\ \mid\ v_1(x)\neq v_2(x)\}} & \text{if } t_1\neq t_2\\ 0 & \text{if } t_1=t_2\end{cases}$$

for $t_1=(v_1,T)$, $t_2=(v_2,T)$, $t_1,t_2\in\mathcal{T}$.

The space $\mathcal{T}$ with this metric becomes a compact space homeomorphic with the Cantor set.

Σ_i^0 (i=1,2,...) denotes the i-th additive Borel class (for i=1 this is the class of open sets), Π_i^0 is the i-th multiplicative Borel class.

3. The hierarchy theorem.

Our construction is based on that from [EHS 66], but there are some essential differences. The main one is that we construct only representatives of Borel classes of finite numbers. The problem of representatives of Borel classes of transfinite numbers (which is connected with the second question stated in the introduction) is left open.

Let us define a mapping $\psi:\mathcal{T}^\omega\to\mathcal{T}$ in the following way : $\psi(t^{(0)},t^{(1)},t^{(2)},\dots)=t=(v,T)$ such that $v(0^k)=0$ and $t_{0^k1} = t^{(k)}$ for k=0,1,2,... . Less formally, the image in ψ of a sequence $t^{(0)},t^{(1)},t^{(2)},\dots$ of trees is a tree which has only zeros on its leftmost path and consecutive right subtrees are equal to $t^{(0)},t^{(1)},t^{(2)},\dots$ respectively (see Fig.1). The mapping ψ is continuous (with respect to the product topology in $\mathcal{T}^\omega$).

Now we construct a sequence of pairs of sets $M_n,A_n\subseteq\mathcal{T}$ (n=1,2,...) such that:

1^0 $M_1=\{t\}$ for some $t\in\mathcal{T}$;

2^0 $A_n=\mathcal{T}-M_n$ for n=1,2,...;

3^0 $M_{n+1}=\psi(A_n^\omega)$ for n=1,2,... .

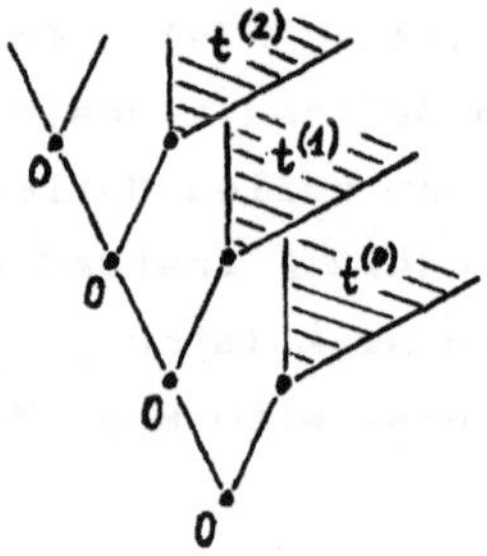

Fig.1

Lemma 1. (see [EHS 66])

If $B \subseteq \mathcal{T}$ is a Borel set of the mutiplicative (additive) class n in $\mathcal{T}$, then there exists a continuous mapping $\varphi: \mathcal{T} \to \mathcal{T}$ such that $\varphi^{-1}(M_n)=B$ (such that $\varphi^{-1}(A_n)=B$).

<u>Proof.</u> The proof is by induction. It should be emphasized that the mapping φ is <u>into</u> and the counterimage of some $X \subseteq \mathcal{T}$ is defined as $\varphi^{-1}(X)=\{t\in\mathcal{T} \mid \varphi(t)\in X\}$. Having φ constructed for some M_n, one can observe that the same mapping φ maps $\mathcal{T}-B$ into A_n, because φ maps the whole space $\mathcal{T}$ into itself and $\varphi^{-1}(M_n)=B$, so $\varphi^{-1}(\mathcal{T}-M_n)=\mathcal{T}-B$.

For n=1 the construction of φ is more or less obvious (if not, see Appendix). Now suppose that the lemma holds for some n. Let $B \subseteq \mathcal{T}$ be a Borel set in Π^0_{n+1}, i.e. $B=B_0 \cap B_1 \cap B_2 \cap \ldots$, where B_i are in Σ^0_n for $i=0,1,2,\ldots$.
By the induction hypothesis, for each B_i there exist a continuous mapping $\varphi_i: \mathcal{T} \to \mathcal{T}$ such that $\varphi_i^{-1}(A_n)=B_i$. The continuous mapping $\varphi: \mathcal{T} \to \mathcal{T}$ defined by the formula $\varphi(t)=\psi(\varphi_i(t) \mid i=0,1,2,\ldots)$ has the property $\varphi^{-1}(M_{n+1})=B$. ■

Lemma 2. (see [EHS 66])

For n=1,2,... M_n is in $\Pi^0_n - \Sigma^0_n$ and A_n is in $\Sigma^0_n - \Pi^0_n$.

<u>Proof.</u> It is enough to prove the lemma for M_n. It is obvious that M_n is in Π^0_n . Suppose that M_n is in Σ^0_n . By Lemma 1 there exists a continuous mapping $\varphi: \mathcal{T} \to \mathcal{T}$ such that $\varphi^{-1}(A_n)= M_n$ (and $\varphi^{-1}(M_n)= A_n$). The mapping φ transforms M_n into a subset of A_n and A_n into a subset of M_n, so the equality $\varphi(t) = t$ does not hold for any $t\in\mathcal{T}$. This contradicts the Brouwer fixed - point theorem which holds in $\mathcal{T}$, homeomorphic with the Cantor set (see [Ku 66] vol. II). ■

Now we shall recall the notion of regular trees . In the literature they are defined in at least three different, but equivalent ways: 1)as such trees $t=(v,T)$ that $v^{-1}(\sigma)$ is a regular set (in the classical sense) for every $\sigma \in \Sigma$; 2)as trees having only finitely many different infinite subtrees; 3)as "definable singletons", i.e. elements of one-element regular sets of trees (see [Th ?] for the equivalence of 1) and 2), and the bibliography, the equivalence of 1) and 3) follows directly from [Ra 72]). We will use the last definition.

Theorem 3.

If $M_1=\{t\}$ for some regular tree t, then all sets M_n and A_n for $n=1,2,...$ are regular. Moreover, they all are accepted by automata with weak conditions.

In order to prove the theorem, we will prove the following

Lemma 4.

If t is a regular tree, then $\{t\}$ is accepted by an automaton with weak conditions, and so is $\mathcal{T}-\{t\}$.

Proof. In the space $\mathcal{T}$ every one-element set is a closed set, so its complement is an open set. According to [Mor 87] and [Mos 87] every regular closed set of trees is accepted by an automaton with no conditions for the run (so it can be treated as a special case of an automaton with weak conditions, where $\mathcal{F}=P(S)$), and every regular open set is also acceptable by an automaton with weak conditions. ■

One can also construct a more "straightforward" proof of the lemma, not using topological properties.

Proof of Theorem 3. The proof is by induction. We have already shown that M_1 and A_1 are accepted by automata with weak conditions. Suppose that for some n the set M_n is accepted by an automaton $\mathfrak{A}_M = \langle \Sigma, S_M, \delta_M, s_{0M}, \mathcal{F}_M \rangle$ and the set A_n by an automaton

$\mathfrak{A}_A = \langle\Sigma, S_A, \delta_A, s_{OA}, \mathcal{F}_A\rangle$. We construct automata $\mathfrak{A}'_M = \langle\Sigma, S'_M, \delta'_M, s'_{OM}, \mathcal{F}'_M\rangle$ and $\mathfrak{A}'_A = \langle\Sigma, S'_A, \delta'_A, s'_{OA}, \mathcal{F}'_A\rangle$ accepting M_{n+1} and A_{n+1} respectively :

1) $S'_M = S_A \cup \{s^*\}$, $s'_{OM} = s^*$, $\delta'_M(s^*,0) = \{(s^*, s_{OA})\}$, $\delta'_M(s^*,1) = \emptyset$,

$\delta'_M(s,\sigma) = \delta_A(s,\sigma)$ for $s \in S_A$, $\sigma \in \Sigma$, $\mathcal{F}'_M = \{\{s^*\} \cup F \mid F \in \mathcal{F}_A \cup \{\emptyset\}\}$;

2) $S'_A = S_M \cup \{\bar{s}, s_N\}$, $s'_{OA} = \bar{s}$, $\delta'_A(\bar{s},0) = \{(\bar{s}, s_N), (s_N, s_{OM})\}$,

$\delta'_A(\bar{s},1) = \{(s_N, s_N)\}$, $\delta'_A(s_N,\sigma) = \{(s_N, s_N)\}$ for $\sigma \in \Sigma$,

$\delta'_A(s,\sigma) = \delta_M(s,\sigma)$ for $s \in S_M$, $\delta \in \Sigma$,

$\mathcal{F}'_A = \{\{\bar{s}\} \cup F \mid F \in \mathcal{F}_M \cup \{\{s_N\}\}\}$. ■

Corollary.

Automata with weak Büchi conditions define an essentially smaller class of sets than automata with weak Muller conditions.

Proof. Automata with weak Büchi conditions can define only Σ^0_2 sets (F_σ sets). We omit details. ■

Appendix.

Given some $t^* \in \mathcal{T}$, $t^* = (v^*, T)$ and a closed set $C \subseteq \mathcal{T}$, we will construct a continuous mapping $\varphi: \mathcal{T} \to \mathcal{T}$ such that $\varphi^{-1}(\{t^*\}) = C$. Of course, for every $t \in C$ we have $\varphi(t) = t^*$. For $t \notin C$ we can always find such $t' \in C$ that $d(t,t') = \min_{\bar{t} \in C} d(t,\bar{t})$ (because C is closed). Let $\tau = (w,T)$ be defined by the formula:

$$w(x) = \begin{cases} v^*(x) & \text{if } \mathrm{length}(x) < \dfrac{1}{d(t,t')} - 1 \\ 1 - v^*(x) & \text{if } \mathrm{length}(x) \geq \dfrac{1}{d(t,t')} - 1 \end{cases} .$$

We take $\varphi(t) = \tau$.

References.

[BL 69] Büchi, J.R., Landweber, L.H., Definability in the monadic second-order theory of successor, Journal of Symbolic Logic 34 (1969), pp.166-170.

[EHS 66] Engelking, R., Holsztyński, W., Sikorski, R., Some examples of Borel sets, Colloquium Mathematicum vol.XV fasc.2 (1966), pp.271-274.

[Ku 66] Kuratowski, K., Topology vol. I and II (the translation of the French edition), Academic Press and Polish Scientific Publishers (1966).

[Mor 87] Moriya, T., Topological characterizations of infinite tree languages, Theoretical Computer Science 52 (1987), pp. 165-171.

[Mos1 80] Mostowski, A.W., Finite automata on infinite trees and subtheories of SkS, Les Arbres en Algebre et en Programmation, 5-eme Colloque de Lille (1980), pp.228-240.

[Mos2 80] Mostowski, A.W., Types of finite automata acceptances and subtheories of SkS, 3rd Symposium on Math. Found. of Comp. Sci., ICS PAS Reports 411 (1980), pp.55-58.

[Mos 87] Mostowski, A.W., Hierarchies of weak monadic formulas for two successors arithmetic, J. Inf. Process. Cybern. EIK 23 (1987), pp.509-515.

[MSS 86] Muller, D.E., Saoudi, A., Schupp, P., Alternating automata, the weak monadic theory of the tree and its complexity, Proc. 13th ICALP (L.Kott ed), Lect.Notes in Comp. Sci.226 (1986), pp.275-283.

[MSW 85] Mostowski, A.W., Skurczyński, J., Wagner, K., Deterministic automata on infinite trees and the Borel hierarchy, Proc. Fourth Hung. Comp. Sci. Conf. (1985), M.Arató, I.Kàtai, L.Varga eds., pp.103-115.

[Ni 85] Niwiński, D., The example of non-Borel set of infinite trees recognizable by a Rabin automaton (in Polish), manuscript, Univ.of Warsaw (1985), 9 pages.

[Ra 69] Rabin, M.O., Decidability of second-order theories and automata on infinite trees, Trans. of Amer. Math. Soc. 141 (1969), pp.1-35.

[Ra 70] Rabin, M.O., Weakly definable relations and special automata, Math.Logic and Found. of Set Theory (1970), pp.1-23.

[Ra 72] Rabin, M.O., Automata on infinite objects and Church's problem, Regional Conference Series in Mathematics 13 (1972), Amer. Math. Soc., Providence,, Rhode Island.

[Sk 86] Skurczyński, J., Automata on infinite trees with conditions for sets of accessible states, unpublished manuscript, Univ. of Gdańsk (1986), 10 pages.

[Th ?] Thomas, W., Automata on infinite objects, in : Handbook of Theoretical Computer Science, North-Holland, to appear.

[Wa 79] Wagner, K., On ω-regular sets, Information and Control 43 (1979), pp.123-177.

PARALLEL GENERAL PREFIX COMPUTATIONS WITH GEOMETRIC, ALGEBRAIC AND OTHER APPLICATIONS

Frederick Springsteel
University of Missouri, Computer Science, Columbia MO 65211, USA;
work partially supported by NSF IRI 8709726.

Ivan Stojmenović
University of Ottawa, Computer Science, Ottawa , Ontario K1N 9B4, Canada;
work partially supported by NSERC.

Abstract

We introduce here a generic problem component and algorithms for it on various parallel models, that captures the most common, difficult "kernel" of many types of problems, e.g. geometric dominance. This kernel involves general prefix computations (GPCs) that can, with insight, be done quickly in parallel and by iterative techniques. GPCs' lower bound complexity of $\Omega(n \log n)$ time is established, and we give optimal solutions on the sequential model in $O(n \log n)$ time, on the CREW PRAM model in $O(\log n)$ time with n processors, on the BSR (broadcasting with selective reduction) model in constant time using n processors, and on mesh-connected computers in $O(\sqrt{n})$ time with $\sqrt{n}$ processors. A solution in $O(\log^2 n)$ time on the hypercube model is given which is the best possible given known sorting limitations. We show that general prefix techniques can be applied to a wide variety of geometric (point set and tree) problems, including triangulation of point sets, two-set dominance counting, ECDF searching, finding two- and three-dimensional maximal points, and the (classical) reconstruction of trees from their traversals. In sum, GPC techniques have many important consequences.

1. INTRODUCTION

In [3,15] several distinct geometric problems are solved separately on the CREW PRAM model (defined in section 3) by combining the merge procedure of [3] with various divide-and-conquer strategies based on merging lists of labeled elements. Instead of solving geometric problems separately, we describe a generic "kernel" technique, General Prefix Computations (GPCs), and parallel algorithms that solve most of these problems as well as others. GPCs cannot be applied to every problem in [3,15], except by further extensions, but the technique is powerful enough to solve efficiently two of the deeper problems considered there: 3-dimensional version of maximal point search and two-set dominance counting. We also demonstrate its efficiency on a variety of other problems, including: triangulating a point set, ECDF searching, standard prefix products, and reconstruction of trees from minimal-length stored strings (parenthesis matching, and counting inversions in a permutation are demonstrated in the full version of the paper).

The key to the wide applicability of GPCs, besides the commonality of general prefixes, is the iterative nature of our techniques for solving the generic problem. Thus, we are able to implement it on various models of computation, parallel as well as sequential. It generalizes the well-known prefix products problem: given an array of "multipliable" elements $f(1), f(2), ..., f(n)$, compute the $n-1$ products $f(1) * f(2) * ... * f(m)$ for $2 \leq m \leq n$, where '$*$' is associative. (cf. [8,13]). While prefix products require only $O(n)$ operations, we will define general prefix computations that require $\Omega(n \log n)$ time on the sequential (uni-processor) model. We will give optimal algorithms for solving it on that model, on the CREW PRAM, BSR and mesh-connected computer parallel models, and give a best known $O(\log^2 n)$ time solution on the hypercube model.

We define the *General Prefix Computations* (GPCs) problem as

> Let $f(m)$ and $y(m)$ be given sequences of elements for integers $1 \leq m \leq n$. There is an arbitrary binary associative operator '$*$' on the f-elements; the y-elements can be compared by a linear order '$<$'. The problem is to compute the sequence of general prefixes: $D_m = f(j_1) * f(j_2) * ... * f(j_k)$ where $j_1 < ... < j_k$ and $\{j_1, ..., j_k\}$ is the set of indices $j < m$ for which $y(j)$ '$<$' $y(m)$, where m is each index from 1 to n.

The GPCs problem is a special formulation of the (very nontrivial) range searching problem, defined as follows: Given a database B, say consisting of 2-tuples of real numbers, that is subject to "orthogonal rectangular" queries $R = [x_1, x_2] \times [y_1, y_2]$, the answer to query R is a real-valued commutative group function '$*$' of point values: $* f(X)$ for all X belonging to both B and R (for example, the number of elements which belong to both B and R).

GPC problem is obtained by setting $R = [-\infty, m] \times [-\infty, y(m)]$ for a given m, $1 \leq m \leq n$, where database B consists of points $(m, y(m))$ for $1 \leq m \leq n$. In other words, GPC problem is to solve the range searching problem for all database points (the form of rectangles is not a restriction since $[x_1,x_2]x[y_1,y_2] = [-\infty,x_2]x[-\infty,y_2] + [-\infty,x_1]x[-\infty,y_1] - [-\infty,x_1]x[-\infty,y_2] - [-\infty,x_2]x[-\infty,y_1]$).

The best known bounds for range searching problem are: query time $O(\log n)$, pre-processing time and space $O(n \log n)$ [10,19]. Applying such more general solutions leads to $O(n \log n)$ time and space for (well-constructed) GPC problems. We reduce utilized space for GPC to $O(n)$, on the sequential model, and present parallel solutions on several models of parallel computation.

We note that while our solutions for the GPCs problem are worst-case optimal, the resulting solutions for (some of the) graph problems solved by it are not necessarily optimal. They are indeed optimal for problems where a lower bound of $\Omega(n \log n)$ is established: triangulating a point set, maximal point searching or two-set dominance counting in 2-D or 3-D sets, and the ECDF searching problem. Also, all the presented solutions are optimal on the mesh-connected computer (MCC), since its best possible time complexity of $O(\sqrt{n})$ is achieved. They are also optimal on a recently introduced BSR (broadcasting with selective reduction) model of computation, where constant time GPC implementation using $O(n)$ processors and space is possible.

For the hypercube model, all presented algorithms are the first known solutions for the given problems, except for another ECDF searching algorithm described by the second author in [16]. In most cases - for example, triangulation, maximal points, two-set dominance, ECDF searching - the obtained solutions are the best possible, in $O(\log^2 n)$ time, given current sorting methods on hypercubes.

The tree reconstruction problem has recently been studied. A $O(n^2)$ iterative method is described in [6]. It was improved in [7] by presenting $O(n \log n)$ solutions. Linear time algorithm is finally achieved by Anderson and Carlsson [2]. Using our method another $O(n \log n)$ tree reconstruction algorithm can be described. Since this method can be parallellized, this gives first parallel tree reconstruction algorithm. On a CREW PRAM the algorithm will work in $O(\log n)$ time with $O(n)$ processors. Another $O(\log n)$ time with $O(n)$ processors CREW PRAM solution can be derived from a technique presented in [5]. However, the reduction of the number of processors to $O(n/\log n)$ as for other problems studied in [5] requires additional investigation, and this remains an open problem.

2. LOWER BOUND

The lower bound argument is based on a linear time transformation of sorting to the generalized prefix computation. Let $x(1),x(2),...,x(n)$ be n real numbers, that we wish to sort. The problem transformation is carried out as follows. Let $y(j)=x(j), f(j)=1$ for $1 \le j \le n$ and $* = +$ be the summation. Then D_m gives the number of elements j such that $x(j) < x(m)$ and $j < m$. Now let $y(j) = x(n-j)$ for $1 \le j \le n$, and f and $*$ defined as before. D_{n-m} now gives the number of elements j such that $x(j) < x(m)$ and $j > m$. The sum of these two numbers for a given m gives the total number of elements in the array x which are less than x_m, i.e. the rank r_m of x_m. Since elements may be equal to each other, an additional step is to be performed in order to get the sorted array $s(1),...,s(n)$ out of $x(1),...,x(n)$ and $r_1,...,r_n$. We introduce an array $b_1,...,b_n$ and set initially $b_j=0$ for $1 \le j \le n$. For $j = 1$ to n do the following: $s(r_j + b(r_j)) = x(j)$ and $b(r_j) = b(r_j)+1$. It is easy to see that $s(1),..., s(n)$ is sorted array $x(1),..., x(n)$.

Since the transform takes only $O(n)$ time, the $\Omega(n \log n)$ lower bound for sorting becames a lower bound for the generalized prefix computation. Note that the proof applies to a specific choice of semigroup operator and total order. Thus, the lower bound may not apply to other operators or orders (for example, settings for prefix products).

3. MODELS OF PARALLEL COMPUTATION

One computational model used in the paper is the synchronous shared memory SIMD (single instruction multiple data) model in which concurrent reads are allowed, but no two processors should attempt to simultaneously write in the same memory location. We henceforth refer to this theoretical model as the CREW PRAM (concurrent read exclusive write parallel random access machine).

Recently Akl and Guenther [1] naturally extended the theoretical CREW and CRCW (concurrent read concurrent write) PRAM models to an even more powerful model called *broadcasting with selective reduction* (BSR). The new model uses only resources already available on the CRCW PRAM. However, some of the restrictions of the CRCW PRAM with respect to memory access are relaxed. As a result, *constant* time solutions to *sorting, parallel prefix* and other important problems on n data can be obtained on a BSR model consisting of (only!) n processors and $O(n)$ memory locations. In addition to constant time concurrent read operation, several processors may simultaneoulsy wish to write into the same memory location. The location will receive, in constant time, an arbitrary associative group value of all the things attempted to be written (for example, their sum, their product, the minimal or maximal value, or a value by a rule of priority). This operation is called *reduction*. One additional instruction supports (simultaneous) *broadcasting* of data from every processor to every memory location. Each processor broadcasts a *tag* and a *datum* . On the "reception" each memory location does a reduction of *selected* datums. More precisely, a memory location c contains three additional fields: *c.red* identifies reduction operator to be used when several datums are written concurrently to this location; *c.sel* and *c.lim* contain selection rule and its limit parameter which is compared with coming tags . The full description is given in [1] together with an excellent insight into CRCW PRAM and motivation for introducing new model.

A mesh-connected parallel computer of size n is a set of n synchronized processing elements (PEs) arranged in a $\sqrt{n} \times \sqrt{n}$ grid. Each PE is connected via bidirectional unit-time communication links to its four neighbors, if they exist. Each processor has a fixed number of registers and can perform standard arithmetic and comparisons in constant time. The standard MCC data movement operations: rotating data within a row (column), sorting, data compression, broadcasting etc. can be performed in $O(\sqrt{n})$ time (cf. [14,17]).

A d-dimensional hypercube is a set of 2^d synchronized processing elements, called nodes, where two nodes i and j $(0 \leq i, j < 2^d)$ are connected by a bidirectional communication link if and only if the binary representations of i and j differ in exactly one bit. The basic data communication techniques on hypercubes are described in [14,16,18].

4. GENERALIZED PREFIX COMPUTATION

In this section we solve GPC problem on sequential and several parallel models of computation. One solution is given which can be applied on all but BSR model. A short and effective algorithm for the BSR model is also described.

By $D(m,S)$ we denote the function D_m restricted on the set of indices S (i.e. $\{m,j_1,...,j_s\}$ is a subset of S). Thus, $D_m = D(m, \{1,..., n\})$. By $Y(S)$ we denote the sorted array of elements $y(j)$ for j in S. Let the rank $B(m,S)$ be the position of element $y(m)$ in the array $Y(S)$ (the element rank with smallest value being 1). Next, let $P(m,S)$ ($E(m,S)$, respectively) be $f(j_1)*f(j_2)*...*f(j_s)$ where $\{j_1, j_2, ..., j_s\}$ is the set of all indices j from S for which $y(j) < y(m)$ ($y(j) \leq y(m)$, respectively) is satisfied (the condition $j < m$ is not required).

If we interpret $y(m)$ as the y-coordinate then i is 'below' j if $y(i) < y(j)$. Also, i is to the left of j if $i < j$. Then $D(m,S)$, $P(m,S)$ and $E(m,S)$ denote the prefix product of points in S which are to the left and below m, below m and not above m, respectively.

The algorithm is best described recursively. Suppose S is divided into two subsets L and R of equal size with $l < r$ for all l in L and r in R. After the recursive calls for L and R in parallel we will have $Y(L)$, $Y(R)$, $D(l,L)$, $D(r,R)$, $P(l,L)$, $P(r,R)$, $B(l,L)$, $B(r,R)$, $E(l,L)$ and $E(r,R)$ for all l in L and r in R. Merge $Y(L)$ and $Y(R)$ to form $Y(S)$. Each element m from S then finds its rank $B(m,S)$ in $Y(S)$. The indices c_r, c'_r (c_l, c'_l) of two elements from $L(R)$ such that the element r(l, resp.) is between them in the sorted array $Y(S)$ are determined by $B(c_r, L) = B(r,S) - B(r,R)$ ($B(c_l, R) = B(l,S) - B(l,L)$, resp.) and $B(c'_r, L) = B(c_r, L) + 1$ ($B(c'_l, R) = B(c_l, R) + 1$, respectively).

Therefore the final result will be obtained directly from the relations (the results for the first and last elements of $Y(S)$, $Y(L)$ and $Y(R)$ can be modified easily):

$D(l,S) := D(l,L)$ for l in L,
$D(r,S) := P(c_r,L)*D(r,R)$ for r in R and $y(c_r) = y(r)$,
$D(r,S) := E(c_r,L)*D(r,R)$ for r in R and $y(c_r) < y(r)$,
$P(l,S) := P(l,L)*P(c_l,R)$ for l in L and $y(c_l)=y(l)$,
$P(l,S) := P(l,L)*E(c_l,R)$ for l in L and $y(c_l) < y(l)$,
$P(r,S) := P(c_r,S)*P(r,R)$ for r in R and $y(c_r)=y(r)$,
$P(r,S) := E(c_r,L)*P(r,R)$ for r in R and $y(c_r) < y(r)$,
$E(l,S) := E(l,L)*E(c_l,R)$ for l in L and $y(c_l) = y(l)$,
$E(l,S) := E(l,L)*P(c'_l,R)$ for l in L and $y(c_l) < y(l) < y(c'_l)$,
$E(l,S) := E(l,L)*E(c'_l,R)$ for l in L and $y(c_l) < y(l) = y(c'_l)$,
$E(r,S) := E(c_r,L)*E(r,R)$ for r in R and $y(c_r) = y(r)$,
$E(r,S) := P(c'_r,L)*E(r,R)$ for r in R and $y(c_r) < y(r) < y(c'_r)$,
$E(r,S) := E(c'_r,L)*E(r,R)$ for r in R and $y(c_r) < y(r) = y(c'_r)$.

Unfolding the recursion yields the iterative solution. Each index m is assigned fields D,P,B and E. Initially $B=1$, $E=f(m)$ while P and D are equal to the identity element for '*'. Then for $i = 0$ to $d-1$ (where d is the smallest integer such that $2^d \geq n$) the algorithm merges consecutive blocks L and R of size 2^i in pairs, finds c_r, c'_r (c_l, c'_l) for each element of R (L, respectively), route the fields P and E from c_r and c'_r (c_l and c'_l) to l (r, respectively) and updates the ranks B and all fields as indicated.

Since merging can be done in $O(n)$ sequential time with $O(n)$ space, an $O(n \log n)$ sequential implementation for generalized prefix computation follows easily. Using the mergesort algorithm [3] the generalized prefix computation can be done in $O(\log n)$ time with $O(n)$ space and $O(n)$ processors on a CREW PRAM model of computation (details are omited since a similar application of mergesort algorithm for 3-D maximal elements and 2-set dominance counting problems is described in [3]). Also, using the optimal merging procedure for mesh computers [17] in $O(\sqrt{n})$ time and other data communication techniques [14], one can implement GPC to run in $O(\sqrt{n})$ time on a MCC. On a hypercube, nodes belonging to a block $L(R)$ at step i (from 0 to $d-1$ where $n= 2^d$) are simply recognized as nodes having i-th coordinate in its binary representation equal to 0 (1, respectively). Using merging, interval broadcast and other data communication techniques, the GPC can be implemented in $O(\log^2 n)$ time on a hypercube.

A short BSR solution can be given as follows. Each processor m broadcasts (in parallel) a tag $y(m)$ and a datum $f(m)$ $(1 \leq m \leq n)$. A memory location m (in parallel for $1 \leq m \leq n$) applies selection operator '<' and limit parameter $y(m)$ to select these datums $f(p)$ for which $y(p)<y(m)$ $(1 \leq p \leq n)$. The reduction operator '*' is then applied on all selected datums $f(p)$. Formal description of the algorithm is given below.

$D_m.red := *$
$D_m.sel := <$
$D_m.lim := y(m)$
BROADCAST $y(m), f(m)$

The presented GPC solution explores the full potential of the BSR model in a very clear way. In fact, the condition that '<' must be a linear order can be relaxed; such a further generalization of GPC technique becames then equivalent to the broadcasting with selective reduction operation on this model. The BSR algorithm clearly runs in constant time using n processors and $O(n)$ space. One can here note that it contradicts to the lower bound for the problem, derived in former section. This is due to the power of the theoretical model. Some of the problems mentioned here (2-D maximal elements, sorting, parallel prefix, convex hull) are already solved in [1] on the BSR with n processors and $O(n)$ space (except the convex hull problem which requires $O(n^2)$ space).

5. ECDF SEARCHING, 2-SET DOMINANCE COUNTING, 2-D AND 3-D MAXIMAL ELEMENTS PROBLEMS

We are given a set $S=\{p_1,..., p_n\}$ of n points in d-dimensional space. A point p_i d-dominates a point p_j iff $p_i[k] > p_j[k]$ for $k=1, 2,..., d$, where $p[k]$ denotes the k-th coordinate of a point p. The 2-dimensional ECDF searching problem consists of computing for each p in S the number $D(p,S)$ of points of S 2-dominated by p. Given two point sets A and B, the 2-set dominance counting problem in two dimensions is the problem of counting for each point p from B the number of points from A that p 2-dominates. The maximal elements problem is the problem of determining points which are dominated by no other point. We can also consider these problems in three dimensions.

In order to solve some of mentioned problems, we sort points by first coordinate and denote the lists sorted in ascending and descending order by $a_1,..., a_n$ and $r_1,..., r_n$, respectively. Then the replacements required to solve the above problems are given in the following table (the result is shown in the last column; min and max denote the minimum and maximum functions, respectively):

	$f(m)$	$y(m)$	*	Result
2-D maximal	$r_m[2]$	m	*max*	max if $r_m[2] > D_m$
ECDF searching	1	$a_m[2]$	+	D_m
2-set dominance	$1(0)$ for a_m in $A(B$, resp.)	$a_m[2]$	+	D_m
3-D maximal	$r_m[3]$	$r_m[2]$	*max*	max if $r_m[3] > D_m$

6. TRIANGULATING A SET OF POINTS IN THE PLANE

Merks [12] presented an optimal parallel algorithm for triangulating an arbitrary set of n points in the plane. The algorithm runs in $O(\log n)$ time using $O(n)$ space and $O(n)$ processors on a CREW PRAM. A parallel divide-and-conquer technique of subdividing a problem into $\sqrt{n}$ subproblems is imployed in [12]. Following the same approach we show that the algorithm of [12] can be simplified by applying the generalized prefix computation technique. The obtained divide-and-conquer will then subdivide a problem into two subproblems (rather than $\sqrt{n}$) still resulting in an optimal algorithm.

We briefly describe the triangulation algorithm of [12]. We exclude degeneracies (they are handled in [12]; for details, including references, also see [12]).

By computing the convex hull of the points, determining the point μ having the lowest y-coordinate, calculatimg the angle β that each point makes at μ in the positive x-direction and sorting according to decreasing β the points are split into subsequences using the extreme points of the convex hull to define boundaries. Now we triangulate each ordered subsequence of points $x_1,\ldots,x_n$ separately by joining them with non-intersecting line segments such that all regions interior to the triangle $\mu x_1 x_n$ are triangles. For each point x_i we calculate the distance ∂_i between x_i and the line segment defined by the two extreme points of the convex hull which determine the containing subsequence. Let us define the following terms:

-the left (right) higher neighbor of a point x_i is the point x_j with j being the largest (smallest) index such that $j < i$ $(j>i)$ and $\partial_i \geq \partial_j$ ($\partial_i > \partial_j$, respectively).

Every point except x_1 and x_n will have precisely three associated connections [12]: a downward connection to μ, a connection to the left higher neighbor and a connection to the right higher neighbor. To compute connections, in [12] the problem is subdivided into $\sqrt{n}$ subproblems of size $\sqrt{n}$ each. We propose a generalized prefix computation technique instead. The index of left neighbor of each point is obtained by replacing $f(j)=j$, $*=max$ and $y(j)=\partial_j-d/(j+1)$, where d is minimal positive difference between any two members of array ∂, obtained after sorting, computing positive differences between neighboring elements and taking the minimum over them. The index of right higher neighbor of each point is determined by reversing sequence x and assigning $*=max$, $f(j)=j$ and $y(j)=\partial_j$. Note that the computation can be done for all subsequences in parallel since by definition each member except boundaries (extreme points) will surely find its higher neighbors within the corresponding subsequence it belongs to.

7. RECONSTRUCTION OF TREES FROM THEIR TRAVERSALS

It is well-known that given the inorder and preorder traversals of a binary tree's nodes, the original binary tree can be reconstructed. This problem has been studied in very recent literature [2,6,7]. Using GPC technique, here we present the first parallel solution to the problem.

It is well-known that any binary tree with n nodes can be represented as a permutation over $1,2,\ldots,n$ called "inorder-preorder sequence" (abreviated "*IP* sequence") [11. As in [6], we first construct *IP* sequence of the tree from its inorder and preorder traversals. Then we build the binary tree from the *IP* sequence. Given any binary tree with n nodes, its *IP sequence* is the numeric sequence output by the following algorithm [11]: label the nodes of the tree as accessed in inorder by the consecutive integers $1,2,\ldots,n$ and output these numeric labels as the nodes are accessed in preorder. For example, let inorder and preorder traversals given as:

IN:	*A*	*B*	*C*	*D*	*E*	*F*	*G*	*H*	*I*	*j*	*K*	*L*	*M*	*N*
	1	*2*	*3*	*4*	*5*	*6*	*7*	*8*	*9*	*10*	*11*	*12*	*13*	*14*
PRE:	*G*	E	B	A	C	D	F	M	*J*	*H*	*I*	L	K	N
	7	*5*	*2*	*1*	*3*	*4*	*6*	*13*	*10*	*8*	*9*	*12*	*11*	*14*

Using the correspondence *A-1, B-2, C-3*, etc., preorder sequence generates *IP* sequence:

7,5,2,1,3,4,6,13,10,8,9,12,11,14. The *IP* sequence can be constructed by sorting both *IN* and *PRE* so that each element can directly learn its indices in both sequences (see also [7]). Our method for reconstructing of binary tree from its *IP* sequence is based on the following lemma which follows directly from the construction of binary search trees.

Lemma. An element $IP[m]$ $(2 \le m \le n)$ will be inserted either as a right child of $IP[l_m]$ (for $l_m > r_m$) or as a left child of $IP[r_m]$ (for $l_m < r_m$) where l_m (r_m) is the index of the greatest (smallest) element of *IP* sequence among elements of *IP* having indices less than m and which is still less (greater, respectively) than $IP[m]$. In other words, $IP[l_m]$ and $IP[r_m]$ are neighboring elements of $IP[m]$ in the sorted list of elements $IP[1],..., IP[m]$.

In our example, $IP[l_m]$, $IP[r_m]$ and parent nodes are as follows:

IP	7	5	2	1	3	4	6	13	10	8	9	12	11	14
$IP[l_m]$	-	-	-	-	2	3	5	7	7	7	8	10	10	13
$IP[r_m]$	-	7	5	2	5	5	7	-	13	10	10	13	12	-
parent	-	7	5	2	2	3	5	7	13	10	8	10	12	13

In order to find the parent for each node we solve the GPC problem. In our case setting $y=f=IP$ and $*=max$ gives $IP[l_m]=D_m$ while setting $y(m)=-IP[m], f=IP$ and $*=min$ gives $IP[r_m]=D_m$. The time complexities for it are the same as for GPC. Now each node is assigned to its parent and the tree is reconstructed. We might like to know the level of each node in the tree in order to complete reconstruction. However, this can be done by applying a list ranking algorithm (each element pointing to its parent) in $O(\log n)$ time on a CREW PRAM (cf. [8]), in $O(\sqrt{n})$ time on a MCC [4] and in $O(\log^3 n)$ time on a hypercube. These time complexities then apply for the complete reconstruction tree algorithm (taking into account time complexities for other steps).

CONCLUSION

We have introduced a general prefix computations problem that on surface seems to be algebraic in nature. However, because the '*' operator on *f*-values can be any associative combinator, and the linear comparator '<' for the associated sequence of y-values can also be very general, we find that it has many applications to geometric cases of dominance, or maximizing, and that there is a linear-time transformation of general sorting to GPCs, giving its lower bound of $\Omega(n \log n)$.

Also, we presented iterative techniques of solving GPCs on various models of computation, and we gave both sequential and parallel algorithms that are (worst-case) optimal time, for each model. For the parallel computation models (CREW PRAM, MCC, BSR, hypercube), most of the geometric problems we considered are solved here in their optimal time on the respective model. Examples of such problems include point set triangulation, 2-D/3-D maximal point searching, two-set dominance counting, and ECDF searching. For all these, a lower complexity bound of $\Omega(n \log n)$ sequential time is proved. On the other hand, linear time sequential algorithms exist for non-general prefix products and binary trees reconstruction problems. Even for them, the GPC solution yields their optimal time on the MCC and BSR models.

It may be of great value to parallel algorithm research to find other applications of GPC techniques, and/or to solve further generalizations of the original problem of prefix products, that help solve other kinds of problems, optimally and in parallel.

REFERENCES

[1] S.G. Akl, G.R. Guenther, Broadcasting with selective reduction, Technical Report No. 88-232, Department of Computing and Information Science, Queen's University, Kingston, Ontario, Canada, August 1988.

[2] A. Anderson and S. Carlsson, Construction of a tree from its traversals in optimal time and space, Comp. Sci. Dept., Lund Univ., Sweden, October 1988.

[3] M.J. Atallah, R. Cole and M.T. Goodrich, Cascading divide-and-conquer: a technique for designing parallel algorithms, *IEEE Symp. Found. Comp. Sci.*, 151-160, 1987.

[4] M.J. Atallah and S.E. Hambrusch, Solving tree problems on a mesh-connected processor array, *Information and Control* 69, 1-3, 168-187, 1986.

[5] I. Bar-On and U. Vishkin, Optimal parallel generation of a computation tree form, *ACM Trans. Program. Lang. Syst.*, 7,2, 384-357, 1985.

[6] H.A. Burgdorff, S. Jajodia, F.N. Springsteel and Y. Zalcstein, Alternative methods for the reconstruction of trees from their traversals, *BIT*, 27,2, 133-140, 1987.

[7] G.H. Chen, M.S. Yu and L.T. Liu, Two algorithms for constructing a binary tree from its traversals, *Inform. Process. Lett.*, 28, 6, 1988.

[8] R. Cole and U. Vishkin, Faster optimal parallel prefix sums and list ranking, *Ultracomputer Note 117*, Comp. Sci. TR 277, February 1987.

[9] F. Dehne and I. Stojmenović, An $O(\sqrt{n})$ algorithm for the ECDF searching problem for arbitrary dimensions on a mesh of processors, *Inform. Process. Lett.*, 28, 2, 67-70, 1988.

[10] H.N. Gabow, J.L. Bentley and R.E. Tarjan, Scaling and related techniques for geometric problems, *ACM Symp. Theory of Computing*, 135-143, 1984.

[11] T. Hikita, Listing and counting subtrees of equal size of a binary tree, *Inform. Process. Lett.*, 17, 225-229, 1983.

[12] E. Merks, An optimal parallel algorithm for triangulating a set of points in the plane, *Int. J. Parallel Programming*, 15, 5, 399-411, 1986.

[13] H. Meijer and S.G. Akl, Optimal computation of prefix sums on a binary tree of processors, *Int. J. Parallel Programming*, 16, 2, 127-136, 1987.

[14] D. Nassimi and S. Sahni, Data broadcasting in SIMD computers, *IEEE Trans. Comput.*, C-30, 2, 101-106, 1981.

[15] J.H. Reif and S. Sandeep, Optimal randomized parallel algorithms for computational geometry, *IEEE Int. Conf. Parallel Processing*, 270-277, 1987.

[16] I. Stojmenović, Computational geometry on a hypercube, *IEEE Int. Conf. Parallel Processing*, 100-103, 1988.

[17] C.D. Thompson and H.T. Kung, Sorting on a mesh-connected parallel computer, *Comm. ACM*, 4, 20, 263-271, 1977.

[18] J.D. Ullman, *Computational aspects of VLSI*, Comp. Sci. Press, Potomac, MD, 1984.

[19] D.E. Wilard, New data structures for orthogonal range queries, *SIAM J. Computing*, 14, 232-253, 1985.

Kolmogorov Complexity and Hausdorff Dimension

Ludwig Staiger
Technische Universität "Otto von Guericke" Magdeburg
Sektion Mathematik
PSF 124 / Magdeburg / DDR - 3010

Introduction

The concept of Kolmogorov or program size complexity of strings was introduced by R.J. Solomonoff, A.N. Kolmogorov and G.J. Chaitin in the sixties. For infinite strings (sequences) this concept is strongly related to P. Martin-Löf's definition of random sequences. Roughly speaking, a sequence is random if almost all of its initial words have a complexity which is close to their length. In other words, random sequences have a relative complexity of 1. Thus one may find it natural to investigate the degree of randomness of a sequence with the help of its relative complexity. For instance, if $x_1x_2 \ldots x_i \ldots$ is a random sequence, the sequences $x_1yx_2y \ldots x_iy \ldots$ or $x_1x_1x_2x_2 \ldots x_ix_i \ldots$ may be considered as random of degree 1/2. Having in mind this form of degree of randomness, i.e. measuring randomness as the amount of information which must be provided on the average in order to specify a particular symbol of that sequence, one is naturally led to the question to which extent this idea consistent is with other measure or information theoretic concepts.

We consider the following two concepts: The first one is called the *entropy* and is also known as *upper Minkowski dimension, upper metric dimension, topological entropy* or, under a convergence condition, as Shannon's *channel capacity*, the second one is the *Hausdorff dimension*.

Either of these, in some sense, measures the size of sets of infinite strings, and it is to expect that sets of large size do also contain complex sequences, whereas sets of small size contain only sequences of limited complexity. For our proposed size measures, however, every maximally complex string ξ is contained in a set $\{\xi\}$ of smallest possible size. It is therefore not possible to obtain upper bounds on the complexity of sequences in sets of small size without any computability constraints on these sets. In order to describe computability constraints it is useful to take advantage of the theory of ω-languages. This theory investigates classes of sets of infinite strings definable by classes of machines or generated via certain operations from classes of languages (sets of finite strings).

Kolmogorov complexity

We start with a brief account on the necessary prerequisites in Kolmogorov or program size complexity, for more detailed information see the book [Sc1] or the survey papers [ZL] and [LV]. The thesis [vL] gives a nice recent survey of the work on random sequences. We conclude this section with a short presentation of our results.

Program size complexity defines the complexity of a finite string to be the length of a shortest program which prints the string. Accordingly, the complexity of an infinite string β is a function $K(\beta/\cdot): \mathbb{N} \to \mathbb{N}$ where $K(\beta/n)$ is the complexity of the initial part of length n of the string β . In this paper we are mainly interested in the first order approximation (i.e. the linear growth) of $K(\beta/\cdot)$. We consider the functions

(0.1) $\underline{\kappa}(\beta) := \liminf_{n\to\infty} K(\beta/n)/n$ and $\kappa(\beta) := \limsup_{n\to\infty} K(\beta/n)/n$,

and we compare the functions

(0.2) $\underline{\kappa}(F) := \sup \{ \underline{\kappa}(\beta): \beta \in F \}$ and $\kappa(F) := \sup \{ \kappa(\beta): \beta \in F \}$

defined for ω-languages to the corresponding entropy H_F or Hausdorff dimension dim F.
Since we are mainly interested in the above mentioned first order approximations, the established in [KS], [LC] and [Sc2] relations between Kolmogorov, Chaitin and other concepts of program complexity prove that the functions $\underline{\kappa}$ and κ do not depend on the particular kind of complexity we use. We therefore (also in view of Theorem 2.5 and Proposition 2.10 below) agree on the following concept of conditional complexity.

(0.3) $K_{\mathfrak{A}}(w\,|\,n) := \inf\{\, |\pi|: \mathfrak{A}(\pi,n) = w \wedge |w| = n \,\}$

Here π denotes a program for the algorithm $\mathfrak{A}$ which under the additional input n outputs the string w of length $|w| = n$.
As usual we consider a complexity function $K := K_{\mathfrak{U}}$, where $\mathfrak{U}$ is an optimal algorithm i.e. for every algorithm $\mathfrak{A}$ there is a constant $c_{\mathfrak{A}}$ such that

(0.4) $K(w\,|\,n) \le K_{\mathfrak{A}}(w\,|\,n) + c_{\mathfrak{A}}$ for all w and n .

After some prerequisites on ω-languages in the next section (more detailed information can be obtained from Chapter XIV of the book [Ei] and the recent survey papers [HR], [S4] or [Th]; [S3] contains a comprehensive study of ω-languages definable by Turing machines.) we begin our study with the derivation of properties of the entropy of ω-languages and of upper bounds on the set-function κ via entropy. The above mentioned computability constraints are specified in terms of recursive (definable by Turing machines) ω-languages, and it is shown that for ω-languages definable by finite automata (regular, or more general finite-state) the maximum growth of $K(\beta/\cdot)$ can be determined more excactly.
Then lower bounds for general ω-languiages are considered. It turns out that the Hausdorff dimension dim F is a general lower bound to $\underline{\kappa}(F)$, and we exhibit examples that there are complexity gaps betweeen $\underline{\kappa}(F)$ and $\kappa(F)$ even for simple recursive ω-languages.
To show that those gaps do not exist for regular ω-languages is the aim of the following part. More precisely, it is shown that the maximum growth of $K(\beta/\cdot)$ in a regular ω-language behaves strongly like the growth of $K(\beta/\cdot)$ for random sequences including P. Martin-Löf's result on complexity dips.
The last part of this paper is devoted to the class of ω-power languages, a class not defined by machines but also exhibiting some regularity in their structure and, therefore, also allowing for a more precise calculation of the functions $\underline{\kappa}$ and κ.

Preliminaries

By $\mathbb{N} = \{ 0,1,2,... \}$ we denote the set of natural numbers. We consider the space X^{ω} of infinite strings (sequences) on a finite alphabet of cardinality $\#X = r$. By X^* we denote the set of finite strings (words) on X, including the *empty word* e. For $w \in X^*$ and $b \in X^* \cup X^\omega$ let $w \cdot b$ be their *concatenation*. This concatenation product extends in an obvious way to subsets $W \subseteq X^*$ and $B \subseteq X^* \cup X^\omega$.[1] As usual we denote subsets of X^* as *languages* and subsets of X^ω as *ω-languages*. For a language W let $W^0 := \{e\}$, $W^* := \bigcup_{i \in \mathbb{N}} W^i$ and by W^ω we denote the set of infinite sequences formed by concatenating words in W. Furthermore $|w|$ is the *length* of the word $w \in X^*$ and $\underline{\ell}(W) := \inf\{|w| : w \in W\}$ denotes the *minimum length* of a word in W.

For a word w and a set $B \subseteq X^* \cup X^\omega$ we call $B/w := \{b : w \cdot b \in B\}$ the *state* of B derived by the word w, and we call a set B *finite-state* provided $\{B/w : w \in X^*\}$ is a finite set. Moreover, $A(B) := \{w : w \in X^* \wedge B/w \neq \emptyset\}$ is the set of all *initial words (prefixes)* of the set $B \subseteq X^* \cup X^\omega$.[1]

Finite-state languages are also known as *regular languages* (cf. [Sa]), whereas *regular ω-languages* are the finite unions of sets of the form $W \cdot V^\omega$ where W and V are regular languages.[2]

Next we mention the following representation of regular ω-languages a proof of which can be found e.g. in Chapter XIV of [Ei]. To this end we say that a language $V \subseteq X^*$ is *prefix-free* provided it does not contain words w, v such that $w \neq v$ and w is a prefix of v, i.e. $w \in A(v)$.

(1.1) **Theorem.** An ω-language $F \subseteq X^\omega$ is regular if and only if there an $n \in \mathbb{N}$ and regular languages $W_i, V_i \subseteq X^*$ such that the languages V_i are prefix-free and

$$F = \bigcup_{i=1}^{n} W_i \cdot V_i^\omega .$$

We consider X^ω as a metric space with the metric ϱ defined by

(1.2) $\quad \varrho(\beta,\xi) := \inf\{ r^{-|w|} : w \in A(\beta) \cap A(\xi) \}$.

Since X is finite, this space is compact. Its open (and simultaneously closed) balls are the sets of the form $w \cdot X^\omega$ $(w \in X^*)$. Thus the *closure* $\mathfrak{C}(F)$ of a subset $F \subseteq X^\omega$ can be described as follows:

(1.3) $\quad \mathfrak{C}(F) = \bigcap_{i \in \mathbb{N}} (A(F) \cap X^i) \cdot X^\omega = \{\beta : A(\beta) \subseteq A(F)\}$

G_δ-sets in (X^ω,ϱ) are characterized as follows (cf. [Ei],[LS]): Let $V^\delta := \{\xi : A(\xi) \cap V \text{ is infinite}\}$, then $E \subseteq X^\omega$ is a G_δ-set if and only if $\exists V (V \subseteq X^* \wedge E = V^\delta)$.

As usual we denote by Σ_1 and Π_1 the classes of languages in the arithmetical hierarchy.

Upper bounds

In this section we derive upper bounds on the complexity of infinite strings ξ in a given ω-language F by means of its entropy H_F. We start with some connections between the entropy of languages H_W and the complexity of words $w \in W$.

To this end let s_W be the *structure function* of the language W which is defined as follows (cf. [Ku])

(2.1) $\quad s_W(n) := \#\{w : w \in W \wedge |w| = n\}$.

---

1)For the sake of brevity, in what follows we shall write $w \cdot B$, $W \cdot b$ and $A(b)$ instead of $\{w\} \cdot B$, $W \cdot \{b\}$ and $A(\{b\})$ respectively.

2)In contrast to the case of languages not every finite-state ω-language is also regular (cf. [S4, Section 5]).

The corresponding *structure generating function* is

(2.2) $\mathfrak{s}_W(t) := \sum_{i\in\mathbb{N}} s_W(i)\cdot t^i$.

The series $\mathfrak{s}_W$ is a positive series and its convergence radius rad W satisfies $r^{-1} \le$ rad W. We define

(2.3) $\mathfrak{s}_W(\text{rad } W) := \sup\{\mathfrak{s}_W(\gamma): \gamma < \text{rad } W\}$ and $\mathfrak{s}_W(\alpha) := \infty$ if $\alpha >$ rad W ,

and consider $\mathfrak{s}_W$ also as a function mapping $[0,\infty)$ to $[0,\infty) \cup \{\infty\}$.

The *entropy* of a language W is defined as follows.

(2.4) $H_W := \limsup_{n\to\infty} n^{-1}\cdot\log_r s_W(n) = -\log_r \text{rad } W$.

For an ω-language $F \subseteq X^\omega$ the notions s_F, $\mathfrak{s}_F$, H_F and rad F are defined as $s_{A(F)}$, $\mathfrak{s}_{A(F)}$, $H_{A(F)}$ and rad A(F) resp. Therefore, the entropy of an ω-language F coincides with the entropy of its closure $\mathfrak{C}(F) = A(F)^\delta$.

First we derive two results for the complexity of finite strings.

(2.5) **Theorem.** If $W \in \Sigma_1 \cup \Pi_1$ then there is a constant c such that for every $w \in W$ it holds $K(w||w|) \le \log_r s_W(|w|) + c$.

Proof. The Σ_1-part of the theorem is Theorem 1.1.i of [dL]. For the Π_1-part define $\mathfrak{A}(\pi,n)$ in the following way: Enumerate $X^*\setminus W$ up to the point when $r^n - r^{|\pi|}$ elements of length n appeared. Then take from the rest the $q(\pi)$th [3] word of length n. If $|\pi| \ge \log_r s_W(n)$ then the above enumeration process terminates. Hence $K_{\mathfrak{A}}(w||w|) \le \lceil\log_r s_W(|w|)\rceil$ for every $w \in W$.□

This result, however, cannot be transferred to the next classes of the arithmetical hierarchy.

(2.6) <u>Example</u>.([Sc1],[vL]). There are sequences $\xi \in X^\omega$ satisfying $A(\xi) \in \Pi_2 \cap \Sigma_2$ and $K(\xi/n) \ge n - o(n)$ [4]. Since $s_{A(\xi)} \equiv 1$, our assertion follows. □

Our Theorem 2.5 leads to the following improvement of Theorem 1 of [S1].

(2.7) **Proposition.** If $A(F) \in \Sigma_1 \cup \Pi_1$ then $\varkappa(F) \le H_F$.

(2.8) **Proposition.** Let $V \in \Sigma_1 \cup \Pi_1$. Then for every $\beta \in V^\delta$ there are infinitely many $n \in \mathbb{N}$ such that $K(\beta/n) \le \log_r s_V(n) + c$ for some constant c .

In Theorem 2 of [S1] for the special case of finite-state ω-languages the following bound stronger than Proposition 2.7 is obtained. Its proof is based on an auxiliary proposition which can be found in [S2].

(2.9) **Proposition.** If $F \subseteq X^\omega$ is a finite-state ω-language for which a language $W \subseteq X^*$ with $\mathfrak{C}(W^\omega) = \mathfrak{C}(F)$ exists then there is a constant c such that for all $n \in \mathbb{N}$ and all $w \in A(F)$ the inequality $|H_F\cdot n - \log_r s_{F/w}(n)| \le c$ is satisfied.□

Remark. The condition $\exists W(W \subseteq X^* \wedge \mathfrak{C}(F) = \mathfrak{C}(W^\omega))$ is equivalent to the following one employed in Corollary 4 of [S2]: $\forall w(w \in A(F) \to \exists v(\mathfrak{C}(F)/w\cdot v \supseteq F))$.

(2.10) **Proposition.**([S1]) Let $F \subseteq X^\omega$ be finite-state. then for every $\xi \in F$ there is a constant c such that $K(\xi/n) \le_{ae} H_F\cdot n + c$.

---

3) By $q(\pi)$ we denote the position of the word π in the lexicographical ordering of the set $\{v: |v| = |\pi| \wedge v \in X^*\}$

4) In what follows we shall use the small-o-notation as well as the following abbreviations $\le_{ae}$ and $>_{io}$ to denote that the corresponding inequalities hold almost everywhere or infinitely often.

Coverings and Hausdorff dimension

An r^{-n}-*covering* of a set $F \subseteq X^\omega$ is a family $(v \cdot X^\omega)_{v\in V}$ of balls such that $\bigcup (v\cdot X^\omega)_{v\in V} = V\cdot X^\omega \supseteq F$ and whose *diameters* diam $v\cdot X^\omega$ do not exceed r^{-n}, i.e. $\underline{\ell}(V) \geq n$.
In Section 14 of [Bi] the general definition of the Hausdorff dimension of a subset of a metric space is given. We recall this definition for our space (X^ω, ϱ). Let

(3.1) $\quad L_\alpha(F;V) := \sum_{v\in V} (\operatorname{diam} v\cdot X^\omega)^\alpha = \sum_{v\in V} r^{-\alpha\cdot|v|} = s_V(r^{-\alpha})$

for a covering $(v\cdot X^\omega)_{v\in V}$ of F and $0 \leq \alpha \leq 1$. Then

(3.2) $\quad L_\alpha(F) := \liminf_{n\to\infty} \{ L_\alpha(F;V) \colon V\cdot X^\omega \supseteq F \wedge \underline{\ell}(V) \geq n \}$

is the α-*dimensional outer measure* of F .
Now, consider $L_\alpha(F)$ for fixed F as a function of α . Then there is an $\alpha_0 \in [0,1]$ such that $L_\alpha(F) = \infty$ if $\alpha < \alpha_0$ and $L_\alpha(F) = 0$ if $\alpha > \alpha_0$. This "change-over" point α_0 is called the *Hausdorff dimension* dim F of the set F , i.e.

(3.3) $\quad \dim F := \sup\{ \alpha \colon L_\alpha(F) = \infty \} = \inf\{ \alpha\colon L_\alpha(F) = 0 \}$.

We observe, that the measures $\underline{\varkappa}$, $\varkappa$, dim and H share some common properties (cf. [Bi], [S2]).
(3.4) **Property.** Let λ be an arbitrary one of the measures $\underline{\varkappa}$, $\varkappa$, dim or H. Then for subsets E, F, $F_i \subseteq X^\omega$ the following identities hold:

(3.4.1) $\quad \lambda(w\cdot F) = \lambda(F)$ and $\lambda(F/w) \leq \lambda(F)$ for $w \in X^*$,

(3.4.2) $\quad \lambda(E \cup F) = \max\{\lambda(E), \lambda(F)\}$,and

(3.4.3) $\quad \lambda(\bigcup_{i\in N} F_i) = \sup\{\lambda(F_i) \colon i \in N\}$,if $\lambda \neq H$.

For nonempty sets of the form W·F the last property implies that $\lambda(W\cdot F) = \lambda(F)$ for $\lambda \neq H$, whereas $H_{W\cdot F} = \max\{H_{A(W)}, H_F\}$ in virtue of $A(W\cdot F) = A(W) \cup W\cdot A(F)$ (cf. [LS]).
Next we draw a connection between the Hausdorff dimension of G_δ-sets and the structure generating function of languages.
(3.4) **Lemma.** If $s_V(r^{-\alpha}) < \infty$ then $L_\alpha(V^\delta) = 0$.
Proof. Define $V^{(i)} := \{ v \colon v \in V \wedge \#(A(v) \cap V) = i-1 \}$. By definition $V^{(i)}\cdot X^\omega \supseteq V^\delta$ and $V = \bigcup_{i\in N} V^{(i)}$ is a disjoint union. Now, the hypothesis guarantees that $L_\alpha(V^\delta; V^{(i)}) = s_{V^{(i)}}(r^{-\alpha})$ tends to zero as i approaches infinity. □
If $s_V(r^{-\alpha}) < \infty$ then in particular $\alpha \geq H_V$. Consequently

(3.5) $\quad \dim V^\delta \leq H_V$.

Lemma 3.4 has in a certain sense a converse.
(3.6) **Lemma.** If $L_\alpha(F) = 0$ then there is a $W \subseteq X^*$ such that $F \subseteq W^\delta$ and $s_W(r^{-\alpha}) < \infty$.
Proof. If $\alpha = 0$ the measure L_α is the counting measure. Hence $F = \emptyset$ and the assertion is obvious. Let $\alpha > 0$. Choose a family $\{W_i \colon i \in N\}$ such that $W_i\cdot X^\omega \supseteq F$ and $L_\alpha(F; W_i) < r^{-i}$. This, in particular, implies $\underline{\ell}(W_i) \geq \alpha^{-1}\cdot i$. Define $W := \bigcup_{i\in N} W_i$. Then $s_W(r^{-\alpha}) \leq \sum_{i\in N} L_\alpha(F; W_i) < \infty$. It remains to show that $W^\delta \supseteq F$. This is obvious, because of $W_i\cdot X^\omega \supseteq F$ and $\underline{\ell}(W_i) \geq \alpha^{-1}\cdot i$ every $\beta \in F$ has arbitrarily long initial words in W.□
Since $s_W(r^{-\alpha}) < \infty$ implies $H_W \leq \alpha$, this leads to the following consequence of Lemma 3.7 and Eqs. (3.3) and (3.6).

(3.8) $\quad \dim F = \inf\{ H_W \colon W^\delta \supseteq F \}$

Now, consider for $F \subseteq X^\omega$ an infinite subset $M \subseteq N$ such that $\left(n^{-1}\cdot \log_r s_F(n)\right)_{n\in M}$ tends to $\liminf_{n\to\infty} n^{-1}\cdot \log_r s_F(n)$ and define $W := \bigcup_{n\in M} A(F) \cap X^n$.

Then $W^{\delta} = \mathfrak{C}(F)$, and Eq. (3.6) proves the following inequality.

(3.9) $\quad \dim \mathfrak{C}(F) \le \liminf_{n\to\infty} \log_r s_F(n) \le H_F$

Our first lower bound is a general one. It is obtained by combining the above relations between the structure function and the Hausdorff dimension with a simple counting argument.

We consider the set $E(\alpha,f) := \{\eta: K(\eta/n) \ge_{ae} \alpha\cdot n - f(n)\}$ of (α,f)-complex sequences. Its complement $X^{\omega}\setminus E(\alpha,f)$ can be described as V^{δ} where $V := \{v: K(v\mid |v|) < \alpha\cdot|v| - f(|v|)\}$. Counting the number of programs π of length $< k$ one obtains

(3.10) $\#\{w: |w| = n \wedge K(w\mid n) < k\} < r^k/(r-1)$.

Hence, $s_V(i) \le 2r^{\alpha\cdot i - f(i)}$ and, consequently $\sum_{i\in\mathbb{N}} r^{-f(i)} < \infty$ implies $\mathfrak{s}_V(r^{-\alpha}) < \infty$. Utilizing Lemma 3.4 we obtain our

(3.11) **Lemma.** Let $F \subseteq X^{\omega}$, and let $f: \mathbb{N} \mapsto \mathbb{N}$ be an arbitrary function such that $\sum_{i\in\mathbb{N}} r^{-f(i)} < \infty$. Then $L_{\alpha}(F) > 0$ implies $F \cap E(\alpha,f) \ne \emptyset$.

Now Lemma 3.11 and Eq. (3.3) yield the following lower bound on $\underline{\varkappa}(F)$.

(3.12) **Corollary.**([R2]) $\forall F(\ F \subseteq X^{\omega} \rightarrow \dim F \le \underline{\varkappa}(F)\ )$.

We conclude this section with examples which show on the one hand that there might be as well large gaps between $\underline{\varkappa}(F)$ and $\varkappa(F)$ as large gaps between $\underline{\varkappa}(\xi)$ and $\varkappa(\xi)$ for a single string $\xi \in F$ even if $A(F)$ is recursive, thus showing that Hausdorff dimension is not an upper bound to $\varkappa$, and on the other hand that the bound of Proposition 2.7 is in several cases very imprecise.

(3.13) Example. Let $F := \prod_{i\in\mathbb{N}} X^{(2i)!}\cdot\{x\}^{(2i+1)!}$ where $x \in X$. Clearly, $A(F)$ is recursive. When we consider F as W with $W := \bigcup_{i\in\mathbb{N}} \prod_{j=1}^{i} X^{(2j)!}\cdot\{x\}^{(2j+1)!}$ Proposition 2.9 proves that $\underline{\varkappa}(F) = 0$. On the other hand, Daley's diagonalization argument [Da] shows that there is a $\beta \in F$ satisfying $\varkappa(\beta) = 1$.□

(3.14) Example.([LS]) Let $E := \{x\}^{\omega} \cup \bigcup_{i\in\mathbb{N}} x^{i}\cdot y\cdot X^{i!}\cdot\{x\}^{\omega}$, where $x,y \in X$ and $x \ne y$. Then one easily verifies that $H_F = 1$ but $\varkappa(F) = 0$.□

Regular ω-languages

Our Example 3.13 shows that Hausdorff dimension is, in general, no suitable upper bound to $\varkappa(F)$. In this section we exhibit a class of ω-languages which behaves quite regular in the sense that dim F is also an upper bound (and, therefore, in view of Proposition 4.3 a tight one) to $\varkappa(F)$.

We start with the following property which is proved in Corollary 7 of [S7].

(4.1) **Property.** If F is finite-state and closed then $\dim F = H_F$.

Next we generalize Theorem 4 of [S6].

(4.2) **Theorem.** If $F \subseteq X^{\omega}$ is regular and $\alpha = \dim F$ then $L_{\alpha}(F) > 0$.

Before we proceed to the proof of the theorem we mention the following properties which can be found e.g. in Section 9 of [LS] or Chapter VIII of [Ei]:

(4.3) $H_{A(V)} = H_V$ if V is a regular language, and

(4.4) $H_V < H_{V^*}$ if V is a regular and prefix-free language.

Moreover, it holds the following.

(4.5) **Lemma.** If V is regular and prefix-free then $L_{\alpha}(V^{\omega}) = L_{\alpha}(\mathfrak{C}(V^{\omega}))$.

Proof. We have $\mathfrak{C}(V^{\omega}) = V^{\omega} \cup V^{*}\cdot E$ where $E := \{\eta : A(\eta) \subseteq A(V)\}$ [LS, Section 5]. Hence, by the above Eqs. (3.6), (4.3), and (4.4): $\dim E \le H_E \le H_{A(V)} = H_V < H_{V^*}$. Moreover, $A(V^{\omega}) \cup \{e\} = A(V^{*})$ implies $H_{V^*} \le H_{V^{\omega}} \le H_{\mathfrak{C}(V^{\omega})} = \dim \mathfrak{C}(V^{\omega})$ by Property 4.1. Thus for $\alpha = \dim \mathfrak{C}(V^{\omega}) > \dim E$ we have $L_{\alpha}(E) = 0$ what in view of the above identity $\mathfrak{C}(V^{\omega}) = V^{\omega} \cup V^{*}\cdot E$ and Eq. (3.4.3) proves our assertion.□

The following consequence is immediate.

(4.6) **Corollary.** If V is regular and prefix-free then $\dim V^{\omega} = \dim \mathfrak{C}(V^{\omega}) = H_{V^{\omega}}$.

Proof of Theorem 4.2. For closed regular ω-languages F the assertion is proved in Theorem 4 of [S6]. According to Theorem 1.1 a regular ω-language F can be represented as a finite union of sets of the form $W\cdot V^{\omega}$ where W and V are regular and V is prefix-free. From Property 3.4 we obtain that there is some $w\cdot V^{\omega} \subseteq F$ such that $\dim w\cdot V^{\omega} = \dim F$. Now, for $\alpha = \dim F$ we have

$$L_{\alpha}(w\cdot V^{\omega}) = r^{-\alpha\cdot|w|}\cdot L_{\alpha}(V^{\omega}) = r^{-\alpha\cdot|w|}\cdot L_{\alpha}(\mathfrak{C}(V^{\omega})) > 0$$

by the above Lemma 4.5 and Theorem 4 of [S6] .□

Moreover, together with Lemma 3.11 we can improve Theorem 6 of [S1] as follows.

(4.7) **Theorem.** Let $F \subseteq X^{\omega}$ be regular, and let $f: \mathbb{N} \mapsto \mathbb{N}$ be a function such that $\sum_{i\in\mathbb{N}} r^{-f(i)} < \infty$. Then there is a $\xi \in F$ such that $K(\xi/n) \ge_{ae} \dim F\cdot n - f(n)$.

As a further consequence we obtain an improvement of the upper bound derived in Proposition 2.10 which together with the preceding theorem gives evidence that in a regular ω-language F the maximum and minimum complexity of a maximally complex sequence ξ differ only slightly

(4.8) **Theorem.** Let F be a regular ω-language. Then for every $\beta \in F$ there is a constant c such that $K(\beta/n) \le_{ae} \dim F\cdot n + c$.

Proof. According to Theorem 1.1 we have some $w\in X^{*}$ and a prefix-free regular language V such that $\beta \in w\cdot V^{\omega} \subseteq F$. Thus, $\beta \in w\cdot\mathfrak{C}(V^{\omega}) = \mathfrak{C}(w\cdot V^{\omega})$. Corollary 4.6 implies $H_{V^{\omega}} = \dim V^{\omega} \le \dim F$, and the assertion follows from Proposition 2.10 .□

In the previous section we gave an example of complexity gaps in (sets of) infinite strings, and in this section we showed that for regular ω-languages there are no such gaps.

In the late sixties P. Martin-Löf and G.J. Chaitin (cf. [LV]) proved that each sequence ξ in the regular ω-language X^{ω} has infinitely many complexity dips $K(\xi/n) \le n - f(n)$ when $f: \mathbb{N} \mapsto \mathbb{N}$ is a recursive function satisfying $\sum_{i\in\mathbb{N}} r^{-f(i)} = \infty$. The proof scetched in [ZL] easily extends to regular ω-languages of the form V^{ω} where $V \subseteq \{w : w\in X^{*} \wedge |w| = k\}$ in the sense that $K(\beta/n) \le_{io} \dim V^{\omega}\cdot n - f(n)$ where f is as above and $\beta \in V^{\omega}$. This has its reason in the regularity of the branching behaviour of the infinite r-nary tree corresponding to $F = V^{\omega}$, more exactly in the fact that for every v, $w \in A(F)$ and for all $n \in \mathbb{N}$ the numbers $s_{F/w}(n)$ and $s_{F/v}(n)$ do not differ too much from each other.

In Proposition 2.9 we have shown that, in particular, for every ω-language of the form $E = w\cdot V^{\omega}$ with $V \subseteq X^{*}$ regular there is a constant c such that $|\log_r s_{E/v}(n) - H_E\cdot n| < c$ holds for all $n \in \mathbb{N}$ and all $v\in A(E)$. Those ω-languages fulfil the above mentioned requirement, and the idea of P. Martin-Löf's proof can be transferred to ω-languages of the above mentioned form $E = w\cdot V^{\omega}$.

This leads to the following result on complexity dips in regular ω-languages.

(4.9) **Theorem.** Let F be a regular ω-language and let $f: \mathbb{N} \mapsto \mathbb{N}$ be a recursive function satisfying $\sum_{i \in \mathbb{N}} r^{-f(i)} = \infty$. Then $K(\xi/n) \leq_{io} \dim F \cdot n - f(n)$ holds for all $\xi \in F$.

ω-power languages

We continue our investigations with considering ω-languages of the special shape W^{ω} where W is some (not necessarily regular) language. It turns out that one can obtain more precise estimates for $\underline{\varkappa}(F)$ and $\varkappa(F)$ in the case of those ω-languages. First we are going to prove an assertion claimed by B.Ya. Ryabko [Ry] concerning the Hausdorff dimension of ω-powers W^{ω}. To this end we quote the theorem of [S5].

(5.1) **Theorem.** Let $W \subseteq X^*$ be an arbitrary language. Then for every $\varepsilon > 0$ there is a finite subset $V \subseteq W$ such that $H_{W^*} - H_{V^*} < \varepsilon$.□

Observe now that for every finite $V \subseteq X^*$ the ω-power V^{ω} is a regular and closed ω-language. Consequently, Property 4.1 and Eq. (4.3) prove $\dim V^{\omega} = H_{V^*}$. Thus, in particular, any finite $V \subseteq W$ satisfies $H_{V^*} = \dim V^{\omega} \leq \dim W^{\omega}$. Applying our theorem, yields $H_{W^*} \leq \dim W^{\omega}$. Conversely, utilizing Eq. (3.5) and the obvious inclusion $W^{\omega} \subseteq (W^*)^{\delta}$, we obtain $\dim W^{\omega} \leq H_{W^*}$ what proves the assertion.

(5.2) $\dim W^{\omega} = H_{W^*}$

Proposition 2.9, Corollary 3.12 and Eq. (5.2) yield the following estimate of $\underline{\varkappa}(W^{\omega})$.

(5.3) **Proposition.** If $W \in \Sigma_1 \cup \Pi_1$ then $\underline{\varkappa}(W^{\omega}) = \dim W^{\omega}$.□

Remark. The Σ_1-part of Proposition 5.3 was proved in a different manner in [Ry].□

Next we give an example which simultaneously shows that neither the bound in Eq. (4.2) is tight in general nor Proposition 5.3 can be extended to higher classes of the arithmetical hierarchy.

(5.4) Example. Consider the same set $A(\xi)$ as in Example 2.6 . From Chapter VIII of [Ei] we know that $\mathfrak{s}_W(r^{-1}) \geq 1$ is a necessary condition for $H_{W^*} = \dim W^{\omega} = 1$. Let $\xi = \prod_{i \in \mathbb{N}} w_i$ where $|w_i| = i!$, and define $W := \{ w_i : i \in \mathbb{N} \}$. One easily verifies $\xi \in W^{\omega}$, $W \in \Pi_2 \cap \Sigma_2$ and $\mathfrak{s}_W(r^{-1}) < 1$. Consequently, $\dim W^{\omega} < 1 = \underline{\varkappa}(\xi) = \underline{\varkappa}(W^{\omega})$.□

We continue this section by showing that for $\varkappa(W^{\omega})$ in Eq. (0.2) the supremum can be substituted by maximum. To this end we mention the following well-known relation between the complexity of a product $w \cdot v$ and the complexity of its factors w and v (cf. [ZL]).

(5.5) $\exists c \forall w \forall v (K(w \cdot v / |w \cdot v|) \geq K(v/|v|) - 2 \cdot \log_r |w| - c)$

Furthermore we need the following theorem which can be proved as Theorem 3 in [S1].

(5.6) **Theorem.** Let $W \subseteq X^*$, and let $(\beta_i)_{i \in \mathbb{N}}$ be a family of elements of $\mathfrak{C}(W^{\omega})$ and $(f_i)_{i \in \mathbb{N}}$ be a family of functions $f_i : \mathbb{N} \mapsto \mathbb{N}$ such that $K(\beta_i/n) >_{io} f_i(n)$ for all $i \in \mathbb{N}$, and for all $k, m \in \mathbb{N}$ the inequality $f_i(n+k) + m \leq_{ae} f_j(n)$ holds whenever $i < j$. Then there is a $\beta \in W^{\omega}$ such that $\forall i (i \in \mathbb{N} \rightarrow K(\beta/n) >_{io} f_i(n))$.□

As an immediate consequence of our theorem we get the announced identity.

(5.7) $\varkappa(W^{\omega}) = \max \{ \varkappa(\beta) : \beta \in W^{\omega} \}$

Utilizing the same idea as in the proof of Theorem 3 of [S1] we can give a more precise (at least for $W \in \Sigma_1$) estimate of the value of $\varkappa(W^{\omega})$.

(5.8) **Lemma.** Let $W \subseteq X^*$. Then $\varkappa(W^{\omega}) \geq H_{W^{\omega}} = \max \{ \dim W^{\omega}, H_{A(W)} \}$.□

As an immediate consequence we obtain from Proposition 2.7 the following identity for languages $W \in \Sigma_1$.

(5.9) $\varkappa(W^{\omega}) = H_{W^{\omega}} = \max\{\dim W^{\omega}, H_{A(W)}\}$ if $W \in \Sigma_1$.

Comparing these results with the corresponding ones, $\underline{\varkappa}(W^{\omega}) \geq \dim W^{\omega}$ or $\underline{\varkappa}(W^{\omega}) \geq \dim W^{\omega}$, for the function $\underline{\varkappa}$ reveals a reason for the possible appearence of complexity gaps in ω-power languages.

References

[Bi] Billingsley, P. , Ergodic Theory and Information. Wiley, New York 1965.

[Da] Daley, R.P., The extent and density of sequences within the minimal-program complexity hierarchies. *J. Comput. System Sci.* **15** (1974), 151 - 163.

[dL] DeLuca, A., On the entropy of a formal language. *In:* Automata Theory and Formal Languages, Proc. 2nd GI Conference (H. Brakhage, Ed.), Lect. Notes Comput Sci. **33**, Springer-Verlag, Berlin 1975, 103 - 109.

[Ei] Eilenberg, S.. Automata, Languages, and Machines. Vol. A, Academic Press, New York 1974.

[HR] Hoogeboom, H.J. and Rozenberg, G., Infinitary languages: Basic theory and applications to concurrent systems. *In:* Current Trends in Concurrency - Overviews and Tutorials (J.W. de Bakker, W.-P. de Roever and G. Rozenberg, Eds.), Lect. Notes Comput. Sci. **224**, Springer-Verlag, Berlin 1986, 266 - 342.

[KS] Katseff, H.P. and Sipser, M., Several results in program size complexity. *Theoret. Comput Sci.* **15** (1981), 291 - 309.

[Ku] Kuich, W., On the entropy of context-free languages. *Inform. and Control* **16** (1970)2, 173 - 200.

[LC] Leung-Yan-Cheong, S.K. and Cover, T., Some equivalences between Shannon entropy and Kolmogorov complexity. *IEEE Trans. Inform. Theory* **IT-24** (1978), 331 - 338.

[LV] Mi, L., and Vitanyi, P.M.B., Two decades of applied Kolmogorov complexity. *In:* Proc. 3rd IEEE Structure in Complexity Conference, 1988.

[LS] Lindner, R. and Staiger, L., Algebraische Codierungstheorie - Theorie der sequentiellen Codierungen. Akademie-Verlag, Berlin 1977.

[Ry] Ryabko, B.Ya., Noiseless coding of combinatorial sources, Hausdoff dimension and Kolmogorov complexity. *Problemy Peredachi Informatsii* **22** (1986) 3, 16 - 26. [Russian]

[Sa] Salomaa, A., Theory of Automata. Pergamon, Oxford 1969.

[Sc1] Schnorr, C.P., Zufälligkeit und Wahrscheinlichkeit. Lect. Notes Math. **218**, Springer-Verlag, Berlin 1971.

[Sc2] Schnorr, C.P., Process complexity and effective random tests. *J. Comput. System Sci.* **7** (1973) 4, 376 - 388.

[S1] Staiger, L., Complexity and entropy. *In:* Mathematical Foundations of Computer Science (J. Gruska and M. Chytil, Eds.), Lect. Notes Comput. Sci. **118**, Springer-Verlag, Berlin 1981, 508 - 514.

[S2] Staiger, L., The entropy of finite-state ω-languages. *Probl. Control and Inform. Theory* **14** (1985) 5, 383 - 392.

[S3] Staiger, L., Hierarchies of recursive ω-languages. *J. Inform. Process. Cybern. EIK* **22** (1986) 5/6, 219 - 241.

[S4] Staiger, L., Research in the theory of ω-languages. *J. Inform. Process. Cybern. EIK* **23** (1987) 8/9, 415 - 439.

[S5] Staiger, L., Ein Satz über die Entropie von Untermonoiden. *Theoret. Comput. Sci.* **61** (1988) (2,3), 279 - 282.

[S6] Staiger, L., Quadtrees and the Hausdorff dimension of pictures. *In:* Proc. GEOBILD '89 (A. Hübler, W. Nagel, B.D. Ripley and G. Werner, Eds), Mathematical Research **51**, Akademie-Verlag, Berlin 1989,

[S7] Staiger, L., Combinatorial properties of the Hausdorff dimension. *J. Statist. Plann. Inference* **22** (1989) , to appear.

[Th] Thomas, W., Automata on infinite objects. Aachener Informatik-Berichte 88-17.

[vL] van Lambalgen, M., Random sequences. Ph.D. Thesis, Univ. of Amsterdam, 1987.

[ZL] Zvonkin, A.K. and Levin, L.A., Complexity of finite objects and the development of the concepts of information and randomness by means of the theory of algorithms. *Russian Math. Surveys* **25** (1970) , 83 - 124.

TREE LANGUAGE PROBLEMS IN PATTERN RECOGNITION THEORY

(Extended abstract)

Magnus Steinby
Department of Mathematics
University of Turku
SF-20500 Turku, Finland

In syntactic pattern recognition patterns are represented as strings, trees, graphs and other constructs familiar from formal linguistics. Here we shall consider only tree representations although some of the issues may be equally relevant in other cases, too. The use of trees in this context can be traced back to the works of FU and BHARGAVA [13] from the early seventies (though SHAW's [28] picture description language published in 1969 could be counted to the genre). Since then there has been a moderate flow of papers suggesting various applications of tree languages in pattern recognition or addressing theoretical questions of the area. The books [18], [25] and [12] are some of the general sources which contain also extensive bibliographies. In the text a few representative more specific references will be mentioned.

In spite of all activity in the field, theoretical progress appears quite modest. Some of the possible reasons for this are linked together forming something of a vicious circle. Firstly, the involvement of the formal language theory community in the work has been almost nonexistent. This is probably largely due to the rather low level of formalization and the conceptual disarray of the area which makes it hard to pinpoint clearly defined theoretical problems to be solved. In fact, most of the papers in the field deal with specific applications using some *ad hoc* formalisms. Frequently the authors' knowledge of tree automaton theory is limited to a superficial familiarity with some of the early basic papers. Therefore, much of the purported theoretical work just rediscovers known facts, expresses in an almost unintelligible manner concepts and results that could be stated and derived more elegantly using existing theory, or, at worst, is simply incorrect. Naturally, this criticism is not intended to apply to everything done in the field.

In this lecture I shall discuss some problems about tree automata, tree languages and tree transformations motivated by pattern recognition theory. The approach taken is purely theoretical and no concrete applications are mentioned, but this could be what is needed at this stage of the development.

Also, considered on this level of generality some of these topics may be of interest even outside the realm of sensory pattern recognition.

The formal framework.

A survey of the examples considered in the literature shows that trees have been used in many different ways to represent patterns. Although it is generally agreed that trees are best suited for representing patterns with hierarchical structures, many of the concrete examples reflect this only in a very vague way. In particular, information of semantic nature is often embedded in structural descriptions in a rather confusing way. Obviously, a general mathematical treatment of the subject is impossible without a precisely defined formal framework in which the basic concepts have clearly stated roles and interpretations.

Assuming that a tree representation is always based on a hierarchical structural decomposition of the pattern, it can be supposed that

(1) each leaf of a tree stands for a pattern primitive named by the label of the leaf, and

(2) each inner node with m immediate descendants is labelled with a symbol representing an m-ary operation which forms a subpattern from the m subpatterns represented by the m maximal subtrees below the node.

Hence patterns are represented by trees in which nodes are labelled by symbols from a ranked alphabet Σ of symbols of pattern primitives and pattern construction symbols. It does not seem to be any real limitation to assume that each symbol has a unique rank.

There are two widely used formal definitions of trees: either they are defined as terms in the sense of universal algebra or as mappings from tree domains to the symbol alphabet. The former definition appears more natural when algebraic tools are used, but tree domains may be more convenient for the implementation of some algorithms. However, the two representations are easily translatable into each other, and the choice between them is largely a matter of taste. We denote the set of all Σ-trees by T_Σ.

Pattern classes are now represented by subsets of T_Σ which are called *Σ-tree languages* or *Σ-forests*. Usually the pattern classes considered are 'regular' enough to be represented by regular forests, *ie* forests which can be recognized by finite tree automata. However, there are natural cases where the greater expressive power of *context-free forests* [26] is needed. For the general theory of tree languages [11,15,33] can be consulted.

Some problems and research topics.

Let us begin with a few classical finite automaton theory type of questions. It has been claimed that the study of the theory of finite tree automata and regular tree languages is just a matter of straightforward generalization, but this is true only partially. Moreover, if systems using regular tree languages are really to be used in pattern recognition, or elsewhere, certain basic tools to deal with them have to be developed. Although proofs of most fundamental results, such as Kleene's theorem, are constructive, they seldom yield any practical algorithms.

One such problem is *the synthesis of pattern classifiers*. Suppose some regular Σ-forests, each representing a pattern class, are given by means of regular tree expressions. The task is to construct a Σ-recognizer which decides to which of the forests a given Σ-tree belongs. It does not seem justified to assume that the forests are necessarily pairwise disjoint. Moreover, this assumption (often made in pattern recognition theory) probably does not simplify the problem at all.

The elegant classical methods by McNaughton and Yamada [24], Gluschkow [16], and Brzozowski [10], and perhaps the combination of these recently presented by Berry and Sethi [7], are natural starting points, but the tree structure will require substantial modifications. These methods yield deterministic recognizers, but one may also ask for a nondeterministic solution with a better upper bound for the number of states in terms of the sizes of the regular expressions. Closely related to this form of the problem is that of *constructing a regular tree grammar* from a regular tree expression.

These problems are relevant in the above forms only if the expressions representing the pattern classes can be formed either directly or through some inference procedure. In this context one may also ask whether there are specification languages better suited for describing regular pattern classes than the language of regular tree expressions. However, this question can be answered only by considering several concrete situations.

The problems of *minimization and reduction of tree recognizers and regular tree grammars* have received some attention in the pattern recognition literature [4,5,21]. The classical minimization method of Aufenkamp and Hohn [3] was generalized to tree automata already in 1968 by Brainerd [8]. However, it seems that the computationally more efficient methods known for ordinary automata, such as Hopcroft's $O(n\log n)$-algorithm [20], have not yet been extended to tree recognizers. Levine [22] proposes a minimization method for tree recognizers based on the idea of derivatives of the forest recognized. Here the foundations of the work could have been derived and consolidated by the use of some tree automaton theory, but even in a formally corrected form the algorithm is hopelessly impractical in any typical case because of the superexponential growth of the number trees to be considered. However, derivatives in the form of unary algebraic functions as they appear in the general theory of minimal recognizers [29,23] may prove useful.

The method presented in [5] merges equivalent nonterminals in a regular tree grammar in normal form (or 'expansive tree grammar'), but this is not quite the same matter as minimizing the grammar. Obviously, this problem is closely related to the problem of minimizing nondeterministic tree recognizers which does not seem to have been considered at all, although for ordinary automata there is an extensive literature on the subject.

Derivatives of regular tree expressions is a topic that should be studied thoroughly. Many related ideas have been used in syntactic pattern recognition, and in particular in connection with methods of inferring regular tree languages from samples [9,21,22,14]. However, the notions suggested actually amount to various types of quotient sets. What would be needed is a theory of formal derivatives of regular tree expressions on a par with Brzozowski's [10] work. Such a theory could also be useful for the syntesis and minimization problems mentioned above. Here the nonlinearity of trees is likely to complicate the generalization from the string case considerably.

The inference of recognizers and grammars from samples is a problem of great practical significance for the usefulness of syntactic pattern recognition, and consequently it has been studied quite extensively also in the case of trees [9,17,14,21,22]. Also here a more systematic and formal approach would be helpful. First of all, it would perhaps make it possible to compare the different methods suggested so far on some more objective grounds, and to understand what the iterative regularities discovered by these methods stand for in the patterns. On the other hand, a commitment to the view of trees as hierarchical pattern representations allows one to systematically look for methods searching for given types of regularities in the samples with a clear understanding of what these mean.

In an attempt [31,32] to create a formal basis for the treatment of *errors in tree representations of patterns* I distinguished two main types of such errors, *local errors* and *structural errors*. The former are assumed to arise from a distortion, a misreading or a misinterpretation of a locally restricted part of the pattern. Structural errors, on the other hand, may affect subpatterns of unlimited size through a structural misinterpretation of a part of the pattern. It turns out that in both cases the possible effects of such errors on the tree representations of patterns can be described by means of a *Σ-algebraic tree transformation* [30]. The main problems considered in [32] were (1) the construction of error-tolerating recognizers and (2) the construction of error-correcting tree transducers. Both problems can be posed in (at least) two natural forms: either an upper limit on the number of errors possible in any given tree is placed in advance or then the possiblity of an unrestricted number of errors is taken into account. If the pattern class is represented by a regular forest, then the constructions can be carried out in all cases. Naturally, one may still look for more efficient algorithms. The *minimum distance correction problem* should also be considered. When can the minimum distance correction be carried out by an frontier-to-root tree transducer? And how should these devices be enhanced for this purpose in the general case?

Barrero, Thomason and Gonzalez [6] considered the effects of simple deletion, insertion and symbol-exchange errors on regular tree expressions. This study could be formalized and extended to cover more general types of errors. Again the cases of limited and unlimited numbers of errors should be distinguished.

In the case of a context-free pattern forest, the problems related to errors are more complicated. First of all, one should find out under which kinds of error transformations the family of context-free forests is closed. It also appears that neither the root-to-frontier tree pushdown automata of Guessarian [19] nor the frontier-to-root version of Schimpf and Gallier [27] are very well suited for the construction of error tolerating recognizers. In fact, it may be that one should look for suitable subfamilies of the context-free forests which are closed under the error transformations and which also allow some convenient type of recognizers, but still substantially increase the expressive power of the regular forests. Aoki and Matsuura [1,2] have studied the least-error problem for insertion, deletion and exchange errors in the case of a certain type of context-free grammars. The exact relationship of these grammars to the usual context-free tree grammars should be clarified and the work could be extended to more general types of errors.

As the last problem area I will consider the question of *representational ambiguity*. In syntactic pattern recognition theory the possibility of multiple representations is usually ignored entirely or then it is confused with grammatical ambiguity. However, it seems quite natural that in a syntactical description system many patterns have more than one fully legitimate representation. In the case of tree representations a possible way to handle such a situation would be to create a list of pairs of terms with variables which describe the basic relationships between the pattern constructions. From this a tree transformation mapping every tree to other trees representing the same pattern could be obtained. It seems that in many cases these transformations would from the mathematical point of view be of the same type as the structural error transformations mentioned earlier, and one could then utilize the error theory. For example, a recognizer for a given regular pattern class could be constructed from an incomplete, but sufficiently representative, regular subclass the same way as one constructs an error tolerating recognizer. A related problem is how to utilize an *a priori* knowledge of the representational ambiguities in the inference process.

References

[1] AOKI, K. and MAZUURA, K.: Syntax-directed least-errors analysis for context-free tree languages for syntactic pattern recognition. - *Systems. Computers. Controls 4, no. 2* (1983), 57-65.

[2] AOKI, K. and MAZUURA, K.: A least-error recognizer for context-free tree languages on the Tai metric. - *ibid 4, no. 6* (1983), 10-18.

[3] AUFENKAMP, D.D. and HOHN, F.E.: Analysis of sequential machines. - *IRE Trans. Electr. Comput. 6* (1957), 276-285.

[4] BARRERO, A. and GONZALEZ, R.C.: Minimization of deterministic tree grammars and automata. - *Proc. IEEE Conf. Decision and Control and the 15th Symp. Adaptive Processes* (Clearwater, Fla., 1976), Inst. Electr. Electron. Engrs., New York (1976), 404-407.

[5] BARRERO, A., GONZALEZ, R.C. and THOMASON, M.G.: Equivalence and reduction of expansive tree grammars. - *IEEE Trans. Pattern Anal. & Mach. Intell. PAMI-3* (1981), 204-206.

[6] BARRERO, A., THOMASON, M.G. and GONZALEZ, R.C.: Regular-like tree expressions. - *Intern. J. Comput. Information Sci. 12* (1983).

[7] BERRY, G. and SETHI, R.: From regular expressions to deterministic automata. - *INRIA Technical Report No. 649*, Rocquencourt, France, 1987.

[8] BRAINERD, W.S.: The minimalization of tree automata. - *Inform. Control 13* (1968), 484-491.

[9] BRAYER, J.M. and FU, K.-S.: A note on the *k*-tail method of tree grammar inference. - *IEEE Trans. Systems Man Cybernetics SMC-7* (1977), 293-300.

[10] BRZOZOWSKI, J.A.: Derivatives of regular expressions. - *J. Assoc. Comput. Mach. 11* (1964), 481-494.

[11] ENGELFRIET, J.: *Tree automata and tree grammars.* - DAIMI FN-10, Inst. Math., Aarhus Univ., Aarhus, 1975.

[12] FU, K.-S.: *Syntactic pattern recognition and applications.* - Prentice-Hall, Englewood Cliffs, N.J., 1982.

[13] FU, K.-S. and BHARGAVA, B.K.: Tree systems for syntactic pattern recognition. - *IEEE Trans. Computers C-22* (1973), 1087-1099.

[14] FUKUDA, H. and KAMATA, K.: Inference of tree automata from sample set of trees. - *Intern. J. Comput. Information Sci. 13* (1984).

[15] GÉCSEG, F. and STEINBY, M.: *Tree automata.* - Akadémiai Kiadó, Budapest, 1984.

[16] GLUSHKOW, V.M.: *Theorie der abstrakten Automaten.* - VEB Deutscher Verlag der Wissenschaften, Berlin, 1963.

[17] GONZALEZ, R.C., EDWARDS, J.J. and THOMASON, M.G.: An algorithm for the inference of tree grammars. - *Intern. J. Comput. Information Sci. 5* (1976), 145-164.

[18] GONZALEZ, R.C. and THOMASON, *Syntactic pattern recognition.* - Addison-Wesley, Reading, Mass., 1978.

[19] GUESSARIAN, I.: Pushdown tree automata. - *Math. Systems Theory 16* (1983), 237-263.

[20] HOPCROFT, J.: An *nlogn* algorithm for minimizing states in a finite automaton. -*Theory of Machines and Computations* (eds. Z. Kohavi and A. Paz), Academic Press, New York and London (1971), 189-196.

[21] LEVINE, B.: Derivatives of tree sets with applications to grammatical inference. - *IEEE Trans. Pattern Anal. Mach. Intell. PAMI-3* (1981), 285-293.

[22] LEVINE, B.: The use of tree derivatives and a sample support parameter for inferring tree systems. - *IEEE Trans. Pattern Anal. Mach. Intell. PAMI-4* (1982), 25-34.

[23] MARCHAND, P.: Construction des algèbres minimales des sous-ensembles des algebres libres. Applications aux reconnaissables. -*Les Arbres en Algébre et en Programmation,* 4éme *Coll. Lille*, Lille (1979), 134-158.

[24] McNAUGHTON, R. and YAMADA, H.: Regular expressions and state graphs for automata. - *IRE Trans. Electron. Computers EC-9* (1960), 39-47.

[25] MICLET, L.: *Structural methods in pattern recognition.* - North-Oxford, London, 1986.

[26] ROUNDS, W.C.: Mappings and grammars on trees. - *Math. Systems Theory 4* (1970), 257-287.

[27] SCHIMPF, K.M. and GALLIER, J.H.: Tree pushdown automata. - *J. Comput. Systems. Sci. 30* (1985), 25-40.

[28] SHAW, A.C.: The formal picture description scheme as a basis for picture processing systems. - *Information Control 14* (1969), 9-52.

[29] STEINBY, M.: Syntactic algebras and varieties of recognizable sets. - *Les Arbres en Algébre et en Programmation,* 4éme *Coll. Lille*, Lille (1979), 226-240.

[30] STEINBY, M.: On certain algebraically defined tree transformations. - *Algebra, Combinatorics and Logic in Computer Science* (Proc. Conf. Györ 1983), North-Holland, Amsterdam 1986, 745-764.

[31] STEINBY, M.: Towards a formal theory of errors in pattern trees. - 2nd *Conf. Automata, Languages and Programming Systems* (Salgótarján, 1988), Department of Mathematics, K. Marx University of Economics, Budapest.

[32] STEINBY, M.: A formal theory of errors in tree representations of patterns, 4th *Workshop on Mathematical Aspects of Computer Science* (Magdeburg, 1988), to appear.

[33] THATCHER, J.: Tree automata: an informal survey. - *Currents in the Theory of Computing* (ed. A. Aho), Prentice-Hall, Englewood-Cliffs, N.J., (1973), 143-172.

The Computational Complexity of Cellular Automata

Klaus Sutner
Stevens Institute of Technology
Hoboken, NJ 07030

Abstract

We study the computational complexity of the evolution of configurations on finite and infinite cellular automata. We will show that a classification of cellular automata suggested by Culik and Yu and based on Wolfram's earlier heuristic classification leads to classes that are Π^0_2-complete and is Σ^0_3-complete.

1 Introduction

The evolution of configurations on infinite as well as finite cellular automata has been studied extensively, see for example [13], [8], [5], [7] and [16]. In this paper we will address a number of combinatorial problems associated with the orbits of configurations on a cellular automaton. The perhaps most natural question is the following:

> **Configuration Reachability Problem** (CREP)
> Given a cellular automaton, a source configuration X and a target configuration Y, does Y occur in the orbit of X?

This problem is well known to be undecidable on infinite one-dimensional cellular automata, see e.g. [1], [3]. The proof hinges on the fact that one-dimensional cellular automata are capable of simulating Turing machines: instantaneous descriptions of a Turing machine are readily expressed as configurations of a one-dimensional cellular automaton. The transition rule of the cellular automaton simulates the transition function of the Turing machine. Since the Halting Problem for Turing machines is undecidable it is also undecidable whether a certain target configuration occurs in the orbit of a source configuration. More precisely, CREP is Σ_1^0-complete for infinite cellular automata of any dimension.

Another question relating to the evolution of configurations that has attracted attention is whether a configuration has a predecessor (or is a Garden-of-Eden, see [9] and [10]).

> **Predecessor Existence Problem** (PEP)
> Given a cellular automaton and a target configuration Y, does Y have a predecessor?

Variations of these problems were also considered in [4], [5] and [16]. With regard to the existence of predecessor configurations a remarkable distinction between one- and higher-dimensional cellular automata occurs. In the one-dimensional case it is decidable whether a given configuration has a predecessor. This follows from the results of Wolfram in [13]: the class of all configurations that have a predecessor on a one-dimensional automaton is a regular language of biinfinite words. Based on this observation it is also possible to develop an algorithm that determines whether a one-dimensional rule is injective or surjective (see [6] and also [2]). On the other hand Yaku shows in [16] that PEP is in general undecidable for two-dimensional cellular automata. In fact, Yaku's argument shows that for any Turing machine $\mathfrak{M}$ there is a rule $\rho_{\mathfrak{M}}$ and a finite configuration Y such that Y has a predecessor under $\rho_{\mathfrak{M}}$ iff Turing machine $\mathfrak{M}$ does not halt on the empty tape. Similarly one can construct a rule $\rho'_{\mathfrak{M}}$ such that Y has a finite predecessor under $\rho_{\mathfrak{M}}$ iff $\mathfrak{M}$ halts on the empty tape. The restricted version of PEP where the predecessor configuration is required to be finite is thus Σ_1^0-complete. By Tychonoff's theorem the space of all configurations with the usual product topology is compact. It follows that the general problem is Π_1^0-complete for infinite two-dimensional cellular automata. (Note that the space of finite configurations fails to be compact; the argument therefore does not apply to the restricted version.)

The orbits of configurations can also be used in an attempt to classify cellular automata as dynamical systems. In [15] Wolfram gave a heuristic classification of one-dimensional cellular automata into four types. His classification is based on "easily observable" characteristics of the behavior of the cellular automaton. In a recent paper by Culik and Yu (see [7]) the authors propose the following formalization of Wolfram's system. A cellular automaton is in CLASS ONE iff all configurations evolve to a fixed point, in CLASS TWO iff all configurations evolve to a periodic configuration and in Class Three iff CREP is decidable for the particular automaton. The fourth class comprises all rules and is trivial. All configurations here are understood to be finite. Culik and Yu show that it is undecidable which class a given cellular automaton belongs to. We will show that the Culik-Yu classes are in fact Π_2^0-complete (Class One and Class Two) and Σ_3^0-complete (Class Three).

By way of contrast CREP as well as PEP become trivially decidable for *finite* cellular automata. However, the computational complexity of these problems varies greatly depending on dimension and rule of the cellular automaton. It is shown in [12] that PEP is complete in **NLOG** for finite one-dimensional cellular automata and **NP**-complete for all dimensions higher than one.

For CREP as stated above it is easy to see that the problem is **PSPACE**-complete even in dimension one. This suggests that in general there is most likely no way of solving CREP other than generating successively all the configurations in the orbit of the source configuration. Also, given $t \geq 0$ and the configuration X, one should not expect to be able to compute $\rho^t(X)$ in polynomial time. Transferring Class Three to finite cellular automata let us call a rule *predictable* iff one can determine in polynomial time whether a configuration occurs in the orbit of another, i.e., iff CREP is in **P** for that rule. Similarly a rule for a finite cellular automaton is *weakly predictable* iff $\rho^t(X)$ can be computed in polynomial time for all configurations X. Typical examples for weakly predictable rules are additive rules: computing $\rho^t(X)$ comes down to a matrix multiplication problem which can be solved in $O(n^3 \cdot \log t)$ steps using the standard algorithm. Weakly predictable rules formalize the notion of computationally reducible rules in [15] and [14]. Note that CREP is in **NP** for all weakly predictable finite cellular automata.

Since the existence of non-predictable or non-weakly-predictable rules is equivalent to the assertion $\mathbf{P} \neq \mathbf{PSPACE}$ it should be expected to be quite difficult to find a description of Class Three for finite cellular automata. In any case, one can show that CREP is **NP**-complete for predictable, two-dimensional rules. For one-dimensional predictable rules one can establish **NP**-completeness for a variation of CREP where the target configuration is only partially specified. For proofs see [12].

2 Culik-Yu Classes

We now show how to pinpoint position of the Culik-Yu classes within the arithmetical hierarchy. Counting quantifiers one easily verifies that CLASS ONE and CLASS TWO occur at level Π_2^0 of the arithmetical hierarchy and CLASS THREE lies at level Σ_3^0. To prove hardness it is convenient to first establish a technical lemma that helps to overcome the major obstacle in the simulation of a Turing machine on a cellular automaton. In recursion theory one is only interested in computations of Turing machines, i.e., sequences of IDs that start with an initial instantaneous description (ID for short) corresponding to some input. In the context of cellular automata, however, one has to contend with the orbits of arbitrary configurations. Due to the fact that as a set of words the collectin of all IDs of a Turing machine is regular, it is not difficult to eliminate configurations that do not correspond to any ID whatsoever. In fact, given the proper coding, a cellular automaton can detect non-IDs in one step. A more serious problem is caused by IDs that do not occur during any computation of the Turing machine. We call such IDs *inaccessible.*

We need to introduce a little more terminology. From now on we will consider only one-dimensional cellular automata. We denote Σ the set of states. A map $X : \mathbf{Z} \to \Sigma$ from the set of all cells to the alphabet is a *configuration* of the cellular automaton. C denotes the space of all configurations. A *local rule* is a map $\rho : \Sigma^N \to \Sigma$ where $N \subset \mathbf{Z}$ is a finite set, called the *basic neighborhood* of the rule. The rule ρ is extended to a *global rule* (also denoted by ρ) $rho : C \to C$ as follows. Given a configuration X define for any cell c the *local configuration* at c, $X_c : N \to \Sigma$, by $X_c(z) := X(c+z)$. Then $\rho(X)(c) := \rho(X_c)$.

As usual let $\mathfrak{M}_e$ and W_e, $e \geq 0$, be an enumeration of all Turing machines and r.e. sets, respectively. In analogy to TOT:= $\{ e \mid W_e = \mathbf{N} \}$ define ALL to be the set of Gödel numbers of Turing machines that halt on *all instantaneous descriptions.* Note that this is a much stronger condition than just requiring the machine to halt on all inputs. ALL is clearly Π_2^0. We will show in a moment that ALL is in fact Π_2^0-complete. Observe that ALL, unlike TOT, fails to be an index set.

For any number $x \geq 0$ we write I_x for the initial ID on machine $\mathfrak{M}_e$ corresponding to input x. $M_e(x) \downarrow$ denotes the fact that $\mathfrak{M}_e$ on I_x halts after finitely many steps:

$$\exists\sigma(I_x \vdash_e^\sigma q_H).$$

Here $I \vdash_e^\sigma J$ indicates that machine number e moves from ID I to J in σ steps. For an ID I of machine $\mathfrak{M}_e$ we write $\mathfrak{M}_e[I]\downarrow$ iff $\mathfrak{M}_e$ started on ID I halts after finitely many steps. Thus an ID I is inaccessible iff

$$\neg\exists x,\sigma(I_x \vdash_e^\sigma I).$$

The set of inaccessible IDs of machine $\mathfrak{M}_e$ is therefore Π_1^0. In fact, it is not hard to see that this set is in general Π_1^0-complete.

Lemma 2.1 ALL, *the set of Gödel numbers of Turing machines that halt on all configurations, is* Π_2^0*-complete.*

Proof. We will show that there is an injective primitive recursive function p such that for all $e \geq 0$:

$W_e = W_{p(e)}$
$\mathfrak{M}_{p(e)}$ halts on all its inaccessible IDs.

To this end we will show that for any Turing machine $\mathfrak{M}_e$ there exists a modified machine $\mathfrak{M}_{e'}$ such that

for any $x \geq 0$: $\mathfrak{M}_e(x)\downarrow \iff \mathfrak{M}_{e'}(x)\downarrow$, and
for any inaccessible ID I of $\mathfrak{M}_{e'}$: $\mathfrak{M}_{e'}[I]\downarrow$.

Moreover, it will be clear from the construction that the index e' can be computed primitive recursively from e. As pointed out above, the class of inaccessible IDs is in general Π_1^0-complete. Thus we cannot effectively eliminate these IDs. We consider instead IDs of $\mathfrak{M}_e$ with an additional tag: the tag contains two numbers x and σ such that the ID in question occurs after σ steps in the computation of $\mathfrak{M}_e$ on input x. A tagged ID (x, σ, I) is *correct* iff indeed $I_x \vdash_e^\sigma I$. Unlike accessibility correctness of tagged IDs is a primitive recursive property and can thus be verified by the Turing machine $\mathfrak{M}_{e'}$. If the tag is correct, machine $\mathfrak{M}_{e'}$ will generate the next ID of $\mathfrak{M}_e$. Also, the counter on the tag will be changed changed to $\sigma + 1$ to preserve correctness. Otherwise the machine $\mathfrak{M}_{e'}$ halts. The process then starts anew.

We now give a more detailed description of machine $\mathfrak{M}_{e'}$. Starting at an initial ID $I_x = q_0 1^x$, machine $\mathfrak{M}_{e'}$ first changes the tape inscription to

$$\#1^x\#\#q_0'1^x\#$$

where q'_0 is the initial state of $\mathfrak{M}_e$. This new configuration of $\mathfrak{M}_{e'}$ represents the tagged ID $(x, 0, I_x)$ of $\mathfrak{M}_e$. Machine $\mathfrak{M}_{e'}$ then cycles through the following three phases.

- Tape Verification
  During this phase $\mathfrak{M}_{e'}$ tests whether its tape contains a tagged ID of $\mathfrak{M}_e$, i.e., an inscription of the form
  $$\#1^x\#1^\sigma\#UqV\# \tag{1}$$
  where $x, \sigma \geq 0, U, V \in \Gamma^*$ and $q \in Q$. If the verification fails $\mathfrak{M}_{e'}$ halts.

- Accessibility Test
  In this phase $\mathfrak{M}_{e'}$ will check whether the tagged ID on its tape is correct. The portion of the tape used during the test are subsequently erased, so the tape inscription will be back to (1) after successful completion of the test. Failure will cause $\mathfrak{M}_{e'}$ to halt.

- Next ID
  We may now assume that the tape contains a correctly tagged ID (x, σ, I). The machine now determines whether I is the halting configuration of $\mathfrak{M}_e$. If so, $\mathfrak{M}_{e'}$ also halts. Otherwise $\mathfrak{M}_{e'}$ computes the next ID and replaces the old ID on its tape by the new one. Furthermore it increments the counter σ to $\sigma + 1$.

This completes the definition of $\mathfrak{M}_{e'}$.

Since on any input $x \geq 0$ machine $\mathfrak{M}_{e'}$ simply simulates machine $\mathfrak{M}_e$, albeit in a very circuitous fashion, we have that $\mathfrak{M}_{e'}$ halts on x iff $\mathfrak{M}_e$ halts on x. Hence $W_{e'} = W_e$ and it remains to show that $\mathfrak{M}_{e'}$ halts on all its inaccessible configurations. So suppose $\mathfrak{M}_{e'}$ is started on an arbitrary ID and performs an infinite computation. The crucial observation is that that $\mathfrak{M}_{e'}$ can perform only finitely many moves before it must enter a tape verification phase (recall that IDs are finitary objects). As $\mathfrak{M}_{e'}$ never halts the verification must be successful, hence the tape inscription must have the form $\#1^x\#1^\sigma\#I\#$ and represents a tagged ID (x, σ, I) of $\mathfrak{M}_e$. Next $\mathfrak{M}_{e'}$ tests accessibility of I (for machine $\mathfrak{M}_e$). Since no failure occurs we must have $I_x \vdash_e^\sigma I$. But then $\#1^x\#1^\sigma\#I\#$ is also accessible for $\mathfrak{M}_{e'}$:

$$I_x = q_0 1^x \vdash_{e'}^{\tau} \#1^x\#1^\sigma\#I\#$$

for some number $\tau \geq \sigma$. Hence machine $\mathfrak{M}_{e'}$ cannot perform an infinite computation on an inaccessible ID.

It now follows that $\mathfrak{M}_e$ halts on all inputs iff $p(e)$ is in ALL. Hence TOT is one-one reducible to ALL. But TOT is well-known to be Π_2^0-complete (see e.g. [11]) and we are done. □

Theorem 2.1 CLASS ONE *and* CLASS TWO *are* Π_2^0*-complete.* CLASS THREE *is* Σ_3^0*-complete.*

Proof. By the lemma we only have to show that ALL, the set of Gödel numbers of Turing machines that halt on all configurations, is reducible to CLASS ONE. To this end we will construct a rule ρ_e for every e such that e is in ALL iff ρ_e is in CLASS ONE. The construction of rule ρ_e is rather standard, we therefore will omit any details and only give a brief description. ρ_e first tests whether configuration X corresponds to an ID of $\mathfrak{M}_e$ (more precisely, whether every isolated non-quiescent part of X corresponds to an ID; there may several such parts). This is possible in one step if one augments the tape alphabet of $\mathfrak{M}_e$ by indicator bits that determine the position of the head relative to the symbol. If the test fails the quiescent configuration is generated in $O(n)$ steps where n is the number of cells in the support of X. Otherwise ρ_e simulates the computation of $\mathfrak{M}_e$ on this ID. If $\mathfrak{M}_e$ ever halts the quiescent configuration is generated, otherwise no stable configuration occurs. Again the map $e \mapsto e'$ is clearly primitive recursive and also injective. Hence ALL $\leq_1$ CLASS ONE and we are through.

A similar technique can be used to show hardness for CLASS TWO and CLASS THREE.

More precisely, we claim that

$$e \in \text{TOT iff } \rho_{p(e)} \text{ is in CLASS TWO}$$
$$e \in \text{REC iff } \rho_{p(e)} \text{ is in CLASS THREE}$$

Here REC := $\{ e \mid W_e \text{ is recursive } \}$ is the collection of indices of recursive r.e. sets. REC is well-known to be Σ_3^0-complete.

To verify our claim first suppose e is in TOT. Then $\mathfrak{M}_{p(e)}$ halts an all its IDs, whence every configuration evolves to the stable configuration X_0 under rule $\rho_{p(e)}$. On the other hand assume e is not in TOT and pick some x such that

$\mathfrak{M}_e$ does not converge on input x. Consider the orbit of configuration

$$X = \#x\#0\#I_x\#$$

under rule $\rho_{p(e)}$. It clearly contains configurations of the form $\#x\#\sigma\#I\#$ for all $\sigma \geq 0$. Hence the orbit of X fails to be periodic and rule $\rho_{p(e)}$ is not in CLASS THREE.

Lastly, for CLASS THREE suppose W_e is recursive and let X and Y be two arbitrary finite configurations of rule $\rho_{p(e)}$. Note that it is decidable whether X corresponds to an accessible ID of $\mathfrak{M}_{p(e)}$. If not, the orbit of X must be finite and we can test whether Y occurs in the orbit by enumerating it. If X is accessible, say, from initial ID I_x, we first test whether x is in W_e. This can be done effectively since W_e is recursive. If indeed x lies in W_e the orbit of X must again be finite and we can test whether Y occurs in it by enumerating it. Otherwise the orbit of X is infinite but the configurations that occur in it are essentially all of the form $\#1^x\#1^\tau\#I\#$. Hence it is easily decidable whether Y lies in the orbit of X: we have to execute at most τ cycles in the computation of $\mathfrak{M}_{p(e)}$.

For the opposite direction note that

$$x \in W_e \quad \text{iff} \quad \mathfrak{M}_e[I_x]\downarrow \quad \text{iff} \quad \mathfrak{M}_{p(e)}[I_x]\downarrow \quad \text{iff} \quad \exists t(\rho^t_{p(e)}(q_0 1^x) = X_0).$$

Thus x is in W_e iff the orbit of I_x contains configuration X_0. Hence W_e is decidable whenever $\rho_{p(e)}$ is in CLASS THREE. □

References

[1] J. Albert and K. Culik II. A simple universal cellular automaton and its one-way and totalistic version. *Complex Systems*, 1(1):1 – 16, 1987.

[2] S. Amoroso and Y. N. Patt. Decision procedures for surjectivity and injectivity of parallel maps for tesselation structures. *Journal of Computers and System Science*, 6:448 – 464, 1972.

[3] A. W. Burks. *Essays on Cellular Automata.* University of Illinois Press, 1970.

[4] U. Golze. Differences between 1- and 2-dimensional cell spaces. In A. Lindenmayer and G. Rozenberg, editors, *Automata, Languages and Development*, pages 369–384. North-Holland, 1976.

[5] F. Green. **NP**-complete problems in cellular automata. *Complex Systems*, 1(3):453–474, 1987.

[6] K. Culik II. On invertible cellular automata. *Complex Systems*, 1(6):1035 - 1044, 1987.

[7] K. Culik II and Sheng Yu. Undecidability of CA classification schemes. *Complex Systems*, 2(2):177–190, 1988.

[8] O. Martin, A. M. Odlyzko, and S. Wolfram. Algebraic properties of cellular automata. *Commun. Math. Phys.*, 93:219–258, 1984.

[9] E. F. Moore. Machine models of self-reproduction. In *Essays on Cellular Automata* [3].

[10] J. Myhill. The converse of moore's garden-of-eden theorem. In *Essays on Cellular Automata* [3].

[11] H. Rogers. *Theory of Recursive Functions and Effective Computability.* McGraw Hill, 1967.

[12] K. Sutner. The complexity of finite cellular automata. Submitted.

[13] S. Wolfram. Computation theory of cellular automata. *Comm. Math. Physics*, 96(1):15–57, 1984.

[14] S. Wolfram. Computer software in science and mathematics. *Scientific American*, 251(3):188–203, 1984.

[15] S. Wolfram. Universality and complexity in cellular automata. *Physica 10D*, pages 1–35, 1984.

[16] T. Yaku. The constructibility of a configuration in a cellular automaton. *Journal of Computers and System Science*, 7:481–496, 1973.

ON RESTRICTED BOOLEAN CIRCUITS

GYÖRGY TURÁN
Automata Theory Research Group of the Hungarian Academy of Sciences
Szeged

and

Department of Mathematics, Statistics, and Computer Science
University of Illinois at Chicago
Chicago

Abstract.

We consider some classes of restricted Boolean circuits: synchronous and locally synchronous circuits, planar circuits, formulas and multilective planar circuits. Bounds are given comparing the computational power of circuits from these classes.

1. Introduction.

There are several restricted classes of Boolean circuits (i.e. acyclic networks of gates computing Boolean functions) studied in complexity theory for which nonlinear (in some cases even exponential) lower bounds are known for explicity defined functions, such as monotone circuits, bounded depth circuits, formulas, synchronous circuits and planar circuits. In this paper we consider the latter two classes and some of their variants.

A circuit is *synchronous* if its nodes (inputs and gates) are divided into levels so that inputs are on level 0, and edges can only go from a level to the next one. The size of the smallest synchronous circuit with fanin ≤ 2, computing f is $C_s(f)$, the synchronous complexity of f. Synchronous circuits were introduced by Harper [10]. There are simple examples such as the sum of two n-bit numbers showing that for some multiple output functions (i.e. functions of the form $f_n : \{0,1\}^n \to \{0,1\}^m$, where $m > 1$, e.g. $n = m$) $C_s(f_n) = \Omega(n \log n)$ but $C(f_n) = O(n)$ (where $C(f_n)$ is the standard circuit complexity), see [28]. Harper and Savage [11] gave $\Omega(n \log n)$ lower bounds for the synchronous complexity of single output functions (i.e. functions of the form $f_n : \{0,1\}^n \to \{0,1\}$, also called predicates or decision problems), such as the Boolean determinant. The question whether synchronous circuits are less powerful than general ones for decision problems is mentioned in Wegener's book [28]. An example of a decision problem with $C_s(f_n) = \Omega(n \log n)$ and $C(f_n) = O(n)$ is given in [25].

Belaga [4], [5], [6], [7] introduced the more general class of *locally synchronous* circuits. Here the circuit is levelled but inputs may occur on any level (though each input on one level only). The locally synchronous complexity of f is denoted by $C_\ell(f)$. Using an interesting proof technique (looking at nonstandard operation modes of a circuit) Belaga proved an $\Omega(n \log n)$ lower bound for the locally synchronous complexity of a multiple output function and showed that for some multiple output function

$C_s(f_n) = \Omega(n \log n)$ and $C_\ell(f_n) = O(n)$ (such an example is the sum). He asked if locally synchronous circuits are less powerful than general ones [6].

In Section 2, it is shown that there is an explicitly defined decision problem with $C_s(f_n) = \Omega(n \log n)$ and $C_\ell(f_n) = O(n)$ and that there is an explicitly defined decision problem with $C_\ell(g_n) = \Omega(n \log n)$ and $C(g_n) = O(n)$ (thus generalizing [25] and answering the question of Belaga). The proofs use the Harper-Savage method and the existence of explicit linear size superconcentrators (Gabber-Galil [9]).

A circuit is ***planar*** if its graph is planar. This class was first studied by Lipton and Tarjan [17]. Savage [22] studied the relationship between planar circuits and VLSI. In particular, he showed that the planar complexity C_p is a lower bound to both measures AT^2 and A^2T. Lipton-Tarjan [17] and Savage [22] gave quadratic lower bounds to the planar complexity of several functions including decision problems such as the predicate corresponding to matrix multiplication. Savage [22] asked if the gap between C_p and C can be as large as quadratic (it is easy to see that it cannot be larger [22]). This question was answered in [25] by giving such an example.

Savage [23] considered an interesting generalization of planar circuits. A circuit is *multilective planar* if its graph is planar but each input may occur several times (thus e.g. formulas are multilective planar circuits). He showed that this model corresponds to VLSI circuits with repeated inputs (lower bounds for this model were given by Kedem-Zorat [14], [15], Kedem [13] and Hochschild [12]. Savage proved $\Omega(n^{4/3})$ lower bounds for the multilective planar complexity C_p^* of multiple output functions such as the shifting function. He also proved an $\Omega(n^{14/13})$ lower bound for a decision problem but the proof of this result is incomplete; it can be completed to yield an $\Omega(n \log n / \log\log n)$ lower bound.

In Sections 3 and 4 we consider planar and multilective planar circuits. In Section 3 an explicitly defined decision problem is given with $C_p(f_n) = \Omega(n^2)$ and formula size $L(f_n) = O(n)$. This generalizes [25] and gives a simpler construction. It is noted that the indirect storage addressing function of Paul with formula size $\Omega(n^2/\log n)$ [21] has planar complexity $O(n)$ and can be computed by a planar circuit of size $O(n \log n)$ even if the inputs are assumed to be located on the outer face. Thus planar circuit complexity and formula size are incomparable complexity measures and therefore multilective planar circuits are a proper generalization of both classes. In Section 4 some lower bounds are given for multilective planar complexity: an $\Omega(n \log n)$ lower bound for a decision problem and $\Omega(n^{3/2})$ lower bounds for some multiple output functions, including the cyclic shifting function. The proof of the $\Omega(n \log n)$ bound uses a lemma of Babai, Pudlák, Rödl and Szemerédi [3] (which is a generalization of a result of Alon and Maass [1]) used in [3] to prove lower bounds for formulas and branching programs.

Some remarks and open problems are mentioned in Section 5.

Due to space limitations we give complete proofs in Section 2 only. A more detailed version of Sections 3 and 4 will be published separately.

For basic definitions and notations we refer to Wegener [28]. We consider circuits over the basis of all ≤ 2-variable functions (with the exception of Theorem 3 all results hold for arbitrary finite complete bases as well). Unless specified otherwise (in Theorems 6 and 7), by a Boolean function f_n we mean a decision problem with variables $x_1, \ldots, x_n$.

A sequence of functions $(f_1, f_2, \ldots)$ is explicity defined in $\bigcup_n f_n^{-1}(1) \in NP$; for every example in this paper $\bigcup_n f_n^{-1}(1) \in P$ holds as well.

2. Synchronous and locally synchronous circuits.

In this section we separate the computational power of synchronous, locally synchronous and general circuits computing decision problems.

Theorem 1. There are explicitly defined functions f_n with

$$C_s(f_n) = \Omega(n \log n) \text{ and } C_\ell(f_n) = O(n).$$

Theorem 2. There are explicitly defined functions g_n with

$$C_\ell(g_n) = \Omega(n \log n) \text{ and } C(g_n) = O(n).$$

First we describe the proof method of Harper and Savage [11].

For a function f with variables $X = \{x_1, \ldots, x_n\}$, $A \subseteq X$ and a truth assignment $\alpha : X \backslash A \to \{0,1\}$, $f|_\alpha$ denotes the subfunction of f on A obtained by fixing variables $x \in X \backslash A$ to $\alpha(x)$. $N(f, A)$ is the number of different subfunctions of f on A and

$$N'(f,a) = \binom{n}{a}^{-1} \sum_{\substack{A \subseteq X \\ |A| = a}} \log N(f, A)$$

is the average of $\log N(f, A)$ taken over the a-element subsets of X.

Lemma 1. (Harper-Savage [11]). If $0 < \delta < 1$, $1 \le a \le n$ and $d = \lfloor \log(\delta(n - a + 1)/(2a + \delta)) \rfloor \le D(f)$, ($D(f)$ is the depth of f), then $C_s(f) \ge (1 - \delta)\ d\ N'(f, a)$. □

Proof of Theorem 1. The function f_n is the universal function of Andreev [2] (an alternative construction is sketched after the proof). Assume n is of the form $2k \cdot 2^k + 2^{2k}$. (For other values of n one can introduce dummy variables. This also holds for all the subsequent constructions except Theorem 4 and will not be mentioned again.) The variables of f_n are $\underline{x}_i = (x_{i,1}, \ldots, x_{i,2^k})$ $(0 \le i \le 2k - 1)$ and $\underline{y} = (y_0, \ldots, y_{2^{2k}-1})$. Put $z_i = x_{i,1} \oplus \ldots \oplus x_{i,2^k}$ $(0 \le i \le 2k - 1)$, $\underline{z} = (z_0, \ldots, z_{2k-1})$ and

$$f_n(\underline{x}_0, \ldots, \underline{x}_{2k-1}, \underline{y}) := SA_{2^{2k}}(\underline{z}, \underline{y})$$

where $SA_{2^{2k}}$ is the "storage access" (or "addressing" function [28]), i.e. $SA_{2^{2k}}(x, y) = y_{\text{bin}\underline{z}}$, $\text{bin}(\underline{z}) = \sum_{i=0}^{2k-1} z_i 2^i$.

To prove the upper bound note that the standard linear size circuit computing SA [28, p. 77] can be turned into a linear size synchronous circuit by repeating the addressing variables on each level (using "delay elements", i.e. gates computing the identity function). If we replace the addressing variables z_i in this circuit by binary trees computing z_i from $\underline{x}_i$, we get a linear size locally synchronous circuit computing f_n.

To prove the lower bound, let $a := \lfloor n^{2/3} \rfloor$.

Lemma 2. $N'(f_n, a) = \Omega(n)$.

Proof. Let us choose a random set A of a variables (with uniform distribution).
We claim

(1) $$Pr(\forall i(0 \le j \le 2k-1)\exists j(1 \le j \le 2^k) : x_{i,j} \in A) = 1 - o(1).$$

For a given i $(0 \le i \le 2k-1)$

$$Pr(\forall j(1 \le j \le 2^k) : x_{i,j} \notin A) = \frac{\binom{n-2^k}{a}}{\binom{n}{a}} < e^{-n^{1/7}}$$

using standard calculations, so the probability in (1) is $\ge 1 - \log n \cdot e^{-n^{1/7}}$, proving the claim.

Now if A satisfies the condition of (1) then

(2) $$\log N(f_n, A) \ge 2^{2k} - n^{2/3} = (1 - o(1)) \cdot n.$$

To show this, let α and α' be two truth assignments which assign 0 to every $x_{i,j} \notin A$ and which differ on some $y_\ell \notin A$. Then there is a truth assignment $\beta : A \to \{0,1\}$ such that for both inputs $\alpha \cup \beta$ and $\alpha' \cup \beta$ the address is ℓ, therefore $f_n|_\alpha \ne f_n|_{\alpha'}$. The number of these α's is $\ge 2^{(2^{2k}-|A|)}$ thus (2) follows. Lemma 2 follows from (1) and (2).
□

To finish the proof of Theorem 1 let $\delta := 1/2$. Then $d = \lfloor \log((1/2)(n - \lfloor n^{2/3} \rfloor + 1)/(2\lfloor n^{2/3} \rfloor + 1/2)) \rfloor \le \log n \le D(f_n)$ and $d = \Omega(\log n)$, i.e. the lower bound follows from Lemmas 1 and 2. □

Another example proving Theorem 1 can be obtained as follows. Let $B = \{b_1, \ldots, b_m\}$, where $0 < b_i < n$ $(i = 1, \ldots, m)$ and put

$$f_B(x_1, \ldots, x_m, y_0, \ldots, y_{n-1}) := y_b,$$

where

$$b = \sum_{i=1}^{m} x_i b_i \pmod{n}.$$

For a fixed k let $r := \lfloor \sqrt{n/k} \rfloor$ and $B := \{i : 1 \le i \le \lfloor \sqrt{kn} \rfloor\} \cup \{ir : 1 \le i \le \lfloor \sqrt{kn} \rfloor + 1\}$. If e.g. $k = \lfloor n^{2/3} \rfloor$, the function f_B can also be used for a proof.

Proof of Theorem 2. We give a general construction which can be used to prove lower bounds to locally synchronous complexity using lower bounds to synchronous complexity.

A directed acyclic graph with k input nodes and k output nodes is a k-superconcentrator if for every $\ell \le k$ and for every set of ℓ inputs and ℓ outputs there are ℓ vertex-disjoint paths connecting these vertices.

Lemma 3. (Gabber-Galil [9]). For every k there are explicitly defined k-superconcentrators with $O(k)$ edges.
□

Definition 1. (The circuit C_n.) Let G_n^1 be a $4n$-superconcentrator with $O(n)$ edges. Select n outputs of G_n^1 and delete all nodes from which there is no path leading to any of these nodes. Let the graph thus obtained be G_n^2 (G_n^2 has $4n$ inputs and n outputs). Form G_n^3 by taking two disjoint copies of G_n^2 and identifying the inputs resp. the outputs of these graphs, In G_n^3 replace nodes having indegree $d > 2$ with binary trees having d leaves and contract edges leading into nodes of indegree 1. Finally we get a graph G_n^4 with $4n$ inputs, n outputs such that every node in G_n^4 has indegree 0 or 2. Replace every node of indegree 2 with a small circuit computing the selection function $(u \wedge w_1) \vee (\bar{u} \wedge w_2)$, where w_1 and w_2 correspond to the two incoming edges and u is a new input. C_n is the circuit obtained this way. The inputs of C_n are $Y = \{y_1, \ldots, y_{4n}\}$ (corresponding to the original inputs of G_n^1) and $U = \{u_1, \ldots, u_m\}$, $m = \theta(n)$ (the programming inputs of the selection circuits). The outputs of C_n are $Z = \{z_1, \ldots, z_n\}$ (corresponding to the original outputs of G_n^2).

Definition 2. (The operator *.) Let $f(x_1, \ldots, x_n)$ be a function and C' be a circuit computing f. Form circuit C^* by identifying the outputs of C_n with the inputs of C'. Then $f^*(y_1, \ldots, y_{4n}, u_1, \ldots, u_m)$ is the function computed by C^*.

Note that by definition f^* has $\theta(n)$ variables and $C(f^*) \leq C(f) + O(n)$.

Lemma 4. For every f, $C_\ell(f^*) \geq \dfrac{\max\limits_{A \subseteq X} \log N(f,A)}{2n} \cdot C_s(f) - n$.

Proof. Let C be a locally synchronous circuit computing f^*. We distinguish two cases.

a) There are $\lceil C_s(f)/2n \rceil$ consecutive levels in C containing altogether $\geq n$ inputs from Y.

Let these levels be $L_a, \ldots, L_{a+t}$, $0 \leq t < \lceil C_s(f)/2n \rceil$ and let $Y' = \{y_{i_1}, \ldots, y_{i_n}\}$ be n variables occurring on these levels. By the concentration property of G_n^2 there is a truth assignment $\alpha : U \to \{0,1\}$ of the programming variables such that $C^*|_\alpha$ computes $f(y_{j_1}, \ldots, y_{j_m})$ for some permutation $(j_1, \ldots, j_n)$ of $(i_1, \ldots, i_n)$. Let α' be an arbitrary extension of α to variables in $U \cup (Y \backslash Y')$. Then $C^*|_{\alpha'}$ computes $f(y_{j_1}, \ldots, y_{j_m})$ and thus the same holds for $C|_{\alpha'}$. The inputs of $C|_{\alpha'}$ are the variables in Y'. (A technical detail is that gates becoming "superfluous" after the substitutions are left in the circuit thus local synchronicity is preserved.)

Now modify $C|_{\alpha'}$ by placing every y_i to level L_a and repeating it using delay elements on all levels $L_{a+1}, \ldots, L_{a+t}$. In this way adding $\leq tn \leq n((C_s(f)/(2n)) + 1) = (C_s(f)/2) + n$ gates we get a synchronous circuit computing f, i.e.

$$|C| \geq |C|_{\alpha'}| \geq C_s(f) - \left(\frac{C_s(f)}{2} + n\right) = \frac{C_s(f)}{2} - n$$

and the lemma follows as $\max\limits_{A \subseteq X} \log N(f,A) \leq n$.

b) Case a) does not hold.

Assume levels $L_0, \ldots, L_{a_0}$ altogether contain $\geq n$ variables from Y and a_0 is the smallest value having this property. As a) does not hold, each level contains $< n$ variables from Y and thus levels $L_0, \ldots, L_{a_0}$ altogether contain $< 2n$ variables from Y. Also, levels $L_{a_0+1}, \ldots, L_{a_0+\lceil C_s(f)/2n \rceil}$ contain $< n$ variables from Y. Thus levels L_j $(j > a_0 + \lceil C_s(f)/2n \rceil)$ altogether contain $> n$ variables from Y.

Let $A := \{x_{i_1}, \ldots, x_{i_r}\} \subseteq X$ be a subset for which $N(f, A)$ is maximal. From the concentration property of the two copies of G_n^2 there is a truth assignment $\alpha : U \to \{0, 1\}$ such that $C^*|_\alpha$ computes $f(y_{j_1}, \ldots, y_{j_n})$ where $j_1, \ldots, j_n$ are different values and

$$x_\ell \in A \Rightarrow y_{j_\ell} \text{ is on a level } > a_0 + \lceil C_s(f)/2n \rceil,$$
$$x_\ell \notin A \Rightarrow y_{j_\ell} \text{ is on a level } < a_0 + 1$$

Let $Y' := \{y_{j_1}, \ldots, y_{j_n}\}$ and let α' be an arbitrary extension of α to variables in $U \cup (Y \setminus Y')$. Then both $C^*|_{\alpha'}$ and $C|_{\alpha'}$ compute $f(y_{j_1}, \ldots, y_{j_n})$.

We claim that each level $L_{a_0+1}, \ldots, L_{a_0+\lceil C_s(f)/2n \rceil}$ contains $\geq \log N(f, A)$ gates in C. If for truth assignments β and β' to variables in Y' on levels $\leq a_0$ the gates of level L_j compute the same values then $f(y_{j_1}, \ldots, y_{j_n})|_\beta = f(y_{j_1}, \ldots, y_{j_n})|_{\beta'}$. Thus if a level L_j ($a_0 + 1 \leq j \leq a_0 + \lceil C_s(f)/2n \rceil$) contains b gates then $N(f, A) \leq 2^b$. Hence the claim follows.

Therefore

$$|C| \geq \left\lceil \frac{C_s(f)}{2n} \right\rceil \max_{A \subseteq X} \log N(f, A)$$

completing the proof of Lemma 4. □

Now let f_m be the function used in Theorem 1 and consider $g_n := f_m^*$, where $n = \theta(m)$ is the number of variables in f_m^*. Then by the remark following Definition 2, $C(g_n) = O(n)$. On the other hand, from Lemmas 2 and 4

$$C_\ell(g_n) = C_\ell(f_m^*) \geq \frac{\Omega(m)}{2n} \cdot C_s(f_m) - n = \Omega(n \log n)$$

which proves Theorem 2. □

3. Planar circuits and formulas.

When defining planar complexity we do not assume that input nodes occur on the outer face (this assumption is used e.g. in [20], but not used in [22]). In Theorem 4, where an upper bound is needed for C_p we mention the slightly weaker upper bound obtained by using this stronger assumption.

For every function f $C(f) \leq C_p^*(f) \leq C_p(f) \leq 6C(f)^2$ (see [28]) and $C_p^*(f) \leq L(f)$ by definition.

Theorem 3. There are explicitly defined functions f_n with

$$C_p(f_n) = \Omega(n^2) \text{ and } L(f_n) = O(n).$$

Theorem 4. There are explicitly defined functions g_n with

$$C_p(g_n) = O(n) \text{ and } L(g_n) = \Omega(n^2/\log n).$$

Also, there is a planar circuit of size $O(n \log n)$ computing g_n with all the input nodes occurring on the outer face.

Sketch of the proof of Theorem 3. A bipartite graph $H = (V_1, V_2, E)$ is an (n, k, d)-expander if $|V_1| = |V_2| = n$, all degrees are $\leq k$ and for every $V \subseteq V_1$, $|\Gamma_V| \geq (1 + d(1 - (|V|/n)))|V|$, where Γ_V is the set of neighbors of V in V_2.

Gabber and Galil [9] showed that there are explicitly defined $(m^2, 5, (2 - \sqrt{3})/4)$-expanders for every m.

Now let $n = m^2$ and construct a graph H_n from H on the vertex set $\{1, \ldots, n\}$ by identifying V_1 and V_2 and deleting loops and multiple edges. Let the edges of H_n be $e_i = (k_i, \ell_i)$ for $i = 1, \ldots, r$ $(r \leq 5n)$ and let

$$f_n(x_1, \ldots, x_n) := \bigvee_{i=1}^{r} (x_{k_i} \wedge x_{\ell_i}).$$

Then $L(f_n) = O(n)$ by definition. The lower bound follows by using the Lipton-Tarjan Planar Separator Theorem [16] as in [17], [22]. □

Sketch of the proof of Theorem 4. We use the "indirect storage access" function of Paul [21] (see also [28, p. 255]). The lower bound is Paul's result [21]. The standard circuits computing g_n can be turned into planar circuits satisfying the requirements of the Theorem. □

4. Multilective planar circuits.

In this section we mention results separating multilective planar circuits and general circuits.

Theorem 5. There are explicitly defined functions f_n with

$$C_p^*(f_n) = \Omega(n \log n) \text{ and } C(f_n) = O(n).$$

The cyclic shifting function $g_n : \{0,1\}^{n+k} \to \{0,1\}^n$, $n = 2^k$ is defined by $g_n(x_0, \ldots, x_{n-1}, y_0, \ldots, y_{k-1}) = (z_0, \ldots, z_{n-1})$, where $z_i = x_{i- \text{bin}(y)}$ for $0 \leq i \leq n-1$ (subtraction is meant $\bmod n$).

Theorem 6. For the cyclic shifting function

$$C_p^*(g_n) = \theta(n^{3/2}) \text{ and } C(g_n) = O(n \log n) \qquad □$$

Theorem 7. There are explicitly defined functions $h_n : \{0,1\}^n \to \{0,1\}^n$ with

$$C_p^*(h_n) = \Omega(n^{3/2}) \text{ and } C(h_n) = O(n). \qquad □$$

The constructions for f_n and h_n use superconcentrators similarly as in Theorem 2.

The lower bound proofs are based on lemmas about partitioning planar graphs into several pieces. Lemma 5 uses a generalization of the Planar Separator Theorem for two cost functions and exact halving, and the existence of a decomposition tree obtained by repeatedly applying the Planar Separator Theorem (see Bhatt-Leighton [8], Savage [23], Ullman [27] for similar results). Lemma 6 follows from Lemma 5 and a result of Babai, Pudlák, Rödl and Szemerédi [3] on the partitioning of trees.

Lemma 5. Let $G = (V, E)$ be a planar graph, $V' \subseteq V$ and $1 \leq r \leq |V'|$. Then there is a partition $(V_1, \ldots, V_r)$ and vertex sets $S_1, \ldots, S_r$ such that

a) $|V' \cap V_i| = \theta(\frac{|V'|}{r})$ for $i = 1, \ldots, r$;
b) $|S_i| = O(\sqrt{|V|})$ and no edge joins V_i to $V \backslash (V_i \cup S_i)$ for $i = 1, \ldots, r$;
c) $|\bigcup_{i=1}^{r} S_i| = O(\sqrt{r|V|})$. □

Lemma 6. Let $G = (V, E)$ be a planar graph, $Z = \{z_1, \ldots, z_s\}$ and $Z' = \{z'_1 \ldots, z'_s\}$ be disjoint sets of labels and $\alpha : V \hookrightarrow Z \cup Z'$ be a (partial) labeling of G such that each label occurs at most k times. Then there are subsets $Z_0 \subseteq Z$, $Z'_0 \subseteq Z'$ and $V^* \subseteq V$ such that

a) $|Z_0| = |Z'_0| \geq s/(9^k)$;
b) $|V^*| = O(k\sqrt{|V|})$;
c) after deleting V^*, none of the remaining components contain labels from both Z_0 and Z'_0. □

5. Some remarks and problems.

1) (Multilective circuits and off-line Turing machines.) Kedem and Zorat [14], [15] noted that allowing the repetition of inputs in the VLSI model is analogous to considering off-line Turing machines, In fact there is a direct connection between these topics. Lemma 5 can be used to improve the $\Omega(n \log n / \log \log n)$ lower bound of Maass, Schnitger and Szemerédi [19] for nondeterministic one-tape off-line Turing machines to $\Omega(n \log n)$. (This application was suggested by W. Maass.)

2) (Complexity and the number of subfunctions.) There is a common feature of several lower bounds for synchronous circuits, planar circuits and formulas: large complexity is implied by the existence of "many" subfunctions, though "many" has different meaning in the different cases. There is a very interesting and apparently less known result of Uhlig [26] (following previous work of Toom [24]). He showed that for every c_1 there is a c_2 such that for every f_n if $\sum_{A \subseteq X} N(f_n, A) \leq c_1 2^n$ then $C(f_n) \leq c_2 n$. Uhlig's proof can be modified to show that the same condition implies $C_s(f_n) \leq c_3 n$, $L(f_n) \leq n^{c_4}$ (for some constants c_3, c_4 also depending on c_1). Thus at least in a very weak sense for these complexity measures there are two-way implications between the properties of having large complexity and having many subfunctions.

3) (Lower bounds for multilective planar circuits.) Considering multilective planar circuits it would be interesting to prove an $\Omega(n^{1+\epsilon})$ lower bound for a decision problem and an $\Omega(n^{(3/2)+\epsilon})$ lower bound for a multiple output function.

Acknowledgement. I am grateful to Wolfgang Maass and Endre Szemerédi for their remarks.

REFERENCES

[1] N. ALON, W. MAASS: Meanders and their application in lower bounds arguments, *J. Comp. Syst. Sci.*, **37**(1988), 118-129.

[2] A. E. ANDREEV: On a method giving larger than quadratic effective lower bounds for the complexity of π-schemes, Vestnik Moscow Univ. Ser. 1 (Math. Mech.), 1986, No. 6, 73-76. (In Russian.)

[3] L. BABAI, P. PUDLÁK, V. RÖDL, E. SZEMERÉDI: Lower bounds to the complexity of symmetric functions, preprint (1986).

[4] E. G. BELAGA: Locally synchronous complexity in the light of the trans-box method, 1. STACS, LNCS **166**(1984), 129-139.

[5] E. G. BELAGA: Constructive universal algebra: an introduction, *Theor. Comp. Sci.*, **51**(1987), 229-238.

[6] E. G. BELAGA: Through the mincing machine with a Boolean layer cake, preprint (1987).

[7] E. G. BELAGA: Through the mincing machine with a Boolean layer cake. Nonstandard computations over Boolean circuits in the lower-bounds-to-circuit-size complexity proving, *Acta Inf.*, **26**(1989), 381-407.

[8] S. N. BHATT, F. T. LEIGHTON: A framework for solving VLSI graph layout problems, *J. Comp. Syst. Sci.*, **28**(1984), 300-343.

[9] O. GABBER, Z. GALIL: Explicit construction of linear size superconcentrators, *J. Comp. Syst. Sci.*, **22**(1981), 407-420.

[10] L. H. HARPER: An $n \log n$ lower bound on synchronous combinational complexity, *Proc. AMS*, **64**(1977), 300-306.

[11] L. H. HARPER, J. E. SAVAGE: Lower bounds on synchronous combinational complexity, *SIAM J. Comp.*, **8**(1979), 115-119.

[12] P. HOCHSCHILD: Multiple cuts, input repetition, and VLSI complexity, *Inf. Proc. Lett.*, **24**(1987), 19-24.

[13] Z. M. KEDEM: Optimal allocation of computational resources in VLSI, **23** FOCS (1982), 379-385.

[14] Z. M. KEDEM, A. ZORAT: Replication of inputs may save computational resources in VLSI, in: H. T. Kung, R. Sproull, G. Steele (eds.): VLSI Systems and Computations, Comp. Sci. Press (Rockwille, Md.), 1981, 52-60.

[15] Z. M. KEDEM, A. ZORAT: On relations between input and communication/computation in VLSI, **22** FOCS (1981), 37-41.

[16] R. J. LIPTON, R. E. TARJAN: A planar separator theorem, *SIAM J. Appl. Math.*, **36**(1979), 177-189.

[17] R. J. LIPTON, R. E. TARJAN: Applications of a planar separator theorem, *SIAM J. Comp.*, **9**(1980), 513-524.

[18] W. MAASS, G. SCHNITGER: An optimal lower bound for Turing machines with one work tape and a two-way input tape, *Structure in Complexity*, LNCS **223**(1986), 249-264.

[19] W. MAASS, G. SCHNITGER, E. SZEMERÉDI: Two tapes are better than one for off-line Turing machines, **19** STOC (1987), 94-100.

[20] W. F. MCCOLL: Planar circuits have short specifications, **2** STACS, LNCS **182** (1985), 231-242.

[21] W. J. PAUL: A 2.5n lower bound on the combinational complexity of Boolean functions, *SIAM J. Comp.*, **6**(1977), 427-443.
[22] J. E. SAVAGE: Planar circuit complexity and the performance of VLSI algorithms, in: H. T. Kung, R. Sproull, G. Steele (eds.): VLSI Systems and Computations, Comp. Sci. Press (Rockwille, Md.), 1981, 61-67. Also: INRIA Report No. 77 (1981).
[23] J. E. SAVAGE: The performance of multilective VLSI algorithms, *J. Comp. Syst. Sci.*, **29**(1984), 243-272.
[24] A. L. TOOM: On the complexity of the realization of binary functions having few "subfunctions", *Problems of Cybernetics*, **18**(1967), 83-90. (In Russian.)
[25] GY. TURÁN: Lower bounds for synchronous circuits and planar circuits, *Inf. Proc. Lett.*, **30**(1989), 37-40.
[26] D. UHLIG: On the relationship between circuit complexity of a Boolean functions and the number of its subfunctions, *Problems of Cybernetics*, **26**(1973), 183-201. (In Russian.)
[27] J. D. ULLMAN: Computational Aspects of VLSI, Comp. Sci. Press (Rockwille, Md.) (1984).
[28] I. WEGENER: The Complexity of Boolean Functions, Wiley-Teubner Series in Comp. Sci. (1987).

The complexity of connectivity problems on context-free graph languages

(Extended Abstract)

Egon Wanke

Universität-Gesamthochschule Paderborn
Fachbereich 17 - Mathematik/Informatik
4790 Paderborn, West Germany

Abstract : We analyze the precise complexity of connectivity problems on graph languages generated by context-free graph rewriting systems under various restrictions. Let L be the family of all context-free graph rewriting systems that generate at least one disconnected resp. connected graph. We show that L is DEXPTIME-complete w.r.t. log-space reductions. If L is finite then L is PSPACE-complete w.r.t. log-space reductions. These results hold true for graph rewriting systems as for example boundary node label controlled (BNLC) graph grammars, hyper-edge replacement systems (HRS's), apex (APEX) graph grammars, simple context-free node label controlled (SNLC) graph grammars, and even for the simple context-free graph grammars introduced by Slisenko in [Sli82].

1 Introduction

Since graph grammars are important tools for describing designs in engineering one of the most interesting question is whether a design system can generate a design that fulfills a certain property. Many authors have discussed the decidability of graph properties on graph languages generated by context-free graph grammars, see [Cou87a,Cou87b,Hab89,HKV87] [LW88a,LW88b,RW86b,Wan89]. In all these references the precise complexity of simple graph properties on possible infinite graph languages remain open, if the graph grammar is given to the input.

In this paper we consider connectivity problems. We investigate the precise complexity of connectivity problems on *simple context-free node label controlled* (SNLC) languages and *boundary node label controlled* (BNLC) languages. BNLC graph grammars are defined in [RW86a]. We use SNLC graph grammars as an example for context-free graph grammars. The notion of *context-free grammars* is well known for strings. There is no such well established notion for rewriting systems generating graphs. We refer to [Cou87b] for a precise definition of context-free graph rewriting. In [Cou87a] BNLC graph grammars are *like* context-free.

The set of graphs generated by a graph grammar Γ is called the *language* $L(\Gamma)$ of Γ. The language $L(\Gamma)$ can be *infinite* ($|L(\Gamma)| = \infty$), *finite* ($|L(\Gamma)| < \infty$) or *single* ($|L(\Gamma)| \leq 1$). If $L(\Gamma)$ is single then Γ may be *deterministic*, i.e. there exists a deterministic graph grammar Γ', such that $L(\Gamma') = L(\Gamma)$. In a deterministic graph grammar each nonterminal object has at most one substitution rule (production). The size of the graph generated by Γ can be exponential in the size of the description of Γ. Therefore, solving graph properties on the single graph in $L(\Gamma)$ is not trivial. If $L(\Gamma)$ is finite then Γ does not allow loops in derivations. In such graph grammars the derivation height is less than or equal to the number of nonterminal objects in Γ. A graph grammar Γ is *linear* if the start graph and each graph on the right hand side of a substitution rule contains only one nonterminal object. For all combinations of these restrictions we show the precise complexity for the question: "Given a SNLC graph grammar or a BNLC graph grammar Γ, does the language $L(\Gamma)$ of Γ contain a disconnected graph?" The results given in this paper are summarized in the following table. All complexities are w.r.t. log-space reductions.

restriction of Γ		complexity for disconnectivity	
$L(\Gamma)$	Γ	SNLC	BNLC
single	not linear	P-complete	NP-complete
finite	linear	NP-complete	NP-complete
	not linear	PSPACE-complete	PSPACE-complete
infinite	linear	PSPACE-complete	PSPACE-complete
	not linear	DEXPTIME-complete	DEXPTIME-complete

We also discuss the problem, whether a given SNLC or BNLC graph grammar Γ can generate a connected graph. This question is always in the same complexity class (resp. in the complementary complexity class if Γ is deterministic). Some of the membership results are already shown in [LW88a,LW88b,Wan89]. All hardness results are shown in this paper.

2 Preliminaries

We use a slightly different notation from [RW86a], because our notation supports the complexity analysis of the procedures that process the graph grammars.

Let Σ be a finite alphabet. Let $\Delta \subseteq \Sigma$ be the set of *terminal* labels and let $\Sigma - \Delta$ be the set of *nonterminal* labels in Σ. A *node labeled graph* $G = (V, E, P, lab, \Sigma)$ *over* Σ is a graph (V, E) whose vertices are labeled by $lab : V \to \Sigma$. In a node labeled graph G over Σ we call a vertex labeled with a terminal label a *terminal vertex* (*nonterminal vertex*, respectively). The vertices in $P \subseteq V$ are called *connection* vertices.

A *node label controlled* (NLC) graph grammar is a system $\Gamma = (G_1, ..., G_k, C, R, \Sigma, \Delta)$ of k node labeled graphs $G_1, ..., G_k$ over Σ, a *connection relation* $C \subseteq \Sigma \times \Sigma$ and a set of *substitution rules* $R \subseteq (\Sigma - \Delta) \times \{1, ..., k\}$. The node labeled graph G_k is the *start graph* of Γ. A *boundary* NLC (BNLC) graph grammar is a NLC graph grammar with a connection relation $C \subseteq \Delta \times \Sigma$. A *simple context-free* NLC (SNLC) graph grammar is a NLC graph grammar with the connection relation $C = \{(e, e) | e \in \Delta\}$.

Let $N(u) = \{v \in V | \{u, v\} \in E\}$ be the *neighborhood* of a vertex u in a graph $G = (V, E)$. In a NLC graph grammar $\Gamma = (G_1, \ldots, G_k, C, R, \Sigma, \Delta)$ a node labeled graph $G = (V, E, P, lab, \Sigma)$ *directly derives* into a node labeled graph J via substitution rule $(e, t) \in R$, denoted by $G \Longrightarrow_{(e,t)} J$, if there exists a nonterminal vertex u in G labeled with e and J is constructed as follows: u is removed from G and substituted by a copy $\overline{G_t} = (V_t, E_t, P_t, lab_t, \Sigma)$ of G_t, such that J is the disjoint union of $G - \{u\}$ and $\overline{G_t}$. The embedding of $\overline{G_t}$ will be done by the addition of edges between a vertex $v \in N(u)$ and a vertex $w \in P_t$ iff $(lab(v), lab_t(w)) \in C$.

The transitive closure of $\Longrightarrow$ is denoted by $\stackrel{*}{\Longrightarrow}$. If $G = (V, E, P, lab, \Sigma)$ is a node labeled graph over Σ we denote the unlabeled graph of G by $unlab(G) = (V, E)$. The language $L(\Gamma)$ is defined by the set of unlabeled graphs of the node labeled graphs over Δ that are derivable from the start graph G_k in Γ, i.e. $L(\Gamma) = \{unlab(G) | G_k \stackrel{*}{\Longrightarrow} G$ and G has no nonterminal vertices$\}$. We consider the graphs in $L(\Gamma)$ as unlabeled since we only use the labeling to derive graphs. In BNLC and SNLC graph grammars edges between nonterminal vertices are redundant. The size of a graph grammar Γ, denoted by $|\Gamma|$, is the size of its description.

$$\Gamma = (G_1, G_2, G_3, \{(a,a), (b,b), (c,c)\}, \{(A,1), (A,2)\}, \{A, a, b, c\}, \{a, b, c\})$$

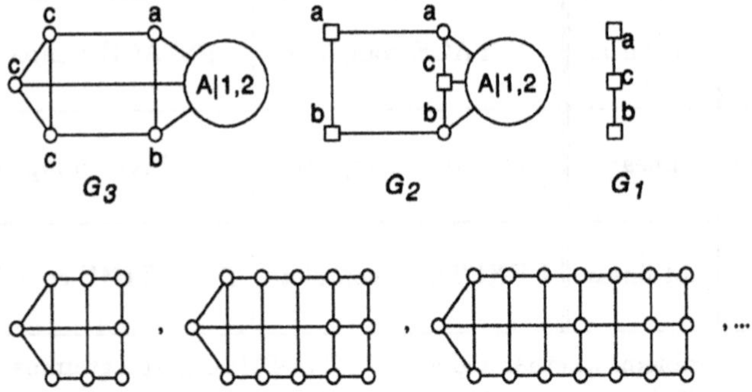

Figure 1: A SNLC graph grammar Γ and some graph in $L(\Gamma)$

Figure 1 shows a SNLC graph grammar Γ and some graphs in $L(\Gamma)$. The start graph is G_3. The letters on the circles represent the labels of the vertices. Upper (lower) case letters denote nonterminal (terminal) labels. The connection vertices are drawn as squares. Nonterminal vertices are drawn as big circles. In the big circle for nonterminal vertex n we write the label $lab(n)$ and/or the right hand side t of each substitution rule $(lab(n), t)$ for n. We allow a node labeled graph to be undirected, directed or mixed (i.e., having directed as well as undirected edges). However, the definition of the edge sets in the node labeled graphs have to correspond with the definition of the new edges created by the embedding.

A number of interesting families of graphs can be generated by BNLC and SNLC graph grammars, see [ELR87,HK87,RW86a] for examples. However, the set of planar graphs and the set of all graphs are not BNLC languages, see [RW87]. All results for SNLC graph grammars shown in this paper also hold true for *hyper-edge replacement systems* [HK87], apex graph grammars [ELR87] and the graph grammar introduced in [Sli82]. Thus, all hardness results for connectivity problems on SNLC graph grammars can be transformed to all these context-free graph grammars. The transformations to these graph grammars are straightforward but not shown here.

3 Connectivity problems on graph languages

Due to space restrictions we only consider unrestricted SNLC graph grammars. The complexity class DEXPTIME=DTIME($2^{\text{poly}(n)}$) is the family of languages that can be recognized deterministically in exponential time where n is the length of the input. Our main result is the following theorem.

Theorem: The following question is DEXPTIME-complete w.r.t. log-space reductions:

Instance: A SNLC graph grammar $\Gamma = (G_1, ..., G_k, C, R, \Sigma, \Delta)$.
Question: Does $L(\Gamma)$ contain a disconnected graph?

Proof: Let L_{SNLC} and L_{BNLC} be the set of SNLC resp. BNLC graph grammars that generate at least one disconnected graph. In [Wan89] it is shown that L_{BNLC} can be recognized deterministically in exponential time. The set L_{SNLC} can also be recognized deterministically in exponential time, because each SNLC language is a BNLC language.

We reduce an arbitrary language $L \in$ DEXPTIME to L_{SNLC}. I.e., we design a log-space transducer T. Given an input x of length n, T writes a SNLC graph grammar Γ_x such that $\Gamma_x \in L_{\text{SNLC}}$ iff $x \in L$. Let M be a 2^m time-bounded deterministic Turing machine accepting L, where m is bounded polynomial in the length n of the input x. Let Q_M be the finite set of states of M, let Σ_M be the finite set of tape symbols of M, and let $\beta_0, ..., \beta_{2^m-1}$ be a computation of M on input x, where each β_i is a *tape description* consisting of exactly 2^m *composite symbols*. Let # be a marker used to denote the ends of the tape descriptions, and let e be a symbol not used in Q_M. In each composite symbol $X = [x, q]$, x is # or a tape symbol of M and q is e or a state of M. In each β_i there is exactly one composite symbol $X = [x, q]$ such that $q \in Q_M$, indicating the head position of M in β_i.

The proof consists of two parts. First we generate a context-free string grammar G_x such that $L(G_x) \neq \emptyset$ iff M accepts x. Then we design a SNLC graph grammar Γ_x using $O(\log(n))$ space such that $\Gamma_x \in L_{\text{SNLC}}$ iff $L(G_x) \neq \emptyset$. The size of G_x is exponential in the length n of the input x. We need G_x only to show the correctness of Γ_x. In [HU79] it is shown that the emptiness problem for context-free string grammars is P-complete. The following lines are similar to the lines in the proof given in [HU79].

<u>The construction of G_x:</u> The nonterminal symbols of G_x are symbols of the form $A_{X,i,t}$, where X is a composite symbol and $0 \leq i, t < 2^m$. The intention is that $A_{X,i,t} \stackrel{*}{\Longrightarrow} w$ for some terminal string w iff X is the ith composite symbol in β_t. The start symbol in G_x is S. The productions of G_x are:

1. $S \longrightarrow A_{X,i,t}$ for all $0 \leq i, t < 2^m$ and all $X = [x, q]$ where q is a final state.
2. Let $f(X, Y, Z)$ be the composite symbol on position i in β_t wherever X, Y, Z occupy the positions $i-1, i, i+1$ in β_{t-1}. Since M is deterministic, $f(X, Y, Z)$ is a unique symbol and is independent of i and t. For each $0 < i < 2^m - 1$, each $0 < t < 2^m$ and each triple X, Y, Z with $W = f(X, Y, Z)$, we have the productions:
$$A_{W,i,t} \longrightarrow A_{X,i-1,t-1} A_{Y,i,t-1} A_{Z,i+1,t-1}.$$
3. $A_{[\#,e],0,t} \longrightarrow \epsilon$ and $A_{[\#,e],2^m-1,t} \longrightarrow \epsilon$ for all $0 \leq t < 2^m$.
4. $A_{X,i,0} \longrightarrow \epsilon$ for $0 < i < 2^m - 1$ iff X is the ith composite symbol in β_0.

An easy induction on t shows that for $0 \leq i < 2^m$, $A_{X,i,t} \stackrel{*}{\Longrightarrow} \epsilon$ iff W is the ith composite symbol in β_t. Of course, no terminal string but ϵ is ever derived from any nonterminal.

Basis: The basis, $t = 0$, is immediate from rule (3) and (4).

Induction: Let $t > 0$. If $i = 0$ or $i = 2^m - 1$, then by rule (3) $A_{W,i,t} \stackrel{*}{\Longrightarrow} \epsilon$ iff W is the composite symbol at position i in β_t, because W must be $[\#, e]$.

If $A_{W,i,t} \stackrel{*}{\Longrightarrow} \epsilon$ and $0 < i < 2^m - 1$, then by rule (2) it must be that for some X, Y, Z, $W = f(X, Y, Z)$ and each of $A_{X,i-1,t-1}, A_{Y,i,t-1}, A_{Z,i+1,t-1}$ derive ϵ. By the inductive hypothesis the composite symbols in β_{t-1} on positions $i-1, i, i+1$ are X, Y, Z, so W is the composite symbol at position i in β_t by the definition of f.

Conversely, if W is the composite symbol at position i in β_t for some $0 < i < 2^m - 1$, then $W = f(X, Y, Z)$, where X, Y, Z are the composite symbols in β_{t-1} on positions $i-1, i, i+1$. By the inductive hypothesis: $A_{X,i-1,t-1} \stackrel{*}{\Longrightarrow} \epsilon$, $A_{Y,i,t-1} \stackrel{*}{\Longrightarrow} \epsilon$, $A_{Z,i+1,t-1} \stackrel{*}{\Longrightarrow} \epsilon$ and thus $A_{W,i,t} \stackrel{*}{\Longrightarrow} \epsilon$.

The construction of Γ_x: Now we will design a SNLC graph grammar Γ_x such that Γ_x can generate a disconnected graph iff $L(G_x) \neq \emptyset$. A m-bit integer i is represented by a graph G_m^i with $m + 2$ connection vertices $0, 1, x_m, ..., x_1$. If the jth bit in i is 0 (is 1) then vertex x_j is connected by an edge with vertex 0 (vertex 1, respectively). Consider two graphs $G_m^{i_1}$, $G_m^{i_2}$. If equal labeled connection vertices are joined by edges then the resulting graph is disconnected iff $i_1 = i_2$. Let Γ_m^i be the SNLC graph grammar that contains only the graph G_m^i.

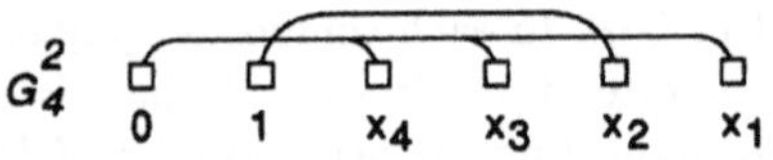

Figure 2: The SNLC graph grammar Γ_4^2

Any graph G_m^i representing a m-bit integer i can be generated by a SNLC graph grammar Γ_m^{rnd}, which consists of three graphs $G_m^{\mathrm{rnd}}, G_0, G_1$, see the example in Figure 3. Each nonterminal vertex in the start graph can be substituted either by G_0 or by G_1. The substitution can specify any coded m-bit integer on the connection vertices of G_m^{rnd}.

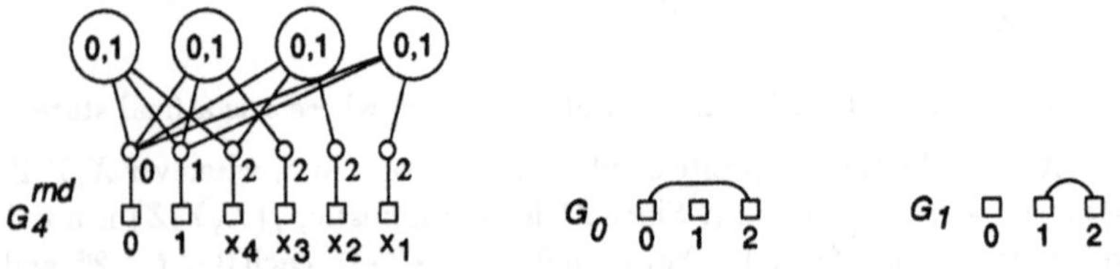

Figure 3: The SNLC graph grammar Γ_4^{rnd}

Now we design a SNLC graph grammar $\Gamma_m^{<c}$ and $\Gamma_m^{>c}$ that tests, whether a given m-bit integer i represented by a graph G_m^i is less than a constant c (greater than a constant c, respectively). Let c, i be two m-bit integers. Then $c < i$ iff there exists an integer l such that the bits

$l+1,\ldots,m$ in c and i are equal, the lth bit in c is 0 and the lth bit in i is 1. Analogously, $c > i$ iff there exists an integer l such that the bits $l+1,\ldots,m$ in c and i are equal, the lth bit in c is 1 and the lth bit in i is 0. This is the idea to construct the SNLC graph grammars $\Gamma_m^{<c}$ and $\Gamma_m^{>c}$, see the example in Figure 4.

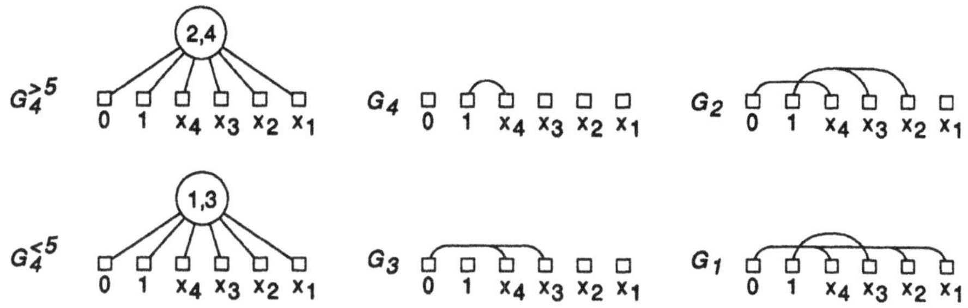

Figure 4: The SNLC graph grammars $\Gamma_4^{<5}$ and $\Gamma_4^{>5}$

The graph $G_m^{<c}$ (graph $G_m^{>c}$) is the start graph for $\Gamma_m^{<c}$ (for $\Gamma_m^{>c}$, respectively). Let G_m^i be a graph representing the m-bit integer i. If we connect the connection vertices of G_m^i with the connection vertices of $G_m^{<c}$ (of $G_m^{>c}$) then $\Gamma_m^{<c}$ ($\Gamma_m^{>c}$) can generate a connected graph iff $i < c$ (iff $i > c$). Here, we connect equal labeled vertices by edges.

Now let us design a SNLC graph grammar Γ_m^+ that adds the constant 1 to a given m-bit integer i represented by a graph G_m^i. Let $i < 2^m - 1$ be a positive m-bit integer. Then there exists a unique integer l such that the lth bit in i is 0 and the bits $1,\ldots,l-1$ in i are 1. This is the idea to construct Γ_m^+, see the example in Figure 5.

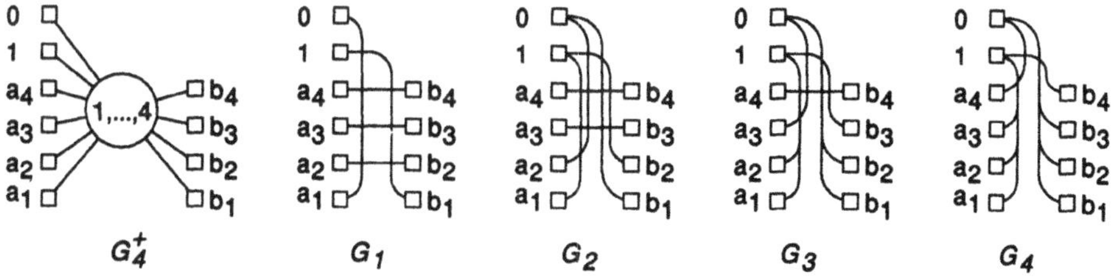

Figure 5: The SNLC graph grammar Γ_4^+

The nonterminal vertex in the start graph G_m^+ can be substituted by one of the graphs G_i for $i \in \{1,\ldots,m\}$. Let $G_m^{i_1}$, $G_m^{i_2}$ be two graphs representing two m-bit integers i_1, i_2. If we connect the connection vertices $0, 1, x_m, \ldots, x_1$ of $G_m^{i_1}$ with the connection vertices in G_m^+ labeled with $0, 1, a_m, \ldots, a_1$ and the connection vertices $0, 1, x_m, \ldots, x_1$ of $G_m^{i_2}$ with the connection vertices in G_m^+ labeled with $0, 1, b_m, \ldots, b_1$ then Γ_m^+ can generate a disconnected graph iff $i_1 = i_2 + 1$. Here, vertex x_i in $G_m^{i_1}$ (in $G_m^{i_2}$) is joined with vertex a_i (vertex b_i) in G_m^+. We can subtract the constant 1 with Γ_m^+ by exchanging the labels $a_m, \ldots, a_1$ with the labels $b_m, \ldots, b_1$ in G_m^+.

We design the SNLC graph grammar Γ_x with the tools defined above. The intention is to represent the rules of G_x by graphs in Γ_x. One graph G_S represents all rules in (1). G_S guesses any integer representation for a final state q, a tape symbol x, a position i and a time t. Let $s = \lceil \log(|\Sigma_M \cup \{\#\}|) \rceil$ and $p = \lceil \log(|Q_M \cup \{e\}|) \rceil$. Each composite symbol X is coded by $k = s + p$ bits. The lower p bits represent the states of M. Assume that the states in Q_M are ordered, such that the first r states in Q_M are final states. Thus, G_S guesses a composite symbol $X = [x, q]$ and verifies whether the lower p bits are less than r.

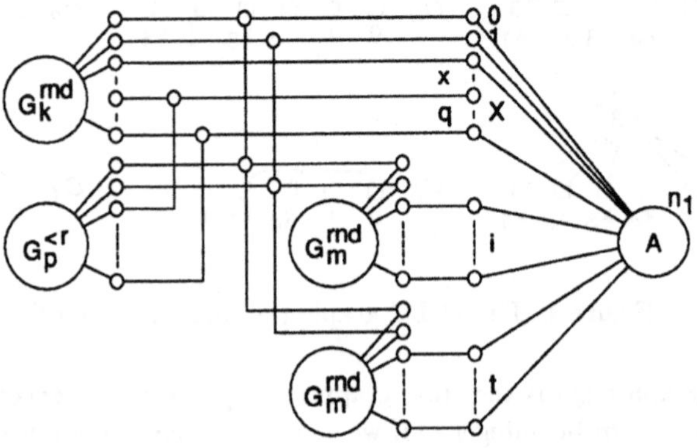

Figure 6: The graph G_S

The graph G_S is the start graph in Γ_x and has one nonterminal vertex n_1 labeled with A, see Figure 6. In the figures we combine vertices and edges in groups. Each nonterminal vertex has two neighboring vertices labeled with 0 and 1. All vertices labeled with 0 (with 1) are connected. Some nonterminal vertices are labeled by graph symbols used for start graphs in SNLC graph grammars defined above. This means, Γ_x contains all graphs of the SNLC graph grammar Γ (disjoint union) and the nonterminal vertex has a substitution rule to the start graph of Γ.

Each rule in (2) is represented by a graph $G_{X,Y,Z,W}$ in Γ_x with $W = f(X, Y, Z)$. Each graph $G_{X,Y,Z,W}$ has three nonterminal vertices n_1, n_2, n_3 labeled with A representing the nonterminal symbols $A_{X,i-1,t-1}, A_{Y,i,t-1}, A_{Z,i+1,t-1}$, see Figure 7. Since f is independent of i and t the number of graphs $G_{X,Y,Z,W}$ is constant.

The graphs representing the rules in (3) and (4) do not contain nonterminal vertices labeled with A. They only represent the composite symbols X on position i in β_t for $i = 0, i = 2^m - 1$ or $t = 0$. Let q_s be the start state of M. Assume that

$$\beta_0 = [\#, e], [0, e], \ldots, [0, e], [x_1, q_s], [x_2, e], \ldots, [x_n, e], [0, e], \ldots, [0, e], [\#, e]$$

and the input $x = x_1, \ldots, x_n$ occupies the positions $a_1, \ldots, a_n$ in β_0 with $0 < a_1$ and $a_n < 2^m - 1$ and $a_i + 1 = a_{i+1}$ for $i = 1, \ldots, n-1$. We design one graph $G_\#$ for the rules in (3). For each $i \in \{1, \ldots, n\}$ we design a graph G_{X_i}, where X_i is the composite symbol on position a_i in β_0. For the remaining positions in β_0 we design a graph $G_=$, see Figure 8. We need one graph $G_\#$ for the end positions of the tape, n graphs G_{X_i} to represent the input x in β_0, and one graph $G_=$ for all other positions in β_0, because all other positions i in β_0 with $0 < i < a_1$ or $a_n < i < 2^m - 1$ contain the same composite symbol $[0, e]$.

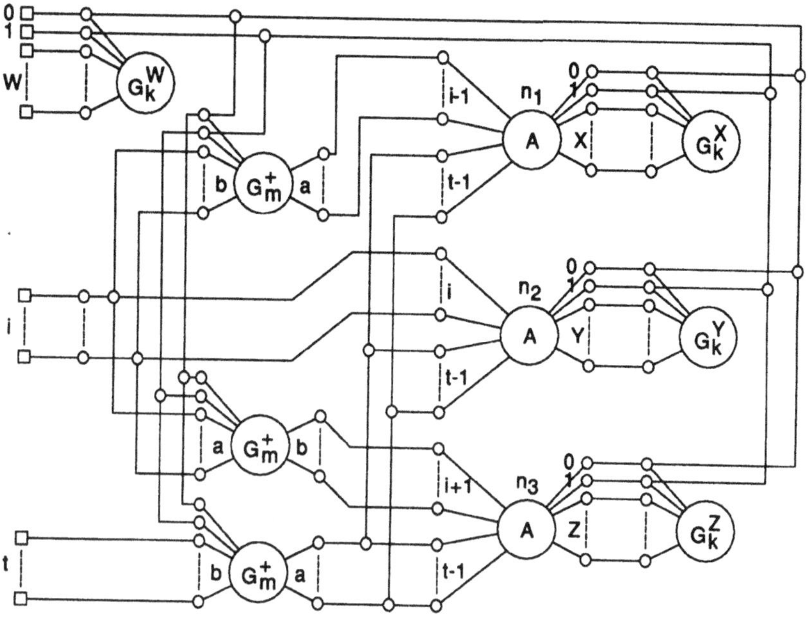

Figure 7: The graphs $G_{\mathrm{X,Y,Z,W}}$

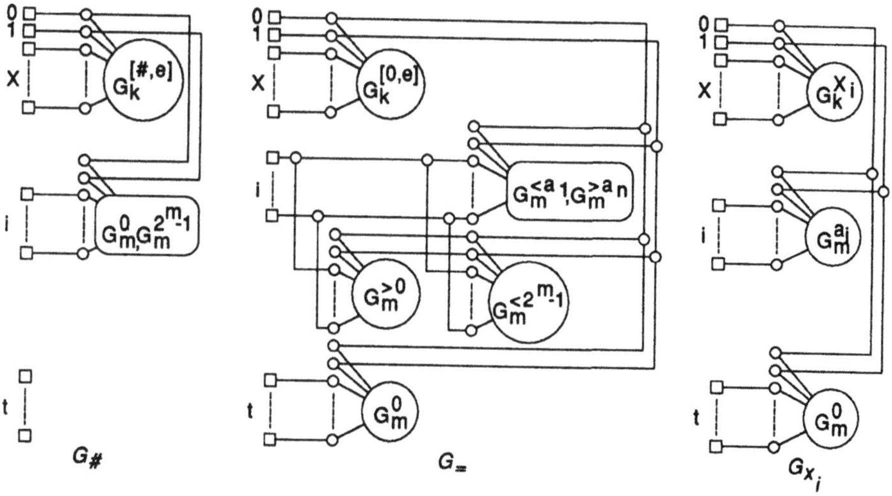

Figure 8: The graphs $G_{\#}, G_{=}$ and G_{X_i} for $i = 1, \ldots, n$

The SNLC graph grammar Γ_x has size $O(m^2 + m \cdot n)$, because Γ_m^+ has size $O(m^2)$, and Γ_x contains $O(n)$ graphs G_{X_i} of size $O(m)$. The rule set R in Γ_x allows to substitute each nonterminal vertex labeled with A with a graph in $\{G_{\mathrm{X,Y,Z,W}}$ for all X, Y, Z, W with $W =$

$f(X,Y,Z)$, $G_\#$, $G_=$, G_{X_i} for $i = 1,\ldots,n\}$. Obviously, Γ_x can be constructed on $O(\log(n))$ space. The correctness of $\Gamma_x \in L_{\text{SNLC}} \iff L(G_x) \neq \emptyset$ follows by the following easy observation. For each sequence P of substitution rules in G_x used to derive a string w we define a sequence Q of substitutions used to derive a graph $G(w)$ in Γ_x. Additionally, we define an one-to-one mapping h between the nonterminals in w and the nonterminal vertices in $G(w)$ labeled with A. Each derivation specified by P can be transformed into a derivation specified by Q. The transformation steps and the one-to-one mapping h are defined in the following table.

Derivation steps in P	Derivation steps in Q
$S \longrightarrow A_{X,i,t}$	$G_{A_{X,i,t}} = G_S$, $h(A_{X,i,t})$ is the nonterminal vertex n_1 in G_S
$A_{W,i,t} \longrightarrow A_{X,i-1,t-1}, A_{Y,i,t-1}, A_{Z,i+1,t-1}$	$h(A_{W,i,t}) \longrightarrow G_{\text{X,Y,Z,W}}$, the nonterminal vertices $h(A_{X,i-1,t-1}), h(A_{Y,i,t-1}), h(A_{Z,i+1,t-1})$ are the nonterminal vertices n_1, n_2, n_3 in $G_{\text{X,Y,Z,W}}$
$A_{[\#,e],0,t} \longrightarrow \epsilon$	$h(A_{[\#,e],0,t}) \longrightarrow G_\#$
$A_{[\#,e],2^m-1,t} \longrightarrow \epsilon$	$h(A_{[\#,e],2^m-1,t}) \longrightarrow G_\#$
$A_{X,i,0} \longrightarrow \epsilon$ with $0 < i < a_1$ or $a_n < i < 2^m - 1$	$h(A_{X,i,0}) \longrightarrow G_=$
$A_{X,i,0} \longrightarrow \epsilon$ with $a_1 \leq i \leq a_n$	$h(A_{X,i,0}) \longrightarrow G_{X_{i-a_1+1}}$

Each derivation $S \stackrel{*}{\Longrightarrow} \epsilon$ in G_x can be transformed into a derivation $G_S \stackrel{*}{\Longrightarrow} G(\epsilon)$ in Γ_x with the transformation steps given above. By the construction of the graphs in Γ_x there exists a derivation $G(\epsilon) \stackrel{*}{\Longrightarrow} G$ in Γ_x such that $G \in L(\Gamma_x)$ and G is disconnected. The derivation $G(\epsilon) \stackrel{*}{\Longrightarrow} G$ verifies the rules used in the derivation $G_S \stackrel{*}{\Longrightarrow} G(\epsilon)$. I.e., a substitution of a nonterminal vertex $n = h(A_{X,i,t})$ in Γ_x is legal iff the vertices in the neighborhood $N(n)$ of n can represent the composite symbol X, the position i, and the time t. Conversely, each graph $G \in L(\Gamma_x)$ that is disconnected can be obtained by the derivations $G_S \stackrel{*}{\Longrightarrow} G(\epsilon)$ and $G(\epsilon) \stackrel{*}{\Longrightarrow} G$ in Γ_x. Since G is disconnected there exists a derivation $S \stackrel{*}{\Longrightarrow} \epsilon$ in G_x. □

The given theorem implies the following: For any $2^{O(n)}$ time-bounded deterministic Turing machine M with given input x of length n one can construct a SNLC graph grammar Γ_x of size $O(n^2)$ such that Γ_x can generate a disconnected graph iff M accepts x. Conversely, if n denotes the size of the description of a SNLC graph grammar Γ, any correct algorithm that determines whether a given SNLC graph grammar Γ can generate a disconnected graph must run for $2^{\Omega(\sqrt{n})}$ steps for infinitely many n's.

A simple modification of the representation of integers shows that the question "Does $L(\Gamma)$ contain a connected graph?" is also DEXPTIME-hard. Here, a m-bit integer i is represented by a graph with $2 \cdot m + 1$ connection vertices $0, x_1, \ldots, x_m, y_1, \ldots, y_m$. If the jth bit in i is 0 (is 1) then vertex x_j (vertex y_j) is connected with vertex 0 by an edge. Now m vertices are connected with 0 and m vertices are isolated. Consider two m-bit integers i_1, i_2 and construct a graph for i_1 and for $2^m - 1 - i_2$. If equal labeled vertices are joined by edges the resulting graph is connected iff $i_1 = i_2$.

References

[Cou87a] B. Courcelle. An axiomatic definition of context-free rewriting and its application to NLC graph grammars. *Theoretical Computer Science*, 55:141–181, 1987.

[Cou87b] B. Courcelle. On context-free sets of graphs and their monadic second-order theory. In H. Ehrig, M. Nagl, A. Rosenfeld, and G. Rozenberg, editors, *Proceedings of Graph-Grammars and Their Application to Computer Science '86*, pages 133–146, LNCS No. 291, Springer Verlag, Berlin/New York, 1987.

[ELR87] J. Engelfriet, G Leih, and G Rozenberg. Apex graph grammars. In H. Ehrig, M. Nagl, A. Rosenfeld, and G. Rozenberg, editors, *Proceedings of Graph-Grammars and Their Application to Computer Science '86*, pages 167–185, LNCS No. 291, Springer Verlag, Berlin/New York, 1987.

[Hab89] A. Habel. Graph-theoretic properties compatible with graph derivations. In J. van Leeuwen, editor, *Proceedings of Graph-Theoretic Concepts in Computer Science, WG '88*, pages 11–29, LNCS No. 344, Springer Verlag, Berlin/New York, 1989.

[HK87] A. Habel and H.J. Kreowski. May we introduce to you: hyperedge replacement. In H. Ehrig, M. Nagl, A. Rosenfeld, and G. Rozenberg, editors, *Proceedings of Graph-Grammars and Their Application to Computer Science '86*, pages 15–26, LNCS No. 291, Springer Verlag, Berlin/New York, 1987.

[HKV87] A. Habel, H.J. Kreowski, and W. Vogler. Compatible graph properties are decidable for hyperedge replacement graph languages. In *Bulletin of the European Association for Theoretical Computer Science No. 33*, pages 55–62, EATCS, October 1987. To appear in Acta Informatica.

[HU79] J.E. Hopcroft and J.D. Ullman. *Introduction to Automata Theory, Languages, and Computation.* Addison-Wesley Publishing Company, Massachusetts, 1979.

[LW88a] T. Lengauer and E. Wanke. Efficient analysis of graph properties on context-free graph languages. In T. Lepistö and A. Salomaa, editors, *Proceedings of ICALP '88*, pages 379–393, LNCS No. 317, Springer Verlag, Berlin/New York, 1988.

[LW88b] T. Lengauer and E. Wanke. Efficient solution of connectivity problems on hierarchically defined graphs. *SIAM Journal of Computing*, 17(6):1063–1080, 1988.

[RW86a] G. Rozenberg and E. Welzl. Boundary NLC graph grammars - basic definitions, normal forms, and complexity. *Information and Control*, 69(1-3):136–167, April/May/June 1986.

[RW86b] G. Rozenberg and E. Welzl. Graph theoretic closure properties of the family of boundary NLC graph languages. *Acta Informatica*, 23:289–309, 1986.

[RW87] G. Rozenberg and E. Welzl. Combinatorial properties of boundary NLC graph languages. *Discrete Applied Mathematics*, 16:59–73, 1987.

[Sli82] A.O. Slisenko. Context-free grammars as a tool for describing polynomial-time subclasses of hard problems. *Information Processing Letters*, 14(2):52–56, 1982.

[Wan89] E. Wanke. *Algorithms for graph problems on BNLC structured graphs.* Technical Report 58, Universität-Gesamthochschule Paderborn, West-Germany, 1989.

CONSTRUCTIVITY, COMPUTABILITY, AND COMPUTATIONAL COMPLEXITY IN ANALYSIS

Klaus Weihrauch
Theoretische Informatik, Fernuniversität D – 5800 Hagen

1 Introduction

Today's analysis has at least two different important branches: Abstract Analysis and Numerical Analysis. Until the end of the last century Analysis was understood mainly constructively. A real function was considered as a "rule" which for any number x determines the value $f(x)$. When more precise foundations became necessary, functions were defined set theoretically: a real function is a single-valued subset $f \subseteq \mathbb{R} \times \mathbb{R}$. Any concrete function f is given by a property formulated in the language of set theory. Generally, such a formulation does not provide a method for computing a function. With the framework of set theory a very rich and powerful abstract analysis has been developed during this century. In Abstract Analysis the existence of numbers, functions, integrals, Fourier series etc. is emphasized while aspects of effictivity are disregarded. Consequently, as a separate branch Numerical Analysis, in which algorithms for determining zeroes, intergrals, solutions of differential equations, etc. are investigated, has been developed.

However, between abstract analysis and numerical analysis there is a gap. We have very elegant definitions and concepts and powerful existence theorems on the one side and a large number of numerical methods and computer programs on the other side. But we cannot sufficiently measure and compare the degrees of constructivity of theorems in abstract analysis, we generally do not understand why certain operations in analysis cannot be performed effictively. We do not really know why some functions are easily computable and others not. Finally, we do not have a satisfactory computational complexity theory for Numerical Analysis in which upper and lower complexity bounds can be defined and investigated.

In the past several partly overlapping, partly competing attempts have been made for closing this effectivity gap between Abstract und Numerical Analysis. Ituitionism (Brouwer [1], Heyting [2], Troelstra [3] et al.) reduces classical logic to "constructive" proofs, whereby proofs of pure existence are avoided. Bishop and Bridges [4] show that "constructive proofs in a narrow sense" suffice to develop the important parts of Analysis. Two other approaches are based on recursion theory. The "Russian approach" uses standard numberings of "computable" real numbers etc. and thus transfers computability from $\mathbb{N}$ to sets of "computable" objects in Analysis (Ceitin [5], Markov [6], Kushner [7], Aberth [8]). The "Polish" approach represents real numbers etc. by sequences of numbers. Computable operators on sequences of numbers yield computable functions on all real numbers, not only on the computable ones (Mazur [9], Grzegorczyk [10], Mostowski [11], Lacombe [12], Klaua [13], Hauck [14]). Scott [15] suggested to embed real numbers into an interval-cpo and thus approximate real numbers by chains of intervals. Finally there are some papers concerning computational complexity of real functions (Brent [16], Ko and Friedman [17], Müller [18, 19]). A detailed discussion of Constructive Mathematic is given by M. Beeson[20]. Although each of the approaches has advantages, none of them has been really accepted by Abstract Analysists or Numerical Analysists.

In this constribution a theory of effectivity is presented which is more powerful than any of the approaches to effectivity in Analysis mentioned above. This "Type 2 theory of effectivity", shortly TTE, admits to close step by step the effectivity gap between Abstract and Numerical Analysis by a concept for constructivity, a concept for computability, and a concept for computational complexity. TTE includes ordinary recursion theory but does not depend on any kind of restricted logic. It can be considered as a consequent extension of the Polish Recursive Analysis.

First some basic definitions of TTE will be introduced and explained. Then it will be shown by examples, how constructivity, computability and computational complexity in analysis can be investigated in TTE. Finally open questions and suggestions for further studies will be given. Details can be found in papers by Kreitz, Müller, and Weihrauch [18, 19, 21 - 30].

2 A Short Outline of TTE

Type 2 Theory of Effectivity, TTE, includes ordinary Type 1 recursion theory. This can be structured as follows:

- Basic set: $\mathbb{N}$ (or Σ^*). A computer can w.l.g. only read and write elements of $\mathbb{N}$ (or Σ^*).
- Partial recursive functions $f :\subseteq \mathbb{N} \to \mathbb{N}$ are defined by Turingmachines. A standard numbering of Turingmachines yields a numbering φ of $P^{(1)}$ satisfying the utm– and the smn–theorem.
- Computability is transferred from $\mathbb{N}$ to other denumerable sets by means of numberings.

- Computational complexity is defined by Turingmachines, random access machines, etc.
- Effectivity of a numbering $\nu :\subseteq \mathbb{N} \to M$ depends on the structure assumed on M and has to be descussed thoroughly.

Type 2 Theory of Effectivity extends Type 1 recursion theory by a formally similar structure.

- Basic set: $B := \mathbb{N}^{\mathbb{N}} = \{(a_0, a_1, \ldots) | a_i \in \mathbb{N}\}$, the set of sequences of numbers.

Although a machine (or man) can never read or write an infinite sequence in finite time, it can read or write such a sequence step by step from the beginning, thus approximating the infinite sequence by longer and longer finite initial parts. This kind of approximation can be described adequately by Baire's topology.

- Baire's topology τ_B on $\mathbb{B}$ is defined by the basis $\{[w] \mid w \in \mathbb{N}^*\}$ where $[w]$ is the "intervall" or "ball" $\{p \in B \mid w$ is an initial part of $p\}$. Continuous functions $\Gamma :\subseteq \mathbb{B} \to \mathbb{B}$ and $\Sigma :\subseteq \mathbb{B} \to \mathbb{N}$ (where on $\mathbb{N}$ the discrete topology is assumed) are interpreted as constructive functions.
- Computable functions $\Gamma :\subseteq \mathbb{B} \to \mathbb{B}$ and $\Sigma :\subseteq \mathbb{B} \to \mathbb{N}$ are defined by Type 2 machines. A Type 2 machine M is a Turingmachine with n infinite inputs tapes for sequences from B, with finitely many work tapes, and one infinite one–way write–only output tape.

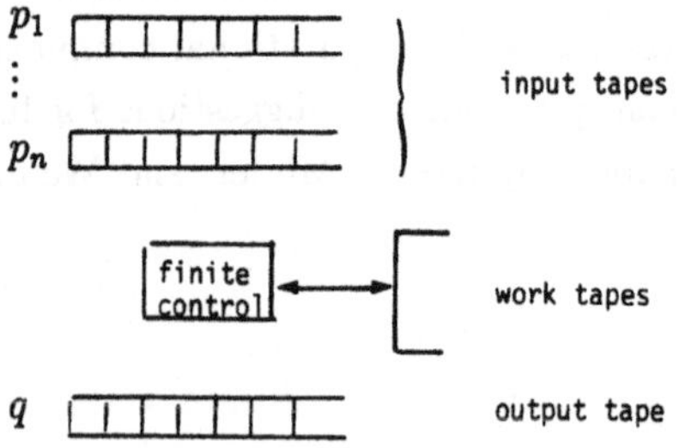

M computes functions $f_M :\subseteq \mathbb{B}^n \to \mathbb{N}$ and $g_M :\subseteq \mathbb{B}^n \to \mathbb{B}$ as follows:

$$f_M(p_1, \ldots, p_n) = k \quad \text{iff } M \text{ with input } (p_1, \ldots, p_n) \text{ halts with } k \text{ on the first position of the output tape.}$$

$$g_M(p_1, \ldots, p_n) = q \quad \text{iff } M \text{ with input } (p_1, \ldots, p_n) \text{ computes forever writing } q \text{ on the output tape.}$$

Every computable function is continuous, thus computability appears as a special case of constructivity. As counterparts of the numbering φ , two "effective" representations $\chi : \mathbb{B} \to [\mathbb{B} \to \mathbb{N}]$ and $\psi : \mathbb{B} \to [\mathbb{B} \to \mathbb{B}]$ of certain continuous functions are introduced.

- Let $M_0, M_1, \ldots$ be a fixed standard numbering of the Type 2 machines with two input tapes. Define

$$\chi < i,p > (q) := f_{M_i}(p,q)$$
$$\psi < i,p > (q) := g_{M_i}(p,q)$$

where $< i,p > := (i, p(0), p(1), \ldots)$. Thus, if $\Gamma = \chi < i,p >$ then p acts as an oracle in the machine M_i. As a result, range $(\chi) = [\mathbb{B} \to \mathbb{N}]$ is the set of all continuous functions $\Gamma :\subseteq \mathbb{B} \to \mathbb{N}$ with open domain, and $[\mathbb{B} \to \mathbb{B}]$ is the set of all continuous functions $\Sigma :\subseteq \mathbb{B} \to \mathbb{B}$ the domain of which is a G_δ set. Both, χ and ψ, satisfy the utm–theorem and the smn–theorem, i.e. they are "effective". As a result, a rich theory of continutity on $\mathbb{B}$ similar to ordinary recursion theory can be developed. Especially, the open subsets of $\mathbb{B}$ correspond to the r.e. subsets and the open and closed subsets of $\mathbb{B}$ correspond to the recurse subsets of $\mathbb{N}$. A subset $X \subseteq \mathbb{B}$ is called r.e. iff $X = dom(\Gamma)$ for some computable function $\Gamma :\subseteq \mathbb{B} \to \mathbb{N}$, and decidable iff X and $\mathbb{B} \setminus X$ are r.e.

Example 1

(1) (Brower's "limited principle of omniscience") The set $X := \{p \in \mathbb{B} \mid 0 \in range(p)\}$ is r.e., hence open, but not closed, hence not decidable. Furthermore, X is complete in τ_B: For any $Y \in \tau_B$ there is some total continuous function $\Gamma : \mathbb{B} \to \mathbb{B}$ with $Y = \Gamma^{-1}(X)$ (shortly $Y \leq_t X$).

(2) (Brower's "lesser limited principle of omniscience") There is no continuous function $\Gamma :\subseteq \mathbb{B} \to \mathbb{N}$ such that for all $p \in \mathbb{B}$ with $p(i) \neq 0$ for at most one $i \in \mathbb{N} : \Gamma(p) \in \{0,1\}$ exists, $\Gamma(p) = 0 \Rightarrow (\forall n)p(2n) = 0$, and $\Gamma(p) = 1 \Rightarrow (\forall n)p(2n,1) = 0$.

(3) The function $T : \mathbb{B} \to \mathbb{N}$ with $T(p) = (0$ if $range(p) = \mathbb{N}, 1$ otherwise) is not continuous.

(4) If q divides a power of p then there is a computable translation from the p –adic into the q–adic representation of real numbers, otherwise there is no continuous translation.

In the above example, the positive answers have the form "there is a computable Γ..." while the negative ones have the form "there is no continuous Γ...". Thus in these cases non–effectivity does not depend on the definition of computability (Church's thesis). These and many other examples suggest to interpret continuity as a very general form of effectivity. Therefore in TTE, "continuous" will be interpreted as "constructive".

- Constructivity and computability are tranfered from $\mathbb{B}$ to other sets M by representations $\delta :\subseteq \mathbb{B} \to M$ (onto).

If $\delta(p) = x$ then p is called "name" of x (w.r.t. δ). Thus, in TTE infinite sequences may act as names. Similar to the case of numberings, the properties open, closed, r.e., and decidable can be transfered from $\mathbb{B}$ to M via the representation δ. Representations δ can be compared by reducibility: $\delta \leq_t \delta' := (\exists \Gamma \in [\mathbb{B} \to B])(\forall p \in dom(\delta))\delta(p) = \delta'\Gamma(p)$. If $\delta \leq_t \delta'$ then δ-names contain at least as much "continuously accessible information" as δ-names.

- Effectivity of a representation depends strongly on the structure assumed on M.

Fixing a representation δ of M is a way of specifying which informations about elements of M are continuously accessible from the names. An important class of structures for which there are "natural" effective representations consists of the T_0-topological spaces with denumerable bases. Important examples are effective cpo's and all the separable metric spaces used in Analysis.

Definition 2

Let (M, τ) be a τ_0-topological space with denumerable basis. A representation $\delta :\subseteq \mathbb{B} \to M$ is called admissible (with final topology τ) iff δ is (τ_B, τ)-continuous and $\delta' \leq_t \delta$ for any (τ_B, τ)-continuous representation δ' of M.

Basic Theorem 3

(1) For any τ_0-space (M, τ) with denumerable basis there is an admissible representation with final topology τ..

(2) For admissible representations δ, δ-continuity is equal to τ-continuity.

Thus, the theory of T_0-spaces with denumerable bases can be embedded into TTE, where, continuity is interpreted as constructivity. An admissible representation for the real line will be given below. We conclude with some examples.

Examples 4

(1) The "enumeration representation" $\mathbb{M} : \mathbb{B} \to P_\omega$ of P_ω is defined by

$$\mathbb{M}(p) := \{i \mid i + 1 \in range(p)\}.$$

$\mathbb{M}$ is admissible w.r.t. the Scott topology on P_ω defined by the basis $\{O_A \mid A \subseteq \mathbb{N} \text{ finite }\}$ where $O_A := \{X \subseteq \mathbb{N} \mid A \subseteq X\}$.

(2) The "characteristic function representation" $M_{cf} : \mathbb{B} \to P_\omega$ of P_ω is defined by

$$\mathbb{M}_{cf}(q) := \{i \mid p(i) = 0\}.$$

$\mathbb{M}_{cf}$ is admissible w.r.t. Cantor's topology (= Scott's interval topology) on P_ω. One easily shows $\mathbb{M}_{cf} \leq_t \mathbb{M}$ and not $\mathbb{M} \leq_t \mathbb{M}_{cf}$.

(3) Let (M, d) be a separable metric space, let $\nu : \mathbb{N} \to A$ be a numbering of a dense subset A of M. The "Cauchy representation" $\delta :\subseteq \mathbb{B} \to M$ is defined by

$$\delta(p) = x :\Leftrightarrow (\forall j > i) d(\nu p(i), \nu p(j)) < 2^{-i} \text{ and } x = lim(\nu p(i)).$$

for all $p \in \mathbb{B}$ and $x \in M$. The representation δ is admissible w.r.t. to the topology induced by the metric d. $dom(\delta)$ consists of the "fast converging" Cauchy sequences on A.

(4) The standard representation ω_R of the open subset of $\mathbb{R}$ is defined by

$$\omega_R(p) := \bigcup \{I_k \mid k \in \mathbb{M}_p\}$$

where $I_{<j,m>} := (\nu_Q(j) - 2^{-m} \nu_Q(j) + 2^{-m})$ and ν_Q is a standard numbering of the rational numbers.

One easily shows $\mathbb{M}_{cf} \leq_t \mathbb{M}$ and not $\mathbb{M} \leq_t \mathbb{M}_{cf}$.

The functions χ and ψ are "effective" representations which are not admissible in the above sense.

3 Constructivity and Computability in Analysis

In this section we apply TTE to some simple questions in Analysis. First, we shall discuss representations of the real numbers. In the past several representations of the real numbers have been proposed (see e.g. Deil [31]). TTE explains clearly why most of them are not natural. Let ν_Q be some standard numbering of the rational numbers, e.g. $\nu_Q < i, j, k >= (i - j)/(k + 1)$. Since the rational numbers are dense in the real line $(\mathbb{R}, \tau_R)$ we may apply Example 4(3) to ν_Q and the distance $(x, y) \to |x - y|$ on $\mathbb{R}$. Let us call the resultant representation δ_C "standard Cauchy representation" of $\mathbb{R}$. Since δ_C is admissible, by the Basic Theorem 3 all the functions like addition, multiplication, division, exponentiation, the trigonometric functions, etc. are continuous w.r.t. δ_C, i.e. constructive. By the choice of ν_Q these functions are even computable w.r.t. δ_C. Roughly speaking, there is a Type 2 machine which for any "fast converging" rational Cauchy sequence with limit x determines a "fast converging" rational Cauchy sequence with limit $exp(x)$, etc.

The most familiar representation of the real numbers is the decimal representation δ_D which can be defined as follows:

$$\delta_D(p) := \nu_Q p(0) + \sum_{i \geq 1} (p(i) \; mod \; 10) \cdot 10^{-i}$$

Although $X \subseteq \mathbb{R}$ is δ_D–open iff $X \in \tau_R$, δ_D is not admissible, especially $\delta_C \leq \delta_D$ is false. As a proof one shows easily, that $x \to 3x$ is not continuous w.r.t. δ_D. Therefore, the decimal representation is unnatural for topological reasons.

Also the usual "naive" Cauchy representation δ_N defined by ($\delta_N(p) = x$ iff $(\nu_Q p(i))_{i \in \mathbb{N}}$ is a Cauchy sequence with limit x) is topologically bad: $X \subseteq \mathbb{R}$ is δ_N–open iff $X = \emptyset$ or $X = \mathbb{R}$. Informally, no finite initial part of p gives any information about $\delta_N(p)$.

Some further representations of $\mathbb{R}$ are discussed in Weihrauch and Kreitz [25, 26]. For studying the real line, only admissible representations with final topology τ_R and especially the "computationally admissible" representation δ_C (or any computationally equivalent one) yield "natural"results. We shall study δ_C–effectivity now.

Proposition 5

No nontrivial property, on $\mathbb{R}$ is δ_C–decidable.

Suppose $X \subseteq \mathbb{R}$ is δ_C–decidable. Then X and $\mathbb{R} \setminus X$ are open. Since the real line is topologically connected, X must be empty or equal to $\mathbb{R}$. At most those properties which correspond to open subsets of $\mathbb{R}$ (or $\mathbb{R}^n$) can be δ_C- r.e. . For example, the sets $\{x \mid x > 0\}, \{x \mid x \neq 0\}$ and $\{(x, y) \mid x \neq y\}$ are δ_C- r.e. . Sets which are δ_C- open but not δ_C- r.e. can be defined easily by using some non recursive set or function.

Classically, if $x \cdot y = 0$ then $x = 0$ or $y = 0$ $(x, y \in \mathbb{R})$. Can we determine effectively a factor which is 0?

Proposition 6

There is no continuous function $\Gamma :\subseteq \mathbb{B}^2 \to \mathbb{N}$ such that for all $p, q \in dom(\delta_C)$ with $\delta_C(p) \cdot \delta_C(q) = 0$ the following holds: $\Gamma(p, q) \in \{0, 1\}$ exists and $(\Gamma(p, q) = 0 \Rightarrow \delta_C(p) = 0)$ and $(\Gamma(p, q) = 1 \Rightarrow \delta_C(q) = 0)$.

The formal proof is easy. Informally, a function Γ with the above property would yield information which does not only depend on finite initial parts of p and q. Hence Γ cannot be continuous. The next examples concern zeroes of continuous functions. Let $C[0; 1]$ be the set of continuous functions $f : [0; 1] \to \mathbb{R}$. With the max–distance $d(f, g) := max\{|f(x) - g(x)| \mid x \in [0; 1]\}$ $C[0; 1]$ becomes a separable metric space. Examples for dense denumerable subsets are the polynomials with rational coefficients, the trigonometric polynomicals with rational coefficients, and Pg = the set of finite polygons with rational edges. Let $\alpha : \mathbb{N} \to Pg$ be some standard numbering of Pg. Let $\delta_\alpha :\subseteq \mathbb{B} \to C[0; 1]$ be the admissible representation derived form α according to Example 4(3). If $\delta_\alpha(p) = f$, then for any i, $B_c(\alpha(i), 2^{-i})$ is a closed ball containing f, where $B_c(g, \varepsilon)$ can be visualized by a band around g with width 2ε. Almost all properties concerning zeroes of continuous functions are not constructive. Let $X_{IV} := \{f \in C[0; 1] \mid f(0) \cdot f(1) < 0\}$, $X_{IVD} := \{f \in$

$X_{IVZ} \mid f^{-1}\{0\}$ is nowhere dense$\}, X_1 := \{f \in C[0;1] \mid f$ has exaclty one zero $\}$.

Theorem 7

(1) The sets $X_N := \{f \in C[0;1] \mid$ has no zero $\}$ and X_{IV} are δ_α–r.e. but not closed, especially not δ_α–decidable.

(2) There is no continuous function $\Gamma :\subseteq \mathbb{B} \to \mathbb{B}$ with $\delta_\alpha(p)\,(\delta_C\Gamma(p)) = 0$ for all $p \in X_{IV}$.

(3) There is a computable function $\Sigma :\subseteq \mathbb{B} \to \mathbb{B}$ with $\delta_\alpha(p)\,(\delta_C\Sigma(p)) = 0$ for all $p \in X_{IVD}$.

(4) There is no $(\delta_\alpha, \delta_C)$–continuous function $F :\subseteq C[0;1] \to \mathbb{R}$ with $f(F(f)) = 0$ for all $f \in X_{IVD}$.

(5) The function $G :\subseteq C[0;1] \to \mathbb{R}$ with $dom(G) = X_1$ and $f(G(f)) = 0$ for all $f \in X_1$. is $(\delta_\alpha, \delta_C)$–computable.

The proofs are not difficult. By (2) the classical intermediate value therorem is not constructive. Under additional conditions it becomes "weakly" computable or even computable (3), (4), (5). Similar properties can be proved for positions of maximal values. Notice, however, that the function $M : C[0;1] \to \mathbb{R}$ with $M(f) = max\{f(x) \mid 0 \leq x \leq 1\}$ is $(\delta_\alpha, \delta_C)$–computable. While determining a zero is not effective in general, "ε–substitutes" for zeroes can be computed:

Proposition 8

There is a computable function such that $|\delta_\alpha(p)\,(\delta_C\Gamma < n,p >)\,| < 2^{-n}$ for all $n \in \mathbb{N}$ and $p \in dom(\delta_\alpha)$ such that $min\{|\delta_\alpha(p)(x)||0 \leq x \leq 1\} < 2^{-n}$.

Again, the proof is easy. While integration is a computable operator, differentiation is not even continuous on $C[0;1]$.

Proposition 9

(1) The function $I : C[0;1] \to C[0;1]$ defined by $I(f)(x) = \int_0^x f(x)dx$ is $(\delta_\alpha, \delta_\alpha)$–computable.

(2) The function $D :\subseteq C[0;1] \to C[0;1]$ with $dom(D) = \{\ f \mid f$ is continuously differentiable $\}$ and $D(f)$ is the derivative of f is not $(\delta_\alpha, \delta_\alpha)$–continuous.

Differently effective kinds of compactness and versions of the Heine–Borel theorem are another instructive example (Kreitz and Weihrauch [21]). Although theorems similar to those above can be proved in other approaches to effectivity in analysis, the distinction between continuity and computability in TTE admits particularily transparent explanations.

4 Computational Complexity

While constructivity (i.e. continuity) and computability can be easily discussed simultaneously, computational complexity requires separate considerations. First, our machine model must be refined.

Realistically, a machine has a finite input/output alphabet T. Therefore, instead of functions $\Gamma :\subseteq \mathbb{B} \to \mathbb{B}$ ($\Sigma :\subseteq \mathbb{B} \to \mathbb{N}$) one should consider functions $\Gamma :\subseteq T^{\mathbb{N}} \to T^{\mathbb{N}}$ ($\Sigma :\subseteq T^{\mathbb{N}} \to T^*$). The formalism explained in Section 2 can be developed for this case equally well. Effectivity can be transferred from $T^{\mathbb{N}}$ and T^* to other sets M (especially to $\mathbb{B}$ and $\mathbb{N}$) by "representations" $\delta :\subseteq T^{\mathbb{N}} \to M$ or "notations" $\nu :\subseteq T^* \to M$. In the case $T = \{0,1\}$, which may be assumed w.l.g., we obtain a Type 2 theory of effectivity on Cantor's space $\{0,1\}^{\mathbb{N}}$. First result of a complexity theory for operators on Cantor's space can be found in Weihrauch and Kreitz [28].

Let $\delta :\subseteq T^{\mathbb{N}} \to \mathbb{R}$ be a representation. Let $f :\subseteq \mathbb{R} \to \mathbb{R}$ be a real function and assume that the Type 2 machine M computes f w.r.t. δ, i.e. $f\delta(p) = \delta g_M(p)$ for all $p \in dom(f\delta)$. As in ordinary Type 1 complexity theory the complexity should (essentially) not depend on the names $p \in \delta^{-1}\{x\}$ but on $x \in \mathbb{R}$ itself. In general, elements x have too many δ–names such that a uniform complexity bound for all $p \in \delta^{-1}\{x\}$ does not exist. There is, however, a special representation ρ of $\mathbb{R}$ such that $\rho^{-1}\{x\}$ in "small"for any $x \in \mathbb{R}$.

Definition 10 *(modified binary representation)*
Let $T := \{0, 1, -1, :\}$. Define a representation $\delta :\subseteq T^{\mathbb{N}} \to \mathbb{R}$ of $\mathbb{R}$ as follows:

$$dom(\rho) := \{a_k a_{k-1} \ldots a_0 : a_{-1} a_{-2} \ldots \mid a_i \in \{0,1,-1\}, a_k \neq 0, a_k a_{k-1} \notin \{1-1, -11\}\}$$

$$\rho\{a_k a_{k-1} \ldots a_0 : a_{-1} a_{-2} \ldots) := \sum_{i=k}^{-\infty} a_i \cdot 2^i.$$

Thus, ρ is a binary representation with positive and negative digits. The three conditions for $a_k a_{k-1}$ guarantee that the integer part of a name cannot become unnecessarily long. The representation ρ is computationally equivalent to our previous representation δ_C of $\mathbb{R}$. Furthermore, $\delta^{-1}(X)$ is a compact subset of $T^{\mathbb{N}}$ for any compact $X \subseteq \mathbb{R}$. Compact subsets of $T^{\mathbb{N}}$ are sufficiently "small" such that uniform complexity bounds exists (see eg. Müller [18, 19]).

Let M be a Type 2 machine which computes a function $f :\subseteq \mathbb{R} \to \mathbb{R}$ w.r.t. to the representation ρ, i.e. $f\rho(p) = \rho g_M(p)$ for all $p \in dom(f\rho)$. For any $p \in dom(f\rho)$. define $TIME(M)(p) : \mathbb{N} \to \mathbb{N}$ by $TIME(M)(p)(n) :=$ "the number of steps which M with input p needs for determining the n–th digit after the binary point of $g_M(p)$". Let

$X \subseteq dom(f)$ be a compact set. Then there is some uniform bound $t : \mathbb{N} \to \mathbb{N}$ with $(\forall p \in \rho^{-1}X)(\forall n) TIME(M)(p)(n) \leq t(n)$. In this case let us say "M computes f on X in time t". The polynomial time computable functions in Ko and Friedman [17] are the polynomial time computable functions in TTE. Also Brent's [16] definition of complexity is in accord with this one. Let $M(n)$ be an upper time bound for integer multiplication, e.g. $M(n) = n \cdot \log n \cdot \log \log n$. Then the following can be shown (Müller [18], Brent [16]).

Proposition 11

(1) On any compact subset $X \subseteq \mathbb{R}^2$, addition is computable in time $0(n)$.

(2) On any compact subset $X \subseteq \mathbb{R}^2$, multiplication is computable in time $0(M(n))$.

(3) On any compact subset $X \subseteq \mathbb{R}$ with $0 \notin X$, $x \to 1/x$ is computable in time $0(M(n))$.

(4) Let f be any trigonometric function or the exponential function. Then on any compact subset $X \subseteq dom(f)$, f can be computed in time $0(M(n) \cdot \log n)$

The time complexity of a real number $a \in \mathbb{R}$ (w.r.t. the representation ρ) can be defined as the computational complexity of the constant function $x \to a$. Hierarchy theorems from ordinary Turingmachine complexity theory can be transferred to the complexity of real numbers and of real functions (Müller [18]).

For certain classes of real functions f, the zeroes of f are in the some complexity class as f (Ko and Friedman [17], Müller [18]). If $t : \mathbb{N} \to \mathbb{N}$ is suficiently large (e.g. $\lambda n \cdot n^3 \in 0(t)$) and "regular" then $x \to \int_0^x f(y)dy$ is computable in $0(n^2 \cdot t(n))$ if $f :\subseteq \mathbb{R} \to \mathbb{R}$ is analytic on $[0;1]$ and computable in time $0(t)$ (Müller [19]).

In addition to $TIME(M)$ we define the input lookahead of a machine M by $ILA(M)(p)(n)$:= "the number of digits after the binary point which M with input p reads for determining the n-th digit after the binary point." Finally we define the dependence of $f : \mathbb{R} \to \mathbb{R}$ at $p \in dom(\rho)$ informally by $DEP(\rho)(p)(n)$:= "the least number of digits after the binary point of p by which $f\rho(p)$ is determined up to an error of 2^{-n}." If M computes f, then in any case $DEP(\rho)(p)(n) \leq ILA(M)(p)(n)$. The machine M computes "online" iff equality holds. It has been shown (Weihrauch [30]) that in any case ("almost –") online machines can be found but that forcing the online property may increase the computation time considerably.

5 Discussion and Open Problems

By means of TTE, five levels of effectivity can be distinguished in Analysis:

(1) Abstract Analysis: $(\forall x)(\exists y)R(x,y)$

(2) Constructivity: $(\exists\Gamma$ continuous$)$ $(\forall p \in Y)R(p,\Gamma(p))$

(3) Computability: $(\exists\Gamma$ computable$)$ $(\forall p \in Y)R(p,\Gamma(p))$

(4) Complexity: $(\exists\Gamma$ easily computable$)$ $(\forall p \in Y)R(p,\Gamma(p))$

(5) Numerical Analysis: "There is a program operating on floating point numbers such that ..."

There is a large number of positive effectivity results in analysis (see e.g. Kushner [7], Bishop and Bridges [4], Beeson [20]). For most of them corresponding theorems of the form "There is a computable ..." can be proved in TTE. In approaches based on Church's thesis negative results have the form "There is no computable ..." and in approaches based on some constructive logic negative results (if they are formulated at all) are of the form "Property P implies the non constructive principle ...". On the other hand, in TTE most of the fundamental negative results in Analysis have the form "There is no continuous ...".

Different kinds of non effectivity are compared in constructive logic approaches in the following way: A is "easier"then B iff A can be proved constructively from B (see e.g. Beeson [20]). As in Type 1 recursion theory, where degrees of non computability can be compared by several kinds of reducibility operators, in TTE it is possible to compare degrees of non continuity by different reducibility operators. For functions $A, B :\subseteq \mathbb{B} \to \mathbb{B}$ (or $\mathbb{B} \to \mathbb{N}$) the strongest form of "topological reducibility" can be defined by $A \leq B$ iff $A(p) = \Sigma < B\Gamma(p), p >$ for continuous functions $\Sigma, \Gamma :\subseteq \mathbb{B} \to \mathbb{B}$. For example, the characteristic function of X in Example 1(1) is equivalent to the function Γ which decides $x = 0$ on $\mathbb{R}$ w.r.t. δ_C (see Prop. 5), and both of them are strictly reducible to any function which translates the standard representation δ_C of $\mathbb{R}$ into the decimal representation. Several other more powerful, hence coarser, reducibilities can be defined for sets and functions, e.g. a kind of tt–reducibility, a kind of Turing–reducibility (by means of flowchart programs) or by the much more general concept of " A is computable in B" (Hinman [32]). Also "A can be proved from B in the constructive logic L" could be expressed in TTE as a special kind of reducibility. Not only to find out the effectivity properties of discontinuity but also to determine the degrees of discontinuity seems to be very important for really understanding the effectivity aspects of Analysis. Not much work has been done until today in this area which promises very interesting results.

For the study of computability in analysis, TTE gives a useful theoretical background. It tells what can be done by a computer (or a Turingmachine). For example the material in books like Bishop and Brides [4] or Kushner [7] should be presented in TTE, although this would require considerable efforts.

As a specialization of computability, TTE covers a powerful complexity Theory for Type 2 objects and functions which is natural down to very small complexity bounds. Until now, only the complexity of real functions and numbers is studied and we have theorems of the form: if $f \in C$ then the zero of f (or the integral of $f, \ldots$) is an element of C' where C and C' are complexity classes. But what is the computational complexity of the intergration operator $I : C[0;1] \rightarrow R$ or $I' : C[0;1] \rightarrow C[0;1]$? As for the real numbers very special representations for $C[0;1]$ (which exist!) must be considered in order to obtain reasonable definitions. We do not yet know, on which spaces there is a reasonable complexity theory. From Approximation Theory we know, that on $C[0;1]$ polynomials can be approximated by rational trigonometric polynomials or by rational polygons and vice versa. We have results about the degrees of polynomials or the numbers of vertices of polygons but not about upper and lower complexity bounds. Finally TTE provides the tools for precisely formulating computational complexity properties in Numerical Analysis.

References

[1] Brouwer, L.E.J.:
Historical background, principles, and methods of intuitionism, South African J.Sc. 49, 139 - 146 (1952)

[2] Heyting, A.:
Intuitionism, an introduction, North-Holland, Amsterdam, 1956 (revised 1972)

[3] Troelstra, A.S.:
Principles of intuitionism, Springer–Verlag, Berlin, Heidelberg, 1969

[4] Bishop,E.; Bridges, D.S.:
Constructive Analysis, Springer-Verlag, Berlin, Heidelberg, 1985

[5] Ceitin, G.S.:
Algorithmic operators in constructive complete separable metric spaces (in Russian), Doklady Akad, Nauk 128, 49 - 52 (1959)

[6] Markov, A.A.:
On constructive mathematics (in Russian), Trudy Mat. Inst. Stektov 67, 8 - 14 (1962)

[7] Kushner, B.A.:
Lectures on constructive mathematical logic and foundations of mathematics, Izdat. "Nauka", Moscow, (1973)

[8] Aberth, O.:
Computable analysis, McGraw-Hill, New York, (1980)

[9] Mazur, S.:
Computable analysis, Rozprawy Matematyczne XXXIII (1963)

[10] Grzegorczyk, A.:
On the definition of computable real continuous functions, Fund.Math. 44, 61 - 71 (1957)

[11] Mostowski, A.:
On computable sequences, Fund.Math. 44, 37 - 51 (1955)

[12] Lacombe, D.:
Quelques procedes de definition en topologie recursive. In: Constructivity in mathematics (A.Heyting, ed.), North–Holland, Amsterdam,1959

[13] Klaua, D.:
Konstruktive Analysis, Deutscher Verlag der Wissenschaften, Berlin, 1961

[14] Hauck, J.:
Berechenbare reelle Funktionen, Zeitschrift f. math. Logik und Grdl. Math. 19, 121 - 140 (1973)

[15] Scott, D.:
Outline of a mathematical theory of computation Science, Proc. 4th Princeton Confernce onInform. Sci., 1970

[16] Brent, R.P.:
Fast multiple precision evaluation of elementary functions, J. ACM 23, 242 - 251 (1976)

[17] Ko, K.; Friedman, H.:
Computational complexity of real functions, Theoret. Comput. Sci. 20, 323 - 352 (1982)

[18] Müller, N.Th.:
Subpolynomial complexity classes of real functions and real numbers. In: Lecture notes in Computer Science 226, Springer–Verlag, Berlin, Heidelberg, 284 - 293, 1986

[19] Müller, N.Th.:
Uniform computational complexity of Taylor series. In: Lecture notes in Computer Science 267, Springer–Verlag, Berlin, Heidelberg, 435 - 444, 1987

[20] Beeson, M.J.:
Foundations of constructive mathematics, Springer–Verlag, Berlin, Heidelberg, 1985

[21] Kreitz,Ch.; Weihrauch,K.:
Compactness in constructive analysis revisited, Informatik–Berichte Nr. 49, Fernuniversität Hagen (1984) and Annals of Pure and Applied Logic 36, 29 - 38 (1987)

[22] Kreitz,Ch.; Weihrauch,K.:
A unified approach to constructive and recursive analysis. In: Computation and proof theory, (M.M. Richter et al., eds.), Springer– Verlag, Berlin, Heidelberg, 1984

[23] Kreitz,Ch.; Weihrauch,K.:
Theory of representations, Theoretical Computer Science 38, 35 - 53 (1985)

[24] Weihrauch, K.:
Type 2 recursion theory, Theoretical Computer Science 38, 17 - 33 (1985)

[25] Weihrauch, K.; Kreitz, Ch.:
Representations of the real numbers and of the open subsets of the set of real numbers, Annals of Pure and Applied Logic 35, 247 - 260 (1987)

[26] Weihrauch, K.:
Computability, Springer–Verlag, Berlin, Heidelberg, 1987

[27] Weihrauch, K.:
On natural numberings and representations, Informatik–Berichte Nr.29, Fernuniversität Hagen, 1982

[28] Weihrauch, K.; Kreitz, Ch.:
Type 2 computational complexity of functions on Cantor's space (to appear)

[29] Weihrauch, K.:
Towards a general effectivity theory for computable metric spaces (to appear)

[30] Weihrauch, K.:
The complexity of online computations of real functions (in preparation)

[31] Deil, Th.:
Darstellungen und Berechenbarkeit reeller Zahlen, Informatik–Berichte Nr.51, Fernuniversität Hagen, 1984

[32] Hinman, P.G.:
Recursion–theoretic Hierachies, Springer–Verlag, Berlin, Heidelberg, , 1978

Vol. 324: M.P. Chytil, L. Janiga, V. Koubek (Eds.), Mathematical Foundations of Computer Science 1988. Proceedings. IX, 562 pages. 1988.

Vol. 325: G. Brassard, Modern Cryptology. VI, 107 pages. 1988.

Vol. 326: M. Gyssens, J. Paredaens, D. Van Gucht (Eds.), ICDT '88. 2nd International Conference on Database Theory. Proceedings, 1988. VI, 409 pages. 1988.

Vol. 327: G.A. Ford (Ed.), Software Engineering Education. Proceedings, 1988. V, 207 pages. 1988.

Vol. 328: R. Bloomfield, L. Marshall, R. Jones (Eds.), VDM '88. VDM – The Way Ahead. Proceedings, 1988. IX, 499 pages. 1988.

Vol. 329: E. Börger, H. Kleine Büning, M.M. Richter (Eds.), CSL '87. 1st Workshop on Computer Science Logic. Proceedings, 1987. VI, 346 pages. 1988.

Vol. 330: C.G. Günther (Ed.), Advances in Cryptology – EUROCRYPT '88. Proceedings, 1988. XI, 473 pages. 1988.

Vol. 331: M. Joseph (Ed.), Formal Techniques in Real-Time and Fault-Tolerant Systems. Proceedings, 1988. VI, 229 pages. 1988.

Vol. 332: D. Sannella, A. Tarlecki (Eds.), Recent Trends in Data Type Specification. V, 259 pages. 1988.

Vol. 333: H. Noltemeier (Ed.), Computational Geometry and its Applications. Proceedings, 1988. VI, 252 pages. 1988.

Vol. 334: K.R. Dittrich (Ed.), Advances in Object-Oriented Database Systems. Proceedings, 1988. VII, 373 pages. 1988.

Vol. 335: F.A. Vogt (Ed.), CONCURRENCY 88. Proceedings, 1988. VI, 401 pages. 1988.

Vol. 336: B.R. Donald, Error Detection and Recovery in Robotics. XXIV, 314 pages. 1989.

Vol. 337: O. Günther, Efficient Structures for Geometric Data Management. XI, 135 pages. 1988.

Vol. 338: K.V. Nori, S. Kumar (Eds.), Foundations of Software Technology and Theoretical Computer Science. Proceedings, 1988. IX, 520 pages. 1988.

Vol. 339: M. Rafanelli, J.C. Klensin, P. Svensson (Eds.), Statistical and Scientific Database Management. Proceedings, 1988. IX, 454 pages. 1989.

Vol. 340: G. Rozenberg (Ed.), Advances in Petri Nets 1988. VI, 439 pages. 1988.

Vol. 341: S. Bittanti (Ed.), Software Reliability Modelling and Identification. VII, 209 pages. 1988.

Vol. 342: G. Wolf, T. Legendi, U. Schendel (Eds.), Parcella '88. Proceedings, 1988. 380 pages. 1989.

Vol. 343: J. Grabowski, P. Lescanne, W. Wechler (Eds.), Algebraic and Logic Programming. Proceedings, 1988. 278 pages. 1988.

Vol. 344: J. van Leeuwen, Graph-Theoretic Concepts in Computer Science. Proceedings, 1988. VII, 459 pages. 1989.

Vol. 345: R.T. Nossum (Ed.), Advanced Topics in Artificial Intelligence. VII, 233 pages. 1988 (Subseries LNAI).

Vol. 346: M. Reinfrank, J. de Kleer, M.L. Ginsberg, E. Sandewall (Eds.), Non-Monotonic Reasoning. Proceedings, 1988. XIV, 237 pages. 1989 (Subseries LNAI).

Vol. 347: K. Morik (Ed.), Knowledge Representation and Organization in Machine Learning. XV, 319 pages. 1989 (Subseries LNAI).

Vol. 348: P. Deransart, B. Lorho, J. Maluszyński (Eds.), Programming Languages Implementation and Logic Programming. Proceedings, 1988. VI, 299 pages. 1989.

Vol. 349: B. Monien, R. Cori (Eds.), STACS 89. Proceedings, 1989. VIII, 544 pages. 1989.

Vol. 350: A. Törn, A. Žilinskas, Global Optimization. X, 255 pages. 1989.

Vol. 351: J. Díaz, F. Orejas (Eds.), TAPSOFT '89. Volume 1. Proceedings, 1989. X, 383 pages. 1989.

Vol. 352: J. Díaz, F. Orejas (Eds.), TAPSOFT '89. Volume 2. Proceedings, 1989. X, 389 pages. 1989.

Vol. 354: J.W. de Bakker, W.-P. de Roever, G. Rozenberg (Eds.), Linear Time, Branching Time and Partial Order in Logics and Models for Concurrency. VIII, 713 pages. 1989.

Vol. 355: N. Dershowitz (Ed.), Rewriting Techniques and Applications. Proceedings, 1989. VII, 579 pages. 1989.

Vol. 356: L. Huguet, A. Poli (Eds.), Applied Algebra, Algebraic Algorithms and Error-Correcting Codes. Proceedings, 1987. VI, 417 pages. 1989.

Vol. 357: T. Mora (Ed.), Applied Algebra, Algebraic Algorithms and Error-Correcting Codes. Proceedings, 1988. IX, 481 pages. 1989.

Vol. 358: P. Gianni (Ed.), Symbolic and Algebraic Computation. Proceedings, 1988. XI, 545 pages. 1989.

Vol. 359: D. Gawlick, M. Haynie, A. Reuter (Eds.), High Performance Transaction Systems. Proceedings, 1987. XII, 329 pages. 1989.

Vol. 360: H. Maurer (Ed.), Computer Assisted Learning – ICCAL '89. Proceedings, 1989. VII, 642 pages. 1989.

Vol. 361: S. Abiteboul, P.C. Fischer, H.-J. Schek (Eds.), Nested Relations and Complex Objects in Databases. VI, 323 pages. 1989.

Vol. 362: B. Lisper, Synthesizing Synchronous Systems by Static Scheduling in Space-Time. VI, 263 pages. 1989.

Vol. 363: A.R. Meyer, M.A. Taitslin (Eds.), Logic at Botik '89. Proceedings, 1989. X, 289 pages. 1989.

Vol. 364: J. Demetrovics, B. Thalheim (Eds.), MFDBS 89. Proceedings, 1989. VI, 428 pages. 1989.

Vol. 365: E. Odijk, M. Rem, J.-C. Syre (Eds.), PARLE '89. Parallel Architectures and Languages Europe. Volume I. Proceedings, 1989. XIII, 478 pages. 1989.

Vol. 366: E. Odijk, M. Rem, J.-C. Syre (Eds.), PARLE '89. Parallel Architectures and Languages Europe. Volume II. Proceedings, 1989. XIII, 442 pages. 1989.

Vol. 367: W. Litwin, H.-J. Schek (Eds.), Foundations of Data Organization and Algorithms. Proceedings, 1989. VIII, 531 pages. 1989.

Vol. 368: H. Boral, P. Faudemay (Eds.), IWDM '89, Database Machines. Proceedings, 1989. VI, 387 pages. 1989.

Vol. 369: D. Taubner, Finite Representations of CCS and TCSP Programs by Automata and Petri Nets. X. 168 pages. 1989.

Vol. 370: Ch. Meinel, Modified Branching Programs and Their Computational Power. VI, 132 pages. 1989.

Vol. 371: D. Hammer (Ed.), Compiler Compilers and High Speed Compilation. Proceedings, 1988. VI, 242 pages. 1989.

Vol. 372: G. Ausiello, M. Dezani-Ciancaglini, S. Ronchi Della Rocca (Eds.), Automata, Languages and Programming. Proceedings, 1989. XI, 788 pages. 1989.

Vol. 373: T. Theoharis, Algorithms for Parallel Polygon Rendering. VIII, 147 pages. 1989.

Vol. 374: K.A. Robbins, S. Robbins, The Cray X-MP/Model 24. VI, 165 pages. 1989.

Vol. 375: J.L.A. van de Snepscheut (Ed.), Mathematics of Program Construction. Proceedings, 1989. VI, 421 pages. 1989.

Vol. 376: N.E. Gibbs (Ed.), Software Engineering Education. Proceedings, 1989. VII, 312 pages. 1989.

Vol. 377: M. Gross, D. Perrin (Eds.), Electronic Dictionaries and Automata in Computational Linguistics. Proceedings, 1987. V, 110 pages. 1989.

Vol. 378: J.H. Davenport (Ed.), EUROCAL '87. Proceedings, 1987. VIII, 499 pages. 1989.

Vol. 379: A. Kreczmar, G. Mirkowska (Eds.), Mathematical Foundations of Computer Science 1989. Proceedings, 1989. VIII, 605 pages. 1989.

Vol. 380: J. Csirik, J. Demetrovics, F. Gécseg (Eds.), Fundamentals of Computation Theory. Proceedings, 1989. XI, 493 pages. 1989.